Suburban Books

40 East Scenic Ave, Richmond, CA 94801

AF892078

Plankton Ecology of the Southwestern Atlantic: From the Subtropical to the Subantarctic Realm [Paperback] [2019] Hoffmey

Order ID: 111-9285533-3685018 (000WR747)
Purchased: 2026-10-22T16:56:46-07:00

Shipping Address:
AMZ APTP ORDER
800 AVONDALE AVE
COLUMBUS, OH 43212-3473

Shipping Method: Standard

Thank you,

Tyson

P.S. If you have any questions or comments, you can contact me at: tyson@bichonbooks.com

Mónica S. Hoffmeyer • Marina E. Sabatini
Frederico P. Brandini • Danilo L. Calliari
Norma H. Santinelli
Editors

Plankton Ecology of the Southwestern Atlantic

From the Subtropical to the Subantarctic Realm

Springer

Editors
Mónica S. Hoffmeyer
CONICET
Instituto Argentino de Oceanografía
Bahía Blanca, Argentina

Frederico P. Brandini
Universidade de São Paulo
Oceanographic Institute
São Paulo, Brazil

Norma H. Santinelli
Laboratorio de Hidrobiología
Universidad de la Patagonia San Juan Bosco
Trelew, Argentina

Marina E. Sabatini
CONICET, INIDEP
Mar del Plata, Argentina

Danilo L. Calliari
Universidad de la Republica
Facultad de Ciencias+CURE
Montevideo, Uruguay

ISBN 978-3-030-08560-5 ISBN 978-3-319-77869-3 (eBook)
https://doi.org/10.1007/978-3-319-77869-3

Printed on acid-free paper

This Springer imprint is published by the registered company Springer International Publishing AG part of Springer Nature.
The registered company address is: Gewerbestrasse 11, 6330 Cham, Switzerland

Preface

Plankton play an indispensable role in maintaining ocean life but are increasingly threatened by multiple natural and human-induced changes to the environment. This advanced textbook explores the intriguing marine plankton communities of the Southwestern Atlantic region along the coast of southern South America. This marine system encompasses low-to-high latitude environments framed by a complex hydrographic scenery, confronting the local plankton communities with the growing challenge of global climate change and other human-induced environmental transformations. This vast and iconic region of the southwestern Atlantic shelf and coasts has been largely under-recognized and under-studied. However, recent years have witnessed a strong and growing interest in the region accompanied by acknowledgment of its high biological productivity.

Editing a contributed volume such as this is a challenging task but not without its rewards. It is with great satisfaction that we have been able to assemble this collection of 25 chapters synthesizing a variety of issues relating to the plankton ecology of the region, written by distinguished colleagues in the field.

Over one hundred (108) experts in plankton ecology and oceanography from Brazil, Uruguay and Argentina have contributed to the book, actually a very high number of authors for the region given the rather small universe of potential contributors. Although the three aforementioned countries together conform most of the Atlantic coastline of South America, there are unfortunately relatively few researchers working on marine sciences. The affiliations of different authors represent well-known universities and scientific institutions: 1 in the USA, 4 in Uruguay, 11 in Brazil and 19 in Argentina, thus ensuring the high quality of the findings presented in all the contributions.

Our team is grateful to Springer for approving our book proposal and enabling this endeavour to reach a successful conclusion. The main goal of the book is to bring to light important up-to-date plankton ecology research on the Southwestern Atlantic shelf and coastal region. The book also provides a comprehensive synthesis of plankton research carried out in recent years in this region and opens up possible new avenues of research for the near future. In short, *Plankton Ecology of the*

Southwestern Atlantic assembles a timely synthesis of the state of the art in the field, setting a new baseline for future research.

The topics covered in the book are organised as follows:

Part I deals with the historical background of plankton ecology and biological oceanography research on the Atlantic shelf and coastal region of South America, focusing on the regional development of the discipline mainly in Brazil and Argentina and mentioning key programmes of scientific cooperation with overseas countries.

Part II consists of five chapters dealing primarily with the oceanographic characterization of the study region. Major circulation patterns over the shelf are described, and the characteristics of several highly productive frontal areas are also portrayed. Inorganic nutrient content in sea water masses, its origin and fate within the region are then addressed, and the associated patterns of chlorophyll a concentration are discussed. Finally, the levels and variability patterns of planktonic primary production in different shelf and coastal systems and the responsible phytoplankton taxonomic groups are depicted.

Part III comprises six chapters dealing with the structure and dynamics of plankton communities over the shelf along various latitudes within the region in relation to water masses, frontal areas and oceanographic processes. Life histories, population structure and trophic conditions of some key zooplankton species are considered. Their link to the distribution areas of exploited fish species over the shelf, oceanographic features and hydroclimatic conditions are also examined.

Part IV contains six chapters dealing with the structure of plankton communities in coastal systems and their spatio-temporal dynamics. The variability patterns of density and biomass are also discussed in relation to natural or anthropogenic stressors. Life histories of some key species are included. Trophic interactions and the effects of major pollutants and eutrophication on plankton communities and individuals are also addressed.

Part V includes two chapters on the taxonomic groups collectively known as "jellyfish", their diversity, abundance and distribution in the region. The characteristics of their occasional blooms, the factors that promote their formation and the biochemical implications for ecosystems are examined mainly from review contributions. Other issues include some key shelf and estuarine species examined from a trophic perspective, thus allowing for the assessment of "jellyfish" as prey for several other marine consumers.

Part VI comprises three chapters dealing with the taxonomy, richness and abundance of phytoplankton toxigenic species (dinoflagellates and diatoms) as well as their genesis and distribution patterns across distinct shelf systems within the region. The characteristics of these species' blooms as well as the environmental factors (natural and human-induced) that appear to trigger and maintain them in particular areas and times are examined. The transfer and accumulation of some toxins through pelagic food webs, either within the first trophic levels (plankton communities) or passed up to large-sized consumers such as whales, are further discussed.

The three chapters of *Part VII* address the effects of sea water warming and other consequences of climate change on the phytoplankton and zooplankton communities of different marine systems within the region. The effects of eutrophication, pollution, and UV radiation on coastal plankton communities and individuals are also considered and the results of experimental approaches presented.

We would like to highlight the invaluable contribution of the long list of authors and reviewers who generously shared their insight and expertise for the editing of this volume and without whose collaboration and hard work this book would not have been possible. The enthusiastic response to our invitation from each and every one of the authors has been truly rewarding to us as editors. Special thanks go to Phyllis Barrantes for her help in polishing the English of some parts of this book. To conclude, we gratefully acknowledge the support provided as in-kind contributions (staff-time, grants and facilities) by research institutions in Brazil, Uruguay and Argentina.

Bahía Blanca, Argentina — Mónica S. Hoffmeyer
Mar del Plata, Argentina — Marina E. Sabatini
São Paulo, Brazil — Frederico P. Brandini
Montevideo, Uruguay — Danilo L. Calliari
Trelew, Argentina — Norma H. Santinelli

A Special Thank You to Our Reviewers

The group of co-editors would like to sincerely thank the reviewers listed below in recognition of their invaluable contribution in reviewing the chapters of this book.

It is by no means common for contributed volumes that individual chapters undergo a peer-review process such as that conducted in the present case. We are highly indebted to our colleagues for contributing to the enhancement of the manuscripts by giving so generously of their time, expertise and resources.

Reviewers

Rut Akselman Cardella. Instituto Nacional de Investigación y Desarrollo Pesquero (INIDEP), Mar del Plata, Buenos Aires, Argentina. E-mail: rutaks@inidep.edu.ar

Paulo C. Abreu. Instituto de Oceanografia, Fundação Universidade Federal de Rio Grande (FURG), Rio Grande, RS, Brazil. E-mail: docpca@furg.br

Gastón O. Almandoz. Facultad de Ciencias Naturales y Museo, Universidad Nacional de La Plata (UNLP), Argentina. E-mail: galmandoz@fcnym.unlp.edu.ar

Ulrich Bathmann. Leibniz-Institute for Baltic Sea Research, Rostock, Germany. E-mail: ulrich.bathmann@io-warnemuende.de

Demetrio Boltovskoy. Facultad de Ciencias Exactas y Naturales, Universidad de Buenos Aires, Buenos Aires, Argentina. E-mail: boltovskoy@gmail.com

Paulo Calil. Instituto de Oceanografia, Fundação Universidade Federal de Rio Grande, Rio Grande, RS, Brazil. E-mail: paulo.calil@gmail.com

Jorge P. Castello. Instituto de Oceanografia, Fundação Universidade Federal de Rio Grande, Rio Grande, RS, Brazil. E-mail: castellojpc@gmail.com

Luciano Chiaverano. Division of Marine Science, University of Southern Mississippi, Hattiesburg, MS, USA. E-mail: luciano.chiaverano@usm.edu

Daniel Conde. Facultad de Ciencias, Universidad de la República, Montevideo, Uruguay. E-mail: vladddcc@gmail.com

Santiago Gassó. GEST/Morgan State University, GFSC/NASA, Greenbelt, MD, USA. E-mail: santiago.gasso@nasa.gov

Elena Guerrero. Institut de Ciències del Mar, Barcelona, Spain. E-mail: eguerrero@icm.csic.es

Máximo Frangopulos. Universidad de Magallanes, Punta Arenas, Chile. E-mail: max.frangopulos@umag.cl

Mónica N. Gil. Centro de Estudio de Sistemas Marinos, CONICET, Puerto Madryn, Chubut, Argentina. E-mail: monicagil@cenpat-conicet.gob.ar

Silvana Halac. Centro de Investigaciones en Ciencias de la Tierra, CONICET, Universidad Nacional de Córdoba, Córdoba, Argentina. E-mail: srhalac@gmail.com

Marcelo P. Hernando. Comisión Nacional de Energía Atómica, Depto. Radiobiología CAC. Gral. San Martín, Buenos Aires, Argentina. E-mail: mphernando09@hotmail.com

Rubén J. Lara. Instituto Argentino de Oceanografía, CONICET – Universidad Nacional del Sur, Bahía Blanca, Argentina. E-mail: rubenlara@iado-conicet.gob.ar

Andrea C. López Cazorla. Instituto Argentino de Oceanografía, CONICET – Universidad Nacional del Sur, Bahía Blanca, Argentina. E-mail: acazorla@criba.edu.ar

María A. Marcoval. Instituto de Investigaciones Marinas y Costeras (IIMyC)/ CONICET-UNMDP, Facultad de Ciencias Exactas y Naturales, Estación Costera J.J. Nágera/ Depto. Acuicultura, Mar del Plata, Argentina. E-mail: marcoval@mdp.edu.ar

Ana M. Martínez. Universidad Nacional del Sur, Bahía Blanca, Argentina. E-mail: anamartinez.quimica@gmail.com

Carmen E. Morales Van de Wingard. Universidad de Concepción, Concepción, Chile. E-mail: camorale@udec.cl

Erik Muxagata. Instituto de Oceanografia, Fundação Universidade Federal de Rio Grande (FURG), Rio Grande, RS, Brazil. E-mail: e.muxagata@gmail.com

Claudia A. Namiki. Instituto Oceanográfico, Universidade de São Paulo, São Paulo, Brazil. E-mail: claunamiki@gmail.com

Marcelo Pájaro. Instituto Nacional de Investigación y Desarrollo Pesquero, Mar del Plata, Argentina. E-mail: mpajaro@inidep.edu.ar

Elbio D. Palma. Instituto Argentino de Oceanografía, CONICET – Universidad Nacional del Sur, Bahía Blanca, Argentina. E-mail: uspalma@criba.edu.ar

Beatriz Reguera Ramírez. Centro Oceanográfico de Vigo, Instituto Español de Oceanografía, Vigo, Spain. E-mail: beatrizreguera@ieo.es

Claudia G. Simionato. CIMA, FCEN, Universidad de Buenos Aires, Buenos Aires, Argentina. E-mail: simionato@cima.fcen.uba.ar

Jefferson Turner. Biology Department, University of Massachusetts Dartmouth, North Dartmouth, MA, USA. E-mail: jturner@umassd.edu

Linda Waters. Instituto Oceanográfico, Universidade de São Paulo, São Paulo, Brazil. E-mail: lindagwen@gmail.com

Cara Wilson. NOAA/NMFS/SWFSC, Environmental Research Division, Monterey, CA, USA. E-mail: cara.wilson@noaa.gov

The original version of the Front Matter was revised.
'A Special Thank You to Our Reviewers' section in the FM was updated with a missed reviewer.

Contents

Contributors

Paulo C. Abreu Instituto de Oceanografia, Universidade Federal do Rio Grande (FURG), Rio Grande, RS, Brazil

Eduardo M. Acha Instituto de Investigaciones Marinas y Costeras (IIMyC), Consejo Nacional de Investigaciones Científicas y Técnicas–Universidad Nacional de Mar del Plata (CONICET – UNMdP), Mar del Plata, Buenos Aires, Argentina

Instituto de Nacional de Investigación y Desarrollo Pesquero (INIDEP), Mar del Plata, Buenos Aires, Argentina

Gastón E. Aguirre Instituto de Biodiversidad y Biología Experimental Aplicada (IBBEA), Consejo Nacional de Investigaciones Científicas y Técnicas–Universidad de Buenos Aires (CONICET – UBA), Buenos Aires, Argentina

Departamento de Biodiversidad y Biología Experimental, Facultad de Ciencias Exactas y Naturales, Universidad de Buenos Aires, Buenos Aires, Argentina

Gastón O. Almandoz División Ficología, Facultad de Ciencias Naturales y Museo, Universidad Nacional de La Plata, La Plata, Argentina

Cecilia Alonso Centro Universitario Regional del Este (CURE), Universidad de la República, Rocha, Uruguay

Julieta C. Antacli Instituto de Diversidad y Ecología Animal (IDEA), Consejo Nacional de Investigaciones Científicas y Técnicas–Universidad Nacional de Córdoba (CONICET – UNC), Córdoba, Argentina

Elena S. Barbieri Laboratorio de Oceanografía Biológica (LOBio), Centro para el Estudio de Sistemas Marinos (CESIMAR), Consejo Nacional de Investigaciones Científicas y Técnicas (CONICET), Puerto Madryn, Argentina

Hugo Benavides Instituto Nacional de Investigación y Desarrollo Pesquero (INIDEP), Mar del Plata, Buenos Aires, Argentina

Anabela A. Berasategui Instituto Argentino de Oceanografía (IADO), Consejo Nacional de Investigaciones científicas y Técnicas–Universidad Nacional del Sur (CONICET–UNS), Bahía Blanca, Buenos Aires, Argentina

Alejandro A. Bianchi Departamento Oceanografía, Servicio de Hidrografía Naval (SHN), Ciudad Autónoma de Buenos Aires, Argentina

Departamento de Ciencias de la Atmósfera y los Océanos, Universidad de Buenos Aires, Ciudad Autónoma de Buenos Aires, Argentina

Demetrio Boltovskoy Facultad de Ciencias Exactas y Naturales, Instituto de Ecología, Genética y Evolución de Buenos Aires (IEGEBA), Consejo Nacional de Investigaciones Científicas y Técnicas–Universidad de Buenos Aires (CONICET – UBA), Buenos Aires, Argentina

Frederico P. Brandini Instituto Oceanográfico, Universidade de São Paulo, Cidade Universitária, São Paulo, Brazil

Luciana Díaz Briz Departamento de Ciencias Marinas, Facultad de Ciencias Exactas y Naturales, Universidad Nacional de Mar del Plata (UNMdP), Mar del Plata, Buenos Aires, Argentina

Daniel Bruno Laboratorio de Ecología, Fisiología y Evolución de Organismos Acuáticos, Centro Austral de Investigaciones Científicas (CADIC), Consejo Nacional de Investigaciones Científicas (CONICET), Ushuaia, Tierra del Fuego, Argentina

Andreana Cadaillón Instituto de Investigación de Hidrobiología, Facultad de Ciencias Naturales y Ciencias de la Salud, Universidad Nacional de la Patagonia San Juan Bosco, Trelew, Argentina

Mariana Cadaveira Instituto de Nacional de Investigación y Desarrollo Pesquero (INIDEP), Mar del Plata, Buenos Aires, Argentina

Danilo L. Calliari Ecología Funcional de Sistemas Acuáticos, Centro Universitario Regional del Este, Universidad de la República, Rocha, Uruguay

Oceanografía y Ecología Marina, Facultad de Ciencias, Universidad de la República, Montevideo, Uruguay

Fabiana L. Capitanio Instituto de Biodiversidad y Biología Experimental Aplicada (IBBEA), Consejo Nacional de Investigaciones Científicas y Técnicas–Universidad de Buenos Aires (CONICET – UBA), Buenos Aires, Argentina

Departamento de Biodiversidad y Biología Experimental, Facultad de Ciencias Exactas y Naturales, Universidad de Buenos Aires, Buenos Aires, Argentina

Mario O. Carignan Instituto Nacional de Investigación y Desarrollo Pesquero (INIDEP), Mar del Plata, Argentina

José I. Carreto Instituto Nacional de Investigación y Desarrollo Pesquero (INIDEP), Mar del Plata, Argentina

Belmiro M. Castro Instituto Oceanográfico, Universidade de São Paulo, São Paulo, Brazil

Georgina D. Cepeda Instituto de Investigaciones Marinas y Costeras (IIMyC), Consejo Nacional de Investigaciones Científicas y Técnicas–Universidad Nacional de Mar del Plata (CONICET – UNMdP), Mar del Plata, Buenos Aires, Argentina

Instituto Nacional de Investigación y Desarrollo Pesquero (INIDEP), Mar del Plata, Buenos Aires, Argentina

Guillermo Cervetto Dirección Nacional de Medio Ambiente. Ministerio de Vivienda, Ordenamiento Territorial y Medio Ambiente, Montevideo, Uruguay

Carlo J. Chazarreta Instituto Argentino de Oceanografía (IADO), Consejo Nacional de Investigaciones científicas y Técnicas–Universidad Nacional del Sur (CONICET–UNS), Bahía Blanca, Buenos Aires, Argentina

Aurea M. Ciotti Universidade de São Paulo, Centro de Biologia Marinha, São Sebastião, São Paulo, SP, Brazil

Augusto C. Crespi-Abril Laboratorio de Oceanografía Biológica (LOBio), Centro para el Estudio de Sistemas Marinos (CESIMAR), Consejo Nacional de Investigaciones Científicas y Técnicas (CONICET), Puerto Madryn, Argentina

Universidad Nacional de la Patagonia San Juan Bosco, Facultad de Ciencias Naturales, Sede Puerto Madryn, Puerto Madryn, Argentina

Valeria C. D'Agostino Centro para el Estudio de Sistemas Marinos (CESIMAR), Consejo Nacional de Investigaciones científicas y Técnicas (CONICET), Puerto Madryn, Chubut, Argentina

Carla M. Derisio Instituto Nacional de Investigación y Desarrollo Pesquero (INIDEP), Mar del Plata, Buenos Aires, Argentina

Ana Dogliotti Instituto de Astronomía y Física del Espacio (IAFE-CONICET), Universidad de Buenos Aires, Facultad de Ciencias Exactas y Naturales, Buenos Aires, Argentina

Marcelo Dottori Instituto Oceanográfico, Universidade de São Paulo, São Paulo, Brazil

María S. Dutto Instituto Argentino de Oceanografía (IADO), Consejo Nacional de Investigaciones científicas y Técnicas–Universidad Nacional del Sur (CONICET–UNS), Bahía Blanca, Buenos Aires, Argentina

Martin D. Ehrlich Instituto de Nacional de Investigación y Desarrollo Pesquero (INIDEP), Mar del Plata, Buenos Aires, Argentina

José L. Esteves Laboratorio de Oceanografía Química y Contaminación de Aguas (LOQyCA), Centro para el Estudio de Sistemas Marinos (CESIMAR), Consejo Nacional de Investigaciones Científicas y Técnicas (CONICET), Puerto Madryn, Chubut, Argentina

Fundación Patagonia Natural, Puerto Madryn, Argentina

Gustavo A. Ferreyra Centro Austral de Investigaciones Científicas (CADIC), Consejo Nacional de Investigaciones Científicas (CONICET), Ushuaia, Tierra del Fuego, Argentina

Carlos A. França Schettini Laboratório de Hidrodinâmica Costeira, Centro de Tecnologia e Geociências, Departamento de Oceanografia, Universidade Federal de Pernambuco (UFPE), Recife, PE, Brazil

Tami M. García Instituto Argentino de Oceanografía (IADO), Consejo Nacional de Investigaciones científicas y Técnicas–Universidad Nacional del Sur (CONICET–UNS), Bahía Blanca, Buenos Aires, Argentina

Maximiliano Garcia Instituto Argentino de Oceanografía (IADO), Consejo Nacional de Investigaciones científicas y Técnicas–Universidad Nacional del Sur (CONICET–UNS), Bahía Blanca, Buenos Aires, Argentina

Gabriel Genzano Departamento de Ciencias Marinas, Facultad de Ciencias Exactas y Naturales, Universidad Nacional de Mar del Plata (UNMdP), Mar del Plata, Buenos Aires, Argentina

Mónica N. Gil Centro para el Estudio de Sistemas Marinos (CESIMAR), Consejo Nacional de Investigaciones Científicas y Técnicas (CONICET), Puerto Madryn, Argentina

Piter D. F. Gomes Instituto de Biologia, Universidade Federal do Rio de Janeiro (UFRJ), Rio de Janeiro, RJ, Brazil

Mónica Gómez-Erache Oceanografía y Ecología Marina, Facultad de Ciencias, Universidad de la República, Montevideo, Uruguay

Rodrigo J. Gonçalves Laboratorio de Oceanografía Biológica (LOBio), Centro para el Estudio de Sistemas Marinos (CESIMAR), Consejo Nacional de Investigaciones Científicas y Técnicas (CONICET), Puerto Madryn, Chubut, Argentina

Leilén Gracia Villalobos Laboratorio de Oceanografía Biológica (LOBio), Centro para el Estudio de Sistemas Marinos (CESIMAR), Consejo Nacional de Investigaciones Científicas y Técnicas (CONICET), Puerto Madryn, Argentina

Raul A. Guerrero Instituto Nacional de Investigación y Desarrollo Pesquero (INIDEP), Mar del Plata, Argentina

Lumi Haraguchi Instituto de Oceanografia, Universidade Federal do Rio Grande (FURG), Rio Grande, RS, Brazil

Institute of Bioscience, Aarhus University, Roskilde, Denmark

Rodrigo D. Hernández-Moresino Laboratorio de Oceanografía Biológica (LOBio), Centro para el Estudio de Sistemas Marinos (CESIMAR), Consejo Nacional de Investigaciones Científicas y Técnicas (CONICET), Puerto Madryn, Chubut, Argentina

Marcelo P. Hernando Departamento Radiobiología CAC, Comisión Nacional de Energía Atómica, Buenos Aires, Argentina

Mónica S. Hoffmeyer Instituto Argentino de Oceanografía (IADO), Consejo Nacional de Investigaciones científicas y Técnicas–Universidad Nacional del Sur (CONICET–UNS), Bahía Blanca, Buenos Aires, Argentina

Facultad Regional Bahía Blanca, Universidad Tecnológica Nacional, Bahía Blanca, Buenos Aires, Argentina

María C. López Abbate Instituto Argentino de Oceanografía (IADO), Consejo Nacional de Investigaciones científicas y Técnicas–Universidad Nacional del Sur (CONICET–UNS), Bahía Blanca, Buenos Aires, Argentina

Vivian Lutz Instituto de Investigaciones Marinas y Costeras (IIMYC), Consejo Nacional de Investigaciones Científicas y Técnicas–Universidad Nacional de Mar del Plata (CONICET - UNMdP), Mar del Plata, Buenos Aires, Argentina

Instituto Nacional de Investigación y Desarrollo Pesquero (INIDEP), Mar del Plata, Buenos Aires, Argentina

María A. Marcoval Instituto de Investigaciones Marinas y Costeras (IIMyC), Consejo Nacional de Investigaciones Científicas y Técnicas–Universidad Nacional de Mar del Plata (CONICET–UMNdP), Buenos Aires, Argentina

Laura Machinandiarena Instituto de Nacional de Investigación y Desarrollo Pesquero (INIDEP), Mar del Plata, Buenos Aires, Argentina

Pedro Vieira Maciel Departamento de Sistemática e Ecologia, Universidade Federal da Paraíba, João Pessoa, PB, Brazil

Gabriela Malanga Facultad de Farmacia y Bioquímica, Fisicoquímica, Universidad de Buenos Aires Ciudad Autónoma de Buenos Aires, Buenos Aires, Argentina

Instituto de Bioquímica y Medicina Molecular (IBIMOL), Consejo Nacional de Investigaciones científicas - Universidad de Buenos Aires (CONICET–UBA), Buenos Aires, Argentina

Marina Marrari Departamento Oceanografía, Servicio de Hidrografía Naval (SHN), Ciudad Autónoma de Buenos Aires, Argentina

Ricardo P. Matano College of Earth, Ocean, and Atmospheric Sciences, Oregon State University, Corvallis, OR, USA

Silvia M. Méndez Dirección Nacional de Recursos Acuáticos (DINARA), Montevideo, Uruguay

Osmar O. Möller Jr Instituto de Oceanografia, Universidade Federal do Rio Grande, Rio Grande, RS, Brazil

Nora G. Montoya Instituto Nacional de Investigación y Desarrollo Pesquero (INIDEP), Mar del Plata, Argentina

André Carrara Morandini Departamento de Zoologia, Instituto de Biociências, Universidade de São Paulo (USP), São Paulo, Brazil

José H. Muelbert Laboratorio de Ecologia do Ictioplâncton, Instituto de Oceanografia, Universidade Federal do Rio Grande (FURG), Rio Grande, RS, Brazil

Renato M. Nagata Instituto de Oceanografia Universidade Federal do Rio Grande (FURG) Rio Grande, Rio Grande, RS, Brazil

Eugenia Nahuelhual Instituto Argentino de Oceanografía (IADO), Consejo Nacional de Investigaciones científicas y Técnicas–Universidad Nacional del Sur (CONICET–UNS), Bahía Blanca, Buenos Aires, Argentina

Lorena Silva do Nascimento Centro de Estudos do Mar, Universidade Federal do Paraná, Paraná, Brazil

Luis F. Niencheski Instituto de Oceanografia, Universidade Federal do Rio Grande (FURG), Rio Grande, RS, Brazil

Miodeli Nogueira Jr. Departamento de Sistemática e Ecologia, Universidade Federal da Paraíba, João Pessoa, PB, Brazil

Clarisse Odebrecht Instituto de Oceanografia, Universidade Federal do Rio Grande (FURG), Rio Grande, RS, Brazil

Lívia Dias Oliveira Departamento de Sistemática e Ecologia, Universidade Federal da Paraíba, João Pessoa, PB, Brazil

Soledad Díaz Ovejero Instituto de Investigación de Hidrobiología, Facultad de Ciencias Naturales y Ciencias de la Salud, Universidad Nacional de la Patagonia San Juan Bosco, Trelew, Argentina

Joanna M. Paczkowska Laboratorio de Oceanografía Biológica (LOBio), Centro para el Estudio de Sistemas Marinos (CESIMAR), Consejo Nacional de Investigaciones Científicas y Técnicas (CONICET), Puerto Madryn, Argentina

Luciano N. Padovani Instituto Nacional de Investigación y Desarrollo Pesquero (INIDEP), Mar del Plata, Buenos Aires, Argentina

Marcelo Pájaro Instituto de Nacional de Investigación y Desarrollo Pesquero (INIDEP), Mar del Plata, Buenos Aires, Argentina

Elbio D. Palma Instituto Argentino de Oceanografía (IADO-CONICET), Consejo Nacional de Investigaciones Científicas y Técnicas and Depatamento de Física, Universidad Nacional del Sur, Bahía Blanca, Argentina

Flavio E. Paparazzo Laboratorio de Oceanografía Química y Contaminación de Aguas (LOQyCA), Centro para el Estudio de Sistemas Marinos (CESIMAR),

Consejo Nacional de Investigaciones Científicas y Técnicas (CONICET), Puerto Madryn, Argentina

Laboratorio de Oceanografía Biológica (LOBio), Centro para el Estudio de Sistemas Marinos (CESIMAR), Consejo Nacional de Investigaciones Científicas y Técnicas (CONICET), Puerto Madryn, Argentina

Daiana Y. Pereyra Instituto Nacional de Investigación y Desarrollo Pesquero (INIDEP), Mar del Plata, Buenos Aires, Argentina

Laura B. Pérez Laboratorio de Biología, Facultad de Ciencias Naturales y Ciencias de la Salud, Universidad Nacional de la Patagonia San Juan Bosco, Comodoro Rivadavia, Argentina

Alberto R. Piola Departamento Oceanografía, Servicio de Hidrografía Naval (SHN), Ciudad Autónoma de Buenos Aires, Argentina

Departamento de Ciencias de la Atmósfera y los Océanos, Universidad de Buenos Aires, Ciudad Autónoma de Buenos Aires, Argentina

María L. Presta Instituto de Biodiversidad y Biología Experimental Aplicada (IBBEA), Consejo Nacional de Investigaciones Científicas y Técnicas–Universidad de Buenos Aires (CONICET – UBA), Buenos Aires, Argentina

Departamento de Biodiversidad y Biología Experimental, Facultad de Ciencias Exactas y Naturales, Universidad de Buenos Aires, Buenos Aires, Argentina

Francisco A. Puente Tapia Departamento de Ciencias Marinas, Facultad de Ciencias Exactas y Naturales, Universidad Nacional de Mar del Plata (UNMdP), Mar del Plata, Buenos Aires, Argentina

Laura Rodríguez-Graña Ecología Funcional de Sistemas Acuáticos, Centro Universitario Regional del Este, Universidad de la República, Rocha, Uruguay

Leonardo R. Rörig Laboratório de Ficologia (LAFIC), Universidade Federal de Santa Catarina (UFSC), Florianópolis, SC, Brazil

Graziela da Rosa Persich Laboratório de Ficologia (LAFIC), Universidade Federal de Santa Catarina (UFSC), Florianópolis, SC, Brazil

Marina E. Sabatini Instituto de Investigaciones Marinas y Costeras (IIMyC), Consejo Nacional de Investigaciones Científicas y Técnicas–Universidad Nacional de Mar del Plata (CONICET – UNMdP), Mar del Plata, Buenos Aires, Argentina

Instituto Nacional de Investigación y Desarrollo Pesquero (INIDEP), Mar del Plata, Buenos Aires, Argentina

Flavia M. P. Saldanha Corrêa Instituto Oceanográfico, Universidade de São Paulo, São Paulo, SP, Brazil

Norma H. Santinelli Instituto de Investigación de Hidrobiología, Facultad de Ciencias Naturales y Ciencias de la Salud, Universidad Nacional de la Patagonia San Juan Bosco, Trelew, Argentina

Betina A. Santos Instituto Nacional de Investigación y Desarrollo Pesquero (INIDEP), Mar del Plata, Buenos Aires, Argentina

Martín Saraceno Centro de Investigaciones del Mar y la Atmósfera, Consejo Nacional de Investigaciones Científicas y Técnicas–Universidad de Buenos Aires, Ciudad Autónoma de Buenos Aires, Argentina

Departamento de Ciencias de la Atmósfera y los Océanos, Universidad de Buenos Aires, Ciudad Autónoma de Buenos Aires, Argentina

Alicia V. Sastre Instituto de Investigación de Hidrobiología, Facultad de Ciencias Naturales y Ciencias de la Salud, Universidad Nacional de la Patagonia San Juan Bosco, Trelew, Argentina

Eliane C. Truccolo Schettini Laboratório de Hidrodinâmica Costeira, Centro de Tecnologia e Geociências, Departamento de Oceanografia, Universidade Federal de Pernambuco (UFPE), Recife, PE, Brazil

Agustín Schiariti Instituto Nacional de Investigación y Desarrollo Pesquero (INIDEP), Mar del Plata, Buenos Aires, Argentina

Instituto de Investigaciones Marinas y Costeras (IIMyC), CONICET, Universidad Nacional de Mar del Plata, Mar del Plata, Buenos Aires, Argentina

Irene R. Schloss Instituto Antártico Argentino, San Martín, Buenos Aires, Argentina

Centro Austral de Investigaciones Científicas (CADIC), Consejo Nacional de Investigaciones Científicas (CONICET), Ushuaia, Tierra del Fuego, Argentina

Universidad Nacional de Tierra del Fuego, Ushuaia, Argentina

Valeria Segura Instituto Nacional de Investigación y Desarrollo Pesquero (INIDEP), Mar del Plata, Buenos Aires, Argentina

Marcio da Silva Tamanaha Laboratório de Ficologia, Universidade do Vale do Itajaí (UNIVALI), Itajaí, SC, Brazil

Miriam E. Solís Laboratorio de Oceanografía Química y Contaminación de Aguas (LOQyCA), Centro para el Estudio de Sistemas Marinos (CESIMAR), Consejo Nacional de Investigaciones Científicas y Técnicas (CONICET), Puerto Madryn, Chubut, Argentina

Gaspar Soria Laboratorio de Oceanografía Biológica (LOBio), Centro para el Estudio de Sistemas Marinos (CESIMAR), Consejo Nacional de Investigaciones Científicas y Técnicas (CONICET), Puerto Madryn, Argentina

Universidad Nacional de la Patagonia San Juan Bosco, Facultad de Ciencias Naturales, Sede Puerto Madryn, Puerto Madryn, Argentina

Mariela L. Spinelli Instituto de Biodiversidad y Biología Experimental Aplicada (IBBEA), Consejo Nacional de Investigaciones Científicas y Técnicas–Universidad de Buenos Aires (CONICET – UBA), Buenos Aires, Argentina

Departamento de Biodiversidad y Biología Experimental, Facultad de Ciencias Exactas y Naturales, Universidad de Buenos Aires, Buenos Aires, Argentina

Virginia Tavano Laboratório de Fitoplâncton e Microorganismos Marinhos, Universidade Federal do Rio Grande, Rio Grande, RS, Brazil

Brenda Temperoni Instituto de Investigaciones Marinas y Costeras (IIMyC), Consejo Nacional de Investigaciones Científicas y Técnicas–Universidad Nacional de Mar del Plata (CONICET – UNMdP), Mar del Plata, Buenos Aires, Argentina

Instituto Nacional de Investigación y Desarrollo Pesquero (INIDEP), Mar del Plata, Buenos Aires, Argentina

Denise Rivera Tenenbaum Instituto de Biologia, Universidade Federal do Rio de Janeiro (UFRJ), Rio de Janeiro, RJ, Brazil

Sebastião Tilbert Departamento de Sistemática e Ecologia, Universidade Federal da Paraíba, João Pessoa, PB, Brazil

Laboratório de Meiofauna, Centro de Educação e Saúde, Universidade Federal de Campina Grande, Cuité, PB, Brazil

Román Uibrig Instituto Argentino de Oceanografía (IADO), Consejo Nacional de Investigaciones científicas y Técnicas–Universidad Nacional del Sur (CONICET–UNS), Bahía Blanca, Buenos Aires, Argentina

Jean L. Valentin Departamento de Biologia Marinha, Instituto de Biologia, Universidade Federal do Rio de Janeiro, Rio de Janeiro, Brazil

Mario Vera Oceanografía y Ecología Marina, Facultad de Ciencias, Universidad de la República, Montevideo, Uruguay

Maria C. Villac Instituto de Biologia, Universidade Federal do Rio de Janeiro (UFRJ), Rio de Janeiro, RJ, Brazil

Fish and Wildlife Research Institute, St. Petersburg, FL, USA

Virginia E. Villafañe Estación de Fotobiología Playa Unión. Consejo Nacional de Investigaciones Científicas y Técnicas, Rawson, Chubut, Argentina

María D. Viñas Instituto de Investigaciones Marinas y Costeras (IIMyC), Consejo Nacional de Investigaciones Científicas y Técnicas–Universidad Nacional de Mar del Plata (CONICET – UNMdP), Mar del Plata, Buenos Aires, Argentina

Instituto Nacional de Investigación y Desarrollo Pesquero (INIDEP), Mar del Plata, Buenos Aires, Argentina

Denise Vizziano Cantonnet Laboratorio de Fisiología de la Reproducción y Ecología de Peces, Instituto de Biología, Facultad de Ciencias/Universidad de la República Oriental del Uruguay, Montevideo, Uruguay

Gabriela N. Williams Laboratorio de Oceanografía Química y Contaminación de Aguas (LOQyCA), Centro para el Estudio de Sistemas Marinos (CESIMAR), Consejo Nacional de Investigaciones Científicas y Técnicas (CONICET), Puerto Madryn, Chubut, Argentina

Part I
Historical Background of Plankton Research in the Region

Overview of the History of Biological Oceanography in the Southwestern Atlantic, with Emphasis on Plankton

Demetrio Boltovskoy and Jean L. Valentin

Abstract The first data on the biological features of the Southwestern Atlantic were the result of European expeditions of the eighteenth to nineteenth and early twentieth centuries. Around the 1920s to 1940s, European-born specialists and their local disciples started playing a central role, and locally produced knowledge grew consistently. Early surveys centered on inventorial and distributional aspects of the flora and fauna, in particular mollusks and fishes, followed by community-level investigations including causal relationships with oceanographic settings (water masses, temperature, salinity, nutrients), red-tide outbursts, prospection of fishing grounds, etc. In Argentina, logistical support for the oceanographic cruises was historically associated with the Naval Hydrographic Service and in Brazil with the Brazilian Navy and the Universities of São Paulo and Rio Grande do Sul. As of 2017, in both these countries, there are >30 teaching and/or research institutions totally or partly dedicated to marine studies. Presently, knowledge of the taxonomy and biogeography of the plankton of the Southwestern Atlantic varies greatly among taxa, but several aspects (e.g., vertical distribution patterns, seasonal and especially multiannual variations, life histories, and many others) have received very little attention. Despite limited financial support and adequate floating platforms and equipment, lack of coordinated efforts, and political turbulences, the scientific output of Argentina, and especially Brazil, has grown in the last two decades, doubling from ~1.2% of the world total in 1996 to ~2.4% in 2016.

Keywords Plankton · Knowledge history · Argentina · Brazil · Southwestern Atlantic

D. Boltovskoy (✉)
Facultad de Ciencias Exactas y Naturales, Instituto de Ecología, Genética y Evolución de Buenos Aires (IEGEBA), Consejo Nacional de Investigaciones Científicas y Técnicas–Universidad de Buenos Aires (CONICET – UBA), Buenos Aires, Argentina

J. L. Valentin
Departamento de Biologia Marinha, Instituto de Biologia, Universidade Federal do Rio de Janeiro, Rio de Janeiro, Brazil

M. S. Hoffmeyer et al. (eds.), *Plankton Ecology of the Southwestern Atlantic*,
https://doi.org/10.1007/978-3-319-77869-3_1

1 Introduction

The present account is an attempt at summarizing the evolution of studies of the plankton of the Southwestern Atlantic starting with the pioneering expeditions of the eighteenth and nineteenth centuries. The task of compiling a fair overview on this subject is complicated by the fact that much of the information has appeared in the so-called gray literature, including institutional reports and journals of restricted distribution.

As we move closer to the present, the body of literature and the number of scientists involved grow substantially, and the information becomes more scattered. Further, while earlier surveys were fairly circumscribed in scope allowing a more or less clear identification of those that dealt specifically with plankton, more recent results are often multidisciplinary where plankton s.s. often plays an accessory role.

Previous efforts at synthesizing this information, especially in the area of general oceanography, ichthyology, and fisheries, have been of much help for tackling this problem (Wüst 1964; López Ambrosioni 1976; López 1976; Servicio de Hidrografía Naval 1976; Ringuelet 1984; Ehrlich and Sánchez 1990; Sánchez 1991; Angelescu and Sánchez 1995, 1997; Brandini et al. 1997; Calliari et al. 2003; Penchaszadeh 2012).

2 The Beginnings (Eighteenth to Early Twentieth Centuries)

Several major seagoing expeditions carried out by various European countries (Britain, France, Spain) since the fifteenth century crossed the Southwestern Atlantic, but they chiefly aimed at expanding their empires and discovering new seaways, lands, and sources of raw materials, rather than centering on scientific research. Nevertheless, some of these voyages contributed descriptions of both terrestrial and marine (mostly fishes) South American flora and fauna. Among the most frequently sampled locations off Argentina were southern Patagonian waters, as well as isolated locations off the Province of Buenos Aires where the expeditions lead by L. A. de Bougainville (1766–1769), J. Cook (1772–1775), L. I. Duperrey (1822–1825), P. P. King (1826–1836), R. Fitzroy (1832–1836), J. C. Ross (1839–1843), R. R. Mayne (1866–1869), and several others collected biological materials subsequently described by P. Commerson, J. G. Forster, J. G. Schneider. R. P. Lesson, G. Cuvier, A. Günther, L. Jenyns, J. Richardson, R.O. Cunningham, and A. d'Orbigny (Ringuelet 1984; Ehrlich and Sánchez 1990). However, systematic studies on the biology of the oceans did not really start until the middle of the nineteenth century (Lalli and Parsons 2007). The first comprehensive effort at investigating the physical, chemical, and biological features of the World Ocean (except for the Arctic) was that of the Challenger expedition (1873–1876), sponsored by the Royal Society (UK). This expedition collected thousands of samples and data from all

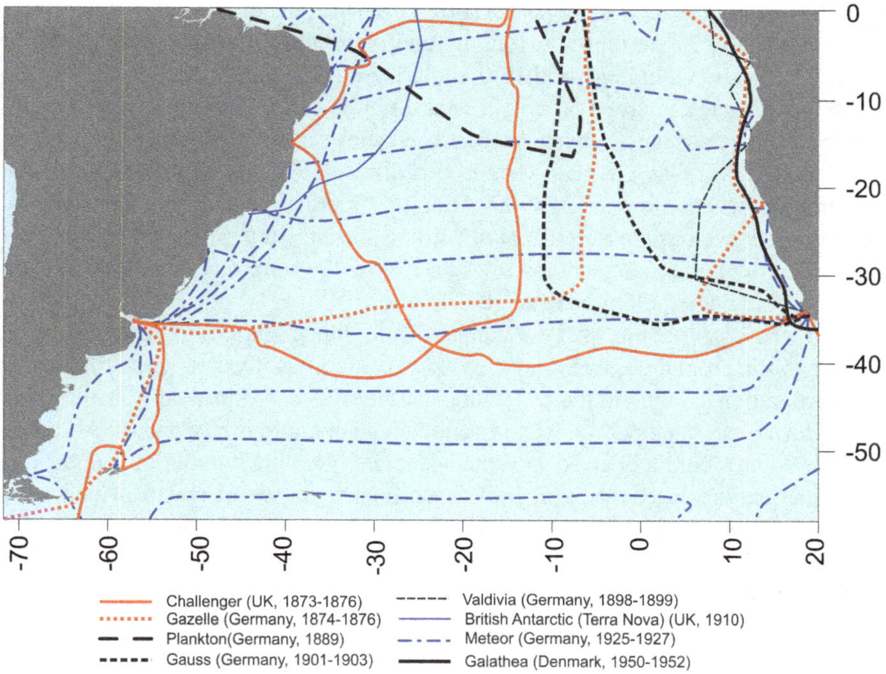

Fig. 1 Major early (1870s to 1950s) oceanographic expeditions that provided the first scientific data from the Southwestern Atlantic

depths of the sea floor and the water column, many of them from the South Atlantic (Fig. 1).

While the results of the Challenger expedition, which continued to be published until 1895 yielding 50 massive volumes authored by 76 scientists, are often considered to mark the birth of modern oceanography, it would be more fair to consider them as the first comprehensive descriptive compendium of the oceans, rather than the cornerstone of our understanding of the processes that govern their functioning. Indeed, many of the most significant mechanisms underlying the structure and dynamics of the physics, chemistry, and biology of the oceans were the result of much earlier efforts (e.g., the Coriolis force in 1835, the Navier-Stokes equations in 1823–1845, the notion of photosynthesis, which started developing around 1600, etc.), whereas others developed after the Challenger expedition, and although they benefited from the vast body of knowledge collected by the Challenger, they were not directly derived from it (e.g., the Ekman transport in 1902, Wegener's plate tectonics and continental drift, in 1912, etc.).

The Challenger inspired many subsequent worldwide voyages, some of which covered parts of the South Atlantic. Among the most prolific in terms of biological information for these waters were the Gazelle (Germany, 1874–1876), the Valdivia (or Deutschen Tiefsee-Expedition; Germany, 1898–1899), Gauss (Germany, 1901–1903), British Antarctic Terra Nova (UK, 1910), Meteor (Germany, 1925–1927), and Galathea (Denmark, 1950–1952) (Fig. 1). The results of these expeditions provided a wealth of data on the taxonomy and distribution of many South Atlantic pelagic animals, including protists (radiolarians, foraminifers), medusae, nemertines, polychaetes, pteropods, crustaceans (amphipods, cumaceans, cladocerans, copepods, mysidaceans, euphausiaceans), chaetognaths, salps, appendicularians, and larval fishes. The German Meteor expedition (1925–1927) was of particular importance because it traversed the Atlantic 13 times performing thousands of soundings, as well as measurements of currents, salinity, temperature, and nutrients and collecting numerous plankton tows (see Wüst 1964, for a comprehensive overview of all major expeditions from 1873 to 1960).

With the construction of the Panama Canal, which started operating in 1914, marine traffic around Cape Horn decreased very significantly, and with it interest in the exploration of the southern Atlantic latitudes also diminished. On the other hand, during the same period, whaling and fishing operations based in the Malvinas and the South Georgia Islands, and somewhat later growing international interest in the Antarctic, increased the number of expeditions to the area, many of which made underway observations and samplings in the Southwestern Atlantic, like the British Scoresby (1927–1932) and Discovery (1931–1935) expeditions (Ehrlich and Sánchez 1990).

3 Biological Oceanography in Argentina

3.1 The Early Years

Until around 1940–1950, the oceanographic activity of the countries bordering the Southwestern Atlantic was limited, but interest in natural history in general, including the vast oceanic realm bordering Argentina's ca. 4000 km-long coast, started in the middle 1800s. Most of these pioneering efforts were devoted to producing inventories of the local flora and fauna (especially mollusks and fishes) and, somewhat later, to addressing fisheries-related issues. Among the specialists involved in these studies were several Europeans, some of which were commissioned by the Argentine government and often appointed to high-ranking positions at several newly created museums and research and teaching centers, as well as Argentines, often trained by the former (G. Burmeister, C. Berg, F. Lahille, L. H. Valette, F. P. Moreno, C. Spegazzini, R. Dabbene, etc.). Reviews of these early contributions to the knowledge of the fishes of Argentina and Uruguay were produced by López (1976) and Ringuelet (1984).

During the first half of the twentieth century, biological oceanography kept expanding in scope and number of specialists, both European and Argentine, often disciples of the former. M. Doello Jurado, the first Argentine naturalist to attempt linking oceanographic processes with faunal traits; J. Frenguelli, a pioneer in the study of diatoms; A. J. Pozzi and E. Mac Donagh (ichthyology); A.R. Carcelles and J. J. Parodiz (malacology); E. L. Holmberg and C. A. Marelli were among the most prolific during this period (Balech 1971, 1976; López 1976). Most of these specialists were associated with the Natural History Museum of Buenos Aires (Penchaszadeh 2012), where their collections were analyzed and stored.

3.2 Logistics and Oceanographic Expeditions: The National Hydrographic Service

In 1879, the Central Hydrographic Office (presently the National Hydrographic Service) was created, and in 1928 the first two oceanographic vessels, San Luis and San Juan, were commissioned. Between 1864 and 1948, the Argentine Navy carried out >60 cruises whose main purpose was charting the poorly known Argentine coast, describing the bottom topography and bottom types and defining coastal tidal regimes. Many of these expeditions had scientists onboard from the Natural History Museums of Buenos Aires and La Plata, as well as other institutions, who performed ancillary collections and observations, chiefly of fishes and mollusks (López Ambrosioni 1976; Ehrlich and Sánchez 1990; Angelescu and Sánchez 1997). By ~1950 these efforts had produced fairly detailed catalogues of marine fishes and some invertebrates, general accounts of the surface circulation and descriptions of some of planktonic groups associated with different water types, rough biogeographic sketches of the Argentine Sea, and general notions on the migrations of commercially valuable fish stocks and oceanographic characterizations of the main fishing areas (Angelescu and Sánchez 1997). Both European-born (Z. Popovici, L. Szidat, V. Angelescu) and Argentine (F. S. Gneri, A. Nani, E. Siccardi, E. Balech) specialists led these studies.

Between 1954 and 1975, the Argentine Navy and the Naval Hydrographic Service conducted >120 oceanographic cruises, almost half of them on the Capitán Cánepa, a Canadian-built ship acquired by Argentina in 1947 which was used for oceanographic work between 1957 and 1972, when it was finally decommissioned. Most of these cruises covered Argentine coastal waters, but several were also conducted in open-ocean areas ranging from the equator to Antarctica and from the coasts of South America to the coasts of Africa (Servicio de Hidrografía Naval 1976). Several of these expeditions were coordinated operations with US research vessels (Vema, Theta). The primary goal of most of them was aimed at the investigation of physical and chemical traits of the areas covered (bathymetry, paleomagnetism, current fields, surface and vertical profiles of temperature, salinity, dissolved oxygen, nutrients, light penetration, etc.), but many also collected bottom and plankton samples and made some seasonal assessments of primary production and recruitment of exploitable mussel, crustacean, and fish stocks, as well as observations on marine bird and mammal populations. Unfortunately, efforts at compiling the vast body of data collected by these cruises have been marginally successful. In 1974, the Argentine Center of Oceanographic Data (CEADO) was created with the aim of compiling, harmonizing, digitizing, storing, and facilitating the accessibility of the thousands of measurements and observations collected throughout many decades in the Southwestern Atlantic and other areas where Argentine oceanographic expeditions were conducted, but the initiative had limited success.

Although a consolidated account of the results of the biological studies derived from these materials is unavailable, much of the information published by Argentine specialists since the 1940s on various planktonic organisms from these waters was based these samples, including diatoms (J. Frenguelli, C. Lange), dinoflagellates and tintinnids (E. Balech), foraminifers (E. Boltovskoy), radiolarians, pteropods and chaetognaths (D. Boltovskoy, J.R. Dadon), benthic mollusk larvae (P. E. Penchaszadeh), pelagic crustaceans and larvae (F. Ramírez, M.D. Viñas, E. Boschi, M. Scelzo), polychaetes (J. M. Orensanz), appendicularians and salps (G. Esnal), fish larvae (J. Ciechomski, M. Ehrlich), etc. Until around 1970–1980, most of these results were published in local or regional journals or institutional reports of limited distribution, in Spanish, and their knowledge outside of the local scientific community was limited. Since the late 1980s, however, the trend to publish in English in international, refereed media grew exponentially, largely due to pressure from funding agencies (in particular CONICET) and major universities, thus strongly enhancing the international visibility of Argentine contributions (see below).

As summarized above, although scientific research was chiefly conducted by personnel affiliated with nonmilitary organizations (see below), operational aspects of oceanographic activities have traditionally been in charge of the Argentine Navy. Presently, the oceanographic vessels Puerto Deseado and the recently acquired El Austral (formerly Sonne), owned by the National Council of Scientific and Technical Research (CONICET), are also operated by the Navy. The Argentine Coast Guard operates a largely overhauled vessel built in 1930 for the Woods Hole Oceanographic Institution and donated to Argentina in 1966 (originally Atlantis, subsequently renamed to El Austral, and later to Dr. Bernardo Houssay), but the history of this ship since it arrived in Argentina has been turbulent, and its contributions to oceanographic studies were limited. No Argentine research center owns and operates a seagoing oceanographic ship, with the exception of the National Institute of Fisheries Investigations and Development (INIDEP), based in Mar del Plata, which has four large fisheries vessels (crewed by civilian personnel): Capitán Cánepa, Oca Balda, E. L. Holmberg, and V. Angelescu (Table 1). These ships operate regularly along the Argentine Sea, and although their primary goal is the assessment of stocks of exploitable living resources, they also perform ancillary observations and samplings which cover a broader range of interests.

3.3 *Institutions*

Table 2 lists the institutions which contributed the most to research in the area of biological oceanography and plankton of the Southwestern Atlantic. Their input, however, has been quite uneven. Some, like the SHN (Naval Hydrographic Service), were chiefly responsible for logistic aspects of cruise implementation and sample/

Table 1 Argentine and Brazilian vessels involved in oceanographic surveys in the Southwestern Atlantic. ARA Argentine Navy (Armada de la República Argentina), CONICET National Council of Scientific and Technical Research (Consejo Nacional de Investigaciones Científicas y Técnicas, Argentina), FURG University of Rio Grande do Sul (Universidade de Rio Grande, Brazil), INIDEP National Institute of Fisheries Investigations and Development (Instituto Nacional de Investigación y Desarrollo Pesquero, Argentina), IO-USP Institute of Oceanography of the São Paulo University (Instituto Oceanografico, Universidade de São Paulo, Brazil), LEPLAC Continental Shelf Research Program (Brazil), PNA Argentine Coast Guard (Prefectura Naval Argentina), REVIZEE Living Economic Resources of the Exclusive Economic Zone program (Brazil), SHN: Naval Hydrographic Service (Servicio de Hidrografía Naval, Argentina)

Name	Operational (as a research vessel) from-to	Operated by	Remarks
Argentina			
Capitán Cánepa	1957–1972	ARA	Probably the most prolific in hydrographic and oceanographic research: 46 oceanographic cruises
Comodoro Laserre	1963–1969	ARA	Three oceanographic cruises
Bahía Blanca (ex San Luis)	1955–1962	ARA	Six oceanographic cruises
Madryn (ex Comodoro Rivadavia, ex San Juan)	1954–1959	ARA	Six oceanographic cruises
Sanaviron	1947–1974	ARA	
Zapiola	1962–1976	ARA	Ten oceanographic cruises
Goyena	1968–1981	ARA	17 oceanographic cruises
Islas Orcadas	1974–1979	ARA	Temporarily operated by ARA in the framework of a US-Argentina agreement (US name: Eltanin). Three oceanographic cruises
Thompson	1975–	ARA	
General San Martin	1954–1982	ARA	Icebreaker, chiefly logistic support to Argentine Antarctic stations
Almirante Irízar	1978–	ARA	Icebreaker, chiefly logistic support to Argentine Antarctic stations
Capitán Cánepa	1979–	INIDEP	Chiefly fisheries monitoring
Oca Balda	1983–	INIDEP	Chiefly fisheries monitoring
E.L. Holmberg	1980–	INIDEP	Chiefly fisheries monitoring
Puerto Deseado	1978-	CONICET-ARA-SHN	
El Austral (ex Sonne)	2017–	CONICET-ARA-SHN	
B. Houssay	1967–	PNA	
V. Angelescu	2017–	INIDEP	Chiefly fisheries monitoring
Brazil			
Alte Saldanha	1964–1990	Brazilian Navy	The first and the most prolific in oceanographic research

(continued)

Table 1 (continued)

Name	Operational (as a research vessel) from-to	Operated by	Remarks
Prof. W. Besnard	1967–2008	IO-USP	260 oceanographic cruises
Alpha-Crusis	2012-	IO-USP	Substitute of NOc Besnard
Atlântico Sul	1978-	FURG	Chiefly fisheries monitoring
Antares	1989-	Brazilian Navy	Support of navy operation and policy. Chiefly for the REVIZEE and LEPLAC programs
Ary Rongel	1994-	Brazilian Navy	Logistic support to high-latitude oceanographic cruises
Cruzeiro do Sul	2008-	Brazilian Navy	Oceanographic research and student formation along the Brazilian coast
Vital de Oliveira	2015-	Brazilian Navy	The most modern vessel of Brazilian navy for monitoring water quality

data collections, while others are host institutions of experts that produced the scientific reports and publications. Among these, a few are restricted to studies of the ocean, but most are multidisciplinary research and teaching centers which host important laboratories that contributed significantly to marine planktonic surveys.

One of the pioneering institutions dedicated specifically to marine studies, in particular biology, was the Institute of Marine Biology (Instituto de Biología Marina – IBM), founded in 1960. The IBM operated as such until 1976 and in 1977 was transferred to the Federal Government and renamed as the National Institute of Fisheries Investigations and Development (Instituto Nacional de Investigación y Desarrollo Pesquero – INIDEP). The IBM was the first large institution, and among the most prolific, to focus on marine sciences, hosting up to ca. 50 resident scientists, many visiting scholars, and a large technical staff (1960–1976). Unfortunately, as many other academic centers, in particular those associated with universities, between the 1960s and the 1970s, the IBM suffered massively from political turmoil and military coups, as result of which many of its researchers emigrated to other countries (Budiansky 1984; Scelzo et al. 2017). On the other hand, the creation of the INIDEP and a significant increase in the budget of this new federal institute whose main lines of research shifted from general marine biology and ecology to the exploration and monitoring of living marine resources boosted its staff and importance very significantly.

A major event for the development of science in general, including marine planktonic studies, was the creation, in 1958, of the National Council of Scientific and Technical Research (Consejo Nacional de Investigaciones Científicas y Técnicas – CONICET). From its beginnings, CONICET implemented the Scientific and the Technical Careers. Basically, this consists in the possibility for researchers based in any public or private research center to apply for a position with CONICET. This application is reviewed by a panel of specially appointed peers, and, if approved, the applicant is accepted as a member of CONICET (in one of presently five positions)

Table 2 Mayor Argentine and Brazilian institutions associated with oceanographic research (notice that most of these host specialists in a wide range of fields). FG, Federal Government. Refer to Fig. 2 for geographic locations

Operational	Name	Acronym	Location(s)	Affiliated with
Argentina				
1812–	Argentine Museum of Natural Sciences Bernardino Rivadavia (Museo Argentino de Ciencias Naturales Bernardino Rivadavia)	MACN	Buenos Aires	CONICET
1874–	University of Buenos Aires, Faculty of Exact and Natural Sciences (Universidad de Buenos Aires, Facultad de Ciencias Exactas y Naturales)	UBA-FCEyN	Buenos Aires	FG, Ministry of Education (Ministerio de Educación)
1879–	National Hydrographic Sevice (Servicio de Hidrografía Naval) (formerly Central Hydrographic Office)	SHN	Buenos Aires	FG, Argentine Navy
1884–	La Plata Museum (Museo de La Plata)	MLP	La Plata (Buenos Aires Prov.)	Buenos Aires Province
1905–	La Plata National University, Faculty of Natural Sciencies and Museum (Universidad Nacional de La Plata, Facultad de Ciencias Naturales y Museo)	UNLP-FCNyM	La Plata (Buenos Aires Prov.)	FG, Ministry of Education (Ministerio de Educación)
1928–	Quequén Port Hidrobiological Station (Estación Hidrobiológica de Puerto Quequén)	EHPQ	Quequén (Buenos Aires Prov.)	MACN, CONICET
1951–	Argentine Antarctic Institute (Instituto Antártico Argentino)	IAA	San Martín (Buenos Aires Prov.)	FG, Ministry of Foreign Affairs, International Commerce and Cult (Ministerio de Relaciones Exteriores, Comercio Internacional y Culto)

(continued)

Table 2 (continued)

Operational	Name	Acronym	Location(s)	Affiliated with
1958–	National Council of Scientific and Technical Research (Consejo Nacional de Investigaciones Científicas y Técnicas)	CONICET	Buenos Aires	FG, Ministry of Science, Technology and Productive Innovation (Ministerior de Ciencia, Tecnología e Innovación Productiva)
1960–1976	Marine Biology Institute (Istituto de Biología Marina) (INIDEP since 1977)	IBM	Mar de Plata (Buenos Aires Prov.)	FG, National Southern University, Univ. of Buenos Aires and Univ. of La Plata
1961–1984	Research Center in Marine Biology (Centro de Investigaciones de Biología Marina)	CIBIMA	Buenos Aires, Puerto Deseado (Santa Cruz Prov.), Ushuaia (Tierra del Fuego Prov.)	FG, University of Buenos Aires, National Institute of Technological Industry (Instituto Nacional de Tecnología Industrial)
1969–	Argentine Institute of Oceanography (Instituto Argentino de Oceanografía)	IADO	Bahía Blanca (Buenos Aires Prov.)	CONICET, Argentine Navy, National Southern Univerity
1970–	National Patagonian Centre (Centro Nacional Patagónico)	CENPAT	Puerto Madryn (Chubut Prov.)	CONICET
1974–	Institute of Marine Biology and Fisheries Admiral Storni (Instituto de Biología Marina y Pesquera Almirante Storni)	IBMPAS	San Antonio Oeste (Río Negro Prov.)	National University of Comahue (FG), Ministry of Production of the Province of Río Negro (Universidad Nacional del Comahue, Ministerio de Producción de la Provincia De Río Negro)
1977–	National Institute of Fisheries Investigations and Development (Instituto Nacional de Investigaciones y Desarrollo Pesquero) (Formerly IBM)	INIDEP	Mar de Plata (Buenos Aires Prov.)	FG, Federal Government, Ministry of Agroindustry (Ministerio de Agroindustria)

(continued)

Table 2 (continued)

Operational	Name	Acronym	Location(s)	Affiliated with
1978–	National University of Mar del Plata, Faculty of Exact and Natural Sciences (Universidad Nacional de Mar del Plata, Facultad de Ciencias Exactas y Naturales)	UNMdP-FCEyN	Mar de Plata (Buenos Aires Prov.)	FG, Ministry of Education (Ministerio de Educación)
1981–	Austral Center of Scientific Research (Centro Austral de Investigaciones Científicas)	CADIC	Ushuaia (Tierra del Fuego Prov.)	CONICET
1996–	National Agency for the Promotion of Science and Technology (Agencia Nacional de Promoción Científica y Tecnológica) (administration, funding)	ANPCyT	Buenos Aires	FG, Ministry of Science, Technology and Productive Innovation (Ministerio de Ciencia, Tecnología e Innovación Productiva)
1997–	Playa Unión Photobiological Station (Estación Fotobiológica Playa Unión)	EFPU	Playa Unión (Chubut Prov.)	CONICET
Brazil				
1946–	University of São Paulo	USP	São Paulo	Educational Secretary of the São Paulo State (Universidade de São Paulo)
1960–	Federal University of Ceará	UFC	Fortaleza	Ministry of Education (MEC) (Universidade Federal do Ceará)
1968–	Federal University of Bahia (Universidade Federal da Bahia)	UFBA	Salvador	Ministry of Education (MEC)
1968–	Federal University of Rio de Janeiro (Universidade Federal do Rio de Janeiro)	UFRJ	Rio de Janeiro	Ministry of Education (MEC)
1970–	Federal University of Rio Grande do Sul (Fundação Universidade Federal do Rio Grande do Sul)	FURG	Rio Grande do Sul	Ministry of Education (MEC)
1977–	University of Rio de Janeiro State (Universidade do Estado do Rio de Janeiro)	UERJ	Rio de Janeiro	Educational Secretary of the Rio de Janeiro State

(continued)

Table 2 (continued)

Operational	Name	Acronym	Location(s)	Affiliated with
1981–	Federal University of Pernambuco (Universidade Federal de Pernambuco)	UFPE	Recife	Ministry of Education (MEC)
1983–	Fluminense Federal University (Universidade Federal Fluminense)	UFF	Niteroi	Ministry of Education (MEC)
1987–	Paulo Moreira Institute for the Marine Research (Instituto de Pesquisa do Mar Almirante Paulo Moreira)	IEAPM	Arraial do Cabo	Brazilian Navy
1992–	Federal University of Paraná (Universidade Federal do Paraná)	UFPR	Curitiba	Ministry of Education (MEC)
1992–	University of Vale do Itajaí (Universidade do Vale do Itajaí)	UNIVALI	Itajaí	Private University
1995–	Federal University of Espírito Santo (Universidade Federal do Espirito Santo)	UFES	Vitoria	Ministry of Education (MEC)
2000–	Federal University of Santa Catarina (Universidade Federal de Santa Catarina)	UFSC	Florianopolis	Ministry of Education (MEC)
2001–	Federal University of Maranhão (Universidade Federal do Maranhão)	UFMA	São Luiz	Ministry of Education (MEC)
2009–	Federal University of Rio de Janeiro State (Universidade Federal do Estado do Rio de Janeiro)	UNIRIO	Rio de Janeiro	Ministry of Education (MEC)

according to his/her qualifications and age. Each category is associated with a salary, which traditionally has been reasonable for Argentine standards and is usually higher than the one earned by the researcher previously. However, because newly incorporated researchers kept working at their original institutions (mostly universities) and had a salary there, CONICET would usually pay only the difference between this original salary and the one offered by CONICET for the corresponding category. Researchers without a paid position elsewhere receive their full salary from CONICET. In addition, CONICET implemented 3–5-year scholarship programs for recent graduates in order to complete their doctoral programs, fellowships for foreign travel and training, research grants, hiring programs, and salaries for technical support personnel, built and maintained many research centers throughout the country, and established partnership programs with several universities.

Fig. 2 Location of major Argentine and Brazilian institutions associated with biological oceanography research (see Table 2 for acronyms and further details). All are presently operational, with the exception of CIBIMA, which was closed down in 1984

Although not without some ups and downs, largely associated with turbulences in Argentine politics, during the last half century, CONICET has grown steadily, both in prestige and influence and in the number of scientists and research centers affiliated with it. The peer-evaluation process that CONICET-associated researchers undergo every 2 years as well as its policy clearly oriented at making Argentine science more visible internationally by publishing in international journals (rather than locally) has very significantly improved both the quality and the output in all branches of science, including biological oceanography (see below).

The National Agency for the Promotion of Science and Technology (Agencia Nacional de Promoción Científica y Tecnológica – ANPCyT), created in 1996, is presently one of the major sources of funding for research and development projects in all areas of science and technology, including grants, loans for startup initiatives, fellowships, and support for joint public-private undertakings. Although more business-oriented than CONICET, its sphere of action largely overlaps those of the latter.

3.4 Foreign Input and International Cooperation

As seen above, the very beginnings of marine studies in Argentina were based on international cooperation insofar as most of the early specialists were Europeans, many of which settled in this country and trained Argentine personnel. As local expertise developed, the direct influence of foreign experts waned, although several of the most salient researchers in the areas of planktology and fisheries that were active until the 1970s–1990s were born abroad (e.g., V. Angelescu, E. Boltovskoy, L. Szidat).

Argentina took part in several large national and international programs. The organization of the International Geophysical Year (1957–1958) marked a period of very active and fruitful international cooperation with the participation of 67 countries (including the USA, the UK, Germany, Japan, Argentina, USSR, and many others) and tens of expeditions in all oceans. The aim of these explorations were chiefly physical, chemical, geological, and, to a lesser extent, biological investigations. In 1963–1964, the Equalant expeditions, in the frame of the International Cooperative Investigation of the Tropical Atlantic (ICITA), sponsored by the Intergovernmental Oceanographic Commission (IOC), involved over a dozen countries. Further examples are the FREPLATA I and II programs (Argentina-Uruguay) and the project Prevention of Coastal Pollution and Management of Marine Biological Diversity (e.g., Boltovskoy 2008), sponsored by the Global Environmental Facility. Although their contributions to physical oceanography and general overviews of the state of knowledge of the Southwestern Atlantic have been considerable, their direct influence on new information in the area of marine plankton was limited.

However, formal participation of Argentina in large multinational projects is obviously a highly biased indicator of the input of foreign specialists and joint surveys with Argentine experts to the knowledge of the Atlantic Ocean around South America. Most international links and cooperation efforts are established on a personal basis, rather than formally under the umbrella of major multinational programs. For example, since around 2000, over 50% of the publications with authors affiliated with Argentine institutions have foreign coauthors as well, chiefly from the USA, Brazil, Spain, Germany, and Chile (Boltovskoy 2015). In 1981, D. Boltovskoy edited a book on marine zooplankton covering methodological aspects and the taxonomy and distribution of 16 zooplanktonic groups. Of the 22 contributing coauthors, 9 were Argentines, and the remaining 13 were from 8 other American and European countries. A more recent edition of a similar work (Boltovskoy 1999b) had 35 coauthors, 6 Argentines and 29 from 15 other countries.

As expected from geographical closeness and shared areas of interest, Argentina has traditionally maintained close ties with Uruguay, and especially with Brazil. Between 1993 and 1999, six cruises by Brazilian vessels were conducted in the framework of the TABIA Brazil-Argentina cooperation program (F. Brandini, PI), collecting data on vertical properties of the water column, nutrients, POC, PON, chlorophyll, plankton (>500 samples), etc. between 30°–60°S and ~50°–60°W. The results of these cruises yielded several doctoral dissertations and many scientific publications. Large programs with other countries and/or international financial support have also been conducted in recent years, like the Argentina-France ARGAU (2000–2005; aimed at assessing CO_2 air-ocean fluxes) and GEF Patagonia (2005–2006), collecting abundant planktonic materials (e.g., Santoferrara and Alder 2009; Carreto et al. 2016).

4 Biological Oceanography in Brazil

The hydrological features of the ~8000 km-long Brazilian coast and continental shelf largely govern the composition and biomass of the plankton and, consequently, biological stocks of commercial interest. According to the limits of the exclusive economic zone (EEZ) defined by the Program REVIZEE (Ministério de Meio Ambiente, Brazil) (2006), based on oceanographic and biological features, the Brazilian coast and its continental shelf can be divided into four large regions: North, Northeast, Central, and Southeast-South.

The northern region, crossed by the equator, is located between the mouths of the Rivers Oiapoque (in the northern hemisphere) and Parnaiba. The pelagic environment is here oligotrophic, and the main source of nutrients is continental drainage by large rivers such as the Parnaiba River and, especially, the Amazon River (Brandini et al. 1997; Ministério de Meio Ambiente (Brazil) 2006).

The Northeast region (from the mouth of Parnaiba River to the city of Salvador) is characterized by the presence of a coastal reef barrier and several oceanic islands (Fernando de Noronha, Atol das Rocas, São Pedro and São Paulo archipelago). This region is influenced by the currents which result from the bifurcation of the South

Equatorial Current – the North Brazil Current, flowing toward the Guianas, and the Brazil Current, heading south. This region is also oligotrophic, with sporadic mineral enrichment limited to the island effect.

In the Central region (from Salvador to Rio de Janeiro), the continental shelf is influenced by the Brazil Current, which transports oligotrophic tropical water to the south. The oligotrophic nature of this region is altered by the effects of the coastal upwelling of Cabo Frio (Valentin 2001) and the presence of vortices, caused by the topographic barrier of the Abrolhos parcel and the Vitória-Trindade seamount chain. These physical processes make hydrography complex in the region (Castro and Miranda 1998) and are responsible for the nutrient enrichment of the surface layers and the concomitant increase in pelagic production, conferring this region the characteristics of a zone of biogeographic transition between the tropical waters to the north and subtropical ones to the south.

In the Southeast region (Rio de Janeiro to the Cape of Santa Marta, in the state of Santa Catarina), oceanographic conditions imposed by the Brazil Current determine the dynamics typical of mid-latitude continental platforms. The southernmost sector of this region, between Cabo de Santa Marta and Uruguay, differs from its northern part due to interactions between the oligotrophic Brazil Current and the nutrient-rich waters of the Malvinas Current and the continental discharge of the Río de la Plata estuary and the Patos Lagoon, which are responsible for its high biological production (Seeliger et al. 1997).

4.1 The Evolution of Planktology in Brazil

The evolution of plankton studies in oceanic and coastal Brazilian waters from 1819 to 1997 was presented in detail by Brandini et al. (1997). The most relevant features are summarized below, appended with more recent information from the last 20 years (1997–2017).

4.1.1 The Early Years

Plankton studies covering Brazilian waters started in the nineteenth century, when international expeditions crossed Brazilian waters and provided taxonomic records, albeit usually with no ecological context: the Challenger expedition (1872–1876), which covered all oceans (except the Arctic) for 3 years; the Plankton Expedition (1889), which collected samples off north and northeastern Brazil; the oceanographic expeditions of the German ship Meteor in the South Atlantic (1925–1927), which included several transects throughout the equatorial and South Atlantic (Fig. 1). These expeditions yielded many publications on the plankton off Brazil (see Brandini et al. 1997 for details).

At the beginning of the twentieth century, the first taxonomic studies on coastal diatoms by local researchers (e.g., Zimmermann 1918, to cite only the last of eight papers published by this author), as well as diatoms, dinoflagellates, and tintinnids collected off the southern coast (Cunha and Fonseca 1918), and on the first record of a harmful algal bloom from Brazil, causing mass fish mortalities in Guanabara Bay (Faria 1914), were published. Faria and Cunha (1917) and Faria et al. (1922) produced the earliest coastal surveys on the microzooplankton and protozoa of the Guanabara Bay and surrounding areas and, later, Paiva Carvalho (1945) on copepods from Santos Bay (state of São Paulo) and the coast of Paraná.

4.1.2 Post-World War II (1945–1960)

After World War II, general interest in the oceans in general, and in the plankton in particular, intensified. Between 1945 and 1960, several pioneering studies on the taxonomy of planktonic organisms from several regions of the Brazilian coast were conducted. The first publications in this period (Oliveira 1946, 1947) studied the composition of both phyto- and zooplankton from the Bays of Sepetiba, Ilha Grande, and Guanabara (State of Rio de Janeiro). The coastal region of the State of São Paulo received the greatest attention after the creation of the Instituto Paulista de Oceanografia. Among these pioneering investigations were those on diatoms, dinoflagellates, and copepods from Cananeia and other coastal regions. Müller-Melchers (1953, 1957) studied the diatoms off the coasts of São Paulo and Rio Grande do Sul, the São Pedro and São Paulo archipelago, and even from the mouth of the Amazon River, based on samples collected during the cruise of the ship Toko Maru. Teixeira and Kutner (1961) contributed data on the diatoms of the Cananeia lagoon system. Paiva Carvalho (1945, 1952) surveyed the zooplankton off the coast of São Paulo, providing the first quantitative estimates.

During this postwar period, the first Brazilian oceanographic cruises were conducted by Navy ships. The first samples of zooplankton from the central region of the Brazilian coast (near the city of Vitória, Espírito Santo State) were collected by the hydrographic ship Rio Branco and analyzed by Vannucci (1949). In the same region, between Vitória and Trindade Island, in the tropical waters of the Brazil Current, cruises of the NOc Almirante Saldanha and the destroyer escort Baependi and the fishing vessel Vega were carried out. The results of these expeditions were numerous qualitative and quantitative analyses of the zooplankton, including hydromedusae (Vannucci 1951a), heteropod mollusks (Vannucci 1951b), tornarian larvae (Bjornberg 1954), Chaetognatha (Vannucci and Hosoe 1952), and Appendicularia (Bjornberg and Forneris 1955). The four oceanographic expeditions conducted by the NOc Almirante Saldanha during this postwar period marked the start of a remarkable growth of Brazilian oceanography in the coming decades.

4.1.3 Years 1960–2017

The second half of the twentieth century was probably the period of greatest impetus in the search for knowledge of the Brazilian sea and its biological components, particularly of the plankton system. Several factors are responsible for this growing trend of oceanography in Brazil and the knowledge of planktonic communities.

4.2 Floating Platforms (Table 1)

In the second half of the twentieth century, the need for large ships for the exploration and exploitation of the Brazilian sea became obvious, since the available knowledge mostly derived from occasional international campaigns, such as those of the Calypso. For more than 20 years (1951–1973), this oceanographic ship, owned by Jacques-Yves Cousteau, sampled the World Ocean, including the Brazilian coast (1961–1962), where it collected numerous plankton samples (Seguin 1965). In July 1999 and June 2000, the research vessel Thalassa of the French Institute for the Exploration of the Sea (IFREMER, France) was leased by Bahia Pesca (Bahia Fishing Company) to survey resources off the central coast of Brazil under the Program REVIZEE. The transformation of the Navy sailboat and school ship Almirante Saldanha into an oceanographic ship by Admiral Paulo Moreira, pioneer of oceanography in Brazil, became the country's first and main floating platform for oceanographic research, carrying out numerous research cruises between 1964 and 1990. In the same period (1967–2008), the University of São Paulo acquired the NOc Prof. W. Besnard, which also participated in many oceanographic cruises, including the Antarctic. It was recently (2012) replaced by the NOc Alpha Crucis and the smaller NOc Alpha Delphini, the first one built in Brazil. In 2008, the NOc Cruzeiro do Sul was commissioned by the Department of Hydrography and Navigation (DHN, Brazilian Navy), carrying out the first transatlantic campaign Brazil-Africa (in 2009) during which phytoplankton and zooplankton were collected. These ships began to fill Brazil's great need for floating platforms for oceanographic research and training of students from numerous undergraduate and postgraduate courses in marine sciences (see Table 2).

4.3 Funding for Oceanographic Research

With the growing interest in the knowledge and exploration of the Brazilian sea, governmental initiatives stimulated development agencies to finance research of the marine environment. In addition to the contributions of the National Research Center of the Ministry of Science and Technology (CNPq/MCTI) and the Foundations for Research Support of the States (FAPs), a decisive step in the evolution of oceanography in Brazil was the creation, in 1974, of the Interministerial

Commission for Sea Resources (CIRM) and its secretariat (SeCIRM), with the objective of elaborating the PSRM (Sectorial Plans for the Resources of the Sea) that establish, every 4 years, research goals and renewable resource forecasts. In 2012, both the VIII PSRM and the work plan for 2012–2015 were approved putting special emphasis on the development of human resources (Krug 2012).

At present, the IX PSRM covers the aims and actions for the 2016–2019 period (www.mar.mil.br/secirm/portugues/psrm.html). In addition, private or mixed economy companies, such as Petrobras, Eletrobrás, and Vale do Rio Doce, provide resources to comply with the requirements of the Brazilian environmental control agency (IBAMA) in terms of environmental studies, for undertaking projects in the marine environment (oil exploration and production offshore, location and operation of nuclear power plants, production and transportation of mining products). An environmental and oceanographic diagnosis of the southern and southeastern regions of Brazil was carried out under the coordination and economic support of Petrobras (Valentin et al. 1994). These initiatives allowed the development of several large-scale research projects and programs, many of which involve the study of plankton in Brazilian marine waters. Among the most important are:

Project Cabo Frio This project, devised and led by the Admiral Paulo Moreira Sea Study Institute (IEAPM), in Arraial do Cabo (RJ), was conceived and carried out by Admiral Paulo Moreira and financed by the Funding Authority for Studies and Projects, (FINEP/MCT). Its main purpose was to study the dynamics of the Cabo Frio upwelling and the use of South Atlantic Central Water (SACW) for aquaculture (Moreira da Silva 1971, 1973). Several years of plankton samplings in the area of this upwelling resulted in many publications (e.g., Valentin 1984, 2001; Valentin et al. 1985, 1987, 1994; Valentin and Coutinho 1990).

Program REVIZEE By ratifying the United Convention on the Law of the Sea in 1988, Brazil accepted the rights and obligations related to the exploitation, use, conservation, and management of living resources in its exclusive economic zone (EEZ). Managed by the Ministry of Environment, the REVIZEE Program was active between 1995 and 2015. Throughout these 10 years, over 300 researchers from ~60 universities and research institutions participated in this program. Over ten oceanographic and fishing vessels were used in the oceanographic campaigns. Many papers have been and are still being published in journals, books, and theses based on the biological materials, including plankton, collected in these campaigns, which covered the entire Brazilian continental shelf in the N-NE regions (Hazin 2009), the central region (Bonecker 2006; Bonecker and de Castro 2006; Tenenbaum 2006; Valentin 2006), and the S-SE regions (Franco et al. 2005; Gaeta and Brandini 2006; Katsuragawa et al. 2006; Lopes et al. 2006a, b; Rossi-Wongtschowski and Madureira 2006).

Project PROABROLHOS The project Productivity, Sustainability, and Utilization of the Abrolhos Bank Ecosystem was financially supported by the MCT/CNPq from 2005 to 2009 and involved around 40 researchers from 14 institutions. Oceanographic cruises were carried out on the vessels Prof. W. Besnard and Atlântico Sul. Its results were published in a special issue of the journal *Continental Shelf Research* (Lopes and Castro 2013).

Project DEPROAS This project, funded by CNPq and FINEP, was aimed at studying the dynamics of the western South Atlantic platform ecosystem, in particular, the impact of the South Atlantic Central Water on biological processes. Its implementation was based on cruises of the NOc Prof. W. Besnard carried out between 2001 and 2003.

Program PELD The Long-Term Ecological Research Program (PELD), funded by CNPq, was initiated in 2000, with a recent contribution from state development agencies (FAPs) and MEC. This ongoing program is used to allocate resources for long-term studies of both marine and coastal ecosystems in the estuary of Patos Lagoon and its adjacent coast, Guanabara Bay, oceanic islands, in the upwelling area of Cabo Frio, in sand dune habitats and coastal lagoons of Northern Rio de Janeiro, and in the Abrolhos Archipelago (Tabarelli et al. 2013).

4.4 Human Resources

The growing number of students interested in the marine sciences led to the creation of numerous courses at the undergraduate and postgraduate levels in different teaching and research institutions. The creation of undergraduate courses in oceanography was decisive for the development of the area in Brazil. Following the first FURG initiative in 1971, nine more oceanography courses were implemented by the following universities: UERJ (1977), UNIVALI (1992), UNIMONTE (1998), UFES (2000), UFPA (2000), UFBA (2004), and UFPR (2004) (see Fig. 2). At the undergraduate level, students have a wide range of marine science programs along the coast of Brazil, including biological oceanography, marine biology, ocean engineering, marine biotechnology, and the possibility of pursuing doctoral degrees in marine plankton in postgraduate programs in zoology, botany, genetics, and ecology at most universities (Table 2).

4.5 International Programs

The engagement of Brazil in international projects grew significantly starting around 1985 and is currently very intense and fostered by UNESCO's Intergovernmental Oceanographic Commission (IOC), whose mission is to encourage marine scientific research. With the decree of January 5, 1994, the MCT was given the function of promoting and coordinating the country's participation in IOC activities related to ocean sciences (Fernandes and Oliveira 2012). New paradigms on the role of planktonic communities in marine ecosystems have emerged in the past 20 years with the implementation of several international research programs, such as JGOFS (Joint Global Ocean Flux Study), an international and multidisciplinary program aimed at understanding the role of the oceans in global carbon and nutrient cycles, and GLOBEC, a study of Global Ocean Ecosystem Dynamics initiated in 1990 by the Scientific Committee on Oceanic Research (SCOR) and UNESCO's IOC.

Due to these favorable factors, the development of marine sciences in Brazil and the country's scientific output, in particular in the area of plankton, have grown both quantitatively and qualitatively over the last decades (Fig. 3). The improvement of sampling methods and laboratory processing also allowed a considerable advance in qualitative and quantitative studies of the various planktonic compartments. Hitherto, studies on the abundance and biomass of protozooplankton and small metazooplankton are rare or even inexistent in Brazil, depending on the region considered (Lopes 2007). Only in the last decade of the twentieth century, publications on both pico- and nanoplankton have increased in several regions of the Brazilian coast (Teixeira and Gaeta 1991), along the southern coast (Odebrecht and Abreu 1995), in São Paulo (Mesquita and Fernandes 1996), in the waters of Guanabara Bay (Gomes et al. 2007; Santos et al. 2007), and the Abrolhos archipelago (Susini-Ribeiro 1999).

5 Knowledge of the Plankton of the South Atlantic

Assessing the degree of knowledge of the plankton of any given geographic area is complicated by the fact that the notion of "knowledge" should ideally encompass not only taxonomic and distributional traits (see reviews in Boltovskoy 1979, 1981; Brandini et al. 1997; Boltovskoy 1999a, b; Boltovskoy et al. 2003, 2005; Miloslavich et al. 2011) but also aspects of the physiology, seasonality, trophic interactions, genetics, population and community dynamics, reproduction, etc. (e.g., Seeliger et al. 1997; Seeliger and Kjerfve 2001). While estimating the sampling coverage and numbers of publications on the taxonomy and distribution of various planktonic groups is in principle feasible (although a major task in itself), assessing the other abovementioned aspects is much more complicated. Indeed, a given species may have never been studied in the South Atlantic, but it might have been the subject of numerous detailed surveys elsewhere, and therefore it is reasonable to assume that at least some of the traits of its South Atlantic populations are not unknown. Thus, the following comments are chiefly centered on our perceived degree of completeness of the inventories and, to a much lesser extent, the distribution patterns of the most important planktonic groups in coastal and offshore waters of the South Atlantic.

Table 3 and Fig. 4 show the approximate numbers of described marine planktonic species and the percentages thereof that have been recorded in the South Atlantic. It should be borne in mind that these figures are approximate, and that they

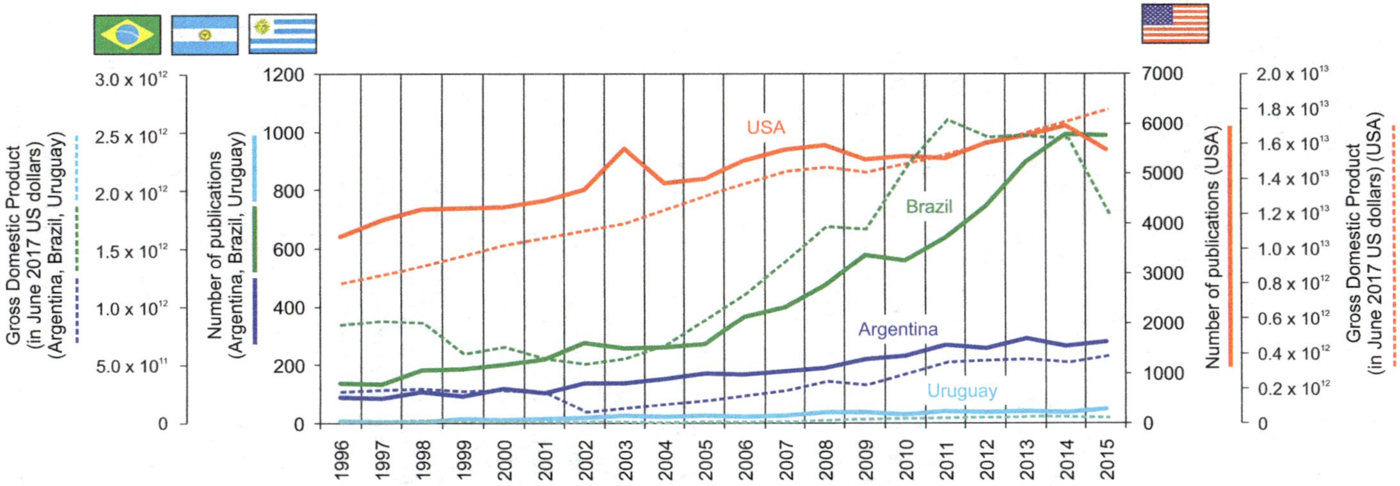

Fig. 3 Numbers of publications on aquatic sciences produced by Argentine, Brazilian, Uruguayan, and US authors in the period 1996–2015 (solid lines) and gross domestic product (in June 2017 US dollars) (broken lines). Notice different scales. Sources: SCIMAGO, Agricultural and Biological Sciences-Aquatic Sciences, and World Bank

Table 3 Approximate numbers of described marine planktonic species in the World Ocean and those recorded in the South Atlantic (0°–60°S, from the coasts of South America to the coasts of Africa). All data refer to morphospecies (rather than genetically distinct species). Excluded are a few planktonic groups (most heterotrophic flagellates, Rotifera, planktonic Nudibranchia, Cumacea), as well as the larval stages of benthic invertebrates (Porifera, Mollusca, Annelida, Crustacea, Phoronida, Bryozoa, Echinodermata, etc.). Data for the phytoplankton are from Sournia et al. (1991) and for zooplankton from Boltovskoy et al. (2003). ND, no data

	No. of species in the World Ocean	No. of species recorded in the South Atlantic	Degree of knowledge in the South Atlantic
Coccolithophorida	650?	70?	Poor
Bacyllariophyta	1600	ND	Fair
Dinoflagellata	1600	ND	Good
Foraminifera	49	39	Good
Acantharia	150	114	Very poor
Radiolaria Polycystina	350	160	Poor
Radiolaria Phaeodaria	350	158	Very poor
Ciliophora (naked ciliates)	150	58	Very poor
Tintinnina	300	151	Fair
Hydromedusae	650	185	Fair
Siphonophorae	190	98	Fair
Scyphozoa	150	38	Poor
Ctenophora	90	20	Poor
Nemertina	97	11	Very poor
Polychaeta	120	61	Fair
Heteropoda	35	26	Fair
Pteropoda	160	91	Fair
Cephalopoda	370	103	Poor
Cladocera	8	8	Fair
Ostracoda	170	120	Good
Copepoda	2000	505	Poor
Mysidacea	700	96	Poor
Amphipoda	400	188	Poor
Euphausiacea	86	61	Fair
Chaetognatha	80	37	Fair
Appendicularia	64	43	Good
Pyrosomatida	8	6	Fair
Doliolida	17	11	Fair
Salpida	45	29	Good
Total	9989	2417	

refer to traditionally defined and described morphospecies only, and therefore may underestimate real values significantly (Mora et al. 2011; Appeltans et al. 2012). Further, they do not include the cryptic species distinguishable by means of molecular tools only, which have been shown to increase the inventories of some taxa

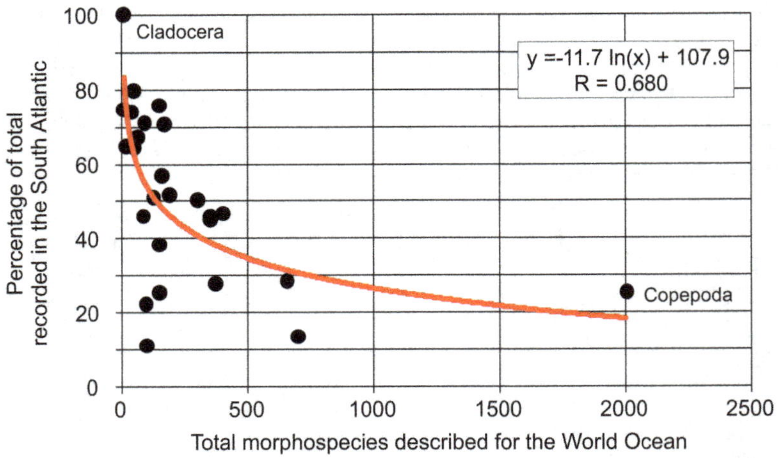

Fig. 4 Approximate numbers of described marine zooplanktonic species in the World Ocean and percentages of the totals recorded in the South Atlantic (0°–60°S, from the coasts of South America to the coasts of Africa). Each data point represents one higher-level taxon (see Table 3 for details)

several-fold (Ishitani et al. 2014; Kulagin et al. 2014; de Vargas et al. 2015; Santoferrara et al. 2015; Morard et al. 2016).

With these limitations in mind, some general trends can be derived from these data (Boltovskoy et al. 2003):

1. Some planktonic taxa are entirely or almost entirely unknown in this area; for most of them there are practically no specialists in South America. Examples are Coccolithophorida, Acantharia, Phaeodarea, Pseudothecosomata, Gymnosomata, Mysidacea, and Hyperiidea.
2. There is very limited knowledge on several groups, such as Polycystina and Doliolida.
3. Siphonophorae, Hydromedusae, Scyphomedusae, Ctenophora, Polychaeta, Ostracoda, Euphausiacea, Chaetognatha, Salpidae, and Appendicularia have been somewhat better but insufficiently studied.
4. A few taxa have been moderately well investigated over the last decades, in particular the Diatomaceae, Dinoflagellata, Foraminifera, and Tintinnina. Their general distribution patterns are fairly well known in coastal and shelf waters but much less so in the open oceanic realm.

As noticed above, the record of a given species in the area does not imply much more than simply the fact that it is present in the South Atlantic. It adds little to its known worldwide distribution range and even less to its potential importance for the dynamics of the ecosystems concerned. Nevertheless, as far as inventorial surveys are concerned, these figures clearly show that different taxa involve different degrees of effort for a reasonable coverage. Some groups comprise a few, easily recognizable species, while others are composed of several thousand different organisms. A few extensive investigations would suffice to provide a general knowledge on the distribution of the Heteropoda (about 35 species in all), while the Copepoda, with >2000 living species, would need an effort many times greater to accomplish the same end. This unbalance is clearly illustrated in Fig. 4, which shows that more speciose groups globally (e.g., Copepoda) are underrepresented in the South Atlantic when compared with the less diversified ones (e.g., Cladocera).

Also in geographic terms, the coverage of the Southwestern Atlantic plankton has been extremely uneven. Southern Ocean waters have been the object of intensive multinational investigations for many years. Both plant and animal plankton have been repeatedly sampled by American, Danish, Russian, Belgian, German, and Argentine expeditions. The limited number of species present in these waters has also contributed to the fact that the Antarctic is by far the best known area in the Southwestern Atlantic. The coastal and shelf waters along eastern South America have been repeatedly sampled by Argentine and Brazilian ships and are second to Antarctica in degree of coverage, with the exception of the equatorial and tropical Brazilian waters, where research efforts have been comparatively scarce, which accounts for the locally meager specific inventories for a region that most probably hosts very high diversity values.

The Central Atlantic is an almost virgin area from the point of view of its plankton. Very few large expeditions collected plankton there (an exception was the German Meteor program, some Soviet Union expeditions from the 1960s to 1980s, and a few US, German, and UK cruises that studied materials collected by ships on their way to and from the Southern Ocean) (e.g., Deevey 1974; Greze 1984; Aiken et al. 2000; Gibb et al. 2000; Wood-Walker et al. 2001). Local cruises (Argentinian, Brazilian, and South African) hardly ever sailed so far from their coasts. Further, the South Atlantic Central Gyre is an oligotrophic area, which therefore does not appeal to sampling programs seeking productive, exploitable waters.

Most of the publications dealing with South Atlantic plankton covered the systematics and qualitative distribution of the organisms. Very few provided quantitative data or aimed at the investigation of more complex problems than relationships between species composition, temperature, and salinity. There is virtually no information on the bathymetric distribution of the plankters, as most samples have been collected from the surface layer or with vertical tows without closing devices. Long-term, seasonal, and especially multiannual monitoring programs have hardly even been undertaken, which very significantly curtails our understanding of seasonal and decadal changes.

6 Trends, Contrasts, and Concluding Remarks

In the Southwestern Atlantic, biological oceanography in general and plankton studies in particular have transited a long and bumpy road. In the two centuries elapsed since the first modest and isolated attempts of the early 1800s at discovering and describing the marine biota off the Atlantic coasts of this subcontinent, the complexity of the issues tackled has grown substantially, generally following the international trend, but when compared with many other oceanic areas, in particular those around Europe, Japan, and North America, the results have been relatively meager.

Several circumstances have conspired against the advancement of scientific knowledge, including regional political instabilities, the excess of ambitious drafts and plans with no or little subsequent implementation due to lack of funding, follow-up actions and/or expertise, inadequate assignment of resources, the scarcity of coordinated efforts at maximizing resources, sampling platforms and expensive equipment, and limited interaction (Ogden et al. 2004; Boltovskoy 2015; Perillo 2015). Political problems have caused major brain drains, especially in Argentina in the 1960s and 1970s, and although a large proportion of these displaced experts moved to other South American countries (Brazil, Venezuela), and several eventually returned to Argentina, many bright scholars ended up settling in elsewhere (Scelzo et al. 2017). Although massive expatriations decreased after political unrest stabilized in the 1980s, local opportunities at pursuing an academic career were and still are marginal when compared with those offered by many of the leading countries in scientific output. Thus, a significant proportion of the brightest young specialists that undertake postdoctoral training programs in these countries end up settling there (Ciocca and Delgado 2017). In contrast to Argentina, Brazil has been more immune to this brain drain, and even during the harshest periods of military dictatorships, the exodus of scientists has been low. The majority of foreign-trained specialists did and presently do return to Brazil, and their input is largely responsible for the growth of local expertise and scientific output in the last decades (Fig. 3).

Support for science, in terms of personnel employed, salaries, and research grants, has been uneven and generally low. With the only exception of Costa Rica, in Latin America, Argentina, Brazil, and Uruguay (in that order) have the highest relative numbers of researchers (as a function of their populations) yet ~4–5 times lower than, for example, the USA and Japan (Ciocca and Delgado 2017). In Argentina, Brazil, and Uruguay, a major hindrance are research grants, which in the best cases are enough to support the day-to-day costs of routine lab and field work, reagents, and some travel, but are rarely sufficient for acquiring, and often even repairing, major scientific equipment. On average, research grants in Argentina are ~10 times lower than those in the USA (Boltovskoy 2015). This situation is aggravated by the fact that most of the supplies and equipment are not produced locally, involving complicated and slow bureaucratic procedures for their acquisition and very substantially increased costs from taxation and the intervention of local distributors (Boltovskoy 2015; Ciocca and Delgado 2017).

The language barrier has also impacted significantly the visibility of scientific work produced in Latin America. Few of the South American journals meet the standards required for inclusion in recognized, international databases, and those that do have low citation rates, especially for articles in Spanish or Portuguese. For example, in the SCOPUS-indexed Argentine ecology journal Ecología Austral, the citation rate of articles in Spanish (4.3) is significantly lower than that for articles in English (6.2) (Boltovskoy 2015).

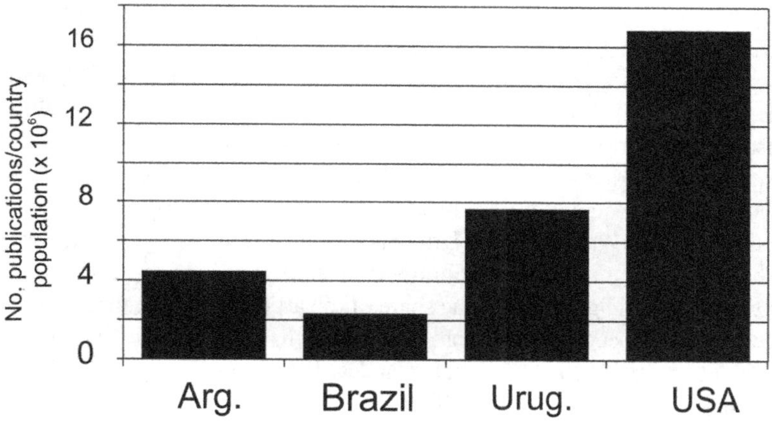

Fig. 5 Ratio number of publications/country population ($\times 10^6$) and means for 1996–2015. Source: SCIMAGO, Agricultural and Biological Sciences-Aquatic Sciences

Figure 3 shows the scientific output of Argentina, Brazil, Uruguay, and, for comparative purposes, the USA (the most prolific worldwide), for the period 1996–2015 in the area of biological aquatic sciences (including marine and freshwater). Clearly, all three Latin American countries lag far behind the USA, which accounts for around 24% of the overall world production in the area of aquatic sciences for this period. However, the growth has been quite uneven. The USA, with the highest initial (1996) values, shows the lowest increase (the mean for 2011–2015 in only 1.25 times higher than that for 1996–2000). Brazil has the highest increase (6), followed closely by Uruguay (4.6), whereas Argentina has a comparative lower figure (2.8). Accounting for the corresponding population growths does not change these figures significantly. Again, the lowest figure is that of the USA (1.2, for increase in publications between 2011 and 2015 with respect to 1996–2000, divided by the increase in population growth for the same time offsets), followed by Argentina (2.4), Brazil (4.1), and Uruguay (4.4).

These figures suggest several interesting trends: (1) In the last 20 years, the USA, with a scientific output ~15–300 times higher than those of the South American countries compared, has maintained its leading role, but the gaps dropped by 30–50%. (2) In Latin America, the production of Brazil has grown more vigorously than those of Argentina and Uruguay, both in absolute terms and when compared with the population growth of each country. (3) In all cases, scientific production parallels closely the gross domestic product of each country (Fig. 3). Interestingly, the strong growth of Brazilian publications, starting around 2005, is preceded by a very noticeably economic growth. (4) For the 1996–2015 time offset, scientific production per capita has been greatest for the USA, followed by Uruguay, Argentina, and Brazil (in that order) (Fig. 5). (5) Between 1996 and 2015, the joint production of Argentina, Brazil, and Uruguay grew from ~6% of that of the USA to ~24%. (6) The international visibility of the local scientific production, as indicated by the numbers of citations per document, remained stable and closely comparable with the world average. However, when compared with the USA (as well as other leading countries), between 1996 and 2015, it increased by ~20%.

Summarizing, despite limited financial support, often low-quality installations and deficient equipment, comparatively low salaries, and a turbulent political history usually with strong repercussions in academic media, the international standard of the scientific production of Argentina, Brazil, and Uruguay has improved over the last two decades. In quantitative terms, Brazil took off vigorously after around 2005, while the growth in Argentina and Uruguay has been slower (Fig. 3). This improvement has occurred roughly in parallel with overall population increase, and especially with economic growth, but the share of the articles by these three countries of the literature produced by those that historically have accounted for >50–60% of overall total (the USA, the UK, Japan, Canada, France, Germany, Australia, China) grew from ~3% in 1996 to 10% in 2015.

Progress has therefore been made, but the pace has been slow. The availability of satellite data and international databases has fostered many important local studies, but these sources of information cannot replace in situ sampling programs, which are scarce, poorly coordinated, and usually limited both in scope and geographically. Much of the cruise-based information collected so far has been ancillary to other activities (in particular, support to Antarctic stations), which has seriously compromised the scientific goals sought. With the increase in public awareness of ecological issues closely associated with sustainable human development, including global warming, species invasions, overexploitation of living resources, and marine pollution, interest in biological oceanography has grown both globally and locally. Hopefully, these threats will change our attitude toward the ocean around us and further foster research activity in South America.

References

Aiken J, Rees N, Hooker S et al (2000) The Atlantic meridional transect: overview and synthesis of data. Prog Oceanogr 45:257–312

Angelescu V, Sánchez RP (1995) A century of oceanographic and fisheries exploration on the continental shelf off Argentina. Helgoländer Meeresun 49:467–487

Angelescu V, Sánchez RP (1997) Exploraciones oceanográficas y pesqueras en el Mar Argentino y la región adyacente del Atlántico Sudoccidental (años 1874–1993). In: Boschi EE (ed) El Mar Argentino y sus recursos pesqueros, INIDEP, Mar del Plata (Argentina), p 11–64

Appeltans W, Ahyong ST, Anderson G et al (2012) The magnitude of global marine species diversity. Curr Biol 22:2189–2202

Balech E (1971) Notas historicas y criticas de la oceanografia biológica argentina. Publicación del Servicio de Hidrografía Naval (Argentina), H1027:1–57

Balech E (1976) Historia de la biología marina en la Argentina. In: Alvarez JA (ed) Evolución de las ciencias en la República Argentina 1923–1972. Sociedad Científica Argentina, Buenos Aires, pp 84–94

Bjornberg TKS (1954) Nota prévia sobre a ocorrência de tornarias na costa sul do Brasil. Arq Mus Par 10:345–348

Bjornberg TKS, Forneris L (1955) Resultados científicos do cruzeiro do "Baependi" e do "Vega" a Ilha da Trindade. Copelata I. Contribuição avulsa do Instituto Oceanográfico 1:1–68

Boltovskoy D (1979) Zooplankton of the South–Western Atlantic. S Afr J Sci 75:541–544
Boltovskoy D (ed) (1981) Atlas del zooplancton del Atlántico Sudoccidental y métodos de trabajo con el zooplancton marino. Instituto Nacional de Investigación y Desarrollo Pesquero (INIDEP), Mar del Plata (Argentina)
Boltovskoy D (1999a) Diversidad y biogeografía del zooplancton del Atlántico Sur. Anales de la Academia Nacional de Ciencias Exactas. Físicas y Naturales (Argentina) 51:111–136
Boltovskoy D (ed) (1999b) South Atlantic zooplankton. Backhuys Publishers, Leiden, pp 1–1705
Boltovskoy D (ed) (2008) Atlas de sensibilidad ambiental de la costa y el Mar Argentino. Secretaría de Ambiente y Desarrollo Sustentable, Buenos Aires
Boltovskoy D (2015) Capítulo 3. Biología de sistemas. In: Depetris P, Rossi R, Tirao J, Charreau E, Cignoli R, Mariscotti M, Vallés E (eds) Estado y perspectivas de las ciencias exactas, físicas y naturales en la Argentina. Academia Nacional de Ciencias (Córdoba) & Academia Nacional de Ciencias Exactas, Físicas y Naturales, Buenos Aires
Boltovskoy D, Correa N, Boltovskoy A (2003) Marine zooplanktonic diversity: a view from the South Atlantic. Oceanol Acta 25:271–278
Boltovskoy D, Correa N, Boltovskoy A (2005) Diversity and endemism in cold waters of the South Atlantic: contrasting patterns in the plankton and the benthos. Sci Mar 69:17–26
Bonecker ACT (ed) (2006) Atlas de zooplâncton da região central da Zona Econômica Exclusiva brasileira. Museu Nacional, Rio de Janeiro
Bonecker ACT, de Castro MS (eds) (2006) Atlas de larvas de peixes da região central da Zona Econômica Exclusiva brasileira. Museu Nacional, Rio de Janeiro
Brandini FP, Lopes RM, Gutseit KS, Spach HL, Sassi R (1997) Planctonologia na plataforma continental do Brasil: Diagnose e revisão bibliográfica. MMA–CIRM–FEMAR (Brazil)
Budiansky S (1984) Army pall over Argentine science. Nature 311:201–204
Calliari D, Defeo O, Cervetto G et al (2003) Marine life of Uruguay: critical update and priorities for future research. Gayana 67:341–370
Carreto JI, Montoya NG, Carignan MO et al (2016) Environmental and biological factors controlling the spring phytoplankton bloom at the Patagonian shelf–break front – degraded fucoxanthin pigments and the importance of microzooplankton grazing. Prog Oceanogr 146:1–21
Castro BM, Miranda LB (1998) Physical oceanography of the western Atlantic continental shelf located between 4°N and 34°S. In: Robinson R, Brink KH (eds) The sea. Wiley, pp 209–251
Ciocca DR, Delgado G (2017) The reality of scientific research in Latin America; an insider's perspective. Cell Stress Chaperones 6:1–6
Cunha AM, Fonseca O (1918) O microplâncton das costas meridionais do Brasil. Mem Inst O Cruz 10:99–103
de Vargas C, Audic S, Henry N et al (2015) Eukaryotic plankton diversity in the sunlit ocean. Science 348:1–11
Deevey GB (1974) Pelagic ostracods collected on Hudson 70 between the equator and 55°S in the Atlantic. Proc Biol Soc Wash 87:351–380
Ehrlich MD, Sánchez RP (1990) Lights and shadows in biological oceanography research in Argentina. A historical review. In: Lenz W, Deacon M (eds) Ocean sciences: their history and relation to man, Deutsche Hydrographische Zeitschrift. Bundesamt für Seeschiffahrt und Hydrographie, Hamburg, pp 481–498
Faria JG (1914) Um ensaio sobre o plâncton, seguido de observações sobre a ocorrência de plâncton monótono, causando mortandade entre peixes na Baía do Rio de Janeiro. Dissertação de Livre Docência, Universidade do Brasil
Faria JG, Cunha AM (1917) Estudos sobre o microplâncton da Baía do Rio de Janeiro e suas imediações. Mem Inst O Cruz 9:68–93
Faria JG, Cunha AM, Pinto C (1922) Estudos sobre os protozoários do mar. Mem Inst O Cruz 15:186–200
Fernandes LPC, Oliveira LL (eds) (2012) O Brasil e o mar no século XXI: Relatório aos tomadores de decisão do País. CEMBRA (Centro de Excelência para o Mar Brasileiro, Base de Hidrografia da Marinha), Niteroi

Franco BC, Muelbert JH, Mata MM (2005) O ictioplâncton da quebra de plataforma da Região Sudeste–Sul do Brasil e sua relação com as condições ambientais. Ministério do Meio Ambiente (Brazil)

Gaeta SA, Brandini FP (2006) Produção primária e fitoplâncton na região entre o Cabo de São Tomé (RJ) e o Chuí (RS). In: Rossi–Wongtschowski CLDB, Madureira LSP (eds) O ambiente oceanográfico da plataforma continental e do talude na região sudeste–sul do Brasil. EDUSP, São Paulo, pp 219–265

Gibb SW, Barlow RG, Cummings DG et al (2000) Surface phytoplankton pigment distributions in the Atlantic Ocean:an assessment of basin scale variability between 50° N and 50° S. Prog Oceanogr 45:339–368

Gomes EAT, Santos VS, Tenenbaum DR, Villac MC (2007) Protozooplankton characterization of two contrasting sites in a tropical coastal ecosystem (Guanabara Bay, RJ). Braz J Oceanogr. https://doi.org/10.1590/S1679-87592007000100004

Greze VN (ed) (1984) Bio–produktzionnaya sistema krupnomashtabnogo okeanicheskogo krugovorota [Bio–productive system of a large scale oceanic gyre]. Naukova Dumka, Kiev (USSR), pp 1–263 In Russian

Hazin FHV (2009) Biomassa fitoplantônica, zooplanctônica, macro–zooplancton, Avaliação espacial e temporal do iclioplancton, estrutura da comunidade de larvas e de peixes e distribuição e abundância do ictioneuston. Programa Revizee – Score NE. Martins & Cordeiro, Fortaleza

Ishitani Y, Ujiié Y, Takishita K (2014) Uncovering sibling species in Radiolaria: evidence for ecological partitioning in a marine planktonic protist. Mol Phylogenet Evol 78:215–222

Katsuragawa M, Muelbert JH, Dias JF (2006) O ictioplâncton na Região entre o Cabo de São Tomé (RJ) e o Chuí (RS). In: Rossi–Wongtschowski CLDB, Madureira LSP (eds) O ambiente oceanográfico da plataforma continental e do talude na região sudeste–sul do Brasil. EDUSP, São Paulo, pp 359–446

Krug LC (ed) (2012) VIII Plano Setorial para os Recursos do Mar –PSRM. Formação de Recursos Humanos em Ciências do Mar: Estado da arte e Plano Nacional de Trabalho 2012–2015. Editora Textos, Pelotas

Kulagin DN, Stupnikova AN, al NTV (2014) Spatial genetic heterogeneity of the cosmopolitan chaetognath Eukrohnia hamata (Möbius, 1875) revealed by mitochondrial DNA. Hydrobiologia 721:197–207

Lalli CM, Parsons TR (2007) Biological oceanography: an introduction. Elsevier Butterworth–Heinemann, Amsterdam

Lopes RM (2007) Marine zooplankton studies in Brazil – a brief evaluation and perspectives. An Acad Bras Cienc. https://doi.org/10.1590/S0001-37652007000300002

Lopes RM, Castro BM (2013) Oceanography, ecology and management of Abrolhos Bank. Cont Shelf Res 70:1–2

Lopes RM, Katsuragawa M, Dias JF et al (2006a) Zooplankton and ichthyoplankton distribution on the southern Brazilian shelf: an overview. Sci Mar 70:189–202

Lopes RM, Montú MA, Gorri C et al (2006b) O zooplâncton marinho entre o Cabo de São Tomé (RJ) e o Chuí (RS). In: Rossi–Wongtschowski CLDB, Madureira LSP (eds) O ambiente oceanográfico da plataforma continental e do talude na região sudeste–sul do Brasil. EDUSP, São Paulo, pp 265–358

López R (1976) La biología marina. In: Alvarez JA (ed) Evolución de las ciencias en la República Argentina 1923–1972. Sociedad Científica Argentina, Buenos Aires

López Ambrosioni N (1976) La oceanografia en la Argentina. In: Alvarez JA (ed) Evolución de las ciencias en la República Argentina 1923–1972. Sociedad Científica Argentina, Buenos Aires

Mesquita HSL, Fernandes AJ (1996) Variação de curta escala temporal de bactérias, picofitoplâncton e nanoheterótrofos na região de Ubatuba –SP, Brasil. Rev Bras Oceanogr 44:47–56

Miloslavich P, Klein E, Diaz JM et al (2011) Marine biodiversity in the Atlantic and Pacific coasts of South America: knowledge and gaps. PLoS One 6:e14631

Ministério de Meio Ambiente (Brazil) (2006) Programa REVIZEE: avaliação do potencial sustentável de recursos vivos na zona econômica exclusiva. Relatório executivo. In: Ministério de Meio Ambiente. Secretaria da Qualidade Ambiental, Brasília

Mora C, Tittensor DP, Adl S (2011) How many species are there on earth and in the ocean? PLoS Biol 9:e1001127

Morard R, Escarguel G, Weiner AKM et al (2016) Nomenclature for the nameless: a proposal for an integrative molecular taxonomy of cryptic diversity exemplified by planktonic foraminifera. Syst Biol 65:925–940

Moreira da Silva PC (1971) Upwelling and its biological effects in southern Brazil. In: Costlow JD (ed) Fertility of the sea. Gordon & Breach, New York, pp 469–474

Moreira da Silva PC (1973) A ressurgencia em Cabo Frio, vol 78. Publicação do Instituto de Pesquisa da Marinha, Rio de Janeiro, pp 1–56

Müller–Melchers FC (1953) New and little known diatoms from Uruguay and the South Atlantic coast. Comunicaciones Botanicas del Museo de Historia Natural de Montevideo, 30:1–11

Müller–Melchers FC (1957) Plankton diatoms of the Toko–Maru voyage (Brazil coast). Bol Inst Oceanogr 8:111–136

Odebrecht C, Abreu PC (1995) Raphidophycean in southern Brazil. Harm Alg News 12:4

Ogden JC, Podestá G, Zingone A et al (2004) Las ciencias del mar en la Argentina. Ciencia Hoy 13:211–228

Oliveira LPH (1946) Estudos sobre o microplâncton capturado durante a viagem do navio oceanográfico Lahmeyer nas baías de Ilha Grande e Sepetiba. Mem Inst Oswaldo Cruz 44:441–488

Oliveira LPH (1947) Distribuição geográfica da fauna e flora da Baía de Guanabara. Mem Inst Oswaldo Cruz 45:709–734

Paiva Carvalho J (1945) Copépodos de Caiobá e Baia de Guaratuba. Arq Mus Par 4:83–116

Paiva Carvalho J (1952) Sobre uma coleção de copepodos, não parasíticos, da Baía de Santos e suas adjacências. Bol Inst Ocean 8:111–136

Penchaszadeh PE (ed) (2012) El Museo Argentino de Ciencias Naturales. Doscientos años. Museo Argentino de Ciencias Naturales, Buenos Aires (Argentina)

Perillo GME (2015) Capítulo 8. Oceanografía. In: Depetris P, Rossi R, Tirao J, Charreau E, Cignoli R, Mariscotti M, Vallés E (eds) Estado y perspectivas de las ciencias exactas, físicas y naturales en la Argentina. Academia Nacional de Ciencias (Córdoba) & Academia Nacional de Ciencias Exactas, Físicas y Naturales, Buenos Aires, pp 307–348

Ringuelet RA (1984) Breve reseña histórica de la ictiología argentina y uruguaya. In: Menni RC, Ringuelet RA, Aramburu RH (eds) Peces marinos de la Argentina y Uruguay. Hemisferio Sur, Buenos Aires, pp 3–8

Rossi–Wongtschowski CLDB, Madureira LSP (eds) (2006) O ambiente oceanográfico da plataforma continental e do talude na região sudeste–sul do Brasil. Editora da Universidade de São Paulo, São Paulo

Sánchez RP (1991) Reseña de las investigaciones sobre ictioplancton marino en Argentina y Uruguay. Atlantica 13:215–231

Santoferrara L, Alder V (2009) Abundance trends and ecology of planktonic ciliates of the South–Western Atlantic (35–63 S): a comparison between neritic and oceanic environments. J Plankton Res 31:837–851

Santoferrara LF, Tian M, Alder VA et al (2015) Discrimination of closely related species in tintinnid ciliates: new insights on crypticity and polymorphism in the genus *Helicostomella*. Protist 166:78–92

Santos VS, Villac MC, Tenenbaum DR et al (2007) Auto– and heterotrophic nanoplankton and filamentous bacteria of Guanabara Bay (RJ, Brazil): estimates of cell/filament numbers versus carbon contents. Braz J Oceanogr 55:133–143

Scelzo MA, Penchaszadeh PE, Castello JP (2017) El Instituto de Biología Marina de Mar del Plata, Argentina (1960–1977). Aportes a su historia. ProBiota (Facultad de Ciencias Naturales y Museo, Universidad Nacional de La Plata), La Plata (Argentina):1–52

Seeliger U, Kjerfve B (eds) (2001) Coastal marine ecosystems of Latin America. Springer, Berlin

Seeliger U, Odebrecht C, Castello JP (eds) (1997) Subtropical convergence environments. The coast and sea in the Southwestern Atlantic. Springer, Berlin
Seguin G (1965) Contribution à la connaissance du plancton des eaux cotières du Brésil (copépodes et amphipodes exceptés) et comparaison avec celui du Sénégal (Campagne de la Calypso, jan–fev 1961). Pelagos (Bulletin de l'Institut Océanographique, Alger) 2:5–44
Servicio de Hidrografía Naval (1976) La Oceanografía en la República Argentina (1952–1975). Armada Argentina. Servicio de Hidrografía Naval, Buenos Aires, pp 5–395
Sournia A, MJ C–D, Ricard M (1991) Marine phytoplankton: how many species in the world ocean? J Plankton Res 13:1093–1099
Susini–Ribeiro SMM (1999) Biomass distribution of pico–, nano– and microplankton on the continental shelf of Abrolhos, East Brazil. ArchFish and MarRes 47:271–284
Tabarelli M, Rocha CFD, Romanowski HP et al (eds) (2013) PELD–CNPq, Dez anos do Programa de pesquisas ecológicas de longa duração do Brasil: achados, lições e perspectivas. Editora Universitária da Universidade Federal de Pernambuco, Recife
Teixeira C, Gaeta S (1991) Contribution of picoplankton to primary production in estuarine, coastal and equatorial waters of Brazil. Hydrobiologia 209:117–122
Teixeira C, Kutner MB (1961) Contribuição para conhecimento das diatomáceas da região de Cananeia. Bol Inst Ocean 3:131–187
Tenenbaum DR (2006) Os Dinoflagelados e os Tintinídeos da região central da Zona Econômica Exclusiva brasileira: guia de identificação. Museu Nacional:1–36
Valentin JL (1984) Spatial structure of the zooplankton community in the Cabo Frio region (Brazil) influenced by coastal upwelling. Hydrobiologia 113:183–191
Valentin JL (2001) The Cabo Frio upwelling system, Brazil. In: Seeliger U, Kjerfve BJN (eds) Ecological studies: coastal marine ecosystems of Latin America. Springer, Berlin, pp 97–105
Valentin JL (ed) (2006) Características hidrobiológicas da região central da Zona Econômica Exclusiva brasileira (Salvador, BA, ao Cabo São Tomé, RJ). IDEAL Gráfica e Editora, Brasilia
Valentin JL, Coutinho R (1990) Modelling maximum chlorophyll in the Cabo Frio (Brazil) upwelling: a preliminary approach. Ecol Model 52:1–11
Valentin JL, Lins DA, Silva NM et al (1985) Les diatomées dans l'upwelling de Cabo Frio (Brésil): liste d'especes et étude écologique. J Plankton Res 7:313–337
Valentin JL, WM M–R, Mureb MA (1987) Sur quelques zooplanctontes abondants dans l'upwelling de Cabo Frio (Brésil). J Plankton Res 9:1195–1216
Valentin JL, Gaeta HL, Spach MA et al (1994) Oceanografia biológica – Plâncton. Diagnostico ambiental oceânico e costeiro das regiões sul e sudeste do Brasil., Rio de Janeiro. PETROBRAS. Rio de Janeiro
Vannucci M (1949) Hydrozoa do Brasil. Boletim Zoologico da Universidade de São Paulo 99:219–265
Vannucci M (1951a) Distribuição dos Hydrozoa até agora conhecidos nas costas do Brasil. Bol Inst Paul Ocean 2:105–124
Vannucci M (1951b) Resultados científicos do cruzeiro do "Baependi" e do "Vega" a Ilha da Trindade. O gênero *Firoloida*, Prosobranchia Heteropoda. Bol Inst Paul Ocean 3:5–30
Vannucci M, Hosoe K (1952) Resultados científicos do cruzeiro do "Baependi" e do "Vega" a Ilha da Trindade. Chaetognatha Bol Inst Paul Ocean 2:73–93
Wood–Walker RS, Kingston KS, Gallienne CP (2001) Using neural networks to predict surface zooplankton biomass along a 50° N to 50° S transect of the Atlantic. J Plankton Res 23:875–888
Wüst G (1964) The major deep–sea expeditions and research vessels 1873–1960. Progr Oceanogr 2:1–52
Zimmermann SJC (1918) Contribuição ao estudo das diatomáceas dos Estados Unidos do Brasil. VIII Contribuição Brotéria 16:113–122

Part II
Regional Settings of Productivity and Nutrient Cycling

Physical Oceanography of the SW Atlantic Shelf: A Review

Alberto R. Piola, Elbio D. Palma, Alejandro A. Bianchi, Belmiro M. Castro, Marcelo Dottori, Raul A. Guerrero, Marina Marrari, Ricardo P. Matano, Osmar O. Möller Jr, and Martín Saraceno

Abstract The continental shelf of the western South Atlantic is characterized by three regions subject to distinct oceanographic regimes. The wide subantarctic shelf, south of approximately 35°S, is occupied by cold, low-salinity waters derived from the Subantarctic Zone and further diluted by the inflow of additional low-salinity waters, primarily from the Magellan Strait. Farther north, the shelf narrows considerably and is subject to the influence of large freshwater discharges and warm-salty intrusions of subtropical waters from the Brazil Current. Intense frontal transitions at various near shore locations and along the shelf break promote vertical circulations that inject nutrients into the upper layer. This nutrient injection leads to

A. R. Piola (✉) · A. A. Bianchi
Departamento Oceanografía, Servicio de Hidrografía Naval (SHN), Ciudad Autónoma de Buenos Aires, Argentina

Departamento de Ciencias de la Atmósfera y los Océanos, Universidad de Buenos Aires, Ciudad Autónoma de Buenos Aires, Argentina
e-mail: apiola@hidro.gov.ar; abianchi@hidro.gov.ar

E. D. Palma
Instituto Argentino de Oceanografía (IADO-CONICET), Consejo Nacional de Investigaciones Científicas y Técnicas and Depatamento de Física, Universidad Nacional del Sur, Bahía Blanca, Argentina
e-mail: uspalma@criba.edu.ar

B. M. Castro · M. Dottori
Instituto Oceanográfico, Universidade de São Paulo, São Paulo, Brazil
e-mail: bmcastro@usp.br; mdottori@usp.br

R. A. Guerrero
Instituto Nacional de Investigación y Desarrollo Pesquero (INIDEP), Mar del Plata, Argentina
e-mail: raul.guerrero@inidep.edu.ar

M. Marrari
Departamento Oceanografía, Servicio de Hidrografía Naval (SHN), Ciudad Autónoma de Buenos Aires, Argentina
e-mail: mmarrari@hidro.gov.ar

M. S. Hoffmeyer et al. (eds.), *Plankton Ecology of the Southwestern Atlantic*,
https://doi.org/10.1007/978-3-319-77869-3_2

enhanced growth of phytoplankton, and, in some regions, to a significant uptake of atmospheric CO_2. While the subantarctic shelf is under the influence of strong westerlies and high-amplitude tides, most of the subtropical shelf undergoes seasonally reversing winds and a micro-tidal regime. The shelf characteristics are also influenced by the offshore circulation, which is dominated by the equatorward flow of cold, nutrient-rich waters of the Malvinas Current in the south and the poleward flow of warm, salty, and oligotrophic waters of the Brazil Current in the north. There is a convergent large-scale mean circulation toward the transition between subantarctic and subtropical shelf waters near 34°S, which is balanced by export of shelf waters to the deep ocean. This article describes the contrasting water masses, frontal features, and circulation patterns of this region.

Keywords Ocean circulation · Water masses · Ocean fronts · Western South Atlantic shelf

1 Introduction

The portion of the continental shelf of eastern South America between Cape Frío (23°S) and Tierra del Fuego (55°S) is characterized by distinct morphological, climatic, and oceanographic features. This 4300-km-long shelf region encompasses three distinct subregions: the Patagonia continental shelf (PS) (38°S–55°S) to the south; the central shelf (CS), which includes portions of the southern Brazil, Uruguay, and northern Argentina shelves and extends from Cape Santa Marta Grande to Cape Corrientes (28°S–38°S); and the South Brazil Bight (SBB) to the north (23°S–28°S). South of 38°S, the continental shelf is a wide plateau subject to intense westerly winds and high tidal variability. The PS is bounded offshore by the

R. P. Matano
College of Earth, Ocean, and Atmospheric Sciences, Oregon State University, Corvallis, OR, USA
e-mail: rmatano@coas.oregonstate.edu

O. O. Möller Jr
Instituto de Oceanografia, Universidade Federal do Rio Grande, Rio Grande, Brazil
e-mail: dfsomj@furg.br

M. Saraceno
Departamento de Ciencias de la Atmósfera y los Océanos, Universidad de Buenos Aires, Ciudad Autónoma de Buenos Aires, Argentina

Centro de Investigaciones del Mar y la Atmósfera, Consejo Nacional de Investigaciones Científicas y Técnicas–Universidad de Buenos Aires, Ciudad Autónoma de Buenos Aires, Argentina
e-mail: saraceno@cima.fcen.uba.ar

cold, nutrient-rich and relatively fresh waters from the Antarctic Circumpolar Current along the southern boundary, which are advected northward by the Malvinas Current (MC) to about 38°S along the western boundary of the Argentine Basin. Farther north, there is the CS, which is 1350 km long and significantly narrower than the other two shelf regions, reaching about 80 km off Mostardas, near 31°16'S. The CS is strongly influenced by the continental discharge of the Río de la Plata (RP) and the Patos/Mirim Lagoon, and by its proximity to the Brazil Current (BC) and the Brazil/Malvinas Confluence, which bound its offshore limit and promote energetic exchanges between shelf and deep ocean waters. The SBB is a 1000-km-long crescent-shaped bay approximately 200 km wide at the center and 70–80 km at its northern and southern boundaries. This region is bounded offshore by the southward-flowing BC, the western boundary current of the Subtropical Gyre, which exerts a significant influence on the shelf circulation, biology, and biogeochemistry. This chapter describes the water mass, chlorophyll *a* and CO_2 flux distributions, and circulation over the continental shelf of these three shelf regions.

2 Data and Methods

For this review we have analyzed historical hydrographic data from holdings at INIDEP, Argentina (http://www.inidep.edu.ar/oceanografia/PERFILES/); CEADO, Argentina (http://www.hidro.gov.ar/ceado/ceado.asp); NOAA's National Centers for Environmental Prediction World Ocean Data (https://www.nodc.noaa.gov/OC5/WOD/pr_wod.html); and Argo profiling float data (http://www.argo.ucsd.edu/). These data were combined to prepare the updated sea surface salinity distribution presented in Fig. 1a. The hydrographic data are also used to prepare seasonal vertical temperature differences in the upper layer. In addition, we analyzed satellite-derived sea surface temperature (SST) and chlorophyll concentration data. We used NOAA's Optimum Interpolated SST version 2.0 with daily 0.25 × 0.25° resolution (Reynolds et al. 2007, available at https://www.ncdc.noaa.gov/oisst) to estimate sea surface temperature gradients presented in Fig. 1b. Surface chlorophyll *a* concentration data are derived from MODIS Aqua collected during 2002–2017. All level 2 data were processed to construct the austral summer climatology of 2 km × 2 km resolution presented in Fig. 1c using the OC3 algorithm (O'Reilly et al. 2000). The data were made available by NASA's Ocean Biology Processing Group at the Goddard Space Flight Center. Only chlorophyll *a* data with concentrations within the 0.02–30 mg m^{-3} range were used in the present analyses. For further details the reader is referred to Marrari et al. (2017). The analysis of surface ocean partial pressure is carried out based on data extracted from the Surface Ocean CO_2 Atlas version 4 (SOCAT, https://www.socat.info/). These data include 130,000 observations collected from 1991 to 2015.

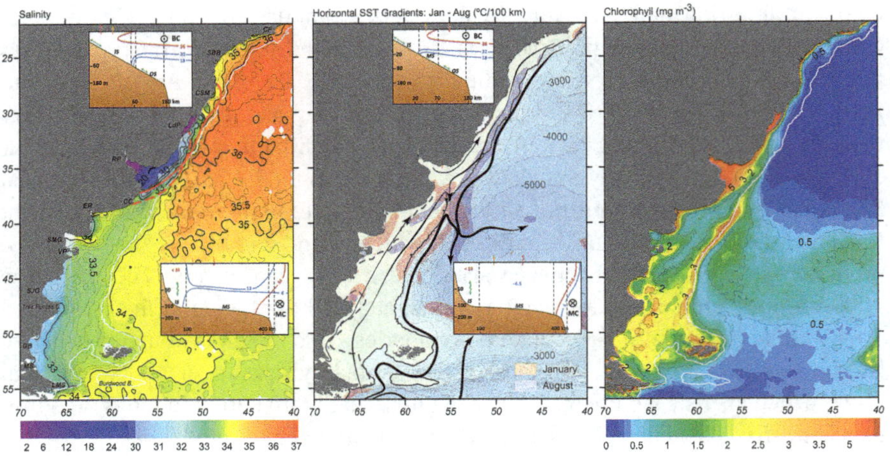

Fig. 1 Climatological sea surface salinity distribution (**a**). The red solid (dashed) contour indicates the austral winter (summer) distribution of the 33.5 isohaline, which marks the offshore edge of Plata Plume Water. CF Cape Frio, SBB South Brazil Bight, CSM Cape Santa Marta Grande, LdP Lagoa dos Patos, RP Río de la Plata, CC Cape Corrientes, ER El Rincón. SMG San Matías Gulf, VP Valdes Peninsula, SJG San Jorge Gulf, GB Grande Bay, MS Magellan Strait, LMS Le Maire Strait. The heavy white line indicates the 200 m isobath. (**b**) Mean summer (orange) and winter (blue) distributions of high sea surface temperature gradients (> 1.5 °C/100 km). The background contours show the bottom topography based on Gebco 30-second-gridded bathymetry. The heavy gray line indicates the 200 m isobath. The insets in a and b display schematic representations of the cross-shelf thermohaline structure in the SBB and the PS. The insets display selected isotherms (light blue) and isohalines (red), surface heat (red wiggly arrow), and freshwater exchanges (orange arrow). The green wiggly arrows indicate vertical mixing. In addition, a schematic surface circulation is shown, adapted from Piola and Matano (2017). (**c**) Mean summer (21 December–21 March) MODIS surface chlorophyll *a* in mg m^{-3} distribution (2002–2017). The heavy white line indicates the 200 m isobath

3 Water Masses and Shelf Fronts

3.1 Patagonian Shelf

In what follows we will use surface salinity, which in this region is not subject to strong seasonal fluctuations, as an identifier of water mass characteristics (e.g., Bianchi et al. 1982; Guerrero and Piola 1997; Piola et al. 2010). Figure 1a presents the climatological near-surface (0–5 meter) salinity distribution based on the analysis of 46,843 observations consisting of bottle, CTD and Argo profiles, hereafter referred to as sea surface salinity (SSS).

The southern portion of the PS is occupied by a modified type of subantarctic waters (SSS < 33.9) that enters the shelf through its southern boundary (e.g., Brandhorst and Castello 1971; Guerrero and Piola 1997). Since the PS is characterized by an excess of evaporation (~2.5 mm day^{-1}, ERA Interim, https://www.ecmwf.int/en/research/climate-reanalysis/era-interim) and only small river runoff, these low-salinity waters reflect the influence of the fresher waters advected from the coastal regions of southern Chile, which are dominated by an excess in precipitation and continental runoff. These low-salinity waters enter the PS via the Magellan Strait (MS, SSS < 32) and the Le Maire Strait (LMS, SSS < 33) (Fig. 1). Slightly saltier waters (SSS < 33.9) enter the PS east of Estados Island. These waters, which are derived from the Cape Horn Current, are diluted along the southern coast of Chile (e.g., Dávila et al. 2002). The low salinity that emanates from the MS and LMS extends northeastward along the coast of Grande Bay, separates from the coast near Cape Tres Puntas (~47°S), and extends to the mid-shelf region at 40°S. This buoyancy plume, which Brandhorst and Castello (1971) called the Patagonian Current, merges with the low-salinity waters derived from the RP near 39°S (Fig. 1a). The Santa Cruz (790 $m^3 s^{-1}$), Negro, and Colorado Rivers (1160 $m^3 s^{-1}$) provide additional freshwater injections that form low-salinity plumes of local significance at 50°S and 40°S (Fig. 1a). Bianchi et al. (1982) defined the coastal waters as those with a SSS < 33.4, shelf waters as 33.4 < SSS < 33.8, and high-salinity shelf water due to local excess evaporation and extended residence time within San Matías Gulf, where SSS > 33.8 (Scasso and Piola 1988; Tonini et al. 2013). Although our SSS analysis is based on a substantially larger number of observations than previously reported (nearly 50% of the data in Fig. 1a were collected after the year 2000), the SSS distribution over the shelf is qualitatively similar to the one described by Bianchi et al. (2005), thus suggesting that the distribution is relatively robust. The most notable differences between our SSS distribution and earlier analyses are subtle: a farther northward extension of the Magellan plume (SSS < 33.6), which merges with the RP plume, and the high-salinity shelf water from San Matías Gulf, which occupies a wider portion of the inner shelf off El Rincón (Fig. 1a). The salty waters derived from the San Matías Gulf extend northeastward during fall–winter and are located east of the gulf's mouth in spring–summer (Lucas et al. 2005). As will be discussed later, the seasonal wind variability is a dominant forcing of the circulation over the shelf north of about 39°S.

The PS hosts a variety of ocean fronts: upwelling, tidal, shelf break, and cold and temperate estuarine fronts, all of which play a significant ecological role (Acha et al. 2004; Sabatini et al. 2004). The most prominent of these fronts is the shelf break front, which marks the limit between the relatively warm and fresh waters on the shelf from the colder and saltier MC waters. From austral spring to autumn, the shelf break front is readily detected by satellite infrared observations (e.g., Martos and Piccolo 1988; Saraceno et al. 2004; Franco et al. 2008; Rivas and Pisoni 2010). During these seasons, the thermal structure of the shelf break front reveals a transition from the stratified shelf waters to the less stratified waters farther offshore (e.g., Romero et al. 2006). The regional extent of the weakening thermal manifestation of the shelf break front in winter is due to weak thermal stratification over the shelf (<0. 5 °C) compared to the summer stratification (>6 °C, Fig. 2).

Near coastal fronts develop primarily in the summer (Rivas and Pisoni 2010, Fig. 1b). These fronts mark the transition between well-mixed coastal waters and stratified mid-shelf waters (Glorioso 1987, Fig. 2). Tidal-induced mixing is one of the main mechanisms generating these fronts, which are characterized by their sharp SST signature (Palma et al. 2004). The most notable of these fronts are located

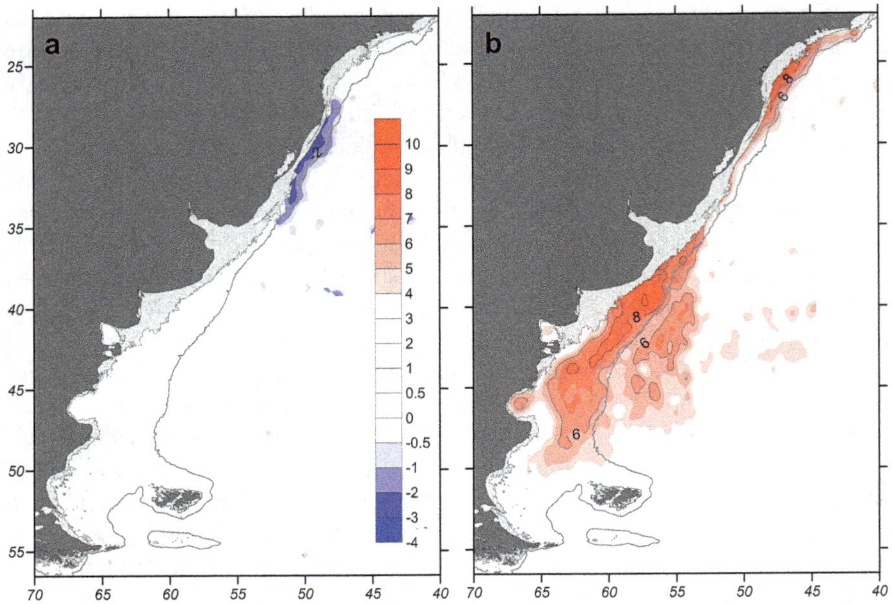

Fig. 2 Temperature difference (°C) between upper layer (10 m) and lower layer (65 m) from historical data collected during (**a**) austral winter (July–September) and (**b**) summer (December–March). Note that only positive temperature differences (T at 10 m > T at 65 m) larger than 4 °C are colored.

at ~47–50°S and along the mouth of San Matías Gulf (~42°S) (Palma et al. 2008; Moreira et al. 2011; Tonini et al. 2013). Tidal mixing is also intense along the coast south of 51°S, but the weak thermal stratification in the mid-shelf region (Fig. 2a and inset in Fig. 1a) leads to weaker thermal gradients. This region, which is characterized by the penetration of the Magellan Strait's buoyant plume and minor contributions from small rivers, is considered a cold estuarine front (Acha et al. 2004). The low-salinity waters, which extend approximately100 km offshore, are moderately warmer/colder than the mid-shelf waters during summer/winter (Rivas and Pisoni 2010). The location of tidal fronts is generally stable although frontal displacements are modulated by the intensity of surface heat flux, the fortnightly tidal cycle, and local mesoscale variability (Pisoni et al. 2015).

3.2 *Central Shelf (CS)*

The water mass structure of the CS is dominated by the presence of subantarctic shelf waters (SASW) with salinities between 33.6 and 33.8 (Piola et al. 2000). This region is narrower than the PS farther south: 180 km at 38°S, 110 km at 34.5°S, and just under 100 km at 31°S. The oceanic circulation over the CS is driven by seasonally reversing winds (e.g., Castro and Miranda 1998; Palma et al. 2008), which have a substantial imprint on the local water mass distribution on account of their influence on the RP plume (e.g., Piola et al. 2000; Simionato et al. 2001, 2010; Palma et al. 2008; Saraceno et al. 2014; Matano et al. 2014; Strub et al. 2015). Thus, the annual mean SSS distribution in the CS should be interpreted with caution. RP, the second largest river of the southern hemisphere, has an average volume discharge of 23,000 $m^3 s^{-1}$ during 1931–2016, although this value has increased to 25,000 $m^3 s^{-1}$during the past few decades (1997–2016) (Borús et al. 2017). The seasonal variations in RP discharge are small, but its interannual variations are significant due to the impact of El Niño on regional rainfall (Depetris et al. 1996; Robertson and Mechoso 1998). The strong Niños of 1983 and 1998, for example, generated discharges larger than 70,000 $m^3 s^{-1}$. These discharges had a significant impact on the SSS and on the biological and biogeochemical characteristics of the continental shelves of Uruguay and southern Brazil (Ciotti et al. 1995). During the austral fall and winter, the RP plume extends beyond the Cape Santa Marta Grande, near 28°S. In late spring and summer, the plume retracts southward, with its northern limit at approximately the mouth of the Patos Lagoon (32°S) and its southern limit near 38°S (Piola et al. 2000, 2005, 2008b; Möller Jr et al. 2008) (Fig. 1a). Events of strong southwesterly winds and large river outflow lead to exceptionally larger northeastward penetrations of the river plume (Piola et al. 2005). Such events exert a strong biological impact in the coastal areas (Stevenson et al. 1998; Proença et al. 2017). Nagai et al. (2014) show that the RP is the major source of terrigenous sediments for shelf areas between 38 and 27°S and only strong northeasterly winds can reverse this northward flow (Zavialov et al. 2002). Lateral mixing between RP waters and Tropical (TW) and South Atlantic Central waters (SACW) of the BC forms the subtropical shelf waters (Piola et al. 2000, 2008a). During the spring and summer seasons, northeasterly winds push the plume waters toward the south and offshore (Möller Jr et al. 2008; Guerrero et al. 2014; Matano et al. 2014), and most of the northern portion of the CS is occupied by subtropical shelf waters and tropical and central waters derived from the BC (see Souza and Robinson 2004). Under these summer conditions, upwelling events are frequently observed south of Cape Santa Marta Grande (Campos et al. 2013). Similarly, coastal bands of negative sea surface temperature anomalies observed along the southern coast of Uruguay are indicative of upwelling associated with strong easterly winds (e.g., Palma et al. 2008; Pimenta et al. 2008; Simionato et al. 2010). These events are also evident as bands of low coastal sea level in satellite-derived sea surface height anomalies and tide gauge data (Saraceno et al. 2014).

Satellite SST data show two distinct shelf break fronts in the region between 45°S and 25°S (Fig. 1b). South of 38°S the shelf break front marks the transition between the SASW and the cold subantarctic waters within the MC (see Sect. 3.1). In the northern region, the shelf break front marks the transition between the warm BC waters and the relatively colder shelf waters. The SST gradient between these regions is most intense during the austral winter (Fig. 1b; Saraceno et al. 2004), when there is a larger northward penetration of cold subantarctic waters flowing from the southern portion of the CS. The opposing flows reinforce the strong thermohaline gradients in the outer shelf (e.g., Möller Jr et al. 2008; Piola et al. 2008a; Matano et al. 2014). The strong cross-shelf temperature gradients can effectively modulate the atmospheric boundary layer and the surface wind field (Mendonça et al. 2017). In winter on the mid- and outer shelf, the cold subantarctic waters combined with the cold Plata plume waters frequently overlie warm-salty waters influenced by the Brazil Current, creating a sharp thermal inversion over the northern portion of CS (Fig. 2a), which has been referred to as the inverted thermocline (Castello and Möller 1977).

The second most remarkable of the shelf fronts in the southwestern Atlantic region—after the Patagonia shelf break front—is the Subtropical Shelf Front (STSF), which marks the subsurface limit between the SASW and the STSW (Piola et al. 2000, 2008a). The STSF extends from the inner shelf at around 32°S nearly to 36°S at the shelf break (see Fig. 10 in Piola et al. 2008a). In winter, the density structure is mostly controlled by the salinity distribution preventing isopycnal mixing across the STSF, while in summer the strong thermocline provides common isopycnal layers, allowing cross-front mixing (Piola et al. 2008a).

3.3 South Brazil Bight (SBB)

The southward-flowing BC sweeps the continental slope of the SBB, although only a small fraction of it intrudes on the outer shelf region (Silveira et al. 2000). This western boundary current transports TW ($T > 20$ °C; $S > 36$) in the mixed layer and SACW ($T < 20$ °C; $35 < S < 36$) in the pycnocline layer (Emilsson 1961). The low-salinity coastal water (CW; $S < 35$) mostly occupies the inner shelf (Castro and Miranda 1998). Castro (2014) identified three regions with distinct thermohaline properties in the cross-shelf direction of the SBB (see insets in Fig. 1a, b): outer shelf (OS), mid-shelf (MS), and inner shelf (IS). The IS is located onshore of a bottom temperature front, also referred to as bottom thermal boundary (BTB), which is defined by the bottom intersection of the 18 °C isotherm. The BTB separates the near-bottom intrusion of SACW from the inshore mixed waters. The IS width varies seasonally between 10–30 km in summer and 40–80 km in winter. The IS waters are usually weakly stratified compared to the MS, except near the medium estuarine systems present in the SBB, and are dominated by relatively low-salinity CW. The MS is located between the BTB and a sharp surface salinity front, also referred to as surface haline boundary (SHB). The SHB is marked by the surface outcrop of the 36 isohaline which determines the transition between TW and inshore mixed waters. Along most of the SBB, the SHB is found ~ 80 km from shore in summer and 130 km in winter.

The volume of each of the three SBB water masses changes seasonally (Cerda and Castro 2014). During autumn, the volume of SACW in the SBB is approximately equal to the volume of TW, but progressively increases, reaching a maximum in summer, when it is almost twice that of the TW. The CW is the product of mixing of TW with small amounts of continental discharges within the SBB. Its volume and physical characteristics change seasonally. In summer, the volume of CW is larger than in the other seasons, but in summer the CW is somewhat saltier

(S > 34.5) than in winter. The main seasonal change in characteristics of the CW is due to northward intrusions in winter of low-salinity waters originating south of the SBB, off the RP and the Patos and Mirim Lagoons (Campos et al. 1995; Stevenson et al. 1998; Lentini et al. 2001; Piola et al. 2000; Acha et al. 2004; Möller Jr et al. 2008; Campos et al. 2013). Interannual variability of the RP intrusions is associated with large outflow events during El Niño (Campos et al. 1999) and, most importantly, with anomalous southwesterlies over the continental shelf (Piola et al. 2005).

Onshore intrusions of nutrient-rich SACW play a significant role in fertilization of the SBB (Castro et al. 2006). There are two main intrusion paths: local, when the oceanic water reaches the MS directly from the shelf break, and remote, when SACW intrudes on the shelf at upstream regions, usually near Cape Frio, and then flows southward along the coast (Cerda and Castro 2014). Intermittent coastal upwelling events of SACW have been observed around Cape Frio, especially in spring and summer, when the region is under the influence of strong and persistent easterly and northeasterly winds (Miranda 1985; Franchito et al. 2008).

Other processes also contribute to the development of upwelling events. Rodrigues and Lorenzetti (2001) showed that the abrupt change in the coastline orientation near Cape Frio favors the effect of wind-induced upwelling. Cerda and Castro (2014) showed that the divergence of the 100 m and 150 m isobaths near Cape Frio favors SACW intrusions toward the coast. Several studies suggested that cyclonic meanders from the BC induce upward motions that inject SACW into the OS (Campos et al. 1995; Cirano and Campos 1996; Campos et al. 2000; Castelão et al. 2004). In addition, the wind stress curl contributes to force local SACW intrusions near Cape Frio (Amor 2004; Castelão and Barth 2006; Castelão 2012). Using numerical models, Palma and Matano (2009) and Matano et al. (2010) showed that the along-shelf pressure gradient near the shelf break, generated by interactions between the BC and the continental slope topography, forces geostrophic currents toward the coast in the bottom boundary layer, elevating the isopycnals toward the sea surface and favoring the onshore penetration of SACW intrusions.

4 Circulation

4.1 Patagonian Shelf (PS)

Numerical simulations suggest that the mean circulation over the continental shelf south of 40°S consists of a broad northeastward flow that intensifies toward the outer shelf (Palma et al. 2008; Combes and Matano 2014). Short-term current records at 43°S indicate mean velocities of 5.9 cm s^{-1} at 17 m and 3.1 cm s^{-1} at 67 m (Rivas 1997) that increase toward the shelf break, reaching values of 40 cm s^{-1} at 10 m depth and 26 cm s^{-1} at 160 m depth (Valla and Piola 2015). This velocity increase is associated with the proximity of the northward-flowing MC (Palma et al. 2008). In the inner shelf, the circulation is modulated by the coastal geometry, which is characterized by the presence of several bays and gulfs. Two of the largest bays are

Grande Bay and the San Jorge Gulf. Numerical models indicate that the circulation in these regions is dominated by relatively weak anticyclonic gyres and poleward coastal currents. South of 49°S, the down shelf development of the Magellan low-salinity plume is highly influenced by tides and wind forcing (Palma and Matano 2012). The circulation features produced by models are in good agreement with geostrophic flows derived from altimeter data at seasonal scales (Ruiz-Etcheverry et al. 2016). The southern shelf exports water northward and eastward in the surface layer that is largely compensated by inflow from the south at all depths and to a lesser extent by subsurface entrainment from the MC (e.g., Combes and Matano 2014). The onshore flow of MC waters is also apparent from the frequent observation of cold-salty subsurface intrusions in the outer shelf (insets in Fig. 1a, b).

Model simulations indicate a moderate strengthening of the northeastward flow during fall and a weakening during spring. North of 48°S, these seasonal variations are mainly driven by the wind forcing over the inner and mid-shelf, while further south, and offshore of the 100 m isobath, the flow is strongly modulated by MC transport variations (Palma et al. 2008). Direct current observations are too short to display a seasonal cycle, but the altimeter-derived currents also present a seasonal pattern. However, in contrast with model results, the intensification and weakening of the altimeter circulation shift to late summer and late winter, respectively (Ruiz-Etcheverry et al. 2016).

The remarkable phytoplankton blooms of the Patagonia shelf break are symptomatic of the upwelling of deep, nutrient-rich waters to the photic zone. It has been postulated and verified using simplified models that the upwelling might be associated with frictionally driven intrusions of the MC onto the shelf (Matano and Palma 2008; Miller et al. 2011). These results have been corroborated by realistic numerical simulations (Palma et al. 2008; Combes and Matano 2014) and observational studies (Valla and Piola 2015). It has also been proposed that the observed variability of the blooms can be modulated by continental shelf waves at seasonal time scales (Saraceno et al. 2005) and by wind variability at intra-seasonal periods (< 15 days, Carranza et al. 2017).

4.2 *Central Shelf (CS)*

The northward mean flow derived from models extends to about 38°S. This flow strengthens during fall and weakens during spring (Palma et al. 2008; Combes and Matano 2014). The relatively weak seasonal variations predicted by models are in fairly good agreement with those from altimeter observations (Strub et al. 2015; Ruiz-Etcheverry et al. 2016). In contrast, farther north, between 38°S and 28°S, models show a general southward flow on the outer shelf, which is largely controlled by the BC. The inner and mid-shelf regions exhibit significant seasonal changes in circulation and in temperature and salinity fields (Palma et al. 2008; Combes and Matano 2014). The circulation in this region is primarily driven by the alongshore component of the wind stress, which generates a positive sea level coastal anomaly; a geostrophically balanced northward-flowing coastal jet during fall and winter; and a depression of sea level with associated southward coastal flow during spring and summer. These seasonal coastal sea level anomalies are also confirmed by satellite altimeter and tide gauge data (Saraceno et al. 2014). The remarkable seasonal changes in surface salinity are associated with redistribution of the large freshwater discharges from the RP derived from in situ and satellite observations (Piola et al. 2000, 2008b; Möller Jr et al. 2008; Guerrero et al. 2014). During the winter months, northward, downwelling favorable winds generate northward currents and onshore Ekman transport (Meccia et al. 2013; Saraceno et al. 2014; Strub et al. 2015) that traps the RP plume near the coast (Palma et al. 2008). Though the northward plume penetrations in winter exhibit significant interannual variations, on average, the plume is positioned near Cape Santa Marta Grande (Fig. 1a). Relaxation of the winter winds allows some of the freshwater to leave the shelf in filaments that extend into the BC (Piola et al. 2008a; Matano et al. 2014), where the low-salinity signal is rapidly diluted by mixing with the salty upper layer carried by the BC. During summer, poleward winds create upwelling and poleward currents, causing freshwater to move south and offshore into the Brazil/Malvinas Confluence (Guerrero et al. 2014; Matano et al. 2014). The RP plume strengthens the inner and mid-shelf currents during fall and early winter. In summer, models show the development of a subsurface northward countercurrent in the mid-shelf, which flows against the prevailing winds (Palma et al. 2008). This countercurrent is presumably responsible for the northward extension of SASW to about 33°S in summer required to sustain the sharp thermohaline gradients across the STSF (Piola et al. 2008a). The summer geostrophic circulation derived from altimetry also shows a southward flow along the inner shelf and a weak northward flow along the outer shelf extending to about 34°S (Saraceno et al. 2014; Strub et al. 2015). Palma et al. (2008) attribute this northward flow to cross-shelf pressure gradients established farther south by the MC that spread northward well beyond the northernmost reach of the MC, as an arrested topographic wave (Csanady 1978).

Off the RP estuary, the CS is characterized by a persistent off-shelf mass flux. Model-derived volume balances of the shelf region between 34°S and 38°S show an annual mean off-shelf transport of ~1.21 Sv (Matano et al. 2014; see also Mendonça et al. 2017). Most of this transport is drawn from SASW from the Patagonian Shelf (1.15 Sv), with very small contributions from the RP discharge and the northern shelf region. The seasonal variations of the off-shelf transport are relatively small and out of phase with the variations of the southern mass flux; it decreases during the winter and increases during the summer (Matano et al. 2014). These results are in agreement with flow patterns inferred using satellite salinity (Guerrero et al. 2014) and the geostrophic velocities derived from altimetry (Strub et al. 2015).

4.3 South Brazil Bight (SBB)

The mean circulation in the middle and outer shelf north of Cape Santa Marta Grande is mostly southwestward and presents significant cross-shelf and seasonal variations in magnitude (Castro and Miranda 1998; Palma and Matano 2009).

Currents in the mid- and outer shelf are stronger in summer and weaker in winter (Matano et al. 2010). These seasonal variations are particularly strong in the inner shelf, where the circulation is driven by baroclinic effects and local winds. Numerical models suggest that the circulation is primarily driven by the poleward flow of the BC in the middle and outer shelf (Palma and Matano 2009), while wind forcing is also important in the inner and middle shelf (Stech and Lorenzzetti 1992). Observations also indicate that the wind stress is a major forcing in the middle shelf region (Dottori and Castro 2009; Dottori and Castro 2018). The influence of remote forcing is manifested in the SST gradients near the shelf break (Campos et al. 2000; Castelão et al. 2004) but is not confined to the surface or to the shelf break, as SACW intrusions and continental shelf waves are additional sources of variability (Castro and Lee 1995; Campos et al. 2010; Filippo et al. 2012; Dottori and Castro 2018). The strongest influence of the BC is observed at depth, where the bottom layer of the outer and middle shelf is permeated by the inflow of SACW throughout the year (Castro and Miranda 1998; Matano et al. 2010). In summer the SACW intrusions extend further onshore (Castro 2014).

The characteristics of the upwelling regime in the SBB are also confirmed by the distribution and variability of passive tracers included in model simulations (Combes and Matano 2014). A large amount of a tracer initially located in the offshore portion of the BC between 25 and 29°S stays offshore and moves south until it meets the MC and leaks offshore in the Brazil/Malvinas Confluence. The remaining part reappears next to the coast as far south as 32°S, transported by the combined action of bottom-layer shelf break upwelling and northeasterly winds (Palma and Matano 2009; Campos et al. 2013). The surface concentration increases at the end of winter when the surface wind stress shifts from downwelling to upwelling favorable and reaches its maximum during spring and summer. A tracer released over the shelf north of ~27°S stays over the shelf but extends just slightly past the 200 m isobath moving south to ~32°S, where it mixes with water from the RP and moves offshore into the Brazil/Malvinas Confluence (Combes and Matano 2014).

5 Surface Chlorophyll *a* Distribution

Satellite observations of chlorophyll *a* distributions reveal the high primary productivity levels of the southwestern South Atlantic shelf (Fig. 1c, see Longhurst 1998; Saraceno et al. 2005; Rivas et al. 2006; Romero et al. 2006; Signorini et al. 2006; Marrari et al. 2013). The largest concentrations are observed close to sediment laden continental discharges such as the RP and along the PS shelf break. Though the RP is a source of nutrients, which may lead to significant blooms in the outer estuary (Calliari et al. 2005; Carreto et al. 2008), global chlorophyll *a* algorithms, such as the one used in this study, frequently overestimate the chlorophyll *a* concentration in these optically complex waters (e.g., Garcia et al. 2005). Thus, the satellite-derived chlorophyll *a* distribution in these regions should be interpreted with caution. South of 38°S, the austral summer chlorophyll *a* distribution suggests a close association between productive regions and semipermanent frontal systems. In austral summer, the shelf break front emerges as an extended band of high concentrations (generally higher than 3 mg m^{-3}, Fig. 1c). In situ chlorophyll *a* estimates in the shelf break front indicate concentrations exceeding 15 mg m^{-3} and are in overall agreement with the high concentrations indicated by satellite observations (e.g., Garcia et al. 2008; Lutz et al. 2010; Carreto et al. 2016). On average, the shelf break bloom initiates in September and remains high (~ 3.5 mg m^{-3}) until March, though there is substantial interannual variability (Saraceno et al. 2005; Romero et al. 2006; Signorini et al. 2006; Piola et al. 2010). In contrast, the spring bloom in mid-shelf regions decays monotonically after it peaks in October–November (e.g., Carreto et al. 1995; Romero et al. 2006). Enhanced vertical circulation and mixing intense enough to overcome the strong summer stratification, such as might be expected close to ocean fronts, appear to be essential to maintain productivity after the spring bloom. Consequently, in summer regions of enhanced surface chlorophyll concentration (Fig. 1c) are closely associated with weakened vertical stratification (Fig. 2b).

The high chlorophyll *a* concentrations associated with the Valdes tidal front reach a spring-summer maximum of ~3.4 mg m^{-3}. This frontal bloom peaks in January and decays monotonically until May (Romero et al. 2006). Chlorophyll blooms are also observed along an extensive band offshore from the 100 m isobath at 85–150 km from shore between 46 and 52°S. This high chlorophyll *a* band is closely associated with the cold estuarine front (see Fig. 1b, c). The highest satellite chlorophyll *a* concentrations in the latter region can exceed 30 mg m^{-3}, and similarly high in situ estimates have been reported off Grande Bay near 51°S (Lutz et al. 2010; Gómez et al. 2011). This extensive mid-shelf region in southern Patagonia blooms in January, and the chlorophyll *a* concentration decays at about the same rate as the Valdes and shelf break fronts (Romero et al. 2006).

In the SBB, onshore intrusions of SACW combined with wind- and BC eddy-induced upwelling are the primary sources of nutrients leading to nearshore blooms (see Sect. 1.3). In addition to wind-induced upwelling (e.g., Campos et al. 2013), the northward intrusions of RP waters in winter appear to play a significant role in fertilizing the CS (e.g., Ciotti et al. 1995). In contrast to the PS, where shelf break upwelling and tidal and wind mixing are permanent sources of nutrients to the upper layer at specific locations, the nutrient sources in the CS and the SBB are modulated by synoptic atmospheric and oceanic processes. The impact of these episodic upwelling events is not readily apparent in long-term surface chlorophyll *a* climatologies, such as the one presented in Fig. 1c. Consequently, in the two northern regions, the satellite-derived chlorophyll *a* concentration is significantly lower than in the PS. Given the seasonality of alongshore component of wind stress in the CS, the wind-induced upwelling is more frequent in austral summer, when the region is under the influence of upwelling-favorable northeasterly winds (see Campos et al. 2013; Ito et al. 2016). The contrasting chlorophyll *a* distributions between the different shelf regions are displayed by the long-term annual mean concentrations in the SBB, CS, and PS, 1.37 ± 0.48, 3.58 ± 0.81, and 2.15 ± 0.39 mg m^{-3}, respectively. The high areal mean chlorophyll *a* estimated over the CS reflects the influence of high concentration associated with the RP plume, even after limiting the calculation to concentrations lower than 5 mg m^{-3}. Though in situ data support a high chlorophyll *a* associated with the RP plume (Garcia et al. 2006; Martinez et al. 2005), the mean value in the CS should be interpreted with caution.

6 Surface Ocean CO_2 Fluxes

Analysis of historical partial pressure of CO_2 (pCO_2) is based on data extracted from the SOCAT.v4 data set. Based on the partial pressure and wind data, and using the Wanninkhof (2014) flux parameterization, the sea-air CO_2 fluxes were estimated in the PS (see Kahl et al. 2017). Preliminary estimates in the PS and the MC regions indicate an oceanic annual averaged CO_2 uptake of -8.1×10^{-3} mol m^{-2} d^{-1} (the minus sign indicates flux from the atmosphere to the ocean). The flux is maximum during the austral spring, with an average of -18.6×10^{-3} mol m^{-2} d^{-1}. These fluxes are among the highest CO_2 uptake per unit area in the world ocean, for instance, doubling the CO_2 sink in the North Sea (Thomas et al. 2004). In particular, the PS shelf break front presents intense CO_2 fluxes into the ocean, with an annual mean of -15×10^{-3} mol m^{-2} d^{-1}. The PS shelf break also presents a large seasonal flux variability, reaching -22.3×10^{-3} mol m^{-2} d^{-1} during spring and -4.8×10^{-3} mol m^{-2} d^{-1} in winter. These new estimates of surface CO_2 fluxes in the PS are higher than previously reported (e.g., Bianchi et al. 2009; Padin et al. 2010; Kahl et al. 2017) due to the substantial increase in observations collected during the past decade. Kahl et al. (2017) suggest that biological processes dominate the CO_2 variability in the shelf break region and that the decreased CO_2 sink during winter is mainly due to the decline in biological activity combined with winter convection. The studies in the PS therefore concluded that vertical stratification plays a key role in CO_2 dynamics.

In contrast with the PS, the subtropical shelf off the Brazilian coast acts as a source of CO_2 to the atmosphere (Ito et al. 2005; Padin et al. 2010). Observations conducted in the SBB in November 1997 and January/February and June 1998 indicate net sea-air CO_2 fluxes between 0.3 and 9.8×10^{-3} mol m^{-2} day^{-1} which decrease to ~0.4 and 3.7×10^{-3} mol m^{-2} day^{-1} over the slope and neighboring deep ocean (Ito et al. 2005). Upwelling events observed primarily in summer and biological regenerative processes are thought to play a key role in the CO_2 dynamics in this region (Ito et al. 2005). Observations in the mid- and outer CS collected during 2000–2008 suggest that in austral spring, most of the region acts as a CO_2 sink, while during the fall, the region is characterized by the sharp transition from atmospheric CO_2 source along the coast of southern Brazil to CO_2 sink off the RP mouth (Padin et al. 2010). More recent observations collected in austral spring 2010 and early summer 2011 in the southern SBB and northern CS indicate that shelf waters are a source of CO_2 to the atmosphere (Ito et al. 2016). In late spring the surface CO_2 fluxes in the northern CS vary between 0.5 and 1×10^{-3} mol CO_2 m^{-2} day^{-1} and decrease offshore. The highest fluxes in this region exceed 3×10^{-3} mol CO_2 m^{-2} day^{-1} in the upwelling region off Cape Santa Marta Grande (Ito et al. 2016). The late summer observations in the northern SBB present much higher CO_2 flux estimates, exceeding 15×10^{-3} mol CO_2 m^{-2} day^{-1} (Ito et al. 2016).

Results from an eddy-resolving, regional ocean biogeochemical model are in good overall agreement with observations, confirming that south of 30° S the western South Atlantic shelf acts as a sink of atmospheric CO_2 (Arruda et al. 2015). The model indicates a weak source of CO_2 to the atmosphere in the SBB. Observed near shore to offshore and meridional pCO_2 gradients are well represented in the simulation. A sensitivity analysis shows that the biological production and solubility are the main processes regulating the model pCO_2, with biological production being particularly important over the continental shelves.

Acknowledgments This chapter summarizes research financed by the Inter-American Institute for Global Change Research (IAI) grant CRN3070 through the US National Science Foundation grant GEO-1128040. This work would not have been possible without the availability of hydrographic data gathered by all participating institutions. We particularly thank Ana Baldoni (INIDEP) and Marcela Charo (SHN) for their efforts in producing and disseminating high-quality data.

References

Acha EM, Mianzán HW, Guerrero RA et al (2004) Marine fronts at the continental shelves of austral South America. Physical and ecological processes. J Mar Syst 44:83–105

Amor CC (2004) Intrusões da Água Central do Atlântico Sul sobre a Plataforma Continental situada entre a Ilha de São Sebastião (SP) e o Cabo de São Tomé (RJ). Dissertation, Universidade de São Paulo

Arruda R, Calil PHR, Bianchi AA et al (2015) Air-sea CO_2 fluxes and the controls on ocean surface pCO_2 variability in the coastal and open-ocean southwestern Atlantic Ocean: a modeling study. Biogeosciences 12:5793–5809

Bianchi AA, Massonneau M, Olivera RM (1982) Análisis estadístico de las características T-S del sector austral de la plataforma continental argentina. Acta Oceanogr Argent 3(1):93–118

Bianchi AA, Bianucci L, Piola AR et al (2005) Vertical stratification and air-sea CO2 fluxes in the Patagonian shelf. J Geophys Res Oceans 110(7):1–10

Bianchi AA, Ruiz-Pino D, Isbert Perlender H et al (2009) Annual balance and seasonal variability of sea-air CO_2 fluxes in the Patagonia Sea: their relationship with fronts and chlorophyll distribution. J Geophys Res 114:C03018. https://doi.org/10.1029/2008JC004854

Borús J, Uriburu Quirno M, Calvo D (2017) Evaluación de caudales diarios descargados por los grandes ríos del Sistema del Plata al estuario del Río de la Plata. Dirección de Sistemas de Información y Alerta Hidrológico. Instituto Nacional del Agua, Ezeiza

Brandhorst W, Castello JP (1971) Evaluación de los recursos de anchoíta (Engraulis anchoita) frente a la Argentina y Uruguay. I. Las condiciones oceanográficas, sinopsis del conocimiento actual sobre la anchoíta y el plan para su evaluación. Proy Des Pesq FAO 29: 63 pp. Mar del Plata

Calliari D, Gómez M, Gómez N (2005) Biomass and composition of the phytoplankton in the Río de la Plata: large-scale distribution and relationship with environmental variables during a spring cruise. Cont Shelf Res 25:197–210

Campos EJ, Goncalves J, Ikeda Y (1995) Water mass characteristics and geostrophic circulation in the South Brazil Bight: summer of 1991. J Geophys Res 100:18537–18550

Campos EJD, Lentini CAD, Miller JL et al (1999) Interannual variability of the sea surface temperature in the South Brazil Bight. Geophys Res Lett 26:2061–2064
Campos EJ, Velhote D, da Silveira IC (2000) Shelf break upwelling driven by Brazil current cyclonic meanders. Geophys Res Lett 27(6):751–754
Campos PC, Möller Jr OO, Piola AR (2013) Seasonal variability and coastal upwelling near Cape Santa Marta (Brazil). J Geophys Res Oceans 118:1420–1433. https://doi.org/10.1002/jgrc.20131
Carranza MM, Gille ST, Piola AR et al (2017) Wind modulation of upwelling at the shelf-break front off Patagonia: observational evidence. J Geophys Res Oceans 122(3):2401–2421
Carreto JI, Lutz VA, Carignan MO et al (1995) Hydrography and chlorophyll-a in a transect from the coast to the shelf-break in the Argentinean Sea. Cont Shelf Res 15:315–336
Carreto JI, Montoya N, Akselman R et al (2008) Algal pigment patterns and phytoplankton assemblages in different water masses of the Río de la Plata maritime front. Cont Shelf Res 28:1589–1606
Carreto JI, Montoya NG, Carignan MO et al (2016) Environmental and biological factors controlling the spring phytoplankton bloom at the Patagonian shelf-break front– degraded fucoxanthin pigments and the importance of microzooplankton grazing. Prog Oceanogr 146:1–21
Castelão RM (2012) Sea surface temperature and wind stress curl variability near a cape. J Phys Oceanogr 42:2073–2087
Castelão RM, Barth JA (2006) Upwelling around Cabo Frio, Brazil: the importance of wind stress curl. Geophys Res Lett 33:L03602. https://doi.org/10.1029/2005GL025182
Castelão RM, Campos EJD, Miller JL (2004) A modelling study of coastal upwelling driven by wind and meanders of the Brazil current. J Coast Res 20(3):662–671
Castello JP, Möller OO (1977) Sobre as condições oceanográficas no Rio Grande do Sul. Atlantica 2:1–119
Castro BM, Lee TN (1995) Wind-forced sea level variability on the southeast Brazilian shelf. J Geophys Res 100(C8):16,045–16,056
Castro BM (2014) Summer/winter stratification variability in the central part of the South Brazil bight. Cont Shelf Res 89:15–26
Castro BM, Miranda LB (1998) Physical oceanography of the western Atlantic continental shelf located between 4°N and 34°S. In: Robinson AR, Brink KH (eds) The sea, vol 11. Wiley, New York, pp 209–251
Castro BM, Lorenzetti JA, Silveira ICA et al (2006) O Ambiente Oceanográfico da Plataforma Continental e do Talude na Região Sudeste-Sul do Brasil, 1st edn. Edusp, p 11–120
Cerda C, Castro BM (2014) Hydrographic climatology of South Brazil Bight shelf waters between São Sebastião (24°S) and Cabo São Tomé (22°S). Cont Shelf Res 89:5–14
Ciotti AM, Odebrecht C, Fillmann G et al (1995) Freshwater outflow and subtropical convergence influence on phytoplankton biomass on the southern Brazilian continental shelf. Cont Shelf Res 15:1737–1756
Cirano M, Campos EJ (1996) Numerical diagnostic of the circulation in the Santos Bight with COROAS hydrographic data. Rev Bras Oceanogr 44(2):105–101
Combes V, Matano RP (2014) A two-way nested simulation of the oceanic circulation in the Southwestern Atlantic. J Geophys Res Oceans 119. https://doi.org/10.1002/2013JC009498
Csanady GT (1978) The arrested topographic wave. J Phys Oceanogr 8:47–62
Dávila PM, Figueroa D, Müller E (2002) Freshwater input into the coastal ocean and its relation with salinity distribution off austral Chile (35–55°S). Cont Shelf Res 22:521–534
Depetris PJ, Kempe S, Latif M et al (1996) ENSO-controlled flooding in the Parana River (1904–1991). Naturwiss 83:127–129
Dottori M, Castro BM (2009) The response of the Sao Paulo continental shelf, Brazil, to synoptic-winds. Ocean Dynam 59:603–614
Dottori M, Castro BM (2018) The role of remote wind forcing in the subinertial current variability in the central and northern parts of the South Brazil Bight. Ocean Dynam, 68: 677, https://doi.org/10.1007/s10236-018-1153-9

Emilsson I (1961) The shelf and coastal waters off southern Brazil. Bol Inst Oceanogr 11(2):101–112
Filippo A, Kjerfve B, Torres AR et al (2012) Low-frequency variability of sea level along the Mid-Atlantic Coast of South America, in 1983. Rev Bras Geofis 30(1):5–14
Franchito SH, Oda TO, Rao VB et al (2008) Interaction between coastal upwelling and local winds at Cabo Frio, Brazil: an observational study. J Appl Meteorol Climatol 47:1590–1598
Franco BC, Piola AR, Rivas AL et al (2008) Multiple thermal fronts near the Patagonian shelf break. Geophys Res Lett 35(2):L02607. https://doi.org/10.1029/2007GL032066
Garcia CAE, Garcia VM, McClain CR (2005) Evaluation of SeaWiFS chlorophyll algorithms in the Southwestern Atlantic and Southern Oceans. Remote Sens Environ 95:125–137
Garcia VMT, Signorini S, Garcia CAE et al (2006) Empirical and semi-analytical chlorophyll algorithms in the South-Western Atlantic coastal region (25–40°S and 60–45°W). Int J Remote Sens 27(8):1539–1562. https://doi.org/10.1080/01431160500382857
Garcia VMT, Garcia CAE, Mata MM et al (2008) Environmental factors controlling the phytoplankton blooms at the Patagonia shelf-break in spring. Deep-Sea Res I 55:1150–1166
Glorioso PD (1987) Temperature distribution related to shelf-sea fronts on the Patagonian shelf. Cont Shelf Res 7(1):27–34
Gómez MI, Piola AR, Kattner G et al (2011) Biomass of autotrophic dinoflagellates under weak vertical stratification and contrasting chlorophyll levels in subantarctic shelf water. J Plankton Res 33(8):1304–1310
Guerrero RA, Piola AR (1997) Masas de agua en la plataforma continental. In: Boschi EE (ed) El Mar Argentino y sus Recursos Pesqueros: Antecedentes Históricos de las Exploraciones en el Mar y las Características Ambientales, vol. 1. Instituto Nacional de Investigación y Desarrollo Pesquero, Mar del Plata, pp 107–118
Guerrero R, Piola AR, Fenco H et al (2014) The salinity signature of the cross-shelf exchanges in the southwestern Atlantic Ocean: satellite observations. J Geophys Res Oceans 119. https://doi.org/10.1002/2014JC010113
Ito R, Schneider B, Thomas H (2005) Distribution of surface pCO_2 and air–sea fluxes in the Southwestern subtropical Atlantic and adjacent continental shelf. J Mar Syst 56:227–242. https://doi.org/10.1016/j.jmarsys.2005.02.005
Ito RG, Garcia CAE, Tavano VM (2016) Net sea-air CO_2 fluxes and modelled pCO_2 in the southwestern subtropical Atlantic continental shelf during spring 2010 and summer 2011. Cont Shelf Res 119(Suppl C):68–84. https://doi.org/10.1016/j.csr.2016.03.013
Kahl LC, Bianchi AA, Osiroff AP et al (2017) Distribution of sea-air CO_2 fluxes in the Patagonian Sea: seasonal, biological and thermal effects. Cont Shelf Res 143:18–28
Lentini CAD, Podestá GG, Campos EJD et al (2001) Sea surface temperature anomalies on the Western South Atlantic from 1982 to 1994. Cont Shelf Res 21:89–112
Longhurst A (1998) Ecological geography of the sea. Academic Press, San Diego
Lucas AJ, Guerrero RA, Mianzán HW et al (2005) Coastal oceanographic regimes of the northern argentine continental shelf (34–43°S). Estuar Coast Shelf Sci 65:405–420
Lutz VA, Segura V, Dogliotti AI et al (2010) Primary production in the argentine sea during spring estimated by field and satellite models. J Plankton Res 32:181–195. https://doi.org/10.1093/plankt/fbp117
Marrari M, Signorini S, McClain CR et al (2013) Reproductive success of the Argentine anchovy, Engraulis anchoita, in relation to environmental variability at a mid-shelf front (Southwestern Atlantic Ocean). Fish Oceanogr 22:247–261
Marrari M, Piola AR, Valla D (2017) Variability and 20-year trends in satellite-derived surface chlorophyll concentrations in large marine ecosystems around South and Western Central America. Front Mar Sci. https://doi.org/10.3389/fmars.2017.00372
Martinez G, Brugnoli E, Hernandez J et al (2005) How valid is the SeaWiFS estimation of chlorophyll-a at the Rio de la Plata estuary and its area of influence? Proc SPIE 5656. Active and Passive Remote Sensing of the Oceans. https://doi.org/10.1117/12.582665
Martos P, Piccolo MC (1988) Hydrography of the Argentine continental shelf between 38 and 42 S. Cont Shelf Res 8:1043–1056

Matano RP, Palma ED (2008) The upwelling of downwelling currents. J Phys Oceanogr 38:2482–2500. https://doi.org/10.1175/2008JPO3783.1
Matano RP, Palma ED, Piola AR (2010) The influence of the Brazil and Malvinas currents on the Southwestern Atlantic shelf circulation. Ocean Sci 6:983–995
Matano RP, Combes V, Piola AR et al (2014) The salinity signature of the cross-shelf exchanges in the southwestern Atlantic Ocean: numerical simulations. J Geophys Res Oceans 119. https://doi.org/10.1002/2014JC010116
Meccia VL, Simionato CG, Guerrero RA (2013) The Río de la Plata estuary response to wind variability in synoptic timescale: salinity fields and salt wedge structure. J Coast Res 29(1):61–77
Mendonça LF, Souza RB, Aseff CRC et al (2017) Regional modeling of the water masses and circulation annual variability at the southern Brazilian continental shelf. J Geophys Res 122:1232–1253
Miller RN, Matano RP, Palma ED (2011) Shelfbreak upwelling induced by alongshore currents: analytical and numerical results. J Fluid Mech 686:239–249. https://doi.org/10.1017/jfm.2011.326
Miranda LB (1985) Correlacão T-S de Massas de água das regiões costeira e oceânica entre o Cabo de São Tomé (RJ) e a Ilha de São Sebastião (SP), Brasil. Bol Inst Oceanogr Univ São Paulo 33(2):105–119
Möller Jr OO, Piola AR, Freitas AC et al (2008) The effects of river discharge and seasonal winds on the shelf off southeastern South America. Cont Shelf Res 28(13):1607–1624
Moreira D, Simionato CG, Dragani W (2011) Modeling ocean tides and their energetics in the North Patagonia gulfs of Argentina. J Coast Res 27(1):87–102
Nagai RH, Sousa SHM, Mahiques MM (2014) The Southern Brazilian shelf. Geol Soc Lond Mem 41:47–54
O'Reilly JE, Maritorena S, O'Brien MC et al (2000) SeaWiFS postlaunch calibration and validation analyses: Part 3. SeaWiFS postlaunch technical report series. In: Hooker SB, Firestone RE (eds) NASA Tech. Memo 2000–206892, 11
Padin XA, Vázquez-Rodríguez M, Castaño M et al (2010) Air-Sea CO_2 fluxes in the Atlantic as measured during boreal spring and autumn. Biogeosciences 7:15871606
Palma ED, Matano RP (2009) Disentangling the upwelling mechanisms of the South Brazil bight. Cont Shelf Res. https://doi.org/10.1016/j.csr.2009.04.002
Palma ED, Matano RP (2012) A numerical study of the Magellan plume. J Geophys Res 117:C05041. https://doi.org/10.1029/2011JC007750
Palma ED, Matano RP, Piola AR (2004) A numerical study of the Southwestern Atlantic shelf circulation: Barotropic response to tidal and wind forcing. J Geophys Res 109:C08014. https://doi.org/10.1029/2004JC002315
Palma ED, Matano RP, Piola AR (2008) A numerical study of the southwestern Atlantic shelf circulation: Stratified Ocean response to local and offshore forcing. J Geophys Res Oceans 113(11):C11010. https://doi.org/10.1029/2007JC004720
Pimenta F, Garvine RW, Munchow A (2008) Observations of coastal upwelling off Uruguay downshelf of the Plata estuary, South America. J Mar Res 66:835–872
Piola AR, Matano RP (2017) Ocean currents: Atlantic western boundary—Brazil current/Falkland (Malvinas) current. Reference module in earth systems and environmental sciences, Elsevier, doi: https://doi.org/10.1016/B978-0-12-409548-9.10541-X
Piola AR, Campos EJD, Möller OO Jr et al (2000) Subtropical shelf front off eastern South America. J Geophys Res 105(C3):6565–6578
Piola AR, Matano RP, Palma ED et al (2005) The influence of the Plata River discharge on the western South Atlantic shelf. Geophys Res Lett. https://doi.org/10.1029/2004GL021638
Piola AR, Möller OO, Guerrero RA et al (2008a) Variability of the subtropical shelf front off eastern South America: winter 2003 and summer 2004. Cont Shelf Res 28:1639–1648
Piola AR, Romero SI, Zajaczkovski U (2008b) Space-time variability of the Plata plume inferred from ocean color. Cont Shelf Res 28:1556–1567
Piola AR, Martínez Avellaneda N, Guerrero RA et al (2010) Malvinas-slope water intrusions on the northern Patagonia continental shelf. Ocean Sci 6(1):345–359

Pisoni JP, Rivas AL, Piola AR (2015) On the variability of tidal fronts on a macrotidal continental shelf, northern Patagonia, Argentina. Deep Sea Res II Top Stud Oceanogr 119:61–68

Proença LAO, Schramm MA, Alves TP, et al (2017) The extraordinary 2016 autumn DSP outbreak in Santa Catharina, South Brazil, explained by large-scale oceanographic processes, Proc 16 Int Conf Harm Algae, Proença LAO, Hallegraeff GM (Eds), Int Soc for the Study of Harmful Algae and Intergovernamental Oceanographic Commission, Brazil, 42–45

Reynolds RW, Smith TM, Liu C et al (2007) Daily high-resolution-blended analyses for sea surface temperature. J Clim 20:5473–5496

Rivas AL (1997) Current meter observations in the argentine continental shelf. Cont Shelf Res 17:391–406. https://doi.org/10.1016/S0278-4343(96)00039-8

Rivas AL, Pisoni JP (2010) Identification, characteristics and seasonal evolution of surface thermal fronts in the Argentinean continental shelf. J Mar Syst 79(1–2):134–143

Rivas AL, Dogliotti AI, Gagliardini DA (2006) Satellite-measured surface chlorophyll variability in the Patagonian shelf. Cont Shelf Res. https://doi.org/10.1016/j.csr.2006.01.013

Robertson AW, Mechoso CR (1998) Interannual and decadal cycles in river flows of southeastern South America. J Clim 11:2570–2581

Rodrigues RR, Lorenzetti JA (2001) A numerical study of the effects of bottom topography and coastline geometry on the southeast Brazilian coastal upwelling. Cont Shelf Res 21:371–393

Romero SI, Piola AR, Charo M et al (2006) Chlorophyll-a variability off Patagonia based on SeaWiFS data. J Geophys Res 111:C05021

Ruiz-Etcheverry LA, Saraceno M, Piola AR et al (2016) Sea level anomaly on the Patagonian continental shelf: trends, annual patterns and geostrophic flows. J Geophys Res Oceans 121:2733–2754. https://doi.org/10.1002/2015JC011265

Sabatini M, Reta R, Matano R (2004) Circulation and zooplankton biomass distribution over the southern Patagonian shelf during late summer. Cont Shelf Res 24:1359–1373

Saraceno MC, Provost C, Piola AR et al (2004) Brazil Malvinas frontal system as seen from 9 years of advanced very high resolution radiometer data. J Geophys Res Oceans 109(5):C05027. https://doi.org/10.1029/2003JC002127

Saraceno MC, Provost C, Piola AR (2005) On the relationship between satellite-retrieved surface temperature fronts and chlorophyll a in the western South Atlantic. J Geophys Res Oceans 110(11):1–16. https://doi.org/10.1029/2004JC002736

Saraceno M, Simionato CG, Ruiz Etcheverry LA (2014) Sea surface height trend and variability at seasonal and interannual time scales in the Southeastern South American continental shelf between 27°S and 40°S. Cont Shelf Res 91:82–94. https://doi.org/10.1016/j.csr.2014.09.002

Scasso LM, Piola AR (1988) Intercambio neto de agua entre el mar y la atmósfera en el Golfo San Matías. Geoacta 15(1):13–31

Signorini SR, Garcia VMT, Piola AR et al (2006) Seasonal and interannual variability of calcite in the vicinity of the Patagonian shelf break (38°S–52°S). Geophys Res Lett 3:L16610. https://doi.org/10.1029/2006GL026592

Silveira ICA, Schmidt ACK, Campos EJD et al (2000) A Corrente do Brasil ao Largo da Costa Leste Brasileira. Rev Bras Oceanogr 48:171–183

Simionato CG, Nuñez MN, Engel M (2001) The salinity front of the Río de la Plata–a numerical case study for winter and summer conditions. Geophys Res Lett 28(13):2641–2644

Simionato CG, Clara Tejedor ML, Campetella C et al (2010) Patterns of sea surface temperature variability on seasonal to sub-annual scales at and offshore the Río de la Plata estuary. Cont Shelf Res 30(19):983–997

Souza RB, Robinson IS (2004) Lagrangian and satellite observations of the Brazilian coastal current. Cont Shelf Res 24:241–262

Stech JL, Lorenzzetti JA (1992) The response of the South Brazil Bight to the passage of wintertime cold fronts. J Geophys Res 97(C6):9507–9520

Stevenson MR, Dias-Brito D, Stech JL et al (1998) How do cold water biota arrive in a tropical bay near Rio de Janeiro, Brazil? Cont Shelf Res 18(13):1595–1612

Strub PT, James C, Combes V et al (2015) Altimeter-derived seasonal circulation on the Southwest Atlantic shelf: 27°–43°S. J Geophys Res Oceans. https://doi.org/10.1002/2015JC010769
Thomas H, Bozec Y, Elkalay K et al (2004) Enhanced open ocean storage of CO_2 from shelf sea pumping. Science 304:1005–1008
Tonini MH, Palma ED, Piola AR (2013) A numerical study of gyres, thermal fronts and seasonal circulation in austral semi-enclosed gulfs. Cont Shelf Res 65:97–110
Valla D, Piola AR (2015) Evidence of upwelling events at the northern Patagonian shelf break. J Geophys Res Oceans 120:7635–7656
Wanninkhof R (2014) Relationship between wind speed and gas exchange over the ocean revisited. Limnol Oceanogr Methods 12(6):351–362
Zavialov PO, Möller OO Jr, Campos EJD (2002) First direct measurements of currents on the continental shelf of southern Brazil. Cont Shelf Res 22:1975–1986. https://doi.org/10.1016/S0278-4343(02)00049-3

Nutrient Transport, Cycles, and Fate in Southern Brazil (Southwestern Atlantic Ocean Margin)

Luis F. Niencheski

Abstract The southwestern Atlantic Ocean margin along the coasts of southern Brazil, Uruguay, and Argentina includes a wide continental shelf which varies in width from about 150 km off southern Brazil to over 500 km along the Patagonian coast of Argentina. The region between 28° and 40° is one of the most biologically productive areas of the World Ocean, and this productivity, perhaps driven mostly by ocean margin processes, extends out across the South Atlantic. Because of this high production, this ocean margin is the largest CO_2 sink in the South Atlantic and is significant on a global scale. Complex interactions of physical, chemical, and biological processes active in this ocean margin control the transport pathways and production in time and space. An appropriate understanding of this system obviously requires interdisciplinary study and information synthesis. This chapter assesses the state of knowledge on ocean margin processes of the southwestern Atlantic Ocean, presents a summary of the present understanding of physical and biogeochemical processes operating in this region and how they are linked, and identifies major areas of uncertainty. This paper provides background information about nutrients in the southwestern Atlantic Ocean margin. Initially we focus on a synthesis of past work and then consider more recent research on nutrients. The chapter emphasizes recent research which considers new nutrient sources to the ocean margin. At the end, the major scientific uncertainties are pointed out to provide a framework for discussion regarding future international, interdisciplinary research in the region.

Keywords Nutrient transport · Southwestern Atlantic · Nutrient sources · Multidiscipline syntheses

L. F. Niencheski (✉)
Instituto de Oceanografia, Universidade Federal do Rio Grande (FURG), Rio Grande, RS, Brazil

M. S. Hoffmeyer et al. (eds.), *Plankton Ecology of the Southwestern Atlantic*,
https://doi.org/10.1007/978-3-319-77869-3_3

1 Introduction

The southwestern Atlantic Ocean region between 28° and 40° is an important part of the global ocean system in terms of ocean circulation and production, but it is of greatest significance to the adjacent South American countries of Brazil, Uruguay, and Argentina. A considerable amount of research has been conducted in this region by South American scientists and by collaborations of foreign scientists with local researchers. Much of this work has been part of local national programs, related to fisheries and other coastal environmental issues, as well as regional programs such as PROSUL (Program to Enhance Cooperation in Science and Technology in South America). International cooperative programs include the World Ocean Circulation Experiment (WOCE), other programs sponsored by the Inter-American Institute for Global Change Research, and a number of North American-South American cooperative programs such as the EcoPlata collaboration between Uruguay, Argentina, and Canada; the NICOP-PLATA collaboration between the US Office of Naval Research, Brazil, Uruguay, and Argentina; and the PATEX (PATagonia Experiment) collaboration between GSFC/NASA, Brazil, and Argentina. Additionally, specific investigator-driven research efforts have brought South American researchers from this area together with North American colleagues to address a number of topics which have both regional and global significance.

Continental shelf processes in this region are influenced by the complex interaction of two major current systems, the Brazil and Malvinas currents, which converge in this region and provide a dynamic environment for cross margin transport and coastal circulation. Material transport to the system from the continent is dominated by a major river system and additional surface runoff sources, but atmospheric transport, groundwater inputs, and coastal water advection through permeable sediments may provide additional important inputs. And, of course, upwelling and intrusions of deep water and offshore transport at the continental shelf edge are important additional processes occurring in this region. The southwestern Atlantic Ocean margin (SWAOM) therefore provides an opportunity to study/understand complex biogeochemical processes which are of interest to a wide range of ocean science disciplines.

2 Shelf Waters

Shelf waters in southern Brazil have been broadly characterized in terms of oceanic water masses and surface inputs from land, and despite the considerable number of nutrient measurements performed in the past four decades in this region (Fillmann 1990; Sales Dias 1994; Niencheski and Fillmann 1997; Del Rosso 2000), the chemical processes and the sources of nutrients in SWAOM are still poorly understood. And part of this is because, with few exceptions, only

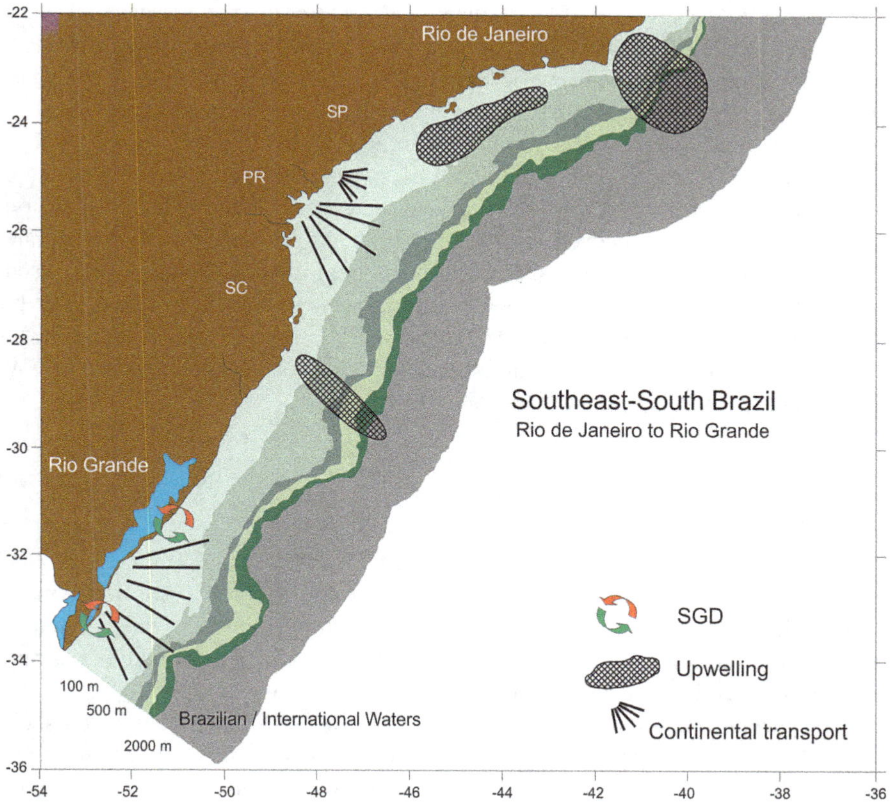

Fig. 1 Terrestrial influences (symbol) submarine groundwater discharge (symbol) and upwelling (symbol) along southern Brazil. (Adapted from Braga and Niencheski 2006)

surface shelf waters concentrations have been the focus of studies. This has limited interpretations of the nutrient association with bottom water masses, mixing, and sources of nutrients in a region where the water chemistry is mainly governed by the mixing of waters of the Brazil and Malvinas currents as they collide (the Brazil-Malvinas Confluence Zone).

A review by Niencheski et al. (1999) indicated that the nutrient characteristics of seawater are well known in regions near Rio de Janeiro, Sao Paulo, and Rio Grande do Sul states. But they point out that coastal studies predominate over those further out on the shelf and slope. More recently studies such as ECOPEL, COROAS, CABO FRIO, PRONEX, and SACC GROUP programs have become more interdisciplinary and systematic and temporally sequenced.

Braga and Niencheski (2006) summarized the existing information for the southern Brazilian continental margin. This included the terrestrial influence from rivers and lagoon inputs and oceanic influences as observed by the presence of upwelled South Atlantic central water and subantarctic shelf water (Fig. 1).

Lopes (2004) suggested that the chemical features of the coastal and shelf waters of southernmost Brazil are mainly influenced by the Brazil current (tropical water, TW), the Malvinas current (subantarctic water, SAW), and subtropical water (STW). In addition to these, the outflow of the Río de la Plata and Patos Lagoon contributes jointly to a coastal water (CW) mass. This study was based on four oceanographic cruises as part of the ECOPEL program (from 1987 to 1991) and a cruise associated with the COROAS program (ocean circulation of the west region of the South Atlantic), carried out in April 1993. Lopes observed higher nitrate concentrations on the continental shelf in the autumn. During the spring increased levels of silicate were observed, while during winter, higher concentrations of nitrate and phosphate were observed on the shelf as a result of the intrusion of SAW. During the summer when lowest nutrient levels occur, the influence of TW on continental shelf is evident.

3 Shelf Break Waters

At the shelf break, upwelling of STW occurs during the spring. This water mass contains high nutrient concentrations and supports new production which is manifested in high fish stocks in the region. Observed heterotrophic conditions during this period result in increased consumption of dissolved oxygen.

Most recent nutrient surveys in SWAOM have focused on the continental shelf (i.e., <200 m), and many of these have included near bottom samples. But most attention has been focused on surficial water and their relation to primary production. Less attention has been given to sources on the shelf or at the shelf break. And even when surveys have extended across the slope, little attention has been paid to bottom waters (i.e., likely nutrient sources). Results from older nutrient surveys (Fig. 2) show considerable silicate enrichment along the shelf break. Clearly the slope water is a likely important source of nutrient supply to the shelf and needs further investigation. This is because very little attention has been paid to bottom water nutrient data in the past.

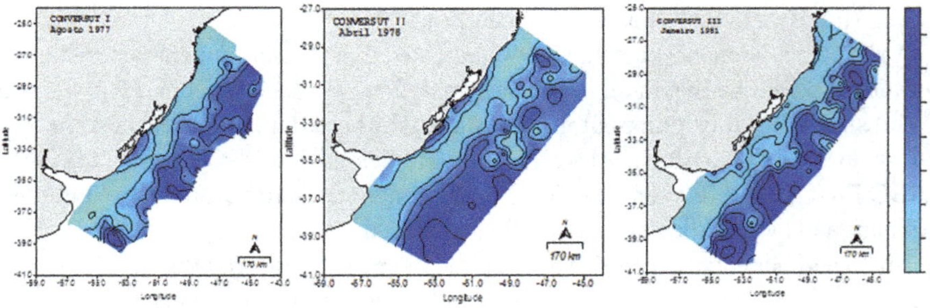

Fig. 2 Silicate concentrations in bottom waters (μM). Data from CONVERSUT program by Brazilian Navy – 1977; 1978 & 1981

4 Coastal Waters

A clear dependence of chemical conditions on oceanographic processes, which create fronts and shelf break upwelling, is evident in the distribution of nutrients on the shelf. Nutrient levels in coastal water (CW), however, are driven by freshwater discharge – from Río de la Plata and Patos Lagoon and the coastal wind stress, all of which depend on weather condition. Because of this, the CW nutrient level also responds to "El Niño" events. But generally, surface inputs such as that of Patos Lagoon outflow have seasonal influences over the local continental shelf. The Patos Lagoon exports dissolved inorganic nitrogen during autumn/winter periods, induced by NE winds.

Braga et al. (2008) (South Atlantic Climate Change Consortium – SACC) presented nutrient distributions over a larger area at continental shelf from 27° 05' S (Brazil) to 39° 31' S (Argentina) during winter 2003 and summer 2004. The relation between nutrient levels and salinity indicates the influences of freshwater discharge over the coastal region and in the front of the Río de la Plata estuary. The low values of salinity associated with high nutrient concentrations clearly suggests the influence of terrestrial inputs, both in summer and in winter, with a larger northward penetration in winter.

Although considerable information exists on nutrient dynamics for the SWAOM, summarized above, scientists in the region view the high-productivity periods in the coastal and shelf waters, as dominantly controlled by river outflow as the major source. Instead, atmospheric transport and submarine groundwater discharge (SGD) of nutrients have only been considered recently.

5 Influence of SGD on Nutrients

During the 1990s research on the variations and interrelation of nutrients in the estuarine region of Patos Lagoon was initiated. The initial aim was to assess the processes occurring within the lagoon (0–30 salinity region) and the role of biological processes (i.e., uptake and remineralization) in the transport and fate of nutrients. In situ benthic flux chamber experiments indicated that remineralization within the sediments dominates the recycling of organic matter and nutrients within the lagoon (Niencheski and Jahnke 2002). But overall, nutrient budgets suggested that the lagoon was a net nutrient trap. From studies of trace metals (Windom et al. 2006), however, it was clear that other processes, specifically surface water-groundwater exchange, may be important to nutrient cycling and transport.

The southern portion of the Brazilian coast is dominated by coastal lagoons formed by sandy barrier spits with small inlets. This coastal configuration is a barrier to the surface flow of freshwater to the sea and accounts for the long water residence time which enhances the lagoon nutrient trapping ability. Significant amount of freshwater from the lagoon has been demonstrated to flow through the permeable

Fig. 3 Schematic representation of SGD on southern Brazil

sands, beneath the barrier spits, where it mixes with seawater (see Figs. 3 and 4; Moore 2008 ; Niencheski et al. 2007, 2010; Niencheski and Windom 2014; Niencheski et al. 2014).

Studies based on nutrient levels in wells from the barrier spit which separates Patos Lagoon from the South Atlantic shelf, along with estimates of SGD, suggest that the following processes are active: products of remineralization of organic detritus accumulated in lagoon sediments are advected through permeable sediments to the oceans, dissolution of biogenic solids and/or solid silicates mobilizes silicate, and phosphate is mobilized from phosphate-rich sediment layers. The resulting SGD has a significantly different composition than that of typical surface water sources. It was pointed out that these processes are not only important to the transport, cycling, and fate of nutrients within the coastal region of southern Brazil but perhaps to coastal areas in general. Niencheski et al. (2007) show that SGD may provide an important flux of nutrients to coastal ocean regions characterized by permeable sediments such as barriers of coastal lagoons. They also suggest that the subterranean transport pathway may be important in explaining the nutrient budgets of coastal lagoons. The authors present a conceptual model of nutrient fluxes in the coastal surface water-groundwater system (Fig. 5). They calculated the nutrient fluxes in the fresh groundwater (the freshwater moving toward the ocean), and these fluxes were compared to those associated with surface freshwater inputs to Patos

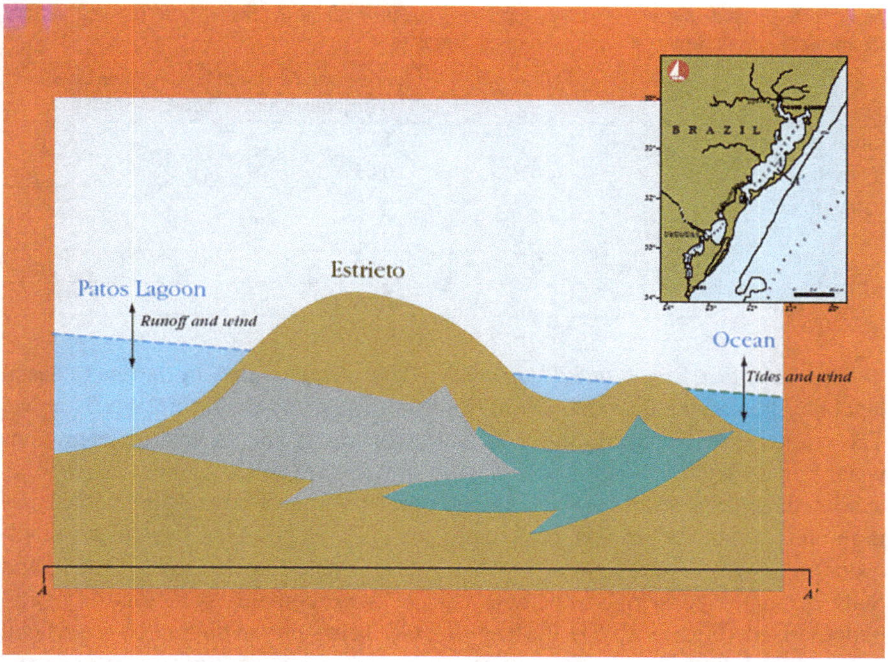

Fig. 4 Schematic representation of groundwater flux in a cross section of Patos Lagoon barrier

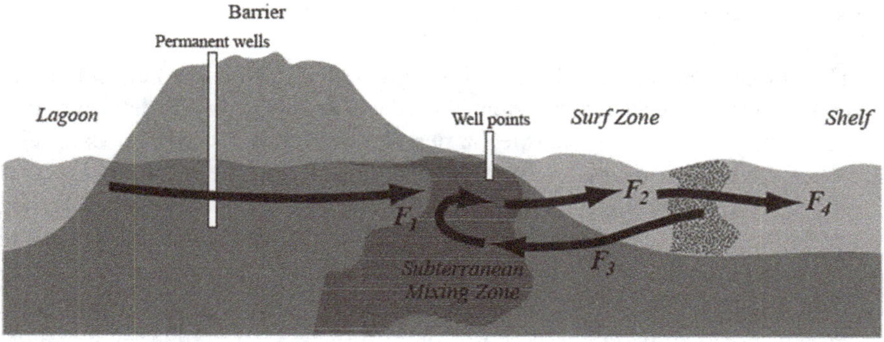

Fig. 5 Conceptual model of nutrient flux in coastal surface water – groundwater systems F1 flux in fresh groundwater, F2 flux associated with SGD, F3 recirculated seawater, F4 transport to the inner shelf, F5 transport offshore

Lagoon. Nitrate is the only nutrient species for which the groundwater flux is considerably less than the surface freshwater flux to Patos Lagoon, but this appears to be compensated by the ammonium flux so that the total nitrogen fluxes are similar (Table 1). Under the assumption that nitrogen limits production in the nearshore or surf zone region, the production supported by the SGD can be calculated. Using this data, the authors estimated that the N flux can support a production rate of about 8 g C m^2 day^{-1} or ca. 3000 g C m^2 $year^{-1}$.

Table 1 Estimated nutrient fluxes to the coast, F_1, in fresh groundwater compared to surface freshwater nutrient inputs to Patos Lagoon (10^6 mol day^{-1})

	F_1	Freshwater nutrient flux to Patos Lagoon (Niencheski and Windom 1994)
Silicate	23.2	13.1
Phosphate	0.25	0.05
Ammonium	1.15	0.25
Nitrite	0.002	0.01
Nitrate	0.086	0.98
Total nitrogen	1.25	1.24

The Southern Brazil and Uruguayan coast is dominated by several coastal lagoons. Of these the Patos and Mirim/Mangueira lagoons are the most well known. This coastal system certainly provides local hydraulic gradients resulting in groundwater flow toward the South Atlantic along ca. 1500 km length of their coastline, as exhibited by water levels in wells placed in the sandy barrier that respond to changing water level in the lagoons. Niencheski et al. (2007), however, show that groundwater transport and seawater cycling through permeable coastal/shelf sediments and SGD along the coast of southern Brazil are more complex than was originally hypothesized (Figs. 3 and 4). Instead of SGD being driven simply by a surficial aquifer which links the lagoon to the ocean, an additional aquifer system(s) underlies this region and may support SGD over a wider region of the shelf. We can speculate that Guarani aquifer or another important regional aquifer may have an even greater geographic influence and may, in fact, provide for nutrient input in SGD on the shelf.

The isotopic composition (d18O, d2H) of groundwater, lagoon water, and seawater collected in a coastal lagoon system in southern Brazil as a tool to define groundwater sources within a permeable barrier that separates the Mirim/Mangueira lagoons from the ocean was described by Schmidt et al. (2011). The isotopic signature of the barrier groundwater differs both from the isotopic composition of lagoon water and seawater. The groundwater is significantly lighter, indicating that the major recharge source to the barrier is precipitation, rather than lagoon water. In the southern and central part of the barrier, no interaction of groundwater with lagoon water was observed. In the northern part, however, lagoon water intrusion into the barrier was identified. This effect is apparently caused by different vegetation types on the barrier influencing subsurface hydrological processes. The central and the southern areas are characterized by natural dune vegetation (high infiltration, low evapotranspiration rates) whereas the northern part is covered by extensive pine tree plantations (low infiltration, high evapotranspiration rates). The presence of pine trees may enhance lagoon and seawater intrusion into the shallow aquifer and decrease fresh submarine groundwater discharge.

Chemistry of southern Brazilian continental shelf (SBCS) waters, between 28.5°S and 34°S, was evaluated in relation to the mixing of thermohaline-defined water masses and concomitant water column processes by Niencheski and Windom. (2014). Data on inorganic nutrient and trace metal (Ba, Cd, Co, Cu, Fe, Mn, U, V, and Zn) concentrations and radium isotopic activities from previous reports and recent analyses were presented and discussed by Niencheski et al. (2014) with the aim of providing a state of knowledge on processes governing their sources, distribution, transport, and fate on the SBS. Nutrient concentrations/distributions on the shelf are controlled mainly by tropical water/subtropical shelf water, seasonally modified as a result of summer upwelling in the northern shelf region, and by Río de la Plata plume water, which is seasonally modified by discharges of submarine groundwater (SGD) and the Patos Lagoon and by the greater northern penetration of the Rio de la Plata plume during winter.^{226}Ra activity varies little across the shelf.^{228}Radium, associated dominantly with SGD, decreases conservatively, with respect to salinity, across shelf transects, converging on a typical ocean end-member activity. The low salinity end-member activity, however, depends on the location of the shelf transect controlled by the variability of coastal SGD. Because SGD is so important to the coastal shelf region, ^{228}Ra activity appears to provide a better tracer of water mass interactions than thermohaline characteristics. Using metal-^{228}Ra relationships, sources, transport, and fate of trace metals are better constrained enabling the following conclusion: the major source of dissolved Co, Mn, and Fe to the shelf is SGD, along the coastal Holocene barrier system, followed by removal as coastal water mixes across the shelf. Cu and Si concentrations are explained as conservative mixtures of three end-members: SGD, surface freshwater discharge, and oceanic. Cd and Zn are largely explained similarly. Vanadium is enriched in coastal waters during the summer presumably due to seasonal remobilization from sediments but exhibits conservative mixing behavior across the shelf. Barium behavior is the most unexpected and is speculated to be the result of biochemical removal in highly productive coastal waters followed by release from decaying phytoplankton detritus as it is advected across the shelf.

6 Potential Nutrient Sources on the Shelf

Silicate is relatively enriched in terrestrial runoff and typically shows a conservative mixing behavior when freshwater mixes with seawater. To determine the sources of nutrient inputs along the coast during different seasons, silicate variations along S-N transect from Chuí (zero km) to Conceição (270 km) (Fig. 6) provide insights.

Data collected along this transect for silicate (Fig. 7) shows that for any time of the year, concentrations near the Albardão region are always high; in some instances, such as spring, they are even higher than the concentrations observed at stations located near the mouth of Patos Lagoon and Chuí (where the input of the Plata should have greatest influence). These results suggest that silicate sources other than the traditional ones (La Plata River and Patos Lagoon) may be responsible for the high silicate concentration which occurs near Albardão.

The map of silicate contours shown in Fig. 8 suggests two areas of high concentrations. But gradients associated with them show no obvious link to the Plata River or Patos Lagoon outlet. Campos et al. (2009) recently reported on the discovery of

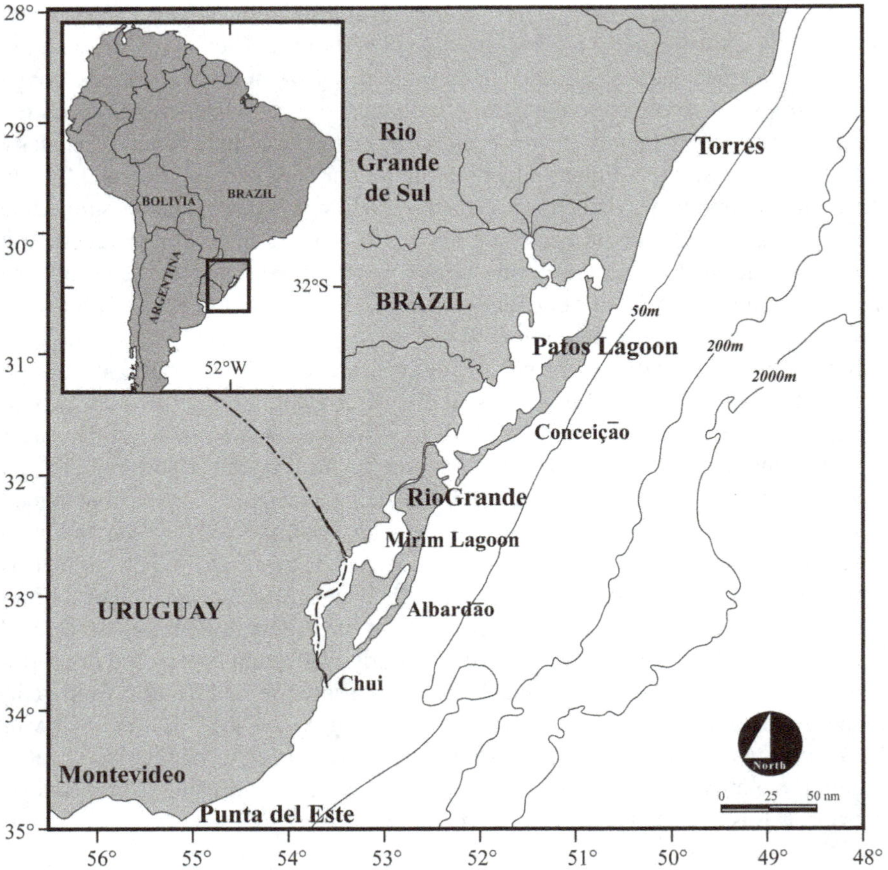

Fig. 6 Southern Brazil and Uruguayan coast

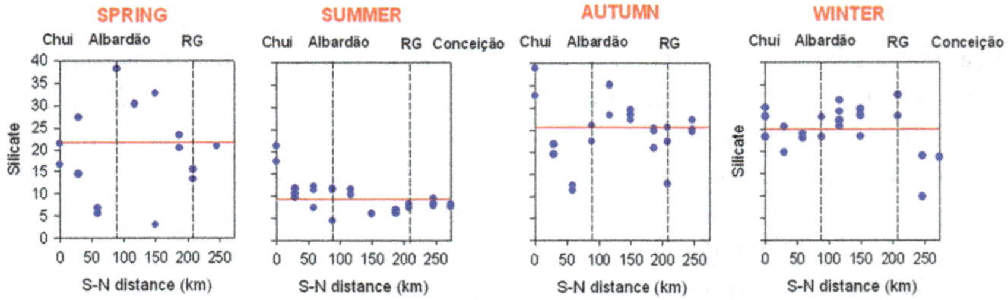

Fig. 7 Distribution of silicate concentrations as μM (down to a 10-m depth) along the coast, from Chuí to Conceição. Adaptado de Attisano et al. (2013)

the Albardão paleochannel which occurs about 35 kms offshore of southern Brazil and Uruguay. It begins near Punta del Este and extends, to approximately 50 km northeast along the coast. This feature reprints an ancient Plata River drainage

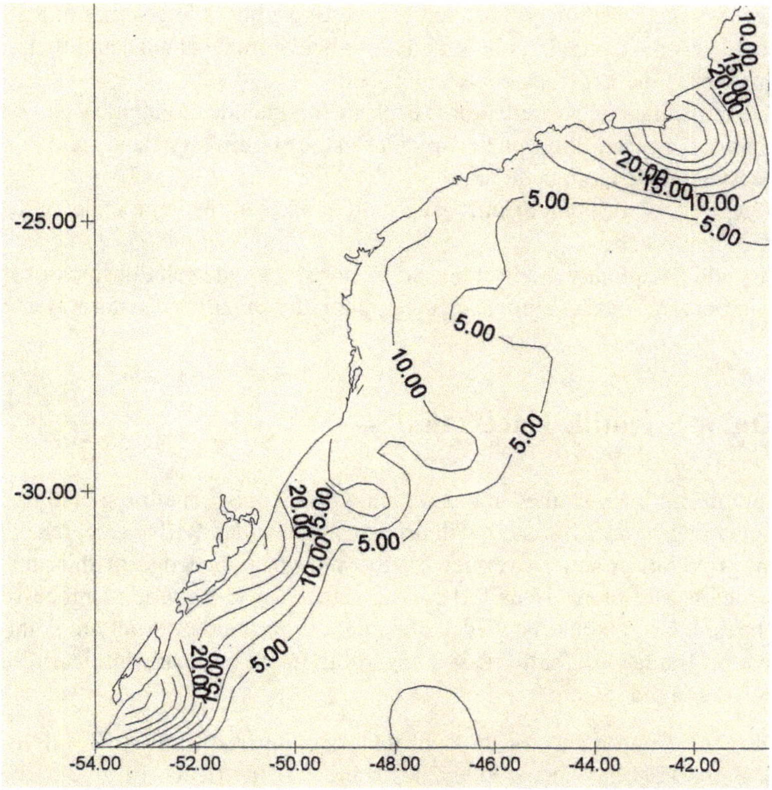

Fig. 8 Surface silicate distribution (autumn) (μM)

channel. SGD controlled by this feature may explain the southernmost area of silicate enrichment.

Although considerable research has been carried out on the southwestern Atlantic Ocean margin, there has been little integration of results across disciplines, and the few recent summaries of research that have been published do not take into account more recent scientific findings in the region. An exception to this was a special issue of *Continental Shelf Research* which was prepared in 2008, to focus on the results of the NICOP-PLATA research on continental shelf oceanography in the region of the Rio de la Plata and Patos Lagoon outflows (Campos et al. 2008).

In 2010, Federal University of Rio Grande (FURG) started the National Institutes of Science and Technology-*Integrated Oceanography and Multiple Uses of the Continental Shelf and the Adjacent Ocean (*Integrated Center of Oceanography – ICO). The main goal of this institute is to understand, quantify, and monitor the changes in the South Atlantic circulation due to changes in the atmospheric circulation and the consequent impacts on coastal processes and on the regional climate. The observational part of this project is conducted in four interlinked components:

(A) Large-scale sub-project, intended to investigate possible changes in the basin-scale South Atlantic circulation due to global climate change and their consequences to the Brazilian continental shelf
(B) A continental edge observational experiment, intended to identify and monitor changes in ocean-atmosphere variables in the Southwest Atlantic and its consequences to the biological pump
(C) A regional-scale component, focused on impacts of the large scale on the shelf and shelf break
(D) A multidisciplinary study intended to investigate the relationships of marine microorganisms and climate changes along the Brazilian coastal waters

7 Major Scientific Uncertainties

The multidisciplinary studies in this region will not only benefit regional researchers in planning future efforts but will also serve the wider marine research community by providing a better overview of the state of knowledge of this important ocean region. But more needs to be done across a wider scope of processes and disciplines. Based on our knowledge, the major uncertainties/issues are pointed out to provide a framework for discussion regarding future international, interdisciplinary research in the region:

- The relative importance of the contributors of surface sources (Rio de la Plata and Patos Lagoon) and groundwater sources of nutrients, in time and space, needs to be understood better.
  - A more detailed geographic distribution of SGD is needed to assess heterogeneity.
- Are there additional SGD nutrient sources on shelf due to larger groundwater systems?
- Understanding the role of slope waters and nutrient sources
- Processes at shelf break on nutrient input to the shelf.

Finally, this paper has a key role to initiate the pursuit of cooperative research projects/programs in the region that articulate with more global research agendas and regional-national collaborations.

References

Attisano KK, Santos IR, Andrade et al. (2013) Submarine groundwater discharge revealed by radium isotopes (Ra–223 and Ra–224) near a paleochannel on the southern Brazilian continental shelf. Braz J Oceanogr 61(3):195–200

Braga ES and Niencheski LFH (2006) Composição das massas de água e seus potenciais produtivos na área entre o Cabo de São Tomé (RJ) e o Chuí (RS). In: O ambiente oceanográfico da plataforma continental e do talude na região sudeste–sul do Brasil. Organizadores Rossi–Wongtschowski, Carmen Lúcia Del Bianco & Madureira Lauro Saint–Pastour, EDUSP, p 161–218

Braga ES, Chiozzini VC, Berbel GBB et al. (2008) Nutrient distributions over the Southwestern South Atlantic continental shelf from Mar del Plata (Argentina) to Itajaí (Brazil): Winter–summer aspects. In: Campos, E.J.D., Piola, AR and RP Matano (Eds.) Synoptic characterization of the Southeastern South American Continental shelf: The NICOP/Plata Experiment. Cont Shelf Res 28(13): 1649–1661

Campos EJD, Piola AR, Matano RP (eds) (2008) Synoptic characterization of the Southeastern South American continental shelf: the NICOP/Plata experiment. Cont Shelf Res 28(13):1551–1692

Campos P, Weigert SC, Madureira LSP (2009) Prospecção hidroacústica do fundo na região do Paleocanal do Albardão – RS – Brasil. Atlântica, Rio Grande. https://doi.org/10.5088/atl.2009.31.1.5

Del Rosso C (2000) Contribuição aos estudos hidroquímicos da plataforma continental sul–sudeste do Brasil. Undergraduate thesis. Oceanologia, FURG. Brazil

Fillmann G (1990) Caracterização química das massas de água da plataforma continental do sul do Brasil. M.Sc. thesis, FURG. 134 p (Brazil)

Lopes MF (2004) Hidroquímica da plataforma sul do Brasil: caracterização e variabilidade sazonal. M.Sc. thesis, FURG (Brazil)

Moore WS (2008) Submarine groundwater discharge. state of knowledge on Southwestern Atlantic ocean margin processes – workshop (SWAOM). Montevideo, November 16–22, 2008. (http://www.skio.usg.edu/meetings/uruguay2008.html)

Niencheski LF and Fillmann G (1997) Chemical characteristics. In: Seeliger U, Odebrecht C, Castello, J.P. (eds.) Ecology of subtropical convergence regions – The coast and sea in the warm–temperate southwestern Atlantic. Chapter: Coastal and marine environments and their biota. Springer, New York, p 96–98

Niencheski LF, Jahnke RA (2002) Benthic respiration and inorganic nutrient release across the sediment water Interface in the estuarine region of Patos Lagoon (Brazil). Aquat Geochem 8:135–152

Niencheski LF, Windom HL (1994) Nutrient flux and budget in Patos Lagoon Estuary. The Science of the Total Environment. 149 (1-2): 53–60

Niencheski LF, Windom HL, Moore WS Brandes J (2010) Nutrient subterranean fluxes to coastal zone of Southern Brazil. American Geophysical Union (AGU). The Meeting of the Americas. Foz do Iguaçu. August 8-12, 2010. Oral presentation at Session H34A

Niencheski LF, Windom H (2014) Chemistry of a surficial aquifer of a large coastal lagoon barrier and its relation to adjacent surface waters of Brazil. J Coast Res. https://doi.org/10.2112/jcoastres–d–13–00175.1

Niencheski LF, Baumgarten MG, Roso et al. (1999) Oceanografia Química – Levantamento Bibliográfico e Identificação do Estado Atual do Conhecimento. Avaliação do Potencial Sustentável de Recursos Vivos na Zona Econômica Exclusiva – REVIZEE. Patrocínio do Ministério do Meio Ambiente, dos Recursos Hídricos e da Amazônia Legal (MMA), Comissão Interministerial para os Recursos do Mar (CIRM), Fundação de Estudos do Mar (FEMAR), p 171 ISBN 85–85966–12–2

Niencheski LF, Windom HL, Moore W et al (2007) Submarine groundwater discharge of nutrients to the ocean along a coastal lagoon barrier, southern Brazil. Mar Chem 106(3–4):546–561

Niencheski LF, Windom HL, Moore WS (2014) Controls on water column chemistry of the southern Brazilian continental shelf. Cont Shelf Res 88:126–139

Sales Dias LM (1994) Caracterização química das águas da plataforma continental sul do Brasil – Verão de 1990 e Outono de 1991. Undergraduate thesis. Oceanologia, FURG, p 55

Schmidt A, Santos IR, Burnett WC et al (2011) Groundwater sources in a permeable coastal barrier: evidence from stable isotopes. J Hydrol 406:66–72

Windom HL, Moore W, Niencheski LF et al (2006) Submarine groundwater discharge: a large, previously unrecognized source of dissolved iron to the South Atlantic Ocean. Mar Chem 102(3–4):252–266

Surface Macronutrient Dynamics of the Drake Passage and the Argentine Sea

Flavio E. Paparazzo and José L. Esteves

Abstract The dynamics of macronutrients on the surface is key for marine life. In this work, we focus on the nitrate, phosphate, and silicate distribution along the Drake Passage and the Argentine Sea. These nutrients have the highest concentration in the south of the Drake Passage because of upwelling of deep waters and inlets of coastal currents. The Antarctic Convergence forms a kind of barrier between water masses, greatly limiting the surface exchange of chemical species to the north. Pacific Ocean waters mixed with surface waters located north of the Polar Front enter the Argentine Sea giving rise to the Patagonian and Malvinas Currents. On their way, primary producers deplete nutrients, and, at a given moment, nitrate reaches limiting concentrations. Two processes locally modify the resulting N-S nutrient gradient: (1) the shelf offshore component receives the contribution of the nutrient-rich Antarctic waters, which move northward along the continental slope through the Malvinas Current; (2) large tidal waves and their interaction with the seabed create seasonal frontal systems that increase the chemical species concentration near the coast. The discharge of the less saline waters of the Magallanes Strait can be observed up to 43°S, but its effect on macronutrients is low. Patagonian rivers present a low flow and seem to make only local contributions. Until now, the fertilization effect of submarine groundwater discharge is unknown and the aeolian dust input is under study. The ice pack coverage in the Drake Passage and the water column stratification in the Argentine Sea govern the seasonal variation. Interannual

F. E. Paparazzo (✉)
Laboratorio de Oceanografía Química y Contaminación de Aguas (LOQyCA), Centro para el Estudio de Sistemas Marinos (CESIMAR), Consejo Nacional de Investigaciones Científicas y Técnicas (CONICET), Puerto Madryn, Argentina

Laboratorio de Oceanografía Biológica (LOBio), Centro para el Estudio de Sistemas Marinos (CESIMAR), Consejo Nacional de Investigaciones Científicas y Técnicas (CONICET), Puerto Madryn, Argentina
e-mail: paparazzo@cenpat-conicet.gob.ar

J. L. Esteves
Laboratorio de Oceanografía Química y Contaminación de Aguas (LOQyCA), Centro para el Estudio de Sistemas Marinos (CESIMAR), Consejo Nacional de Investigaciones Científicas y Técnicas (CONICET), Puerto Madryn, Argentina

M. S. Hoffmeyer et al. (eds.), *Plankton Ecology of the Southwestern Atlantic*,
https://doi.org/10.1007/978-3-319-77869-3_4

differences have been associated with ENSO-like events, but information is not enough to draw conclusions. Due to the importance of knowing the nutrient dynamics to understand the biological processes of the region, multidisciplinary studies focusing on this topic should be promoted.

Keywords Macronutrients · Currents · Fronts · Drake Passage · Argentine Sea

1 Introduction

The study of macronutrient dynamics on the surface is key to understand the relationship between marine chemistry and primary production. The products of rock weathering, decay of organic material, glacial action, and volcanic activity are the major sources of nutrients to the sea. Instead, biogeochemical processes, such as incorporation by living organisms and sedimentation, control the removal. To determine the particular patterns of distribution, mixing and advective processes coupled to transport by currents and frontal gradients must be considered.

In this work, we focus on nitrate, phosphate, and silicate seeking to understand their surface dynamics along the Drake Passage (DP) and the Argentine Sea (AS) (Fig. 1a). First, an analysis of the permanent dynamics is developed followed by the main seasonal and interannual variations. Next, we show an estimate of nutrient concentration (based on literature). Finally, the general conclusion is presented.

2 Permanent Surface Nutrient Dynamics

In deep Antarctic waters, the nutrient concentration is high (Levitus et al. 1993). The Antarctic shelf break front (ASBF – Fig. 1c) is the primary site for the renewal of Antarctic surface waters and the deepest layers of the world ocean (Jacobs 1991). There, upwelling, advection, and convection processes transport the deep nutrient-rich water mass to the surface. Once emerged, surface and intermediate currents distribute the nutrients through the Antarctic Circumpolar Current (ACC – Fig. 1b) and its derivations (Barker et al. 2007).

Near the Antarctic Peninsula, the ACC interacts with shelf waters facilitating the lateral transport of shelf-derived components such as iron into high-macronutrient offshore regions. Lateral fluxes overwhelm vertical inputs and exports from the water column (Dulaiova et al. 2009). The elevated trace metal concentrations near the 25 de Mayo Island (Shetland del Sur) correspond to the flow path of the Mar de la Flota Current (MFC – Fig. 1b). This enriched plume can also be seen exiting the Mar de la Flota Sea, between the Clarence and Elefante Islands into the Ona Basin, where it mixes with high macronutrient–low trace metal concentration waters from the ACC (Hatta et al. 2013). Nutrients enter the Ona Basin from the continental shelf through advection along an isopycnic, resulting in an iron concentration peak

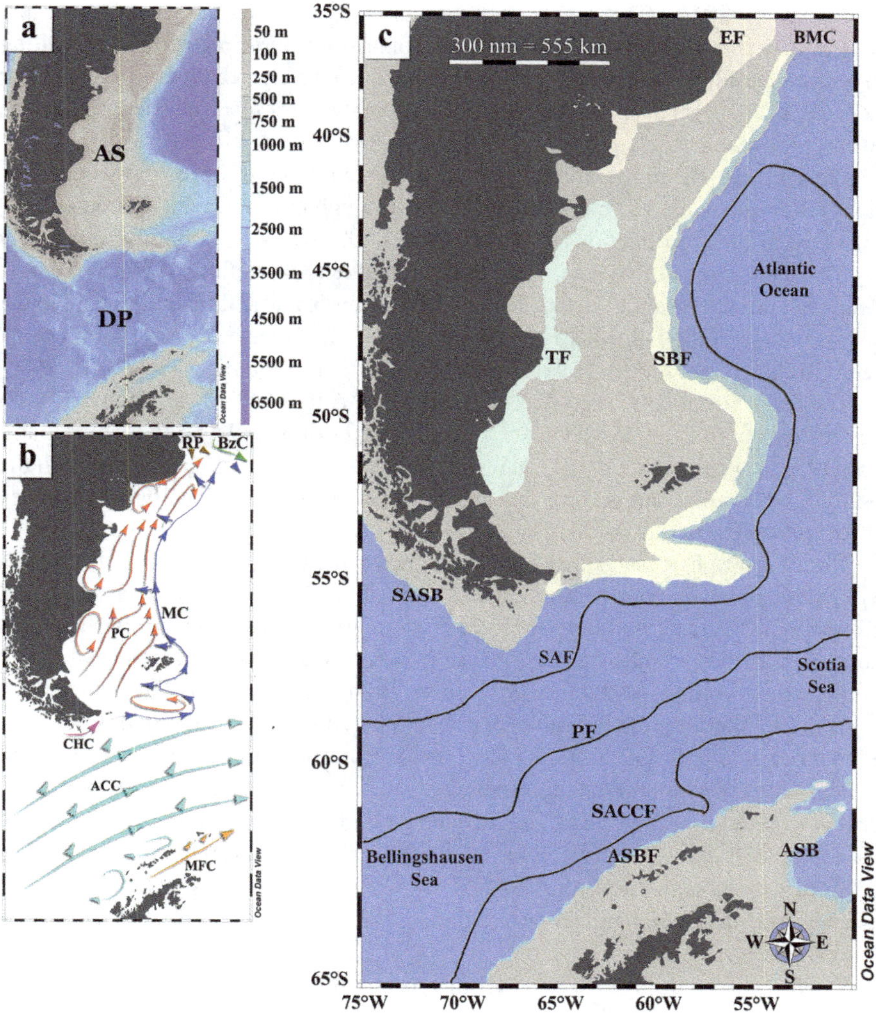

Fig. 1 Main environmental factors regulating surface nutrient distribution (**a**) Topography. AS Argentine Sea, DP Drake Passage. (**b**) Currents (Modified after Palma et al. 2008). RP Rio de la Plata, BzC Brazil Current, MC Malvinas Current, PC Patagonian Current, CHC Cape Horn Current, ACC Antarctic Circumpolar Current, MFC Mar de la Flota Current. (**c**) Fronts (Modified after Acha et al. 2004; Sprintall 2003). EF estuarine front, BMC Brazil-Malvinas Confluence, TF tidal fronts, SBF shelf break front, SASB South American shelf break, SAF Subantarctic Front, PF Polar Front, SACCF Southern Antarctic Circumpolar Current Front, ASBF Antarctic shelf break front, ASB Antarctic shelf break. The Ocean Data View software (Schlitzer 2017) was used for mapping

(Frants et al. 2013). The mixing between the iron-poor waters of the ACC and iron-rich waters on the shelf through horizontal transport and vertical upwelling processes could be responsible for the enhanced primary productivity in this sector (Zhou et al. 2010). This process increases the variability of macronutrients along the SE sector of the DP.

Entering the DP to the north, macronutrients continue to have high concentrations (Brandini et al. 2000; Kim et al. 2004), although micronutrients become limiting of the primary production (e.g., Smith et al. 2008). The DP exhibits exceptionally high concentrations of silicate, which might be due to the inflow of North Atlantic deep water in the Atlantic sector (Berger 2007). This is particularly significant in the southern half of the DP, where a large gradient in the silicate concentration can be seen (Tréguer and Jacques 1992; Paparazzo et al. 2016).

The Polar Front (PF – Fig. 1c), also known as Antarctic Convergence, acts as a surface barrier that divides the water mass properties of both the northern and southern DP (Paparazzo et al. 2016). The front intensity significantly correlates with the bottom topography, suggesting that the front intensifies over shallow bathymetry (Freeman and Lovenduski 2016).

Nitrate, phosphate, and silicate present a decreasing trend from south to north with a significant lower concentration than south of the PF (Paparazzo et al. 2016). However, all the macronutrients meet the phytoplankton needs, and the silicate of this zone is particularly higher in the DP than in other longitudes (e.g., Brzezinski et al. 2005), never limiting the primary production (Brandini et al. 2000; Kim et al. 2004).

In the southern tip of the South American coast, nutrient concentrations decrease even more although it is not possible to speak of low concentration because it continues being abundant for the requirement of the primary producers. Simultaneously, primary production increases. This explains a higher consumption of nutrients by phytoplankton due to better conditions for its growth within the continental shelf north of the Subantarctic Front (SAF – Fig. 1c), where there is a decrease in the mixing layer depth and an increase in temperature and micronutrients such as iron (Klunder et al. 2014).

Toward the south of the Tierra del Fuego province, three masses of water converge (Garzón et al. 2016): (1) high macronutrient oceanic water derived from the northern DP, (2) intermediate macronutrient water from the Southeast Pacific transported by the Cape Horn Current (CHC – Fig. 1b), and (3) low macronutrient water from the mouth of the Beagle Channel. The mixture of these three bodies of water enters to the AS through the Le Maire Strait and joins the Magallanes Strait discharge, originating the Patagonian Current (PC – Fig. 1b).

In the AS, primary producers consume macronutrients while the PC moves toward the north, generating a marked superficial gradient. In the case of nitrate, it reaches limiting concentrations for primary production during the stratification season (Paparazzo et al. 2010; Song et al. 2016). However, frontal contributions with different characteristics generate a significant increase of macronutrients from both the west and east sides of the AS, breaking the continuity along the gradient.

In the western AS, the effect of tidal currents generates the formation of tidal fronts (TF – Fig. 1c) (e.g., Glorioso 2000) whose consequences have been studied by several authors (e.g., Acha et al. 2004; Romero et al. 2006; Rivas et al. 2006). During thermal stratification periods, tidal fronts are an important source of nutrients for the AS surface waters (Acha et al. 2004; Romero et al. 2006; Paparazzo et al. 2010). The main tidal front zones were located off the Valdés Peninsula, near Cape Blanco, and in a region stretching south of 50 °S.

In the external sector of the AS, subantarctic waters coming from the Malvinas Current (MC; Fig. 1b) enter from the eastern margin, along the shelf break front (SBF; Fig. 1c) (Matano and Palma 2008). It is a key source of nutrient-rich waters to the AS. The subantarctic cold waters of this stream not only flow through the slope but also through the bottom of the continental shelf fertilizing beyond the frontal sector (Anderson and Kaltin 2001). In the interaction zone between the shelf and the slope, gradients of high salinity (Fedúlov et al. 1990), temperature (Piola and Falabella 2009) and biological activity were observed (Acha et al. 2004). Many studies have discussed the fertilization generated by the MC toward the AS (Brandhorst and Castello 1971; Carreto et al. 1995; Brandini et al. 2000; Romero et al. 2006).

To the north of the AS, the combined effects of the freshwater discharge from the Rio de la Plata river (RP –1b) (Depetris and Paolini 1991) and winds (Piola et al. 2005) contributed to the modification of nutrient distributions over the continental shelf. The nitrate concentration increased near the continental shelf break (Carreto et al. 2007), while silicate and phosphate established an anticorrelation with salinity, showing to be associated with the freshwater of the RP (Braga et al. 2008). Provost et al. (1996) identified four types of fronts in the upper layer over the Brazil–Malvinas Confluence region (BMC; Fig. 1c), highlighting the great dynamics of this sector.

Little is currently known about other local sources that contribute to the fertilization of the AS. The lower salt waters of the Magallanes Strait (MS) cover a large surface area, reaching 43 °S (Palma and Matano 2012; Strub et al. 2015). In this way, its influence on numerous processes in the AS is evident. However, although the macronutrient records at the MS nearest sector of the AS show non-limiting concentrations (e.g., Decembrini et al. 2014), it is not proved that the nutrient intake of the MS extends beyond its mouth. In addition, surface and submarine groundwater flows from Patagonian rivers (Colorado, Negro and Santa Cruz) could contribute significant amounts of nutrients to the coast and probably play a significant role in maintaining the rich Patagonian productivity (Depetris et al. 2005).

Fertilization could also occur throughout the contribution of sediments transported by the intense winds coming from the west sector of the Patagonian region. If these sediments were rich in nutrients, they would constitute an additional contribution to primary producers. Iron fertilization through the wind transport of volcanic ash from an eruption event reaches the Southern Ocean at the south of Africa (Simonella et al. 2015). Simultaneously with the present work, a series of experiments and field measurements indicate a possible supply of nitrate and silicate through aeolian dust transport (Paparazzo et al. unpublished data). Such a source of nutrients could be of great relevance, especially when the local chemical conditions limit primary production (Crespi Abril et al. chapter "Perspective: Continental Inputs of Matter into Planktonic Ecosystems of the Argentinean Continental Shelf, the Case of Atmospheric Dust").

3 Seasonal Surface Nutrient Dynamics

The variations in the sea ice distribution and the freshwater inputs from glacial and melting sea ice are the dominant influences on the biogeochemical processes in the Antarctic coastal systems (Ducklow et al. 2013). The extent of the ice field surrounding Antarctica in winter reduces to a fifth in summer (Arrigo et al. 1997); this exposes a large extension of sea surface to the interaction with strong winds. In comparison with the whole continent, there are no large expanses of ice in the north of the Antarctic Peninsula (Fig. 2). In consequence, the seasonal pack ice of the DP is limited to a narrow sector surrounding the Shetland del Sur Islands, leaving ice-free surface in the area of greater upwelling (Martinson et al. 2008). This leads to delaying the drop in the surface concentration of macronutrients during summer (Smith et al. 2008). The DP shows a higher concentration of macronutrients in winter as compared to summer (e.g., Munro et al. 2015). However, detailed studies showed nitrate and phosphate concentrations somewhat lower in fall than in summer (Paparazzo et al. 2016). In early fall, the ice cover begins to grow and macronutrients decrease. However, along the Antarctic coast, this process occurs simultaneously with the drop in temperature and the increase of macronutrients by overall mixing of the water column. Consequently, the minimum of macronutrients in the southern half of the DP would take place in early fall. Although it is an area with a relative low fluctuation of the ice cover, the DP is definitely an ecosystem that strongly depends on it.

The seasonal variability in the location of the PF reaches more than 100 km southward in March with respect to February (Paparazzo et al. 2016). There is a slight trend for the PF to being farthest south in austral summer and farthest north in austral spring, and the spatial changes among years would be strongly correlated to both zonal wind stress and wind stress curl in the southeast Pacific Ocean (Sprintall 2003). Because of the PF migration, sectors of different nutrient concentrations extend or reduce (Fig. 2).

The stability of the water column is the key process to understand the seasonality on the surface macronutrients concentration in the AS (Fig. 2). In winter, strong vertical mixing leads to relatively high nutrient concentrations throughout the water column (Brandhorst and Castello 1971; Carreto et al. 1995). However, as of mid-August, the water column begins to stratify north of 40°S, beginning a process toward the south that ends up generating the complete surface stratification of the AS at the end of September (Rivas et al. 2006). The stability in the photic layer leads to a great phytoplankton bloom during the beginning of spring (Romero et al. 2006). As the summer progresses and the stratification strengthens, plankton consumes the surface nutrients until it reaches limiting concentrations, especially nitrate (Paparazzo et al. 2010). Here, the tidal fronts play a very important role allowing the recirculation of bottom water to the surface. From the other side, in the eastern sector of the platform, the shelf break front takes a preponderant role (Acha et al. 2004). This front provides nutrients to sustain high phytoplankton biomass levels during the warm season (García et al. 2008). In the north of the AS, the front moves

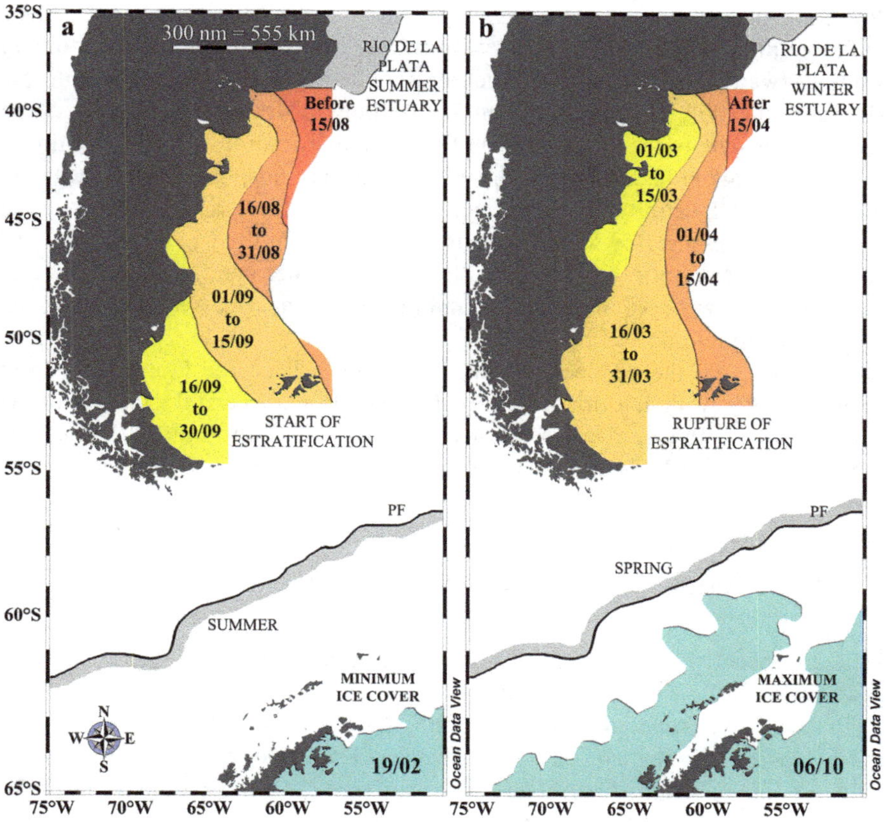

Fig. 2 Seasonal factors regulating surface nutrient distribution. (**a**) Higher summer effect of the Rio de la Plata estuary over the shelf, progressive dates of the start of stratification along the Argentine Sea from late winter to early spring, the southern location of the Polar Front in summer, and the minimum Antarctic ice cover in February. (**b**) Lower winter effect of the Rio de la Plata estuary over the shelf, progressive dates of the rupture of stratification along the Argentine Sea from late summer to early fall, the northern location of the Polar Front in spring, and the maximum Antarctic ice cover in October (Modified after Piola et al. 2008; Sprintall 2003, NASA Earth observatory https://earthobservatory.nasa.gov). The Ocean Data View software (Schlitzer 2017) was used for mapping

offshore during summer and onshore during spring and autumn (Carreto et al. 1995). As of March, the rupture of the stratification in the AS occurs (Rivas et al. 2006), and the deepwater nutrients recirculate on the surface (Paparazzo et al. 2010), reestablishing the concentration observed during the winter. At the same time, there is no stability of the water column for plankton, and primary production decreases (Romero et al. 2006).

To the north of 40°S, there is a high seasonal variability due to estuary coverage (Fig. 2). Intrusions of cold subantarctic waters are consequences of a northward expansion of mixtures of the Rio de la Plata waters in late fall and a slower retraction of the plume during spring–summer (Palma et al. 2008).

Primarily in response to variations in the along-shore wind stress, the Rio de la Plata plume presents large northeastward penetrations in late fall–winter and southeastward retractions along the continental shelf in late spring–summer (Piola et al. 2005, 2008). Such combined effects of freshwater discharge and winds contribute to the modification of nutrient distributions over the continental shelf (Braga et al. 2008). Dissolved silicate and phosphate show a negative correlation with salinity and remarkable features of the terrestrial inputs in the winter period associated with low-salinity waters from the Rio de la Plata outflow that contribute to local fertilization. On the other hand, dissolved nitrate shows the key presence of upwelling and regeneration processes at some points near the continental shelf break.

Offshore, near the continental slope during the winter, the Malvinas Current grows stronger while the Brazil Current transport decreases and the latitude of the BMC moves northward. During the austral summer, a southward displacement of the latitude of the BMC occurs (Matano et al. 1993).

4 Interannual Surface Nutrient Dynamics

Unlike seasonality, processes that affect the macronutrient concentration along the years did not follow predictable patterns.

In the DP, the Southern Annular Mode (SAM) and El Niño Southern Oscillation (ENSO) and its counterpart (La Niña) take a lead role in the variation between years (Ducklow 2008; Stammerjohn et al. 2008, Martinson et al. 2008; Smith et al. 2008). The greater influx of warmer Upper Circumpolar Deep Water (UCDW) from the ACC is a possible consequence of the strengthening of southward winds over the Southern Ocean, often quantified by the movement of the SAM to a more positive state. These atmospheric changes could induce stronger upwelling (Waugh et al. 2013). In the north of the Antarctic Peninsula, intense winds drive an Ekman northward transport and a consequent Ekman pumping, increasing the nutrient concentration. The SAM is strongly involved in the direct atmospheric changes affecting the area, along with the El Niño and La Niña via teleconnections with the tropical Pacific.

Weak southerly winds are the atmospheric response to "El Niño" in the DP, fact that generates the delay of ice pack recoil during spring, particularly offshore. Consequently, it also decreases the surface friction and reduces the upwelling of nutrient-rich waters. On the other hand, during La Niña strong northwesterly winds are the response from spring to fall. This leads to a quick retreat of the offshore ice and a pileup of onshore ice. In addition, a strong flow of warm moist air occurs over the Antarctic Peninsula (Massom et al. 2006). The sum of these conditions leads to a quick and extensive retreat of the ice field (Turner et al. 2002). In consequence, the water column mixing is intensified and the upwelling of deep water strengthened (Martinson et al. 2008), leading to an increase in macronutrient concentration in surface waters.

Somewhat north, during El Niño the ACC fronts displace northward, resulting in the reduced mixing of coastal and oceanic waters. During La Niña, the ACC fronts displace southward increasing the mixing between coastal and oceanic waters (Ducklow et al. 2013). Besides, a weakening of the surface ACC flow in the Drake Passage associated with ENSO events was suggested (Lenn et al. 2007).

In the southern AS, all major Patagonian rivers show a similar pattern at ENSO-like frequency range; however, other factors exert a greater effect on their flows. During the last years, probably due to anthropic causes, the Negro River has decreased its annual discharge significantly. In the same way, due to global warming, by the influence of glacier melt water, the Santa Cruz River shows an increasing discharge trend (Pasquini and Depetris 2007). These changes can have major consequences in the coming years.

Some authors have suggested that the MC intensity is related to the ACC intensity in the DP (e.g., Vivier and Provost 1999). As mentioned above, this would mean a weakening in intensity associated with ENSO events and a decrease in the injection of nutrients into the AS.

In the northern AS, the hydrological variability of the Río de la Plata tributaries seems to be linked simultaneously to long-term tendencies and interannual teleconnections of ENSO (Boulanger et al. 2005). ENSO usually triggers anomalously high precipitation in most areas of the Río de la Plata drainage basin (e.g., Boulanger et al. 2005). Nevertheless, in relation to the concentration of surface nutrients, the distribution of the plume plays a preponderant role. At interannual time scales, the northernmost Rio de la Plata plume penetrations in winter are associated with more intense and persistent northeastward wind stress. In contrast, in years characterized by weaker northeastward wind stress, the winter plume only reaches half of that distance (Piola et al. 2008). In other words, increases on discharge led to a greater influence of the Rio de la Plata plume over the continental shelf although during ENSO events it did not penetrate farther north because of the unfavorable northerly wind pattern (Gonzalez-Silvera et al. 2006).

With respect to the confluence, if during the ENSO events the flow of the MC is weakened, the consequence should be a southward movement of the BMC.

5 Estimated Surface Nutrient Concentration

The sum of the abovementioned processes generates a localized distribution of macronutrients. Thirteen zones can be chemically characterized based on existing information (Fig. 3):

1. The seasonal ice zone (SIZ) localized between the Antarctic shelf break and the northern limit of the ice pack
2. The Permanently Open Ocean Zone (POOZ) localized between the northern limit of the ice pack and the Polar Front
3. The Polar Front zone (PFZ) localized between the Polar Front and the Subantarctic Front
4. The transitional subantarctic zone (TSAZ) localized between the Subantarctic Front and the South American shelf break

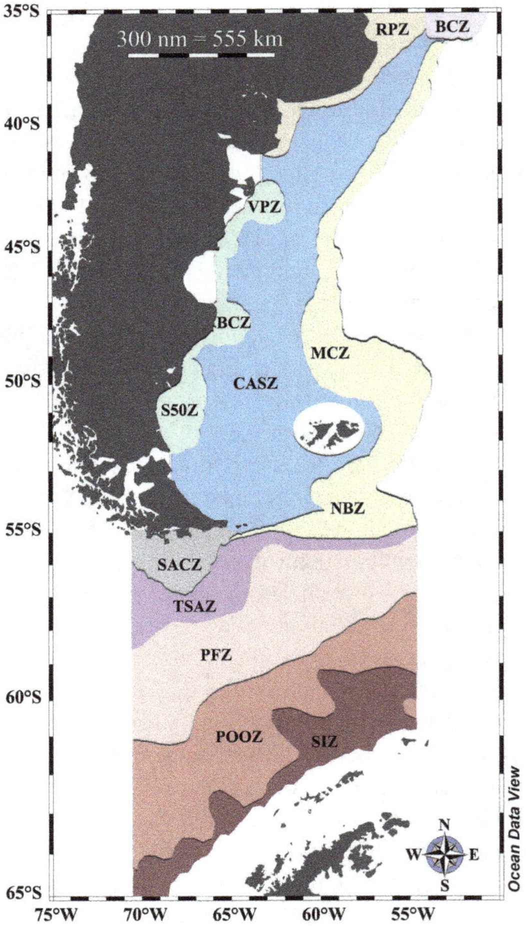

Fig. 3 Zones determined from considering the main processes that govern surface nutrient concentration in the Southwestern Atlantic. SIZ seasonal ice zone, POOZ Permanently Open Ocean Zone, PFZ Polar Front zone, TSAZ transitional subantarctic zone, SACZ South American coastal zone, NBZ Namuncurá bank zone, CASZ central Argentine Sea zone, MCZ Malvinas Current zone, S50Z south of 50°S zone, BCZ Blanco Cape zone, VPZ Valdés Peninsula zone, RPZ Rio de la Plata zone, BzCZ Brazil Current zone. The Ocean Data View software (Schlitzer 2017) was used for mapping

5. The South American coastal zone (SACZ) localized between the South American shelf break and the southern continental tip
6. The Namuncurá bank zone (NBZ)
7. The central Argentine Sea zone (CASZ) covering the shelf from the Estados Island to the Rio de la Plata estuary
8. The Malvinas Current zone (MCZ) covering the influence area of this current
9. The south of 50°S zone (S50Z), a combined estuarial and tidal frontal zone
10. The Cape Blanco zone (BCZ)
11. The Valdés Peninsula zone (VPZ), BCZ and VPZ both tidal frontal areas
12. The Rio de la Plata zone (RPZ), the northern estuarial limit of the AS
13. The Brazil Current Zone (BzCZ), out of our study area but important in the overall context

For these zones, ranges of surface macronutrient concentration were summarized from data obtained during field surveys (Table 1). These databases with historical

Table 1 Macronutrient concentration ranges over the study area. Ranges were established on the basis of published data and organized so that they summarize the magnitude of macronutrient concentration in surface waters across the zones depicted in Fig. 3. Please note that available information is quite scarce compared to other regions of the world ocean. SIZ Seasonal Ice Zzone, POOZ Permanently Open Ocean Zone, PFZ Polar Front Zone, TSAZ Transitional Subantarctic Zone, SACZ South American Coastal Zone, NBZ Namuncurá Bank Zone, CASZ Central Argentine Sea Zone, MCZ Malvinas Current Zone, S50Z South of 50°S Zone, BCZ Blanco Cape Zone, VPZ Valdés Peninsula Zone, RPZ Rio de la Plata Zone, BzCZ Brazil Current Zone

Zone	Nitrate (μM)	Phosphate (μM)	Silicate (μM)	Seasonality	References
SIZ	20–30	1.5–2.0	40–60	High due to variability of the ice pack	Tréguer and Jacques (1992) Lara et al. (2010) Paparazzo (2011) García et al. (2014) Paparazzo et al. (2016)
POOZ	20–25	1.5–1.8	15–50	High due to variability of the ice pack	Tréguer and Jacques (1992) Lara et al. (2010) Paparazzo (2011) García et al. (2014) Paparazzo et al. (2016)
PFZ	15–20	1.2–1.8	5–20	Medium	Tréguer and Jacques (1992) Lara et al. (2010) Paparazzo (2011) García et al. (2014) Paparazzo et al. (2016)
TSAZ	12–18	1.1–1.6	5–15	Medium	Brandhorst and Castello (1971) Piola and Falabella (2009) Lara et al. (2010) Paparazzo (2011) García et al. (2014)
SACZ	9–15	0.8–1.4	1–5	Medium	Brandhorst and Castello (1971) Piola and Falabella (2009) Lara et al. (2010) Paparazzo (2011) García et al. (2014)

(continued)

Table 1 (continued)

Zone	Nitrate (µM)	Phosphate (µM)	Silicate (µM)	Seasonality	References
NBZ	0–18	0.5–1.6	0–15	Insufficient available information	Brandhorst and Castello (1971) Paparazzo (2011)
CASZ	0–18	0.5–1.4	0–10	High due to thermal stratification	Brandhorst and Castello (1971) Piola and Falabella (2009) Lara et al. (2010) Paparazzo (2011)
MCZ	5–20	0.5–2.0	0.9–15	Medium	Brandhorst and Castello (1971) García et al. (2008) Carreto et al. (2007) Carreto et al. (2016)
S50Z	6–14	0.9–1.4	2–10	Low/medium	Paparazzo et al. (2010) Paparazzo (2011) Sabatini et al. (2012)
BCZ	6–8	0.5–1.4	2–8	Low/medium	Brandhorst and Castello (1971) Paparazzo et al. (2010) Paparazzo (2011)
VPZ	2–6	1.0–1.5	1–10	Low/medium	Brandhorst and Castello (1971) Carreto et al. (2007) Paparazzo et al. (2010) Paparazzo (2011)
RPZ	0–4	0.4–0.8	5–30	Medium	Brandhorst and Castello (1971) Braga et al. (2008) Piola and Falabella (2009)
BzCZ	0–3	0.1–0.5	1–3	Medium	Brandhorst and Castello (1971) Braga et al. (2008) García et al. (2014)

information have contributed to depict global pictures (e.g., WOA13 – García et al. 2014), and some works have covered almost the entire study area (e.g., Lara et al. 2010; Paparazzo 2011). In addition, some publications added precision in the spatial and temporal interpretation of the DP (e.g., Tréguer and Jacques 1992; Paparazzo et al. 2016) and the AS (e.g., Brandhorst and Castello 1971; Braga et al. 2008; García et al. 2008; Piola and Falabella 2009; Paparazzo et al. 2010).

6 Conclusions

In this work, some of the numerous processes that affect the macronutrient concentration in the complex environment under study were evidenced. Currents modulated by topography, tides, and winds, play a relevant role for transporting nutrients while fronts generate abrupt changes in the surface conditions. In addition, the ice pack coverage in the DP and the water column stratification in the AS govern the seasonal oscillation. Interannual events like ENSO produce variations in currents and rivers. Moreover, wind conditions have effects over the stacking of ice packs, the intensity and direction of currents, and the extension of low-salinity plumes. However, despite the great advances that have taken place in recent years, there is still much to know. Available information in macronutrients is not sufficient for a detailed analysis in the water column, neither for interpreting long-term temporal changes nor for explaining localized fertilization events, among others.

What is clear in terms of what is known is that ocean life (and earth life) depends on the surface dynamics of nutrients, and this can only be understood by intensifying a multidisciplinary scientific approach.

Acknowledgments An important part of this study was based on our work in the framework of a cooperative research program (ARGAU, Programme de Coopération avec la ARGentine pour l'étude de l'océan Atlantique AUstral) between the Laboratoire de Biogéochimie et Chimie Marines at the Université Pierre et Marie Curie in Paris (France), the Instituto Antártico Argentino and the Servicio de Hidrografía Naval Argentina from 2001 to 2004. We would like to thank I. Schloss for allowing us to participate in the program, as well as V. Alder for her valuable help in the interpretation of oceanographic information.

References

Acha EM, Mianzan HW, Guerrero RA et al (2004) Marine fronts at the continental shelves of austral South America: physical and ecological processes. J Mar Syst 44(1):83–105

Anderson LG, Kaltin S (2001) Carbon fluxes in the Arctic Ocean—potential impact by climate change. Polar Res 20(2):225–232

Arrigo KR, Worthen DL, Lizotte MP et al (1997) Primary production in Antarctic Sea ice. Science 276(5311):394–397

Barker PF, Filippelli GM, Florindo F et al (2007) Onset and role of the Antarctic circumpolar current. Deep Sea Res II Top Stud Oceanogr 54(21):2388–2398

Berger WH (2007) Cenozoic cooling, Antarctic nutrient pump, and the evolution of whales. Deep Sea Res II Top Stud Oceanogr 54:2399–2421

Boulanger JP, Leloup J, Penalba O et al (2005) Observed precipitation in the Paraná-Plata hydrological basin: long-term trends, extreme conditions and ENSO teleconnections. Clim Dyn 24(4):393–413

Braga ES, Chiozzini VC, Berbel GB et al (2008) Nutrient distributions over the southwestern South Atlantic continental shelf from mar del Plata (Argentina) to Itajaí (Brazil): winter–summer aspects. Cont Shelf Res 28(13):1649–1661

Brandhorst W, Castello JP (1971) Evaluación de los recursos de anchoíta (*Engraulis anchoita*) frente a la Argentina y Uruguay: 1. las condiciones oceanográficas, sinopsis del conocimiento actual sobre la anchoíta y el plan para su evaluación. Contr Inst Biol Mar, Mar del Plata, Argentina 166:1–63

Brandini FP, Boltovskoy D, Piola A et al (2000) Multiannual trends in fronts and distribution of nutrients and chlorophyll in the southwestern Atlantic (30–62°S). Deep Sea Res I Oceanogr Res Pap 47:1015–1033

Brzezinski MA, Jones JL, Demarest MS (2005) Control of silica production by iron and silicic acid during the Southern Ocean Iron experiment (SOFeX). Limnol Oceanogr 50(3):810–824

Carreto JI, Lutz VA, Carignan MO et al (1995) Hydrography and chlorophyll a in a transect from the coast to the shelf-break in the Argentinian sea. Cont Shelf Res 15(2/3):315–336

Carreto JI, Carignan MO, Montoya NG, Cucchi Colleoni AD (2007) Ecología del fitoplancton en los sistemas frontales del Mar Argentino. In: Sánchez RP, Bezzi SI (eds) El Mar Argentino y sus recursos pesqueros 5. El ecosistema marino. Publicaciones Especiales INIDEP, Mar del Plata, p 11–31

Carreto JI, Montoya NG, Carignan MO et al (2016) Environmental and biological factors controlling the spring phytoplankton bloom at the Patagonian shelf-break front–degraded fucoxanthin pigments and the importance of microzooplankton grazing. Prog Oceanogr 146:1–21

Decembrini F, Bergamasco A, Mangoni O (2014) Seasonal characteristics of size-fractionated phytoplankton community and fate of photosynthesized carbon in a subantarctic area (Straits of Magellan). J Mar Syst 136:31–41

Depetris PJ, Paolini JE (1991) Biogeochemical aspects of South American rivers: the Paraná and the Orinoco. Biogeochemistry of major world. Rivers 74:105–125

Depetris PJ, Gaiero DM, Probst JL et al (2005) Biogeochemical output and typology of rivers draining Patagonia's Atlantic seaboard. J Coast Res 21(4):835–844

Ducklow HW (2008) Long-term studies of the marine ecosystem along the West Antarctic peninsula. Deep Sea Res II Top Stud Oceanogr 55:1945–1948

Ducklow HW, Fraser WR, Meredith MP et al (2013) West Antarctic peninsula: an ice-dependent coastal marine ecosystem in transition. Oceanography 26(3):190–203

Dulaiova H, Ardelan MV, Henderson PB et al (2009) Shelf-derived iron inputs drive biological productivity in the southern drake passage. Global Biogeochem Cycles 23:GB4014

Fedúlov PP, Remeslo AV, Burykin SN et al (1990) Variabilidad de la Corriente de Malvinas. Frente Marítimo 6:121–127

Frants M, Gille ST, Hewes CD et al (2013) Optimal multiparameter analysis of source water distributions in the southern drake passage. Deep Sea Res II Top Stud Oceanogr 90:31–42

Freeman NM, Lovenduski NS (2016) Mapping the Antarctic polar front: weekly realizations from 2002 to 2014. Earth Syst Sci Data 8(1):191–198

García VM, García CA, Mata MM et al (2008) Environmental factors controlling the phytoplankton blooms at the Patagonia shelf-break in spring. Deep Sea Res I Oceanogr Res Pap 55(9):1150–1166

García HE, Locarnini RA, Boyer TP et al (2014) World Ocean Atlas 2013, vol 4 Dissolved inorganic nutrients (phosphate, nitrate, silicate). Levitus S (ed), Mishonov A (tech ed) NOAA Atlas NESDIS 76

Garzón JC, Martínez AM, Barrera F et al (2016) The Pacific-Atlantic connection: biogeochemical signals in the southern end of the Argentine shelf. J Mar Syst 163:95–101

Glorioso P (2000) Patagonian shelf 3D tide and surge model. J Mar Syst 24:141–151

Gonzalez-Silvera A, Santamaria-del-Angel E, Millán-Núñez R (2006) Spatial and temporal variability of the Brazil-Malvinas confluence and the La Plata plume as seen by SeaWiFS and AVHRR imagery. J Geophys Res Oceans 111(C6)

Hatta M, Measures CI, Selph KE et al (2013) Iron fluxes from the shelf regions near the south Shetland Islands in the drake passage during the austral-winter 2006. Deep Sea Res II Top Stud Oceanogr 90:89–101

Jacobs SS (1991) On the nature of the Antarctic slope front. Mar Chem 35:9–24

Kim D, Shim J, Kim KT et al (2004) Distribution of total CO2, nutrients, chlorophyll *a* in the Scotia Sea, during austral summer. Ocean Pol Res 26(3):401–414

Klunder MB, Laan P, De Baar HJW et al (2014) Dissolved Fe across the Weddell Sea and drake passage: impact of DFe on nutrient uptake. Biogeosciences 11:651–669

Lara RJ, Alder V, Franzosi CA et al (2010) Characteristics of suspended particulate organic matter in the southwestern Atlantic: influence of temperature, nutrient and phytoplankton features on the stable isotope signature. J Mar Syst 79:199–209

Lenn YD, Chereskin TK, Sprintall J et al (2007) Mean jets, mesoscale variability and eddy momentum fluxes in the surface layer of the Antarctic circumpolar current in drake passage. J Mar Res 65(1):27–58

Levitus S, Conkright ME, Reid JL et al (1993) Distribution of nitrate, phosphate and silicate in the world oceans. Prog Oceanogr 31:245–273

Martinson DG, Stammerjohn SE, Iannuzzi RA et al (2008) Western Antarctic peninsula physical oceanography and spatio-temporal variability. Deep Sea Res II Top Stud Oceanogr 55:1964–1987

Massom RA, Stammerjohn SE, Smith RC et al (2006) Extreme anomalous atmospheric circulation in the West Antarctic peninsula region in austral spring and summer 2001/2, and its profound impact on sea ice and biota. J Clim 19(15):3544–3571

Matano RP, Palma ED (2008) On the upwelling of Downwelling currents. J Phys Oceanogr 38:2482–2500

Matano RP, Schlax MG, Chelton DB (1993) Seasonal variability in the southwestern Atlantic. J Geophys Res Oceans 98(C10):18027–18035

Munro DR, Lovenduski NS, Stephens BB et al (2015) Estimates of net community production in the Southern Ocean determined from time series observations (2002–2011) of nutrients, dissolved inorganic carbon, and surface ocean pCO2 in drake passage. Deep Sea Res II Top Stud Oceanogr 114:49–63

Palma ED, Matano RP (2012) A numerical study of the Magellan plume. J Geophys Res Oceans 117(C5)

Palma ED, Matano RP, Piola AR (2008) A numerical study of the southwestern Atlantic shelf circulation: Stratified Ocean response to local and offshore forcing. J Geophys Res Oceans 113(C11)

Paparazzo FE (2011) Distribución espacio–temporal de nutrientes en el Mar Argentino, Pasaje Drake y Península Antártica. Tasa de incorporación por fitoplancton. Tesis doctoral, Facultad de Ciencias Exactas y Naturales, Universidad de Buenos Aires

Paparazzo FE, Bianucci L, Schloss IR et al (2010) Cross-frontal distribution of inorganic nutrients and chlorophyll *a* on the Patagonian continental shelf of Argentina during summer and fall. Rev Biol Mar Oceanogr 45(1):107–119

Paparazzo FE, Alder VA, Schloss IR et al (2016) Spatial and temporal trends in the distribution of macronutrients in surface waters of the drake passage. Ecol Austral 26(1):27–39

Pasquini AI, Depetris PJ (2007) Discharge trends and flow dynamics of south American rivers draining the southern Atlantic seaboard: an overview. J Hydrol (Amst) 333(2):385–399

Piola A, Falabella V (2009) Principales características oceanográficas y físicas. In: Falabella V, Campagna C, Croxall J (eds) Atlas del Mar Patagónico. Especies y espacios. Wildlife Conservation Society and ByrdLife International, Buenos Aires, pp 56–76

Piola AR, Matano RP, Palma ED et al (2005) The influence of the Plata River discharge on the western South Atlantic shelf. Geophys Res Lett 32:L01603

Piola AR, Möller OO, Guerrero RA et al (2008) Variability of the subtropical shelf front off eastern South America: winter 2003 and summer 2004. Cont Shelf Res 28(13):1639–1648
Provost C, Garçon V, Falcon LM (1996) Hydrographic conditions in the surface layers over the slope-open ocean transition area near the Brazil-Malvinas confluence during austral summer 1990. Cont Shelf Res 16(2):215–221, 219–235
Rivas AL, Dogliotti AI, Gagliardini DA (2006) Seasonal variability in satellite-measured surface chlorophyll in the Patagonian shelf. Cont Shelf Res 26(6):703–720
Romero SI, Piola AR, Charo M et al (2006) Chlorophyll-a variability off Patagonia based on SeaWiFS data. J Geophys Res Oceans 111(C5)
Sabatini ME, Akselman R, Reta R et al (2012) Spring plankton communities in the southern Patagonian shelf: hydrography, mesozooplankton patterns and trophic relationships. J Mar Syst 94:33–51
Schlitzer, R (2017) Ocean data view. http://odv.awi.de
Simonella LE, Palomeque ME, Croot PL et al (2015) Soluble iron inputs to the Southern Ocean through recent andesitic to rhyolitic volcanic ash eruptions from the Patagonian Andes. Global Biogeochem Cycles 29(8):1125–1144
Smith RC, Martinson DG, Stammerjohn SE et al (2008) Bellingshausen and western Antarctic peninsula region: pigment biomass and sea-ice spatial/temporal distributions and interannual variability. Deep Sea Res II Top Stud Oceanogr 55:1949–1963
Song H, Marshall J, Follows MJ et al (2016) Source waters for the highly productive Patagonian shelf in the southwestern Atlantic. J Mar Syst 158:120–128
Sprintall J (2003) Seasonal to interannual upper-ocean variability in the drake passage. J Mar Res 61(1):27–57
Stammerjohn SE, Martinson DG, Smith RC (2008) Sea ice in the western Antarctic peninsula region: Spatio-temporal variability from ecological and climate change perspectives. Deep Sea Res II Top Stud Oceanogr 55:2041–2058
Strub PT, James C, Combes V et al (2015) Altimeter-derived seasonal circulation on the Southwest Atlantic shelf: 27°–43° S. J Geophys Res Oceans 120(5):3391–3418
Tréguer P, Jacques G (1992) Dynamics of nutrients and phytoplankton, and fluxes of carbon, nitrogen and silicon in the Antarctic Ocean. Polar Biol 12:149–162
Turner J, Harangozo S, Marshall G et al (2002) Anomalous atmospheric circulation over the Weddell Sea, Antarctica, during the austral summer of 2001/02 resulting in extreme sea ice conditions. Geophys Res Lett 29(2160):1–4
Vivier F, Provost C (1999) Direct velocity measurements in the Malvinas current. J Geophys Res Oceans 104(C9):21083–21103
Waugh DW, Primeau F, De Vries T et al (2013) Recent changes in the ventilation of the southern oceans. Science 339(6119):568–570
Zhou M, Zhu Y, Dorland RD et al (2010) Dynamics of the current system in the southern drake passage. Deep Sea Res I Oceanogr Res Pap 57(9):1039–1048

Perspective: Continental Inputs of Matter into Planktonic Ecosystems of the Argentinean Continental Shelf—the Case of Atmospheric Dust

Augusto C. Crespi-Abril, Elena S. Barbieri, Leilén Gracia Villalobos, Gaspar Soria, Flavio E. Paparazzo, Joanna M. Paczkowska, and Rodrigo J. Gonçalves

Abstract Land-derived dissolved and particulate matter (allochthonous matter) affect pelagic ecosystems by changing factors which include light penetration, nutrient availability, substrate concentration, and in general, biogeochemical cycles in the ocean. In a context of growing anthropogenic impact, this material may not only increase its load but also carry toxic substances. Riverine runoff is the most studied mechanism of particulate matter input from the continent to the sea in the southern region of South America where the continental shelf is widest (e.g., Atlantic Patagonia). However, there are other sources of particulate matter which are not affected by rivers in this semiarid region: aeolian material. Winds in this region (notably the Southern Hemisphere westerlies) are the only way continental

A. C. Crespi-Abril · G. Soria
Laboratorio de Oceanografía Biológica (LOBio), Centro para el Estudio de Sistemas Marinos (CESIMAR), Consejo Nacional de Investigaciones Científicas y Técnicas (CONICET), Puerto Madryn, Argentina

Universidad Nacional de la Patagonia San Juan Bosco, Facultad de Ciencias Naturales, Sede Puerto Madryn, Puerto Madryn, Argentina
e-mail: crespi@cenpat-conicet.gob.ar; soria@cenpat-conicet.gob.ar

E. S. Barbieri · L. G. Villalobos · J. M. Paczkowska · R. J. Gonçalves (✉)
Laboratorio de Oceanografía Biológica (LOBio), Centro para el Estudio de Sistemas Marinos (CESIMAR), Consejo Nacional de Investigaciones Científicas y Técnicas (CONICET), Puerto Madryn, Argentina
e-mail: barbieri@cenpat-conicet.gob.ar; gracia@cenpat-conicet.gob.ar

F. E. Paparazzo
Laboratorio de Oceanografía Biológica (LOBio), Centro para el Estudio de Sistemas Marinos (CESIMAR), Consejo Nacional de Investigaciones Científicas y Técnicas (CONICET), Puerto Madryn, Argentina

Laboratorio de Oceanografía Química y Contaminación de Aguas (LOQyCA), Centro para el Estudio de Sistemas Marinos (CESIMAR), Consejo Nacional de Investigaciones Científicas y Técnicas (CONICET), Puerto Madryn, Argentina
e-mail: paparazzo@cenpat-conicet.gob.ar

M. S. Hoffmeyer et al. (eds.), *Plankton Ecology of the Southwestern Atlantic*,
https://doi.org/10.1007/978-3-319-77869-3_5

aeolian material (atmospheric aerosols or "dust") can reach not only the shelf but even further onto oceanic HNLC (high nutrient–low chlorophyll) regions of the Atlantic Southern Ocean. This potential impact of Patagonian dust beyond the continental shelf attracts the attention of the global climate community, and at the same time, it opens questions about the potential effects of dust in coastal waters. According to previous work and ongoing studies, deposited particles can have significant impacts in the chemical and biological components in the euphotic zone. However the effects of this airborne material in plankton communities of South America are largely unknown, mostly due to the lack of in situ studies and observations. Since the events of dust mobilization, transport, and deposition are expected to increase (due to climate change) and interact with other global change factors such as warming and more intensive land use, the influence of dust input may become more prominent for coastal and oceanic regions of southern South America in the next decades.

Keywords Aeolian dust · Atmospheric deposition · Particulate matter · Southwest Atlantic

1 Introduction

The continental shelf of Argentina is one of the largest in the world and sustains fisheries of high economic value. Factors which support coastal productivity include frontal regions (e.g., tidal, shelf-break, upwelling, and estuarine fronts) and input of nutrients and particles from continental sources into the adjacent seawater (e.g., riverine and underwater discharge, coastal runoff, atmospheric deposition) (Acha et al. 2004, 2008; Gassó et al. 2010a). This input results in a continuous supply of material into coastal and oceanic areas, thus affecting the pelagic realm, planktonic (and ultimately benthic) organisms. In general, dissolved and particulate matter affect pelagic ecosystems by changing factors which include light penetration, nutrient and detritus availability, substrate concentration, and in general, biogeochemical cycles in coastal and oceanic areas. In addition, nutrients and particles may also carry pollution from human activities or make the aquatic environment more prone to the occurrence of harmful algal blooms.

The most conspicuous source of land-derived material into the coast of Argentina is the Río de la Plata, the most important river in terms of discharge and area of influence. Due to its large size and location, the Río de la Plata has traditionally been the focus of extensive studies. We focus here on a second, less studied source: atmospheric transport. Dominant winds in Patagonia transport material eastward, adding particles into the Patagonian Atlantic continental shelf (PACS), one of the broadest in Earth, and into the oceanic HNLC (high nutrient–low chlorophyll) areas of the Southern Ocean. We present here a brief discussion of the current knowledge, compare with study cases in other regions of the world, and offer some thoughts on how this could affect pelagic ecosystems within PACS and beyond. When reading the text, at times one can ask: is this relevant for Patagonia? For many cases, the answer is "unknown, but likely." Simply, there is a lack of basic data, thus our poor understanding of even basic questions about sources of emission, let alone the overall effects on planktonic communities. However, these phenomena have been studied in different regions of the world where the same components are present (i.e., wind and mobilizable soil). Therefore the point of this perspective is to call into attention the importance of continental dust deposition on planktonic organisms and, by using analogies with other regions, try to infer possible scenarios likely to be important in Patagonia as well. The quantification of its effects will probably remain unknown until more studies are completed and integrated with different fields like marine biology, remote sensing, geology, and urbanization/land management.

2 Particle Flux from Land to Oceans

It is estimated that the South Atlantic Ocean receives a total sediment load of 400×10^6 t yr^{-1} and 240×10^6 t yr^{-1} of dissolved load from rivers (Milliman 2001). In terms of sediment volume, the Río de la Plata is the second largest river in South America. Its drainage area covers about 20% of the South American continental area (Acha et al. 2008). The turbidity front in the innermost part of the estuary is partially formed by transported riverine suspended material, with contributions from resuspended material by tidal stirring. This turbidity restricts photosynthesis (by blocking sunlight penetration) even in presence of high nutrients (Nagy et al. 2002), which makes this environment suitable for detritus-based food webs (Acha et al. 2008; Derisio et al. 2014).

Concentrations of suspended matter range from 100 to 300 mg L^{-1} (Framiñan and Brown 1996). Due to its large size and location, the Río de la Plata has been extensively studied, and the reader may refer to reviews elsewhere (e.g., Acha et al. 2008). South of the Río de la Plata, several rivers discharge along the Argentine coastline, but their contribution is smaller. Each year all the main rivers in Patagonia combined export an estimated average of 2×10^6 t of suspended particulate matter into the South Atlantic Ocean, which represents only ~2% of the total annual export by the Río de la Plata (Gaiero et al. 2003).

3 Dust

Another mechanism by which land-derived material can reach coastal and oceanic regions is aeolian "dust", or more precisely atmospheric transport of mineral aerosols. We will use a broad definition of "dust" as terrestrial sediment, sized < 100 μm, which is transported in an aeolian suspension. We focus here on more steady dust sources and do not consider cases of deposition from volcanic eruptions (due to ash ejection into the atmosphere or remobilization from the soil immediately after eruptions). In spite of being more transient, these can contribute mineral dust significantly larger than other continental sources (Gaiero et al. 2003; Duggen et al. 2010). We use the terms "dust" and "mineral aerosols" and "atmospheric inputs" interchangeably, although more rigorous definitions are used in specific studies (McTainsh and Strong 2007; Després et al. 2012). The composition and size distribution of aerosol particles depend on the type of source and emission (see below). Atmospheric dust includes mixtures of various mineral species with different shapes, coatings, and mixing states (Gassó et al. 2010a) and may include also biogenic aerosols (Jaenicke 2005; Burrows et al. 2009; Després et al. 2012). These particles can be transported very long distances (Gillette 1981) and can impact areas hundreds (if not thousands) of kilometers downwind from the source (Gassó et al. 2010a). It is estimated that, globally, 2000 t of dust are emitted yearly into the atmosphere, 25% of which is deposited to the ocean (Shao et al. 2011).

4 Sources

Several regions have been studied as dominant sources of mineral aerosols, such as the arid regions of North Africa and the Arabian Peninsula (dust belt) among others (Washington et al. 2003; Mahowald et al. 2005). As an example, 240 ± 80 t of dust are transported annually from Africa to the Atlantic Ocean, 140 ± 40 t are deposited in the Atlantic Ocean (Yu et al. 2015). Dust sources in the Southern Hemisphere are considerably less active than those in the Northern Hemisphere, and the most active regions are located in Argentina, southern Africa, and Australia (Maher et al. 2010). Total atmospheric dust inputs to the oceans is ~450 t yr^{-1}, among which about 4% settles into the South Atlantic Ocean (Jickells et al. 2005). Source areas are usually soils with erodible material, dry climate, and strong winds. In present-day southern South America, the major dust source areas are located in a continuous N–S band of arid and semiarid terrains ("Diagonal árida," arid diagonal) extending from the coastal regions of Perú to Patagonia. Three main persistent source areas stand out: Patagonia, central–western Argentina, and the Puna/Altiplano plateau. These areas were continuously active over the last several glacial cycles (Gili and Gaiero 2014). Not only it has extensive dry areas prone to wind erosion, Patagonia is located in the zonal latitudes of the strongest (in yearly average) oceanic winds in the world, the Southern Hemisphere westerlies (also known as "Roaring Forties"). The importance of present-day Patagonian sources has been acknowledged in global remote sensing studies (Prospero et al. 2002; Shao et al. 2011). Patagonia emission sources are linked to thousands of small enclosed basins, and the area is strongly influenced by the mentioned westerlies (Maher et al. 2010). Dust deposition rates at the Patagonian coasts are estimated to be in average about ~5 g m^{-2} yr^{-1} (Gaiero et al. 2003, 2007; Maher et al. 2010). It has been estimated that Patagonian sediments are supplied to the South Atlantic shelf in approximately equivalent amounts from the atmosphere (~30×10^6 t yr^{-1}) and coastal erosion (~40×10^6 t yr^{-1}) with much less coming from the rivers (~2×10^6 t yr^{-1}) (Gaiero et al. 2003). Yet, the contribution from the atmosphere has been suggested to be even larger (Simonella et al. 2015). These particles can travel ~1800 km eastward from Patagonia (Gassó and Stein 2007; Gassó et al. 2010b). In many cases, dust events in Patagonia yield lower signals in global studies due in part to the scarcity (Bullard et al. 2016) and intermittency (Gassó et al. 2010b) of observations and also to cloud cover (Johnson et al. 2011). However, northern Patagonia has been identified as an important area of dust source for the Atlantic Ocean (Johnson et al. 2010) that is continuously emitting dust (Crespi-Abril et al. 2018).

5 Effects on Primary Production

Airborne particles impact surface of the sea without the typical salinity-driven stratification found in rivers; thus particles deposited on the sea surface immediately start undergoing concomitant marine processes, including iron solubilization (Duce and Tindale 1991; Bonnet et al. 2005; Knippertz and Stuut 2014). Major oceans are exposed to the transport and deposition of mineral dust, which alters the biogeochemical cycles of a number of nutrients, especially nitrogen (Baker et al. 2017) and iron (Mahowald et al. 2005, 2009). In some cases, the atmospheric input of nutrients is even of the same magnitude as riverine inputs (Guieu et al. 1991; Gallisai et al. 2014) or even higher (Ridgwell 2002). In global terms dust provides a rather small fraction of the total iron input to the oceans (river discharge is still the major coastal source); however it is disproportionately important in some areas of the ocean, especially where availability of the micronutrient iron might limit productivity (Jickells et al. 2005). This happens usually in oligotrophic regions (Bonnet et al. 2005) (or high nutrient–low chlorophyll) areas but may be more important than previously thought due to the input of phosphorus and nitrogen (Herut et al. 1999; Pulido-Villena et al. 2010). In addition, a lower concentration of particles from the atmosphere (as compared to riverine input) may have an unexpected effect, as it has been observed that solubility of iron decreases with the increasing amount of particles introduced following a power law (Bonnet and Guieu 2004). Similarly, lower density of air versus water may contribute to this nonlinear effect as iron solubility also changes with particle size (Baker and Jickells 2006). Thus, dust effect might be important in some regions of the PACS which may be nitrogen-limited (Paparazzo et al. 2017). This means in turn that, via input of micro- and macronutrients, mineral dust can potentially affect biological activity in the upper mixed layer of the sea. This has been studied in other regions where it was found that dust can influence positively or negatively phytoplankton population (Paytan et al. 2009; Gassó et al. 2010a; Maher et al. 2010) and change food web structure (Bonnet et al. 2005; Lekunberri et al. 2010). Regarding iron interactions with dust, the most affected plankton species will be those with specific nutrient requirements (Boyd and Doney

2003; Jickells et al. 2005). Just to mention one of the examples, Gallisai et al. (2014) argue that aeolian material from Sahara desert can have ecologically relevant effects on plankton growth in the Mediterranean Sea. The situation may be different in coastal waters off Patagonia, as the influence of particulate material may not be as strong as in the oligotrophic Mediterranean. However, the conditions would be very different beyond the continental shelf, where dust deposition may enhance plankton growth in HCLN waters of the Southern Atlantic Ocean. Thus, what is known in other regions is still unknown here, but hundreds of studies allow us to speculate and foresee that similar scenarios may be happening with Patagonian dust. The magnitude however remains uncertain and leaves many open questions which would need local-regional studies to address them.

In South America, when studying atmospheric deposition of nutrients to the Atlantic Ocean, Patagonia was detected as one of the main sources (Baker et al. 2003). Large areas in Patagonia have dust sources, as it has a dry steppe that covers ~40% of the surface area (Gaiero et al. 2003), is going through a process of desertification (Del Valle et al. 1998; Mazzonia and Vazquez 2009), and exhibit dust-related events in different areas along the year (Fig. 1). In this sense, in several places of the world, it has been demonstrated that heavy human activities on the land surface have led to severe dust episodes in the last century (Gill 1996; Mahowald et al. 2009). The lack of studies relating land management and desertification in Patagonia prevents us to draw solid conclusions. There are simply no data; thus even basic knowledge of this relationship is still poor. However on the basis of several other cases, we speculate that it could very well be the case in Patagonia, where the main factors are present and acting year-round: strong, persistent winds, dry soil, and poor land management (due to urbanization and sheep grazing). Thus using analogies with other regions, we argue that desertification can gradually increase the surface area which can act as source of aeolian material in Patagonia. The extent and effects of these particles can be expected not only in the PCAS region but also far away from the sources. For example, dust from north Patagonia may reach HNLC waters of the Southern Ocean and Antarctica in ~30–96 h (Gassó and Stein 2007; Johnson et al. 2011). Using remote sensing (MODIS) data of dust events, it has been shown that satellite-detectable dust plumes can cover significant areas (an area larger than 400,000 km^2 was covered through a total of six events) over the PACS, even beyond the limits of the continental shelf (Fig. 1, Crespi-Abril et al. 2016). Much of dust particles reach altitudes higher than 5000 m above sea level while being transported (Gassó and Stein 2006; Johnson et al. 2011).

6 Effects of Global Change

Global changes are affecting natural process and biogeochemical cycles worldwide, and Patagonia is not an exception. However, the lack of historical records of dust emissions in Patagonia does not allow determining potential long-term changes in dust fluxes due to global changes or to address which are the process and factors that

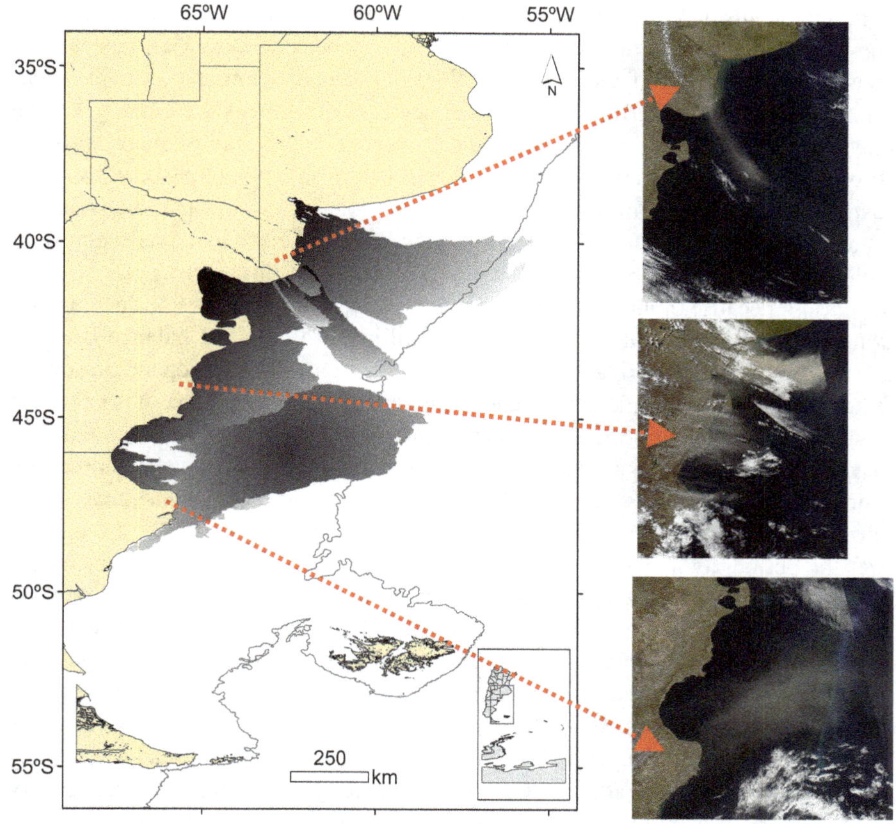

Fig. 1 The map on the left (modified from Crespi-Abril et al. 2016) shows the area of influence of selected dust events over the continental shelf of Argentina. The shaded areas represent the extension of dust plumes into the adjacent ocean. The dust events were obtained from MODIS satellite images of three events of aeolian dust: 23 Jan 2009 (top-right panel), 28 Mar 2009 (middle-right panel) and 24 Jan 2010 (bottom-right panel). Images from NASA/GSFC, Rapid Response website (https://earthdata.nasa.gov)

modulate dust emissions into the PCAS and most importantly how changes on those factors will affect dust emission. However, some possible scenarios can be inferred to occur in Patagonia by considering similarities with other well-studied cases elsewhere.

As anthropogenic impacts grow steadily world-wide, events of atmospheric aerosols may not only increase its load and frequency, but aeolian particles may also carry toxic substances. As a source of dust material, Patagonia is under the influence of several regional (e.g., human population growth, land use, desertification, field fires) and global changes (warming, changes in atmospheric circulation and winds, changes in precipitation regimes) which may enhance atmospheric dust depositions. First, human activities and poor land management may result in more areas becoming exposed to erosion and thus mobilization of particles to be carried away by dominant winds. For example, it has been reported in 2011 that a region located in coastal Atlantic Patagonia (Viedma, 41°S–63°W) became active as dust source "possibly due to a combination of poor livestock management and drought conditions" (Johnson et al. 2011). Actually, in some environments with heavy human intervention, atmospheric aerosol may be regarded as an anthropogenic product (Prospero 1996). Urbanization and use of rivers are also a potential source of more suspended particles in the atmosphere. It would probably also mean more sediment going into river flow. Interestingly, it has been observed elsewhere that while sediment loads of rivers may be increasing, actual sediment flux to the ocean may be decreasing because of another human impact: increased river diversion (e.g., irrigation and flood protection through levees) (Milliman 2001). The balance between these two inputs (i.e., aeolian vs riverine origin) will most likely depend on several factors including economic growth and industrialization of the region. It is unknown how this will develop in the Patagonian region, although it has been hypothesized that in developed nations rivers will carry less sediment due to dams' construction. Using the "Three Gorges Dam" in China as an extreme example, the sediment flux into the estuary decreased by 85×10^6 t yr^{-1} (31%) in comparison to a non-dam case over the period 2003–2005 (Yang et al. 2007). It is likely that, due to dams, rivers will also carry less amounts of dissolved nutrients and at different ratios (e.g., in cases where dams and coastal lagoons act as silica sinks; Humborg et al. 2000; Turner et al. 1998). In developing countries, poor land management (e.g., related to livestock, water management, extractive activities, etc.) will increase dissolved and particulate loads, while wind-borne dust is probably increasing everywhere (Milliman 2001). As mentioned above, these are phenomena already occurring somewhere else, and we argue that may be happening in Patagonia at ecologically relevant scales. Secondly, global warming may have two effects. On the one hand, atmospheric circulation is changing, and dominant winds in Patagonia have overall intensified over the past decades (Thompson and Solomon 2002; Thompson et al. 2011). It is unclear whether and how this would affect dust transport and deposition in coastal and oceanic regions off South America and even Antarctic waters, but an increased frequency of dust events is certainly a possibility to consider, since extreme weather events (droughts, rainfalls, storms, heat waves, field fires) are expected to increase (Easterling 2000; Coumou and Rahmstorf 2012). The effects of these potentially more frequent dust events on aquatic environments may be modulated by interactions with other global change factors. In continental waters, for example, it has been found that recurring nutrient inputs, which would mimic the effects of dust events, in combination with high solar ultraviolet radiation (UVR) may affect cell size and composition of the microbial aquatic community (Cabrerizo et al. 2017). In the same line, the joint impact of solar UVR and Saharan dust inputs resulted in a strong change toward autotrophic metabolism in oligotrophic areas (Cabrerizo et al. 2016). On the other hand, the effects of mineral dust on primary production can be more pronounced in conditions of stratification and nutrient limitation. In this sense, global warming is expected to strengthen ocean stratification (Manabe and Stouffer 1993; Sarmiento et al. 1998; Le Quéré et al. 2003 but see also Russell et al. 2006) thus resulting in a longer exposure of plankton communities to atmospheric influence. Unlike in the northern Argentina marine area, which is largely influenced by the Río de la Plata, seasonal stratification in waters of PACS is controlled by temperature (Lucas et al. 2005), and thus changes in thermal stratification may also have important consequences for these coastal and shelf pelagic regions.

Finally, one further element of dust research should be taken into account as human impacts spread on global land and ocean, and that is the biogenic component of atmospheric aerosols. This includes bacteria (Burrows et al. 2009), virus, fungi, and many other materials of biological origin (Després et al. 2012). Biogenic aerosols have traditionally been underestimated or ignored; however in some marine areas, cellular material and proteins compose up to 25% of the atmospheric aerosol (Jaenicke 2005; Jaenicke et al. 2007). Negative impacts on marine life can be expected in some cases from these biogenic aerosols, as well as from other mineral components such as toxic copper particles (Shinn et al. 2000; Paytan et al. 2009).

7 Conclusions

Studies of atmospheric transport of nutrients, inorganic particles, and biogenic aerosols are fields with some history in world research. Nevertheless, when looking for dedicated studies in Patagonia, the lack of relevant observations and data is striking. The area is "a vast and remote region holding a rich variety of past environmental records but a small number of meteorological stations" (Garreaud et al. 2012), and so most studies rely on remote sensing information and numerical simulations. Rather recently, dust research is regaining interest in the region with focus on present-day dust events, and there is a need for more in situ studies and different approaches aiming to understand how marine organisms are adapted to current emission levels and to predict how marine plankton will respond to future changes.

Acknowledgments This work was supported by CONICET (PIP 6447-2016 to R.J.G.) and FONCYT (PICT-2015-1837 to A. C.-A.).

References

Acha EM, Mianzan H, Guerrero R et al (2004) Marine fronts at the continental shelves of austral South America: physical and ecological processes. J Marine Sys 44(1):83–105

Acha EM, Mianzan H, Guerrero R et al (2008) An overview of physical and ecological processes in the Rio de la Plata Estuary. Cont Shelf Res 28:1579–1588. https://doi.org/10.1016/j.csr.2007.01.031

Baker AR, Jickells TD (2006) Mineral particle size as a control on aerosol iron solubility. Geophys Res Lett 33:L17608. https://doi.org/10.1029/2006GL026557

Baker AR, Kelly SD, Biswas KF et al (2003) Atmospheric deposition of nutrients to the Atlantic Ocean. Geophys Res Lett 30(24):2296. https://doi.org/10.1029/2003GL018518

Baker AR, Kanakidou M, Altieri KE et al (2017) Observation- and model-based estimates of particulate dry nitrogen deposition to the oceans. Atmos Chem Phys Discuss:1–38. https://doi.org/10.5194/acp-2016-1123

Bonnet S, Guieu C (2004) Dissolution of atmospheric iron in seawater. Geophys Res Lett 31:L03303. https://doi.org/10.1029/2003GL018423

Bonnet S, Guieu C, Chiaverini J et al (2005) Effect of atmospheric nutrients on the autotrophic communities in a low nutrient, low chlorophyll system. Limnol Oceanogr 50:1810–1819. https://doi.org/10.4319/lo.2005.50.6.1810

Boyd PW, Doney SC (2003) The impact of climate change and feedback processes on the ocean carbon cycle. In: Fasham MJR (ed) Ocean biogeochemistry — the role of the ocean carbon cycle in global change. Springer, Berlin, pp 157–193

Bullard JE, Baddock M, Bradwell T et al (2016) High-latitude dust in the earth system. Rev Geophys 54. https://doi.org/10.1002/2016RG000518

Burrows SM, Elbert W, Lawrence MG et al (2009) Bacteria in the global atmosphere–part 1: review and synthesis of literature data for different ecosystems. Atmos Chem Phys 9:9263–9280. https://doi.org/10.5194/acp-9-9263-2009

Cabrerizo MJ, Medina-Sánchez JM, González-Olalla JM et al (2016) Saharan dust inputs and high UVR levels jointly alter the metabolic balance of marine oligotrophic ecosystems. Sci Rep 6:35892. https://doi.org/10.1038/srep35892

Cabrerizo MJ, Medina-Sánchez JM, Dorado-García I et al (2017) Rising nutrient-pulse frequency and high UVR strengthen microbial interactions. Sci Rep 7:43615. https://doi.org/10.1038/srep43615

Coumou D, Rahmstorf S (2012) A decade of weather extremes. Nat Clim Chang 2:491–496. https://doi.org/10.1038/nclimate1452

Crespi Abril AC, Montes AMI, Williams GN et al (2016) Uso de sensores remotos para la detección de eventos de transporte eólico de sedimentos hacia ambientes marinos en Patagonia. Meteor-Forschung 41:33–47

Crespi-Abril AC, Soria G, De Cian A et al (2018) Roaring forties: an analysis of a decadal series of data of dust in Northern Patagonia. Atmos Environ 177:111–119

Del Valle HF, Elissalde NO, Gagliardini DA et al (1998) Status of desertification in the Patagonian region: assessment and mapping from satellite imagery. Arid Soil Res Rehabil 12:95–121. https://doi.org/10.1080/15324989809381502

Derisio C, Braverman M, Gaitán E et al (2014) The turbidity front as a habitat for *Acartia tonsa* (Copepoda) in the Río de la Plata, Argentina-Uruguay. J Sea Res 85:197–204. https://doi.org/10.1016/j.seares.2013.04.019

Després VR, Huffman JA, Burrows SM et al (2012) Primary biological aerosol particles in the atmosphere: a review. Tellus Ser B Chem Phys Meteorol 64:15598. https://doi.org/10.3402/tellusb.v64i0.15598

Duce RA, Tindale NW (1991) Atmospheric transport of iron and its deposition in the ocean. Limnol Oceanogr 36:1715–1726. https://doi.org/10.4319/lo.1991.36.8.1715

Duggen S, Olgun N, Croot P et al (2010) The role of airborne volcanic ash for the surface ocean biogeochemical iron-cycle: a review. Biogeosciences 7:827–844

Easterling DR (2000) Climate extremes: observations, modeling, and impacts. Science 289:2068–2074. https://doi.org/10.1126/science.289.5487.2068

Framiñan MB, Brown OB (1996) Study of the Río de la Plata turbidity front, part 1: spatial and temporal distribution. Cont Shelf Res 16:12591269–12671282

Gaiero DM, Probst J-L, Depetris PJ et al (2003) Iron and other transition metals in Patagonian riverborne and windborne materials: geochemical control and transport to the southern South Atlantic Ocean. Geochim Cosmochim Acta 67:3603–3623. https://doi.org/10.1016/S0016-7037(03)00211-4

Gaiero DM, Brunet F, al PJ-L (2007) A uniform isotopic and chemical signature of dust exported from Patagonia: rock sources and occurrence in southern environments. Chem Geol 238:107–120. https://doi.org/10.1016/j.chemgeo.2006.11.003

Gallisai R, Peters F, Volpe G et al (2014) Saharan dust deposition may affect phytoplankton growth in the Mediterranean Sea at ecological time scales. PLoS One 9:e110762. https://doi.org/10.1371/journal.pone.0110762

Garreaud R, Lopez P, Minvielle M et al (2012) Large-scale control on the Patagonian climate. J Clim 26:215–230. https://doi.org/10.1175/JCLI-D-12-00001.1

Gassó S, Stein AF (2007) Does dust from Patagonia reach the sub-Antarctic Atlantic Ocean? Geophys Res Lett 34:L01801. https://doi.org/10.1029/2006GL027693

Gassó S, Grassian VH, Miller RL (2010a) Interactions between mineral dust, climate, and ocean ecosystems. Elements 6:247–252. https://doi.org/10.2113/gselements.6.4.247

Gassó S, Stein A, Marino F et al (2010b) A combined observational and modeling approach to study modern dust transport from the Patagonia desert to East Antarctica. Atmos Chem Phys 10:8287–8303. https://doi.org/10.5194/acp-10-8287-2010

Gili S, Gaiero DM (2014) South American dust signature in geological archives of the Southern Hemisphere. Pages Mag 22:78

Gill TE (1996) Eolian sediments generated by anthropogenic disturbance of playas: human impacts on the geomorphic system and geomorphic impacts on the human system. Geomorphology 17:207–228. https://doi.org/10.1016/0169-555X(95)00104-D

Gillette DA (1981) Production of dust that may be carried great distances. Geol Soc Am Spec Pap 186:11–26. https://doi.org/10.1130/SPE186-p11

Guieu C, Martin JM, Thomas AJ et al (1991) Atmospheric versus river inputs of metals to the Gulf of Lions. Mar Pollut Bull 22:176–183. https://doi.org/10.1016/0025-326X(91)90467-7

Herut B, Krom MD, Pan G et al (1999) Atmospheric input of nitrogen and phosphorus to the Southeast Mediterranean: sources, fluxes, and possible impact. Limnol Oceanogr 44:1683–1692. https://doi.org/10.4319/lo.1999.44.7.1683

Humborg C, Conley DJ, Rahm L et al (2000) Silicon retention in river basins: far-reaching effects on biogeochemistry and aquatic food webs in coastal marine environments. AMBIO J Hum Environ 29:45. https://doi.org/10.1639/0044-7447(2000)029[0045:SRIRBF]2.0.CO;2

Jaenicke R (2005) Abundance of cellular material and proteins in the atmosphere. Science 308:73–73. https://doi.org/10.1126/science.1106335

Jaenicke R, Matthias-Maser S, Gruber S (2007) Omnipresence of biological material in the atmosphere. Environ Chem 4:217. https://doi.org/10.1071/EN07021

Jickells TD, An ZS, Andersen KK et al (2005) Global iron connections between desert dust, ocean biogeochemistry, and climate. Science 308:67–71. https://doi.org/10.1126/science.1105959

Johnson MS, Meskhidze N, Solmon F et al (2010) Modeling dust and soluble iron deposition to the South Atlantic Ocean. J Geophys Res Atmos 115:D15202. https://doi.org/10.1029/2009JD013311

Johnson MS, Meskhidze N, Kiliyanpilakkil VP et al (2011) Understanding the transport of Patagonian dust and its influence on marine biological activity in the South Atlantic Ocean. Atmos Chem Phys 11:2487–2502. https://doi.org/10.5194/acp-11-2487-2011

Knippertz P, Stuut J-BW (eds) (2014) Mineral Dust. Springer, Dordrecht

Le Quéré C, Aumont O, Monfray P et al (2003) Propagation of climatic events on ocean stratification, marine biology, and CO2: case studies over the 1979–1999 period. J Geophys Res Oceans 108:3375. https://doi.org/10.1029/2001JC000920

Lekunberri I, Lefort T, Romero E et al (2010) Effects of a dust deposition event on coastal marine microbial abundance and activity, bacterial community structure and ecosystem function. J Plankton Res 32:381–396. https://doi.org/10.1093/plankt/fbp137

Lucas AJ, Guerrero RA, Mianzan HW et al (2005) Coastal oceanographic regimes of the northern argentine continental shelf (34?43?S). Estuar Coast Shelf Sci 65:405–420. https://doi.org/10.1016/j.ecss.2005.06.015

Maher BA, Prospero JM, Mackie D et al (2010) Global connections between aeolian dust, climate and ocean biogeochemistry at the present day and at the last glacial maximum. Earth-Sci Rev 99:61–97. https://doi.org/10.1016/j.earscirev.2009.12.001

Mahowald NM, Baker AR, Bergametti G et al (2005) Atmospheric global dust cycle and iron inputs to the ocean. Glob Biogeochem Cycles 19:GB4025. https://doi.org/10.1029/2004GB002402
Mahowald NM, Engelstaedter S, Luo C et al (2009) Atmospheric iron deposition: global distribution, variability, and human perturbations. Annu Rev Mar Sci 1:245–278. https://doi.org/10.1146/annurev.marine.010908.163727
Manabe S, Stouffer RJ (1993) Century-scale effects of increased atmospheric C02 on the ocean–atmosphere system. Nature 364:215–218. https://doi.org/10.1038/364215a0
Mazzonia E, Vazquez M (2009) Desertification in Patagonia. In: Latrubesse EM (ed) Natural hazards and human-exacerbated disasters in Latin America, Develop earth surf process, vol 13, pp 351–377
McTainsh G, Strong C (2007) The role of aeolian dust in ecosystems. Geomorphology 89:39–54. https://doi.org/10.1016/j.geomorph.2006.07.028
Milliman JD (2001) River inputs. In: Steele J, Thorpe S, Turekian K (eds) Encyclopedia of ocean sciences (1st ed., pp 2419–2427). New York (USA): Academic Press. Retrieved from http://linkinghub.elsevier.com/retrieve/pii/B012227430X00074X
Nagy GJ, Gómez-Erache M, López CH et al (2002) Distribution patterns of nutrients and symptoms of eutrophication in the Rio de la Plata River estuary system. In: Orive E, Elliott M, de Jonge VN (eds) Nutrients and eutrophication in estuaries and coastal waters, Proc 31st Symp Estuar coastal Sci Assoc (ECSA). Springer, Dordrecht, pp 125–139
Paparazzo FE, Williams GN, Pisoni JP et al (2017) Linking phytoplankton nitrogen uptake, macronutrients and chlorophyll- a in SW Atlantic waters: the case of the Gulf of san Jorge, Argentina. J Mar Syst 172:43–50. https://doi.org/10.1016/j.jmarsys.2017.02.007
Paytan A, Mackey KRM, Chen Y et al (2009) Toxicity of atmospheric aerosols on marine phytoplankton. Proc Natl Acad Sci 106:4601–4605. https://doi.org/10.1073/pnas.0811486106
Prospero JM (1996) Saharan dust transport over the North Atlantic Ocean and Mediterranean: an overview. In: Guerzoni S, Chester R (eds) The impact of desert dust across the Mediterranean. Springer, Dordrecht, pp 133–151
Prospero JM, Ginoux P, Torres O et al (2002) Environmental characterization of global sources of atmospheric soil dust identified with the Nimbus 7 Total ozone mapping spectrometer (TOMS) absorbing aerosol product. Rev Geophys 40:1002. https://doi.org/10.1029/2000RG000095
Pulido-Villena E, Rérolle V, Guieu C (2010) Transient fertilizing effect of dust in P-deficient LNLC surface ocean. Geophys Res Lett 37:L01603. https://doi.org/10.1029/2009GL041415
Ridgwell A (2002) Dust in the earth system: the biogeochemical linking of land, air, and sea. Philos Trans R Soc Lond A 360:2905–2924
Russell JL, Dixon KW, Gnanadesikan A et al (2006) The Southern Hemisphere Westerlies in a warming world: propping open the door to the deep ocean. J Clim 19:6382–6390. https://doi.org/10.1175/JCLI3984.1
Sarmiento JL, Hughes TMC, Stouffer RJ et al (1998) Simulated response of the ocean carbon cycle to anthropogenic climate warming. Nature 393:245–249. https://doi.org/10.1038/30455
Shao Y, Wyrwoll K-H, Chappell A et al (2011) Dust cycle: an emerging core theme in earth system science. Aeolian Res 2:181–204. https://doi.org/10.1016/j.aeolia.2011.02.001
Shinn EA, Smith GW, Prospero JM et al (2000) African dust and the demise of Caribbean coral reefs. Geophys Res Lett 27:3029–3032. https://doi.org/10.1029/2000GL011599
Simonella LE, Palomeque ME, Croot PL et al (2015) Soluble iron inputs to the Southern Ocean through recent andesitic to rhyolitic volcanic ash eruptions from the Patagonian Andes. Global Biogeochem Cycles 29:1125–1144. https://doi.org/10.1002/2015GB005177
Thompson DWJ, Solomon S (2002) Interpretation of recent southern hemisphere climate change. Science 296:895–899
Thompson DWJ, Solomon S, Kushner PJ et al (2011) Signatures of the Antarctic ozone hole in southern hemisphere surface climate change. Nat Geosci 4:741–749. https://doi.org/10.1038/ngeo1296
Turner RE, Qureshi N, Rabalais NN et al (1998) Fluctuating silicate: nitrate ratios and coastal plankton food webs. Proc Natl Acad Sci 95:13048–13051

Washington R, Todd M, Middleton NJ et al (2003) Dust-storm source areas determined by the Total ozone monitoring spectrometer and surface observations. Ann Assoc Am Geogr 93:297–313. https://doi.org/10.1111/1467-8306.9302003

Yang SL, Zhang J, Xu XJ (2007) Influence of the three gorges dam on downstream delivery of sediment and its environmental implications, Yangtze River. Geophys Res Lett 34:L10401. https://doi.org/10.1029/2007GL029472

Yu H, Chin M, Bian H et al (2015) Quantification of trans-Atlantic dust transport from seven-year (2007–2013) record of CALIPSO lidar measurements. Remote Sens Environ 159:232–249

Overview on Primary Production in the Southwestern Atlantic

Vivian Lutz, Valeria Segura, Ana Dogliotti, Virginia Tavano, Frederico P. Brandini, Danilo L. Calliari, Aurea M. Ciotti, Virginia F. Villafañe, Irene R. Schloss, Flavia M. P. Saldanha Corrêa, Hugo Benavides, and Denise Vizziano Cantonnet

Abstract Photosynthesis is the fundamental process by which autotrophs produce organic matter to sustain the biosphere using basic elements (i.e., CO_2 and H_2O) and solar irradiance as energy source. Marine phytoplankton provides near half of the global primary production (PP), being at the base of most marine trophic webs and playing an important role in the cycling of atmospheric CO_2. Therefore, it is crucial to estimate and understand the relationships between environmental conditions and PP rates in the global ocean. There are scarce field estimations of PP in the southern hemisphere and in the Southwestern Atlantic in particular. Hence, global estimates are generally made using indirect methods, such as satellite or biogeochemical models, which should be validated and adjusted with field data to produce reliable results.

In this section we synthesize the available information, assembling recent field PP estimations obtained by research groups from Argentina, Brazil, and Uruguay. We evaluate the insights derived from this integrated dataset on the spatial and temporal dynamics of the phytoplankton production in the Southwestern Atlantic. In addition, a general view of the spatial-temporal variation in PP at a regional scale using a simple satellite PP model is presented. Finally, we offer perspectives and recommendations for future studies.

V. Lutz (✉)
Instituto de Investigaciones Marinas y Costeras (IIMyC), Consejo Nacional de Investigaciones Científicas y Técnicas–Universidad Nacional de Mar del Plata (CONICET – UNMdP), Mar del Plata, Buenos Aires, Argentina

Instituto Nacional de Investigación y Desarrollo Pesquero (INIDEP), Mar del Plata, Buenos Aires, Argentina
e-mail: vlutz@inidep.edu.ar

V. Segura · H. Benavides
Instituto Nacional de Investigación y Desarrollo Pesquero (INIDEP), Mar del Plata, Buenos Aires, Argentina
e-mail: vsegura@inidep.edu.ar; hbenavid@inidep.edu.ar

M. S. Hoffmeyer et al. (eds.), *Plankton Ecology of the Southwestern Atlantic*,
https://doi.org/10.1007/978-3-319-77869-3_6

Keywords Primary production · In situ measurements · Satellite-production models · Phytoplankton · Southwestern Atlantic

1 Introduction

Phytoplankton photosynthesis was the fundamental process that allowed the evolution of oxygenic life on the planet around 3500 million years ago and still now contributes about half of the organic production on Earth (Sarmiento and Bender 1994; Longhurst et al. 1995). At present we are facing pronounced changes in CO_2 concentration in the atmosphere, in great part due to anthropogenic causes,

A. Dogliotti
Instituto de Astronomía y Física del Espacio (IAFE-CONICET), Universidad de Buenos Aires, Facultad de Ciencias Exactas y Naturales, Buenos Aires, Argentina
e-mail: adogliotti@iafe.uba.ar

V. Tavano
Laboratório de Fitoplâncton e Microorganismos Marinhos, Universidade Federal do Rio Grande, Rio Grande, RS, Brazil
e-mail: docvmtg@furg.br

F. P. Brandini · F. M. P. Saldanha Corrêa
Instituto Oceanográfico, Universidade de São Paulo, São Paulo, SP, Brazil
e-mail: fsalcorr@usp.br; brandini@usp.br

D. L. Calliari
Grupo Ecología Funcional de Sistemas Acuáticos, Centro Universitario Regional del Este (CURE), Universidad de la República, Rocha, Uruguay
Facultad de Ciencias, Oceanografía y Ecología Marina, Universidad de la República, Montevideo, Uruguay
e-mail: dcalliar@fcien.edu.uy

A. M. Ciotti
Universidade de São Paulo, Centro de Biologia Marinha, São Sebastião, São Paulo, SP, Brazil
e-mail: ciotti@usp.br

V. F. Villafañe
Estación de Fotobiología Playa Unión. Consejo Nacional de Investigaciones Científicas y Técnicas, Rawson, Chubut, Argentina
e-mail: virginia@efpu.org.ar

I. R. Schloss
Instituto Antártico Argentino, San Martín, Buenos Aires, Argentina
Centro Austral de Investigaciones Científicas (CADIC)/Universidad Nacional de Tierra del Fuego, Ushuaia, Tierra del Fuego, Argentina
e-mail: ireschloss@gmail.com; irene.schloss@cadic-conicet.gob.ar

D. Vizziano Cantonnet
Laboratorio de Fisiología de la Reproducción y Ecología de Peces, Instituto de Biología, Facultad de Ciencias/Universidad de la República Oriental del Uruguay, Montevideo, Uruguay
e-mail: vizziano@gmail.com

which are promoting changes in the world climate. In order to understand how the carbon cycle is being affected by these changes, we need to produce a better accounting of the rates of photosynthesis both on land and in the ocean.

Marine photosynthesis, which is responsible for an important fraction of CO_2 absorption (Takahashi 2004; Sabine et al. 2004), has been estimated in the field using a variety of methods since the beginning of the twentieth century (e.g., Gaarder and Gran 1927; Steeman Nielsen 1952). Most measurements have been performed in the waters of the northern hemisphere during cruises carried out by laboratories mainly from the United States (USA), Canada, Europe, and Japan, with much less estimations in those of the southern hemisphere, especially in the Southwestern Atlantic. One of the few pioneer studies on PP in this region (covering the Argentine shelf and shelf-break, Drake Passage, and Antarctica) was the collaborative program *Productivity* between "Texas A&M University" (USA) and the "Servicio de Hidrografía Naval" (Argentina) carried out during the 1960s (El-Sayed 1967), which showed a large spatial and seasonal variability in production.

The scarcity of data and the increasing need to evaluate changes in the carbon cycle at a global scale have led to the use of satellite (Platt and Sathyendranath 1988; Behrenfeld and Falkowski 1997; Behrenfeld et al. 2005) and biogeochemical (Buitenhuis et al. 2013) models to estimate PP in a more synoptic way. Nevertheless, these models have varying degrees of reliability and should be validated and readjusted using actual field estimations in order to provide a more robust representation of PP in different regions of the ocean (Carr et al. 2006; Buitenhuis et al. 2013; Bouman et al. 2018).

Here a series of field estimations of PP carried out by research groups from Brazil, Uruguay, and Argentina from 1982 to 2009 on the Southwestern Atlantic Ocean are reviewed. This information is spread throughout time and space; hence, to have a synoptic view of the PP in the whole area, we also run a simple satellite model for the year 2004.

1.1 Brief Background on Techniques to Estimate Field Primary Production

Due to the complexity involved in the PP estimations, several revisions have been made about the different analytic techniques (e.g., Vernet and Smith 2007; Regaudie-de-Gioux et al. 2014), as well as on the different mathematical models used to arrive at the final production rate estimates (Behrenfeld and Falkowski 1997; Platt and Sathyendranath 2009). In order to introduce the biological processes involved in the different methods for measuring PP, and since the scarcity of data demands more field measurements, we offer below a brief description of the main approaches available.

There are three main types of field estimations of PP based in the following: (1) oxygen evolution, (2) carbon assimilation, and (3) variable fluorescence. All of them involve exposing seawater samples to natural or artificial light for a period of time and measuring some aspect of the photosynthetic process; the first two measure the amount of product (oxygen release) and reactive (carbon assimilation) involved in the photosynthesis chemical reaction, while the last involves estimating the quantic performance of the "photosynthetic machinery."

There is a range of particular methods within each of these techniques, each one susceptible to different bias. Techniques 1 and 2 require incubating seawater samples for time periods ranging from 2 to 24 h, which convey associated artifacts generically known as "bottle effects" that include grazing, possible nutrient depletion, bacterial growth, and changes in mixing conditions, among others (Vernet and Smith 2007). Another issue to consider is that, according to the incubation time, the estimated rates can represent different production stages: for short periods (about <4 h) the values represent all production without losses, which is known as 'gross primary production (GPP)'; for long incubations (up to 24 h) only the amount of products remaining after consumption (mainly by phytoplankton due to their own metabolic processes), known as net primary production (NPP). There are also uncertainties regarding whether in some cases measurements could represent intermediate levels between GPP and NPP (Vernet and Smith 2007).

1. *Oxygen evolution.* The first estimations of photosynthetic rates were conducted following changes in the oxygen concentration (Gaarder and Gran 1927) in samples of seawater enclosed in transparent bottles exposed to natural light or an artificial light source simulating the radiation prevailing at the site and depth of sample collection, with a dark bottle for evaluation of respiration rate (through oxygen consumption). This technique is still in use, with improvements in the method of oxygen detection (e.g., micro-Winkler, optodes). There is also a variant of the oxygen technique using isotopic ^{18}O (e.g., Bender et al. 1987).
2. *Carbon assimilation.* This technique developed in the 1950s is based on measuring the assimilation of carbon (C), after a period of incubation, by a sample that was previously inoculated with sodium bicarbonate marked either with the radioactive ^{14}C (Steeman Nielsen 1952; Platt and Jassby 1976) or with the stable mass isotope ^{13}C (Hama et al. 1983). This is up to the present the most popular technique to estimate field PP. Therefore, a great variety of methods of differing complexity have been developed from this technique. These vary in the source of light used for the incubation: (a) actual in situ incubations (where samples from different depths are inoculated with C isotopes and deployed back to be incubated at their original depth), (b) on deck or "simulated in situ incubations" (where samples are wrapped with neutral light-attenuating meshes and incubated in a container, with circulating water, under sunlight), and (c) use of what is known as production (P) versus irradiance (E) or "P&E incubation box" (with circulating water, providing artificial light of different intensities). Further complexity is introduced by accounting or not for variations in the spectral composition of light, including in some cases the effect of ultraviolet radiation (UV).
3. *Variable fluorescence (Fv).* This is to date the only technique that does not require sample incubation; and some instruments allow recording of water

column in situ profiles of *Fv*. This technique is based on measuring the fluorescence emitted by phytoplankton in a sample when exposed to short flashes of light of different intensities; hence, it is called "variable fluorescence." Some of the instruments developed are the "pump and probe," the "pulse amplitude modulated (PAM)," and the "fast repetition rate fluorometer (FRRF)." Nevertheless, the estimation of actual PP from this technique is not trivial, and important considerations have to be taken into account: (a) variable fluorescence only measures the activity of photosystem II (PSII), which is not always proportional to the functioning of the whole photosynthetic apparatus due to uneven distribution of chlorophyll-a between PSI and PSII and to mechanisms of non-photochemical quenching (Geider and Osborne 1992; Lutz et al. 1998; Johnsen and Sakshaug 2007), and (b) to transform this fluorescence signal into the actual amount of photosynthetic products (usually carbon incorporation) requires the use of physiological factors (e.g., the amount of chlorophyll-a molecules per reaction center in PSII) that are kept fixed for all calculations. Currently, these are based on few laboratory determinations, while it is known that they can have a wide variation according to phytoplankton type and physiological state (Kolber and Falkowski 1993; Sugget et al. 2004).

Another important distinction in the way the mentioned techniques are applied is whether a natural seawater sample is incubated at a single light intensity (similar to that at which it was exposed at sea) or if the sample is incubated at a gradient of light intensities. In the first case, the information retrieved is the rate of production per unit volume at a given site and time, i.e., the instantaneous production (p) (mg C m^{-3} h^{-1}). In the second case, one can reconstruct the response of P at different irradiances E, what is known as a "*P&E* curve" from where a set of photosynthetic parameters can be derived: α, the slope of the curve at low irradiances, and P_m, the maximum production at saturating irradiance. In this case, the parameters can be used to calculate p for that place and time, but also under the assumption that these parameters would be representative of the physiological responses of phytoplankton in the area and season, they can be used in satellite or biogeochemical models to extrapolate production information at larger spatial and temporal scales. All approaches imply, however, that reliable measurements of irradiance at the visible range (photosynthetic available radiation – PAR), both incident and in the water column, are performed, except in the case of remote sensing or model approaches.

It is also relevant to mention that phytoplankton growth mediates the link between the carbon and nitrogen cycling in the oceans because nitrogen must be assimilated concomitantly with carbon in the approximate average molar proportion of 106 moles of carbon to 16 moles of nitrogen, known as the Redfield ratio (Redfield 1934) in order to build up organic matter. Nitrogen is available in the euphotic zone in the form of oxidized nitrogen compounds such as nitrate (usually) and nitrite or as reduced forms such as ammonium and urea. Dugdale and Goering (1967) defined the primary production supported by physical injection of nitrate into the euphotic zone (e.g., upwelling) as "new production" in contrast to the "regenerated production" based on ammonium and urea derived from community excretion or microbial regeneration of organic matter in the euphotic zone. A production system dominated by new production tends to export the excess of particulate carbon to deeper waters, feeding the long-term carbon reservoir through the oceanic biological pump, whereas regenerated production mediates the short-term carbon cycling in surface layers. Hence, it is important to know the relative contribution of new to total production (*f*-ratio; Eppley and Peterson 1979) at the regional scale such as the Southwestern Atlantic in order to better access its role in the global carbon cycle. The usual approach to do so is to measure the incorporation of tracer amounts of a 99% enriched ^{15}N solutions of nitrate (as $Na^{15}NO_3$ or $K^{15}NO_3$) and ammonium (as $^{15}NH_4Cl$ or $[^{15}NH_4]_2\ SO_4$), simultaneously spiked into water samples to be incubated, for which the natural concentration of both nitrate and ammonia are known (Dugdale and Wilkerson 1986). Unfortunately, estimates of *f*-ratio have been rarely applied in the Southwestern Atlantic.

1.2 Brief Description of the PP Satellite Model

Remote sensing of ocean color is an ideal tool to assess PP on regional and global scales, since it offers good spatial and temporal coverage providing daily estimations of the phytoplankton biomass (as indexed by Chl *a* concentration), attenuation coefficient, and photosynthetically available radiation (PAR). Algorithms in use today range from very simple and purely empirical, such as a simple relationship between Chl *a* and PP, to highly complex models based on plant physiology, in which many variables are resolved with depth and with the spectral irradiances (some are listed in Behrenfeld and Falkowski 1997; Carr et al. 2006). The main limitation of this technique is that satellites only "see" the upper layer (first optical depth) of the ocean and that other accessory information, like the photosynthetic parameters and biomass profile parameters, cannot be directly derived from remote sensors.

In this study, a relatively simple model (Platt and Sathyendranath 1988) was used to estimate daily water-column-integrated PP (PP_{ZT}) in the whole area using satellite and in situ information. This spectrally integrated model assumes uniform biomass vertical profiles and a sinusoidal irradiance distribution during the day. The PP at depth z and time t is calculated using.

$$\mathrm{PP}_{ZT} = \mathrm{Chl}a \times P_m^{\ B} \int_{t_1}^{t_2}\int_{0}^{\infty} 1 - \exp\left[-\left(\alpha^B \times E(z,t) / P_m^{\ B}\right)\right] dtdz$$

where α^B and $P_m^{\ B}$ are the Chl *a*-normalized *P&E* curve photosynthetic parameters and $E(z,t)$ is PAR irradiance vertical profile given by $E(Z,T) = E(0,t) \times \exp(-K_d \times z)$ where $E(0)$ is PAR incident on the surface and K_d is the diffuse attenuation coefficient for PAR.

In order to estimate seasonal PP maps for the study area (between 57°–23°S and 70°–43° W), level 3 (SMI) monthly composites of MODIS-Aqua sensor for 2004 were downloaded from the NASA Ocean Color website (http://oceancolor.gsfc.nasa.gov). The following Ocean Biology Processing Group (OBPG) standard data products at 4 km resolution were obtained: Chl *a* (OCI algorithm; Hu et al. 2012), PAR (http://oceancolor.gsfc.nasa.gov/DOCS/seawifs_par_wfigs.pdf), and K_d (490) (P.J. Werdell, June 2009) from which K_d (PAR) was obtained using Morel et al. (2007).

In order to assign the photosynthetic parameters in a per pixel basis, the approach of dividing the area into static biogeochemical provinces proposed by Longhurst et al. (1995) was used. The study area encompasses three biomes defined by Longhurst, i.e., the Trade and Westerly Winds and the Coastal biomes. For pixels located in the first two biomes, parameters were taken from the literature since no in situ data were available (Table 3). However, the in situ data gathered in this study and knowledge of the area allowed us to divide the coastal domain into three zones, namely, the North, Central, and South Shelf. These zones were divided latitudinally at 30°S and 37.5°S and delimited to the west by the 1000 m isobath. All the available parameters averaged within each zone and season (Table 3) were used to estimate monthly mean PP maps for 2004 and then averaging the corresponding months to derive the seasonal maps for summer (January–February–March), fall (April–May–June), winter (July–August–September), and spring (October–November–December).

2 Results and Discussion

2.1 Field Primary Production

We assembled a database of recent field PP estimations obtained by research groups in the region. The types of techniques used, as well as references of publications with details on the procedures and particular descriptions for each dataset, are provided in Table 1. The common variable most easily retrievable from the different studies was the instantaneous production, *p*, at the surface according to the available solar irradiance. In cases where production was estimated by measuring changes in oxygen concentration in samples incubated in light/dark bottles, the values were converted into carbon using the corresponding factors (Schloss et al. 2007). Incubations were made during a given time period (3 to 24 h) and recalculated on a per hour basis. We plotted the positions of the sites where field estimations were made (Fig. 1). In some of these studies, *P&E* curves have been carried out; thus photosynthetic parameters are available for those sites (identified in Fig. 1). In this figure we have included also positions of data not fully available yet (samples chemically analyzed but not mathematically processed), in order to have a first glance of the geographic distribution of all potential information for the region.

Table 1 Information regarding the data gathered from the participating research groups at three shelf zones: South (>37.5°S), Central (<37.5°S and > 30°S), and North (<30°S). See the text for explanation on this zonal division

Shelf zones	Year	*n* data	PP method	Data status	Research Group/Publications
South	1994	4	^{14}C (P&E curve)	Processed – unpublished	INIDEP-COCOAS, Argentina (Benavides unpublished data)
	2005–2006	71	^{13}C (P&E curve)	Published Available in doi:10.1594/ PANGAEA.874087	GEF, Argentina (Lutz et al. 2010; Segura et al. 2013; Dogliotti et al. 2014; Bouman et al. 2018)
	2008–2014	5	^{13}C (P&E curve)	Unpublished data	DiPlaMCC EPEA cruises, INIDEP, Argentina (Segura and Lutz unpublished data)
	2008–2014	11	^{13}C (P&E curve)	Unpublished data	Crustacean fisheries cruises, INIDEP, Argentina (Segura unpublished data)
	2008–2014	68	^{13}C (P&E curve)	Not available yet	Various cruises. Argentina (Segura and Lutz unpublished data)
	2002–2004	20	Oxygen evolution	Processed –published	ARGAU, Argentina (Schloss et al. 2007)
	1999–2003	39	^{14}C	Processed –published	EFPU, Argentina (Villafañe et al. 2004a, b)
	2004	4	^{14}C	Processed –published	PATEX, Brazil (Garcia et al. 2008)
Central	2003	5	^{14}C (P&E curve)	Processed –published	OAS, Uruguay (Calliari et al. 2009)
	1988–1990	24	^{14}C	Processed –published (ECOPEL 2) Unpublished data. (ECOPEL 3).	ECOPEL 2, Brazil (Ciotti et al. 1992) ECOPEL 3, Brazil (Ciotti et al. unpublished data)
	1990–1993	12	^{14}C ^{14}C (P&E curve)	Unpublished data. (ECOPEL-4). Unpublished data. (COROAS 1)	ECOPEL 4, Brazil (Tavano and. Odebrecht unpublished data) COROAS 1, Brazil (González-Silvera 1994)
	2010–2014	27	^{13}C (P&E curve)	Not available yet	Various cruises, Argentina. (Segura and Lutz unpublished data)
North	1982–1983	10	^{14}C ^{14}C(P&E curve)	Processed –published	SUESTE, Brazil. (Brandini 1988, 1990)
	1991–1993	2	^{14}C	Unpublished data. (COROAS 1)	COROAS 1, Brazil (González-Silvera 1994)
	1994–1997	11	^{14}C (P&E curve)	Processed – -published	OPISS, Brazil (Saldanha-Correa and Gianesella 2008)

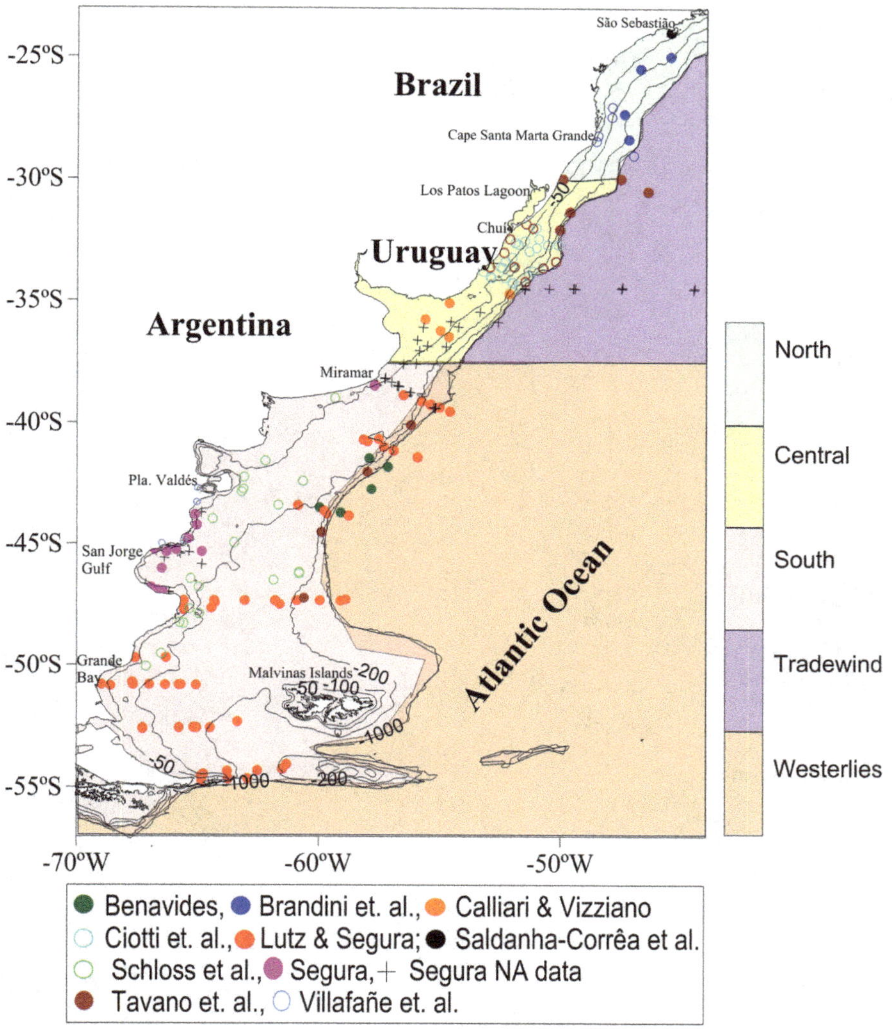

Fig. 1 Locations where field primary production estimations have been performed. The positions are identified with different colors according to the different research groups (identified by name of PI). Filled symbols mark sites for where photosynthetic parameters are available. Crosses are used for sites where production samples have been chemically analyzed, but final data is not yet available

A total of 211 values of PP were gathered and separated by seasons (summer, 70; fall, 32; winter, 42; spring, 67) regardless of the year of collection (1982–2009). Figure 2 shows the distribution of p (mg C m^{-3} h^{-1}) at the surface for each season.

A first glance at these maps (Fig. 2) evidences the scarcity of field data in the region; higher sampling frequencies took place in both summer and spring and lower ones in fall and winter. It becomes also apparent that the higher production values occurred in spring and summer, normally the growing phytoplankton seasons,

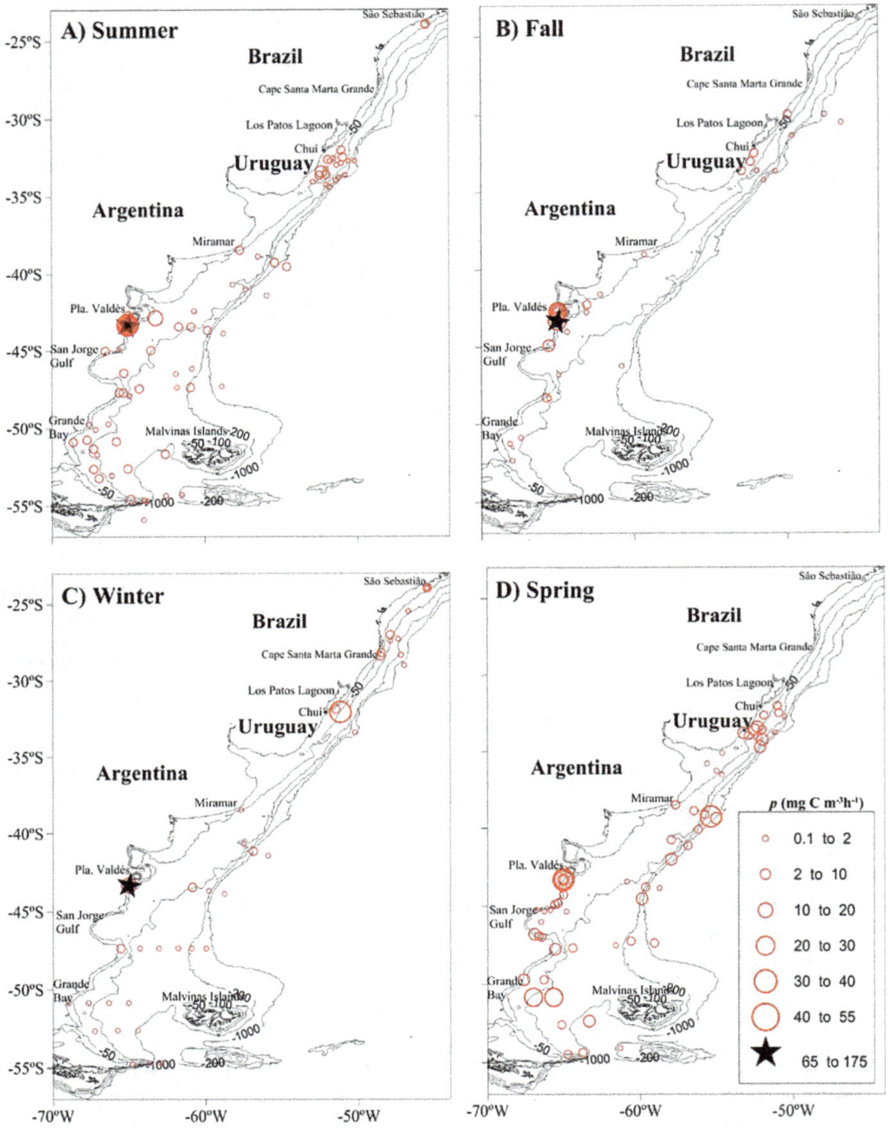

Fig. 2 Distribution of instantaneous production rates at the surface, p (mg C m^{-3} h^{-1}), during the different seasons in the Southwestern Atlantic. Size-coded circles show the range of values at the sites where field estimations were performed

with the exception of a few coastal sites in Patagonia (close to Valdés Peninsula, subjected to high input of nutrients mostly of anthropogenic/eolic origin), which showed very high values also in fall/winter. Within the richer seasons, it is possible to recognize some areas of higher PP, associated to the shelf-break (>55 mg C m^{-3} h^{-1} spring) and Grande Bay (>30 mg C m^{-3} h^{-1} spring) in the Argentine shelf and the southeast of the Brazilian shelf (>19 mg C m^{-3} h^{-1} spring).

Table 2 Distribution of mean instantaneous production values, *p* (mg C m^{-3} h^{-1}), by zone and season; number of data and standard deviations are also shown. Values on shadowed cells were the ones used for the statistical description

		South		Central		North	
		All	Without extremes	All	Without extremes	All	Without extremes
Summer	n	53	27	12		5	
	mean	**7.08**	**2.80**	**1.69**		**4.30**	
	SD	12.52	3.69	1.64		1.09	
Fall	n	21	10	9		2	
	mean	**20.87**	**1.81**	**1.71**		**1.60**	
	SD	31.78	2.79	1.76		1.48	
Winter	n	26	23	3	2	13	
	mean	**11.64**	**1.39**	**15.04**	**2.95**	**2.94**	
	SD	36.36	1.48	21.93	2.93	4.46	
Spring	n	48		17		2	
	mean	**9.41**		**7.02**		**1.54**	
	SD	12.77		6.39		0.25	

We analyzed the variations in the field PP data within three shelf zones (Table 2): South (>37.5°S), Central (<37.5°S and > 30°S), and North (<30°S). The limits among these shelf zones were established based on dominating oceanographic features: (South) the Patagonian shelf and shelf-break, mainly dominated by tidal forcing and upwelling along the Malvinas Current; (Central) the Uruguayan and South Brazilian shelves, the first under strong influence of the Brazil-Malvinas confluence and both dominated by continental discharge of the Río de la Plata, particularly in fall-winter; and (North) the eastern Brazilian coast, mainly dominated by the South Brazilian Bay system (Garcia et al. 2008; Calliari et al. 2009; Brandini 1988).

By excluding a few extreme values, corresponding to very coastal sites (probably due to high input of nutrients mostly of anthropogenic/eolic origin), it is possible to observe that the highest average PP occurs in the South in spring ($p = 9.41 \pm 12.77$ mg C m^{-3} h^{-1}; range, 0.215–55.400 mg C m^{-3} h^{-1}), while the lowest occurs in the same area in winter (average without extremes $p = 1.39 \pm 1.48$ mg C m^{-3} h^{-1}; range with extremes, 0.374–175.044 mg C m^{-3} h^{-1}). Although interannual variability is expected, and this dataset is limited and uneven, a preliminary analysis indicates that seasonal PP does not follow the same temporal pattern in the three zones. Average seasonal PP values decreased in the following order, for each zone: in the South from spring, summer, fall to winter; in the Central from spring, winter, fall to summer; and in the North from summer, winter, fall to spring.

For the analysis of the variability of the normalized photosynthetic parameters (α^B and P_m^B), 102 pairs were available within the region. They were divided by zone and season following the biogeochemical zones proposed and used in the satellite model here implemented; since for some cases there was no field data available, a pair of more generic parameters was taken from Longhurst et al. (1995) (indicated in red in Table 3).

Regarding field data alone, the two parameters showed a high variability among zones and seasons. However, it is noticeable that P_m^B reaches higher values in both the Central and North zones than in the South, although the result for the latter is more robust, due to a much higher number of data points ($n = 86$) as compared to the first two zones ($n = 8$). This integrated field dataset, the largest gathered for the region so far, provides a general view of the heterogeneous distribution, including specific hot spots, of PP in the Southwestern Atlantic. On the other hand, it should be kept in mind that each dataset was obtained through specific projects having different aims. While the goal of some studies was to follow the production dynamics in a local area or a time series at a given site, others aimed to investigate particular physiological aspects of production by different types of phytoplankton under specific light regimes. A brief review of the main conclusions from these individual studies is provided below.

2.2 Brazil

SUESTE (Primary Production in the Southeast Brazilian Shelf – SBS) The shelf ecosystem off southeastern and southern Brazil, between Cabo de São Tomé and Chuí, are dominated by oligotrophic conditions driven by the Tropical Water (TW) brought from the northeast by the Brazil Current, on top of colder and nutrient-rich South Atlantic Central Water (SACW). Overall, the regenerative production dominates over new production, even nearshore and out of the influence of local upwelling or continental runoff where the pool of ammonia tends to be higher than nitrate due to excretion of coastal and estuarine biota. Incubations with ^{15}N uptake experiments (Metzler et al. 1997) reported higher phytoplankton uptakes of reduced forms of nitrogen (ammonia, urea) throughout the shelf and in the offshore oligotrophic waters, with f-ratios (i.e., the contribution of nitrate-based production relative to total production, sensu Eppley and Peterson 1979) ranging from 0.16 to 0.3. Nevertheless, the geographic extension of the main water masses and boundaries between them varies remarkably among seasons due to changes in the wind pattern, and the *f*-ratio tends to increase episodically during upwelling of the nutrient-rich SACW reaching f-ratios values up to 0.86 in offshore regions and 0.26 in the coastal zone (Metzler et al. 1997).

Table 3 Values of the normalized photosynthetic parameters [α^B mg C (mg Cl *a*)$^{-1}$ h^{-1} (Wm^{-2})$^{-1}$ and P_m^B mg C (mg Chl *a*)$^{-1}$ h^{-1}] for each of the different shelf zones considered in this work (North, Central, and South) identified within the Coastal biome and the main offshore biomes according to Longhurst (Trade and Westerly Wind Biomes). Values in red were taken from Longhurst et al. (1995). NA: Not available

Biome	Province	Season	N	α^B		P_m^B	
				Mean	SD	Mean	SD
Coastal	North-shelf	Total	8	0.199	0.170	4.463	2.581
		Summer	NA	0.140	NA	3.600	NA
		Fall	2	0.143	0.093	4.055	1.26
		Winter	6	0.224	0.186	4.582	2.996
		Spring	NA	0.110	NA	2.800	NA
	Central-shelf	Total	8	0.050	0.042	6.310	4.726
		Summer	NA	0.140	NA	3.600	NA
		Fall	3	0.0763	0.051	3.925	2.480
		Winter	NA	0.120	NA	3.000	NA
		Spring	5	0.034	0.032	7.742	5.402
	South-shelf	Total	86	0.105	0.130	2.708	2.412
		Summer	30	0.065	0.030	2.169	1.797
		Fall	NA	0.240	NA	5.000	NA
		Winter	23	0.053	0.012	1.004	0.288
		Spring	33	0.177	0.188	4.394	2.679

(continued)

Table 3 (continued)

	Shelf total	**Total**	**102**	**0.108**	**0.131**	**3.128**	**2.829**
		Summer	30	0.066	0.030	2.160	1.798
		Fall	5	0.096	0.074	3.997	1.855
		Winter	29	0.089	0.105	1.740	1.960
		Spring	38	0.158	0.182	4.835	3.267
Trade winds	South atlantic gyral (SATL)	Summer	NA	0.058	NA	2.930	NA
		Fall	NA	0.058	NA	2.930	NA
		Winter	NA	0.058	NA	2.930	NA
		Spring	NA	0.058	NA	2.930	NA
Westerly winds	South subtropical convergence (SSTC) subantarctic water ring (SANT)	Summer	NA	0.065	NA	2.290	NA
		Fall	NA	0.056	NA	2.420	NA
		Winter	NA	0.092	NA	3.980	NA
		Spring	NA	0.103	NA	4.880	NA

Intrusions of SACW may be enhanced by cyclonic meanders and eddies originated in the western wall of the Brazil Current (Campos et al. 1995; Castro and Miranda 1998; Castro et al. 2006; Calado et al. 2010) increasing new production in the outer shelf (Gaeta et al. 1999). A diatom-dominated deep chlorophyll maximum layer develops regularly from October to March usually between 1 and 10% of surface irradiance as a consequence of these intrusions (Brandini et al. 2014). Metzler et al. (1997) estimated f-ratios >0.8 at the base of the euphotic zone in the mid- and outer-shelf sections, which confirms the importance of eukaryotic autotrophs as dominant primary producers under the fertilization of the intrusion. While the deep intrusions of SACW enhance PP in summer, mass fertilization at the Subtropical Shelf Front (sensu Piola et al. 2000) as a consequence of the large freshwater outflow from the Río de la Plata and, to a lesser extent, the Patos Lagoon keeps high productivity and phytoplankton biomass in terms of Chl *a*, even in winter time. Internal waves may also fertilize the base of the euphotic zone. The contribution of internal waves to the annual PP of the SBS is a gap to be solved if modeling and prediction of PP estimations and prediction are to be applied. The integrated PP over the euphotic zone in the SBS out of the influence of coastal upwelling ranges from <0.01 to 2.67 g C m^{-2} d^{-1}. It is hard to define a clear seasonal or geographic pattern other than remarkable differences between the more productive inner-shelf regions and the mid- and outer-shelf areas, even during intrusions or shelf-break upwelling.

Especially during summer periods, upwelling may occur in specific coastal areas of the SBS such as in Cabo de São Tomé and Cabo Frio, both on the northern coast of the state of Rio de Janeiro off São Sebastião Island (São Paulo state) and in Cabo de Santa Marta Grande, off Santa Catarina state. Systematic investigations on the biological response to upwelling in the plankton system off Cabo Frio (23°S, 42°W) began in the 1950s (Emilsson 1961). Gonzalez-Rodrigues et al. (1992) measured photosynthetic rates ranging from 2 to 14 mg C m^{-3} h^{-1} during upwelling. Seasonal studies (Gonzalez-Rodrigues 1994) on the photosynthetic characteristics of phytoplankton reported values of P_m^B ranging from 0.5 to 15.3 mg C (mg Chl *a*)$^{-1}$ h^{-1}. Satellite maps of surface temperature distribution in the SBS during the summer (see Part 2.5, this volume) reveal the geographical extension of the upwelling in Cabo Frio, affecting the fertility of the mid-shelf and hundreds of kilometers southward (Lorenzetti and Gaeta 1996).

OPISS (Oceanografia da Plataforma Interna de São Sebastião) Samplings were conducted at a fixed station in the southern portion of São Sebastião Island (Lat. 23°58' S – Long. 45°29.09' W) over the 46 m isobath, in summer of 1994 and spring 1997, respectively, for 72 and 78 h. During summer the higher *p* rates were detected at surface (2.54 to 5.06 mg C m^{-3} h^{-1}), under high-light condition. Secondary maxima were detected between 10 and 25 m depth, associated to the deep Chl *a* maximum. Picoplankton contributed on average with 64% of the total PP, being more important in deeper layers. In spring time, the euphotic zone was twice shallower than in summer. Maximum *p* value was 9.02 mg C m^{-3} h^{-1}. Integrated PP ranged from 12.13 to 50.79 mg C m^{-2} h^{-1}, being picoplankton responsible for 40 to 58% of the total (Saldanha-Corrêa and Gianesella 2008).

ECOPEL (Estudo do Ecossistema Pelágico do Extremo sul do Brasil) These cruises were performed over the shelf between Chui and Cabo de Santa Marta Grande. During the spring (ECOPEL 2), the mid-continental shelf showed the highest hourly depth-integrated PP, varying from 44.1 to 76.7 mg C m^{-2} h^{-1}, which was positively related to Chl *a* values. Maximum surface values were observed adjacent to Patos Lagoon, of 19.66 mg C m^{-3} h^{-1}. During the summer (ECOPEL 3), consistent with low nutrients and Chl *a* values, integrated PP was lower, with an average of 29.1 mg C m^{-2} h^{-1}. During the winter (ECOPEL 4) integrated production rates were relatively high, mostly between 8.5 and 17.2 mg C m^{-2} h^{-1}.

However, at one particular site, the integrated production reached a much higher value, 247 mg C m^{-2} h^{-1}. This inner shelf coastal site was associated with very low salinity (28.7) and high concentrations of surface nutrients, indicating a strong influence of continental water discharge. This corroborates the strong effects of mainly the Rio de La Plata plume, but also locally of the Patos Lagoon, on the South Brazilian coast, particularly in winter.

COROAS-1 (Circulação Oceânica na Região Oeste do Atlântico Sul) During the COROAS-1 cruise (fall), *P&E curves* were performed, and results indicated that phytoplankton populations were mainly acclimated to high light levels, since there was no photo-inhibition even at irradiances up to 500 W m^{-2}. The highest value of PP was associated with the lowest temperature.

2.3 Uruguay

FEMCIDI-OAS (Organization of American States grant) Some PP estimations were performed during spring (November) on the continental shelf off Uruguay as part of a research project aimed at tuning remote-sensing estimations for the area (Martinez et al. 2005; Calliari et al. 2009). This is a hydrographically complex area under the influence of four distinct water masses: Subtropical Water (STW), Subantarctic Water (SAW) carried by the northward flowing Malvinas Current, Tropical Water (TW) advected from the north by the Brazil Current, and Río de la Plata Waters (RPW), which spread over large areas of the shelf off Uruguay and southern Brazil as a thin surface lens (Piola et al. 2000). Primary production at the surface was low-moderate over most of the area. Maximum PP was measured at (a) mid-shelf stations, where SAW reached the euphotic depth, and (b) a station associated with an oceanographic front between STW and TW. No clear coupling was observed between surface Chl *a* concentration and PP. Different phytoplankton assemblages were found at different sites. Small phytoplankton prevailed at low biomass in mid-shelf and other shelf-break stations, while large diatoms and dinoflagellates were representative at other mid-shelf stations. Loss process (i.e., grazing) may explain in part the uncoupling between the observed production and biomass levels: assemblages of small phytoplankton are tightly controlled by micro-sized protozoan predators which prevent biomass buildup; in turn, assemblages dominated by larger cells are more able to escape of grazing control by metazoan herbivores (Kiørboe 1993; Calbet and Landry 2004) allowing for biomass accumulation. Thus, while oceanographic conditions appear as important drivers of PP rates, community structure may play a key role for determining biomass levels attained.

2.4 Argentina

ARGAU (Programme de coopération avec la ARGentine pour l'étude del'océan Atlantique AUstral) In this study PP data collected during 2002–2004 was analyzed in relationship to the type of phytoplankton present and the levels of CO_2 uptake. Average production for these years, when including an important phytoplankton bloom in the Grande Bay area in 2003, was 45.4 ± 9.6 mg C m^{-3} d^{-1}. In the coastal-most stations, although high PP was measured, the area behaved as a CO_2 source to the atmosphere, indicating that other processes leading to CO_2 production were more important than PP. In addition, significant differences in net community production and CO_2 sink were observed when phytoplankton assemblages were dominated by diatoms or by flagellates. In the first case, GPP was significantly higher ($p < 0.05$) and twice the value of the latter. Moreover, a significant correlation was evident between Chl *a*, a proxy for phytoplankton biomass, and the difference in partial pressure of CO_2 between the ocean and the atmosphere, in the diatom-dominated stations, while this relationship was not evident when the assemblages were dominated by phytoflagellates, which was the case in 64% of the studied stations (Schloss et al. 2007).

EFPU (Estación de Fotobiología de Playa Unión) A series of field studies of PP were performed at inner coastal sites from 1999 to 2002 by the group of the EFPU. In this case the main aim was to look at the effect of solar ultraviolet radiation (UVR, 280–400 nm) on phytoplankton production. UVR is a well-known stressor of different physiological processes in phytoplankton, particularly inhibiting PP (Villafañe et al. 2003; Häder et al. 2014). To evaluate the extent of this inhibition, studies were performed at several bays south from Valdés Peninsula in Patagonia, Bahía Engaño, Bahía Nueva, and Bahía Camarones (Helbling et al. 2001, 2005; Villafañe et al. 2004a, b). These sites have high PP favored by the high concentrations of nutrients (either from anthropogenic or eolic origin). In addition, during some calm periods (promoting higher stratification), phytoplankton growth is highly favored and blooms can develop (mainly during winter time) as it occurs in the Bahía Engaño area. A time series of 26 estimations carried out at this site throughout an annual cycle showed variations in *p* from 2.11 to 175.04 mg C m^{-3} h^{-1}. Surface UVR-induced inhibition of PP was as high as 60% during spring (Villafañe et al. 2004a, b). On the other hand, under low-light conditions, mainly during winter, UVR could be used for photosynthesis (Barbieri et al. 2002). Overall, a high variability was determined in the responses that were mainly dependent on the taxonomic composition, the previous light history of the organisms, their acclimation capacity, and the depth of the mixed layer as well as the intensity of mixing (Barbieri et al. 2002; Helbling et al. 2005). The inhibitory effects of UVR in these Patagonian coastal sites may have important consequences for the productivity in higher trophic levels; hence, more experimental studies should consider this effect to better validate remote sensing algorithms.

PATEX (PATagonian EXperiment) During the first project cruise (PATEX-1), the study was conducted along a visible (from remote sensing) Chl *a* patch on the shelf-break front in the Patagonian region during spring 2004. High PP rates were associated with phytoplankton communities mainly dominated by relatively large centric diatoms, reaching a maximum of 7800 mg C m^{-2} d^{-1}. Phytoplankton biomass was closely and positively related with O_2 saturation, implying that PP was actively driving O_2 variations. Phytoplankton growth in the region was presumably fueled by upwelling along the front, causing high biomass accumulation, a consistent feature observed by ocean color remote sensing, particularly in spring (Garcia et al. 2008). These blooms certainly cause a significant drawdown of CO_2, as has been demonstrated for the Patagonian shelf and shelf-break (Bianchi et al. 2005, 2009).

GEF (Global Environment Facility Grant) PP was estimated, by *P&E* experiments, during three extensive cruises covering the Argentine shelf and shelf-break area in 2005–2006. The highest production values were found in spring (GEF-1) associated with frontal areas (Lutz et al. 2010). Along the shelf-break, where phytoplankton was dominated by the diatom *Thalassiosira* cf *oceanica* (Sabatini et al. 2012), integrated production reached up to 5470 mg C m^{-2} d^{-1}, while at the Grande Bay front, where a bloom of the dinoflagellate *Prorocentrum minimum* occurred (Gómez et al. 2011; Sabatini et al. 2012), integrated production showed values of 1272 mg C m^{-2} d^{-1}. During the late summer (GEF-2), production was considerably lower, mean 298.2 mg C m^{-2} d^{-1}, than in spring; the lowest rates were recorded during the winter cruise (GEF-3), with a mean value of 183.4 mg C m^{-2} d^{-1}. Although there was a positive correlation between surface Chl *a* concentration and *p* for all stations in the three cruises ($r_s = 0.61$, $p < 0.05$), there was a high data dispersion, indicating that other factors, such as the phytoplankton composition and their physiological state, played an important role in regulating the production rates. An analysis was performed to discriminate photosynthetic and bio-optical phytoplankton types (PBPT) using field measurements from these three cruises (Segura et al. 2013). A combination of the diverse taxa and their physiological flexibility, according to a cluster analysis, revealed the presence of 11 PBPT. Variations in both the photosynthetic and bio-optical properties did not always follow the expected trend according to cell size; e.g., a PBPT composed of a mixture of taxonomic groups with prevalence of micro-phytoplankton was characterized by high α^B values. Therefore, the assumption of representative photosynthetic parameters for different phytoplankton types to model primary production in this region would be quite challenging. This fact was further investigated in a study comparing different satellite models to estimate primary production in the area, adjusted by using the estimated field photosynthetic parameters (Dogliotti et al. 2014). Although different approaches were tried to assign these parameters according to expected oceanographic features, variability was too high. Finally, the selected model (Platt and Sathyendranath 1988) was run using average seasonal parameters for the whole area.

COCOAS VI (Corrientes de Contorno del Atlántico Sudoccidental) P&E experiments were carried out at a few sites around the Argentine shelf-break front at the end of summer. Notoriously, data from the site closest to the front showed parameter values one order of magnitude higher than the remaining sites (P_m^B = 8.66 mg C (mg Chl *a*)$^{-1}$ h^{-1}; α^B = 0.16 mg C (mg Cl *a*)$^{-1}$ h^{-1} (Wm^{-2})$^{-1}$). This discrepancy may be due to a change in phytoplankton composition and/or their physiological state, since Chl *a* concentrations were within the same range of those found at the other sites.

DiPlaMCC (Dinámica del Plancton Marino y Cambio Climático) As part of this INIDEP project dedicated to the time series study of environmental variables and plankton components, a series of *P&E* experiments were carried out at the coastal station EPEA (Estación Permanente de Estudios Ambientales) located off Miramar. A subset of available data corresponding to spring-summer 2008–2009 (November, December, and January) showed that a special high PP event occurred in December 2008, with *p* values of 214.61 mg C m^{-3} h^{-1}, three times larger than the values in November. This event has been associated to the upwelling of subantarctic waters from the mid-shelf (Negri et al. 2010).

Crustacean Fisheries This INIDEP project is dedicated to the survey of *Patagonian red shrimp* stock in San Jorge Gulf and adjacent area, and during some of the cruises *P&E* experiments were performed. Preliminary results from 11 estimations taken during spring 2008 showed variations in *p* from 0.21 to 12.03 mg C m^{-3} h^{-1}.

Although an effort has been made to synthesize most of the field estimations of primary production in the region, we are aware that there are some datasets from other projects that for different reasons (e.g., difficulty in retrieving old data) could not be included in this compilation. There are also some recent studies, whose data are not included here (due to difficulties to compare their variable fluorescence values with the other data), aimed to look at some specific photo-physiological aspects of phytoplankton. One of them was carried out at *Araça Bay* and focused on the following variations in primary production in a shallow bay and at the adjacent deeper channel on scales of days to months and their relationship to phytoplankton size classes. High values of Chl *a*-normalized primary production, up to 2.59 mg C (mg Chl)$^{-1}$ h^{-1}, were found during pulses of nutrient input into the bay, when the community was co-dominated by the ultraplankton and microplankton fractions (Giannini and Ciotti 2016).

3 Satellite Estimation of Primary Production

In order to have a larger-scale view and the spatial-temporal variability of the daily water-column-integrated PP [$PP_{Z,T}$ (mg C m^{-2} d^{-1})] in the whole region, seasonal composites for the year 2004 were generated using a simple model (Platt and Sathyendranath 1988) applied to satellite-derived monthly composites of standard products (like Chl *a*, K_d(490), and PAR) from the MODIS-Aqua sensor and using mean field photosynthetic parameters for each province (Table 3, Fig. 3).

These satellite estimations of PP confirm the general description attained from the scattered field values: PP seems higher in summer/spring and lower in fall/

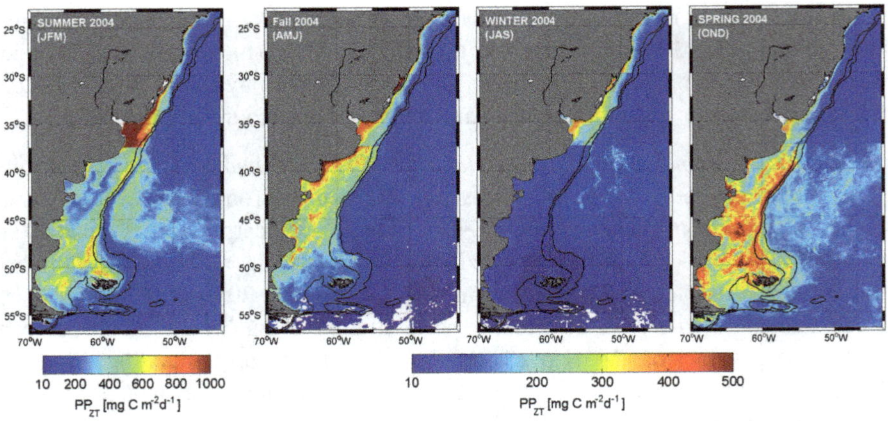

Fig. 3 Seasonal primary production composites for 2004 obtained using a simple model (Platt and Sathyendranath 1988), MODIS-Aqua data, and averaged field photosynthetic parameters (when available) for the Biomes and zones defined in the text and indicated in Table 3. Note that the scale for summer has a higher range than for the rest of the seasons

winter; the areas with higher values during the rich seasons are located along the shelf-break, in the Patagonian gulfs, and the southeast coast of Brazil. Furthermore, these maps allow following with more detail the effect of well-known oceanographic features, such as the front on the shelf-break. During spring and mainly summer at this area, nutrient inputs from the Malvinas current favor a phytoplankton bloom on the outer Argentinean shelf (where stratification maintains the cells in the upper illuminated layer), while the core of the current remains with relatively low values (possibly due to turbulence and the lack of chelating metals compounds), and some spillover of nutrients carried by the retroflection and eddies from the current allows a moderate production to spread through the Westerly Wind Biome (South Subtropical Convergence Province – SSTC).

Conspicuous abrupt "artificial" changes in PP are observed between the static biogeochemical zone boundaries. These can be caused by several factors, such as the high variability in the photosynthetic parameters among provinces evidenced by the field data collected in this study and the use of PP parameters from the literature, some of which are assumed to be the same for all seasons (see Table 3). In turn, the Central Shelf Province is heavily influenced by the discharge of the Río de la Plata, carrying huge loads of sediments and dissolved organic matter (Dogliotti et al. 2016; Negri et al. 2016), which interfere with the Chl *a* detection by remote sensing, causing anomalous PP estimates. Questionably high PP values were estimated in this region, which is most likely an artifact due to the failure of standard Chl *a* algorithms in these optically complex waters which has already been shown in previous works (Armstrong et al. 2004; Martinez et al. 2005; Garcia et al. 2005, 2006; Coló-Gianini et al. 2013). Different approaches can be used to overcome the unrealistic straight boundaries resulting from the applied method, such as using environmental proxies that can be retrieved from remote sensors (e.g., Bouman et al. 2005), and the use of archived datasets, either by dynamically partitioning the study area into biogeochemical regions (Devred et al. 2007) or rearranging them according to Chl *a*, SST, and day of the year, referred to as the nearest-neighbor method (Platt et al. 2008). However, they require a larger database of field measurements to characterize their variability and to improve the assignment of representative parameters on a per pixel basis. Another way to reduce the unrealistic boundaries between provinces is to average all the available parameters within the whole coastal domain, i.e. the North, Central, and South Shelf data together. In doing so during a trial exercise, a smoother map was produced. However, an overestimation of the PP in the South was evident; possibly due to the effect of including parameters from the Central/North zones in the average (corresponding probably to smaller cells acclimated to low nutrients and high solar irradiance). Therefore, these maps with "artificial" abrupt changes highlight the importance of, on the one hand, counting with more field data on photosynthetic parameters and on the other hand the need to improve atmospheric correction and Chl *a* algorithms in these nearshore sites (extreme case 2 waters) to be able to produce more realistic satellite-based production maps.

4 Final Remarks

In the present context of climate change, there is a growing interest in following the dynamics of marine PP and its influence on the carbon cycle. The Southwestern Atlantic appears as one of the most productive areas of the world ocean according to satellite estimations. Nevertheless, available information about PP in this region is scarce, in comparison with information available for other areas of the world ocean, e.g., Northwestern Atlantic (Devred et al. 2007), the Southern Ocean, such as the Weddell Sea, and the West Antarctic Peninsula (Moreau et al. 2015).

The spatial distribution of the available data is mostly close to the coast. There is a fair amount of specialized production studies being performed close to the coast or in mesocosms, which highlight the capacity of research groups to carry out these analyses. This reflects on one hand the limitation to have access to ship time to develop this type of production experiments and probably also the complex logistic and costs of performing these estimations in extensive cruises. In addition, most of the studies here presented were carried out during the spring and summer (and less in fall), but winter studies are rare (El-Sayed 1967; Segura et al. 2013). *A first recommendation is to increase the efforts to make PP estimations covering the shelf and shelf-break, especially during fall and winter.*

Here an effort was made to retrieve the variable most readily comparable among the individual sets, which was the production at the surface per unit volume. As explained in the background section, differences in the values of production may arise due to the methods used in the estimations. Nevertheless, considering the larger errors that have been detected in the validation of some PP satellite models (Dogliotti et al. 2014) and the uncertainty in the results from biogeochemical models, the utility of a set of field data like the one here gathered resides in providing a first-order comparison to results obtained from modeling exercises. On the other hand, it is unfortunate that at the time of applying a satellite model of PP for the whole shelf, there are very few sets of photosynthetic parameters for this region, which hinders the use of more adequate approaches. *A second recommendation is to perform more P&E curves to retrieve sets of field photosynthetic parameters, in order to be able to improve satellite and biogeochemical models used to estimate carbon budgets at regional and global scale.*

The high variability observed in the values of production and photosynthetic parameters from the individual sets points to the well-known fact that coastal and shelf zones are more active and heterogeneous than the open ocean. A high spatial and seasonal variability in the phytoplankton composition, and their physiological condition, is driven by pronounced changes in the light and nutrient environment, as well as in the grazing pressure. Since a high biological production occurs in these coastal shelf zones, which provide important ecosystem services (e.g., food provision, regulation), more attention needs to be paid to the variability of PP and factors driving it in specific areas of interest. It seems alright to pool values from very different sets to retrieve bulk information, e.g., an average value of total carbon uptake for the whole region, but these average values will be far from reality when applied to answer a specific question at a local area (e.g., regarding fisheries, effect of eutrophication). The value of time series of observations at local sites becomes then clear; local effects of upwelling (in the Brazilian coast), the influence of high input of nutrients and UVR (in the Patagonian coast), or extreme events (in the north Argentinean coast) could only be detected by periodic sampling. Events at such high frequency are at the scale that biological activity occurs at those places. *A third recommendation is to enhance the existing, and start new, time series of biogeochemical observations, which include primary production estimations.*

A crucial factor regulating the rate of PP is the types of phytoplankton present in a place. There has been throughout the decades a shift from detailed taxonomic studies on phytoplankton species composition toward the estimations of what are known as "bulk properties" of phytoplankton (e.g., Chl *a*, pigments, PP). Nevertheless, there is nowadays a renewed appreciation that not all phytoplankton are the same and that the health of an ecosystem depends on the different functions that the different groups play in it (e.g., Le Quéré et al. 2005). One of the functions, of current global importance, is the role of phytoplankton removing atmospheric CO_2 and "sequestering" it in the deep ocean; as it is expected, large cells of the micro-phytoplankton (e.g., large diatoms) contribute more to the export of carbon. A recent review by Tilstone et al. (2017) highlights the importance of the South Atlantic, especially the South Subtropical Convergence Province, in exporting carbon, thanks to the higher proportion of PP contributed by the micro-phytoplankton. Schloss et al. (2007) reported the importance of large phytoplankton cells in the CO_2 uptake for the Argentine shelf. Several other studies mentioned in this review showed also that the rate of PP varied with the type of phytoplankton present and that Chl *a* alone is not a good indicator of PP in this region. *A fourth recommendation that follows is the importance of associating information on phytoplankton composition to the estimations of primary production.*

Acknowledgments Primary production estimations are labor intensive, hence, the authors wished to thank all the many colleagues that have collaborated throughout the years to produce the results here shown for the region. The financial support from all the institutions involved and the grants held by all the authors are acknowledged. We want to thank the reviewers of the work for their useful comments. This is INIDEP contribution # 2120 .

References

Armstrong RA, Gilbes F, Guerrero R et al (2004) Validation of SeaWiFS–derived chlorophyll for the Rio de la Plata estuary and adjacent waters. Int J Remote Sens 25:1501–1505

Barbieri ES, Villafañe VE, Helbling EW (2002) Experimental assessment of UV effects upon temperate marine phytoplankton when exposed to variable radiation regimes. Limnol Oceanogr 47:1648–1655

Behrenfeld MJ, Falkowski PG (1997) Photosynthetic rates derived from satellite–based chlorophyll concentration. Limnol Oceanogr 42:1–20

Behrenfeld MJ, Boss E, Siegel D et al (2005) Carbon–based ocean productivity and phytoplankton physiology from space. Glob Biogeochem Cycles 19:GB1006. https://doi.org/10.1029/2004GB002299

Bender ML, Grande K, Johnson K et al (1987) A comparison of four methods for determining planktonic community production. Limnol Oceanogr 32:1085–1098. https://doi.org/10.4319/lo.1987.32.5.1085

Bianchi AA, Bianucci L, Piola A et al (2005) Vertical stratification and air–sea CO2 fluxes in the Patagonian shelf. J Geophys Res 110:C07003. https://doi.org/10.1029/2004JC002488

Bianchi AA, Ruiz Pino D, Isbert Perlender HG et al (2009) Annual balance and seasonal variability of sea–air CO2 fluxes in the Patagonian Sea: their relationship with fronts and chlorophyll distribution. J Geophys Res 114:C03018. https://doi.org/10.1029/2008JC004854

Bouman H, Platt T, Sathyendranath S, Stuart V (2005) Dependence of light saturated photosynthesis on temperature and community structure. Deep-Sea Res I 52:1284–1299

Bouman HA, Platt T, Doblin M et al (2018) Photosynthesis–irradiance parameters of marine phytoplankton: synthesis of a global data set. Earth Syst Sci Data. https://doi.org/10.5194/essd–2017–40

Brandini FP (1988) Hydrography, phytoplankton biomass and photosynthesis in shelf and oceanic waters off southeastern Brazil during autumn (May/June 1983). Bol Inst Oceanogr 36:63–72

Brandini FP (1990) Primary production and phytoplankton photosynthetic in the southeastern Brazilian coast. Bol Inst Oceanogr 38:147–159

Brandini FP, Nogueira M, Simião M et al (2014) Deep chlorophyll maximum and plankton community response to oceanic bottom intrusions on the continental shelf in the South Brazilian bight. Cont Shelf Res 89:61–75. https://doi.org/10.1016/j.csr.2013.08.002

Buitenhuis ET, Hashioka T, LeQuéré C (2013) Combined constraints on global ocean primary production using observations and models. Glob Biogeochem Cycles 27:847–858. https://doi.org/10.1002/gbc.20074

Calado L, Silveira ICA, Gangopadhyay A et al (2010) Eddy–induced upwelling off Cape São Tomé (22 S, Brazil). Cont Shelf Res 30:1181–1188

Calbet A, Landry MR (2004) Phytoplankton growth, microzooplankton grazing, and carbon cycling in marine systems. Limnol Oceanogr 49:51–57

Calliari D, Brugnoli E, Ferrari G et al (2009) Phytoplankton distribution and production along a wide environmental gradient in the South–West Atlantic off Uruguay. Hydrobiologia 620:47–61

Campos EJD, Gonçalves JE, Ikeda Y (1995) Water mass characteristics and geostrophic circulation in the South Brazil bight – summer of 1991. J Geophys Res 100:18537–18550

Carr ME, Friedrichs MAM, Schmeltz M et al (2006) A comparison of global estimates of marine primary production from ocean color. Deep-Sea Res II 53:741–770
Castro BM, Miranda LB (1998) Physical oceanography of the western Atlantic continental shelf located between 4N and 34 S. In: Robinson AR, Brink KH (eds) The sea, vol 11. Wiley, New York, pp 209–251
Castro BM, Brandini FP, MAS P–V et al (2006) Multidisciplinary oceanographic processes on the Western Atlantic continental shelf located between 4°N and 34°S. In: Robinson AR, Brink KH (eds) The sea, vol 14. Wiley, New York, pp 259–293
Ciotti AM, Odebrecht C, Moller Jr O (1992) South Brazilian continental shelf: chlorophyll–a, primary production, and the relationship between abiotic and biotic parameters. In: Falkowski PG, Woodhead AD, Vivirito K (eds) Primary productivity and biogeochemical cycles in the sea. Springer, Boston, pp 507–508
Coló Giannini MF, Ciotti AM (2016) Parameterization of natural phytoplankton photo–physiology: effects of cell size and nutrient concentration. Limnol Oceanogr 61:1495–1512
Coló Giannini MF, Garcia CAE, Tavano VM et al (2013) Effects of low–salinity and high–turbidity waters on empirical ocean colour algorithms: an example for Southwestern Atlantic waters. Cont Shelf Res 59:84–96
Devred E, Sathyendranath S, Platt T (2007) Delineation of ecological provinces using ocean colour radiometry. Mar Ecol Prog Ser 346:1–13
Dogliotti AI, Lutz VA, Segura V (2014) Estimation of primary production in the southern Argentine continental shelf and shelf–break regions using field and remote sensing data. Remote Sens Environ. https://doi.org/10.1016/j.rse.2013.09.021
Dogliotti AI, Ruddick K, Guerrero R (2016) Seasonal and inter–annual turbidity variability in the Río de la Plata from 15 years of MODIS: El Niño dilution effect. Est Coastal Shelf Sci 182:27–39
Dugdale RC, Goering JJ (1967) Uptake of new and regenerated forms of nitrogen in primary productivity. Limnol Oceanogr 12(2):196–206
Dugdale RC, Wilkerson FP (1986) The use of ^{15}N to measure nitrogen uptake in eutrophic oceans: experimental conditions. Limnol Oceanogr 31:673–689
El–Sayed SZ (1967) On the productivity of the Southwest Atlantic Ocean and the waters West of the Antarctic Peninsula. In: Schmitt W, Llano GA (eds) Biology of the Antarctic seas III, Antarctic research series, vol 11. Am Geophys Society, Washington, pp 15–47
Emílsson I (1961) The shelf and coastal waters off southern Brazil. Bol Inst Oceanogr 11(2): 101–112. https://doi.org/10.1590/S0373-55241961000100004
Eppley RW, Peterson BJ (1979) Particulate organic matter flux and planktonic new production in the deep ocean. Nature 282(5740):677–680. https://doi.org/10.1038/282677a0
Gaarder T, Gran HH (1927) Investigations of the production of plankton in the Oslo Fjord. Rapp P v Reun Cons Int Explor Mer 42:1–48
Gaeta SA, Ribeiro SMS, Metzler PM et al (1999) Environmental forcing on phytoplankton biomass and primary productivity of the coastal ecosystem in Ubatuba region, Southern Brazil. Rev Bras Oceanogr 47(1):11–27
Garcia CAE, Garcia VMT, McClain CR (2005) Evaluation of SeaWiFS chlorophyll algorithms in the southwestern Atlantic and Southern oceans. Remote Sens Environ 95:125–137. https://doi.org/10.1016/j.rse.2004.12.006
Garcia VMT, Signorini S, Garcia CAE et al (2006) Empirical and semi–analytical chlorophyll algorithms in the southwestern Atlantic coastal region (25–40°S and 60–45°W). Int J Remote Sens 27(8):1539–1562. https://doi.org/10.1080/01431160500382857
Garcia VMT, Garcia CAE, Mata MM et al (2008) Environmental factors controlling the phytoplankton blooms at the Patagonia shelf–break in spring. Deep-Sea Res I 55:1150–1166
Geider RJ, Osborne BA (1992) Algal photosynthesis: the measurement of algal gas exchange. In: Dring MJ, Melkonian M (eds) Current phycology, vol 2. Chapman and Hall, New York, p 256
Gómez MI, Piola A, Kattner G et al (2011) Biomass of autotrophic dinoflagellates under weak vertical stratification and contrasting chlorophyll levels in subantarctic shelf waters. J Plankton Res 33:1304–1310

Gonzalez–Rodriguez E (1994) Yearly variation in primary productivity of marine phytoplankton from Cabo Frio (RJ, Brazil) region. Hydrobiologia 294(2):145–156
Gonzalez–Rodriguez E, Valentin JL, Andrè DL et al (1992) Upwelling and downwelling at Cabo Frio (Brazil): comparison of biomass and primary production. J Plankton Res 14(2):289–306
González–Silvera A (1994) Modelos Semi–Analíticos para estimar la Producción Primária del Fitoplâncton a través de Sensoriamento Remoto: Una aplicación en el âmbito regional. MSc Dissertation, Federal University of Rio Grande
Häder DP, Villafañe VE, Helbling EW (2014) Productivity of aquatic primary producers under global climate change. Photochem Photobiol Sci 13:1370–1392
Hama T, Miyazaki T, Ogawa Y et al (1983) Measurement of photosynthetic production of a marine phytoplankton population using a stable 13C isotope. Mar Biol 73:31–36
Helbling EW, Buma AGJ, de Boer MK et al (2001) In situ impact of solar ultraviolet radiation on photosynthesis and DNA in temperate marine phytoplankton. Mar Ecol Prog Ser 211(1):43–49
Helbling EW, Barbieri ES, Marcoval MA et al (2005) Impact of solar ultraviolet radiation on marine phytoplankton from Patagonia. Photochem Photobiol 81:807–818
Hu C, Lee Z, Franz B (2012) Chlorophyll *a* algorithms for oligotrophic oceans: a novel approach based on three–band reflectance difference. J Geophys Res 117:C01011. https://doi.org/10.1029/2011JC007395
Johnsen G, Sakshaug E (2007) Biooptical characteristics of PSII and PSI in 33 species (13 pigment groups) of marine phytoplankton, and the relevance for pulse–amplitude–modulated and fast–repetition–rate–Fluorometry. J Phycol 43:1236–1251
Kiørboe T (1993) Turbulence phytoplankton cell size and the structure of marine pelagic food webs. Adv Mar Biol 29:1–72
Kolber Z, Falkowski P (1993) Use of active fluorescence to estimate phytoplankton photosynthesis in situ. Limnol Oceanogr 38:1646–1665
Le Quéré C, Harrison PH, Prentice C et al (2005) Ecosystem dynamics based on plankton functional types for global ocean biogeochemistry models. Glob Change Biol 11:2016–2040. https://doi.org/10.1111/j.1365–2486.2005.1004.x
Longhurst A, Sathyendranath S, Platt T et al (1995) An estimate of global primary production in the ocean from satellite radiometer data. J Plankton Res 17:1245–1271
Lorenzzetti JA, Gaeta SA (1996) The Cape Frio upwelling effect over the South Brazil Bight northern sector shelf Waters: a study using AVHRR images. ISPRS Archives, vol XXXI, PartB7:448–453
Lutz VA, Sathyendranath S, Head EJH et al (1998) Differences between in vivo absorption and fluorescence excitation spectra in natural samples of phytoplankton. J Phycol 34:214–227
Lutz VA, Segura V, Dogliotti AI et al (2010) Primary production in the Argentine sea during spring estimated by field and satellite models. J Plankton Res 32:181–195
Martínez G, Brugnoli E, Hernández J et al (2005) How valid is the SeaWiFS estimation of chlorophyll–a at the Río de la Plata estuary and its area of influence? In: Frouin R, Kawamura H, Pan D (eds) Active and passive remote sensing of the oceans proc of SPIE, vol 5656. https://doi.org/10.1117/12.582665
Metzler PM, Glibert PM, Gaeta AS et al (1997) New and regenerated production in the South Atlantic off Brazil. Deep-Sea Res I 44(3):363–384
Moreau S, Mostajir B, Bélanger S et al (2015) Climate change enhances primary production in the Western Antarctic Peninsula. Glob Chang Biol. https://doi.org/10.1111/gcb.12878
Morel A, Claustre H, Antoine D, Gentili B (2007) Natural variability of bio-optical properties in Case 1 waters: attenuation and reflectance within the visible and near-UV spectral domains, as observed in the South Pacific and Mediterrranean waters. Biogeosciences 4:913–925
Negri RM, Akselman R, Carignan MO et al (2010) Plankton community and environmental conditions during a mid shelf waters intrusion and upwelling at the EPEA atation (Argentina). In: Abstracts of the meeting of the Americas AGU, Foz do Iguazu, Brazil, 8–10 August 2010
Negri RM, Mollinari G, Carignan M et al (2016) Ambiente y Plancton en la Zona Común de Pesca Argentino–Uruguaya en un escenario de cambio climático (marzo 2014). Rev Frente Mar 24:251–316

Piola A, Campos EJD, Môller Jr OO et al (2000) Subtropical shelf front off Eastern South America. J Geophys Res 107:6565–6578
Platt T, Jassby AD (1976) The relationship between photosynthesis and light for natural assemblages of coastal marine phytoplankton. J Phycol 12:421–430
Platt T, Sathyendranath S (1988) Oceanic primary production: estimation by remote sensing at local and regional scales. Science 241:1613–1620
Platt T, Sathyendranath S (2009) Light and marine primary production. Seibutsu Kenkyusha, Tokyo, p 174
Platt T, Sathyendranath S, Forget M–H et al (2008) Operational mode estimation of primary production at large geographical scales. Remote Sens Environ 112:3437–3448
Redfield AC (1934) On the proportions of organic derivatives in sea water and their relation to the composition of plankton. In: Daniel RJ (ed) James Johnstone memorial volume. University Press of Liverpool, Liverpool, pp 177–192
Regaudie–de–Gioux A, Lasternas S, Agustí S et al (2014) Comparing marine primary production estimates through different methods and development of conversation equations. Front Mar Sci. https://doi.org/10.3389/fmars.2014.00019
Sabatini ME, Akselman R, Reta R et al (2012) Spring plankton communities in the southern Patagonian shelf: hydrography, mesozooplankton patterns and trophic relationships. J Mar Syst 94:33–51
Sabine CL, Feely RA, Gruber N et al (2004) The oceanic sink for anthropogenic CO2. Science 305:367–371
Saldanha–Corrêa FMP, Gianesella SMF (2008) Produção Primária e Fitoplâncton. In: Pires–Vanin AMS (ed) Oceanografia de um ecossistema subtropical. EDUSP, São Paulo, pp 223–251
Sarmiento JL, Bender M (1994) Carbon biogeochemistry and climate change. Photos Res 39: 209–234. https://doi.org/10.1007/bf00014585
Schloss IR, Ferreyra GA, Ferrario ME et al (2007) Role of plankton communities in pCO2 sea–air variation in the southwestern Atlantic Ocean. Mar Ecol Prog Ser 332:93–106
Segura V, Lutz VA, Dogliotti AI et al (2013) Phytoplankton functional types and primary production in the Argentine sea. Mar Ecol Prog Ser 491:15–31
Steeman Nielsen E (1952) The use of radioactive carbon (^{14}C) for measuring production in the sea. J Cons Perm Int Explor Mer 18:117–140. https://doi.org/10.1093/icesjms/18.2.117
Sugget DJ, MacIntyre HL, Geider RJ (2004) Evaluation of biophysical and optical determinations of light absorption by photosystem II in phytoplankton. Limnol Oceanogr Methods 2:316–332
Takahashi T (2004) The fate of industrial carbon dioxide. Science 305:352–353
Tilstone GH, Lange PK, Misra A et al (2017) Micro–phytoplankton photosynthesis, primary production and potential export production in the Atlantic Ocean. Prog Oceanogr. https://doi.org/10.1016/j.pocean.2017.01.006
Vernet M, Smith RC (2007) Measuring and modeling primary production in marine pelagic ecosystems. In: Fahey TJ, Knapp AK (eds) Principles and standards for measuring primary production. Oxford University Press, New York. https://doi.org/10.1093/acprof:oso/9780195168662.003.0009
Villafañe VE, Sundbäck K, al FFL (2003) Photosynthesis in the aquatic environment as affected by UVR. In: Helbling EW, Zagarese HE (eds) UV effects in aquatic organisms and ecosystems. Comprehensive series in photochemical and Photobiological sciences. The Royal Society of Chemistry, Cambridge, pp 357–397
Villafañe VE, Barbieri ES, Helbling EW (2004a) Annual patterns of ultraviolet radiation effects on temperate marine phytoplankton off Patagonia, Argentina. J Plankton Res 26:167–174
Villafañe VE, Marcoval MA, Helbling EW (2004b) Photosynthesis versus irradiance characteristics in phytoplankton assemblages off Patagonia (Argentina): temporal variability and solar UVR effects. Mar Ecol Prog Ser 284:23–34

Part III
Plankton of Shelf and Boundary Systems

Phytoplankton Assemblages of the Subtropical South West Atlantic: Composition and Dynamics in Relation to Physical and Chemical Processes

Frederico P. Brandini

Abstract Marine ecosystems respond first to physical-chemical changes in the upper euphotic layers through its phytoplankton community, an assemblage of unicellular eukaryotic and prokaryotic photoautotrophs growing at the same time and spatial scales of hydrodynamic processes. Temporal and spatial environmental changes are followed by physiological adaptations of opportunistic species that dominate the phytoplankton assemblage at a certain time and location. This chapter aims to describe the composition and distribution of the phytoplankton in shelf waters of the subtropical Southwestern Atlantic Ocean, one of the poorly studied areas of the global ocean. It includes regional scale frontal systems caused by thermal upwelling and estuarine plumes together with a variety of meso- and submesoscale hydrodynamic processes that potentially fertilize the euphotic zone and phytoplankton dynamics in shelf systems. Cause-effect relationships of phytoplankton composition and distributional patterns in relation to environmental properties rely on published information which are geographically limited and seasonally fragmented. Nevertheless, the chapter gives a general overview of particular ecological features of the subtropical Brazilian province and its environmental drivers. Despite the recent progress of biomolecular techniques in facilitating taxonomic identification in order to expand the knowledge of phytoplankton dynamics and composition in the South West Atlantic, a joint regional effort is necessary to better understand the baselines of the ecosystem functioning provided by phytoplankton organisms. This will be necessary for modeling future scenarios of regional marine ecosystem services facing the global climatic changes expected in the next decades.

Keywords Phytoplankton · Composition · Dynamics · South Brazil Bight

F. P. Brandini (✉)
Instituto Oceanográfico, Universidade de São Paulo, São Paulo, Brazil
e-mail: brandini@usp.br

M. S. Hoffmeyer et al. (eds.), *Plankton Ecology of the Southwestern Atlantic*,
https://doi.org/10.1007/978-3-319-77869-3_7

1 Introduction

Phytoplankton is an assemblage of photoautotrophic cells suspended in the water column, found in all aquatic ecosystems. They are composed of prokaryotic and unicellular microalgae belonging to various taxonomic groups. This community forms a 3D "planetary lawn" growing in the upper euphotic layers of the global ocean, subjected to all scales of turbulence and currents. Phytoplankton cells are the main primary producers of the ocean, contributing approximately half of the annual net photosynthetic carbon uptake in the biosphere (Field et al. 1998) and, ultimately, the chemical energy that fuels the vast majority of marine ecosystems. Phytoplankton is also responsible for the vertical flux of particulate organic matter that sequesters enormous carbon reservoirs in deeper layers, an important step in the global carbon cycle. Its abundance and diversity drive marine biogeochemical processes and provide important oceanic ecosystem services, such as renewal of at least half the atmospheric oxygen and the maintenance of global temperature by absorbing atmospheric CO_2.

Light is the most limiting factor for growth of phytoplankton because it is available in less than 2% of the ocean by volume. The thickness of the euphotic zone, where net accumulation of phytoplankton biomass takes place, varies according to the concentration of suspended particles. Phytoplankton dynamics changes within specific space and time scales, being less abundant in light-limited turbid estuarine waters. Phytoplankton cells need more than light to grow and accumulate. Macronutrients and trace elements (e.g., nitrogen, phosphorus, silicon, and iron) are also essential for the buildup of net organic matter accumulated in phytoplankton cells. The abundance and/or relative concentration of specific nutrients per unit area of the euphotic zone determines the taxonomic composition of phytoplankton assemblages. The seasonal light regime and nutrient dynamics in the euphotic layer are dependent on latitude, ocean topography, and wind-driven turbulence. Hence, phytoplankton composition and distributional patterns co-vary with these environmental properties along cross-shelf gradients and oceanic fronts.

Phytoplankton size and diversity shape the structure of the marine ecosystem (Kiørboe 1993; Marañón 2009). Understanding the organization and dynamics of phytoplankton communities is the stepping stone to better assess the energy flux in pelagic food webs. Micro-sized (>20 μm) diatoms with a low surface to volume ratio are well-adapted to nutrient-rich upwelling systems where they provide energy for a short copepod and/or euphausiid food chain (Cushing 1989). The ballast of their siliceous exoskeleton (frustule) makes them sink rapidly during physiological stress, transferring organic matter to the benthic shelf ecosystem. In contrast, small pico-and nano-sized cells have a much higher surface to volume ratio and are more adapted to low-nutrient conditions. Most of the organic matter produced in the euphotic zone by cyanobacteria and pico-sized eukaryotes, the main primary producers in tropical oligotrophic oceans (Partensky et al. 1999; Flombaum et al. 2013), is recycled primarily within the microbial food web (sensu Azam et al. 1983). Pico- and nano-plankton thus contribute less to the vertical transport of organic matter and oceanic biological pump yet keep a steady-state balance between atmospheric carbon uptake and microbial respiration in the upper ocean layers at tropical and subtropical latitudes, covering approximately 70% of the oceanic area. They also play an important role in the global biogeochemical cycles of atmospheric gases.

2 Community Composition and Dynamics

Three hydrographically different latitudinal bands occupy the SWA: (i) the subtropical province off the Brazilian margin, (ii) the temperate shelf subjected to the La Plata river plume, and (iii) the subantarctic Patagonian shelf. Shelf morphology, seasonal changes in continental runoff, and tidal and estuarine fronts affect distributional patterns of phytoplankton across the southeastern shelf (Fig. 1). The background

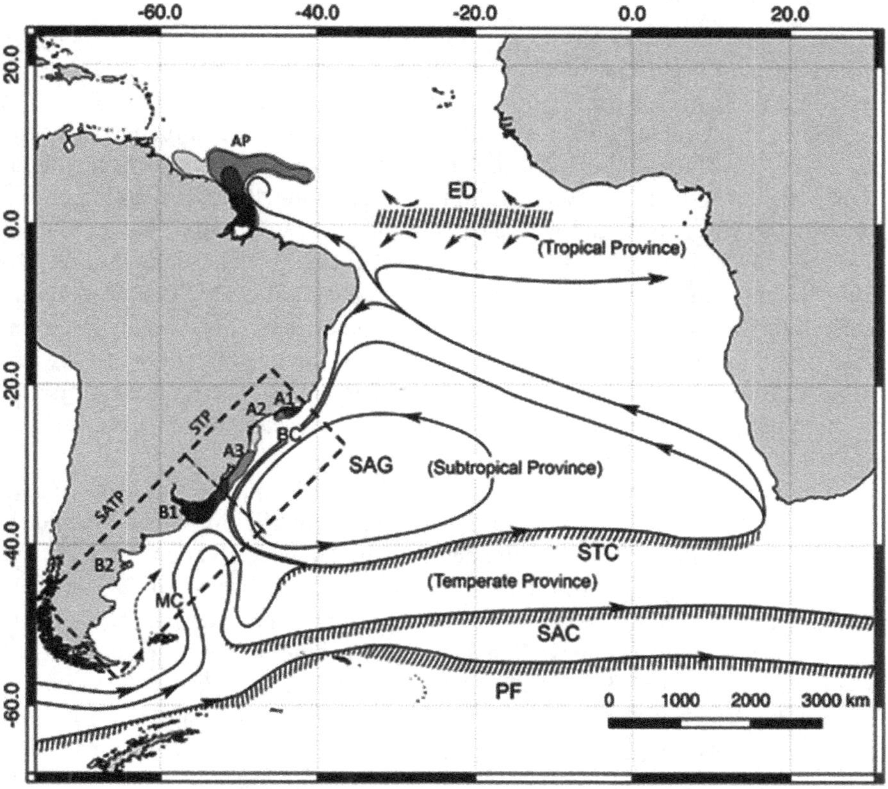

Fig. 1 Map of the South Atlantic Ocean showing hydrographic provinces and frontal systems of the western oceanic and continental South American continent, illustrating main hydrodynamic processes in shelf areas. The dot-delimited area encompasses the subantarctic-temperate province (SATP) and the subtropical province (STP). A1-Cape Frio upwelling, A2-wide shelf off the South Brazil Bight, A3-southernmost shelf affected by the La Plata plume, B1-temperate province affected by the Rio de la Plata estuary, B2-Subantarctic Patagonian Shelf, ED-Equatorial Divergence, AP-Amazon Plume, SAG-Subtropical Atlantic Gyre, MC-Malvinas (=Falkland) Current, STC-Subtropical Convergence, PF-Polar Front

oligotrophic scenario prevails along the whole Brazilian continental margin due to the accumulation of equatorial waters driven by westerly winds. West boundary currents (the North Brazil and Brazil currents) spread over the north and south continental shelves in this system, with the plankton adapted to nutrient-poor conditions, dominated by the cyanobacteria of the genus *Synechococcus* and *Prochlorococcus,* and classes of smaller eukaryotes, except in inshore and inner-shelf areas where the development of micro-phytoplankton, mostly diatoms, is stimulated by nutrient-rich estuarine plumes and coastal upwelling. Hereafter, phytoplankton characteristics will be described taking into account differences in hydrodynamics and geomorphologic features of a selected cross-shelf transect representative of these contrasting subtropical plankton systems.

2.1 The Subtropical Province

The STP extends from 23 to 34°S; it includes the South Brazil Bight (SBB) between Cape Frio and Cape of Santa Marta Grande and the southernmost section of the Brazilian continental margin directly affected by the La Plata river plume during the winter (Fig. 1). The shelf is occupied by three water masses: (i) the warm and oligotrophic tropical water (TW: > 25°C, salinity >36, chlorophyll biomass <0.5 mg m^{-3}); (ii) the cold, less saline, and nutrient-rich South Atlantic Central Water (SACW: 10–20°C, 35–36, Redfield ratio = 13.6; Castro et al. 2006), which dominate, respectively, the upper mixed layers and the bottom layers of mid-shelf (50–100 m), outer-shelf (100–200 m), and shelf-break (>200 m) areas; and (iii) coastal water that mostly occupies the inner shelf (<50 m depth) and is directly affected by freshwater drainage. Boundaries between them form hydrographic fronts that have a marked effect on plankton composition and dynamics on both seasonal and spatial scales, which will be described in more detail below.

Multiple sources of nutrients may fertilize the shelf euphotic zone, driven by a very complex hydrodynamic regime. Their relative contribution to the total nutrient load and ultimately the development of phytoplankton varies spatially and seasonally, creating an episodic "patchwork" of a diatom-dominated new production system. Fertilization of the euphotic zone is not restricted to the nearshore, subjected to coastal upwelling and estuarine plumes. Onshore intrusions of the deeper nutrient-rich SACW take place on the mid-shelf (50–100 m) during summer that balance the Ekman transport of surface waters away from the coast driven by the persistence of east-northeast winds (Castro et al. 2006). Cyclonic eddies and Brazil Current meanders over the shelf-break also enrich the lower half of the euphotic zone, replacing the pico-sized cell assemblages with micro-sized as well as chain-forming diatoms and, to a lesser extent, eukaryotic micro flagellates that may flourish at the base of the euphotic zone in outer-shelf areas (Brandini et al. 2014; Bergo et al. 2017). Mass fertilization of the southernmost regions takes place in austral winter (June–August) due to lateral advection of the La Plata river plume (Brandini 1990a; Campos et al. 2008; Piola et al. 2008), remarkably changing the phytoplankton composition in inner- and mid-shelf waters (Brandini 1988).

The phytoplankton community of the STP is by far the most studied along the Brazilian margin. Pioneer studies on the composition and distribution of micro-sized (>20 μm) phytoplankton in open shelf waters began in the early 1960s during hydrographic cruises by the Brazilian Navy. During these early cruises, *Ceratium tripos*, *Dictyocha fibula*, and *Distephanus speculum* were reported for the first time as indicators of water bodies (Barth and Castro 1964, 1965). In particular, the silicoflagellates were associated with subantarctic waters from the La Plata plume, which was later examined in more detail during the winter season (Brandini 1988).

Overall, cross-shelf differences in phytoplankton composition are clearly distinguished and partitioned through their characteristic diatoms, cyanobacteria, and picoeukaryotes. Biomolecular techniques have recently reported the numeric dominance of cyanobacteria in the surface layers over the mid- and outer-shelf regions (Alves et al. 2015; Ribeiro et al. 2016; Bergo et al. 2017) where nutrient concentrations are low due to the proximity of the oligotrophic warm waters of the Brazil Current. *Synechococcus* and a diverse community of flagellated picoeukaryotes are more abundant at the surface and in deep layers of the euphotic zone across the shelf. Toward the shelf-break, this community tends to be gradually replaced by *Prochlorococcus* under the extremely oligotrophic conditions of the TW (Fig. 2).

This background picoplanktonic community, typical of oligotrophic systems, is replaced in the nearshore by a highly diverse coastal phytoplankton assemblage, particularly in those areas affected by upwelling or estuarine systems, such as off Guanabara, Santos, Cananéia, Paranaguá, and São Francisco do Sul (Valentin et al. 1987; Brandini et al. 1997; Odebrecht and Djurfeldt 1996; Fernandes and Brandini 2004). The dominant taxonomic groups usually found on plankton samples in the nearshore areas are (mostly) diatoms and flagellated cells of the nano- and micro-sized categories belonging to the groups Dinophyceae, Haptophyceae, Cryptophyceae, Prasinophyceae, and Chlorophyceae (Fernandes and Brandini 2004; Tenenbaum et al. 2004; Villac et al. 2008). As part of the GloBallast program network (www.globallast.imo.org), Tenenbaum et al. (2004) identified 271 species of eukaryotic phytoplankton in Sepetiba Bay, southeastern Rio de Janeiro, including 160 diatoms, 76 dinoflagellates, 15 Chlorophyceae, 2 Euglenophyceae, 2 silicoflagellates, 1 Raphidophyceae, 1 Zygnematophyceae, and 4 cyanobacteria. Villac et al. (2008) updated that species list along with one for the coast of São Paulo state; this was based on a detailed review of the literature about phytoplankton composition surveyed in the last 100 years. They identified 572 taxa, dominated by diatoms (82%) and followed by dinoflagellates (16%).

Tidal and geostrophic currents spread these diatom-rich waters along the inner shelf, enriched further south with estuarine diatoms coming from the Itajai River and Patos Lagoon. *Skeletonema costatum* has a notable high frequency and density

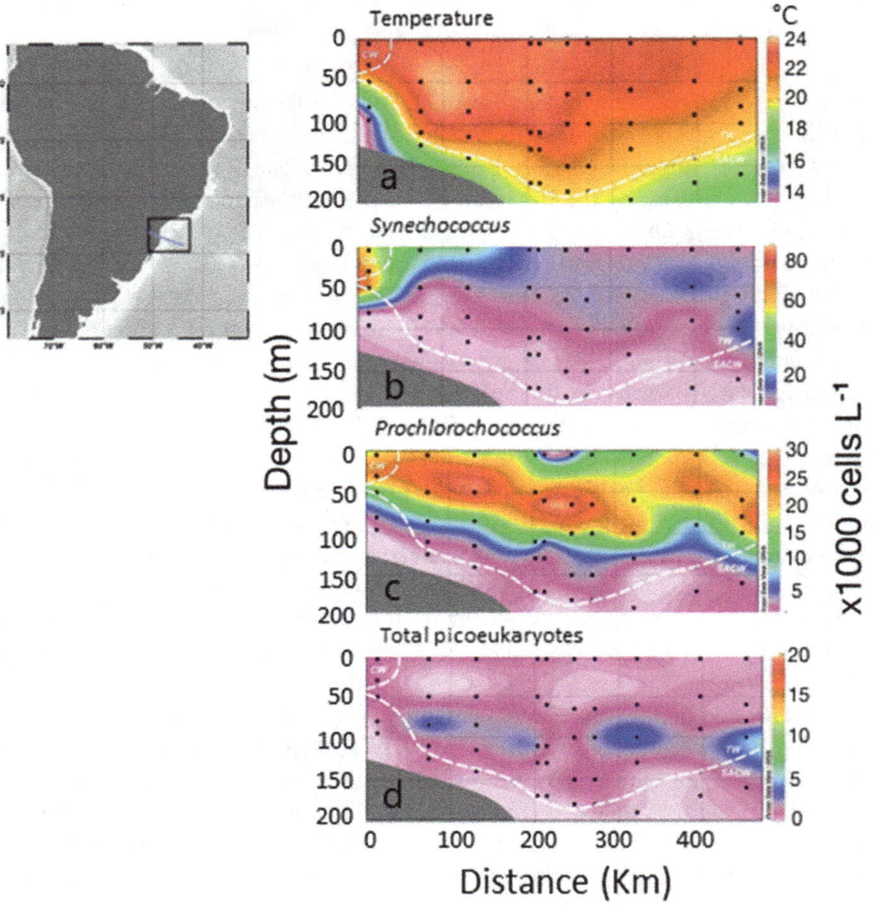

Fig. 2 Cross-shelf-break distribution of temperature and pico-phytoplankton density (in cells 10^{-3} L^{-1}) distribution in the subtropical Brazilian shelf shown in the left inserted map. (**a**) Temperature (°C), (**b**) *Synechococcus*, (**c**) *Prochlorococcus*, (**d**) picoeukaryotes). (Redrawn from Ribeiro et al. 2016)

in phytoplankton assemblages of these estuarine systems between November and March. The worldwide ubiquity of this species is explained by its preference for less saline and nutrient-rich estuarine waters, a common situation in southeast Brazil during rainy seasons when continental runoff is maximum (Rörig et al. 1998). Their presence is therefore not uncommon among the phytoplankton of the nearshore open waters along the STP, being advected out of the estuaries by tidal currents. Another feature of phytoplankton dynamics in the southernmost shelf of the STP, particularly along the coast of Santa Catarina and Rio Grande do Sul, is the seasonal blooms of *Asterionellopsis* in surf zones of sandy beaches (Odebrecht et al. 2010, 2014; Franco et al. 2016). This is common in late summer and early spring seasons and seems to be associated with the enhancement of turbulence driven by the passage of cold atmospheric fronts.

2.1.1 Phytoplankton off the Cape Frio Upwelling System

Coastal upwelling is unusual in western boundary systems. However, due to the particular topography combined with the persistence of N–NE winds and the Coriolis effect, surface waters off Cape Frio (23°S–42°W) are transported offshore, causing deep oceanic nutrient-rich SACW to surface very near the coast (Valentin 2001). During upwelling, the diatom-based primary production increases remarkably up to 14 mg C m^{-3} h^{-1} (Gonzalez-Rodriguez 1994). The upwelling front moves fast at approximately 50 cm sec^{-1}, being detected 400 km southward in mid-shelf regions off the SBB (Lorenzetti and Gaeta 1996). Moreira-Filho (1965) reported the first inventory of diatoms in Cape Frio as a mixed assemblage of 90 species of neritic and oceanic diatoms; *Asterolampa marylandica* and *Odontella sinensis* were indicators of the tropical water transported by the Brazil Current, and *Asteromphalus hookeri* indicated the presence of subantarctic waters. Other common species during upwelling were *Rhizosolenia alata, Bacteriastrum hyalinum, Odontella mobiliensis, Chaetoceros affinis, Chaetoceros compressus, Rhizosolenia setigera, Stephanopyxis turris, Stephanopyxis palmeriana, Thalassionema nitzschioides, Synedra tabulata, Rhizosolenia calcar-avis, Skeletonema costatum, Coscinodiscus gigas*, and *Coscinodiscus jonesianus*. Later studies (Macedo and Valentin 1974; Macedo et al. 1975) reported a sudden shift in phytoplankton composition and densities from the low point of 5×10^3 cells L^{-1} to 600×10^3 cells L^{-1} soon after the upwelling initiate. Two decades later, Valentin et al. (1985) updated the species list, reporting a 226-diatom inventory that revealed Cape Frio is possibly the most important hot spot of diatoms in Brazilian coastal waters. They also reported high densities of *Hemiaulus sinensis* in the upper warm water above the thermocline near the upwelling plume. Although the phytoplankton composition and its size-class distribution in terms of chlorophyll concentration have been determined in the Cape Frio upwelling system (Guenther et al. 2008), a more comprehensive investigation of the community size range is necessary for modeling of fisheries production, supporting management of marine resources in this important Brazilian ecosystem.

2.1.2 Phytoplankton Composition and Dynamics on the Wide Shelf off the South Brazil Bight

South of Cape Frio's upwelling and further offshore, the wind field, shelf morphology, and dynamics of the Brazil Current jointly control seasonal changes to the hydrographic regime; this affects the composition and dynamics of phytoplankton on the continental shelf off the SBB. There are important nutrient-rich frontal boundaries on this wide (*ca* 200 km) shelf system along the Brazilian margin, where phytoplankton tends to accumulate physically or through growth rate enhancement (Brandini et al. 1989). The most important fertilization mechanism is the summer intrusion of the SACW (see review by Castro et al. 2006). Similar to the Cape Frio upwelling, N–NE winds are more frequent and persist during summertime, moving surface waters away from the coast due to the Coriolis effect. Deep SACW moves shoreward to balance the Ekman transport of surface waters seaward, increasing physical gradients and the stability of a more fertile euphotic zone. Oceanic intrusions cause strong stratification in mid-shelf sections in late summer periods (Castro 2014). With the relaxation of N–NE winds, bottom layers of intrusions are tidally mixed; dinoflagellates, coccolithophorids, and silicoflagellates accumulate in tidal shelf fronts between the tidally mixed inner shelf and the stratified oceanward side of the front (Brandini et al. 1989, 2014), similar to what has been described for tidal fronts over shelf regions in the North Atlantic (Kiørboe 1993; Holligan et al. 1984; Sharples et al. 2001).

Ascension of the thermocline/nutricline caused by an intrusion fertilizes the lower euphotic layers, in which a phytoplankton community adapted to low light then grows over the mid-shelf. The deep chlorophyll maximum layer (DCML) is a conspicuous feature in the mid- and outer shelf of the SBB, with shade-adapted phytoplankton mostly composed by diatoms and nanoeukaryotes (Brandini et al. 2014). In an early investigation, Brandini et al. (1989) discussed the development of this subsurface maximum of chlorophyll accumulation through three simultaneous processes: (i) photoadaptation with the increase of intracellular chlorophyll concentration without increase of biomass, (ii) *in situ* net production with an increase in cell density, and (iii) a decrease in the rate of cellular sedimentation. In the lower layers of the euphotic zone, light becomes a limiting factor, so the intracellular concentration of chlorophyll increases to compensate for the low light intensity. Subsurface photo adaptation of picoplankton in oceanic areas was reported recently in the shelf-break off the South Brazil Bight without an increase of biomass; the intracellular carbon concentration was the same as in the phytoplankton populations in the mixing zone above the thermocline/nutricline (Ribeiro et al. 2016). In contrast, photo adaptation of diatoms in neritic waters off the bight increases not only the intracellular concentration of chlorophyll but also the photosynthetic efficiency, i.e. the rate of photosynthesis per unit of chlorophyll (Brandini 1990b). Consequently, more carbon can be fixed by photosynthesis and the cells multiply. Brandini et al. (1989) found higher concentrations of dissolved oxygen in layers just above the DCML, confirming the photosynthetic activity in these layers.

Phytoplankton at Subsurface Layers: Seasonal Dynamics Across the Shelf

Only a few studies have addressed the phytoplankton composition of the DCML in particular, so a specific assemblage at a particular shelf location or season cannot be associated with the DCML. Identifying a clear annual pattern in that phytoplankton diversity is necessary for a better understanding of the energy flux in the shelf ecosystem, taking into account its importance for the recruitment of commercial pelagic fish and for the vertical flux of particulate matter to the benthic ecosystem. Odebrecht and Djurfeldt (1996) performed size fractionation of phytoplankton chlorophyll in the open shelf off Cape Santa Marta during spring. Their results suggested that

<20 μm cells dominate the phytoplankton of the upper mixed layer while cells >20 μm, mainly those of the genus *Coscinodiscus* and *Thalassiosira*, dominate the DCML and contribute up to 85% of the chlorophyll stock in the water column.

In 2005–2006, a seasonal cross-shelf survey of the plankton system was conducted off Santa Catarina, central SBB, to study the structure and dynamics of the phytoplankton in relation to the intrusion of the SACW (Brandini et al. 2014). Spatial trends of community properties showed statistically significant differences between the coastal and the oceanic sides of the transect within each survey and among seasons. Species richness, diversity, and percent contribution of coastal species to total diatoms all decreased oceanward; the exception was January, when these properties did not change significantly at the surface or in the DCML compared to coastal assemblages (Fig. 3).

Along the same transect, chlorophyll concentrations varied from <0.1 to 6.2 μg L^{-1} with a maximum at depths that differed according to the isobath. Peak concentrations tended to be deeper toward the outer shelf, associated with the volume of SACW in the bottom layers and the potential of fertilization by this water mass at the base of the euphotic zone (Fig. 4). An interesting feature, probably related to physical forcing, was the formation of two independent nuclei of higher phytoplankton concentrations along the thermocline: one between 20 and 100 m, mostly dominated by large diatoms, and the other located in deeper layers near the shelf-break dominated by nano-planktonic diatoms and flagellates adapted to lower light intensities. This was confirmed by Simião (2010) who studied the phytoplankton composition along the same sampling transect, reporting a total of 103 species of diatoms, including 30 morphotypes of pennate and 8 morphotypes of centric species. Overall, the specific composition of the DCML varied sharply toward the outer shelf between all sampling surveys. This early and late summer assemblage was mainly composed of cosmopolitan tropical and subtropical groups of centric and pennate species dominated by *Aulacoseira granulata* (60%), *Coscinodiscus* sp. (40%), *Cyclotella styllorum* (51%), *Cylindrotheca closterium* (46%), *Diploneis* cf. *bombus* (56%), *Pseudo-nitzschia* sp. (42%), *Thalassionema nitzschioides* (58%), *Thalassiosira* cf. *decipiens* (51%), and *Hemiaulus indicus* (36%). The overall composition and the relative contribution of dominant cells were similar in the upper mixed layer above the DCML, though at lower density. Subsurface patches of diatoms, usually contributing 29–99% to total phytoplankton density, were conspicuous across the shelf. In January 2006, the intrusion was at its maximum shoreward penetration with a very thick DCML between the 50- and 80-m isobaths, being dispersed in the inner-shelf area in late summer (March) due to the tidal stirring of the bottom layers.

During this 2-month period of maximum intrusion, the diatom composition in the DCML varied by altering the species dominance across the shelf. Unusual features deserving further investigation are the presence of the invasive species *Coscinodiscus wailesii* and presence and (occasional) dominance of the "freshwater" *Aulacoseira granulata* across the whole shelf. *C wailesii* was first identified in the coastal area of Paraná state by Procopiak et al. (2006). The samples with the

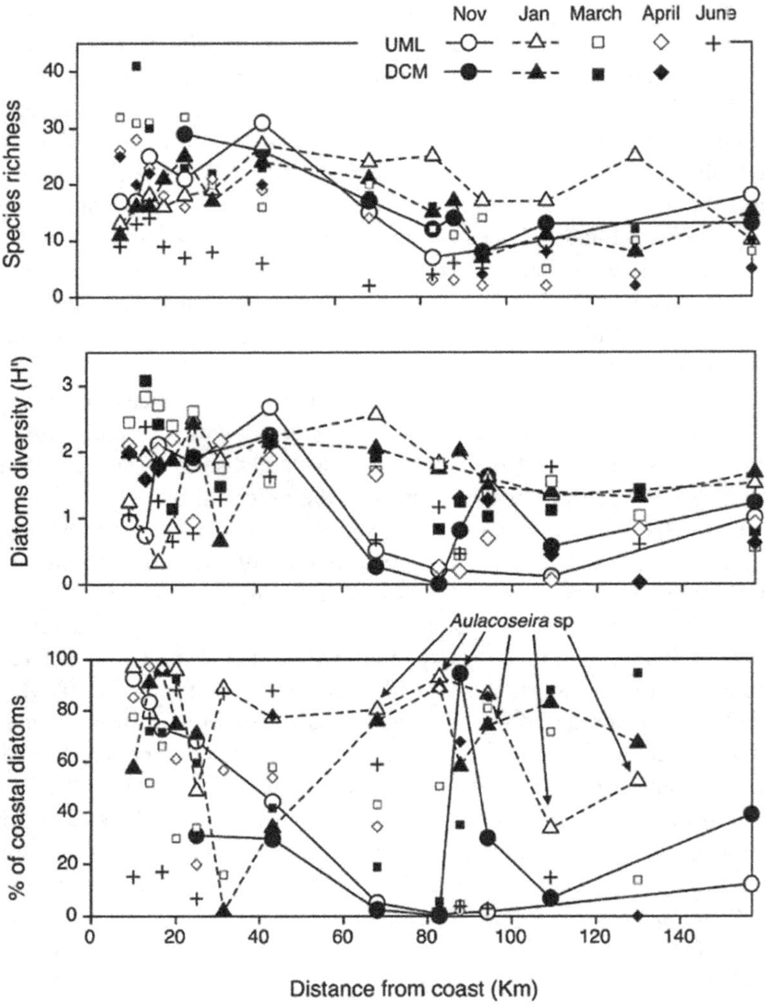

Fig. 3 Seasonal trends in the cross-shelf distribution of diatom community properties in the South Brazil Bight between November 2005 and June 2006 (UML upper mixed layer, DCML deep chlorophyll maximum layer). (Redrawn from Brandini et al. 2014)

smallest cell densities and largest carbon biomass (determined according to Utermöhl 1958) that were collected in late summer (March) were all related to the presence of *C. wailesii*. The same co-occurrence of low densities and high concentrations of chlorophyll was also associated with the occurrence of *C. wailesii* by Fernandes and Brandini (2004), who examined the seasonal variation of diatoms on the shelf off Paraná.

Aulacoseira granulata is a well-known freshwater diatom, which apparently tolerates high salinity environments. It commonly dominates the plankton community in the Paraná River hydrographic basin and in the estuaries of southern

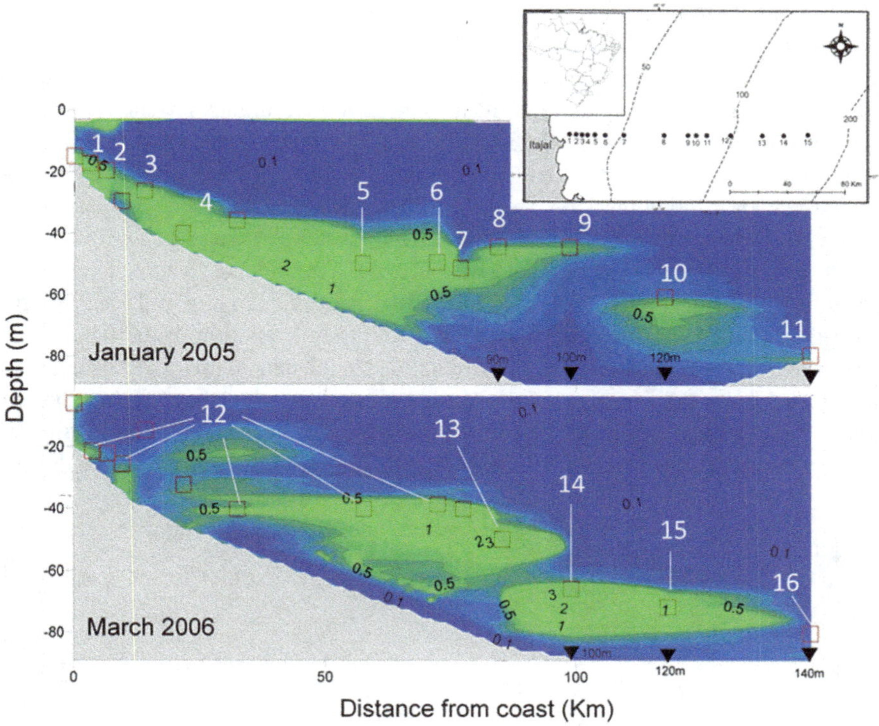

Fig. 4 Vertical distribution of chlorophyll and the main diatoms in the DCM along a cross-shelf transect in the SBB. Utermöhl samples were obtained at 15 stations (upper map) during two late summer surveys. The numbers along the chlorophyll profiles (lower panels) indicate dominance of diatoms as follows: 1-81% *Coscinodiscus*; 2-50% *H. indicus*, 12% *pennate diatoms*; 3-61% *C. wailesii*, 21% *Coscinodiscus*; 4-80–97% *Coscinodiscus* spp; 5-28% *C. wailesii*, *22% Coscinodiscus* spp, 10% *G. striata*, 2% *Aulacoseira granulata*; 6-24% *G. striata*, 22% *Chaetoceros* spp., 13% *Coscinodiscus* spp; 11% *P alata*, 10% *Bacteriastrum* sp; 7-25% *Coscinodiscus* sp., 21% *Rhizosolenia* sp., 18% *Thalassiosira*, 11% *P. alata*, 2% *Aulacoseira*; 8-96% *Coscinodiscus*, 3% *Aulacoseira*; 9-71% *Coscinodiscus*, 16% *P. alata*; 10-66% *G. striata*, 34% centric diatoms; 11-90% *Coscinodiscus* spp, 9% *Aulacoseira*; 12-80–95% *Coscinodiscus wailesii*, *Coscinodiscus* spp; 13-54% *Aulacoseira granulata*; 14 *Paralia sulcata*, *Actinoptychus*, *Coscinodiscus* spp., *Cyclotella* sp; 15-73% *Thalassiosira* sp. <20; 16-32% *Pleurosigma* sp. 21% centric diatom. (Redrawn from Brandini et al. 2014; diatoms were examined by Simião 2010)

Brazil during rainy seasons (O'Farrell et al. 1996; O'Farrell and Podlejski 2001; Odebrecht et al. 2005; Procopiak et al. 2006). Empty frustules of this species contribute <2% to the total frustules identified in slope sediments of Argentina and the southernmost Brazilian continental margin (Romero and Hensen 2002). Resting cells may be transported northward by the La Plata river plume during winter periods, adapted to shade and the high nutrient conditions of the thermocline/nutricline layer next to the top of the SACW intrusion. This species may therefore be a good indicator of the effect of the La Plata plume over the neritic plankton system off the SBB.

The Diatom Hydrodynamic Loop

The coastal diatom assemblage described in the previous section provides an important seeding stock to colonize and eventually guarantee higher productivity areas elsewhere on the wide shelf off the SBB. In 2002, a seasonal cross-shelf survey program in the central part of the SBB provided evidence of a diatom cell circulation driven by a combination of physical and physiological processes. Brandini et al. (2014) argue that this "diatom hydrodynamic loop" takes place between the inner- and mid-shelf areas off the SBB (Fig. 5). The balance between offshore-driven surface waters and onshore bottom intrusions, tidal stirring in the nearshore, and the physiological stress of diatoms due to nitrate limitation in the outer-shelf region are jointly responsible for this circulation. In the near shore, diatoms are kept in suspension by tidal stirring; if the northeast wind persists, they are transported along with surface waters toward the mid-shelf where nutrients, mainly nitrate, become limited under the influence of warm oligotrophic waters of the Brazil

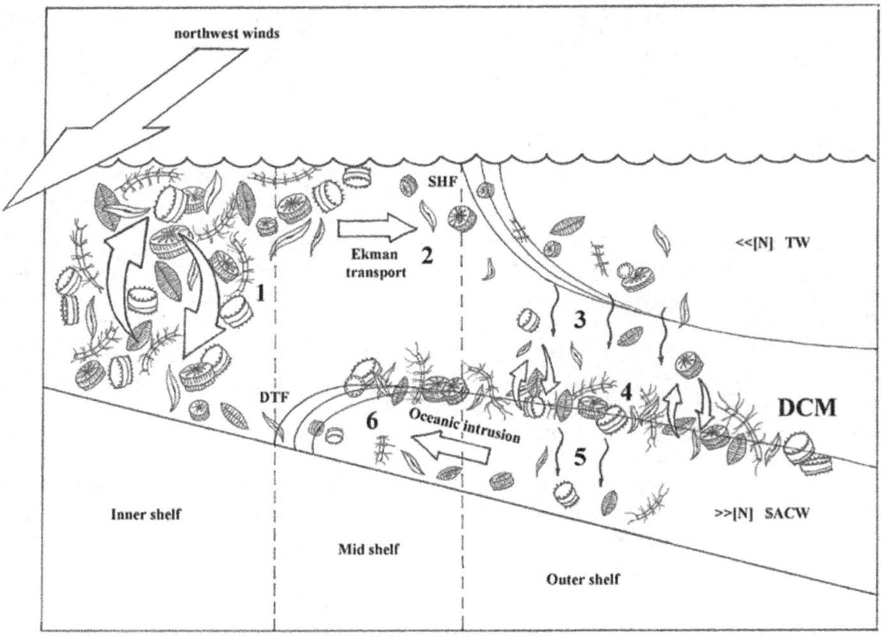

Fig. 5 The diatom loop across the shelf off the South Brazil Bight: 1-diatoms are resuspended by tidal stirring on the inner shelf; 2-northeast winds and the Coriolis effect caused Ekman transport of the surface community toward oligotrophic waters in the mid-shelf; 3-due to nutrient limitation, diatoms lose buoyancy and are forced downward by the convergence of saltier tropical water (TW) near mid-shelf fronts; 4-under better nutritional conditions at the top of the oceanic intrusion, diatoms adapt to low light intensities and start growing again, contributing to the formation of the DCML on the mid-shelf; 5-the increase of cell densities causes light-limited cells to sink and resting cells to accumulate in the dark sediments; 6-resting cells are moved coastward by summer intrusions; DTF-deep thermal front; SHF-surface haline front; SACW-South Atlantic Central Water; DCM-deep chlorophyll maximum. (Redrawn from Brandini et al. 2014)

Current. Because of nutrient limitation, cells become heavier due to the ballast of silica frustules and tend to sink faster. The convergence of the saltier TW enhances their sinking rate. Sinking cells reach the nutrient-rich pycnocline in the top of the SACW, grow, and then accumulate in the lower half of the euphotic zone, contributing to the formation of the DCML in the mid-shelf. Below this layer, light-limited cells sink faster and accumulate in sediments as resting cysts where they can be transported shoreward by the oceanic intrusions during the next summer seasons. A similar mechanism may occurs in eastern upwelling systems but on a much larger geographical scale. This loop is important to maintain perpetual seeding stocks of diatoms in shelf regions where nutrients are usually more abundant.

Trichodesmium Blooms

Another important feature of the phytoplankton dynamics on the shelf of the SBB is the seasonal accumulation of the globally ubiquitous cyanobacteria *Trichodesmium* sp. It was first detected in the northeast of Brazil when causing toxicity in the surf spray of sandy beaches known as "Tamandaré fever" (after Satô et al. 1963). They are common in equatorial waters off the northeast during dry seasons, being pushed away from the coast by continental runoff during the rainy season (Medeiros et al. 1999). Surface blooms have been reported in the nearshore and over the shelf-break off the South Brazil Bight given calm winds and oligotrophic conditions (Brandini 1988; Gianesella-Galvão et al. 1995; Rörig et al. 1998; Siqueira et al. 2006; Detoni et al. 2016). Trichomes do not accumulate, yet filaments are commonly identified in plankton samples during winter; these are restricted to the northern half of the SBB, pushed northward by the Subtropical Shelf Front (SSF, *sensu* Piola et al. 2008; Brandini 1988). The life cycle of this important diazotroph in the SWA has gaps that need further investigation. Queiroz et al. (2004) reported trichomes of this species in sediments from 18 meters off Paraná state; they were probably not fixing molecular nitrogen because higher nitrate concentrations near the coast stop the activity of nitrogenase, the enzyme responsible for molecular nitrogen fixation. It is unclear how these coastal populations were advected from distant offshore oligotrophic waters to the nearshore (Fig. 6).

Winter Changes due to the La Plata River Plume

Besides local nutrient-rich continental discharges, which are enhanced during rainy seasons (Ciotti et al. 1995), inner- and mid-shelf areas of the southern half of the bight are also subjected to mass fertilization of the La Plata plume in winter seasons. Signatures of the river plume arrive already poor in nitrate yet carry large phosphate loads (Brandini 1990a; Muelbert et al. 2008). Taking into account its continental origin, the plume must be enriched with trace elements (e.g., Fe). Besides its importance for photosynthesis and respiration, Fe is essential for nitrogen fixation by diazotrophic cyanobacteria and may explain the frequency of *Trichodesmium* in

Fig. 6 *Trichodesmium* bloom at the western Brazilian shelf-break that was dominated by the Brazil Current in January 2017. (Source: The author)

mid- and outer-shelf waters of the SBB (Detoni et al. 2016). The SSF is revealed in satellite images as consisting of cold-water tongues moving progressively alongshore toward the northeast, carrying a high volume of chlorophyll-rich La Plata plume (Fig. 7, higher panel). The multiyear pattern of mean chlorophyll obtained from satellite data in the southern Brazilian shelf revealed higher mean concentrations associated with the advance of the SSF in wintertime. Two peak concentrations coincided with the El Niño of 2007–2008 and 2016–2017 (Fig. 7, lower panel), when high altitude jetties crossing over the South America continent are intensified, hampering the atmospheric fronts in middle latitudes. This increases the precipitation over the La Plata hydrographic basin and (hence) the input of nutrients over the southern Brazilian shelf (Ciotti et al. 1995).

Surface phytoplankton composition and spatial distribution change remarkably during the advection of the plume. Brandini (1988, 1990a) has reported remarkable changes in the surface distributional patterns of environmental properties and dominant eukaryotic phytoplankton between the southernmost Brazilian and Uruguayan shelf province, during the winter of 1983 (Fig. 8). Out of the cold La Plata plume, either to the north or oceanward from the plume front, different subtropical and temperate species of eukaryotic phytoplankton depicted individual distribution patterns limited by the 20 °C isotherm. The bulk of diatoms was concentrated in the cold waters of the plume. The dominant genera in the plume were *Thalassiosira*, followed by *Leptocylindrus* and *Skeletonema*, and dinoflagellates mostly dominated by *Ceratium lineatum* (Fig. 8m), a cosmopolitan species of temperate and subpolar waters (http://www.algaebase.org). The abundance of *C. lineatum* of the La Plata plume reaching southern Brazilian waters may support the argument of lateral

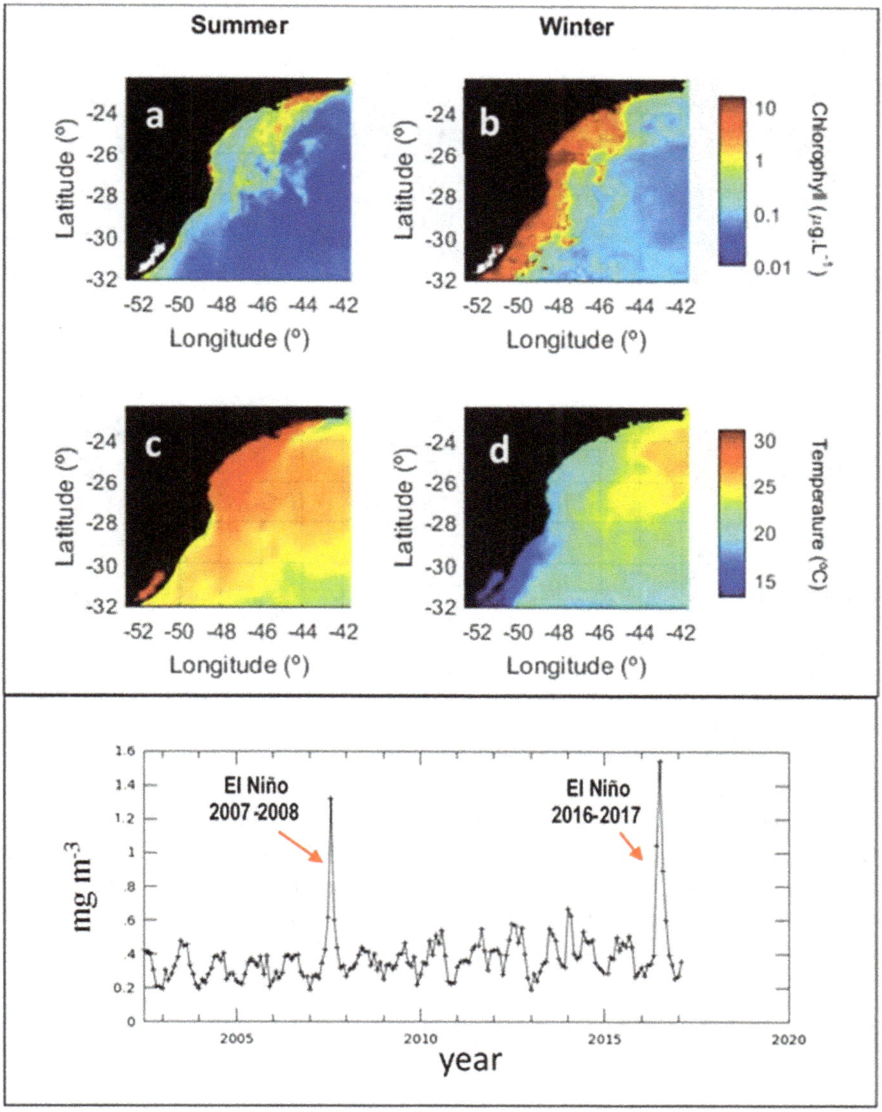

Fig. 7 Upper panel: satellite images of surface temperature and chlorophyll during summer (**a**, **c**) and winter (**b**, **d**) in the subtropical Brazilian shelf provinces. (From MODIS_Aqua_L3_CHLA_Daily_4km_R at http://oceancolor.gsfc.nasa.gov. Lower panel: Multiyear variation of the mean chlorophyll concentration in the same area showing two conspicuous peaks of biomass related to the El Niño of 2007–2008 and 2016–2017)

advection of chlorophyll-rich La Plata plume during the El Niño instead of enrichment of the shelf with nutrient-rich drainage. The genera *Chaetoceros*, *Thalassionema*, *Thalassiothrix*, *Bacteriastrum*, *Coscinodiscus*, and *Nitzschia* were represented by species with more affinity for the warmer (>20 °C) and saltier (>35) waters found in the north and offshore side of the plume. Small coccolithophorids

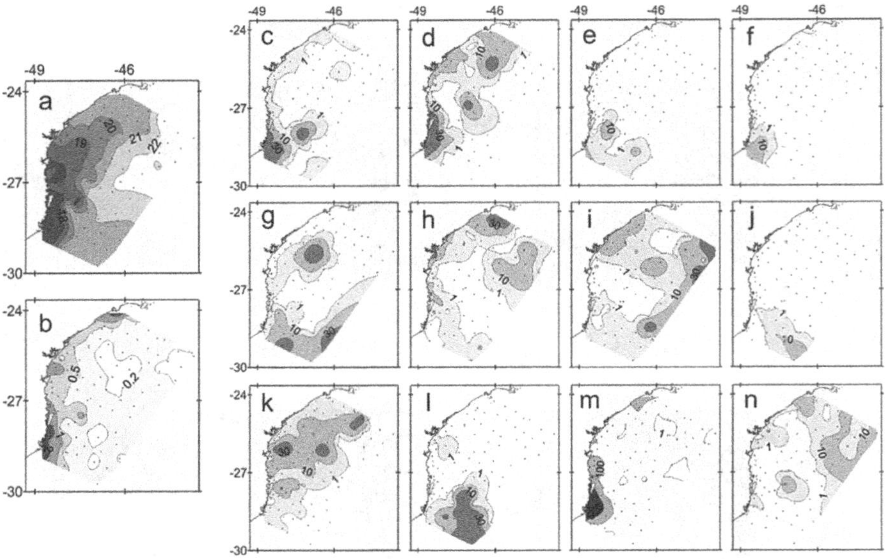

Fig. 8 Winter (July, 1982) geographic distribution of sea surface chlorophyll-a concentration (μg L^{-1}) and dominant phytoplankton species (10^3 cells L^{-1}) in the South Brazil Bight. (**a**) Temperature (°C), (**b**) chlorophyll *a* (mg m^{-3}), (**c**) total diatoms (10^3 cells L^{-1}), (**d**) *Thalassiosira* spp (%), (**e**) *Leptocylindrus* sp. (%), (**f**) *Skeletonema costatum* (%), (**g**) *Chaetoceros* spp (%), (**h**) *Thalassionema nitzschioides* (%), (**i**) *Thalassiothrix* sp. (%), (**j**) *Bacteriastrum* sp. (%), (**k**) *Coscinodiscus* spp. (%); (**l**) *Nitzschia* sp. (%), (**m**) *Ceratium lineatum* (cells L^{-1}), (**n**) coccolithophores (10^3 cells L^{-1}). (Redrawn from Brandini 1988)

were frequent out of the plume, mostly dominated by an unidentified species with general size (4–6 μm) and morphology similar to the cosmopolitan *Gephyrocapsa oceanica*, followed by *Calciosolenia murrayi.* Among the prokaryotes, trichomes of *Trichodesmium* were common out of the plume, yet never abundant.

3 Summary and Final Remarks

The SWA has complex hydrodynamics associated with important western boundary currents of different chemical features that affect remarkably the phytoplankton composition. Particularly the Southeastern Brazilian subtropical shelves are generally oligotrophic, dominated by nutrient-poor warm waters brought by the Brazil Current. Understanding the spatial and temporal patterns of phytoplankton diversity and distribution in relation to environmental forces is the stepping stone for modeling ecosystem services in this wide shelf province of the Southwest Atlantic Ocean. This includes modeling of oceanic primary production, lateral advection and vertical flux of organic matter to the benthic system, long-term dynamics of deep water carbon reservoirs (i.e., oceanic biological pump), and regional ecosystem-based fishery management and human health.

In particular, the geographic area occupied by the warm waters of the subtropical Brazilian province may be regarded as a biological *hot spot* of phytoplanktonic cells, responding at once to complex and variable physical oceanographic processes that may potentially fertilize the euphotic zone and/or mix different phytoplankton communities. Yet its phytoplankton is one of the least studied among the tropical and subtropical shelf regions. Large-scale physical processes like oceanic intrusions of deep nutrient-rich waters, cyclonic eddies (and meandering) of the Brazil Current, and mass fertilization of large freshwater plumes of the La Plata river impart a well-defined seasonal pattern that may be modeled in terms of phytoplankton abundance. To do so, however, many ecological and taxonomic gaps must be filled. Studies on community diversity and individual species dynamics that were reported to once dominate in some locations and seasons are still not enough to allow prediction. A better assessment of taxonomic diversity, abundance, and cell-size structure is crucial for understanding other oceanographic processes directly linked to the energy flow at low trophic levels, hence the initial oceanic steps of global carbon cycling.

The answers on how future anthropogenic impacts and global climate changes may affect the structure and function of regional marine ecosystems certainly rely on phytoplankton resilience to changes of environmental properties on a large geographic scale. A joint regional collaboration between Brazilian, Uruguayan, and Argentinean scientists will be helpful to address this issue and the best track toward elucidation. That collaboration would be a promising step forward to better assess marine resources and environmental sustainability for future generations of southwest Atlantic countries.

References

Alves N Jr, Meirelles P, Santos EO et al (2015) Microbial community diversity and physical-chemical features of the Southwestern Atlantic Ocean. Arch Microbiol 197:165–179

Azam F, Fenchel T, Field JG et al (1983) The ecological role of water-column microbes in the sea. Mar Ecol Prog Ser 10:257–263

Barth R, Castro MLRO (1964) Sobre silicoflagelados como indicadores biológicos na costa sul do Brasil. Notas Técnicas Inst Pesq Mar 14:1–22

Barth R, Castro MLRO (1965) Observações biogeográficas em *Ceratium tripos* (Dinoflagellata). Notas Técnicas Inst Pesq Mar 21:1–12

Bergo NM, Signori CM, Amado AM et al (2017) The partitioning of carbon biomass among the pico- and nanoplankton community in the South Brazilian Bight during a strong summer intrusion of South Atlantic Central Water. Front Mar Sci 4:1–12

Brandini FP (1988) Composição e distribuição do fitoplâncton na Região Sueste e suas relações com as massas d'água (Operacão Sueste I – Inverno/1982). Ciência e Cultura, São Paulo 40:334–341. (in Portuguese)

Brandini FP (1990a) Hydrography and characteristics of the phytoplankton in shelf and oceanic waters off Southeastern Brazil during winter (July/August 1982) and summer (February/March 1984). Hydrobiology 196:111–148

Brandini FP (1990b) Producão primária e características fotossintéticas do fitoplâncton na Região Sueste do Brasil. Bol Inst Oceanogr 38(2):147–159. (in Portuguese)

Brandini FP, Moraes CLB, Thamn CAC (1989) Shelf break upwelling, subsurface maxima of chlorophyll and nitrite, and vertical distribution of subtropical nano- and microplankton community off Southeastern Brazil. In: Brandini FP (ed) Memórias do III Encontro Brasileiro de Plâncton. Editora da UFPR, Curitiba, pp 47–55

Brandini FP, Lopes RM, Gutseit KS et al (1997) Planctonologia na plataforma continental do Brasil. Diagnose e revisão bibliográfica. Ministério do Meio Ambiente e da Amazônia Legal – IBAMA, Brasília (in Portuguese)

Brandini FP, Nogueira M, Simião MS et al (2014) Deep chlorophyll maximum and plankton community response to oceanic bottom intrusions on the continental shelf in the South Brazilian Bight. Cont Shelf Res 89:61–75

Campos EJD, Piola AR, Matano RP et al (2008) A synoptic characterization of the Southwest Atlantic shelf under influence of the Plata River and Patos Lagoon outflows. Cont Shelf Res 28(13):1551–1555

Castro BM (2014) Summer/winter stratification variability in the central part of the South Brazil Bight. Cont Shelf Res 89:15–23

Castro BM, Brandini FP, Pires Vanin AMS et al (2006) Multidisciplinary oceanographic processes on the western Atlantic continental shelf between 4°N and 34° S. In: Robinson AR, Brink K (eds) The sea: the global Coastal Ocean: interdisciplinary regional studies and syntheses. Harvard University Press, Cambridge, MA, pp 259–293

Ciotti AM, Odebrecht C, Fillmann G et al (1995) Fresh-water outflow and subtropical convergence influence on phytoplankton biomass on the Southern Brazilian continental shelf. Cont Shelf Res 15:1737–1756

Cushing DH (1989) A difference in structure between ecosystems in stongly sratified waters and in those that are only weakly stratified. J Plankton Res 11:1–3

Detoni AD, Aurea C, Paulo C et al (2016) *Trichodesmium* latitudinal distribution on the shelf-break in the southwestern Atlantic Ocean during spring and autumn. Global Biogeochem Cycles 30. https://doi.org/10.1002/2016GB005431

Fernandes LF, Brandini FP (2004) Diatom associations in shelf waters off Parana State, Southern Brazil: annual variations in relation to environmental factors. Rev Bras Oceanogr 52(1):19–34

Field C, Behrenfeld M, Randerson J et al (1998) Primary production of the biosphere: integrating terrestrial and oceanic components. Science 281(5374):237–240

Flombaum P, Gallegos JL, Gordillo RA et al (2013) Present and future global distributions of the marine Cyanobacteria *Prochlorococcus* and *Synechococcus*. Proc Natl Acad Sci USA 110:9824–9829

Franco AOR, They NH, Canani LGC et al (2016) Asterionellopsis tropicalis (Bacillariophyceae): a new tropical species found in diatom accumulations. J Phycol 52:888–895

Gianesella-Galvão SMF, Costa MPF, Kutner MBB (1995) Bloom of *Oscillatoria* (*Trichodesmium*) erythraea (Ehr.) Kutz in coastal waters of the Southwest Atlantic. Publ Esp Inst Oceanogr 11:133–140

Gonzalez-Rodriguez E (1994) Yearly variation in primary productivity of marine phytoplankton from Cabo Frio (RJ, Brazil) region. Hydrobiologia 294:145–156

Guenther M, Gonzalez-Rodriguez E, Carvalho WF et al (2008) Plankton trophic structure and particulate organic carbon production during a coastal downwelling-upwelling cycle. Mar Ecol Prog Ser 363:109–119

Holligan PM, Balch WM, Yentsch CM (1984) The significance of subsurface chlorophyll, nitrite and ammonium maxima in relation to nitrogen for phytoplankton growth in stratified waters of the Gulf of Maine. J Mar Res 42:1051–1073

Kiørboe T (1993) Turbulence, phytoplankton cell size, and the structure of pelagic food webs. Adv Mar Biol 29:1–72

Lorenzetti JA, Gaeta SA (1996) The Cape Frio upwelling effect over the South Brazil Bight northern sector shelf waters: a study using AVHRR images. Int Arch Photogramm Remote Sens 31(B7):448–453

Macedo FE, Valentin JL (1974) O plâncton na ressurgência de Cabo Frio (Brasil). III – Primeiras observações sobre o microfitoplâncton (período de 04/02 a 16/04/73). Publ Inst Pesq Marinha 84:1–10

Macedo FE, Tenenbaum DR, Valentin JL (1975) O plâncton na ressurgência de Cabo Frio (Brasil). VI Composição florística e suas variações nas águas da estação fixa oceânica. Publ Inst Pesq Marinha 87:1–9

Marañón E (2009) Phytoplankton size structure. In: Steele JH, Turekian KK, Thorpe SA (eds) Encyclopedia of ocean science. Academic, Oxford, pp 445–452

Medeiros C, Macêdo SJ, Feitosa FAN et al (1999) Hydrography and phytoplankton biomass and abundance of North-East Brazilian Waters. Arch Fish Mar Res 47(2/3):133–151

Moreira-Filho H (1965) Contribuição ao estudo das diatomáceas da região de Cabo Frio (Estado do Rio de Janeiro – Brasil). An Acad Brasil Ciênc 37:231–238. (in Portuguese)

Muelbert JH, Acha EM, Mianzan H et al (2008) Biological, physical and chemical properties at the subtropical shelf front zone in the SW Atlantic continental shelf. Cont Shelf Res 28:1662–1673

Odebrecht C, Djurfeldt L (1996) The role of nearshore mixing on the phytoplankton size structure off Cape Santa Marta Grande, Southern Brazil (spring 1989). Arch Fish Mar Res 43(3):13–26

Odebrecht C, Abreu PC, Möller OO et al (2005) Drought effects on pelagic properties in the shallow and turbid Patos Lagoon, Brazil. Estuaries 28(5):675–685

Odebrecht C, Bergesh M, Rörig LR et al (2010) Phytoplankton interannual variability at Cassino Beach, Southern Brazil (1992–2007), with emphasis on the surf zone diatom *Asterionellopsis glacialis*. Estuaries Coast 33:570–583

Odebrecht C, Campbell E, Du Preez D, Abreu PC (2014) Surf zone diatoms: a review of the drivers, patterns and the role in sandy beaches food chains. Estuar Coast Shelf Sci 150:24–35

O'Farrell G, Podlejski TA (2001) Morphological variability of *Aulacoseira granulata* (Ehr.) Simonsen (Bacillariophyceae) in the lower Paraná River (Argentina). Limnology 2(2):65–71

O'Farrell G, Izaguirre I, Vinocur A (1996) Phytoplankton ecology of the lower Paraná River (Argentina). Arch Hydrobiol Suppl 115(1):75–89

Partensky F, Blanchot J, Vaulot D (1999) Differential distribution and ecology of *Prochlorococcus* and *Synechococcus* in oceanic waters: a review. Bull Inst Océanogr 19:457–475

Piola AR, Möller OO, Guerrero R et al (2008) Variability of the subtropical shelf front off eastern South America: winter 2003 and summer 2004. Cont Shelf Res 28:1639–1648

Procopiak LK, Fernandes LF, Moreira-Filho H (2006) Diatomáceas (Bacillariophyta) marinhas e estuarinas do Paraná, Sul do Brasil: lista de espécies com ênfase em espécies nocivas. Biota Neotropica 6(3):1–27. (in Portuguese)

Queiroz RLV, Brandini FP, Pellizari FM (2004) Dynamics of microalgal communities in the water-column/sediment interface of the inner shelf off Parana State, southern Brazil. Rev Bras Oceanogr 52(3/4):183–194

Ribeiro CG, Santos AL, Dominique M et al (2016) Pico and nanoplankton abundance and carbon stocks along the Brazilian Bight. PeerJ 4:2587–2016

Romero O, Hensen C (2002) Oceanographic control of biogenic opal and diatoms in surface sediments of the Southwestern Atlantic. Mar Geol 186:263–280

Rörig LR, Yunes JS, Kuroshima KNS et al (1998) Studies on the ecology and toxicity of *Trichodesmium* spp blooms in Southern Brazilian coast waters. Harmful Algae 1:22–25

Satô S, Paranagua M, Eskinazi E (1963) On the mechanism of red tide of Trichodesmium in Recife, northeastern Brazil, with some considerations of the relation to human disease Tamandare fever. Trab Inst Oceanogr Univ Recife 5(6):7–49

Sharples J, Moore MC, Rippeth TP et al (2001) Phytoplankton distribution and survival in the thermocline. Limnol Oceanogr 46:486–496

Simião MS (2010) Máximos subsuperficiais de clorofila-a na plataforma Sudeste-Sul brasileira. Master's Dissertation, Univ Federal do Paraná (in Portuguese)

Siqueira A, Kolm HE, Brandini FP (2006) Offshore distribution patterns of the cyanobacterium *Trichodesmium erythraeum* Ehrenberg and associated phyto- and bacterioplankton in the Southern Atlantic Coast (Paraná, Brazil). Braz Arch Biol Technol 49(2):323–337

Tenenbaum DR, Villac MC, Viana SC et al (2004) Phytoplankton identification Atlas–Sepetiba Bay, Brazil. GloBallast Monogr Ser 16, Rio de Janeiro/London

Utermöhl H (1958) Zur Vervolkommung der quantitativen Phytoplankton. Methodik Mitt Int Ver Theor Angew Limnol 9:1–38

Valentin JL (2001) The Cabo Frio upwelling system, Brazil. In: Seeliger U, Kjerfve B (eds) Coastal marine ecosystems of Latin America. Springer, Berlin, pp 97–105
Valentin JL, Lins da Silva NM, Bastos CTBT (1985) Les diatomées dans l'upwelling de Cabo Frio (Brésil): list d'espèces et étude écologique. J Plankton Res 7(3):313–337
Valentin JL, Andre DL, Jacob SA (1987) Hydrobiology in the Cabo Frio (Brazil) upwelling: two-dimensional structure and variability during a wind cycle. Cont Shelf Res 7:77–88
Villac MC, Noronha VAPC, Pinto TO (2008) The phytoplankton biodiversity of the coast of the state of São Paulo, Brazil. Biota Neotrop 8(3):151–173

Community Structure and Spatiotemporal Dynamics of the Zooplankton in the South Brazilian Bight: A Review

Miodeli Nogueira Jr. and Frederico P. Brandini

Abstract The South Brazilian Bight (23–28.5°S; SBB) is a typical western boundary current system with a wide shelf and which has high ecologic and economic importance, supporting nationally important fishing grounds. Away from the coastal areas, the planktonic production is mostly controlled by regenerative processes prevailing in the oligotrophic Tropical Water (TW) of the Brazil Current, enriched by the advection of the nutrient-rich South Atlantic Central Water (SACW). About 790 zooplanktonic invertebrate species have been recorded, what is an underestimate considering that most meroplanktonic and some holoplanktonic taxa are largely unstudied. An inshore-offshore gradient is clear. Zooplankton abundance and biomass are typically higher in more coastal areas, diminishing offshore as the influence of the oligotrophic TW increases, while the diversity has the inverse tendency. The cold and nutrient-rich SACW intrusions certainly are the most relevant mesoscale physical feature over the shelf of the SBB, increasing considerably the primary production and subsequently zooplankton abundance and production. Indeed, available data suggest that the intrusions of the SACW and their strength are an important factor influencing both seasonal and interannual variability in zooplankton diversity, biomass, abundance, production, and size-spectra. While the distributional patterns of most of the dominant groups in relation to the main water masses are relatively well-known, little is known about the life strategy, trophic interactions, physiological responses, and the impact of the main physical processes on these populations. In this review, we also emphasize the need for process-oriented studies along SBB with spatiotemporal scales relevant to the main physical events in order to better understand the zooplankton dynamics and their role in the regional fishery production.

Keywords Zooplankton · Composition · Dynamics · South Brazilian Bight

M. Nogueira Jr. (✉)
Departamento de Sistemática e Ecologia, Universidade Federal da Paraíba, João Pessoa, Paraíba, Brazil

F. P. Brandini
Instituto Oceanográfico, Universidade de São Paulo, Cidade Universitária, São Paulo, Brazil
e-mail: brandini@usp.br

M. S. Hoffmeyer et al. (eds.), *Plankton Ecology of the Southwestern Atlantic*, https://doi.org/10.1007/978-3-319-77869-3_8

1 Introduction

The South Brazilian Bight (23–28.5°S, SBB) is a typical western boundary current system with a wide shelf that may reach up to ca. 200 km wide (Fig. 1). It is part of a transitional zone between subtropical and temperate domains (Lohrenz and Castro 2005; Longhurst 2006). The SBB hosts a large biodiversity and is the most productive Brazilian continental shelf region, supporting important fishing grounds with fish catches >200,000 tons/year (MPA 2011), emphasizing its economic and ecologic importance. The main targeted species is the Brazilian sardine *Sardinella brasiliensis* (MPA 2011). This species is primarily planktivorous (e.g., Dias et al. 2004), as is the case of many other economically important fish species from the SBB (e.g., Kurtz and Matsuura 2001; Soares and Vazzoler 2001) and elsewhere. Even though several adult fish do not eat plankton directly, the survivorship of their larvae and consequently their recruitment rates

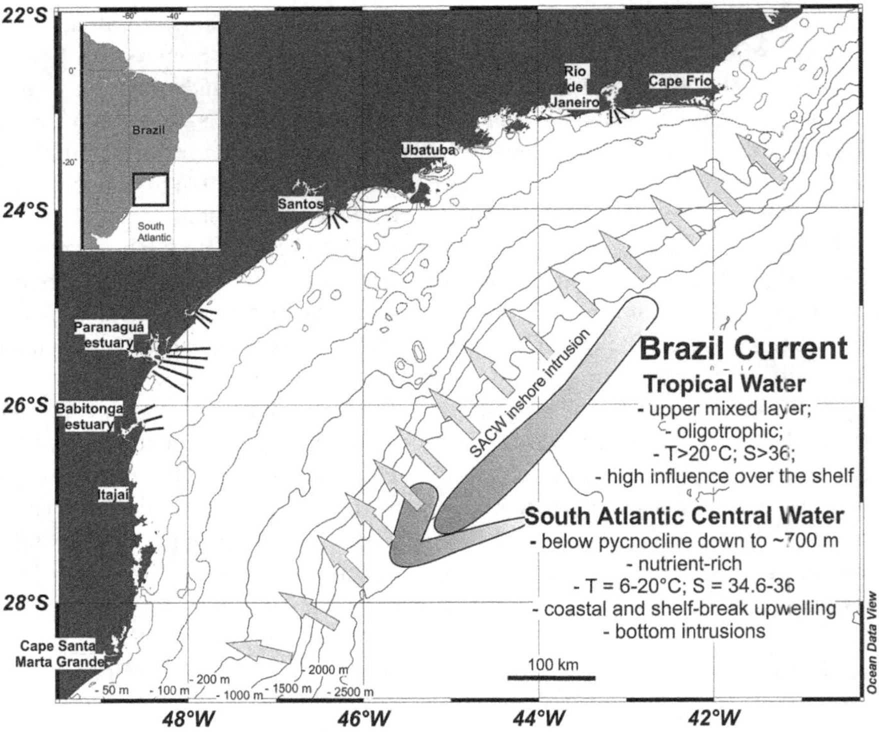

Fig. 1 Map of the South Brazilian Bight showing the main oceanographic features, such as the Brazil Current (large arrow) with the Tropical Water in the upper layer and the South Atlantic Central Water below along with its wind-driven inshore intrusions, and the main sources of continental drainage (solid lines). The Coastal Water is not shown, but it occupies mainly the shallow shelf (<50 m isobath) and has a variable signature with low salinity (<34) because of the mixing of continental runoff and the other two water masses. Generated using Ocean Data View 4.7.10 software (Schlitzer 2017)

are greatly dependent on the availability of their planktonic food (Matsuura et al. 1992; Dias et al. 2004). Thus, the knowledge of zooplankton variability is important to understand the ecosystem functioning and its ecological patterns as well as to fisheries management, providing important information about the energy flow in marine communities (Montú et al. 1997). In the present study, we provide synoptic information on diversity, structure, and temporal variability of invertebrate zooplankton from the SBB in relation to the main water masses and oceanographic processes taking place over the shelf (Fig. 1).

2 Hydrography and Circulation Patterns

Several authors have described the general physical oceanographic characteristics of the SBB (e.g., Lohrenz and Castro 2005; Castro et al. 2005, 2006; Cerda and Castro 2014; Brandini et al. 2007, 2014). The overall circulation over the shelf follows the worldwide pattern of western boundary current systems, as part of the South Atlantic Anticyclonic Subtropical Gyre. The Brazilian Current (BC) originates at *ca* 10°S and flows southward near the continental slope through all the SBB until it meets the Malvinas Current, forming the Brazil–Malvinas Confluence Zone southward the SBB, between 33°S and 38°S, where a complex frontal system with cyclonic eddies and water column stabilization support high biological production (e.g., Brandini et al. 2000). On the SBB, the upper mixed layer in mid- to outer-shelf is dominated by warm (>20 °C), saline (>36), and oligotrophic Tropical Water (TW) of the BC (Figs. 1, 2, and 3). This water mass is formed as a consequence of the intense solar radiation and excess of evaporation, characteristic of tropical and subtropical oceans. Below the TW is the South Atlantic Central Water (SACW), formed by subsidence at the Subtropical Convergence, with temperatures varying from 6 to 18 °C and salinities from 34.6 to 36. The Coastal Water (CW) with lower salinities (<34) results from the mixing of TW and/or SACW with local continental drainage, and it occupies most of the shallow shelf. It is characterized by high seasonal and geographical physical-chemical fluctuations according to geomorphology, local precipitation and river discharge (Castro et al. 2006).

Most of the SBB's shelf area is dominated by one of these water masses or a mixture of them (Figs. 2 and 3); CW tends to dominate the inner shelf (Fig. 3a, b), and TW with the SACW below dominate the outer-shelf and offshore (Fig. 3c, d). A stable thermocline separates the TW and SACW, while the surface haline front (SHF) and the deep thermal front (DTF) separate the CW and TW and CW/TW and SACW, respectively (Castro et al. 2005, 2006; Brandini et al. 2014). These boundaries are dynamic and change their position mainly driven by the wind regime pattern which tends to be seasonal (Fig. 2).

During spring-summer northeasterly winds tend to dominate and CW is pushed offshore by surface Ekman transport with SHF reaching more than 100 km offshore (Fig. 2). This results in the onshore intrusions of the cold SACW (and the consequent

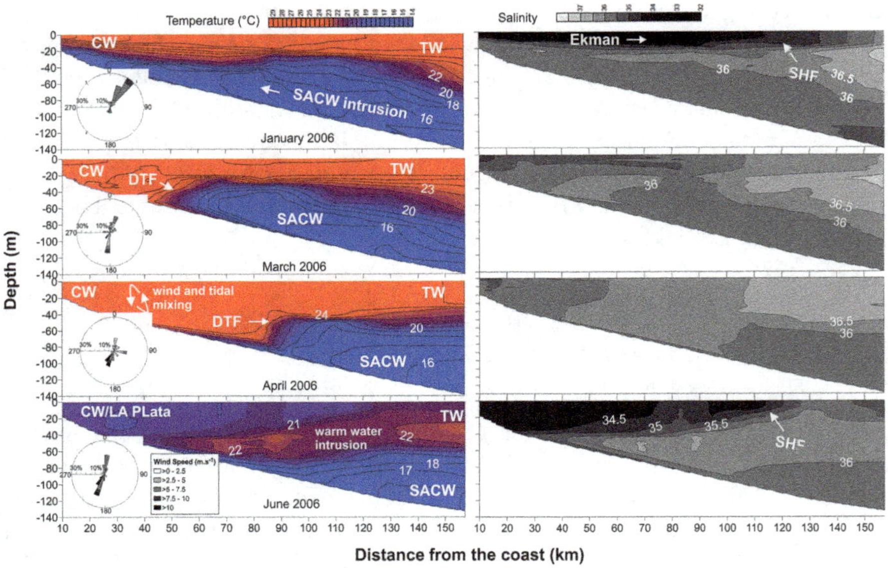

Fig. 2 Seasonal cross-shelf vertical profiles of temperature and salinity off Itajaí, between January and June 2006. Inserted in the temperature profiles are the wind fields off Santa Catarina. CW coastal water, TW tropical water, SACW South Atlantic Central Water, DTF deep thermal front, SHF surface haline front. (Modified from Brandini et al. (2014) and Nogueira Júnior et al. (2014))

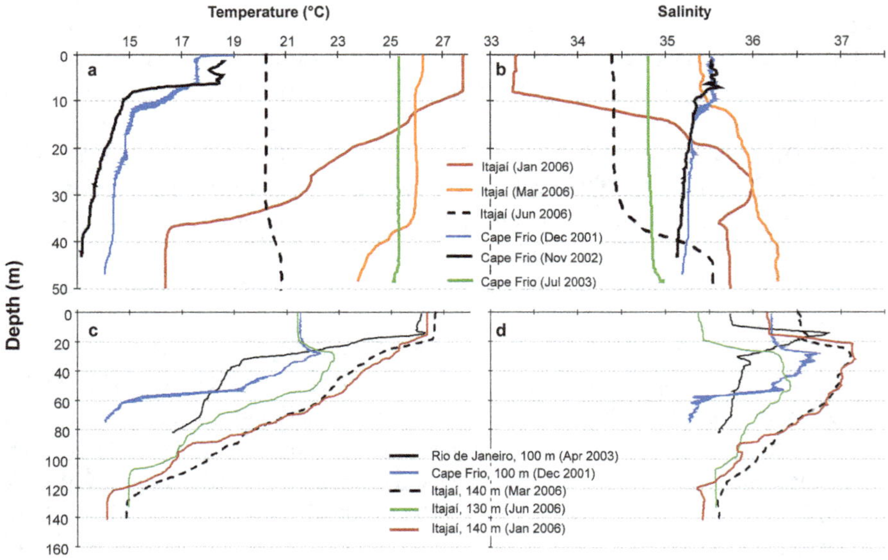

Fig. 3 South Brazilian Bight representative vertical profiles of temperature and salinity from different seasons and locations. (**a**, **b**) at 50 m depth, (**c**, **d**) at 100–140 m depth. (Data from Cape Frio and Rio de Janeiro are derived from DEPROAS cruises (IO–USP) and data from Itajaí taken from PROPEL cruises (CEM–UFPR Brandini et al. 2014; Nogueira Júnior et al. 2014))

displacement of the DTF inshore) which provide important nutrient input and also enhance light conditions over the shelf by increasing the ratio between the depth of the euphotic and upper mixing zone, thus increasing considerably phytoplankton growth and production (Brandini et al. 2007, 2014). The particularities of the intrusions may change interannually and with latitude throughout SBB, depending mostly on wind conditions and interactions with local topography, respectively. For instance, in some locations with relatively shorter shelf such as Cape Frio and Cape of Santa Marta Grande, the SACW may reach the surface (Matsuura 1986; Castro et al. 2005, 2006; Cerda and Castro 2014; see Fig. 3), while in most other sites of the SBB, the intrusion fertilizes the euphotic zone and may reach the inner shelf areas (Fig. 2 and 3), but due to the larger continental shelf and gentle slope, it typically does not reach the surface (Brandini et al. 2007, 2014; Cerda and Castro 2014; Nagata et al. 2014).

These differences are clear comparing, for instance, the vertical temperature and salinity profiles of shallow waters (50 m) off Cape Frio and Itajaí (Fig. 3a, b). In the former, in December 2001 and November 2002, temperature was low (17–19 °C) in the surface, reaching <15 °C below 20 m, while the salinity was nearly homogeneous vertically 35.2–35.5 (Fig. 3b), denoting the SACW influence through the water column. Off Itajaí in January 2006, water column was very stratified both in temperature and salinity in the 50 m isobath (Fig. 3a, b); the lower temperature near the bottom indicates the influence of the SACW only in this depth stratum, whereas the lower salinities in the surface layer indicate the high influence of CW in the upper layer, with the SHF positioned offshore (Fig. 2).

During winter, shelf waters tend to overturn vertically (Fig. 2) due to tidal circulation and Ekman convergence of outer-shelf waters driven by southerly winds, and therefore the SHF and DTF reverse, being placed inshore and offshore, respectively. Also, the winter predominance of southwest winds shifts the offshore BMCZ northward, transporting subantarctic and less saline waters (T = 4–15 °C, S = 33.5–34) from the Argentinian shelf, along with a coastal branch greatly dominated by the continental drainage of the La Plata River basin which may reach the southernmost shelf areas of the SBB (Brandini 1990; Castro et al. 2005, 2006). For instance, temperature and salinity were vertically homogeneous both off Cape Frio and Itajaí shallow waters during winter (Fig. 3a, b), in both cases highly influenced by the CW but colder and less saline off Itajaí due to influence of the La Plata Plume (Brandini et al. 2014).

Offshore, seasonal and latitudinal variations are less pronounced with the typical tropical vertical profile of temperature and salinity with warm (>20 °C) and saline (>36) waters in the upper layer (Fig. 3c, d), typically under TW influence, and cold waters of the SACW below through the year. Relatively lower temperatures and salinity in the surface layer off Itajaí at the 130 m isobath during winter is resultant from the influence of the Plata Plume River (Brandini et al. 2014). The meandering of the Brazil Current along the continental slope and cyclonic eddies may lead to shelf-break upwellings of the SACW (Castro et al. 2005, 2006).

3 Zooplankton from South Brazilian Bight

3.1 Diversity and Association with Water Masses

About 790 invertebrate zooplanktonic species have been recorded at SBB. This number certainly is an underestimate, considering that (i) all meroplanktonic taxa and some holoplanktonic ones such as ctenophores, amphipods, euphausiaceans, turbellarians, polychaetes, and ostracods are largely unstudied and that (ii) there is a continuous discovery of new species (e.g., Bersano and Boxshall 1994; Nogueira Júnior et al. 2013) or records for the area (e.g., Campos and Vega-Pérez 2004; Carvalho and Bonecker 2008; Nogueira Júnior et al. 2016).

As observed in other pelagic coastal and oceanic systems, copepods are the most abundant and diverse taxa, representing ~33% of all invertebrate zooplanktonic species recorded at SBB (Fig. 4). Copepods are followed by cnidarians (~30%) and other crustaceans (mero + holoplankton = ~19%). However, the potential diversity of larval benthic crustaceans, polychaetes, and many other invertebrate taxa, such as mollusks and echinoderms, is gigantic. Each of these taxa have thousands of known species worldwide, most of them with planktonic larvae but which have not been described yet and are particularly understudied at the SBB. Amphipods are also locally understudied, only analyzed in details by Lima and Valentin (2001a).

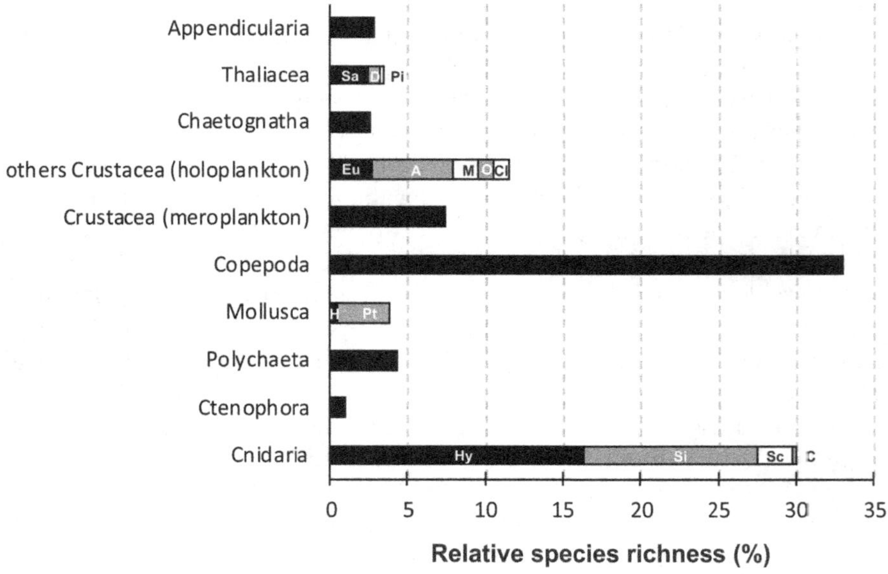

Fig. 4 Percentage of zooplankton species recorded at South Brazilian Bight according to main taxonomic groups. (Data from Lopes et al. (2006) complemented with recent new species and records up to 2015). Total number of species = 790. Hy hydromedusae, Si Siphonophorae, SC Scyphozoa, C Cubozoa, Pt Pteropoda, H Heteropoda, Eu Euphausiacea, A Amphipoda, M Mysidacea, O Ostracoda, Cl Cladocera, Sa Salpida, D Doliolida, Pi Pyrosomatida

The diversity of planktonic ostracods recorded (Fig. 4) is quite low (7 species) when compared to the 124 species recorded in the whole South Atlantic (Angel 1999). This taxon is locally understudied, and publications with specific or at least generic identifications of planktonic marine ostracods from the SBB are rare (Rocha 1983; Valentin 1980), although they may reach high abundances of up to 7600 ind m^{-3} (Valentin et al. 1987). Regardless, many planktonic ostracod species often dwell close to the bottom or live at depths greater than 200 m with considerably fewer species occurring over the shelf (Deevey 1974). The groups with lesser numbers of species are the pyrosomes and cubomedusae (2 species each). These taxa are also understudied in Brazil (Carvalho and Bonecker 2008; Nogueira Júnior and Haddad 2008), but unlike meroplanktonic larvae, amphipods, and ostracods, pyrosomes and cubomedusae are not very diverse, with only six species each recorded in whole South Atlantic (Esnal 1999; Carvalho and Bonecker 2008; Gershwin and Gibbons 2009; Oliveira et al. 2016).

3.1.1 Cnidarians

Among hydromedusae, the holoplanktonic *Liriope tetraphylla* is the most ubiquitous, being the commonly dominant species in the shallow to mid-shelf under CW influence with densities up to 930 ind m^{-3} along with the occurrence of many other meroplanktonic, usually less abundant, species (Vannucci 1963; Nagata et al. 2014; Nogueira Júnior et al. 2014). High biomass of the scyphomedusa *Lychnorhiza lucerna* occurs in shallow coastal areas (<15 m), with recorded wet weight up to 6.4 kg km^{-2} (Nogueira Júnior and Haddad 2017). Besides very abundant, this large jellyfish (>30 cm in bell diameter) is edible and may be commercially exploited as exportation product for human consumption, especially to Japan and China (Nagata et al. 2009), and may serve as shelter for other organisms such as young fishes, crabs, and shrimps (Nogueira Júnior and Haddad 2005; Martinelli Filho et al. 2008). Medusae from TW of the Brazilian Current are typically holoplanktonic like *Aglaura hemistoma*, *Solmundella bitentaculata*, and *Rhopalonema velatum*, the former commonly being the most abundant, with densities offshore typically not exceeding 10 ind m^{-3} (Vannucci 1957, 1963; Nogueira Júnior et al. 2014, 2015). In upwelling areas of Cape of Santa Marta Grande, the meroplanktonic hydromedusae *Olindias sambaquiensis* and *Rhacostoma atlanticum* are extremely abundant, with the latter representing up to 68% of the macrozooplankton carbon biomass (Mianzan and Guerrero 2000), both also occurring in areas dominated by CW (Vannucci 1951a; Nogueira Júnior and Haddad 2017). The holoplanktonic *Solmaris corona* is also found with relative high densities (up to 84 ind m^{-3}) associated with SACW in offshore regions (Nogueira Júnior et al. 2014).

Siphonophores are usually more oceanic, with only a few species like *Muggiaea kochii*, *Diphyes bojani*, and *Nanomia bijuga* occurring abundantly in coastal areas. High diversity and abundance of siphonophores is found typically offshore in the upper layer of the BC. *M. kochii* tend to dominate under CW influence, where they may reach abundances of 140 colonies per cubic meter, while *D. bojani* and

Abylopsis spp. (mainly *A. tetragona*) in mid- to outer-shelf with densities typically <25 colonies per 10 cubic meters, and mainly *Chellophyes appendiculata* and *Eudoxoides spiralis* in oceanic areas under direct influence of BC (Cordeiro and Montú 1991; Nogueira and Oliveira 1991; Nagata et al. 2014; Nogueira Júnior et al. 2014, 2015). Other common siphonophores associated with TW offshore include *Enneagonum hyalinum*, *Eudoxoides spiralis*, *Eudoxoides mitra*, *Lensia* spp., and *Sulcoleolaria* spp., commonly in lower densities.

3.1.2 Ctenophores

Since only a few species of ctenophores (mainly smaller forms of *Beroe* spp.) may be routinely fixed and preserved with success (but see Sullivan and Gifford 2009; Engel-Sorensen et al. 2009), studies on ctenophores are generally scant. This is especially true for Brazilian ecosystems where these ctenophores, although common, are among the least-studied marine animals (Oliveira et al. 2007, 2016). Only 13 species of planktonic ctenophores have been recorded thus far (Oliveira et al. 2016). *Mnemiopsis leidyi* and *Beroe ovata* are common and occasionally abundant in coastal and estuarine areas (Oliveira et al. 2007; Nogueira Júnior et al. 2018). In offshore areas over the shelf abundances are low (<1 ind m^{-3}), and small-sized (<10 mm high) *Beroe* sp. perform diel vertical migration through the thermocline (Nogueira Júnior et al. 2015).

3.1.3 Mollusks

Gastropod and bivalve larvae may be quite abundant on some situations, mostly in the shallow shelf dominated by CW and off estuaries influenced by river plumes with densities peaking around 1000–2000 ind m^{-3} (Valentin et al. 1987; Codina 2003). Among pteropods, *Creseis virgula* dominates the shelf TW, while *Limacina trochiforme* characterizes oceanic TW (Resgalla and Montú 1994; Resgalla 2008). *Creseis acicula* and *Limacina* spp. are quite abundant at Cape Frio with densities up to 13,600 ind m^{-3} (Valentin et al. 1987). *Cavolinia inflexa*, *Clio pyramidata*, and *Hyalocylis striata*, contrary, are widely distributed (Montú et al. 1997). Heteropods are commonly less abundant with dominance of *Atlanta* spp. in the northern SBB (Valentin et al. 1987). The heteropod *Firoloida desmarestia* was formally recorded only by Vannucci (1951b) near Trindade Islands. It is not abundant (<0.1 ind 10 m^{-3}) but is moderately common offshore in the upper layer under TW influence and can be considered an indicator of this water mass offshore (MNJ, unpublished data).

3.1.4 Crustaceans

Due to the high diversity found at TW, 150–200 copepod species may be found in a typical cross-shelf transect and offshore areas (Lopes et al. 2006). Most of the coastal species are small-sized. The dominant copepods under CW influence typically include *Acartia lilljeborgi*, *Paracalanus quasimodo*, *Parvocalanus crassirostris*, *Pseudodiaptomus acutus*, *Euterpina acutifrons*, *Labidocera fluviatilis*, *Ditrichocorycaeus amazonicus*, *Oncaea waldemari*, *Oithona hebes*, and *Oithona oswaldocruzi*, and these species may reach thousands and up to tenths of thousands ind m^{-3} (Valentin et al. 1987; Eskinazi-Sant'Anna and Björnberg 2006; Miyashita et al. 2009; Brandini et al. 2014; Melo Júnior et al. 2016). *Lubbockia squillimana* and *Paraeucalanus sewelli* are typically oceanic but may also occur in shallow waters, probably due to TW and SACW influences (Miyashita et al. 2009). Many of the coastal species are restricted to estuaries and the inner shelf, disappearing under influences of oceanic oligotrophic waters (Sartori and Lopes 2000).

Copepods associated with the TW, such as *Acrocalanus longicornis*, *Calanopia americana*, *Clausocalanus furcatus*, *Farranula gracilis*, *Haloptilus*, *Mecynocera*, and *Candacia* among others, are commonly more abundant in the upper mixed layer (Björnberg 1963, 1981; Lopes et al. 1999; Brandini et al. 2014). Among these, *C. furcatus* commonly is the dominant species; this small-sized copepod is typically oceanic and known to be well-adapted to thrive successfully in oligotrophic waters, producing more offspring under lower food concentrations (Mazzochi and Paffenhöfer 1998; Brandini et al. 2014). Species typical from the SACW include *Heterorhabdus*, *Euatideus*, *Temeropia*, *Haloptilus* spp., *Centropages violaceus*, *Ctenocalanus vanus*, and *Calanoides carinatus* (Valentin 1989). Most of this species disappear shortly after upwelling or mixing events, except for the two latter which persist and are commonly used as indicator of SACW presence on the shelf (Björnberg 1963, 1981; Valentin 1984a, b, 1989). *C. vanus* typically dominate the mid-shelf during SACW intrusion periods, usually accumulating in the deep chlorophyll maximum layer, but its populations may persist throughout the seasons (Brandini et al. 2014).

Cladocerans are essentially neritic, with higher concentrations near the coast (Muxagata and Montú 1999), being particularly common in the upper mixed layer reaching densities up to ca. 2000 ind m^{-3} (Rocha 1982; Valentin et al. 1987; Miyashita et al. 2011; Domingos-Nunes and Resgalla 2012; Brandini et al. 2014). Parthenogenetic reproduction is the main reproductive strategy of SBB cladoceran populations allowing a rapid populational growth under favorable conditions, and thus population variations are accentuated, and abundance pulses over the shelf are commonly observed (Resgalla and Montú 1993; Miyashita et al. 2011). In shallow coastal waters, *Pseudevadne tergestina* and *Penilia avirostris* are the most common and abundant, particularly under stratified conditions due to bottom intrusions of the SACW (Rocha 1982; Valentin 1984a, b; Miyashita et al. 2010, 2011; Brandini et al. 2014). *Pleopsis polyphemoides* is a CW indicator, frequently associated with estuarine waters (Valentin et al. 1987; Resgalla and Montú 1993; Miyashita et al. 2011), and when associated with *Evadne nordmanni* and *Pleopsis schmackeri* differentiates thermal conditions of summer and winter, respectively, on South Brazilian Shelf (Resgalla 2008). *Evadne spinifera* characterizes the oceanic TW (Resgalla and Montú 1993; Resgalla 2008) but also may occur in the inner shelf in low abundance and frequency (Miyashita et al. 2011). *Podon intermedius* is a cold-water species, typically associated with SACW intrusions (Rocha 1982; Miyashita et al. 2011).

Most mysids tend to be more common in coastal and estuarine areas (Murano 1999). They may reach high concentrations at surf zone of sandy beaches, where *Metamysidopsis elongata atlantica* may reach high densities (up to 3500 ind m^{-3}) and biomass, dominating the zooplankton (Bersano 1994; Ávila et al. 2009). In contrast, euphausiids are more common over the shelf-break and oceanic areas, especially the adult forms of some dominant species such as *Euphausia similis*, *E. recurva*, *E. lucens*, and *Thysanoessa gregaria* (Montú et al. 1997). Species like *Euphausia recurva*, *E. americana*, *E. hemigiba*, *Stylocheiron* spp., *Thysanopoda* spp., and *Nematobrachion flexipes* are typical of warm waters and indicate the influence of BC over the shelf (Lansac-Tôha 1991).

Two species of holoplanktonic decapods of the genus *Lucifer* are common, *L. faxoni*, most common, is typically more coastal, while *L. typus* occurs in areas of mixed CW/TW in salinities higher than 36 (Brandão et al. 2015), although both species also occur on TW offshore until near the Mid-Atlantic Ridge (~15°W, Marafon-Almeida et al. 2016). Hyperiids are understudied in the SBB, and the unique comprehensive survey found 40 species and densities up to ca. 160 ind m^{-3}, with *Lestrigonus bengalensis* and *Simorhynchotus antennarius* dominating, the former typical from the CW and usually absent from salinities >36 of the TW, and both also may occasionally be associated with siphonophores and/or salps (Lima and Valentin 2001a, b).

3.1.5 Chaetognaths

Chaetognaths are important predators and usually considered good water mass indicators. *Parasagitta friderici* is associated with waters of lower salinities (Nogueira Júnior et al. 2018) and considered typical of CW, where it commonly is the most abundant species, along with *Parasagitta tenuis* reaching densities usually smaller than 250 ind m^{-3} (Almeida Prado 1961, 1968; Liang 2002); however at Cape Frio chaetognath densities up to 2600 ind m^{-3} have been reported, and *Flaccisagitta enflata* is typically the dominant species (Valentin et al. 1987). This later species also typically dominates the TW over the shelf along with the less abundant *Ferosagitta hispida*, *Serratosagitta serratodentata*, *Flaccisagitta hexaptera*, *Pterosagitta draco*, and *Krohnitta pacifica* which are typical of oceanic TW, while *Krohnitta subtilis*, *Decipisagitta decipiens*, and *Pseudosagitta lyra* are typical from SACW (Almeida Prado 1968; Liang and Vega-Pérez 1994, 2002; Montú et al. 1997; Liang 2002; Resgalla 2008).

3.1.6 Appendicularians

A little more than 20 species of appendicularians have been recorded over the SBB, with *Oikopleura longicauda* being the dominant species (Tundisi 1970; Sinque 1982), particularly in the upper mixed layer where it may reach densities up to ca. 1100 ind m^{-3} (Miyashita and Lopes 2011). *O. fusiformis* also reaches high densities at CW, and *Fritillaria pellucida* is more abundant at SACW intrusions. Other species such as *O. rufescens*, *F. borealis*, and *F. pellucida* are typical of shelf and slope waters, and *O. albicans*, *O. cophocerca*, and *F. formica* are predominantly from oceanic waters (Montú et al. 1997). Appendicularians have high secondary production on SBB inner shelf equaling to 77% of the copepod production (Miyashita and Lopes 2011).

3.1.7 Thaliaceans

The salp *Thalia democratica* is the most ubiquitous species, being widely distributed over shelf, coastal, and offshore areas (Tavares 1967; Valentin and Monteiro-Ribas 1993; Amaral et al. 1997; Nogueira Júnior et al. 2015), occasionally reaching estuarine waters in salinities down to 21 (Nogueira Júnior et al. 2018). This opportunist species has high asexual reproduction rates (Esnal and Daponte 1999), forming huge aggregates that may exert negative impact over the recruitment of pelagic fishes (Matsuura et al. 1992; Katsuragawa et al. 1993; Resgalla et al. 2001) by depleting resources from the water column. The doliolids *Doliolum nationalis* and *Dolioletta gegenbauri* are the most common in the CW and TW over the shelf (Tavares 1967), with abundances up to 1000 ind m^{-3} (Valentin et al. 1987), with gonozooids being particularly abundant in the upper mixed layer and phorozooids over the deep chlorophyll maximum layer (Nogueira Júnior et al. 2015). High biomass of *Salpa fusiformis* over the shelf can be observed associated with the SACW intrusions (Nogueira Júnior and Brandini, unpublished data). Other species like *Brooksia rostrata*, *Ihlea punctata*, *Riteriella retracta*, and *Cyclosalpa polae* may be found over the shelf at much smaller concentrations (Amaral et al. 1997; Esnal and Daponte 1999), the former typically associated with TW (Nogueira Júnior and Brandini, unpublished data).

3.2 Abundance and Biomass Distribution

An inshore-offshore gradient in the zooplankton abundance, biomass and diversity is typical over the whole SBB. While abundance and biomass are higher in more coastal areas diminishing offshore as the influence of the oligotrophic TW increases, the diversity has the inverse tendency (Katsuragawa et al. 1993; Vega-Pérez 1993; Lima and Valentin 2001a; Resgalla et al. 2001; Lopes et al. 2006; Brandini et al. 2014). Although data from different studies may vary considerably due to latitudinal, interannual differences and also due to different sampling strategies such as sampling frequency and mesh size (e.g., Miyashita et al. 2009), generally higher zooplankton abundance and biomass are associated with CW, decreasing in other water masses independently of the season, and higher diversity is associated with the TW (e.g., Valentin 1984a; Valentin and Monteiro-Ribas 1993; Lopes et al. 2006; Brandini et al. 2014; Nogueira Júnior et al. 2014, 2015).

For instance, zooplankton biovolume and copepod biomass of up to 1.7 ml m^{-3} and 100 mg C m^{-3}, respectively, have been reported from shallow waters off Cape Frio, while offshore the values were <0.1 ml m^{-3} (Nogueira and Oliveira 1991) and <10 mg C m^{-3} (Dias et al. 2015). Off Ubatuba, biovolumes also are commonly >1 ml m^{-3} and up to 6 ml m^{-3} in shallow waters and much lower (<0.5 ml m^{-3}) in offshore areas (>100 m isobaths, Vega-Pérez 1993; Codina 2003). A high variability has been reported through different years, with high concentrations (up to 38 ml m^{-3}) occasionally also found in mid- to the outer-shelf mostly due to the massive presence of salps (Katsuragawa et al. 1993) which are typically oceanic (see above). Similarly, further south of the SBB, average biovolumes between 1977 and 1990 reached >1 ml m^{-3} in coastal areas and typically lower values (<0.25 ml m^{-3}) offshore, with differences more pronounced during summer and autumn (Resgalla et al. 2001). Copepod biomass off Ubatuba may reach up to ca. 100 mg DW m^{-3}, but more commonly is <30 mg DW m^{-3} (Melo Júnior et al. 2016), with summer averages around 7.3 mg C m^{-3} (Miyashita et al. 2009). Zooplankton abundances over the shallow shelf are typically >2000 ind m^{-3}, and concentrations of tenths of thousands ind m^{-3} have commonly been reported with up to ca. 100,000 ind m^{-3}, while offshore of the 100 m isobath densities are typically <1000 ind m^{-3}, and commonly lower, occasionally reaching up to 1700 ind m^{-3} (Vega-Pérez 1993; Schettini et al. 1998; Valentin 1984a, b; Valentin et al. 1987; Sartori and Lopes 2000; Miyashita et al. 2009; Brandini et al. 2014; Dias et al. 2015).

A few exceptions to these general patterns of decreasing abundance and increasing diversity toward offshore include (i) the tendency of some taxa to have higher diversity and abundance in coastal and estuarine waters, mostly those somehow associated to the seafloor such as meroplanktonic hydromedusae, mysids, and decapod larvae (Nogueira Júnior 2012; Nogueira Júnior et al. 2014; Miyashita and Calliari 2014; Brandão et al. 2015; Nogueira Júnior et al. 2018), and (ii) some typically oceanic taxa such as euphausiaceans, salps, and siphonophores tend to have higher abundances over the mid- to outer-shelf and shelf-break (Valentin and Monteiro-Ribas 1993; Amaral et al. 1997; Esnal and Daponte 1999; Gibbons et al. 1999; Nogueira Júnior et al. 2014).

Horizontally, higher zooplankton concentrations are usually associated with areas under continental drainage influence, like off Guanabara, Cananéia, and Paranaguá Bays (Vannucci and Almeida Prado 1959; Resgalla et al. 2001; Lopes et al. 2006; Nagata et al. 2014; Brandão et al. 2015). Areas subjected to the upwelling (Valentin 1984a, b, 1989; Valentin and Monteiro-Ribas 1993) or bottom intrusions (Miyashita et al. 2009, 2011; Brandini et al. 2014) of the SACW also typically tend to have high seasonal zooplankton concentrations. The dynamics of the SACW movements certainly is the most relevant mesoscale physical feature over the shelf of the SBB (Castro et al. 2005, 2006; Cerda and Castro 2014) impacting biological communities (Miyashita et al. 2009, 2011; Brandini et al. 2014). Either coastal and shelf-break upwelling or bottom intrusion of the SACW causes enrichment over the shelf and shelf-break (Brandini et al. 1997, 2014), increasing considerably the primary production and subsequently zooplankton abundance, biomass, and production (Guenther et al. 2008; Miyashita et al. 2009, 2011; Brandini et al. 2014; Marcolin et al. 2015).

Off Cape Frio the relation of SACW upwelling with zooplankton dynamics was relatively well-studied in different spatiotemporal scales (e.g., Valentin 1984a, b, 1989; Valentin et al. 1987; Valentin and Monteiro-Ribas 1993; Lopes et al. 1999; Guenther et al. 2008). In the downwelling phase, a microbial structure prevails with dominance of pico- and nanophytoplankton and high contribution of bacterioplankton and microzooplankton. During upwelling, microplankton, mostly diatoms, and mesozooplankton, mostly copepods, are the main producers and consumers, respectively, characterizing an herbivorous, more productive, food web (Guenther et al. 2008). During downwelling, mesozooplankton is dominated mostly by *Paracalanus parvus*, *Temora turbinata*, and *Pseudevadne tergestina*, while during the upwelling the two latter species decrease in abundance, while zooplankton biomass is dominated by opportunistic herbivores including *Paracalanus parvus*, *Oikopleura longicauda*, *Thalia democratica*, *Calanoides carinatus*, and *Ctenocalanus vanus*, and the main carnivore is the chaetognath *Flaccisagitta enflata*. These herbivores can achieve high growth rates induced by pulses of autotrophic microplankton biomass, typical of upwelling areas (Valentin 1984a, b, 1989; Valentin et al. 1987; Guenther et al. 2008). After maximum period, there is the appearance of herbivores, predators, and eventually detritivores (*Ctenocalanus vanus*, *Creseis acicula*, *Penilia avirostris*, *Doliolum* spp., *Eucalanus* spp.), detritivores and omnivores (*Temora stylifera*, *Centropages furcatus*, *Euterpina acutifrons*, *Microsetella* sp., *Conchoecia* spp., and *Oithona plumifera*), and carnivores like *Flaccisagitta enflata*, siphonophores, and copepods of the families Euchaetidae, Candacidae, and Pontellidae (Valentin 1989). The zooplankton biomass in this area is high, with mean annual dry weight of 66 mg m^{-3}, typically <50 mg m^{-3} during the downwelling phase and up to >200 mg m^{-3} associated to the SACW upwelling (Valentin 1984b; Valentin et al. 1987).

In the central and south parts of the SBB, copepod abundance and biomass is considerably higher during SACW intrusion periods, apparently due to enhanced food for the dominants *Oncaea waldemari*, *Oithona plumifera*, and *Ctenocalanus vanus*, the two latter clearly vertically concentrated at the deep chlorophyll maximum layer (Katsuragawa et al. 1993; Vega-Pérez 1993; Miyashita et al. 2009; Brandini et al. 2014). Off São Paulo shallow shelf, the SACW intrusion contributed more to the increase in copepod production than the plume of the small Santos Bay (Miyashita et al. 2009). Similarly, the cladocerans *Penilia avirostris*, *Pseudevadne tergestina*, and *Evadne spinifera* typically increase their abundance in inner-to mid-during intrusions of the SACW (Miyashita et al. 2011; Brandini et al. 2014), as well as abundance and biomass of appendicularians, dominated mostly by *Oikopleura longicauda* (Miyashita et al. 2011). Differently, coastal hydromedusae tend to have reduced distribution and abundance during SACW intrusion (Nogueira Júnior et al. 2014). Although SACW intrusions support a general increase in zooplankton abundance and biomass, it has been suggested that the high dominance of small copepods and cladocerans indicat the intrusions do not last long enough to sustain the development of large organisms, with longer generation times (Marcolin et al. 2015).

The dynamics of water mass changes and fertilization processes in the region are wind-driven which are typically seasonal (e.g., Castro et al. 2005, 2006; Brandini et al. 2014; Cerda and Castro 2014) leading to seasonal changes in the zooplankton assemblages of the SBB. The general tendency is that zooplankton abundance and biomass over the shelf peaks between December and March, associated with periods of the SACW intrusion, with minimum values during autumn-winter when the water column is vertically mixed and SACW influence is restricted to offshore (Matsuura et al. 1980, Resgalla et al. 2001; Miyashita et al. 2009, 2011; Miyashita and Lopes 2011; Brandini et al. 2014; Marcolin et al. 2015; Melo Júnior et al. 2016). This seasonal pattern is recurrent and has been observed for different taxonomic groups such as copepods (Lopes et al. 1999; Sartori and Lopes 2000; Brandini et al. 2014), decapod larvae (Fehlauer and Freire 2002; Koettker and Freire 2006; Marafon-Almeida et al. 2008), heteropod mollusks (Resgalla and Montú 1994), hydromedusae (Vannucci 1963), siphonophores (Nogueira Júnior et al. 2014), and ichthyoplankton (Matsuura 1996; Godefroid et al. 1999). Multi-year data suggest that the intrusions of the SACW and their strength over the inner shelf is an important factor influencing both seasonal and interannual variability in zooplankton diversity, abundance, biomass, and size-spectra (Katsuragawa et al. 1993; Vega-Pérez 1993; Marcolin et al. 2015; Melo Júnior et al. 2016).

Although these SACW intrusions are the most important mesoscale physical forcing over the SBB typically leading to increased zooplankton abundance and biomass during warm stratified periods, other seasonal patterns can also be found including the absence of seasonal variations on chaetognaths off Santos and Cananéia (Almeida Prado 1968), and winter peaks as is the case of larval decapod abundance at Guanabara Bay entrance (Fernandes et al. 2002), copepod biomass in the São Sebastião Channel (Eskinazi-Sant'anna and Björnberg 2006), and occasionally off Ubatuba (Melo Júnior et al. 2016), and hydromedusae in the inner shelf of the central SBB (Nagata et al. 2014). These are usually related to complex hydrodynamic and meteorological processes that commonly occur during winter. It is not clear whether these winter increments represent an advection of shelf populations, concentrating in the inner shelf, an actual populational increase (Nagata et al. 2014; Melo Júnior et al. 2016), or a combination of both. During winter, the typically lower rainfalls decrease the volume of coastal water and carry shelf waters toward the coast, enhanced by the prevalent southeasterly winds and cold fronts, causing coastward advection and accumulation of the zooplankton (Nagata et al. 2014). Also, the cold fronts may lead to pycnocline erosion and particle resuspension (Gaeta et al. 1999; Castro et al. 2006), what could provide additional food sources for the plankton community growth (Melo Júnior et al. 2016). The combination and interaction of these factors, along with intra-regional variations in biotic factors such as food and predator type and concentration, may lead to irregular and unpredictable seasonal patterns, particularly for groups with complex life cycles (Nagata et al. 2014).

Smaller-scale variations (hours, days, weeks) also are substantial and may be wider than seasonal changes, particularly in coastal areas under strong tidal influence, but have been less studied locally apart from a few studies inside estuaries (e.g., Ara 2002; Liang et al. 2003; Mendes et al. 2012), or in the inner shelf focused on the diel vertical migration (e.g., Moreira 1973, 1976a; Sinque 1982). The causes of these small-scale temporal fluctuations seem to be more linked to exchange of water masses, populations transfer, and trophic factors (Valentin 1989). Offshore over the shelf, a more physically stable environment, abundance, and community structure remained relatively constant throughout 48 h of sampling (Nogueira Júnior et al. 2015); however patch formation and horizontal variation have not been addressed.

These general spatiotemporal trends of the SBB zooplankton dynamics described above are comparable to other western boundary current systems, such as the South Atlantic Bight (Atkinson 1977; Atkinson et al. 1984; Paffenhöfer 1985; Paffenhöfer et al. 1984, 1994, 1995; Coston-Clements et al. 2009) and the Kuroshio Current System (Chern et al. 1990; Houng-Yung and Yuh-ling 1992; Qiu 2001) which are subjected to seasonal, wind-driven oceanic intrusions and/or surface salinity fronts, whose dynamics largely control plankton diversity, composition, and abundance, except that the SBB shelf tends to be larger (up to ca. 200 km; Castro et al. 2005, 2006; Lohrenz and Castro 2005). Zooplankton abundances and composition of dominant species and/or genera from SBB also are similar to these other ecosystems, with tenths of thousands ind m^{-3} commonly found and up to >100,000 ind m^{-3} (Paffenhöfer et al. 1984, 1987, 1994, 1995; Toda 1989; Nakata et al. 2000; Kâ and Hwang 2011; Lo et al. 2012).

4 Final Remarks

The zooplankton community of the SBB is structured according to the water masses that shelter the dominant taxonomic groups, tending to have higher densities in more coastal areas and higher diversity in more offshore waters, under the influence of the TW. Onshore bottom intrusion of cold oceanic waters (SACW) over the shelf, although understudied, is one of the processes that most affect the planktonic communities over the SBB, hence affecting the nourishment of pelagic fish stocks, such as the Brazilian sardines and anchovies.

Taxonomy is among the most studied issue for the main zooplankton groups. Great gaps still exist, however, for the meroplankton as a whole, and some sporadically abundant holoplanktonic taxa such as ctenophores, ostracods, polychaetes, amphipods, euphausiids, and thaliaceans. Effort in this direction is being carried out with the descriptions of the larval stages of several decapods (e.g., Fransozo and Bertini 2003; Negreiros-Fransozo et al. 2003; Barros-Alves et al. 2013; Pantaleão et al. 2013; Alves et al. 2016) and the publication of an identification key for Brazilian ctenophores (Oliveira et al. 2007), for instance. The connection of zooplanktonic associations with different water masses is satisfactorily addressed, being one of the few ecological aspects relatively well-studied locally. We can predict with considerable confidence which are the main species in each major water mass. However, it is still needed to intensify sampling efforts at the outer-shelf and shelf-break zones, especially associated with the frontal eddies and shelf-break upwellings associated with the BC meanders, as already pointed out a decade ago (Lopes et al. 2006).

Main gaps on marine zooplankton knowledge off Brazil, pointed out 20 years ago (Brandini et al. 1997), are still hampering a broader knowledge of the plankton community off the SBB. We still need (i) detailed analyses of the meroplankton taxonomy and spatiotemporal distributional patterns over the shelf; (ii) more studies on larval dispersion and recruitment in relation to meteorological, hydrographical, and biological behaviors; (iii) simultaneous studies coupling planktonic communities with the dominant oceanographic processes, including relevant vertical resolution and higher sampling frequency throughout the seasonal cycles; (iv) to know the small- and medium-scale temporal variations (hours, days, weeks) and the influence of physical and biological processes on this variability; and (v) detailed knowledge of population dynamics and life history of all taxa, as well as determination of metabolic rates and trophic interactions.

The general lack of studies on metabolism, physiology, and survivorship (but see, e.g., Moreira 1976b, 1978; Moreira and Vernberg 1978; Miyashita and Calliari 2016) of the main species along with the scattered data on secondary production (e.g., Miyashita et al. 2009; Miyashita and Lopes 2011; Melo Júnior et al. 2013) and trophic interactions (e.g., Vega-Pérez and Liang 1992; Liang and Vega-Pérez 1995; Vega-Pérez et al. 1996) does not allow adequate extrapolation of trophic models and energy flow through food chains. These are crucial information to understand the basic structure and functioning of the SBB ecosystem and, ultimately, shift the usual target species to an ecosystem-oriented fisheries management in one of the most important fishing zones among the Brazilian regional seas.

Acknowledgments The present study was initially written as a partial requirement for qualification in the PhD Program in Zoology, UFPR. We thank Dr. Carolina Freire and Paulo Lana for their encouragement and critical reading of the first version of the text and to Dr. Guilherme Bersano Filho for providing literature and suggestions to improve the text. We thank also Dr. Demetrio Boltovskoy for his valuable comments and Dr. Jefferson Turner for his suggestions and help with English editing.

References

Almeida Prado MS (1961) Distribuição dos Chaetognatha no Atlântico Sul Ocidental. Bol Inst Oceanogr 12(4):15–49

Almeida Prado MS (1968) Distribution and annual occurrence of Chaetognatha off Cananéia and Santos coast (São Paulo, Brazil). Bol Inst Oceanogr 17(1):33–55

Alves FRA, Pantaleão JAP, Barros-Alves SP et al (2016) First zoeal stage of the crab *Domecia acanthophora* (Desbonne, in Desbonne and Schramm, 1867) (Decapoda, Brachyura) and revision of the larval morphology of superfamily Trapezioidea. Nauplius 24:e2016021

Amaral WJA, Montú MA, Gloeden IM (1997) Salpidae (Thaliacea) da plataforma continental do extremo sul do Brasil: composição, distribuição e abundância. Atlantica 19:51–66

Angel MV (1999) Ostracods. In: Boltovskoy D (ed) South Atlantic Zooplankton. Backhuys Publishers, Leiden, pp 815–868

Ara K (2002) Temporal variability and production of *Temora turbinata* (Copepoda: Calanoida) in the Cananéia Lagoon estuarine system, São Paulo, Brazil. Sci Mar 66(4):399–406

Atkinson LP (1977) Modes of gulf stream intrusions into the South Atlantic Bight shelf waters. Geophys Res Lett 4(12):583–586

Atkinson LP, O'Malley PG, Yoder JA et al (1984) The effect of summertime shelf break upwelling on nutrient flux in Southeastern United States continental shelf waters. J Mar Res 42:969–993

Ávila TR, Pedrozo CS, Bersano JG (2009) Variação temporal do zooplâncton da Praia de Tramandaí, Rio Grande do Sul, com ênfase em Copepoda. Iheringia, Sér Zool 99(1):18–26

Barros-Alves AP, Alves DFR, Bolla EA et al (2013) First zoeal stage of ? *Cataleptodius parvulus* (Fabricius, 1793) and *Xanthodius denticulatus* (White, 1848) (Decapoda: Brachyura): larval evidences and systematic position. Zootaxa 3731(1):234–242

Bersano JG (1994) Zooplâncton da zona de arrebentação de praias arenosas situadas ao sul de Rio Grande, RS. Primavera de 1990 e verão de 1991. Dissertation, Fundação Universidade Rio Grande

Bersano JG, Boxshall GA (1994) Planktonic copepods of the genus *Oncea* (Poecilostomatoida: Onceaidae) from the waters off Southern Brazil. Nauplius 2:29–41

Björnberg TKS (1963) On the marine free-living copepods off Brazil. Bol Inst Oceanogr 13(1):3–142

Björnberg TKS (1981) Copepoda. In: Boltovskoy D (ed) Atlas del zooplâncton del atlantico sudoccidental y métodos de trabalho con el zooplâncton marino. INIDEP, Mar del Plata, pp 587–679

Brandão MC, Garcia CAE, Freire AS (2015) Large-scale spatial variability of decapod and stomatopod larvae along the South Brazil Shelf. Cont Shelf Res 107:11–23

Brandini FP (1990) Hydrography and characteristics of the phytoplankton in shelf and oceanic waters off southeastern Brazil during winter (July/August 1982) and summer (February/ March 1984). Hydrobiologia 196:111–148

Brandini FP, Lopes RM, Gutseit KS et al (1997) Planctonologia na plataforma continental do Brasil. Diagnose e revisão bibliográfica. Ministério do Meio Ambiente e da Amazônia Legal – IBAMA, Brasília

Brandini FP, Boltovskoy D, Piola A et al (2000) Multiannual trends in fronts and distribution of nutrients and chlorophyll in the Southwestern Atlantic (30–62° S). Deep Sea Res Part 1 Oceanogr Res Pap 47(6):1015–1033

Brandini FP, Silva AS, Silva E et al (2007) Sources of nutrients and seasonal dynamics of chlorophyll in the inner shelf off Paraná State – South Brazil Bight. J Coast Res 23(5):1131–1140

Brandini F, Nogueira M Jr, Simião M et al (2014) Deep chlorophyll maximum and plankton community response to oceanic bottom intrusions on the continental shelf in the South Brazilian Bight. Cont Shelf Res 89:61–75

Campos MAG, Vega-Pérez LA (2004) First records of *Fritillaria venusta* Lohmann, 1896 (Tunicata: Appendicularia) in South Brazil Bight. Oceánides 19(1):11–17

Carvalho PF, Bonecker SL (2008) Tunicata, Thaliacea, Pyrosomatidae, *Pyrosomella verticillata* (Neumann, 1909): first record from the Southwest Atlantic Ocean. Check List 4(3):272–274

Castro BM, Brandini FP, Pires-Vanin AM et al (2005) Multidisciplinary oceanographic processes on the Western Atlantic continental shelf between 4°N and 34°S. In: Robinson R, Brink KH (eds) The sea, vol 14. Harvard College, Harvard, pp 259–293

Castro BM, Lorenzzetti JA, Silveira ICA et al (2006) Estrutura termohalina e circulação na região entre o Cabo de São Tomé (RJ) e o Chuí (RS). In: Rossi-Wongtschowski CLB, Madureira LSP (eds) O ambiente oceanográfico da plataforma continental e do talude na região Sudeste-Sul do Brasil. Editora da Universidade de São Paulo, São Paulo, pp 11–120

Cerda C, Castro BM (2014) Hydrographic climatology of South Brazil Bight shelf waters between Sao Sebastiao (24°S) and Cabo Sao Tome (22°S). Cont Shelf Res 89:5–14

Chern C-S, Wang J, Wang D-P (1990) The exchange of Kuroshio and East China Sea shelf water. J Geophys Res 95(9):16017–16023

Codina JCU (2003) Dinâmica sazonal das larvas de moluscos na plataforma continental interna do Paraná – Brasil. Dissertation Universidade Federal do Paraná

Cordeiro TA, Montú MA (1991) Distribuição dos Siphonophorae: Calycophorae (Cnidaria) em relação às massas de água ao largo dos estados do Paraná e Santa Catarina, Brasil (28° – 31°S). Nerítica 6(1–2):107–126

Coston-Clements L, Waggett RJ, Teste PA (2009) Chaetognaths of the United States South Atlantic Bight: distribution, abundance and potential interactions with newly spawned larval fish. J Exp Mar Biol Ecol 373:111–123

Deevey GB (1974) Pelagic ostracods collected on Hudson 70 between the equator and 55°S in the Atlantic. Proc Biol Soc Wash 87:351–380

Dias JF, Clemmensen C, Ueberschär B et al (2004) Condition of the Brazilian sardine *Sardinella brasiliensis* (Steindachner, 1879) larvae in the São Sebastião inner and middle continental shelf (São Paulo, Brazil). Braz J Oceanogr 52(1):81–87

Dias CO, Araujo AV, Vianna SC et al (2015) Spatial and temporal changes in biomass, production and assemblage structure of mesozooplanktonic copepods in the tropical South-West Atlantic Ocean. J Mar Biol Assoc UK 95(3):483–496

Domingos-Nunes R, Resgalla C Jr (2012) The zooplankton of Santa Catarina continental shelf in Southern Brazil with emphasis on Copepoda and Cladocera and their relationship with physical coastal processes. Lat Am J Aquat Res 40(4):893–913

Engel-Sorensen K, Andersen P, Holmstrup M (2009) Preservation of the invasive ctenophore *Mnemiopsis leidyi* using acidic Lugols's solution. J Plankton Res 31(8):917–920

Eskinazi-Sant'Anna EM, Björnberg TKS (2006) Seasonal dynamics of mesozooplankton in Brazilian coastal waters. Hydrobiologia 563:253–268

Esnal GB (1999) Pyrossomatida. In: Boltovskoy D (ed) South Atlantic Zooplankton. Backhuys Publishers, Leiden, pp 1401–1408

Esnal GB, Daponte MC (1999) Salpida. In: Boltovskoy D (ed) South Atlantic Zooplankton. Backhuys Publishers, Leiden, pp 1423–1444

Fehlauer KH, Freire AS (2002) Occurrence of decapods larvae, specially *Xiphopenaeus kroyeri* (Penaeidae) in the shallow shelf of Paraná. Nauplius 10(1):37–45

Fernandes LD, Bonecker SLC, Valentin J (2002) Dynamic of decapod crustacean larvae on the entrance of Guanabara Bay. Braz Arch Biol Technol 45:491–498

Fransozo A, Bertini G (2003) The first zoeal stage of *Lithadia rotundata* A. Milne Edwards, 1880 (Brachyura: Leucosiidae: Ebaliinae) obtained in the laboratory. Nauplius 10(1):55–59

Gaeta S, Ribeiro SMS, Metzler PM et al (1999) Environmental forcing on phytoplankton biomass and primary productivity of the coastal ecosystem in Ubatuba region, southern Brazil. Braz J Oceanogr 47:11–27

Gershwin L, Gibbons MJ (2009) *Carybdea branchi*, sp. nov., a new box jellyfish (Cnidaria: Cubozoa) from South Africa. Zootaxa 2088:41–50

Gibbons MJ, Spiridonov VA, Tarling GA (1999) Euphausiacea. In: Boltovskoy D (ed) South Altantic Zooplankton. Backhuys Publishers, Liden, pp 1241–1279

Godefroid RS, Hofstaetter M, Spach HL (1999) Larval fish in the surf one f Pontal do Sul beach, Paraná, Brazil. Rev Bras Zool 16(4):1005–1011

Guenther M, Gonzalez-Rodriguez E, Carvalho WF et al (2008) Plankton trophic structure and particulate organic carbon production during a coastal downwelling-upwelling cycle Mar Ecol Prog Ser 363:109–119

Houng-Yung C, Yuh-ling C (1992) Quantity and quality of summer surface net zooplankton in the Kuroshio current-induced upwelling northeast of Taiwan. Terr Atmos Ocean Sci 3(3):321–334

Kâ S, Hwang J-S (2011) Mesozooplankton distribution and composition on the northeastern coast of Taiwan during autumn: effects of the Kuroshio current and hydrothermal vents. Zool Stud 50(2):155–163

Katsuragawa M, Matsuura Y, Suzuki K et al (1993) O ictioplâncton ao largo de Ubatuba, SP: composição, distribuição e ocorrência sazonal (1985–1988). In: Pires-Vanin AMS (ed) Estrutura e função do ecossistema de plataforma continental do Atlântico Sul brasileiro. Publicação Especial do instituto Oceanográfico de São Paulo 10:85–121

Koettker AG, Freire AS (2006) Spatial and temporal distribution of decapod larvae in the subtropical waters of the Arvoredo Archipelago, SC, Brazil. Iheringia, Sér Zool 96(1):31–39

Kurtz FW, Matsuura Y (2001) Food and feeding ecology of Brazilian sardine (*Sardinella brasiliensis*) larvae from the Southeastern Brazilian Bight. Rev Bras Oceanogr 49(1/2):61–74

Lansac-Tôha FA (1991) Chaves de identificação e diagnose das espécies de Euphausiacea (Crustacea) encontradas ao largo da costa sul do Brasil (Lat. 22°S–29°). Rev Bras Biol 51(3):623–638

Liang TH (2002) Chaetognath seasonality and distribution at surface layers off Ubatuba, South Brazil Bight coastal waters. Oceánides 17(2):103–111

Liang TH, Vega-Pérez LA (1994) Studies on chaetognaths off Ubatuba region, Brazil. I. Distribution and abundance. Bol Inst Oceanogr 42(1/2):73–84

Liang TH, Vega-Pérez LA (1995) Studies on Chaetognaths off Ubatuba region, Brazil. II. Feeding habitats. Bol Inst Oceanogr 43(1):35–48

Liang TH, Vega-Pérez LA (2002) Distribution, abundance and biomass of Chaetognaths off São Sebastião region, Brazil in February 1994. Rev Bras Oceanogr 50:1–12

Liang TH, Ara K, Miranda LB et al (2003) On the variability of the chaetognath *Sagitta friderici* Ritter-Z`ahony at Cananéia Lagoon Estuarine system, São Paulo, Brazil. Hydrobiologia 510:91–102

Lima MCG, Valentin JL (2001a) Preliminary results to the holistic knowledge of the Amphipoda Hyperiidea faunal composition off the Brazilian coast. J Plankton Res 23:469–480

Lima MCG, Valentin JL (2001b) New records of Amphipoda Hyperiidea in association with gelatinous zooplankton. Hydrobiologia 448:229–235

Lo WT, Kang P-R, Hsieh H-Y (2012) Siphonophores from a transect off Southern Taiwan between the Kuroshio current and South China Sea. Zool Stud 51(8):1354–1366

Lohrenz SE, Castro BM (2005) Eastern Ocean boundaries pan-regional overview. In: Robinson AR, Brink H (eds) The sea, vol 14. Harvard College, Harvard, pp 3–20

Longhurst AR (2006) Ecological geography of the sea, 2nd edn. Academic Press, San Diego

Lopes RM, Brandini FP, Gaeta SA (1999) Distribution patterns of epipelagic copepods off Rio de Janeiro (SE Brazil) in summer 1991/1992 and winter 1992. Hydrobiologia 411:161–174

Lopes RM, Montú MA, Gorri C et al (2006) In: Rossi-Wongtschowski CLB, Madureira LSP (eds) O ambiente oceanográfico da plataforma continental e do talude na região Sudeste-Sul do Brasil. Editora da Universidade de São Paulo, São Paulo, pp 265–358

Marafon-Almeida A, Souza-Conceição JM, Pandolfo PSV (2008) Distribuição e abundância de larvas de três espécies de peneídeos (Decapoda) na plataforma continental interna adjacente à Baía da Babitonga, Sul do Brasil. Panam. J Aquat Sci 3(3):340–350

Marafon-Almeida A, Pereira JB, Fernandes LFL (2016) Distribution of the species of *Lucifer* Thompson, 1829 in the subtropical South Atlantic between parallels 20° and 30°S. Braz J Oceanogr 64(3):217–226

Marcolin CR, Gaeta S, Lopes RM (2015) Seasonal and interannual variability of zooplankton vertical distribution and biomass size spectra off Ubatuba, Brazil. J Plankton Res 37:808–819

Martinelli Filho JE, Stampar S, Morandini AC et al (2008) Cleaner shrimp (caridea: palaemonidae) associated with scyphozoan jellyfish. Vie et Milieu 58(2):133–140

Matsuura Y (1986) Contribuição ao estudo da estrutura oceanográfica da região sudeste entre Cabo Frio (RJ) e Cabo de Santa Marta Grande (SC). Cienc Cult 38(8):1439–1450

Matsuura Y (1996) A probable cause of recruitment failure of the Brazilian sardine *Sardinella aurita* population during the 1974/75 spawning season. Afr J Mar Sci 17:29–35

Matsuura Y, Nakatami K, Tamassi STJ (1980) Distribuição sazonal do zooplâncton, ovos e larvas de peixes na região Centro-Sul do Brasil (1975–1977). Bol Inst Oceanogr 29(2):231–235

Matsuura Y, Spach HL, Katsuragawa M (1992) Comparison of spawning patterns of the Brazilian sardine (*Sardinella brasiliensis*) and anchoita (*Engraulis anchoita*) in Ubatuba region, South Brazil, during 1985 through 1988. Bol Inst Oceanogr 40(1/2):101–115

Mazzochi MG, Paffenhöfer GA (1998) First observations on the biology of *Clausocalanus furcatus* (Copepoda, Calanoida). J Plankton Res 20(2):331–342

Melo Júnior M, Miyashita LK, Silva NJ et al (2013) Reproductive traits of *Euterpina acutifrons* in a coastal area of Southeastern Brazil. Mar Ecol 34:363–372

Melo Júnior M, Marcolin CR, Miyahita LK et al (2016) Temporal changes in pelagic copepod assemblages off Ubatuba, Brazil. Mar Ecol 37(4):877–890

Mendes F, Figueiredo GM, Valentin JL (2012) Reproduction and structure of the population of the Chaetognath *Parasagitta friderici* in Guanabara Bay (Brazil) based on short term sampling. An Acad Bras Cienc 84:103–111

Mianzan HW, Guerrero RA (2000) Environmental patterns and biomass distribution of gelatinous macrozooplankton. Three study cases in the South-Western Atlantic Ocean. Sci Mar 64(1):215–224

Miyashita LK, Calliari D (2014) Revision of mysids (Crustacea: Peracarida: Mysida) zoogeographical distribution in the South-West Atlantic (0° to 40°S). Mar Biodiv Rec 7(e5):1–13

Miyashita LK, Calliari D (2016) Distribution and salinity tolerance of marine mysids from a subtropical estuary, Brazil. Mar Biol Res 12(2):133–145

Miyshita LK, Pompeu M, Gaeta SA, Lopes RM (2010) Seasonal contrasts in abundance and reproductive parameters of *Penilia avirostris* (Cladocera, Ctenopoda) in a coastal subtropical area. Mar Biol 157, 2511–2519

Miyashita LK, Lopes RM (2011) Larvacean (Chordata, Tunicata) abundance and inferred secondary production off southeastern Brazil. Estuar Coast Shelf Sci 92:367–375

Miyshita LK, Gaeta SA, Lopes RM (2011) Life cycle and reproductive traits of marine podonids (Cladocera, Onychopoda) in a coastal subtropical area. J Plankton Res 33, 779–792

Miyashita LK, Melo Junior M, Lopes RM (2009) Estuarine and oceanic influences on copepod abundance and production of a subtropical coastal area. J Plankton Res 31:815–826

Montú M, Gloeden IM, Duarte AK, Resgalla C Jr (1997) Zooplankton. In: Seeliger U, Odebrecht C, Castello JP (eds) Subtropical convergence environments. The coast and sea in the Southwestern Atlantic. Springer, Berlin, pp 110–114

Moreira GS (1973) On the diurnal vertical migration of hydromedusae off Santos Brazil. Publ Seto Mar Biol Lab 20:537–566

Moreira GS (1976a) Sobre a migração vertical diária do plâncton ao largo de Santos, estado de São Paulo, Brasil. Bol Inst Oceanogr 25:55–76

Moreira GS (1976b) Laboratory studies on the copepod *Euterpina acutifrons* (Dana). An Acad Bras Ciênc 47:109–111

Moreira GS (1978) A preliminary laboratory study on the salinity and temperature tolerances of some medusae from the São Paulo coast, Brazil. Bol Inst Oceanogr 27(2):45–55

Moreira GS, Vernberg WB (1978) Synergistic effects of environmental variables on the metabolism of the copepod *Euterpina acutifrons* from two different areas off the coast of the state of São Paulo, Brazil. Fish Bull 76(2):449–455

MPA (2011) Boletim estatístico da pesca e aquicultura. Ministério de Pesca e Aquicultura, Brasília

Murano M (1999) Mysidacea. In: Boltovskoy D (ed) South Altantic Zooplankton. Backhuys Publishers, Leiden, pp 1099–1140

Muxagata E, Montú M (1999) Os cladocera da plataforma continental sudeste brasileira: distribuição, densidade e biomassa (inverno de 1995). Nauplius 7:151–172

Nagata RM, Haddad MA, Nogueira M Jr (2009) The nuisance of jellyfish to shrimp trawls in central part of Southern Brazilian Bight, from the perspective of artisanal fishermen. Pan Am J Aquat Sci 4(3):312–325

Nagata RM, Nogueira Júnior M, Brandini FP et al (2014) Spatial and temporal variation of planktonic cnidarian density in subtropical waters of the Southern Brazilian Bight. J Mar Biol Assoc UK 94:1387–1400

Nakata H, Kimura S, Okazaki Y et al (2000) Implications of meso-scale eddies caused by frontal disturbances of the Kuroshio current for anchovy recruitment. ICES J Mar Sci 57(1):143–152

Negreiros-Fransozo ML, González-Gordillo JI, Fransozo A (2003) First larval stage of *Exhippolysmata oplophoroides* (Holthuis, 1948) (Decapoda, Caridea, Hippolytidae) obtained in laboratory. Nauplius 10(1):67–71

Nogueira Júnior M (2012) Gelatinous zooplankton fauna (Cnidaria, Ctenophora and Thaliacea) from Baía da Babitonga (southern Brazil). Zootaxa 3398:1–21

Nogueira Júnior M, Haddad MA (2005) *Lychnorhiza lucerna* Haeckel (Scyphozoa, Rhizostomeae) and *Libinia ferreirae* Brito Capello (Decapoda, Majidae) association in southern Brazil. Rev Bras Zool 22(4):908–912

Nogueira Júnior M, Haddad MA (2008) The diet of Cubomedusae (Cnidaria, Cubozoa) in Southern Brazil. Braz J Oceanogr 56(3):157–164

Nogueira Júnior M, Haddad MA (2017) Seasonal distribution, abundance and biomass of large medusae in subtropical coastal Brazil. In: Mariottini GL (ed) Jellyfish, ecology, distribution patterns and human interactions. Nova Science Publihsers, New York, pp 3–25

Nogueira Júnior M, Rodriguez C, Mianzan HW et al (2013) Description of a new hydromedusa from the southwestern Atlantic Ocean, *Bougainvillia pagesi* sp. nov. (Cnidaria, Hydrozoa, Anthoathecata). Mar Ecol 34:113–122

Nogueira Júnior M, Brandini FP, Codina JCU (2014) Distribution of planktonic cnidarians in response to South Atlantic central water intrusion in the South Brazilian Bight. Cont Shelf Res 89:93–102

Nogueira Júnior M, Brandini FP, Codina J (2015) Diel vertical dynamics of gelatinous zooplankton (Cnidaria, Ctenophora and Thaliacea) in a subtropical stratified ecosystem (South Brazilian Bight). PLoS One 10(12):e0144161–e0144128

Nogueira Júnior M, Brandini FP, Haddad MA (2016) First record of the hydromedusa *Aequorea macrodactyla* (Leptothecata: Aequoreidae) in Brazilian waters. Mar Biodivers 46:737–742

Nogueira Júnior M, Silva Nascimento L, Maciel PV, Tilbert S, Oliveira LD, (2018) Diversity, species composition and assemblage dynamics of estuarine gelatinous and semi-gelatinous zooplankton from Brazil. In: Hoffmeyer M, et al. (eds.), Plankton Ecology of the Southwestern Atlantic, chapter 18, this volume

Nogueira CR, Oliveira SR Jr (1991) Siphonophora from the coast of Brazil (17° to 24°S). Bol Inst Oceanogr 39(1):6–69

Oliveira OMP, Mianzan H, Migotto AE et al (2007) Chave de identificação dos Ctenophora da costa brasileira. Biota Neotrop 7(3):341–350

Oliveira OM, Miranda TP, Araujo EM et al (2016) Census of Cnidaria (Medusozoa) and Ctenophora from South American marine waters. Zootaxa 4194(1):1–256

Paffenhöfer G-A (1985) The abundance and distribution of zooplankton on the Southeastern shelf of the United States. Coast Estuar Sci 2:104–117

Paffenhöfer G-A, Wester BT, Nicholas WD (1984) Zooplankton abundance in relation to state and type of intrusions onto the southeastern United States shelf during summer. J Mar Res 42:995–1017

Paffenhöfer G-A, Sherman BK, Lee TN (1987) Summer upwelling on the Southeastern continental shelf of the USA during 1981. Abundance, distribution and patch formation of zooplankton. Prog Oceanogr 19:403–436

Paffenhöfer G-A, Atkinson LP, Lee TN et al (1994) Variability of particulate matter and abundant zooplankton off the Southeastern United States during spring of 1984 and 1985. Cont Shelf Res 14(6):629–654

Paffenhöfer G-A, Atkinson LP, Lee TN et al (1995) Distribution and abundance of thaliaceans and copepods off the Southeastern USA during winter. Cont Shelf Res 15(2/3):255–280

Pantaleão J, Terossi M, Costa RC, Mantelatto FL (2013) First zoeal stage of the partner shrimp *Periclimenes paivai* Chace, with remarks on the genus *Periclimenes* O.G. Costa (Caridea, Palaemonidae). Zootaxa 3750(4):367–374

Qiu B (2001) Kuroshio and Oyashio currents. In: Encyclopedia of ocean science. Academic, New York, pp 1413–1425

Resgalla C Jr (2008) Pteropoda, Cladocera, and Chaetognatha associations as hydrological indicators in the Southern Brazilian shelf. Lat Am J Aquat Res 36(2):271–282

Resgalla C Jr, Montú MA (1993) Cladóceros marinhos da plataforma continental do Rio Grande do Sul–Brasil. Nauplius 1:63–79

Resgalla C Jr, Montú MA (1994) Distribuição espacial e temporal de Pteropoda Euthecosomata (Mollusca, Gastropoda) a plataforma do Sul do Brasil. Atlânttica 16:99–123

Resgalla C Jr, de La Rocha C, Montú MA (2001) The influence of Ekman transport on zooplankton biomass variability off Southern Brazil. J Plankton Res 23(6):641–650

Rocha CEF (1982) Distribution of the marine cladocerans (Crustacea, Branchiopoda) off Santos, Brazil. Bol Zool Univ S Paulo 7:155–169

Rocha CEF (1983) Distribution of *Euconchoecia chierchiae* G.W. Muller (Crustacea, Ostracoda) off Santos, Brazil. Bol Inst Oceanogr 6:53–61

Sartori LP, Lopes RM (2000) Seasonal variability of pelagic copepods assemblages on the inner continental shelf off Paraná, Brazil. Nauplius 8:79–88

Schettini CA, Kuroshima N, Pereira J et al (1998) Oceanographic and ecological aspects of the Itajaí-açu River plume during a high discharge period. An Acad Bras Cienc 70(2):335–351

Schlitzer R (2017) Ocean data view. http://www.odv.awi.de. Accessed 15 Aug 2017

Sinque C (1982) Distribuição vertical de Appendicularia (Tunicata) ao largo de Santos, Brasil (24°16′08"S e 46°00′04"W). *Oikopleura*. Arq Biol Tecnol 26(1):61–73

Soares ISH, Vazzoler AEAM (2001) Diel changes in food and feeding activity of sciaenid fishes from the South-Western Atlantic, Brazil. Rev Bras Biol 61(2):197–216

Sullivan LJ, Gifford DJ (2009) Preservation of the larval ctenophore *Mnemiopsis leidyi* A. Agassiz (Ctenophora, Lobata). J Plankton Res 31(8):921–926

Tavares DQ (1967) Occurrence of doliolids and salps during 1958, 1959 and 1960 off the São Paulo coast. Bol Inst Oceanogr 16(1):87–97

Toda H (1989) Surface distributions of copepods in relation to regional upwellings around the Izu Islands in summer of 1988. J Oceanogr Soc Jpn 45:251–257

Tundisi TM (1970) On the seasonal occurrence of Appendicularians in waters off the coast of São Paulo state. Bol Inst Oceanogr 19:13–144

Valentin J (1980) L'écosystème d'upwelling à Cabo Frio (Brésil). Analyse en composantes principales du plancton à une station fixe. Oceanol Acta 3(1):43–50

Valentin JL (1984a) Spatial structure of the zooplankton community in the Cabo Frio region (Brazil) influenced by coastal upwelling. Hydrobiologia 113:183–199

Valentin JL (1984b) Analyse des paramètres hydrobiologiques dans la remontée de Cabo Frio (Brésil). Mar Biol 82:259–276

Valentin JL (1989) A dinâmica do plâncton na ressurgência de Cabo Frio–RJ. In: Brandini FP (ed) Memórias do III Encontro Brasileiro de Plâncton. Editora da UFPR, Curitiba, pp 25–35

Valentin JL, Monteiro-Ribas WM (1993) Zooplankton community structure on the east-southeast Brazilian continental shelf (18–23°S latitude). Cont Shelf Res 13(4):407–424

Valentin JL, Monteiro-Ribas WM, Mureb MA (1987) Sur quelques zooplanctontes abondants dans l'upwelling de Cabo Frio (Brésil). J Plankton Res 9(6):1195–1216

Vannucci M (1951a) Hydrozoa e Scyphozoa existentes no instituto paulista de oceanografia. Bol Inst Oceanogr 2(1):69–100

Vannucci M (1951b) Resultados científicos do cruzeiro do "Baependi" e do "Vega" à ilha da Trindade. O gênero *Firoloida*, Prosobranchia Heteropoda. Bol Inst Paul Oceanogr 2(2):73–89

Vannucci M (1957) On Brazilian hydromedusae and their distribution in relation to different water masses. Bol Inst Oceanogr 8(1–2):23–109

Vannucci M (1963) On the ecology of Brazilian medusae at 25°Lat. S. Bol Inst Oceanogr 13(1):143–184

Vannucci M, Almeida Prado MS (1959) Sobre as coletas de plâncton na II e IV viagens no N/Oc. "Almirante Saldanha". Contrib Avulsas Inst Oceanogr 1:1–16

Vega-Pérez LA (1993) Estudo do zooplâncton da região de Ubatuba, Estado de São Paulo. Publção esp Inst Oceanogr 10:65–84

Vega-Pérez LA, Liang TH (1992) Feeding of a pelagic chaetognath, *Sagitta friderici* Ritter-Zahony off Ubatuba region (São Paulo). Bol Inst Oceanogr 40(1, 2):93–100

Vega-Pérez LA, Ara K, Liang TH et al (1996) Feeding of the planktonic shrimp *Lucifer faxoni* Borradaile (Crustacea: Decapoda) in the laboratory. Revta Bras Oceanogr 44(1):1–8

Zooplankton Communities of the Argentine Continental Shelf (SW Atlantic, *ca.* 34°–55°S), An Overview

Georgina D. Cepeda, Brenda Temperoni, Marina E. Sabatini, María D. Viñas, Carla M. Derisio, Betina A. Santos, Julieta C. Antacli, and Luciano N. Padovani

Abstract A profuse literature related to the ecology of the Argentine continental shelf has been produced in the last four decades, documenting its biological richness and high productivity. Distinctive environmental characteristics define particular systems along and across the shelf, which in all cases are inhabited by mammal, bird, fish and cephalopod species in all life history stages, either as spawning, mating, nursery or juvenile grounds or just for adult feeding. At the productive base of these systems, zooplankton certainly plays a crucial role. This paper reviews the available information on zooplankton diversity and ecology for this huge region in the Southwest Atlantic Ocean, with the focus primarily upon copepods and

---

Dedicamos este artículo al Dr. Fernando César Ramírez, nuestro mentor y de tantos otros zooplanctólogos a través de las generaciones. Para él, nuestro orgulloso y enorme reconocimiento por sus logros, y admiración por el entusiasmo siempre presente que continúa contagiándonos cada día

We proudly dedicate this article to our mentor Dr. Fernando César Ramírez, whose professional achievements and never-ending enthusiasm still inspire many generations of Argentinian zooplanktologists.

G. D. Cepeda (✉) · B. Temperoni · M. E. Sabatini · M. D. Viñas
Instituto de Investigaciones Marinas y Costeras (IIMyC), Consejo Nacional de Investigaciones Científicas y Técnicas–Universidad Nacional de Mar del Plata (CONICET – UNMdP), Mar del Plata, Buenos Aires, Argentina

Instituto Nacional de Investigación y Desarrollo Pesquero (INIDEP), Mar del Plata, Buenos Aires, Argentina
e-mail: gcepeda@inidep.edu.ar; btemperoni@inidep.edu.ar; marsab@inidep.edu.ar; mdvinas@inidep.edu.ar

C. M. Derisio · B. A. Santos · L. N. Padovani
Instituto de Nacional de Investigación y Desarrollo Pesquero (INIDEP), Mar del Plata, Buenos Aires, Argentina
e-mail: cderisio@inidep.edu.ar; bsantos@inidep.edu.ar; lucianopadovani@inidep.edu.ar

J. C. Antacli
Instituto de Diversidad y Ecología Animal (IDEA), Consejo Nacional de Investigaciones Científicas y Técnicas–Universidad Nacional de Córdoba (CONICET – UNC), Córdoba, Argentina

M. S. Hoffmeyer et al. (eds.), *Plankton Ecology of the Southwestern Atlantic*,
https://doi.org/10.1007/978-3-319-77869-3_9

secondarily on hyperiid amphipods and euphausiids. We describe general aspects of biogeographic zonation and diversity for the entire shelf, with emphasis on key zooplankton species. Then, we consider the structure and dynamics of the communities in relation to water masses, frontal areas and the overall circulation, specifically for (i) the northern shelf (34°–41°S), (ii) the Valdés frontal system over the northern Patagonian shelf (41°–45°S) and (iii) the southern Patagonian shelf (47°–55°S). We finally go over the open questions and prospects for the future work on zooplankton in the region.

Keywords Copepods · Hyperiid amphipods · Euphausiids · Diversity patterns · Community patterns · Southwest Atlantic Ocean

1 Introduction

From the Rio de la Plata south to Tierra del Fuego, the Argentine continental shelf (ACS) is one of the longest and widest submarine coastal plains in the world. With a surface of approximately 1,000,000 km^2, this region includes almost the whole of the extensive and highly productive Patagonian Shelf Large Marine Ecosystem (Sherman and Duda 1999), which is also termed the Southwest Atlantic Shelf Province based on productivity patterns at the regional scale (Longhurst 2007). The Argentine shelf is among the richest and most diverse ocean areas, and it is unique in many ways. It is a habitat for schools of fish and squids of high economic and ecological value (hakes, anchovy, toothfish, shortfin squid and others; e.g. Sánchez and Bezzi 2004), and it also includes the migration corridors and breeding grounds of many charismatic marine mammals and birds (whales, dolphins, elephant seals and large sea lions, penguins, albatross and other birds; e.g. Falabella et al. 2009). Supporting these varied top-predator populations, the shelf 's zooplankton communities are key intermediate components of the pelagic food web, transferring energy and matter from phytoplankton to the upper trophic levels.

A great deal of research into the zooplankton communities of the ACS has been carried out since the mid-1960s, coinciding with the beginning of large-scale fisheries off Argentina. Main fishery targets include small and medium pelagic species: *Engraulis anchoita* and *Scomber colias*, demersal hakes *Merluccius hubbsi* and *Macruronus magellanicus* and squid *Illex argentinus*. For all of those, copepods, hyperiid amphipods and euphausiids represent major food resources, as shown by a collection of local trophic studies which will be addressed later on. Stock assessment surveys have historically provided a cooperative platform for opportunistic zooplankton work. While assuring ship time, facilities and technical support for ecological studies, the zooplankton sampling strategies and sampling gear selected usually have had to be accommodated to the overall cruise designs and schedules. As a result, our historical dataset is largely biased toward the spring–summer period and to the mesozooplankton fraction (0.2–20 mm), with consequent underestimation of both the larger and faster-moving macrozooplankton and the microzooplankton fraction less than 100 μm.

Studies over the past few decades in the ACS have created through faunal inventories and provided reasonable knowledge of zooplankton spatial patterns in relation to the major regional hydrographic features. However, this valuable body of information, gathered in different areas at different times, has never been examined in one piece. Therefore, our goal here is to assemble all the available data on the zooplankton diversity and ecology for this huge region in the Southwest Atlantic. Because of their abundance and relevance for the ACS pelagic food webs, we will focus primarily upon copepods and secondarily on hyperiid amphipods and euphausiids. Cladocerans (Viñas et al. 2007), hyperiid amphipods (Viñas et al. 2016) and euphausiids (Ramírez 2016) have been reviewed to some extent before.

In this review, we first address general aspects for the entire shelf, such as the overall biogeography and diversity from an updated species inventory, including "summary cards" for a few key species. Next, we summarize the structure and dynamics of the zooplankton communities in relation to water masses, frontal areas and general circulation, in particular for (i) the northern shelf system, (ii) the Valdés frontal system over the northern Patagonian shelf and (iii) the southern Patagonian shelf system (Fig. 1). Finally, the information gaps and prospects for future work on zooplankton in the region are highlighted. Piola et al. (2018) have dealt with the physical oceanography of the subtropical and subantarctic shelves of the western South Atlantic, which includes our study area. We refer readers to that article for further details on the physical background of our studies.

2 Biogeographic Zonation and Species Diversity

The three major systems along the ACS that we are addressing in this review roughly agree with three of the 16 Marine Ecoregions proposed for temperate South America by Spalding et al. (2007). They based their biogeographic scheme on distinct oceanographic features of coastal and shelf areas (e.g. temperature regimes, freshwater influx) that can be recognized in the ACS. We agree with them that a broad range of species distributions are associated, probably causally so, with readily discerned physical patterns. Hence, those physical criteria allow our data to be cross-referenced to more general, regional biogeographic classifications.

Worldwide, *ca.* 7000 species of marine zooplankton have been described. More than a third of those occur in the South Atlantic Ocean, including 505 species of Copepoda, 188 of Amphipoda and 61 of Euphausiacea (Boltovskoy et al. 2003). Particularly in the Southwest Atlantic Ocean (SWAO, 0°–55°S; 30°–70°W), approximately 430 species of copepods have been recorded (Björnberg 1981; Bradford-Grieve et al. 1999; Neumann-Leitão et al. 2008; Magalhães et al. 2009), 109 of hyperiid amphipods (Vinogradov 1999) and 31 of euphausiids (Mauchline and Fisher 1969; Mauchline 1980).

A closer look at the copepod composition over the ACS indicates the presence of 101 species/taxa representing 23.48% of total copepod species reported in the SWAO (Table 1); 95 of those occur over the northern shelf, whereas 26 and 24 species are

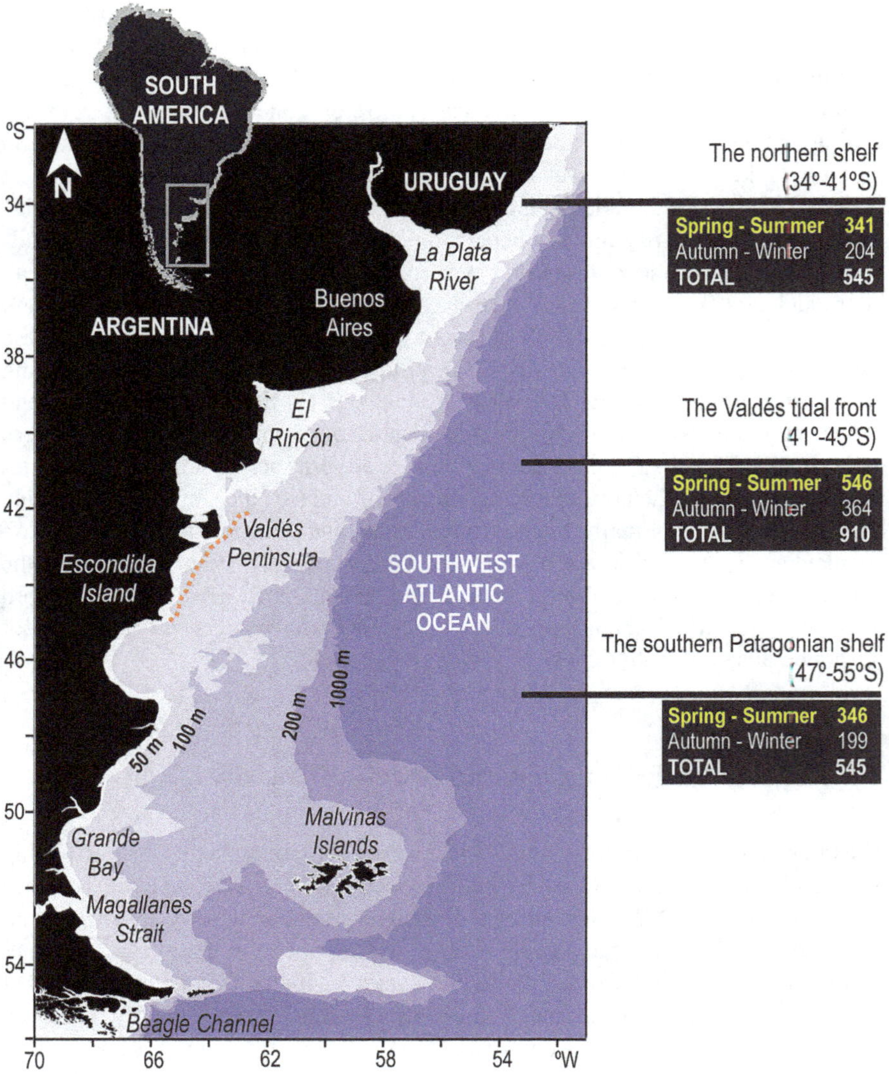

Fig. 1 Study area in the Southwest Atlantic Ocean. Mesozooplankton sampling effort is expressed as number of sampled stations during the spring–summer and autumn–winter periods for the three reviewed systems along the Argentine continental shelf. The orange dashed line represents the interannual average position of the Valdés tidal front

found over the northern and southern Patagonian shelves, respectively. A number of 20 species are widely distributed over the three systems, and 71 occur only over the shelf between 34°S and 41°S, whereas the same number of species (3) are found exclusively either over the northern or southern Patagonian shelves. In a broad sense, these patterns accord with the latitudinal gradients of species richness globally (e.g. Valentine 2009) and of some zooplankton groups particularly (e.g. McPherson 2002; Rombouts et al. 2009).

Table 1 Checklist of copepod species in the three systems along the Argentine continental shelf

Order–family–*species*	ACS system		
	34°–41°S	41°–45°S	47°–55°S
Calanoida			
Calanidae (8)			
1. *Calanoides carinatus*	●	●	
2. *Calanoides patagoniensis*			●
3. *Calanus australis*	●	●	●
4. *Calanus simillimus*	●	●	●
5. *Neocalanus tonsus*	●	●	●
6. *Neocalanus gracilis*	●		
7. *Nannocalanus minor*	●		
8. *Undinula vulgaris*	●		
Paracalanidae (9)			
9. *Calocalanus plumulosus*	●		
10. *Calocalanus pavoninus*	●		
11. *Calocalanus* sp.			●
12. *Calocalanus pavo*	●		
13. *Paracalanus parvus*	●	●	●
14. *Paracalanus aculeatus*	●		
15. *Parvocalanus scotti*	●		
16. *Parvocalanus crassirostris*	●		
17. *Delibus* sp.	●		
Mecynoceridae (1)			
18. *Mecynocera clausi*	●		
Eucalanidae (8)			
19. *Eucalanus elongatus*	●		
20. *Pareucalanus attenuatus*	●		
21. *Pareucalanus langae*	●		
22. *Rhincalanus gigas*		●	
23. *Rhincalanus nasutus*	●	●	●
24. *Subeucalanus monachus*	●		
25. *Subeucalanus pileatus*	●		
26. *Subeucalanus longiceps*	●	●	●
Clausocalanidae (6)			
27. *Clausocalanus brevipes*	●	●	●
28. *Clausocalanus furcatus*	●	●	●
29. *Clausocalanus laticeps*	●	●	●
30. *Clausocalanus arcuicornis*	●		

(continued)

Table 1 (continued)

Order–family–*species*	ACS system		
	34°–41°S	41°–45°S	47°–55°S
31. *Ctenocalanus vanus*	●	●	●
32. *Drepanopus forcipatus*	●	●	●
Aetideidae (3)			
33. *Aetideus armatus*	●	●	
34. *Gaetanus tenuispinus*	●		
35. *Euchirella rostrata*	●		
36. *Undeuchaeta plumosa*	●		
Euchaetidae (2)			
37. *Euchaeta marina*	●		
38. *Paraeuchaeta barbata*	●		
Scolecitrichidae (4)			
39. *Scolecithricella minor*	●		
40. *Scolecithrix danae*	●		
41. *Scottocalanus securifrons*	●		
42. *Scaphocalanus magnus*	●		
Metridinidae (5)			
43. *Pleuromamma robusta*	●	●	●
44. *Pleuromamma gracilis*	●		
45. *Pleuromamma xiphias*	●		
46. *Pleuromamma abdominalis*	●		
47. *Metridia lucens*	●	●	●
Centropagidae (4)			
48. *Centropages brachiatus*	●	●	●
49. *Centropages calaninus*	●		
50. *Centropages bradyi*	●		
51. *Centropages furcatus*	●		
Heterorhabdidae (1)			
52. *Heterorhabdus austrinus*	●		
Augaptilidae (3)			
53. *Haloptilus longicornis*	●		
54. *Haloptilus oxicephalus*	●		
55. *Haloptilus spiniceps*	●		
Candaciidae (5)			
56. *Candacia longimana*	●		
57. *Candacia pachydactila*	●		
58. *Candacia cheirura*		●	
59. *Candacia simplex*	●		
60. *Candacia aethiopica*	●		

Order–family–*species*	ACS system		
	34°–41°S	41°–45°S	47°–55°S
Lucicutiidae (1)			
61. *Lucicutia flavicornis*	●		
Temoridae (2)			
62. *Temora stylifera*	●		
63. *Temora turbinata*	●		
Acartiidae (3)			
64. *Acartia tonsa*	●	●	●
65. *Acartia negligens*	●		
66. *Acartia danae*	●		
Pontellidae (3)			
67. *Pontella patagoniensis*	●		
68. *Labidocera fluviatilis*	●		
69. *Pontellina plumata*	●		
Cyclopoida			
Oithonidae (5)			
70. *Oithona nana*	●	●	●
71. *Oithona helgolandica*	●	●	●
72. *Oithona atlantica*	●	●	●
73. *Oithona flemingeri*	●		
74. *Oithona tenuis*	●		
Harpacticoida			
Euterpinidae (1)			
75. *Euterpina acutifrons*	●	●	●
Ectinosomatidae (3)			
76. *Microsetella norvegica*	●	●	●
77. *Microsetella rosea*	●		
78. *Macrosetella gracilis*	●		
Aegisthidae (1)			
79. *Aegisthus mucronatus*	●		
Clytemnestridae (1)			
80. *Goniopsyllus rostratus*	●		●
Poecilostomatoida			
Corycaeidae (8)			
81. *Ditrichocorycaeus amazonicus*	●		
82. *Onychocorycaeus pacificus*	●		
83. *Onychocorycaeus giesbrechtii*	●		
84. *Corycaeus furcifer*	●		
85. *Corycaeus speciosus*	●		
86. *Urocorycaeus lautus*	●		

(continued)

Table 1 (continued)

Order–family–*species*	ACS system		
	34°–41°S	41°–45°S	47°–55°S
87. *Farranula gracilis*	●		
88. *Farranula concinna*	●		
Clausidiidae (1)			
89. *Hemicyclops thalassius*	●		
Peltiidae (1)			
90. *Eupelte minuta*		●	
Oncaeidae (5)			
91. *Triconia antarctica*	●	●	
92. *Oncaea venusta*	●		
93. *Conaea rapax*	●		
94. *Lubbockia squillimana*	●		
95. *Lubbockia aculeata*	●		
Sapphirinidae (5)			
96. *Sapphirina angusta*	●		
97. *Sapphirina metallina*	●		
98. *Sapphirina opalina*	●		
99. *Copilia mirabilis*	●		
100. *Copilia quadrata*	●		
Siphonostomatoida			
101. *Siphonostomatoida* sp.			●
Number of species/taxa	**95**	**26**	**24**

Species are listed by order and family. In brackets, total count of species/taxa per family. Source data from 64 to 500 μm nets available from all the literature reviewed from 1966 to date. Overall nomenclatural update follows World Register of Marine Species (WoRMS, 2017)

Distribution patterns of zooplankton, in general, and of copepods, in particular, typically occur in agreement with the environmental variations determined by water masses and major currents (e.g. Boltovskoy 1999). Thus, the highest species richness recorded between 34°S and 41°S, that is, 94% of the ACS total copepod species, is mainly a result of the mixing of subtropical species intruding with waters of the Brazil Current, typical subantarctic species and species arriving with the Rio de la Plata influx (Table 1). This is in agreement with the transitional patterns described by Boltovskoy and Correa (2008), namely, euryhaline, cryophilic and thermophilic species coming together. Our present data show that the south of 41°S species richness is much lower, in close relation with intense hydrographic features that dictate the almost exclusive presence of subantarctic species. That is true, although the presence of subtropical species has been mentioned occasionally (Boltovskoy and Correa 2008). The trends suggest that penetration of subantarctic species northward is much stronger than expatriation of subtropical organisms southward. Except for unusual expatriates, subtropical species do not occur south of 41°S (Table 1).

When considering macrozooplankton, species diversity for hyperiid amphipods is relatively low in the ACS; only 11 species are found, representing 10% of all hyperiid species in the SWAO (Ramírez and Viñas 1985; Padovani 2013). For euphausiids, on the contrary, the proportion of SWAO species in the ACS is similar to that of copepods (26%) since 8 of the 31 species in the SWAO are typically reported (Ramírez 1971, 1973, 2016) (Table 2). Among the three ecosystems along the ACS, the latitudinal gradients of species richness for both amphipods and euphausiids are not so evident as that of copepods. However, more species northward becomes evident when sampling at latitudes north of the ACS, which adds more subtropical species. Greater richness has also been described toward the outer reaches of the ACS, where the occurrence of cold-water species increases (Padovani 2013). Trends across the shelf are not evident from Table 2, because our analysis was focused on the latitudinal arrangement of the three systems.

Themisto gaudichaudii strongly dominates hyperiid abundance over the ACS, whereas species such as *Primno macropa, Cyllopus magellanicus* and *Hyperiella antarctica* occur mostly over the outer shelf (Padovani 2013 and references therein). Among euphausiids, Ramírez (1971, 1973) showed that *Euphausia lucens* is the most widely distributed species along the entire ACS and also in cold waters of the Malvinas Current. *Nematoscelis megalops, Thysanoessa gregaria* and *Euphausia vallentini* are found mainly over the Patagonian shelf south of 44°S and in colder, more oceanic waters northward. *Euphausia similis*, another cryophile species, distributes mainly in oceanic waters of the Malvinas Current, while *Stylocheiron affine* and *Stylocheiron longicorne* are recorded in subtropical oceanic waters of the Brazil Current.

The general species richness patterns described here are based on typical, morphological classifications that have historically supported zooplankton research along the ACS. These estimates may change if DNA-based approaches are coupled to this traditional identification technique, possibly leading to new understanding of global, regional and local zooplankton biodiversity (for review, see Bucklin et al. 2010 and references therein).

Table 2 Checklist of hyperiid amphipod and euphausiid species in the three systems along the Argentine continental shelf

(Sub)order–family–species	ACS system		
	34°–41°S	41°–45°S	47°–55°S
Hyperiidea			
Hyperiidae (4)			
1. *Themisto gaudichaudii*	●	●	●
2. *Hyperia medusarum*			●
3. *Hyperiella antarctica*	●	●	●
4. *Hyperoche medusarum*	●		●
Phrosinidae (1)			
5. *Primno macropa*	●	●	●
Vibiliidae (3)			
6. *Vibilia antarctica*	●	●	●
7. *Vibilia armata*	●	●	●
8. *Cyllopus magellanicus*	●	●	●
Paraphronima (1)			
9. *Paraphronima gracilis*			●
Phronimidae (1)			
10. *Phronima sedentaria*		●	●
Lestrigonidae (1)			
11. *Lestrigonus bengalensis*	●		
Number of hyperiid species	**8**	**7**	**10**
Euphausiacea			
Euphausiidae (8)			
1. *Euphausia lucens*	●	●	●
2. *Euphausia similis*	●		●
3. *Euphausia recurva*	●		
4. *Euphausia vallentini*	●	●	●
5. *Nematoscelis megalops*	●	●	●
6. *Thysanoessa gregaria*	●	●	●
7. *Stylocheiron affine*	●		
8. *Stylocheiron longicorne*	●		
Number of euphausiid species	**8**	**4**	**5**

Species are listed by order/suborder and family. In brackets, total count of species/taxa per family. Source data from 64 to 500 μm nets available from all the literature revised reviewed from 1966 to date

From the currently up-to-date species inventory for the ACS, we identified a few key taxa (Figs. 2, 3, 4 and 5), mainly because of their high abundances, latitudinal ranges, year-round occurrence, biogeographic significance and roles in their local food webs. Some of these species exhibit circumglobal distributions, while others are endemic in the Southern Hemisphere or have a restricted location off southern

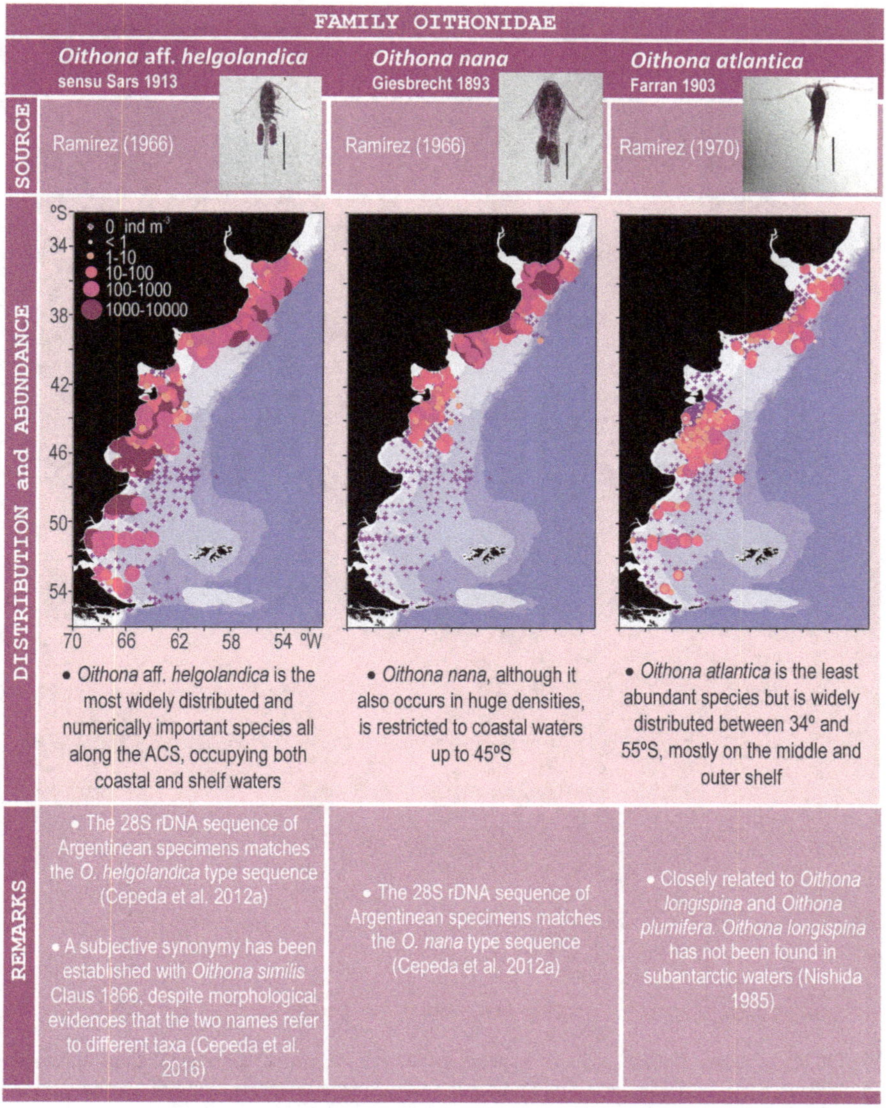

Fig. 2 Key species of Oithonidae present over the Argentine continental shelf during the spring–summer period. The taxonomic sources used for species identification and specific remarks are indicated on the basis of a full literature review. The distribution and abundance plots are based on the quantitative dataset of copepod species that was assembled during the present review. Bars represent 400 μm for *O.* aff. *helgolandica*, 200 μm for *O. nana* and 600 μm for *O. atlantica*

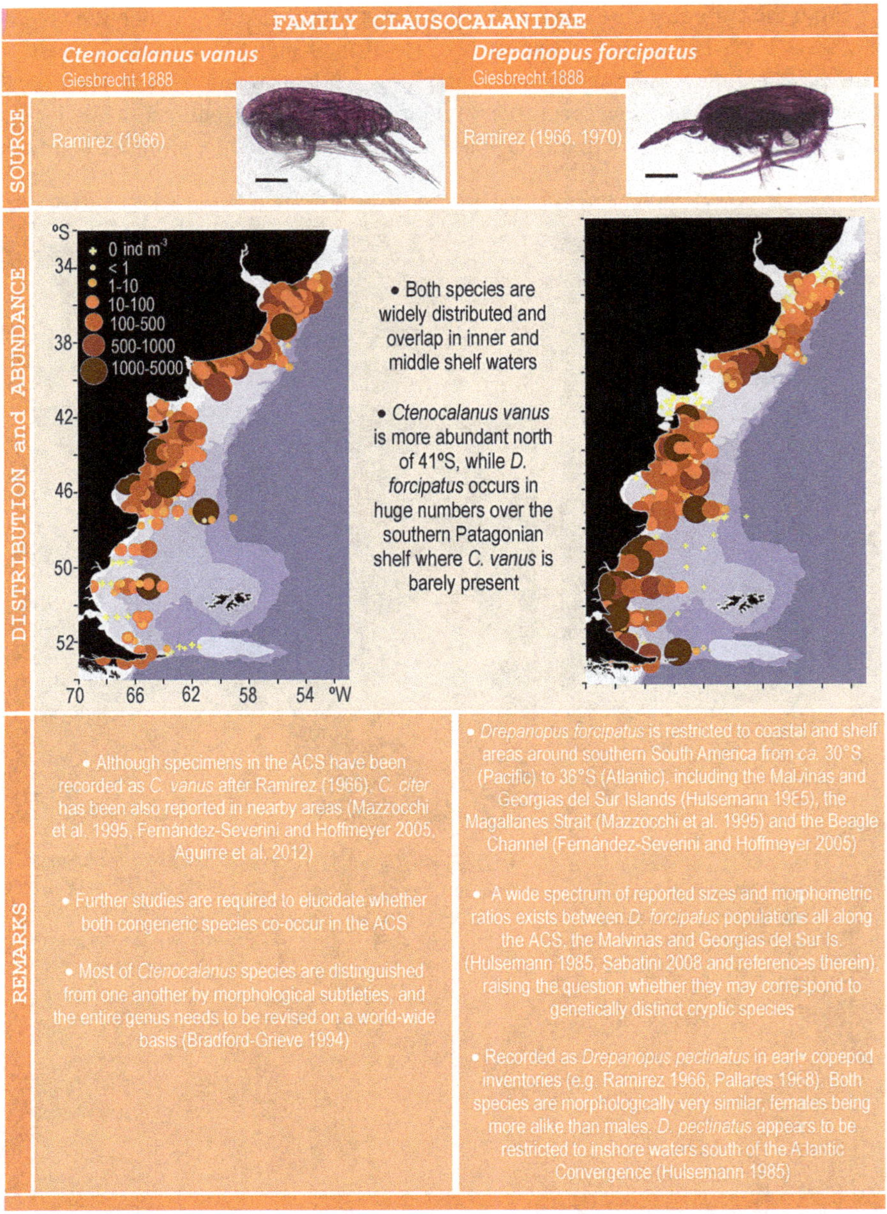

Fig. 3 Key species of Clausocalanidae present over the Argentine continental shelf during the spring–summer period. The taxonomic sources used for species identification and specific remarks are indicated on the basis of a full literature review. The distribution and abundance plots are based on the quantitative dataset of copepod species that was assembled during the present review. Bars represent 200 μm for both species

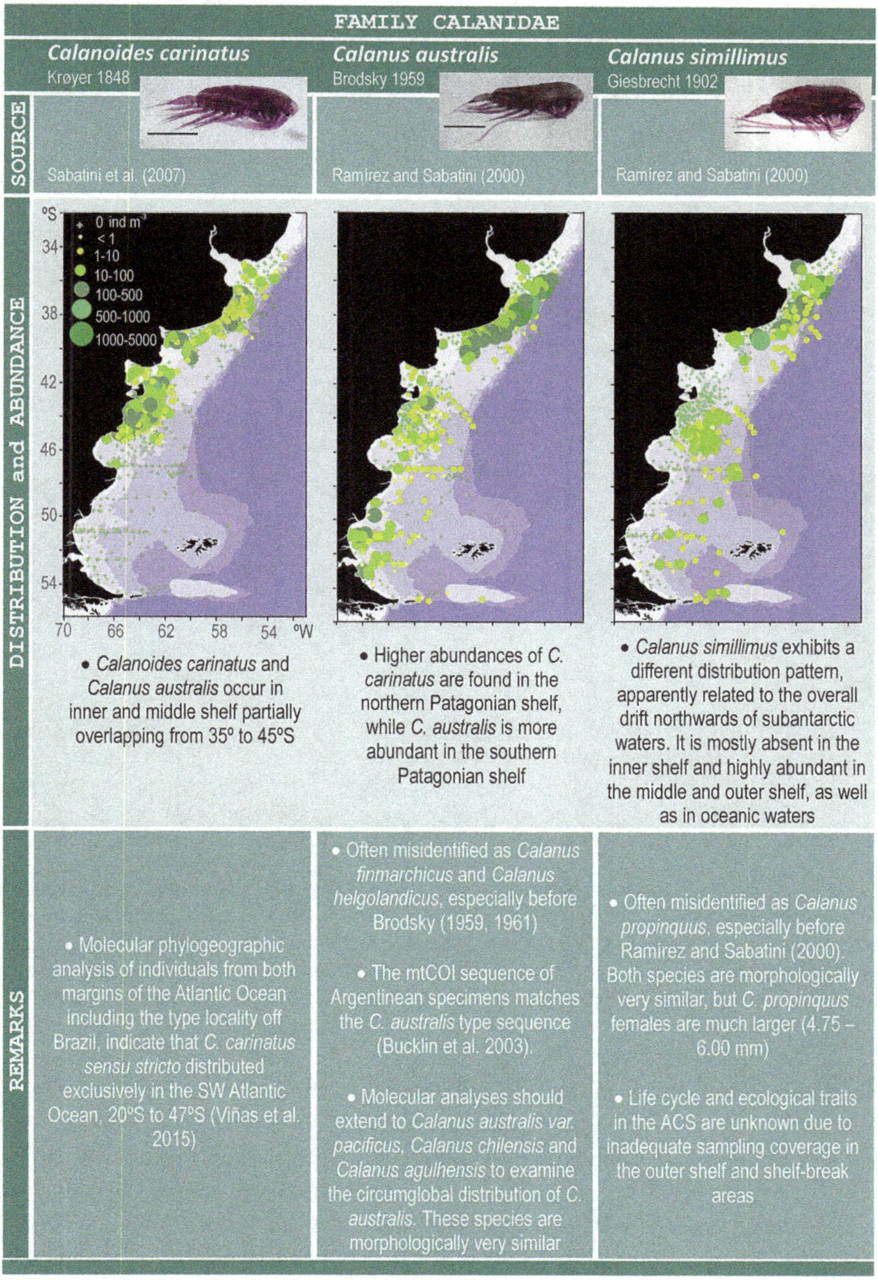

Fig. 4 Key species of Calanidae present over the Argentine continental shelf during the spring–summer period. The taxonomic sources used for species identification and specific remarks are indicated on the basis of a full literature review. The distribution and abundance plots are based on the quantitative dataset of copepod species that was assembled during the present review. Bars represent 1 mm for the three species

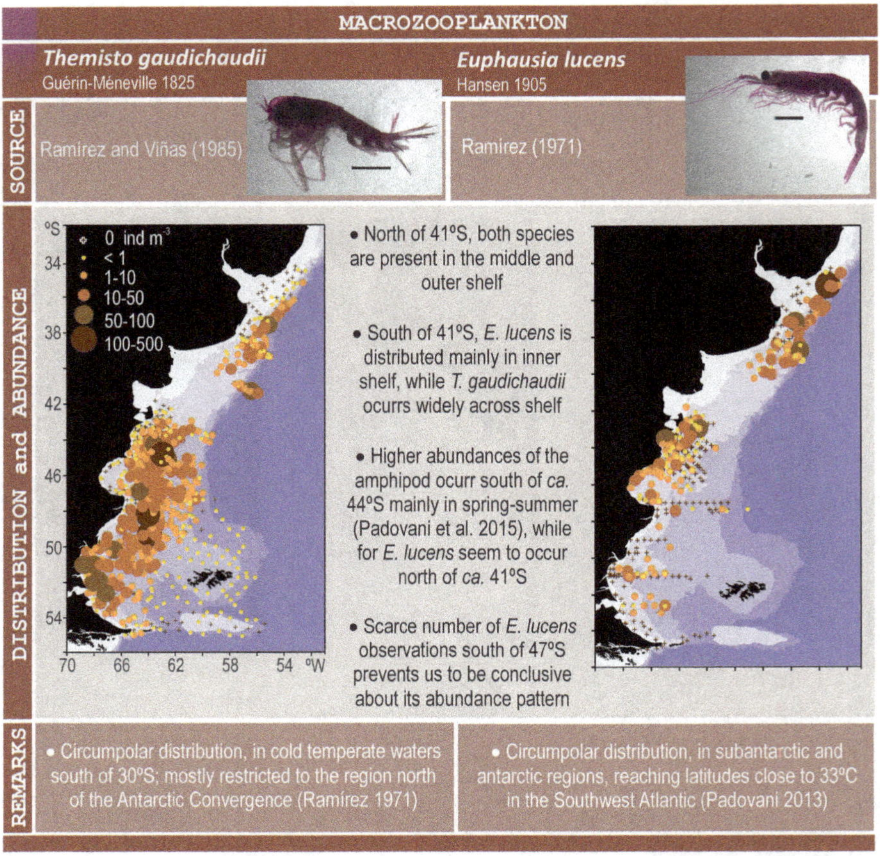

Fig. 5 Key species of macrozooplankton present over the Argentine continental shelf during the spring–summer period. The taxonomic sources used for species identification and specific remarks are indicated on the basis of a full literature review. The distribution and abundance plots are based on the quantitative dataset of hyperiid amphipod and euphausiid species that was assembled during the present review. Note that this dataset is different from that of copepods. Bars represent 2 mm for both species

South America. Among the small-sized copepods, three cyclopoids of the family Oithonidae are considered keystone species, including *Oithona* aff. *helgolandica*, *Oithona nana* and *Oithona atlantica* (Fig. 2), with three markedly different distributions. Medium-sized key copepods are represented by two clausocalanids, *Ctenocalanus vanus* and *Drepanopus forcipatus*, which are very abundant along the ACS (Fig. 3). In the largest copepod size fraction, three species of Calanidae, namely, *Calanoides carinatus*, *Calanus australis* and *Calanus simillimus*, are major components of the zooplankton communities along the ACS (Fig. 4). They often dominate in terms of biomass, with lipid-storing, fifth-stage copepodids outnumbering adults (e.g. Antacli et al. 2014a). Among macrozooplankton, *T. gaudichaudii* and *E. lucens* are the most important and widely distributed species, overlapping along the ACS (Fig. 5).

3 Structure and Dynamics in Relation to Water Masses, Frontal Areas and Circulation

Distributions of planktonic organisms depend strongly on currents and water masses. In view of that, the complex and diverse hydrographic features occurring along the ACS provide heterogeneity of habitats, water-type interfaces, more and less productive systems and hydrographic discontinuities where mixing of ecological assemblages occurs. Historically, the abundance, distribution and relations among zooplankton species have been studied in relation to the habitat's heterogeneity along the ACS. A robust body of published information allows the following update for the three systems spanning the ACS.

3.1 The Northern Shelf Ecosystem (34°–41°S)

The oceanographic conditions in this latitudinal range are highly complex and variable. Shelf waters of subantarctic origin flow in from the south, large rivers add freshwater runoff, and two distinct western boundary currents converge: the warm and salty, southward flowing Brazil Current ($S \geq 36$), and the cold and relatively fresh, northward flowing Malvinas Current (chapter "Physical Oceanography of the SW Atlantic Shelf: A Review").

The Rio de la Plata estuary, at ~35°–36°S, contributes substantially to the coastal oceanography of this system. Its great discharge has a weak seasonal signal with a maximum in winter and a minimum in summer. These estuarine waters are distributed with seasonal variations over the adjacent shelf, influenced mainly by the prevailing winds of the region (Piola et al. 2018). At ~39°–41°S, the less intense "El Rincon" estuarine system is established by freshwater from the Negro and Colorado Rivers. The shelf area between these two estuarine systems is occupied by subantarctic shelf waters (SASW, S 33.4–34) (chapter "Physical Oceanography of the SW Atlantic Shelf: A Review").

Despite the characteristic spatio-temporal hydrographical variability of this system, studies of its zooplankton communities, mainly dominated by copepods, show a seasonally persistent across-shelf zonation delimiting contiguous but distinct faunistic areas (FA), that is, an estuarine FA in the low-salinity surface plume of the Rio de la Plata discharge (Plata plume water (PPW) $S < 33.3$); inner-shelf (depths <50 m) and middle-shelf (~50–100 m) FAs, both associated with SASW but separated by a seasonal thermal front; and an outer-shelf FA (>100 m), strongly influenced by subantarctic water of the Malvinas Current flowing along the shelf-break front (Fernández Aráoz et al. 1991; Santos and Ramírez 1991; Ramírez and Santos 1994; Cepeda et al. 2012b; Viñas et al. 2013a and references therein). It is worth highlighting that the geographical extension of each FA is strongly dependent on the distribution of water masses.

For each FA a set of representative species occurs frequently. The balance among their degrees of tolerance to the environmental variability, their ecological preferences, the productive features of each inhabiting system as well as the relationships among species determine the contribution of each species to defining the assemblages. The eurytopic *Acartia tonsa* is the most typical species of the estuarine FA (Viñas et al. 2013a and references therein). Although this species is found along the inner shelf as far as 51°S, it is the absolutely dominant copepod at the turbidity maximum of the inner Rio de la Plata estuary (Derisio et al. 2014a). A gradual decrease in the relative abundance of *A. tonsa* can be observed across shelf coinciding with a reduction of Rio de la Plata influence (Fig. 6a). Other small-sized but less abundant euryhaline species, including *Corycaeus amazonicus*, *Temora turbinata*, *Temora stylifera*, *Hemicyclops thalassius* and *Oncaea* spp., can occur in these nutrient-rich waters of reduced salinity (Cepeda et al. unpublished). Large amounts of suspended material, detritus and dissolved substances promote high concentrations of cyanobacteria, bacterioplankton and ciliates constituting a well-developed microbial food web (e.g. Carreto et al. 2007; Negri et al. 2016), making available a proper and varied food supply for these copepods (e.g. Derisio et al. 2014a).

Usually, a great proportion of small-sized copepod species, particularly *O. nana*, *Paracalanus parvus*, *Microsetella norvegica* and *Euterpina acutifrons*, composes the copepod assemblage of the inner-shelf FA (Fig. 6a). These species in general and especially *O. nana* dominate the inner-shelf mesozooplankton community numerically throughout the year (Viñas et al. 2013b), reproducing continuously (Temperoni et al. 2011). This vertically homogeneous and regenerative system exhibits relatively lower phytoplankton abundance than the contiguous middle-shelf FA (Carreto et al. 1995). Even though microphytoplankton are, on average, the main carbon contributors over the year, there is a remarkable picophytoplankton abundance (> 50% of the total phytoplankton biomass) in summer and early autumn (Silva et al. 2009; Viñas et al. 2013b). The dominance and intense reproduction of the microbial filter feeders *O. nana* and *Paracalanus* spp. observed during this period suggest that the system provides adequate food sources for the development of these small copepod species that typically feed upon small particles. High abundances of ciliates and flagellates have also been reported in this system (Silva et al. 2009).

Medium-sized and larger copepods, including *C. vanus*, *D. forcipatus*, *C. carinatus*, *C. australis* and *C. simillimus*, along with the cyclopoids *O.* aff. *helgolandica* and *O. atlantica*, commonly occur in the middle-shelf FA (Fig. 6a). These herbivorous, filter-feeding calanoids likely exploit the well-defined phytoplankton maxima occurring in this system during spring, when diatoms dominate (Carreto et al. 1995). In fact, an absolute dominance of diatoms was observed in the diet of adult females of *C. vanus* and *D. forcipatus*. More variety, including also dinoflagellates, silicoflagellates, tintinnids and microcrustaceans, characterizes the diets of *C. carinatus* and *C. australis* (Santos B. unpublished data).

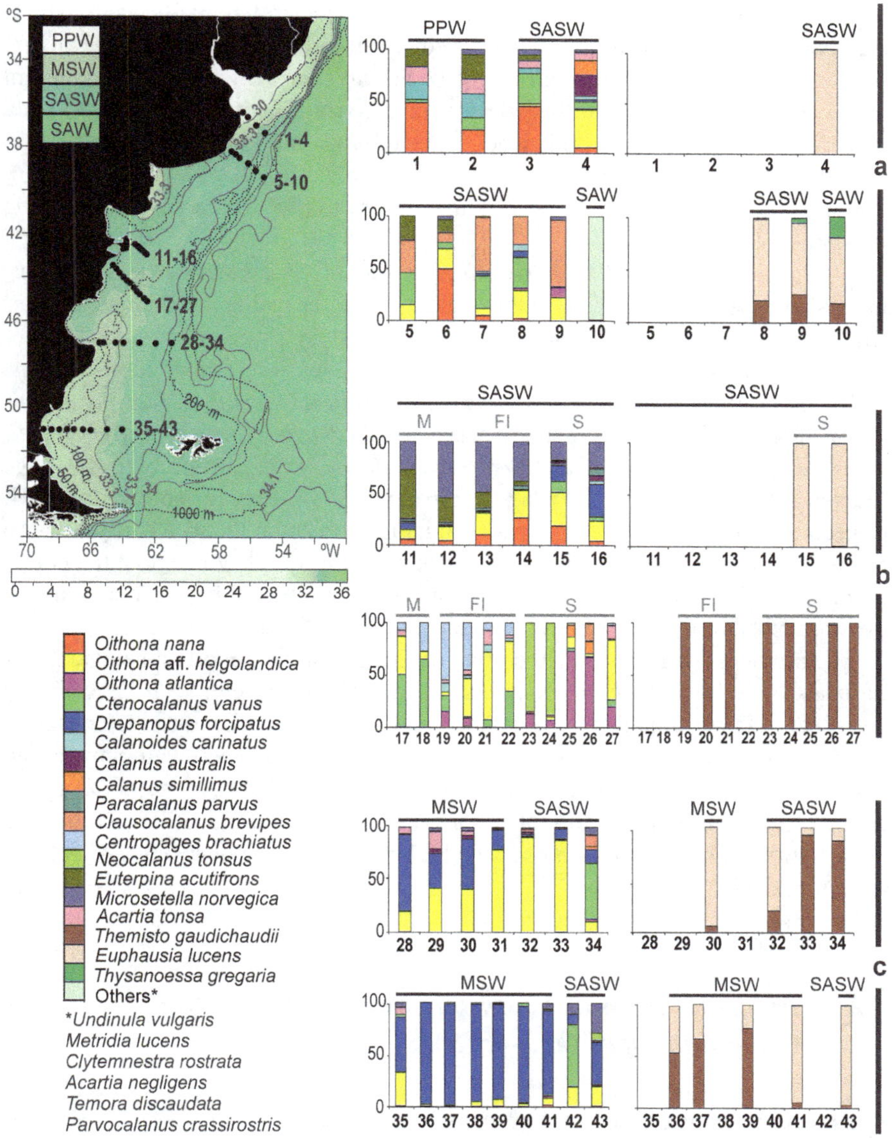

Fig. 6 Relative abundance of copepod and macrozooplankton species along across-shelf transects (**a**) the northern shelf (34°–41°S), (**b**) the Valdés frontal system over the northern Patagonian shelf (41°–45°S) and (**c**) the southern Patagonian shelf (47°–55°S). The map shows the horizontal distribution of annual average surface salinity (modified from Baldoni et al. (2015)) with the corresponding water masses: PPW Plata plume water, MSW Magallanes Strait water, SASW subantarctic shelf water, SAW subantarctic shelf water of the Malvinas Current. Symbols: M well-mixed waters, FI frontal interface and S stratified represent the three sectors of the Valdés tidal front

The cyclopoids *O.* aff. *helgolandica* and *O. atlantica* preferentially occur in middle-shelf waters (e.g. Ramírez 1970; Cepeda et al. 2012b; Viñas et al. 2013a). The relatively higher egg production rates of *O.* aff. *helgolandica* recently observed in saltier shelf waters than in freshened water inshore confirm this preference (Cepeda et al. 2015). In this sector, both species can rely on small phytoplankton such as dinoflagellates and ultraplanktonic forms (Carreto et al. 2007; Negri et al. 2016). These cyclopoids would thus occupy a different trophic niche from that of the medium-to-large calanoids.

The macrocrustaceans *T. gaudichaudii*, *E. lucens* and *T. gregaria*, in lesser proportions than copepods, also occur in this middle-shelf area and in the contiguous outer shelf (Viñas et al. 2013b and references therein) (Fig. 6a). A mix of cryophilic large-sized copepod species including *Subeucalanus longiceps*, *Rhincalanus nasutus*, *Rhincalanus gigas*, *Clausocalanus laticeps*, *Scolecithricella minor*, *Neocalanus tonsus* and *Metridia lucens* and subtropical species like *Undinula vulgaris*, *Mecynocera clausi*, *Eucalanus elongatus* and *Scolecithrix danae* occurs in the outer-shelf FA (Santos and Ramírez 1991; Fernández Aráoz et al. 1994; Ramírez and Santos 1994, Cepeda et al. unpublished). This species mixture appears in the data partly because of the commonly adopted, oblique zooplankton net tows. They integrate the cooler subantarctic waters of the Malvinas Current in deeper levels of the water column and the warmer subtropical waters of the Brazil Current in the upper levels.

The northern shelf system is the principal habitat of the important pelagic fishes of the ACS, i.e. anchovy *E. anchoita* and Atlantic chub mackerel *S. colias* (Sánchez and Bezzi 2004). Calanidae, Oithonidae and Clausocalanidae species constitute dominant prey in the diets of adult *E. anchoita*, along with hyperiid amphipods and euphausiids (Angelescu 1982; Padovani et al. 2011). *Calanoides carinatus* and *T. gaudichaudii* are the main prey of adult *S. colias* (Viñas et al. 1999) and juvenile demersal fishes such as *M. hubbsi*, which also prey upon euphausiids (Angelescu and Prenski 1987).

3.2 The Valdés Frontal Ecosystem over the Northern Patagonian Shelf (41°–45°S)

A distinctive characteristic of the northern Patagonian shelf is the seasonal development of a major tidal front that becomes a highly productive ecosystem (e.g. Carreto et al. 2007). The front starts developing during austral spring as a thermocline originates offshore, and it persists until autumn when stratification weakens. The strongest offshore stability is reached in summer. Formation of the front is controlled by tidal currents and winds, enhanced by inshore headlands. The system develops within a single water mass of typical SASW (Piola et al. 2018), and it is characterized by a strong horizontal temperature gradient as the thermocline intersects the surface and bottom layers, separating vertically homogeneous waters inshore from stratified layering offshore. Consequently, three different sectors can be distinguished across the shelf: well-mixed waters nearshore, the frontal interface and stratified waters offshore. The average position of the front estimated over a decade follows an overall NE–SW alignment along the bathymetry, although exhibiting strong interannual variability (Ehrlich et al. 2000; Pisoni et al. 2015). There is also latitudinal variability in the intensity of the frontal signal both at the surface and the bottom (Sabatini and Martos 2002). Maximum development occurs off the Valdés Peninsula. The variability makes this front an intensely dynamic biological system.

The Valdés tidal front was first described in relation to a harmful red tide event that caused the deaths of some fishermen (Carreto et al. 1981a). Since then much research has been conducted, with several studies addressing crustacean zooplankton issues (Santos and Ramírez 1995; Viñas and Ramírez 1996; Sabatini and Martos 2002; Spinelli et al. 2012; Derisio et al. 2014b; Temperoni et al. 2014). These studies show that zooplankton communities – mainly dominated by copepods – vary year to year, depending on both physical and biological forcing. In the northern area of the system, where the three sectors are typically well developed, a close inspection of the frontal structure along two transects across the shelf shows that species lists of medium- and large-sized zooplankton (150 and 300 μm nets) are about the same in all three sectors, while their relative abundances are variable. Higher numbers are found at the frontal interface and in stratified waters, leading to different assemblages (Sabatini and Martos 2002; Derisio et al. 2014b; Temperoni et al. 2014). The community structure is strongly dependent on the physics of the system which has large interannual variability. Overall, in well-mixed waters and at the frontal interface, calanoid copepods are mainly represented by *D. forcipatus*, *C. vanus, P. parvus*, *C. australis* and *C. carinatus*, while in stratified waters the cyclopoids *O.* aff. *helgolandica* and *O. atlantica* are dominant. *Calanoides carinatus* is progressively replaced offshore by *C. simillimus* and *C. australis.* Most of these dominant calanoid copepods are known to be mainly herbivorous (Santos B. unpublished data), *C. carinatus* and *C. vanus* being indicators of highly productive systems (Sabatini and Martos 2002). Adults of *T. gaudichaudii* and *E. lucens* are relatively more abundant offshore (Fig. 6b).

The array of fine and coarse nets (67 + 300 μm) reveals a somewhat different community pattern, with the prevalence of smaller species, such as the harpacticoids *M. norvegica* and *E. acutifrons* in the mixed sector and the cyclopoids *O. nana* and *O.* aff. *helgolandica* added to those among the dominants at the frontal interface and a more diverse community – with a larger contribution of calanoids – predominating in the stratified waters (Fig. 6b). In the southern part of the system off Escondida Island, the front usually occurs very close to shore, and the mixed and frontal sectors are much narrowed as a result. There, copepod abundance either increases from less-stratified waters close to shore to strongly stratified waters seawards or is higher in the frontal interface, where *D. forcipatus*, *C. vanus* and *C. carinatus* are largely dominant (Sabatini and Martos 2002; Temperoni et al. 2014).

Latitudinal differences in the physical forcing (mainly tidal dissipation and winds) approaching either the northern or southern boundaries of the system may lead to contrasting plankton communities. A prevalence of relatively smaller copepods is recorded northwards, independent of the mesh size used for sampling. Closer to 41°S, a high proportion of matter and energy seems to be channelled through a microheterotrophic food web. Approaching 44°S larger copepods become dominant, and a more classical herbivorous food web may be expected (Sabatini and Martos 2002). Changes in copepod abundances are also observed throughout the summer (December to February–March), with decreasing numbers from spring to autumn as a consequence of the increased stability of the system (Temperoni et al. 2014).

Particular hydrometeorological conditions appear to strongly affect both the physical and biological structures of the system. Certain conditions of stability, nutrient availability and radiation favour the occurrence of harmful red tide events (Carreto et al. 1981a), which may negatively affect copepod populations. Characteristics of these harmful algal species such as toxicity, production of inhibitory substances and luminescence may cause the episodes of copepod exclusion occasionally reported in the northern area (e.g. Santos and Ramírez 1995). On the other hand, anomalous warming events in the system have led to massive blooms of gelatinous zooplankton (ctenophores) and of the phagotrophic dinoflagellate *Noctiluca*, with the concurrent predatory diminishment of copepod populations (Sabatini and Martos 2002 and references therein).

Predation is also imposed on zooplankton communities in this system by early stages of *E. anchoita* and *M. hubbsi*, which have their spawning and nursery grounds closely associated with the Valdés tidal front. Larvae of both fish species include species of Calanidae and Clausocalanidae in their diets (Ciechomski and Weiss 1974; Viñas and Ramírez 1996; Viñas and Santos 2000; Temperoni and Viñas 2013), whereas juveniles and adults prey mainly upon *E. lucens* and *T. gaudichaudii* (Angelescu and Prenski 1987; Belleggia et al. 2014).

The extent of the water column stratification is likely the main factor responsible for the zooplankton patterns across and along the Valdés tidal front, although other physical and biological processes, such as transport, retention, species-specific life history strategies and trophic relationships, surely also condition each species success in this complex system (Derisio et al. 2014b and references therein). For example, the frontal interface appears to represent a breeding area for copepods, since maximum nauplius abundance (Viñas and Ramírez 1996; Derisio et al. 2014b) and highest calanoid egg production rates (Sabatini and Martos 2002; Derisio 2012) have been reported there, as well as higher cyclopoid egg production rates and copepodid abundance (Cepeda 2013).

3.3 The Southern Patagonian Shelf Ecosystem (47°–55°S)

Circulation over the southern shelf is driven by large freshwater inflows, high tidal amplitudes, the predominance of westerly winds and the strong influence of the Malvinas Current flowing northwards along the shelf-break (Piola et al. 2018 and references therein). Most of the system is dominated by a distinct, low-salinity, surface plume ($S < 33.4$) associated with the discharge from the Magallanes Strait (MSW). The outer shelf is much more influenced by the colder, more saline

($S > 34$) subantarctic water (SAW) of the Malvinas Current along the seaward boundary. Separating these two, the SASW extends over the middle shelf. The SASW is a relatively fresh variant of subantarctic water that is injected into the region through Le Maire Strait and along the shelf break, after passing around Hornos Cape (Piola et al. 2018).

A variety of fronts occurs over this portion of the ACS: tidal mixing fronts, estuarine and river plumes, water mass convergences and a shelf-break front (Piola et al. 2018). Modelling experiments suggest a cross-shelf circulation due to the dynamic interaction of fronts, which probably includes the inflow of nutrient-rich waters from offshore (Piola et al. 2018). In turn, a major source of the waters on the southern ACS shelf and in the offshore zone is found at the depth in the southeast Pacific Ocean. These waters, rich in both nitrate and iron, are primed to support new (as opposed to regenerated) productivity along the entire Patagonian shelf (Song et al. 2016).

The hydrographic diversity described above translates into habitat heterogeneity for plankton communities in terms of both nutrient and prey availability, eventually promoting dissimilar food web structures. Water masses over the southern Patagonian shelf have distinctive mesozooplankton assemblages that persist in all seasons (Sabatini et al. 2012, 2016). The dissimilarity among assemblages associated with specific water masses actually increases with increasing geographical distance on the ACS. Correspondingly, the MSW and SAW mesozooplankton communities are clearly different, while boundaries between the MSW and the SASW assemblages are less definite. Because the SASW and SAW water masses are so closely related in origin, their associated communities are very alike (Fig. 6c). Species abundance differences resulting from distinctive life histories and population development seem to be more important in defining the assemblages than simply the presence or absence of particular species. The differences between the MSW and SASW mesozooplankton communities relate to changes in the relative dominance of a few shared species, rather than to fully distinctive taxonomic compositions, for example, average abundances of shared taxa are often considerably lower in the SASW. The middle-shelf SASW community may be characterized as an ecotone assemblage, i.e. to some extent related to the MSW community, but also with strong contributions of species more fully developed in the SAW. With varying abundances through the seasons, major components for the MSW are adults and late copepodids of *D. forcipatus*, copepodids C5 and adult females of *C. australis* and the amphipod *T. gaudichaudii.* The SASW is characterized by copepodids C4–5 of *D. forcipatus*; females and late copepodids of *C. vanus*, *Clausocalanus brevipes* and *C. simillimus*; the cyclopoids *O.* aff. *helgolandica* and *O. atlantica*; *T. gaudichaudii*; and euphausiid juveniles. Most of the latter species are also typical of the SAW assemblage. Epipelagic seasonal migrants such as *N. tonsus* C5, *Subeucalanus longiceps* and *M. lucens* are also recorded, though in low numbers, over the outer shelf near the slope (Ramírez 1981; Ramírez and Sabatini 2000). Because of the wide extent of the continental shelf off southern Patagonia, the SAW community has only minor representation in the area that has been historically surveyed.

The significant contributions of small-sized species, such as *Oithona* aff. *helgolandica* and *Microsetella norvegica*, have only recently become evident from sampling with fine-mesh nets (Antacli et al. 2010, 2014b). In terms of both numerical abundance and occurrence, these copepods occupy, respectively, the second and third places in the community after *D. forcipatus*, which is consistently the most conspicuous component of the mesozooplankton in the southern ACS system, by far outnumbering any other species across all seasons. *Ctenocalanus vanus* appears as the fourth most numerous species, mostly concentrated in offshore waters, while the relative numerical importance of the larger *C. australis* seems to be less than established previously using coarser nets (Fig. 6c).

The seasonal development of plankton communities over the southern Patagonian shelf is typical of cold temperate regions, with a clear seasonal signal in mesozooplankton abundance long after the spring phytoplankton bloom (Carreto et al. 1981b; Ramírez 1981; Sabatini and Álvarez Colombo 2001; Sabatini et al. 2016). Most copepod species show different population structures in the northerly and southerly areas of the southern Patagonian shelf. North versus south differences also apply to the feeding and reproductive activities of the *D. forcipatus* and *C. australis* populations in particular (Antacli et al. 2014a; Sabatini et al. 2016). This is likely as much due to the importance of temperature as to food. In addition to seasonal warming and cooling, the temperature has a clear latitudinal gradient along the southern Patagonian shelf, colder southward (Sabatini et al. 2004).

Total mesozooplankton abundance increases about 2.5-fold from the beginning of spring to late summer and then decreases at least four orders of magnitude in winter. In all seasons, copepods represent >70–80% of all mesozooplankton over most of the southern Patagonian shelf system. Although seasonal differences in abundance are striking, the spatial distribution of mesozooplankton is largely similar across seasons, with relatively higher concentrations occurring mainly in Grande Bay and its surroundings. Biomasses of primary producers appear to be locally enhanced in that area throughout spring and summer (Lutz et al. 2010; Dogliotti et al. 2014), and large mesozooplankton biomasses are recurrently recorded by the end of the productive season (Sabatini and Álvarez Colombo 2001; Sabatini et al. 2004; Sabatini 2008; Antacli et al. 2014b). The influences of the Malvinas Current and strong tidal forcing on the local frontal dynamics seem to strengthen in Grande Bay and are likely generating a nutrient-rich, recirculating area (Sabatini et al. 2004; Piola et al. 2018), where a significant proportion of production can be retained at temporal scales that allow for the development of abundant primary and secondary producers (Sabatini et al. 2016).

Less work has been conducted at the population level, with the focus on a few key species: the medium- and large-sized copepods *D. forcipatus* and *C. australis* and the amphipod *T. gaudichaudii*. After high production during spring and early summer, late summer marks the beginning of a less productive season on the southern Patagonian shelf (e.g. Lutz et al. 2010 and references therein), with the prevalence of a microbial trophic web (e.g. Antacli et al. 2014a) and, thus, food-limiting conditions for copepods. By then the populations of the two dominant copepods are largely lipid-storing C4 and C5 copepodids, and the species distribute differently in the water column. The bulk of *D. forcipatus* population is concentrated in the upper layers down to *ca.* 50 m, whereas *C. australis* are deeper, even close to the bottom. Sampling results suggest that while *D. forcipatus* is still feeding in the upper water column, the bulk of the *C. australis* population may be entering a period of arrested metabolism at depth and thus is not feeding or scarcely so (Sabatini 2008). Actually, late summer adult females and late copepodids of *D. forcipatus* and *C. australis* all display overall low feeding activity, although it is relatively greater in the former species. Also, reproductive activity is much reduced, yet both populations show some ongoing spawning (Antacli et al. 2014a). Diets of adult females during late summer, based on gut-content analysis, indicate for *D. forcipatus* opportunistic feeding on the smaller but more abundant particles in the environment. Gut contents of *C. australis* suggest relatively more ingestion of autotrophic prey, particularly large diatoms. The ability to ingest small food particles could provide *D. forcipatus* an important advantage over the larger copepods and perhaps is a reason for its overwhelming numerical abundance in some areas of the southern Patagonian shelf (Antacli et al. 2014a).

The hyperiid *T. gaudichaudii* reaches its highest biomass in this system (Sabatini and Álvarez Colombo 2001; Padovani et al. 2015), where its populations develop on the bases of very high concentrations of copepods as food and adequate water temperatures. Very large biomasses of this amphipod have been particularly reported from the Grande Bay area over the inner and mid-shelf, mainly in summer–autumn (Sabatini and Álvarez Colombo 2001). The local population has a more extended reproductive period (spring–summer), a larger number of cohorts (two main ones) and smaller size at maturity (around 10 mm) than other *T. gaudichaudii* populations at similar latitudes. These adaptations would favour larger abundances, placing the southern Patagonian shelf system among the most favourable for the development of this amphipod anywhere in its global distribution range (Padovani et al. 2015). *T. gaudichaudii* strongly supports the community of planktivorous fish and squid in the area (Ivanovic and Brunetti 1994; Padovani et al. 2012). Due to this trophic relevance, *T. gaudichaudii* has been postulated as a "wasp-waist" species, channelling the energy flow in a short and efficient food chain, role similar to that of krill in Antarctic waters.

4 Gaps, Open Questions and Prospects for Future Work

As shown in this review, exhaustive taxonomic work, based chiefly on traditional morphological techniques and coupled with a (or the) strong understanding of the association between zooplankton species and hydrographic conditions, sustains our current knowledge of their communities in the ACS. However, there are many important issues remaining poorly understood or altogether unexplored.

Taxonomic uncertainties on the specific identification of the few animals appearing to be keystone species affect this synopsis, with the possibility of similar situations for many other species along the ACS. In this sense, there is a strong need to combine morphological and genetic studies and to engage several other disciplines (e.g. physiology, behavioural ecology) in what is nowadays known as integrative taxonomy. Thus, we could acquire more precise and unequivocal species definitions. Under the current global warming, it has become practically mandatory to determine the number of species and their distribution limits in order to evaluate further possible biogeographical changes. Without a proper description of biodiversity and its functioning over time, it is difficult to ensure appropriate ecosystem management. In this regard, a major secondary product of this review has been the construction of a quantitative data set on copepod species along the entire ACS, which is now suitable for further, in-depth analysis of diversity.

Furthermore, little to nothing is known about the population dynamics for most of the key species. Future directions should focus on the details of zooplankton life cycles, such as generation times, growth rates, reproductive biology, secondary production as well as life cycle strategies (i.e. dormancy, resting eggs).

Vast information exists about the importance of copepods, hyperiid amphipods and euphausiids as main prey for higher trophic levels. However, nutritional studies focused on the nutritional quality of these species as foods are still scarce. Development is currently in progress of biochemical tools to establish the energetic density available in the lipid content of each major zooplanktonic prey, as well as their contents of long-chain polyunsaturated fatty acids, including the essential lipids EPA and DHA.

Integrated studies of the processes influencing the pelagic ecosystem are also lacking. Specifically, it is necessary to quantify energy flow through zooplankton communities, as well as to assess their functional diversity (i.e. how organisms interact with each other) for each system. No attention has been paid yet to ecosystem modelling for the ACS, which in the future will be extremely useful for predicting how upcoming cyclic or non-cyclic changes will affect its distinct zooplankton communities.

Although not mentioned in detail in this article, several zooplankton time series based mainly on abundance data are currently being collected along the ACS. Such time series are essential to assess the long-term effect of climate changes on planktonic communities. However, abundance data alone are often insufficient for understanding regime shifts and their consequences for higher trophic levels. More information is needed on biomass and size spectra. In the short term, the addition of modern semi-automatic analysis methods, currently in progress, will improve our capability to fill this gap. These types of field studies should be associated with experimental work directed to assess the effects of environmental variables, singly and together, on zooplankton dynamics.

Acknowledgements We would like to express our gratitude to the countless scientists, technicians and research vessel crews of INIDEP and CONICET who have for 50 years helped build our present understanding of zooplankton communities in the ACS. Their effort, enthusiasm and expertise are the basis of the results reported here. We would also like to acknowledge the contributions by many colleagues who, for reasons of space limitations, have not been referenced in these pages. We are grateful in particular to Gustavo Álvarez Colombo, Roxana Di Mauro, Nora Fernández Aráoz and Marina Marrari for kindly providing their data for the zooplankton datasets assembled throughout this review. Special thanks must be given to the heads of the several assessment groups at INIDEP for receiving and helping plankton researchers on board their stock assessment cruises. Also, the facilities at INIDEP and numerous grants from CONICET and the Universidad Nacional de Mar del Plata have been essential for the completion of our work. This is INIDEP contribution N°2099.

References

Aguirre GE, Capitanio FL, Lovrich GA (2012) Seasonal variability of mesozooplankton in coastal sub-Antarctic waters (Beagle Channel). Mar Biol Res 8(4):341–353

Angelescu V (1982) Ecología trófica de la anchoíta del Mar Argentino (Engraulidae, *Engraulis anchoita*). Parte II. Alimentación, comportamiento y relaciones tróficas en el ecosistema. Ser Contrib Inst Nac Invest Desarr Pesq 409, 83 pp

Angelescu VA, Prenski LB (1987) Ecología trófica de la merluza común del Mar Argentino (Merlucciidae, *Merluccius hubbsi*). Parte 2. Dinámica de la alimentación analizada sobre la base de las condiciones ambientales, la estructura y las evaluaciones de los efectivos en su área de distribución. Ser Contrib Inst Nac Invest Desarr Pesq 561, 205 pp

Antacli JC, Hernández D, Sabatini ME (2010) Estimating copepods' abundance with paired nets: implications of mesh size for population studies. J Sea Res 63:71–77

Antacli JC, Sabatini ME, Silva RI (2014a) Feeding and reproductive responses of the copepods *Drepanopus forcipatus* and *Calanus australis* to ambient food limitation during late summer over the southern Patagonian shelf (Argentina, 47°–55°S). Braz J Oceanogr 62:295–314

Antacli JC, Hernández D, Sabatini ME (2014b) First report on the contribution of small sized species to the copepod community structure of the southern Patagonian shelf (Argentina, 47°–55°S). Sci Mar 78:17–26

Baldoni A, Molinari GN, Reta R et al (2015) Atlas de temperatura y salinidad de la plataforma continental del Atlántico Sudoccidental: períodos cálido y frío. Instituto Nacional de Investigación y Desarrollo Pesquero, Mar del Plata

Belleggia M, Figueroa DE, Irusta G et al (2014) Spatio-temporal and ontogenetic changes in the diet of the Argentine hake *Merluccius hubbsi*. J Mar Biol Assoc 94(8):1701–1710

Björnberg T (1981) Copepoda. In: Boltovskoy D (ed) Atlas del Zooplancton del Atlántico Sudoccidental y Métodos de Trabajo con el Zooplancton Marino. Publicaciones Especiales INIDEP, Mar del Plata, pp 587–679

Bradford-Grieve JM (1994) Pelagic Calanoid Copepoda: Megalanidae, Calanidae, Paracalanidae, Mecynoceridae, Eucalanidae, Spinocalanidae, Clausocalanidae. NZ Oceanogr Inst Mem 102:1–160

Bradford-Grieve JM, Markhaseva EL, Rocha CEF et al (1999) Copepoda. In: Boltovskoy D (ed) South Atlantic Zooplankton. Backhuys Publishers, Leiden, pp 869–1098

Boltovskoy D (ed) (1999) South Atlantic Zooplankton. Backhuys Publishers, Leiden

Boltovskoy D, Correa NM, Boltovskoy A (2003) Marine zooplanktonic diversity: a view from the South Atlantic. Oceanol Acta 25:271–278

Boltovskoy D, Correa NM (2008) Zooplancton: biogeografía y diversidad. In: Estado de Conservación del Mar Patagónico y Áreas de Influencia. Publicación del Foro, Puerto Madryn. Available at: http://www.marpatagonico.org

Brodsky KA (1959) On the phylogenetic relationship of certain species of *Calanus* (Copepoda) from the northern and southern hemispheres. Zool J Acad Sci URSS 38:1537–1553

Brodsky KA (1961) Comparison of *Calanus* species (Copepoda) from the southern and northern hemispheres. New Zeal Ocean Inst Contr 95:1–22

Bucklin A, Frost BW, Bradford-Grieve J et al (2003) Molecular systematic and phylogenetic assessment of 34 calanoid copepod species of the Calanidae and Clausocalanidae. Mar Biol 142:333–343

Bucklin A, Nishida S, Schnack-Schiel S et al (2010) A census of zooplankton of the global ocean. In: McIntyre AD (ed) Life in the World's oceans. Diversity, distribution, and abundance. Blackwell Publishing, Oxford, pp 247–262

Carreto JI, Lasta ML, Negri RM et al (1981a) Los fenómenos de marea roja y toxicidad de moluscos bivalvos en el Mar Argentino. Contr Inst Nac Invest Desarr Pesq 399. 89 pp

Carreto JI, Negri RM, Benavides HR (1981b) Fitoplancton, pigmentos y nutrientes. Resultados campañas III y IV de B/I "Shinkai Maru" 1978. In: Angelescu V (ed) Campañas de investigación pesquera en el Mar Argentino, años 1978 y 1979. Ser Contrib Inst Nac Invest Desarr Pesq 399, 89 pp

Carreto JI, Lutz VA, Carignan MO et al (1995) Hydrography and chlorophyll *a* in a transect from the coast to the shelf-break in the Argentinean Sea. Cont Shelf Res 15:315–336

Carreto JI, Carignan MO, Montoya NG et al (2007) Ecología del fitoplancton en los sistemas frontales del Mar Argentino. In: Sánchez RP, Bezzi SI (eds) El Mar Argentino y sus recursos pesqueros, 5. El ecosistema marino. Publicaciones Especiales INIDEP, Mar del Plata, pp 11–31

Cepeda GC (2013) Sistemática molecular, distribución y dinámica poblacional de las especies de *Oithona* (Copepoda, Cyclopoida) en diversos sectores costeros del Mar Argentino. Dissertation, Universidad Nacional de Mar del Plata

Cepeda GC, Blanco-Bercial L, Bucklin A et al (2012a) Molecular systematic of three species of *Oithona* (Copepoda, Cyclopoida) from the Atlantic Ocean: comparative analysis using 28S rDNA. PLoS One 7(4):1–7

Cepeda GC, Di Mauro R, Martos P et al (2012b) A section-sampling design to assess copepods and cladocerans distribution patterns in a highly complex hydrographic region (SW Atlantic Ocean, 34–41°S): insights for a new time series? Braz J Oceanogr 60(3):381–390

Cepeda GC, Di Mauro R, Hozbor MC et al (2015) Spatial variation in life-history traits of *Oithona* spp. in a shallow temperate estuarine system (Río de la Plata, south-west Atlantic) during spring. Mar Freshw Res 66(9):795–804

Cepeda GC, Sabatini ME, Scioscia CL et al (2016) On the uncertainty beneath the name *Oithona similis* Claus, 1866 (Copepoda, Cyclopoida). Zookeys 552:1–15

Ciechomski JD, Weiss G (1974) Estudios sobre la alimentación de larvas de la merluza *Merluccius merluccius hubbsi* y de la anchoita *Engraulis anchoita* en el mar. Physis 33(86):199–208

Derisio C (2012) El rol del frente de mareas de Península Valdés en el control de la comunidad zooplanctónica. Dissertation, Universidad Nacional de Mar del Plata

Derisio C, Braverman M, Gaitán E et al (2014a) The turbidity front as a habitat for *Acartia tonsa* (Copepoda) in the Río de la Plata, Argentina–Uruguay. J Sea Res 85:197–204

Derisio C, Alemany D, Acha EM et al (2014b) Influence of a tidal front on zooplankton abundance, assemblages and life histories in Península Valdés, Argentina. ICES J Mar Sci 139:475–482

Dogliotti AI, Lutz VA, Segura V (2014) Estimation of primary production in the southern Argentine continental shelf and shelf-break regions using field and remote sensing data. Remote Sens Environ 140:497–508

Ehrlich MD, Martos P, Madirolas A et al (2000) Causes of spawning pattern variability of anchovy and hake on the Patagonian shelf. ICES CM 2000/N: 06, 13 pp

Falabella V, Campagna C, Croxall J (eds) (2009) Atlas of the Patagonian Sea. Species and spaces. Wildlife Conservation Argentina, Buenos Aires and BirdLife International, Cambridge

Fernández Aráoz NC, Pérez Seijas GM, Viñas MD et al (1991) Asociaciones zooplanctonicas de la Zona Común de Pesca argentino-uruguaya en relación con parámetros ambientales. Primavera 1986. Frente Marit 8:85–99

Fernández Aráoz N, Santos B, Ramírez F (1994) Análisis ecológico de la distribución de los copépodos planctónicos de una campaña de primavera en la zona común de pesca. Frente Marit 15(A) 412: 133–140

Fernández-Severini MD, Hoffmeyer M (2005) Mesozooplankton assemblages in two bays in the Beagle Channel (Argentina) during January 2001. Sci Mar 69(Suppl. 2):27–37

Hulsemann K (1985) Two species of *Drepanopus* Brady (Copepoda, Calanoida) with discrete ranges in the southern hemisphere. J Plankton Res 7:909–925

Ivanovic ML, Brunetti NE (1994) Food and feeding of *Illex argentinus*. Antarct Sci 6:185–193
Longhurst AR (2007) Toward an ecological geography of the sea. In: Longhurst AR (ed) Ecological geography of the sea. Academic, London, pp 1–17
Lutz VA, Segura V, Dogliotti AI et al (2010) Primary production in the Argentine Sea during spring estimated by field and satellite models. J Plankton Res 32:181–195
Magalhães A, Leite ND, Silva JGS et al (2009) Seasonal variation in the copepod community structure from a tropical Amazon estuary, Northern Brazil. An Acad Bras Ciênc 81(2):187–197
Mauchline J, Fisher LR (1969) The biology of euphausiids. Adv Mar Biol 7:1–421. Academic, London
Mauchline J (1980) The biology of mysids and euphausiids. Adv Mar Biol 18:1–681. Academic, London
Mazzochi MG, Zagani G, Ianora A et al (1995) Copepods. In: Guglielmo L, Ianora A (eds) Atlas of marine zooplankton, Straits of Magellan. Springer, Berlin, pp 166–170
McPherson E (2002) Large-scale speciesrichness gradients in the Atlantic Ocean. Proc R Soc Lond B 269:1715–1720
Negri RM, Molinari G, Carignan M et al (2016) Ambiente y Plancton en la Zona Común de Pesca Argentino–Uruguaya en un escenario de cambio climático (marzo, 2014). Frente Marit 24:251–316
Neumann-Leitão S, Sant'anna EME, Gusmão LMO et al (2008) Diversity and distribution of the mesozooplankton in the tropical Southwestern Atlantic. J Plankton Res 30(7):795–805
Nishida S (1985) Taxonomy and distribution of the family Oithonidae (Copepoda: Cyclopoida) in the Pacific and Indian Oceans. Bull Ocean Res Inst Univ Tokyo 20:1–167
Padovani LN (2013) Biodiversidad y ecología de los anfípodos hiperideos del Mar Argentino y aguas adyacentes: *Themisto gaudichaudii*, una especie clave. Dissertation, Universidad Nacional de Mar del Plata
Padovani LN, Viñas MD, Pájaro M (2011) Importance of the Rio de la Plata estuarine front (Southwest Atlantic Ocean) in the feeding ecology of Argentine anchovy, *Engraulis anchoita* (Cupleiformes, Cupleidae). Lat Am J Aquat Res 39:205–213
Padovani LN, Viñas MD, Sánchez F et al (2012) Amphipods-supported food web: *Themisto gaudichaudii*, a key food resource for fishes in southern Patagonian Shelf. J Sea Res 67(1):85–90
Padovani LN, Viñas MD, Sabatini ME et al (2015) Dinámica poblacional de *Themisto gaudichaudii*, una especie clave en la trama trófica de la plataforma patagónica austral. Rev Invest Desarr Pesq 26:69–88
Pallares RE (1968) Copépodos marinos de la Ría Deseado (Santa Cruz, Argentina). Contribución sistemático-ecológica. Servicio de Hidrografía Naval H.1024:1–125
Piola AR, Castro BM, Guerrero RA et al (2018). Overview on water masses, fronts and circulation of the subtropical and subantarctic shelves of the western South Atlantic. In: Hoffmeyer M (ed) Plankton ecology of Atlantic South America. From the subtropical to the subantarctic realm. Springer, Heidelberg
Pisoni JP, Rivas AL, Piola AR (2015) On the variability of tidal fronts on a macrotidal continental shelf, Northern Patagonia, Argentina. Deep-Sea Res II 119:61–68
Ramírez FC (1966) Copépodos Cyclopoidos y Harpacticoidos del plancton de Mar del Plata. Buenos Aires: Asociación Argentina de Ciencias Naturales. Physis 26(72):285–292
Ramírez FC (1970) Copépodos planctónicos del sector patagónico: resultados de la campaña 'Pesquería XI. Physis 29:473–476
Ramírez FC (1971) Eufáusidos de algunos sectores del Atlántico Sudoccidental. Physis 30(81):385–405
Ramírez FC (1973) Eufáusidos de la expedición oceanográfica "Walther Herwig" 1966. Buenos Aires: Asociación Argentina de Ciencias Naturales. Physis 32(84):105–114
Ramírez FC (1981) Zooplancton y producción secundaria. Parte I. Distribución y variación estacional de los copépodos. In: Angelescu V (ed) Campañas de investigación pesquera realizadas en el Mar Argentino por los B/I "Shinkai Maru" y "Walter Herwig" y el B/P "Marburg", años 1978 y 1979. Resultados de la parte Argentina. Ser Contrib Inst Nac Invest Desarr Pesq 383:202–212

Ramírez FC (2016) Eufáusidos. In: Boschi EE (ed) El Mar Argentino y sus recursos pesqueros, 6. Los crustáceos de interés pesquero y otras especies relevantes en los ecosistemas marinos. Publicaciones Especiales INIDEP, Mar del Plata, pp 15–28

Ramírez FC, Viñas MD (1985) Hyperiid amphipods found in Argentine shelf waters. Physis A 43(104):25–37

Ramírez FC, Santos BA (1994) Análisis del zooplancton de la plataforma bonaerense, en relación con algunas variables ambientales: Campañas Transección de 1987. Frente Marít 15(A):141–156

Ramírez FC, Sabatini ME (2000) The occurrence of Calanidae species in waters off Argentina. Hydrobiologia 439:21–42

Rombouts I, Beaugrand G, Ibañez F et al (2009) Global latitudinal variations in marine copepod diversity and environmental factors. Proc R Soc B 276:3053–3062

Sabatini ME (2008) Life history trends of copepods *Drepanopus forcipatus* (Clausocalanidae) and *Calanus australis* (Calanidae) in the southern Patagonian shelf (SW Atlantic). J Plankton Res 30:981–996

Sabatini M, Álvarez Colombo G (2001) Seasonal pattern of zooplankton biomass in the Argentinian shelf off Southern Patagonia (45°–55°S). Sci Mar 65(1):21–31

Sabatini ME, Martos P (2002) Mesozooplankton features in a frontal area off northern Patagonia (Argentina) during spring 1995 and 1998. Sci Mar 66:215–232

Sabatini M, Reta R, Matano R (2004) Circulation and zooplankton biomass distribution over the southern Patagonian shelf during late summer. Cont Shelf Res 24:1359–1373

Sabatini ME, Ramírez FC, Bradford-Grieve J (2007) Redescription of *Calanoides carinatus* (Krøyer, 1848) (Copepoda, Calanoida, Calanidae), with a discussion on the status of related species. Invertebr Syst 21:341–364

Sabatini M, Akselman R, Reta R et al (2012) Spring plankton communities in the southern Patagonian shelf: hydrography, mesozooplankton patterns and trophic relationships. J Mar Syst 94:33–51

Sabatini ME, Reta R, Lutz V et al (2016) Influence of oceanographic features on the spatial and seasonal patterns of mesozooplankton in the southern Patagonian shelf (Argentina, SW Atlantic). J Mar Syst 157:20–38

Sánchez RP, Bezzi SI (eds) (2004) El Mar Argentino y sus recursos pesqueros. Tomo 4. Los peces marinos de interés pesquero. Caracterización biológica y evaluación del estado de explotación. Publicaciones Especiales INIDEP, Mar del Plata

Santos BA, Ramírez FC (1991) Variación estacional de las asociaciones de copépodos en una transecta de la plataforma argentina. Bolm Oceanogr São Paulo 39(1):71–86

Santos BA, Ramírez FC (1995) Distribución y abundancia de copépodos en el sistema frontal de Península Valdés durante florecimientos fitoplanctónicos. Thalassas 11:133–142

Sherman K, Duda AM (1999) An ecosystem approach to global assessment and management of coastal waters. Mar Ecol Prog Ser 10:271–287

Silva RI, Negri RM, Lutz V (2009) Summer succession of ultraphytoplankton at the EPEA coastal station (Northern Argentina). J Plankton Res 31:447–458

Song H, Marshall J, Follows MJ et al (2016) Source waters for the highly productive Patagonian shelf in the southwestern Atlantic. J Mar Syst 158:120–128

Spalding MD, Fox HE, Allen GR et al (2007) Marine ecoregions of the world: a Bioregionalization of coastal and shelf areas. Bioscience 57(7):573–583

Spinelli ML, Pájaro M, Martos P et al (2012) Potential zooplankton preys (Copepoda and Appendicularia) for *Engraulis anchoita* in relation to early larval and spawning distributions in the Patagonian frontal system (SW Atlantic Ocean). Sci Mar 76(1):39–47

Temperoni B, Viñas MD, Diovisalvi N et al (2011) Seasonal production of *Oithona nana* Giesbrecht, 1983 (Copepoda: Cyclopoida) in temperate coastal waters off Argentina. J Plankton Res 33(5):729–740

Temperoni B, Viñas MD (2013) Food and feeding of Argentine hake (*Merluccius hubbsi*) larvae in the Patagonian nursery ground. Fish Res 148:47–55

Temperoni B, Viñas MD, Martos P et al (2014) Spatial patterns of copepod biodiversity in relation to a tidal front system in the main spawning and nursery area of the Argentine hake *Merluccius hubbsi*. J Mar Syst 139:443–445

Valentine JW (2009) Overview of marine biodiversity. In: Witman JD, Roy K (eds) Marine macroecology. University of Chicago Press, Chicago, pp 3–28

Vinogradov G (1999) Amphipoda. In: Boltovskoy D (ed) Zooplankton of the Southwestern Atlantic. Backhuys, Leiden, pp 1141–1240

Viñas MD, Santos BA (2000) First-feeding of hake (*Merluccius hubbsi*) larvae and prey availability in the North Patagonian spawning area – comparison with anchovy. Arch Fish Mar Res 48:242–254

Viñas MD, Ramírez FC (1996) Gut analysis of first-feeding anchovy larvae from Patagonian spawning area in relation to food availability. Arch Fish Mar Res 43:231–256

Viñas MD, Sánchez F, Marrari M et al (1999) Zooplancton, hidrografía y ecología trófica de la caballa (*Scomber japonicus*) en el área de El Rincón (39–41°S). Paper presented at the 8th COLACMAR, Trujillo, Perú, 17–21 October 1999

Viñas MD, Ramírez FC, Santos B et al (2007) Spatial and temporal distribution patterns of Cladocera in the Argentine Sea. Hydrobiologia 594:59–68

Viñas MD, Marrari M, Di Mauro R et al (2013a) El zooplancton del hábitat reproductivo de la población bonaerense de anchoíta (*Engraulis anchoita*), con especial énfasis en crustáceos. Rev Invest Desarr Pesq 23:125–144

Viñas MD, Negri RM, Cepeda GC et al (2013b) Seasonal succession of zooplankton in coastal waters of the Argentine Sea (Southwest Atlantic Ocean): prevalence of classical or microbial food webs. Mar Biol Res 9(4):371–382

Viñas MD, Blanco-Bercial L, Bucklin A et al (2015) Phylogeography of the copepod *Calanoides carinatus* s.l. (Krøyer) reveals cryptic species and delimits *C. carinatus* s.s. distribution in SW Atlantic Ocean. J Exp Mar Biol Ecol 468:97–104

Viñas MD, Álvarez Colombo G, Padovani L (2016) Anfípodos Hipérideos. In: Boschi EE (ed) El Mar Argentino y sus recursos pesqueros, 6. Los crustáceos de interés pesquero y otras especies relevantes en los ecosistemas marinos. Publicaciones Especiales INIDEP, Mar del Plata, pp 29–39

WoRMS Editorial Board (2017) World register of marine species. Available from http://www.marinespecies.org at VLIZ. doi:https://doi.org/10.14284/170. Accessed 25 Sept 2017

Ecological Role of Common Appendicularian Species from Shelf Waters Off Argentina

Fabiana L. Capitanio, Mariela L. Spinelli, María L. Presta, Gastón E. Aguirre, Guillermo Cervetto, Marcelo Pájaro, and Carla M. Derisio

Abstract Appendicularians generally comprise a significant fraction of mesozooplanktonic tunicates in marine environments. Their eggs, trunks, and houses are important food supply to large copepods, chaetognaths, ctenophores, and larvae and adults of engraulids. They are semelparous and hermaphrodites (except *O. dioica*) organisms, with a short and temperature-dependent life cycle. In this chapter, we discuss the seasonal dynamics of appendicularians, comparing life strategies of dominant species at distinct coastal environments of the Southwest Atlantic Ocean. *O. dioica*, *O. fusiformis*, *Appendicularia sicula*, and *Fritillaria borealis* are common coastal species in the southwestern Atlantic. Total abundance, biomass, and house production of *O. dioica* and *A. sicula* were higher during spring and summer. *O. dioica* and *A. sicula* bloomed during summer with temperatures between 17 and 20 °C. *O. fusiformis* appeared occasionally during summer and fall but in very low densities. *Fritillaria borealis* prefers subantarctic and Antarctic cold (<11 °C) and salty waters. The contribution of appendicularians to the zooplankton secondary production had been underestimated. Here we emphasized the role of appendicularians as extraordinary producers of carbon and

F. L. Capitanio (✉) · M. L. Spinelli · M. L. Presta · G. E. Aguirre
Instituto de Biodiversidad y Biología Experimental Aplicada (IBBEA), Consejo Nacional de Investigaciones Científicas y Técnicas–Universidad de Buenos Aires (CONICET – UBA), Buenos Aires, Argentina

Departamento de Biodiversidad y Biología Experimental, Facultad de Ciencias Exactas y Naturales, Universidad de Buenos Aires, Buenos Aires, Argentina
e-mail: capitani@ege.fcen.uba.ar

G. Cervetto
Dirección Nacional de Medio Ambiente. Ministerio de Vivienda, Ordenamiento Territorial y Medio Ambiente, Montevideo, Uruguay
e-mail: guillermo.cervetto@mvotma.gub.uy

M. Pájaro · C. M. Derisio
Instituto Nacional de Investigación y Desarrollo Pesquero (INIDEP), Mar del Plata, Buenos Aires, Argentina
e-mail: mpajaro@inidep.edu.ar; cderisio@inidep.edu.ar

M. S. Hoffmeyer et al. (eds.), *Plankton Ecology of the Southwestern Atlantic*,
https://doi.org/10.1007/978-3-319-77869-3_10

macroscopic aggregates in planktonic ecosystems, as it has been shown by several studies at the northern hemisphere and herein for the southern SW Atlantic Ocean.

Keywords Tunicates · Seasonal cycles · Carbon flux · Vertical migration · Southwest Atlantic

1 Introduction

Pelagic appendicularians or "larvaceans," the smallest organisms among the tunicates, inhabit mainly coastal than oceanic environments. They are able to occasionally bloom which is enhanced by their capacity to filter and concentrate a wide range of food particles, such as nano-picoplankton, bacteria, and colloidal organic material, using a renewable mucopolysaccharide structure known as "house" (Flood et al. 1992). Due to their high grazing rates, dense appendicularian populations may deplete available food in a few days representing a single-step shunt between small food preys and large predators (Touratier et al. 2003).

In the Southwest Atlantic Ocean, the general zoogeography of appendicularians has been reviewed by Esnal (1999) mainly focused on the morphology of the following dominant species and their distributional patterns: (i) *Oikopleura dioica* an euryhaline circumglobal species typical of coastal environments, including brackish estuaries and bays; (ii) *Oikopleura fusiformis* an eurythermic and widely distributed neritic and oceanic species, scarcely tolerant to estuarine waters; (iii) *Appendicularia sicula* a small-bodied species inhabiting warm and temperate coastal waters; and (iv) *Fritillaria borealis* a cosmopolitan species which prefers subantarctic and Antarctic waters. In this review, life-history strategies and ecological relevance of these four common coastal southwestern Atlantic species were assessed given an overview of their sizes at maturity, gonadal development, distribution patterns, and trophic role at distinct coastal environments between 34° and 55°S which are the Río de la Plata and El Rincón estuaries, the *Estación Permanente de Estudios Ambientales* (EPEA coastal station), the Valdés Peninsula tidal front, and the Beagle Channel. The responses of appendicularian populations to environmental conditions and gradients are discussed mainly focused on their ecological impact on these coastal ecosystems.

2 Life Cycle and Sizes at Maturity of Common Oikopleurids and Fritillarids

The appendicularian life cycle is very simple, with direct development and a unique reproductive event. They are protandric hermaphrodites (except *O. dioica*), and the fertilized eggs hatch into a tadpole-like juvenile stage. The first feeding house is inflated within hours (Trøedsson et al. 2002), and the development rate is rapid compared with other tunicate groups. Within oikopleurids, both *O. fusiformis* and *O. dioica* have a typical ovoid trunk with the incipient gonads ventrally positioned. In the former species, the caecum is noticeable visible in juvenile animals dorsally extended, and when animal matures, both the ovary and testis surround the stomach laterally. In *O. dioica* – the only species with separate sexes – the ovary or testis is next to the gut. Mature gonads are turgid, the ovary has spherical cells, and the testis shows a fine-grained texture. In juvenile stages of *A. sicula*, a small dorsal ovotestis can be observed which originates ovary and testis as the development proceeds. The ovary is spherical, and the testis expands posteriorly when the animal is fully mature (Aguirre et al. 2006). Differing from all appendicularians, this species exhibits a peculiar characteristic which is the lack of anus. This exceptional "blind gut" explains the typical pear-shaped aspect caused by the enlargement of the rectum and the accumulation of undigested fecal material (Brena et al. 2003). Conversely, in *F. borealis*, the body trunk is elongated, and the gonads are symmetrically arranged along a longitudinal axis. In juveniles, gonads are not yet differentiated. In mature specimens, the testis is elongated behind a spherical ovary, and in later stages, an increase in the volume of gonads accompanied by lateral extensions of the genital sac is observed. As all appendicularians are semelparous when animal matures, the last house is abandoned, and the body wall is ruptured to allow the eggs release causing the animal death (Fig.1a).

Temperature is a key driver in regulating the sizes at maturity of appendicularians, even though other factors such as quantity and quality of food may be considered. Generally, when temperature decreases, generation time and maximum trunk length increase. At the EPEA station, the population of *O. dioica* begins and ends the maturation at larger sizes when temperatures are lower (between 750 and 1400 μm at 11 °C) contrasting to the summer population (between 450 and 900 μm at 20 °C). Only data of sizes at maturity during the autumn-winter period (around 12 °C) in the case of *O. fusiformis* and during summer (around 21 °C) in the case of *A. sicula* are available, these species reaching sizes of 2000 μm and 450 μm, respectively. Likewise, in *F. borealis* the same size-temperature-dependent pattern was observed. At the Beagle Channel ecosystem, sizes at maturity were slightly larger at lower temperatures, while at the EPEA station, this difference was more noticeable with sizes decreasing from 1300 to 600 μm when temperature increased from 10 to 17 °C (Fig. 1b).

3 Spring Spatial Assemblages of Species at the Río de La Plata Estuary and Their Contribution to the Carbon Flux

Coastal water masses may show great variability as they can be modified by the continental discharge of rivers. In the SW Atlantic Ocean, differences in salinity distribution could be attributed to diverse factors including ENSO (El Niño Southern Oscillation) events. During ENSO years, high precipitation anomalies are associated with high discharges of tributary rivers. The intensity of wind stress increases

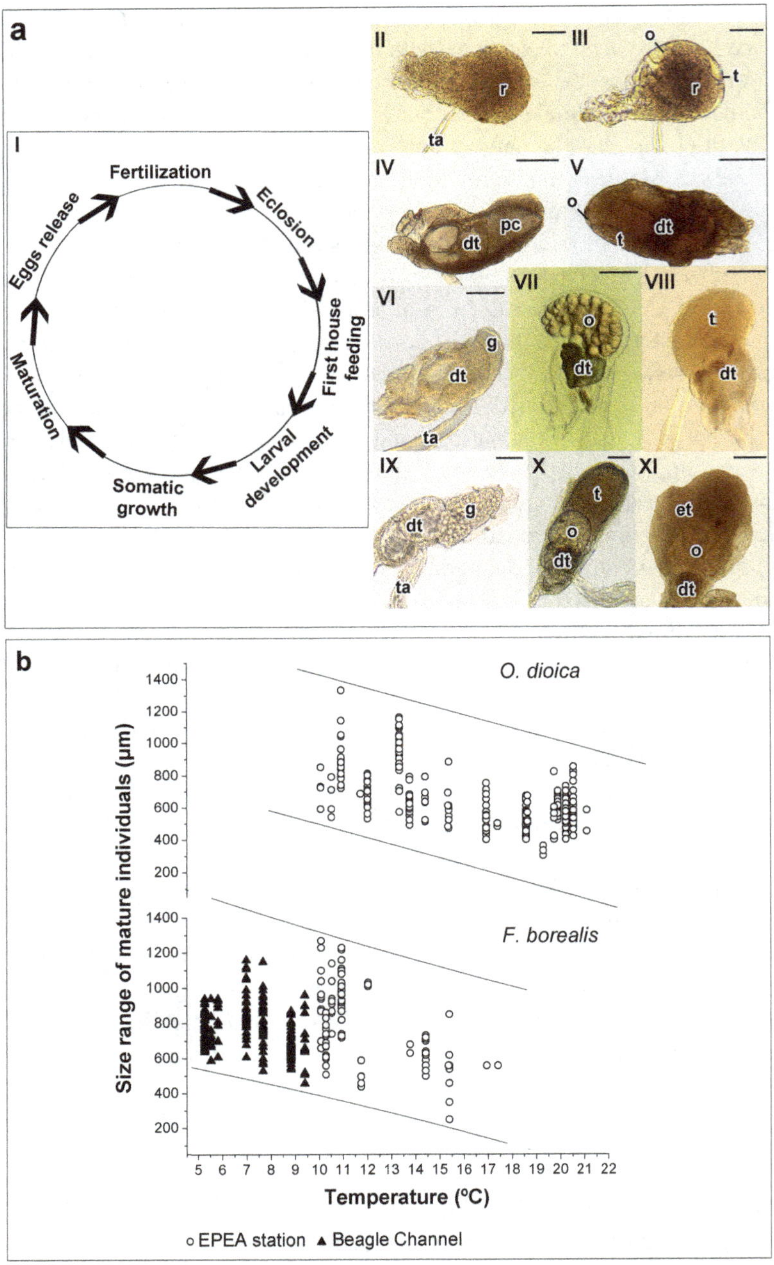

Fig. 1 (**a**) Generalized life cycle of appendicularians and sizes at maturity of (i), juvenile and mature stages of *Appendicularia sicula* (ii, iii), *Oikopleura fusiformis* (iv, v), *O. dioica* (vi–viii), and *Fritillaria borealis* (ix–xi). (**b**) Body sizes (µm) at maturity of *O. dioica* and *F. borealis* as a function of temperature (°C). dt digestive tract, et expanded testis, g gonad, o ovary, p postcardial caecum, r rectum, ta tail, t testis. Bar scales: (ii) 50 µm, (iii) 62.5 µm, (iv) 54.5 µm, (v) 167 µm, (vi) 65 µm, (vii–viii) 164 µm, (ix) 55 µm, (x) 106 µm, (xi) 178.5 µm. Data corresponding to EPEA station (Estación Permanente de Estudios Ambientales, 38°28'S, 57°41'W) and Beagle Channel (ca 55°S, 68°W) from Capitanio et al. (2008), Aguirre et al. (2012) and Presta et al. (2015)

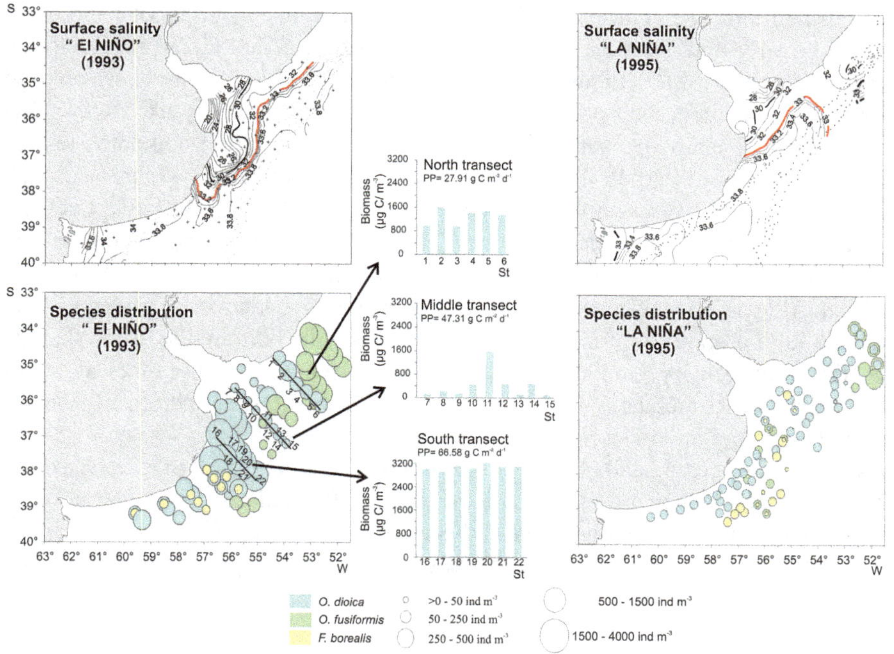

Fig. 2 Surface salinity distribution and spring spatial abundance of *Oikopleura dioica*, *O. fusiformis*, and *Fritillaria borealis* during a typical "El Niño" (ENSO, El Niño Southern Oscillation) and "La Niña" years at the Río de La Plata estuary. For the El Niño 1993, total biomass (µg C m^{-3}) and pellet production (PP: g C $m^{-2}d^{-1}$) of appendicularians were indicated at three (northern, middle, and southern) transects along the estuary. (Data modified from Spinelli et al. (2009))

and the La Plata River spread offshore, contrasting to La Niña years characterized by a decrease in precipitation, a reversal in the wind fields and high salinity waters within the estuary (Piola et al. 2005).

The outflow variability of the La Plata River plume could be reflected on the nearshore ecosystem as it has been documented by Calliari et al. (2004) and Acha et al. (2008), among others. Salinity distribution in a typical El Niño (1993) and La Niña (1995) years and its effect on the appendicularian community were analyzed during springtime. Spatial distribution of species markedly differed between both spring events (Fig. 2). During El Niño year, the La Plata River plume was toward the southeast reaching 38.5°S, and the estuarine external limit marked by the isohaline of 33 showed an extension toward the north. On the contrary, during La Niña year, no river plume was observed at the middle estuary. The intrusion of stenohaline marine species, such as *F. borealis* and *O. fusiformis*, in the estuary characterized La Niña year. Lower densities of species, including *O. dioica*, were also reported (Spinelli et al. 2009). *Oikopleura dioica* was present at all salinity ranges both in El Niño and La Niña years, but highest densities were found during the former year, particularly at the south of estuary (around 4000 ind m^{-3}). Similarly, *O. fusiformis* was very abundant during El Niño year mainly at the north of estuary. This pattern could probably be associated with the discharge of nutrient-enriched waters of the La Plata River enhancing higher chlorophyll-*a* concentration.

Furthermore, the contribution in biomass at this estuary during El Niño event could be very high as it was shown in the biomass gradient along three selected transects (see Fig. 2). Particularly at the southern one, *O. dioica* biomasses reached 3200 μg C m^{-3}, and their pellet production estimated from the trunk length-pellet volume relationship was very high (mean: 66.58 g C m^{-2} d^{-1}). Oikopleurids are often mediators of carbon flux as they produce compact pellets which have high sinking rates. At fixed samples, the flow of pellets through the digestive tract can easily be observed in an orderly sequence, and an active feeding can be inferred by counting the number of pellets inside their guts, which would be up to three in *O. dioica*. Although pellets may be ingested by other zooplankters or degraded by bacteria, a high proportion of them could rapidly reach the bottom due to the shallow waters of this estuary. On the other hand, it is known that the house renewal rate of appendicularians increases linearly when salinity decreases, an increment from 46 to 53 houses per animal corresponds to a 30–25 salinity decline (Sato et al. 2001). As a consequence, the daily house production would be increased in estuarine environments such is the case of the La Plata estuary. Particularly, the contribution of appendicularians to the carbon flux in this estuary would be significantly enhanced during El Niño events due to higher biomasses, higher pellet production, and higher houses renewal.

4 Seasonal Cycles of Appendicularians at Temperate and Subantarctic Coastal Environments

Appendicularians seasonality and community succession at two distinct coastal environments of the Argentine Sea were studied. The EPEA station (38°28’S-57°41’W, 48 m depth) is a temperate coastal system characterized by high salinity waters (33.7–34.2) over the year. The advection of low salinity waters from the La Plata River sporadically occurs during warm months. A seasonal stratification of temperature with a strong thermocline during summer-autumn months is followed by a vertical water column mixing during winter (Lucas et al. 2005; Lutz et al. 2006). At this station, a short-term study (biweekly to monthly samples, except for some skipped months) of appendicularian species annual succession was conducted along 2000–2002 years.

Total abundance, biomass, and production of appendicularians were noticeably higher during spring and summer months. Highest abundance was recorded in February 2002 (19,213 ind m^{-3}) and highest biomass (2321 μg C m^{-3}) and somatic and houses production (1872 μg C m^{-3} d^{-1} and 4405 μg C m^{-3} d^{-1} respectively) in January 2001. Surface temperature varies between 10 and 21 °C and salinity between 33.2 and 34.3. During 2000–2001 period, higher concentrations of chlorophyll-a were registered for winter months. No data are available for 2001–2002 period. *Oikopleura dioica* appeared regularly and was the dominant species peaking mainly at temperatures between 17 and 20 °C. During summer 2002, an extraordinary bloom of *A. sicula* replaced *O. dioica* niche persisting until autumn. Autotrophic ingestion rate (0.17 μg C ind^{-1} d^{-1}) and clearance rate (5.0 ml ind^{-1} d^{-1})

of mature *A. sicula* were high revealing their ecological impact on this temperate system (Aguirre et al. 2006). *Oikopleura fusiformis* appeared occasionally during summer and fall months but in very low densities. *Fritillaria borealis* was more abundant during winter months associated with temperatures below 11 °C (Fig. 3). The abovementioned highest contribution in biomass during summer corresponds to the predominance of weighted matures *O. dioica*. At the EPEA station, Viñas et al. (2013) pointed out a contrasting seasonal pattern in phytoplankton composition highlighting the predominance of pico-phytoplankton in summer and of micro-phytoplankton in winter months. Thus, the seasonal succession of appendicularians at this coastal station would be regulated by the seasonality of temperature and the phytoplankton size fraction composition. It is worth mentioning that production of *O. dioica* houses is twofold above their secondary production; an overall somatic secondary production of 76 mg C m^{-3} and a house production of 123 mg C m^{-3} were estimated for 2000–2001 year (Capitanio et al. 2008) which is in concordance with other reports for temperate seas (e.g., Uye and Ichino 1995). Even more in tropical waters, secondary production of appendicularians is at least 50% higher than that of copepods if houses production is considered (Hopcroft and Roff 1995).

Otherwise, the Beagle Channel is a high latitude ecosystem at the southern extreme of Argentina (*ca* 55°S, 68°W) which connects the Pacific and Atlantic Oceans. It represents the southernmost limit of the Magellan region, belonging to the subantarctic neritic domain. Their waters exhibit coastal-estuarine characteristics with surface salinities <32. The estuarine dynamic is governed by seasonal precipitations and ice melting as well as tidal flows from the west to east (Balestrini et al. 1998; Isla et al. 1999). Zooplankton samples and environmental data were collected seasonally at different coastal areas of the Beagle Channel along with a longitudinal gradient from west to east.

Appendicularians exhibited a very strong seasonal cycle with their abundance following the changes in the environmental conditions (Fig. 4). Maximum abundances were observed in spring in coincidence with the phytoplankton bloom (mean chlorophyll-*a* concentration of 1.32 $mg.m^{-3}$) and warmer temperatures (around 9 °C). These abundances were highly contrasting with those found in summer and autumn when the lowest chlorophyll-*a* concentrations were registered. Conversely to the EPEA station, *F. borealis* was the dominant species in the Beagle Channel, generally representing more than 90% of the appendicularian abundance. This species was followed very far by *O. fusiformis* in spring and *O. dioica* in the remaining seasons.

The spatial distribution of *F. borealis* during spring when it represented almost 10% of the total meta-zooplankton of the Beagle Channel (Aguirre et al. 2012) revealed the existence of an increasing abundance gradient from the west to east, which seemed to be strongly linked to spatial differences in salinity conditions. Temperature and chlorophyll-*a* concentration did not exhibit a significant spatial variability. Freshwater discharge in the west area is remarkable causing a dilution effect being the water exchange with the rest of the channel limited (Isla et al. 1999). Both *F. borealis* abundance and salinity values were minimal in the estuarine complex of the west area. In the eastern most sites (11 and 12), *F. borealis* bloomed (>5000 $ind.m^{-3}$) which is consistent with the lower influence of glacier runoff.

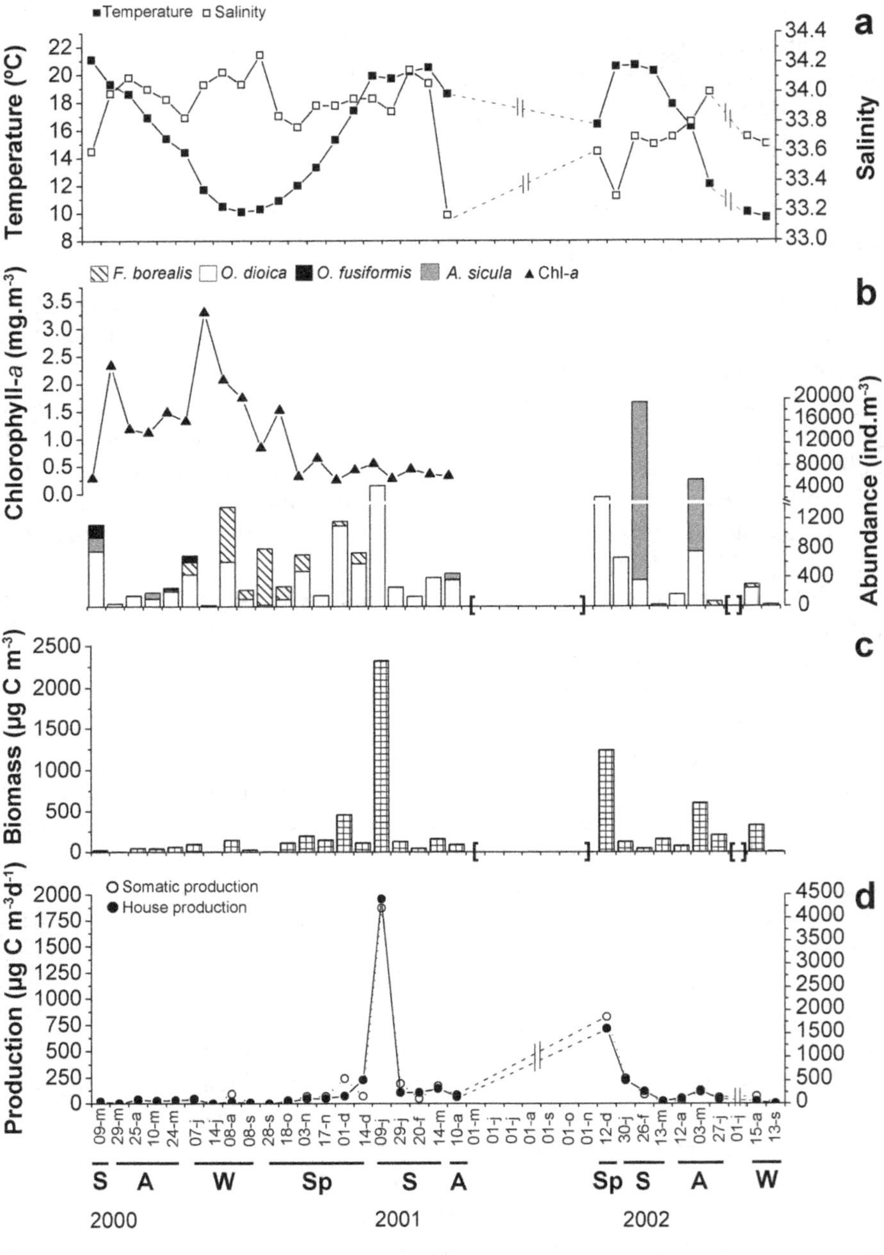

Fig. 3 Seasonal cycle of appendicularians at the EPEA (Estación Permanente de Estudios Ambientales, 38°28'S, 57°41'W) coastal station over 2000–2002 years. Seasonal variation in surface temperature (°C) and salinity (**a**); chlorophyll-*a* (mg m^{-3}) concentration and abundances (ind. m^{-3}) of *Oikopleura dioica*, *O. fusiformis*, *Appendicularia sicula*, and *Fritillaria borealis* (**b**); total biomass (µg C m^{-3}) of appendicularians (**c**); and somatic and house production (µg C m^{-3}d^{-1}) (**d**). (Data obtained from Capitanio et al. (2008). Somatic production and house production were estimated from Hopcroft and Roff (1995), López-Urrutia et al. (2003) and Tomita et al. (2003))

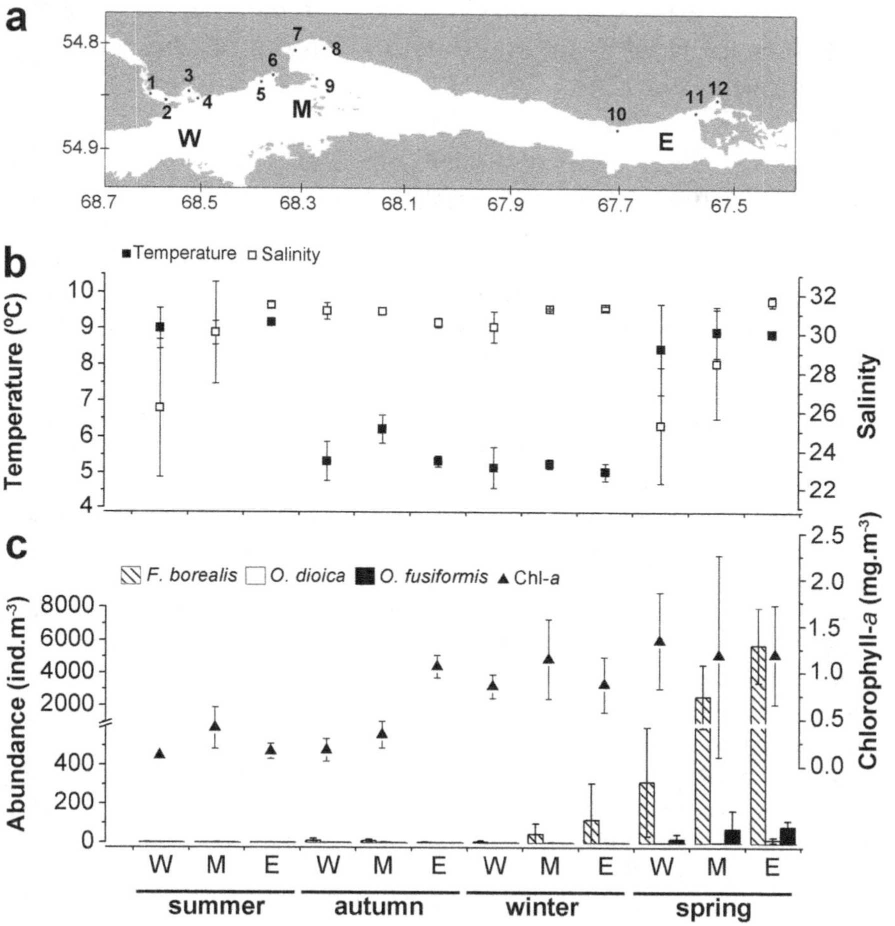

Fig. 4 Seasonal cycle of appendicularians at the Beagle Channel during 2005–2006 years. Sampling sites at west, middle, and east areas (**a**), seasonal variation of surface temperature and salinity (**b**) and chlorophyll-*a* concentration (mg m^{-3}), and *Fritillaria borealis*, *Oikopleura dioica*, and *O. fusiformis* abundances (ind.m^{-3}) (**c**). W west (1–4 stations), M middle (5–9 stations), E east (10–12 stations). (Data modified from Aguirre et al. (2012))

Comparing both temperate and subantarctic coastal environments, a strong geographical influence on the species composition was detected as *O. dioica* and *A. sicula* bloomed at the EPEA temperate station and *F. borealis* did it at the Beagle Channel. Temperature seasonality is not significant at this high latitude ecosystem, and primary production is strongly limited by reduced radiation availability during autumn and winter months (Almandoz et al. 2011). Thus, the optimal conditions to make an opportunistic bloom would be reduced to a shorter time lapse forbidding the succession of appendicularians along the year; this is opposite to what happens at the temperate EPEA station where a seasonal succession of species occurs.

5 Diel Vertical Distribution of *Oikopleura dioica* at the Valdés Peninsula Tidal Front and Potential Causes of Their Migration Patterns

During springtime, the spatial distribution of *O. dioica* was studied at the Valdés Peninsula tidal frontal system. Frontal systems contribute to the control of several aspects of the ecology of zooplankton communities such as species spatial and vertical distributions, among others (Lee et al. 2005). This front is developed during spring and summer, extending from Valdés Peninsula at 42–45°S. Circulation occurs at two-layer flows, the upper layer is directed toward the northeast, and the bottom layer moves more slowly in the opposite direction (Palma et al. 2008; Álvarez Colombo et al. 2011).

Spatial and diel vertical patterns in the abundance and population structure of *O. dioica* at this front are shown in the day and night profiles (Fig. 5a). Higher abundances were found in transitional and stratified stations associated with the highest concentrations of chlorophyll-*a* and a marked thermocline. The analysis of appendicularian samples collected at two depth layers revealed differences in their vertical distribution pattern. Higher densities were found in the upper layers of transitional and stratified waters, but similar densities were detected at both layers of the homogeneous waters (Fig. 5b). At these waters, only juvenile animals were detected both during the day and the night. However, at transitional and stratified waters, mature specimens were well represented mainly in the upper layer at night (up to 90% in stratified stations).

Diel vertical migration (DVM) is considered a central factor in shaping the vertical fluxes of organisms in the water column and their trophic interactions. It is commonly referred to as a trade-off between feeding opportunities and predation risk induced by changes in light intensity (Cohen and Forward 2009; Ringelberg 2010). Thus, while the light is the proximate cause of DVM of zooplankton, influencing the vertical extension and time of migration (Ringelberg and Van Gool 2003), the distribution of food and predation risk are regarded as the major drivers for their DVM (Neilson and Perry 1990; Sutton and Hopkins 1996). As appendicularians have only one reproductive episode during their life cycle, it would be considered more advantageous for their development to spawn in favorable conditions. At this front, Spinelli et al. (2015) remarked a high percentage of well-fed appendicularians mainly in the upper layers of transitional and stratified waters, which are defined as actively feeding animals with three pellets inside their tracts (López-Urrutia and Acuña 1999). The predominance of mature animals in the upper layer during the night may be related to a reproductive strategy to release their eggs on surface as it was demonstrated in culture experiments (Lombard et al. 2009). Also, larger appendicularians, including mature ones, are more prone to adjust their own depth in the water column (Lombard 2006). This may explain the reverse diel vertical migration of larger animals to the lower layers in order to avoid predators.

Taking these considerations, a schematic model of the possible causes of DVM of appendicularians at frontal and stratified waters off Valdés Peninsula was

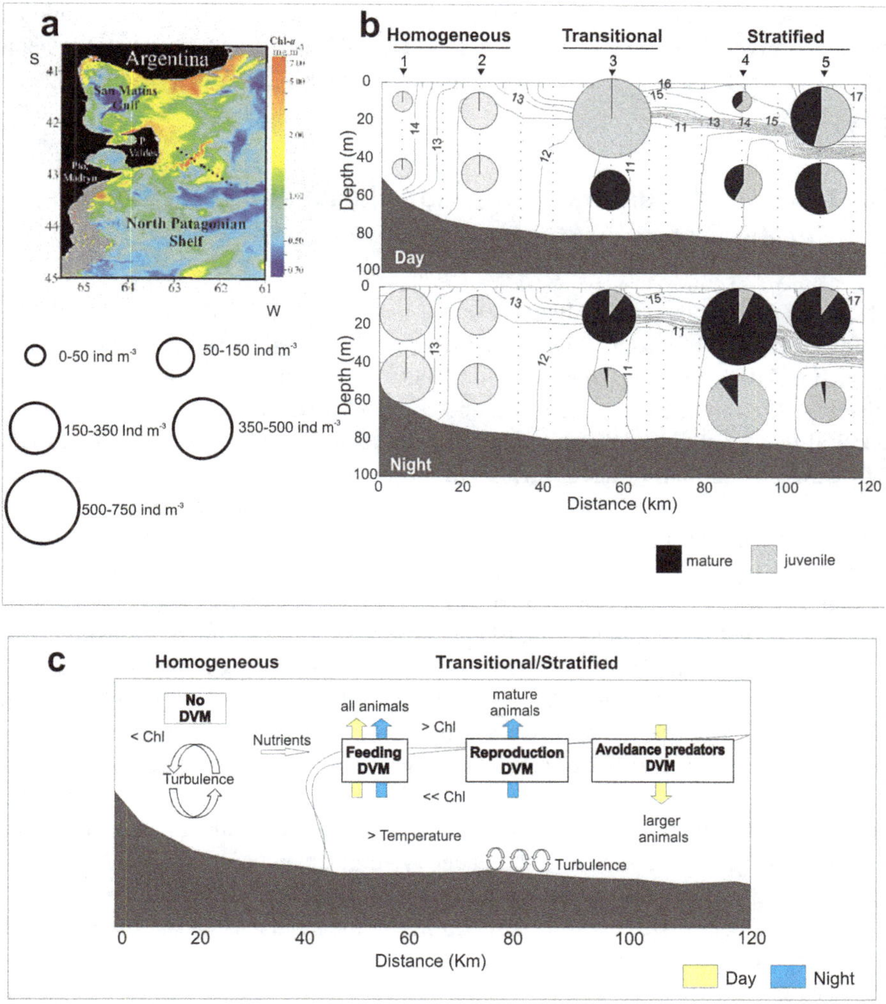

Fig. 5 Diel vertical *Oikopleura dioica* distribution at the Valdés Peninsula (42°–45°S) tidal frontal system. (**a**) Satellite chlorophyll concentration (mg m^{-3}) along a transect crossing the front. (**b**) Day and night vertical distribution of mature and juvenile *O. dioica* at homogeneous, transitional, and stratified waters. (Data modified from Spinelli et al. (2015)). (**c**) Schematic diagram of possible life-history strategies of *O. dioica* at the different sectors of Valdés Peninsula front. DVM diel vertical migration

postulated (Fig. 5c). The potential drivers of *O. dioica* DVM suggest different life-history strategies to be successful dwellers inhabiting this system which are (i) a reproduction DVM strategy which implies the migration of matures *O. dioica* to upper layers mainly at night, (ii) a feeding *O. dioica* DVM strategy to upper layers at day and night times, and (iii) an avoidance predators DVM strategy of larger *O. dioica* animals to lower layers at daytime, even though we cannot determine if DVM of predators respond directly to *O. dioica* diel patterns. No DVM is postulated in the homogeneous waters given that appendicularians, mainly juvenile stages, would be certainly mixed in the water column (Koski et al. 2007). Thus, at this frontal system, *O. dioica* would be retained through their DVM, favoring its life cycle.

6 Common Predators of Appendicularians at SW Atlantic Coastal Waters

It is well known that appendicularians are an important food supply worldwide, being both their trunks and their houses commonly preyed by several pelagic fishes (e.g., mackerel, tuna, herrings, capelin, flatfish) and many invertebrates such as chaetognaths, ctenophores, medusae, and copepods (Purcell et al. 2005).

Within the small zooplanktivorous fishes, the anchovy *Engraulis anchoita* is a key economically important species in the Argentine Sea with two distinct populations, north of 40°S and between 40°S and 47°S, inhabiting the SW Atlantic waters. Usually, during springtime, adults migrate from oceanic to coastal reproductive habitats to lay their eggs. They feed on meso- and macro-zooplankters including appendicularians. This was noticed by the stomach content analysis of 853 anchovy adults collected at the La Plata and El Rincón estuaries (north of 40°S) and at Valdés Peninsula tidal front (42–44°S) along a 4-year spring sampling program when great amounts of *O. fusiformis* and *O. dioica* were identified in their guts (Capitanio et al. 1997, 2005). Generally, copepods are the main diet of anchovies, followed by appendicularians. Comparing fishing trawls with plankton samples collected simultaneously at the same depth, large number of appendicularians in the stomach contents coincided with their high densities in the water column. In the same way, appendicularians were scarce in the anchovy stomachs when their densities in the plankton were low (Table 1). Remarkably, the proportion of juveniles and matures of both oikopleurid species were also similar in the stomachs and in the plankton, excepting for one campaign at the Río de La Plata estuary where all ingested animals were matures contrasting with the distribution pattern detected in plankton tows. Thus, adult anchovies filter all larvacean stages around the water column, including occupied and non-occupied breeding mature specimens when these are grouped near the surface for spawning events.

In Fig. 6 trunk length distribution of appendicularians at the EPEA station, El Rincón and Valdés Peninsula fronts are plotted in relation to the abundance of their main potential predators (e.g., copepods, chaetognaths, *E. anchoita* larvae, and ctenophores) prevailing in each one of these systems. At the EPEA station, a predominance of largely sized appendicularians was detected during spring-summer, whereas all size-classes were found during autumn-winter period. A strong size structure regulation of *O. dioica* by highly abundant large copepods (up to 45,000 ind.m^{-3}) could be occurring during spring-summer as only large larvaceans were found. The suppression of eggs and newly enclosed *O. dioica* (up to 0.2 mm TL) would be caused mainly by the predation pressure of *Paracalanus* spp., *Ctenocalanus vanus*, and *Calanoides carinatus*. Also, first-feeding anchovy

Table 1 Percentage (%) of food items in the stomach contents of adults of *Engraulis anchoita* anchovy collected during 1993–1996 campaigns at the Rio de la Plata and El Rincón estuaries and Valdés Peninsula tidal front

			Stomach contents *E. anchoita* adults										
Anchovy stock	Year	Catch	Ap	Co	Cl	Fe	Am	Pb	Pc	Eu	Sa	Ap (% mature)	Ap density sample
Rio de La Plata estuary	1993	6	66.93	33.05	0.02	0	0	0	0	0	0	100	17,500
		1	13.99	0	0	55.94	0.30	0	0	0.30	29.47	100	100
		5	92.53	3.40	1.33	0.31	0	2.43	0	0	0	100	7500
	1994	5	65.21	32.23	2.31	0.25	0	0	0	0	0	40	200
		9	2.10	68.02	4.60	0.99	0	0	0.23	0	0	30	11–100
	1995	7	18.94	58.18	5.11	14.58	0	0	3.18	0	0	30	11–100
		11	26.15	68.02	4.60	0.99	0	0	0.23	0	0	25	11–100
	1996	4	71.73	21.47	1.28	5.16	0	0	0.37	0	0	20	101–200
		7	16.09	61.01	16.11	6.653	0	0.07	0.07	0	0	13	11–100
El Rincón estuary	1994	15	4.42	91.38	0.11	4.08	0	0	0	0	0	100	10
		16	19.56	71.90	1.23	7.31	0	0	0	0	0	70	101–200
	1995	6	45.66	6.96	19.44	0.19	0	27.76	0	0	0	30	200
	1996	19	24.07	58.57	16.05	0.23	0.05	0.95	0.07	0	0	50	11–101
Valdés Peninsula	1993	1	0.29	94.18	5.14	0.29	0.07	0.02	0	0	0	100	300
		4	2.99	68.08	28.74	0.19	0	0	0	0	0	100	350
		6	9.41	87.67	0	0.01	0.03	0	0	2.87	0	100	1800
	1994	7	2.58	97.35	0	0.07	0	0	0	0	0	12.5	101–200
		8	36.68	61.94	1.18	0.16	0	0	0.03	0	0	10	11–100
	1995	9	73.37	20.92	4.85	0.84	0.02	0	0	0	0	10	200
		12	30.71	61.23	5.79	0	0	0	0.63	1.64	0	12.5	11–100
	1996	3	78.19	13.58	2.54	2.19	0	0	3.50	0	0	5	11–100
		5	29.98	66.02	1.54	2.42	0	0.03	0	0	0	10	11–100

Data modified from Capitanio et al. (1997, 2005)

For each catch % of each item was indicated. *Ap* Appendicularia, *Co* Copepoda, *Cl* Cladocera, *Fe* eggs of fish, *Am* Amphipoda, *Pb* postlarva of bivalve, *Pc* postlarvae of crustacean, *Eu* Euphausiacea, *Sa* Salpidae. Data of % of mature appendicularians in the stomach of anchovies and density of appendicularians (ind m^{-3}) simultaneously collected in plankton samples were included

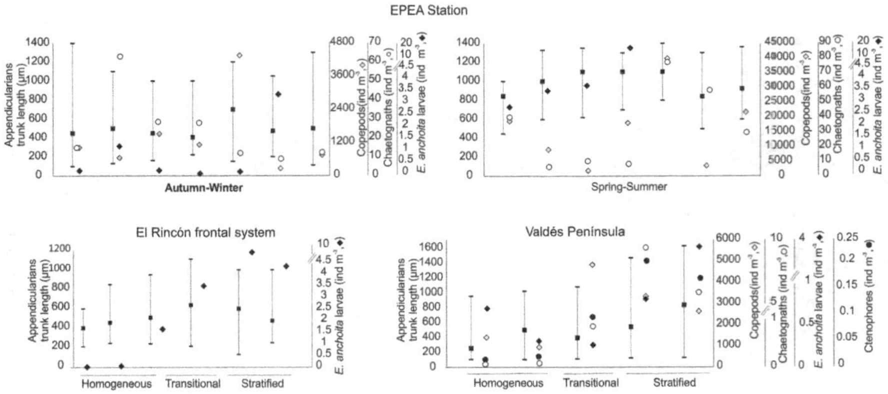

Fig. 6 Appendicularian trunk length distribution at both autumn-winter and spring-summer periods at the EPEA station, El Rincón and Valdés Peninsula frontal systems in relation to the abundance of their main potential predators (copepods, chaetognaths, *Engraulis anchoita* larvae, and ctenophores) at these systems. (Data of predators were obtained from Daponte et al. (2004), Capitanio et al. (2008), Pájaro et al. (2009), and Spinelli et al. (2013, 2015))

larvae were dominant during this period contributing to the top-down control of small-sized larvaceans. At this station, Sato et al. (2011) analyzed the gut content of first-feeding larvae exhibiting small prey items such as dinoflagellates, tintinnids, diatoms, eggs, nauplii, and copepodites of several copepod species together with non-identified appendicularians. Regarding other potential predators, even though chaetognaths (e.g., *Sagitta friderici*, now *Parasagitta friderici*) are present year-round, juveniles are better represented during summer and adults during winter (Daponte et al. 2004). Several examples are reported worldwide, mentioning *O. dioica*, *O. labradoriensis*, *O. longicauda*, and *O. fusiformis* larvacean species in the diets of *S. hispida*, *S. enflata*, *S. elegans*, *S. setosa*, *Eukrohnia hamata*, and *P. friderici* chaetognaths (Purcell et al. 2005). It is known that the effect of predation on appendicularians increase with chaetognath sizes. Usually, older chaetognaths attack free-swimming appendicularians detecting the tail beating which is more perceptible when they are not feeding inside their houses. Thus, at the EPEA station, *P. friderici* predation pressure would be higher during autumn-winter months.

At both transitional and stratified waters of El Rincón and Valdés Peninsula frontal systems, suitable environments for *O. dioica* reproduction, higher density of potential predators was detected. Particularly, high densities of anchovy larvae were positively correlated with high densities of *O. dioica* at El Rincón front during summer (Spinelli et al. 2013). Small larvaceans (<600 μm mean trunk length) were dominant at this front suggesting a higher predation on larger animals, although no data of other potential predators are available to delve into prey-predator interactions. At Valdés Peninsula, large copepods (i.e., *Drepanopus forcipatus*, *Ctenocalanus vanus*, *Calanoides carinatus*, and *Centropages brachiatus*) were present reaching up to 5000 ind m^{-3} at transitional stations (Derisio 2012). These potential predators, together with mature *S. friderici* chaetognaths, anchovy larvae, and the ctenophore *Mnemiopsis leidyi*, would be regulating all sizes of larvacean populations at this front (Spinelli et al. 2015).

Finally, it is known that several predators would prey on abandoned houses of appendicularians (Purcell et al. 2005). For example, Nishibe et al. (2015) reported copepod species of the genus *Oncaea* would perform an important role in the degradation of appendicularian houses. Plankton samples collected at the Beagle Channel revealed that the spring bloom of *F. borealis* was correlated with higher densities of *Oncaea curvata* (Aguirre et al. 2012) suggesting that probably a selective predation would be occurring. At the SW Atlantic Ocean, more studies describing the feeding behavior of different predators on discarded houses – a significant source of the marine snow in coastal ecosystems – are still required.

7 Main Remarks

Appendicularians have successfully colonized coastal environments due to their rapid adaptation and response to environmental changes. From a bottom-up perspective, their ability to feed on a wide range of food particles allows them to survive at highly variable food conditions present in coastal ecosystems. Thermal dependence of appendicularians is an outstanding characteristic of worldwide pelagic tunicates having a positive effect on gonadal maturation time. At the SW Atlantic Ocean, several appendicularian species (*Oikopleura dioica*, *O. fusiformis*, *Fritillaria borealis*, and *Appendicularia sicula*) show a twofold increase in body trunk length when temperature approximately decreases by half. This was observed at subantarctic (Beagle Channel) and temperate (EPEA station) coastal environments when sizes and maturity of species were compared.

Among environmental variables, salinity could be the main factor driving species spatial assemblages if coastal areas associated with river discharges or glacier runoff were considered. At the Río de La Plata estuary, seasonal fluctuations in the plume discharges are strongly reflected on species spatial composition. The contribution of *O. dioica*, the most euryhaline and abundant neritic species, to the carbon flux could be enhanced due to the significant proportion of fecal pellets and discarded houses which could reach and be quickly recycled on the seabed of this shallow estuary.

Historically, the contribution of appendicularians to the zooplankton secondary production had been underestimated. While copepod biomass would be up to ten times higher than that of appendicularians during the same period, copepod growth rate hardly achieves one-third of those of appendicularians. Thus, it is emphasized the role of appendicularians as extraordinary producers of carbon and macroscopic aggregates in planktonic ecosystems, as it has been shown by several studies at the northern hemisphere and herein for the southern SW Atlantic Ocean.

From a top-down perspective, houses of appendicularians would be functioning as a refuge from potential visual predators especially when houses are newly secreted because of their transparency. However, predation risk would be more complex. Firstly, the risk would be increased with the age of houses as they would be more exposed to be selected by visual predators, thus increasing the risk of dying before reproducing. However, another alternative would be that appendicularians escape from their houses without affecting the recruitment of new generations since a slight turbulence can trigger the spawning of mature houseless animals. Thus, attending the variety of mechanisms and the diversity of predators that can be involved, top-down regulation would be the main control of appendicularian populations at coastal systems. As an evidence of this type of regulation, diel vertical migration of matures *O. dioica* is reported to avoid predation strategy at stratified waters of the Valdés Peninsula tidal front.

References

Acha EM, Mianzan H, Guerrero R et al (2008) An overview of ecological processes in the Río de la Plata estuary. Cont Shelf Res 28:1579–1588

Aguirre GE, Capitanio FL, Viñas MD et al (2006) Gonadal development, allometric growth and ecological impact of *Appendicularia sicula* (Appendicularia: Fritillariidae) from the South-Western Atlantic Ocean. J Mar Biol Assoc 86:1215–1220

Aguirre G, Capitanio F, Lovrich G et al (2012) Seasonal variability of metazooplankton in coastal sub-Antarctic. Mar Biol Res 8:341–353

Almandoz GO, Hernando MP, Ferreyra GA et al (2011) Seasonal phytoplankton dynamics in extreme southern South America (Beagle Channel, Argentina). J Sea Res 2:47–57

Álvarez Colombo G, Dato C, Macchi GJ et al (2011) Distribution and behavior of argentine hake larvae: evidence of a biophysical mechanism for self-recruitment in northern Patagonian shelf waters. Cienc Mar 37:633–657

Balestrini C, Manzella G, Lovrich G (1998) Simulación de corrientes en el Canal Beagle y Bahía Ushuaia, mediante un modelo bidimensional. Servicio de Hidrografía Naval. Informe Técnico 98:1–58

Brena C, Cima F, Burighel P (2003) The highly specialized gut of *Fritillaridae* (Appendicularia: Tunicata). Mar Biol 143:57–71

Calliari D, Cervetto G, Castiglioni R (2004) Summer time herbivory and egg production by *Acartia tonsa* at the Montevideo coast – Rio de La Plata. Ophelia 58:115–128

Capitanio F, Pájaro M, Esnal GB (1997) Appendicularians (Chordata, Tunicata) in the diet of anchovy *Engraulis anchoita* in the Argentine sea. Sci Mar 61:9–15

Capitanio F, Pájaro M, Esnal GB (2005) Appendicularians: an important food supply for the Argentine anchovy *Engraulis anchoita* in coastal waters. J Appl Ichthyol 21:414–419

Capitanio FL, Curelovich J, Tresguerres M et al (2008) Seasonal cycle of appendicularians at a coastal station (38°28′S, 57°41′W) of the SW Atlantic Ocean. Bull Mar Sci 82:171–184

Cohen JH, Forward RB (2009) Zooplankton diel vertical migration a review of proximate control. Oceanogr Mar Biol 47:77–110

Daponte MC, Capitanio FL, Nahabedian DE et al (2004) *Sagitta friderici* Ritter–Zahony (Chaetognatha) from South Atlantic waters. Abundance, population structure and life–cycle. ICES J Mar Sci 61:680–686

Derisio C (2012) El rol del frente de marea de Península Valdés en el control de la comunidad zooplanctónica. Tesis doctoral, Univ de Mar del Plata, Mar del Plata, 134 pp

Esnal GB (1999) Appendicularia. In: Boltovskoy D (ed) Zooplankton of the South Atlantic Ocean. Leiden, Backhuys, pp 1375–1399

Flood PR, Deibel D, Morris C (1992) Filtration of colloidal melanin from seawater by planktonic tunicates. Nature 355:630–632

Hopcroft R, Roff J (1995) Zooplankton growth rates: extraordinary production by the larvacean *Oikopleura dioica* in tropical waters. J Plankton Res 17:205–220

Isla F, Bujalesky G, Coronato A (1999) Procesos estuarinos en el Canal Beagle, Tierra del Fuego. Rev Asoc Geol Argent 54:307–318

Koski M, Møller EF, Maar M et al (2007) The fate of discarded appendicularian houses: degradation by the copepod, *Microsetella norvegica,* and other agents. J Plankton Res 29:641–654

Lee O, Nash RDM, Danilowicz BS (2005) Small–scale spatio–temporal variability in ichthyoplankton and zooplankton distribution in relation to a tidal–mixing front in the Irish Sea. ICES J Mar Sci 6:1021–1036

Lombard F (2006). Étude expérimentale et modélisation de l'écophysiologie de l'appendiculaire *Oikopleura dioica*. Doctoral Thesis. Université Pierre et Marie Curie, Paris VI. 284 pp

Lombard F, Renaud F, Sainsbury C et al (2009) Appendicularian ecophysiology I. Food concentration dependent clearance rate, assimilation efficiency, growth and reproduction of *Oikopleura dioica*. J Mar Syst 78:606–616

López-Urrutia A, Acuña JL (1999) Gut throughput dynamics in the appendicularian *Oikopleura dioica*. Mar Ecol Prog Ser 191:195–205

López-Urrutia A, Acuña JL, Irigoyen X et al (2003) Food limitation and growth in temperate epipelagic appendicularians (Tunicata), Mar Ecol Prog Ser 252:143-157

Lucas AJ, Guerrero RA, Mianzan MW et al (2005) Coastal oceanographic regimes of the northern Argentine continental shelf (34–43°S). Estuar Coast Shelf Sci. https://doi.org/10.1016/j.ecss.2005.06.015

Lutz V, Subramaniam A, Negri RM et al (2006) Annual variations in bio–optical properties at the ¨Estación Permanente de Estudios Ambientales (EPEA)¨ coastal station, Argentina. Cont Shelf Res 26:1093–1112

Neilson JD, Perry RI (1990) Diel vertical migrations of marine fishes an obligate or facultative process. Adv Mar Biol 26:115–168

Nishibe Y, Takahashi K, Ichikawa T et al (2015) Degradation of discarded appendicularian houses by oncaeid copepods. Limnol Oceanogr 60:967–976

Pájaro M, Diaz MV, Leonarduzzi E et al (2009). Abundancia de huevos y larvas de anchoíta en la Estación Permanente de Estudios Ambientales (EPEA) en el período 2000–2007. Variación estacional e interanual. Informe de Investigación INIDEP N°28/09, 14 pp

Palma ED, Matano RP, Piola AR et al (2008) A numerical study of the Southwestern Atlantic shelf circulation: Stratified Ocean response to local and offshore forcing. J Geophys Res 113:1–22

Piola AR, Matano RP, Palma ED et al (2005) The influence of the Plata river discharge on the western South Atlantic shelf. Geophys Res Lett. https://doi.org/10.1029/2004GL021638

Presta ML, Hoffmeyer MS, Capitanio FL (2015) Population structure and maturity stages of *Fritillaria borealis* (Appendicularia, Tunicata): seasonal cycle in Ushuaia Bay (Beagle Channel). Braz J Oceanogr 63:279–288

Purcell JE, Sturdevant MV, Galt CP (2005) A review of appendicularians as prey of invertebrate and fish predator. In: Gorsky G (ed) Response of marine ecosystems to global change: ecological impact of Appendicularians. Editions Scientifiques, Paris, pp 360–435

Ringelberg J (2010) Diel migration of zooplankton in lakes and oceans: causes explanations and adaptive significance. Springer, London

Ringelberg J, Van Gool E (2003) On the combined analysis of proximate and ultimate aspects in diel vertical migration (DVM) research. Hydrobiologia 491:85–90

Sato R, Tanaka J, Ishimaru T (2001) House production by *Oikopleura dioica* (Tunicata, Appendicularia) under laboratory conditions. J Plankton Res 28:415–423

Sato NE, Hernández D, Viñas MD (2011) Feeding habits of *Engraulis anchoita* (Hubbs & Marini, 1935) larvae in coastal waters off Buenos Aires Province, Argentina. Lat Am J Aquat Res 39:6–24

Spinelli M, Martos P, Esnal G et al (2009) Appendicularian assemblages and their space–time variability off the La Plata River, SW Atlantic Ocean. Estuar Coast Shelf Sci 85:97–106
Spinelli ML, Guerrero R, Pájaro M et al (2013) Distribution of *Oikopleura dioica* (Tunicata, Appendicularia) at a coastal frontal system (39°– 41°S) of the SW Atlantic Ocean in the spawning area of *Engraulis anchoita* anchovy. Braz J Oceanogr 61:141–148
Spinelli M, Derisio C, Martos P et al (2015) Diel vertical distribution of the larvacean *Oikopleura dioica* in a north Patagonian tidal frontal system (42°–45°S) of the SW Atlantic Ocean. Mar Biol Res 6:633–643. https://doi.org/10.1080/17451000.2014.978338
Sutton TT, Hopkins TL (1996) Trophic ecology of the stomiid (Pisces: Stomiidae) fish assemblage of the eastern Gulf of Mexico: strategies, selectivity and impact of a top mesopelagic predator. Mar Biol 127:179–192
Tomita M, Shiga N, Ikeda T (2003) Seasonal occurrence and vertical distribution of appendicularians in Toyama Bay, southern Japan Sea. J Plankton Res 25(6):579–589
Touratier F, Carlotti F, Gorsky G (2003) Individual growth model for the appendicularian *Oikopleura dioica*. Mar Ecol Prog Ser 248:141–163
Trøedsson C, Bouquet CJM, Aksnes DL et al (2002) Resource allocation between somatic growth and reproductive output in the pelagic chordate *Oikopleura dioica* allows opportunistic response to nutritional variation. Mar Ecol Prog Ser 243:83–91
Uye S, Ichino S (1995) Seasonal variations in abundance, size composition, biomass and production rate of *Oikopleura dioica* (Fol) (Tunicata: Appendicularia) in a temperate eutrophic inlet. J Exp Mar Biol Ecol 189:1–11
Viñas MD, Negri RM, Cepeda GD et al (2013) Seasonal succession of zooplankton in coastal waters of the Argentine sea (Southwest Atlantic Ocean): prevalence of classical or microbial food webs. Mar Biol Res 9:371–382

Ichthyoplankton Associated to the Frontal Regions of the Southwestern Atlantic

Eduardo M. Acha, Martin D. Ehrlich, José H. Muelbert, Marcelo Pájaro, Daniel Bruno, Laura Machinandiarena, and Mariana Cadaveira

Abstract The study region covers a wide latitudinal range, from 22°S to 55°S, and a bathymetric scope ranging from the coast to the shelf-break and adjacent region to a depth of ca. 1000 m. Hydrography of the region is extremely complex. Oceanographic structures are diverse, including several estuaries, water masses, wind systems, tidal regimes, and two major oceanic currents. A significant portion of the biological production is related to the existence of several marine fronts, characterized by different forcing, and temporal and spatial scales. Fish species that inhabit the region are diverse, covering a range of life strategies. Among those fishes that produce eggs and/or planktonic larvae, there are estuarine, shelf and oceanic forms, and species characteristic of tropical/subtropical or subantarctic waters. Pelagic species are more important in the northern part, while demersal species dominate the southernmost area. The region looks like an intricate mosaic of opportunities for life cycle closure of distinct species. Marine fronts seem to be the preferred areas to establish spawning and nursery grounds by fishes, and current knowledge allow to identify the elements of Bakun's triad for several of those fronts.

This is INIDEP Contribution N° 2103

E. M. Acha (✉)
Instituto de Investigaciones Marinas y Costeras (IIMyC), Consejo Nacional de Investigaciones Científicas y Técnicas–Universidad Nacional de Mar del Plata (CONICET – UNMdP), Mar del Plata, Buenos Aires, Argentina

Instituto de Nacional de Investigación y Desarrollo Pesquero (INIDEP), Mar del Plata, Bueos Aires, Argentina
e-mail: macha@inidep.edu.ar

M. D. Ehrlich · M. Pájaro · L. Machinandiarena · M. Cadaveira
Instituto de Nacional de Investigación y Desarrollo Pesquero (INIDEP), Mar del Plata, Bueos Aires, Argentina
e-mail: mehrlich@inidep.edu.ar

J. H. Muelbert
Laboratorio de Ecologia do Ictioplâncton, Instituto de Oceanografia, Universidade Federal do Rio Grande (FURG), Rio Grande, RS, Brazil
e-mail: docjhm@furg.br

D. Bruno
Laboratorio de Ecología, Fisiología y Evolución de Organismos Acuáticos, Centro Austral de Investigaciones Científicas (CADIC), Consejo Nacional de Investigaciones Científicas (CONICET), Ushuaia, Tierra del Fuego, Argentina

M. S. Hoffmeyer et al. (eds.), *Plankton Ecology of the Southwestern Atlantic*,
https://doi.org/10.1007/978-3-319-77869-3_11

Keywords Ichthyoplankton · Fronts · Larvae retention · Bakun's triad · Southwest Atlantic Ocean

1 Introduction

The study region corresponds to the Southeast South American shelf large marine ecosystem (SSASLME) (Bisbal 1995). It comprises the continental shelf of Southeast Brazil, Uruguay, and Argentina and the steep continental slope to a depth of approximately 1000 m. The shelf is relatively narrow at its northern end (approximately 70 km at Cape Frio, Brazil) and widens progressively to the south, where it reaches a maximum width of about 850 km. This broad submarine terrace constitutes the largest continental shelf in the southern hemisphere. Biological production is high and sustains several large fisheries. A significant portion of the primary production is there related to the existence of several types of marine fronts, characterized by distinct forcing, and temporal and spatial scales (e.g., Acha et al. 2004).

A highly diverse and abundant ichthyofauna inhabits this region (e.g., Castello 1997; Cousseau 1997). Demersal and benthic forms dominate species richness and abundances. Notwithstanding, the richness of pelagic species is relatively high in South Brazil (Castello 1997), while in the Patagonian region (south to 46°S) just a couple of pelagic species are common (Cousseau and Perrotta 1998). Species composition and abundance show a strong seasonal signal. Most species exhibit large home ranges, and their populations undergo extensive migrations, seasonally vacating and reoccupying specific subregions within their respective ranges. Marine fishes select the environment in which their eggs will incubate, so most of them have a tendency to undertake extensive movements to specific breeding sites (e.g., Secor 2015). Spawning activities usually take place in relatively small areas inside the species' geographical range, which are associated with well-defined and geographically predictable or stable oceanographic systems (Secor 2015).

Marine fronts have been widely reported as preferred spawning grounds for fishes (Bakun 1996; Acha et al. 2015). Although fronts are diverse in forcing and spatiotemporal scales, most of them seem to accomplish the requirements of the "fundamental triad hypothesis" (Bakun 1996) that identify suitable spawning habitats for fishes: (i) nutrient enrichment processes, (ii) concentration of food particles, and (iii) retention of eggs and larvae within a favorable habitat.

Notwithstanding the complex oceanographic scenario in the region, fronts can be arranged in five zones according to their location, main forcing, key physical variables, seasonality, and enrichment mechanisms (Acha et al. 2004): (i) the South Brazil upwelling zone, (ii) the temperate estuarine zone, (iii) the Patagonian Current zone, (iv) the Patagonian tidal zone, and (v) the shelf-break zone. Frontal zones result in an adequate framework to organize and study the ichthyoplankton information existing for the area.

2 The South Brazil Upwelling Zone

Two upwelling systems occur between 22°S and 28°S related to prominent capes in South Brazil: Cape Frio (CF) and Cape Santa Marta Grande (CSM) (Fig. 1). The Cape Frio upwelling system (CFUS) is present throughout the year north of Rio de Janeiro (22°S, 43°W) (Fig.1). The temperature variability around CF has been associated with Ekman transport caused by NE and SW winds (Miranda 1982). The high frequency and persistence of the prevailing NE winds are the main factor promoting the rise of South Atlantic Central Water (SACW) (Rodrigues and Lorenzzetti 2001; Mazzini and Barth 2013), but meanders and eddies in the Brazil Current as well as local topography and coast line are also important (Coelho-Souza et al. 2012). Wind and wind stress curl variability influence the development of upwelling and downwelling cycles (Castelão and Barth 2006). Upwelling around CF has been reported in winter (Moser and Gianesella-Galvão 1997), but the most frequent and intense events are observed in summer and spring (Coelho-Souza et al. 2012).

The upwelling events of CF are short-lived and constrained to the coast and marked by a temperature decrease from 23 to 15 °C. The upwelling pulses of cold waters last for 15 days or less (Gonzalez Rodriguez 1994; Lopes et al. 2006) and are limited to occur within a 5-km-wide coastal strip. At that offshore distance, isotherms descend sharply, forming a practically vertical frontal plane. From the continental slope to the shore, the upwelling contributes to the formation of an intermediate layer between warm surface waters of the Brazil Current and deeper cold water (Valentin et al. 1987a).

Upwelling events rise nutrient concentration in the CF region and promote maximum phytoplankton concentration when zooplankton is dominated by opportunistic herbivores and carnivores. After that, there is a succession of facultative herbivores, predators, detritivores, and omnivores (Coelho-Souza et al. 2012). Since CF region is influenced by coastal, tropical, and subtropical waters (Valentin 1984; Pereira et al. 2008), the composition and abundance of bacterioplankton, phytoplankton, and zooplankton and community changes correlate with upwelling events (Valentin et al. 1987b; Cury et al. 2011).

It is suggested that drift currents carry planktonic larvae out to the open sea which later return as adults to coastal surface waters (Valentin 2001). *Engraulis anchoita* spawning in summer is related to the cold upwelled waters of CF, and in winter, it occurs mainly offshore in the neritic region, when a mid-water stability zone is formed (Bakun and Parrish 1991). Thus, displacement of the spawning area seems to be related to water column stability (Matsuura and Kitahara 1995).

During summer, fish egg densities are higher at the CF upwelling region (Lopes et al. 2006; Macedo-Soares et al. 2014). Intensity and frequency of summer upwelling events can influence the location of the Brazilian sardine (*Sardinella brasiliensis*) spawning areas (Matsuura 1998; Lopes et al. 2006). Fish larvae abundance is higher near the shelf at CF (Macedo-Soares et al. 2014). In summer, *E. anchoita* spawning is mainly associated with the upwelled colder waters (Matsuura et al. 1992; Matsuura and Kitahara 1995), and larvae show enhanced nutritional condition during SACW upwelling (Dias et al. 2016). Pulsing upwelling events may

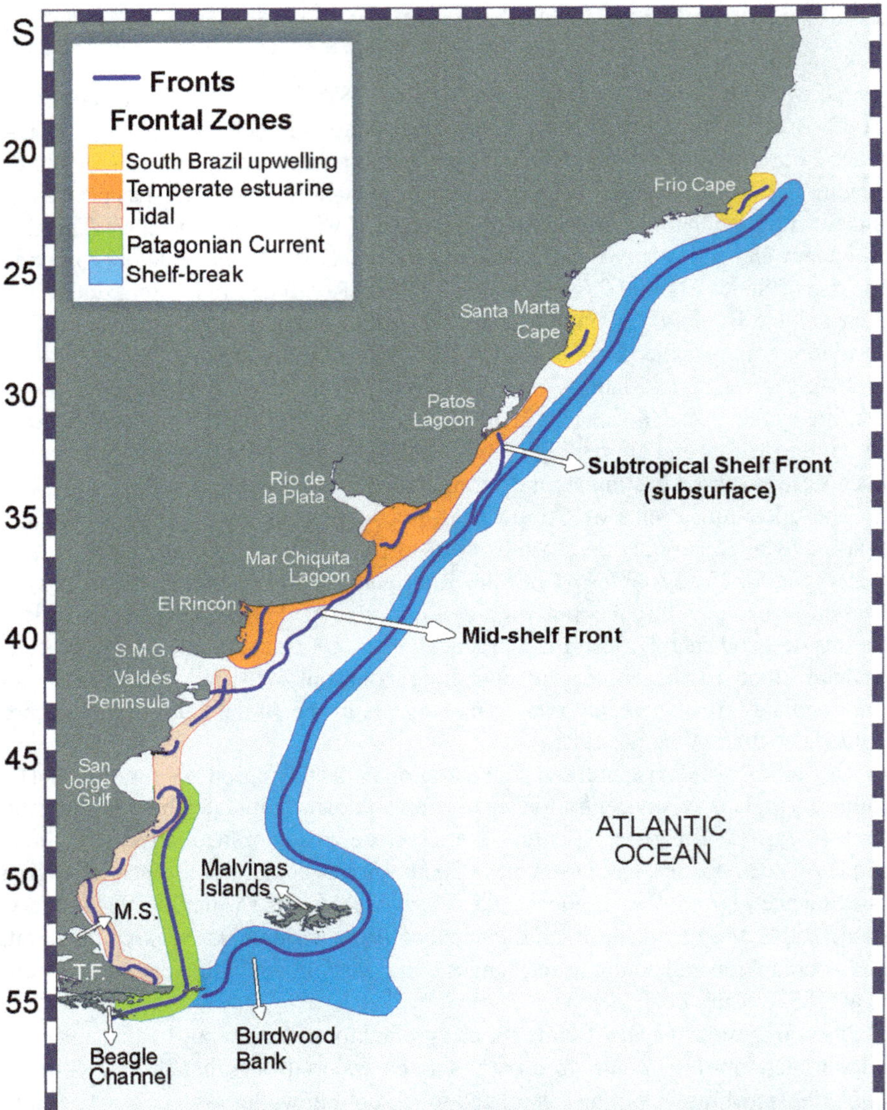

Fig. 1 Frontal zones in the Southwestern Atlantic. (Modified from Acha et al. (2004)). M.S. Magellan Strait, S.M.G. San Matias Gulf, T.F. Tierra del Fuego

explain broadscale structuring of ichthyoplankton assemblages (Moraes et al. 2012). During spring, a shelf assemblage dominated by larvae of small pelagic fishes, such as *S. brasiliensis*, *E. anchoita*, and *Trachurus lathami*, is associated with the enrichment of shallow water by the SACW upwelling (Katsuragawa et al. 2014).

Cape Santa Marta Grande (CSM) is marked by a change in coastal orientation at 28°36'S and 48°48'W (Fig. 1). North of the CSM, the coast follows an N-S orientation, while farther south, it is oriented NE-SW. At CSM, the continental shelf is relatively narrow (~100 km), and the 50 m isobath is only 5 km away from the coast, while farther north and south, the shelf width increases. In this region, upwelling of SACW is the result of NE winds (Katsuragawa and Matsuura 1992; Campos et al. 2013). Changes associated with the alternation of NE and SW winds due to the passage of cold fronts lead to pulses of stratification during upwelling and downwelling events during spring and summer. Modeling studies suggest that upwelling results from the synergy between shelf-break upwelling and the regional shelf circulation induced by the northeasterly winds, both of which are modulated by changes in the coastline orientation and shelf width (Campos et al. 2013). Nutrient injection into the euphotic zone is driven by bottom turbulence that results in a subsurface chlorophyll maximum (Odebrecht and Djurfeldt 1996). High zooplankton biomass present in this area has been associated with the influence of cold SACW (Resgalla et al. 2001; Lopes et al. 2006). However, studies have reported both, high and low zooplankton biovolume and fish eggs and larvae abundance in the region (Matsuura 1998; Freitas and Muelbert 2004), and have suggested these changes are related to the intensity of the upwelling events (Freitas and Muelbert 2004). Intense spawning of the Brazilian sardine has been registered in the offshore region of the CSM upwelling (Matsuura 1998).

E. anchoita is present in the region, but its abundance seems not to be related to the upwelling. In the winter, the highest abundance of eggs and larvae is confined to the shore areas, but larvae were also present toward offshore. Eggs and larvae were mainly associated with waters from subtropical shelf waters and Plata plume waters. During summer, *E. anchoita* larvae are present but in low abundances and not associated to any particular water mass (Torquato and Muelbert 2015). Larvae from CSM have greater growth rate than larvae from the Uruguay and Argentine coasts (Castello and Vasconcellos 1995). Few or if any starved larvae have been found in the region (Sieg 1998), but a high frequency of poor condition larvae may indicate the influence of a dynamic environment (Castello and Vasconcellos 1995).

3 The Temperate Estuarine Zone

The temperate estuarine zone extends from the south of Cape Santa Marta (Brazil), about 30°S, to the Negro river mouth (41°S) (Argentina), encompassing the estuarine fronts generated by the Patos Lagoon, the Río de la Plata and El Rincón, and several minor coastal lagoons.

3.1 The Patos Lagoon

The Patos Lagoon is located at 32°S in South Brazil (Fig. 1). It is a warm subtropical river-dominated chocked lagoon, with a significant relationship between total rainfall at the hydrographic basin and annual river discharge (Odebrecht et al. 2010). About 80% of the lagoon is a freshwater system, but at its southern limit, the lagoon connects to the ocean by means of a deep (15 m) and narrow inlet (800 m). There, brackish waters and fringing marshes comprise an estuarine ecosystem of about 1000 km^2. Tidal energy is scarce due to amphidromic conditions presented at 30–32°S (Odebrecht and Castello 2001). As a result, the main forcing controlling water dynamics and salinity distribution is regional precipitation and wind patterns (northeasterly winds promote flushing, while the southerly winds force saltwater into the estuary) (Odebrecht and Castello 2001; Odebrecht et al. 2010). The mean annual discharge is 2000 $m^3 s^{-1}$ (Vaz et al. 2006), but peaks of 12,000–25,000 $m^3 s^{-1}$ have been observed during El Niño events (Fernandes et al. 2002). Seasonal means range from 700 $m^3 s^{-1}$ in summer to 3000 $m^3 s^{-1}$ in spring (Möller et al. 2001).

The variable freshwater outflow produces a plume with a bulge extending from 10 up to 50 km offshore and a width of approximate 33 km (Burrage et al. 2008). This plume is characterized by salinities from 12 to 27 near the coast with a sharp salinity front of 2–3 contrast near the Lagoon entrance. The Patos plume is only weakly affected by earth rotation, being dominated instead by weather-scale surface wind and bottom stress. In summer, the Patos plume is typically absent or only weakly developed. During winter, the Patos plume is found embedded in the hydrographic and current fields of the larger-scale Plata plume (see below). The Plata plume will modify the environment of the embedded Patos plume by enhancing along-shelf advection and increasing ambient vertical and horizontal shear (Burrage et al. 2008). Spawning activity of fishes is scarce inside the lagoon. Most of them spawn at sea, near the inlet. Fish eggs, larvae, and juveniles are transported into the lagoon by deep channel currents, and retention and survival of these organisms in the estuarine region are dependent on the water exchange and prevailing winds (Muelbert and Weiss 1991; Odebrecht et al. 2010, Sinque and Muelbert 1997c).

Despite several studies on ichthyoplankton of the Patos Lagoon estuary, there is no information of fish eggs and larvae related to the estuarine front. Several fish species in South Brazil are dependent on the Patos Lagoon as a nursery ground, especially *Micropogonias furnieri*, *Lycengraulis grossidens*, *Brevoortia pectinata*, and *Mugil* spp. (Muelbert and Weiss 1991; Sinque and Muelbert 1997c). The transport of fish eggs and larvae is related to variations in local circulation patterns of coastal currents (Muelbert and Weiss 1991). Most of this variability is determined by the dynamics of freshwater outflow from the lagoon and wind-induced penetration of seawater into the estuary (Costa et al. 1988). With increase in freshwater outflow, recirculation cells provide a mechanism for concentration and retention of eggs and larvae in the adjacent region to the Patos Lagoon entrance (Busoli 2001). During periods of low freshwater influence, shelf circulation is predominantly northward, and the area becomes unfavorable for larval retention. Anomalous periods of freshwater outflow, linked to ENSO events, have a profound effect on the diversity of young fish in the Patos Lagoon estuary (Garcia et al. 2001).

Eggs and larvae of at least 29 fish species are found in the estuarine and adjacent coastal waters (Sinque and Muelbert 1997c). Estuarine resident fish are represented by the eggs and larvae of the flatfish *Achirus garmani* and by larvae of Atherinidae, *Syngnathus folletti*, Blenniidae, Gobiidae, and Gobiesocidae. These species are not that abundant (0.4–3.3%) and are usually found in shallow inland estuarine regions. A large number of marine species depend on the estuary for their development, such as *B. pectinata*, *L. grossidens*, and *M. furnieri*. They are present in egg and larval stages and together account for 88% of the eggs and 66% of the larvae found in the estuary. A less abundant group, which is present only in the larval stage, includes many commercially important species, like *Macrodon atricauda*, *Menticirrhus* spp., *Paralonchurus brasiliensis*, and the flatfishes *Paralichthys* sp. and *Symphurus jenynsi*. *Trichiurus lepturus*, present as egg and larvae, can be included in this group. Freshwater species, mainly *Parapimelodus nigribarbis*, use the estuary opportunistically during periods of intense freshwater runoff. A last group of species, composed of larvae of marine origin, is occasionally present under conditions of saltwater intrusion and includes larvae of *Anchoa marinii*, *Ophichthus gomesii*, *Urophycis brasiliensis*, *Porichthys porosissimus*, *Epinephelus* sp., *Prionotus punctatus*, *Cynoscion* spp., *Umbrina canosai*, and *Peprilus paru* (Table 1).

A complex extension of the Brazil-Malvinas Confluence over the shelf defines a thermohaline subsurface front, called the Subtropical Shelf Front (STSF) (Fig. 1), between subtropical and subantarctic shelf waters (Piola et al. 2000). The front is located near the 50 m isobath, at 32°S, and extends southward toward the shelf-break close to 36°S. The front, covered by a low-salinity surface layer generated by the discharge of the Rio de la Plata and the Patos Lagoon, is not evident on the surface (Muelbert et al. 2008). Variability in the front position is the result of the dynamics of water masses in the region caused by the seasonal expansion and reversion of the Plata plume waters along the coastal zone. In winter, the low-salinity plume can be up to 100 km wide, and, during summer, the low-salinity plume is reduced, and a strong thermocline is present, and the STSF is marked and more intrusive toward the coast (Muelbert et al. 2008).

The region shows high variability in physical, chemical, and biological properties at seasonal and spatial scales. During winter, an inner-shelf salinity front and a mid-shelf STSF present high nutrients and surface Chl-*a* concentration in the coastal region. During summer, diluted waters extend over the shelf to join the STSF in the upper layer, and the concentration of inorganic nutrients and surface Chl-*a* decreases in shallow waters. *E. anchoita* is the only fish larvae common to both seasons, but with low abundances in summer.

Merluccius hubbsi is present in coastal waters during winter, indicating the influence of waters of southern origin over the coast. At this season, fish larvae of 19 species or taxonomic categories were found (Table 1). *E. anchoita* is the most abundant fish larvae, followed by *M. hubbsi*. The highest abundance of *E. anchoita* eggs was associated to the coastal salinity front. An inshore-frontal assemblage occupies from the innermost to the offshore boundary of the STSF, and an offshore group is

Table 1 Species or taxonomic categories reported as ichthyoplankton for the zones analyzed

Species	South Brazil upwelling zone	Temperate estuarine zone	Patagonian tidal zone	Patagonian Current zone	Shelf-break zone
Acanthistius brasilianus	–	+	+	–	–
Achirus garmani	–	+	–	–	–
Agonopsis chiloensis	–	–	+	+	–
Anchoa marinii	–	+	–	–	–
Argyropelecus aculeatus	–	–	–	–	+
Astronesthidae	–	–	–	–	+
Atherinidae	–	+	+	–	–
Balistes sp.	–	+	–	–	–
Bassanago albescens	–	–	+	–	–
Bathylagus antarcticus	–	–	–	–	+
Bathylagus gracilis	–	–	–	–	+
Bathylagus tenuis	–	–	–	–	+
Blenniidae	–	+	–	–	–
Bregmaceros atlanticus	–	+	–	–	–
Bregmaceros cantori	–	–	–	–	+
Brevoortia aurea	–	+	–	–	–
Brevoortia pectinata	–	+	–	–	–
Careproctus pallidus	–	–	–	+	–
Coelorhynchus fasciatus	–	–	–	–	+
Conger orbignianus	–	–	+	–	–
Congiopodus peruvianus	–	–	+	–	–
Cynoscion guatucupa	–	+	–	–	–
Cynoscion spp.	–	+	–	–	–
Diplodus argenteus	–	–	+	–	–
Dissostichus eleginoides	–	–	–	–	+
Engraulis anchoita	+	+	+	–	+
Epinephelus sp.	–	+	–	–	–
Etropus sp.	–	+	–	–	–
Genypterus blacodes	–	–	+	–	–
Gobiesocidae	–	+	–	–	–
Gobiidae	–	+	–	–	–
Gobiosoma parri	–	+	–	–	–
Gonostoma atlanticum	–	–	–	–	–

(continued)

Table 1 (continued)

Species	South Brazil upwelling zone	Temperate estuarine zone	Patagonian tidal zone	Patagonian Current zone	Shelf-break zone
Gymnoscopelus spp.	–	–	–	–	+
Harpagifer bispinis	–	–	–	+	–
Helcogrammoides cunninghami	–	–	+	–	–
Helicolenus dactylopterus	–	–	–	–	+
Hippocampus punctulatus	–	+	–	–	–
Hoplunnis tenuis	–	–	–	–	+
Hygophum spp.	–	–	–	–	+
Hypleurochilus fissicornis	–	+	–	–	–
Idiacanthus atlanticus	–	–	–	–	+
Krefftichthys anderssoni	–	–	–	–	+
Lampanyctus spp.	–	–	–	–	+
Lepdophanes guentheri	–	–	–	–	+
Lestidiops jayakari	–	–	–	–	+
Lycengraulis grossidens	–	+	–	–	–
Macrodon atricauda	–	+	–	–	–
Macrourus whitsonia	–	–	–	–	+
Maurolicus muelleri	–	–	+	–	+
Maurolicus stehmanni	–	–	–	–	+
Melamphaes sp.	–	–	–	–	+
Menticirrhus americanus	–	+	–	–	–
Menticirrhus spp.	–	+	–	–	–
Merluccius hubbsi	–	+	+	–	–
Micromesistius australis	–	–	–	–	+
Micropogonias furnieri	–	+	–	–	–
Mugil spp.	–	+	–	–	–
Mullus argentinus	–	+	–	–	–
Myctophidae	–	–	–	–	+
Myctophum spp.	–	–	–	–	+
Notolepis coatsi	–	–	–	–	+
Notoscopelus spp.	–	–	–	–	+

(continued)

Table 1 (continued)

Species	South Brazil upwelling zone	Temperate estuarine zone	Patagonian tidal zone	Patagonian Current zone	Shelf-break zone
Nototheniidae	–	–	–	–	+
Odontesthes sp.	–	+	–	–	–
Ophichthus gomesii	–	+	–	–	–
Pagrus pagrus	–	+	–	–	–
Paralepididae	–	–	–	–	+
Paralepis atlantica	–	–	–	–	+
Paralichthys spp.	–	+	–	–	–
Paralonchurus brasiliensis	–	+	–	–	–
Parapimelodus nigribarbis	–	+	–	–	–
Parona signata	–	+	–	–	–
Patagonotothen cornucola	–	–	–	+	–
Patagonotothen sp.	–	–	+	–	–
Patagonotothen tessellata	–	–	–	+	–
Peprilus paru	–	+	–	–	–
Percophis brasiliensis	–	+	+	–	–
Phosichthyidae	–	–	–	–	+
Platanichthys platana	–	+	–	–	–
Pleuronectiformes	–	–	+	–	–
Porichthys porosissimus	–	+	–	–	–
Prionotus nudigula	–	–	+	–	–
Prionotus punctatus	–	+	–	–	–
Prionotus spp.	–	+	–	–	–
Protomyctophum spp.	–	–	–	–	+
Pseudopercis semifasciata	–	–	+	–	–
Salilota australis	–	–	–	+	–
Sardinella brasiliensis	+	–	–	–	+
Scomber colias	–	–	+	–	+
Scopelarchidae	–	–	–	–	–
Scopelosaurus sp.	–	–	–	–	+
Scorpaenidae	–	–	–	–	+
Sebastes oculatus	–	–	+	–	–
Serranus auriga	–	+	–	–	–

(continued)

Table 1 (continued)

Species	South Brazil upwelling zone	Temperate estuarine zone	Patagonian tidal zone	Patagonian Current zone	Shelf-break zone
Sprattus fuegensis	–	–	–	+	+
Sternoptychidae	–	–	–	–	+
Stromateus brasiliensis	–	+	+	–	–
Symbolophorus rufinus	–	–	–	–	+
Symphurus jenynsi	–	+	–	–	–
Symphurus plagiusa	–	+	–	–	–
Syngnathidae	–	–	+	–	–
Syngnathus folleti	–	+	–	–	–
Synodontidae	–	–	–	–	+
Synodus sp.	–	+	–	–	–
Trachurus lathami	+	+	+	–	+
Trichiurus lepturus	–	+	–	–	–
Umbrina canosai	–	+	–	–	–
Urophycis brasiliensis	–	+	–	–	–
Urophycis sp.	–	–	–	–	+
Valencienellus tripunctulatus	–	–	–	–	+
Vinciguerria nimbaria	–	–	–	–	+
Xystreuris rasile	–	+	+	–	–
Zoarcidae	–	–	–	–	+

located farther east. *E. anchoita* and *M. hubbsi* characterized the inshore group. The offshore group showed a heterogeneous composition, and Myctophidae and *Gonostoma atlanticum* are the discriminating species (Muelbert et al. 2008).

The presence of the STSF in summer separates an inshore from an offshore group of species. Inshore waters are occupied by larvae from fish known to spawn in the coastal region, among them *M. furnieri* and other Sciaenidae (Sinque and Muelbert 1997b). The frontal group contained a mixture of larvae of coastal origin such as *T. lepturus* (Sinque and Muelbert 1997b) and fish of oceanic provenance such as *Bregmaceros atlanticus* (Matsuura et al. 1993). During summer, 24 species or taxonomic categories of fish larvae were reported. Mean abundance was much lower than in winter, and the most abundant species were *T. lepturus* and *B. atlanticus*. Abundance was equally distributed from coastal to offshore waters and above and below the thermocline (Muelbert et al. 2008).

Inshore, frontal, and offshore groups of fish larvae are formed during summer. The inshore assemblage occupies the water column of the coastal region and is characterized by larvae of *E. anchoita*, *Etropus* sp., and *T. lepturus*. The frontal assemblage matched with the upper layer of the salinity front and is represented by *Balistes* sp., Pleuronectiformes, and *B. atlanticus*. The offshore group is associated with a region stratified both in temperature and salinity and is marked by the presence of *Synodus* sp. (Muelbert et al. 2008) (Table 1).

3.2 The Río de la Plata Estuary

The Río de la Plata is an extensive and shallow coastal plain estuary located at 35°S (Fig. 1). It receives a mean freshwater discharge of 22,000 $m^3\ s^{-1}$ (Framiñan and Brown 1996). The estuary is characterized by a semipermanent salt-wedge regime, fast reaction to atmospheric forcing, low tidal amplitude, and weak seasonality in river discharge (Guerrero et al. 1997). At larger temporal scales, however, the system is under the effects of large runoff fluctuations, primarily associated with the El Niño Southern Oscillation (ENSO) cycle (e.g., Acha et al. 2012b). During El Niño events, an excess of rainfall is observed in the Plata basin and a shortage during La Niña years, which produce positive and negative discharge anomalies, respectively. A strong vertical stratification characterizes the estuary: fresher waters flow seaward on the surface, while denser shelf water intrudes along the bottom, taking the shape of a salt wedge (Guerrero et al. 1997). In the inner estuary, the flocculation of suspended matter at the tip of the salt wedge, and resuspension of sediment due to tidal current friction at the bottom, forms a turbidity front (Framiñan and Brown 1996). On the continental shelf, the influence of the Plata manifests as a plume of diluted waters that extends north eastward along a coastal strip of 1300 km (Piola et al. 2005) merging to the Patos Lagoon discharge.

The Plata estuary shows an intense reproductive activity of fishes. At least 64% ($n = 21$) of the teleosts that occur in this estuary spawn also inside it (Berasategui et al. 2004). Most abundant species, such as *Brevoortia aurea* and *M. furnieri*, concentrate to spawn near the tip of the salt wedge, at the bottom salinity-turbidity front (BSF/TF) (Acha et al. 2008). Spawning activity at this specific region has been related to retention mechanisms for eggs and larvae, driven by the natural wind variability acting over bathymetric features (Simionato et al. 2008). Modeling results predict that eggs and/or larvae retention along the BSF/TF would be affected by extreme river discharge events (high runoff promotes lower retention and vice versa) and by the spawning location (more upriver spawning promotes higher retention and vice versa). The recruitment variability of *M. furnieri* has been attributed to the interannual river discharge fluctuations acting on the retention properties of the estuary (Acha et al. 2012b).

High zooplankton abundances occur at the BSF/TF that would provide an abundant food supply for fish larvae. The highest observed densities of the dominant copepod *Acartia tonsa* (more than 10,000 ind m^{-3}) were mostly associated with this front (Derisio et al. 2014b). The main preys reported for fish larvae in the estuary are adults and juvenile copepods, mainly of *A. tonsa* and *Paracalanus* sp., while the smaller fish larvae (preflexion stage) mainly feed on copepod nauplii and eggs (Machado et al. 2017).

Fish larvae also occur all along the salt wedge, arranged into assemblages related to different environments (Berasategui et al. 2004). The freshwater assemblage is composed by catfishes (order Siluriformes) larvae, while *M. furnieri* and *B. aurea* characterize the BSF/TF group. The assemblage of the upper/middle estuary is composed by *Gobiosoma parri* and *Paralonchurus brasiliensis*, which occupied the region immediately offshore the BSF/TF. The middle/lower estuary assemblage occurs at the outermost region. Several species of this group are coastal or shelf spawners, with the estuary representing a marginal reproductive area. This assemblage shows the highest diversity and is composed by larvae of *E. anchoita*, *Cynoscion guatucupa*, *Hypleurochilus fissicornis*, *Peprilus paru*, *Prionotus* spp., *Paralichthys* spp., *Symphurus plagiusa*, *Trachurus lathami*, *Macrodon atricauda*, *Trichiurus lepturus*, and *Hippocampus punctulatus*. *Pogonias cromis* spawns also inside the estuary (Acha et al. 2008) though their larvae have not been recorded (Table 1).

Some small rivers discharge into the Plata estuary, where fish eggs and larvae also occur. For example, in the Solis river on the Uruguayan coast, larvae of *P. paru*, *M. furnieri*, Engraulidae, *Platanichthys platana*, *Odontesthes* sp., *B. aurea*, *H. fissicornis*, and *P. orbignyanus* have been reported (Machado et al. 2017). Several of the species that are estuarine spawners in the Plata estuary (*M. furnieri*, *B. aurea*, *Anchoa marinii*, *M. atricauda*, *P. brasiliensis*) are marine spawners in the Patos Lagoon, and they colonize then the lagoon as late larvae or juveniles to establish their nursery grounds (Weiss 1981; Sinque and Muelbert 1997a).

To the north and to the south of the Plata estuary, there are several small coastal lagoons, permanently or intermittently connected to the sea, in which marine ichthyoplankton has been detected. In the estuarine regions of the Rocha and Castillos coastal lagoons (Uruguay), unidentified fish eggs and larvae of *Anchoa marinii*, *M. furnieri*, *C. guatucupa*, *Syngnathus folleti*, *Odontesthes* sp., *B. aurea*, *and Lycengraulis grossidens* were recorded (Machado et al. 2011). The occurrence of spawning females of *M. furnieri* was reported in Rocha coastal lagoon from November to March (Vizziano et al. 2002). Mar Chiquita is a coastal lagoon located on the Argentine side, to the south of the Plata estuary. Most fish species recorded at the surf zone near the inlet and at the inner areas of the lagoon are presumably originated in marine spawning, such as *A. marinii*, *B. aurea*, *C. guatucupa*, *G. parri*, *H. fissicornis*, *M. atricauda*, *Menticirrhus americanus*, *M. furnieri*, *and P. orbignyanus*. At sea, in the mouth surroundings, larvae of *E. anchoita*, *P. paru*, *Porichthys porosissimus*, *Stromateus brasiliensis*, and *Umbrina canosai* were also detected (Bruno et al. 2014).

All these coastal lagoons are microtidal estuaries, <1 m tidal height. The main forcing controlling larvae recruitment into these lagoons is the wind, acting over different scales. At Mar Chiquita lagoon, relatively strong onshore winds aid the larvae to reach the coast and be accumulated in the surf zone, near the lagoon inlet; on the contrary, offshore winds hinder larvae recruitment into the lagoon (Bruno et al. 2014; Bruno and Acha 2015). At shorter time scales, sea breeze (which develops during diurnal hours) blows in onshore direction and aids larvae to penetrate into the lagoon. During the night, sea breeze diminishes, and the incoming tide becomes a more relevant factor for larvae entering into the system (Bruno, unpublished data).

3.3 The El Rincón System

The coastal regime called El Rincón (39–41°S, depth < 40 m, Fig. 1) is characterized by vertical homogeneity due to tidal forcing, and a coastal front that separates diluted coastal waters by the Negro and Colorado rivers discharge (960 $m^3 s^{-1}$ total average discharge), and shelf waters. Salinity gradient is increased by the pouring into the shelf of high-salinity waters from the San Matías Gulf. This front, oriented north-south, shows weak seasonality. Bathymetry and the mean shelf circulation contribute in maintaining frontal structure (Guerrero and Piola 1997; Lucas et al. 2005). Toward the east of the front, shelf waters are stratified in temperature during the warmer months (Martos and Piccolo 1988; Lucas et al. 2005).

El Rincón is a multispecies spawning ground for coastal fishes, several of them of commercial interest. Based on adult studies, reproductive activity of *M. furnieri*, *C. guatucupa*, *Parona signata*, *Percophis brasiliensis*, *Paralichthys patagonicus*, *Xystreuris rasile*, *Mullus argentinus*, *Serranus auriga*, *P. porosissimus*, and *Acanthistius brasilianus* has been documented (Acha et al. 2012a; Macchi and Acha 1998). Studies on ichthyoplankton (Cassia and Booman 1985; Acha et al. 2012a; Hoffmeyer et al. 2009) extend the above list of species to include *A. marinii*, *Prionotus* sp., *B. aurea*, *E. anchoita*, *Pagrus pagrus*, *Peprilus paru*, *Stromateus brasiliensis*, and *Odontesthes argentinensis* (Table 1). In the outer, deeper zone of El Rincón, *E. anchoita* spawn between late spring and early summer. Highest abundances of eggs and larvae have been recorded between October and November near the coast, but the bulk of the spawning seems to be associated to the mid-shelf front located to the east, in deeper waters (see below) (Pájaro 1998; Pájaro et al. 2009).

Due to a combination of the coastline shape and the northward-flowing shelf waters, a recirculation cell would exist in this area (Piola and Rivas 1997; Auad and Martos 2012). Eggs and larvae could be retained by that gyre. Modeling exercises show a gradient in flow intensity across the cell that would favor an upwelling (Auad and Martos 2012) able of injecting nutrients into the photic layer and to promote primary production. Chl-*a* field measurements show the highest values very close to the coast (Hoffmeyer et al. 2009), in a region in which satellite images are not reliable. El Rincon presents high densities of small copepods, nauplii, and eggs, which constitute adequate prey for fish larvae (Marrari et al. 2004).

4 The Tidal Zone

The tidal zone extends from north of Valdés (42°S) to Staten Island (55°S). The Valdés tidal front is the largest and best known of those fronts, but some minor and less studied tidal fronts exist southward, generated by the high levels of tidal energy dissipation along the Patagonian coast (Palma et al. 2004b). We include into this zone also the Bonaerense mid-shelf front, which is not a true tidal front but its vertical structure and the ichthyoplankton species resemble those of the tidal fronts.

4.1 Valdés Peninsula Tidal Front

The Valdés Peninsula tidal mixing front (Fig. 1) is a mesoscale (100–1000 km) thermal front observed in spring and summer, which defines the boundary between stratified offshore waters and a coastal, vertically mixed water mass (Carreto et al. 1986). Surface warming during spring and summer induces the stratification of shelf waters, and the mixing of the coastal water is forced by vertical shear induced by tidal currents at particular topographic shoals southeast and northeast of the peninsula (Carreto et al. 1986). Wind stress on the surface layer also contributes to the formation and maintaining of the homogeneous side of the front. The structure of the front is maintained until autumn when stratification of shelf waters decays.

The front is highly productive during spring and summer; nutrient enrichment is largely dependent on intensified vertical mixing in the homogeneous side of the front (Carreto et al. 1986). Zooplankton of several size fractions occur at different sectors of the front. Microzooplankton, mostly copepod eggs and nauplii, mainly occupy the transitional zone (between mixed coastal waters and stratified shelf waters), in coincidence with the maximum Chl-*a* values (Viñas and Ramírez 1996; Derisio et al. 2014a). High abundance of early stages of zooplankton constitutes a valuable trophic offer for fish larvae such as *E. anchoita*, *Merluccius hubbsi*, and *Genypterus blacodes*.

E. anchoita spawning peaks in December at this region. The tidal front seems to be the axis for the distribution of the spawning shoals (Pájaro 1998; Hansen et al. 2001). *M. hubbsi* also spawns related to the front (Ehrlich and Ciechomski 1994; Álvarez-Colombo et al. 2011), but peak spawning occurs in January (Pájaro et al. 2005). Larvae of both species are mainly observed in the stratified sector, but while *M. hubbsi* larvae occurred at the bottom layer, *E. anchoita* larvae were more frequent in the upper layer (Viñas and Santos 2000).

Variability in phytoplankton abundance as Chl-*a* concentration would likely have an influence on the survival of *E. anchoita* and *M. hubbsi* larvae via changes in the production of adequate prey (Viñas and Ramírez 1996; Temperoni and Viñas 2013; Marrari et al. 2016). Small copepod species usually dominate in coastal waters and produce eggs and nauplii in the optimum size range of prey for first-feeding *E. anchoita* larvae (Viñas and Ramírez 1996). *M. hubbsi* larvae prey also on adults of smaller copepod species, mainly individuals <2 mm in length (Viñas and Santos 2000; Temperoni and Viñas 2013).

Although *E. anchoita* and *M. hubbsi* eggs and larvae dominate the region, ichthyoplankton of other species have been also observed. In the stratified waters, at depths around 100 m, leptocephalus larvae of congers *Conger orbignianus* and

Bassanago albescens occur in low densities. These larvae occurred at the end of spring, when the surface water temperature reaches 14–16 °C (Ciechomski et al. 1979). Between spring and summer, high densities of *Maurolicus muelleri* were reported between 41°S and 43°S (Ciechomski et al. 1979) (Table 1).

4.2 Isla Escondida and San Jorge Gulf Tidal Fronts

The continental shelf between 43°S and 47°S at depths lesser than 100 m constitutes the main spawning ground of the Patagonian stocks of the Argentine *M. hubbsi* and also for the *E. anchoita*. Several tidal fronts, very close to the coast, characterize this region (Palma et al. 2004b). The San Jorge Gulf (Fig. 1) is a semi-enclosed basin with a rather deep connection to the open shelf. Vertical mixing by wind and tides and the entrance of southern diluted waters from the Magellan Strait (see below) characterize the northern and southern gulf's extremes, generating frontal systems. Those fronts produce nutrient enrichment and retention mechanisms of ecological importance (Guerrero and Piola 1997; Carreto et al. 2007). Most of eggs and larvae in this region correspond to *E. anchoita*, followed by *M. hubbsi*. The copresence of eggs of both species is common; usually *E. anchoita* eggs are located at the mixed upper layer, while *M. hubbsi* eggs are located within and below the thermocline (Ehrlich 1998).

High concentrations of *M. hubbsi* eggs and larvae occur between 43°S and 45°S in coastal and mid-shelf waters. The location of the spawning area is well documented and matches the bottom expression of a tidal front located parallel to the coastline at ca. 80 m depth (Pájaro et al. 2005; Macchi et al. 2010). Larvae attain retention making daily vertical migrations in a region where flux is vertically structured, so early larvae remain near the spawning ground. In the following months, they are slowly advected southwestward by bottom currents to their nursery and settlement locations, particularly the San Jorge Gulf, considered being the main nursery ground (Álvarez-Colombo et al. 2011, 2014).

M. hubbsi larvae in this region feed mainly on Calanoid copepodites (1–2 mm in length) but also on small-sized adult copepods (<1 mm) such as *Calanoides carinatus* and *Drepanopus forcipatus*. Larvae would feed on small prey at the onset of feeding and replace them with larger prey as they grow (Viñas and Santos 2000; Temperoni and Viñas 2013).

Although *M. hubbsi* and *E. anchoita* larvae are dominant, ichthyoplankton of other 17 species or taxonomic categories were found in lower abundances: *Genypterus blacodes*, *Sebastes oculatus*, *Congiopodus peruvianus*, *Prionotus nudigula*, *Agonopsis chiloensis*, *Helcogrammoides cunninghami*, *T. lathami*, *Diplodus argenteus*, *Patagonotothen* sp., *P. brasiliensis*, *Pseudopercis semifasciata*, *Stromateus brasiliensis*, *Xystreuris rasile*, and larvae of Pleuronectiformes, Atherinidae, and Syngnathidae (Table 1). Most of these species were regularly caught along several years showing the stability of the assemblage.

4.3 *The Bonaerense Mid-Shelf Front*

Along the shelf of the Buenos Aires province (37–42°S) and close to the 50 m isobath, a mid-shelf front (MSF) is established during the spring and summer months (Fig. 1). The vertical and horizontal distribution of properties allows identifying a cross-shelf zonation: a coastal homogenous zone (depth lesser than 40–50 m), separated by a shallow sea front from the mid-shelf regime (Martos and Piccolo 1988; Lucas et al. 2005; Romero et al. 2006). This front has the same orientation than that of Valdés Peninsula, and likewise it is formed during the warmer months. However, this is not a true tidal front, tidal amplitudes are by far smaller than in the Patagonia region, and mixing is mainly due to the winds (Lucas et al. 2005).

A main population of *E. anchoita* occurs south of 34–41°S. Spawning of this stock is extensive and takes place throughout the year, but maximum reproductive activity occurs during spring, dominating the regional ichthyoplankton abundance. In October and November, the peak of the spawning takes place in the coastal-shelf area. However, *E. anchoita* eggs and newly hatched larvae are distributed widely over the entire region. Highest egg and larvae abundances (>1000 ind m^{-2}) were found placed along the 40–60 m isobaths in relation to the mid-shelf front (Pájaro 1998). The timing of initiation, duration, and magnitude of the spring phytoplankton bloom at the front rule larval survival, suggesting that phytoplankton dynamics strongly influence recruitment (Marrari et al. 2013). Small copepod species usually dominate in coastal waters and produce eggs and nauplii in the optimum size range of prey for first-feeding *E. anchoita* larvae (Viñas et al. 2002).

Although *E. anchoita* eggs and larvae are dominant, ichthyoplankton of other species also occur in the region such as *Scomber colias*, *Prionotus nudigula*, *Acanthistius brasilianus*, and *Stromateus brasiliensis*, mainly during spring and summer (Ciechomski et al. 1979; Sánchez and Ciechomski 1995) (Table 1).

5 The Patagonian Current Zone

A diluted plume, originated in the discharge of Magellan Strait and those waters entering the continental shelf through the Le Maire Strait, spreads on the shelf due to the effects of diffusion and advection of the Ekman layer caused by the prevailing westerlies (e.g., Acha et al. 2004) (Fig. 1). The plume is vertically homogeneous due to tide mixing and wind stress. It is traced 200 km offshore (100 m depth) and 800 km northward, reaching the southern extreme of the San Jorge Gulf. At this point, the plume meets a coastal tidal front (see above). Circulation models have shown that local wind field generates a large anticlockwise circulation cell in the Grande Bight at 50–52°S (Palma et al. 2004a, b). Close to the coast, along the Grande Bight and Tierra del Fuego, those models predict the occurrence of tidal fronts. Satellite and field measurements persistently show a regional maximum of surface Chl-*a* in the Grande Bay (Romero et al. 2006; Sabatini et al. 2016), and a maximum in primary production has been estimated

(Lutz et al. 2010). Consequently, copepod biomasses are high, peaking at the end of the austral summer, after the phytoplankton bloom (Sabatini et al. 2016).

The Fuegan spratt (*Sprattus fuegensis*) spawns near the coast of Tierra del Fuego, mostly in the Beagle Channel where the largest densities were registered, but also in the Le Maire Strait and to the north of Mitre Peninsula (Sánchez et al. 1995, 1997). Larvae occur in the same stations that eggs, but also in large densities all along the Fuegan Atlantic coast off the Magellan Strait, and to a minor extent in the coastal region of Santa Cruz province, in correspondence to the tidal fronts described for the area by modeling (Palma et al. 2004a, b). The large anticlockwise circulation cell advects the larvae to the Patagonian coast, where they are transported by the southern coastal waters, to the warmer nursery grounds at the south extreme of San Jorge Gulf. A southward coastal countercurrent from San Jorge Gulf, a probable consequence of a slackening and offshore displacement of the Magellan discharge, may facilitate retention in the coastal habitat. The bulk of 0-group sprat from metamorphosis onward occurs in association with southern coastal waters, mostly at bottom depths below 50 m.

The Beagle Channel is part of the Fuegian system of fjords and canals. It extends in an E-W direction for 200 km approximately, connecting the Atlantic and the Pacific Oceans. Strong winds from the SW are dominant, and currents in the channel are strong, especially along its middle portion, running from the west to east (Balestrini et al. 1998). Large forests of the giant kelp *Macrocystis pyrifera* (Linnaeus) characterize the coasts of the channel providing refuge for fish larvae. Though ichthyofaunal diversity is modest, reproductive strategies are varied. *Harpagifer bispinis*, *Patagonotothen tessellata*, and *P. cornucola* spawn in the rocky intertidal where the parents, both for the former species and only males for the two latter, guard the eggs until hatching. *Careproctus pallidus* extrude eggs through an anteriorly positioned ovipositor into the branchial chambers of king crabs (*Lithodes santolla*), and *Salilota australis* and *S. fuegensis* spawn free planktonic eggs. Free larvae of all those species were recorded in the plankton of the Beagle Channel, together with those of *Agonopsis chiloensis* whose eggs are unknown (Bruno et al. in press).

Fish larvae at the Beagle Channel are present throughout the year. Regardless of the reproductive strategy, larvae are more abundant at the edge of the giant kelp forests than in open waters (Bruno et al. in press). The presence of giant kelp forests mainly reduces water currents (Wu et al. 2017) and concentrates plankton (Jones et al. 1994) allowing fish larvae to be retained in an area of abundant preys diminishing their dispersion (Wu et al. 2017).

5.1 The Shelf-Break Zone

The open ocean circulation is dominated by the opposite flow of the Brazil (subtropical) and the Malvinas (subantarctic) currents. Both currents flow along the shelf-break and meet, in average, at 36°S. In this area, referred to as the Brazil-Malvinas Confluence, the two flows turn offshore in a series of large amplitude meanders.

5.1.1 The Brazil Current Shelf-Break System

Cyclonic eddies of the Brazil Current (BC) (Fig. 1) are common mesoscale processes on the mid- and outer shelf throughout the year (Silveira et al. 2000). The BC flows southward along the continental slope near the shelf edge, where vertical current shear and bottom topography contribute to the development of meandering and eddies (Campos et al. 1995). During summer, eddy-derived waters are pushed toward the inner shelf with SACW intrusions, but during winter, eddies are short-lived and geographically restricted (Lopes et al. 2006). Shelf-break upwelling of the SACW is enhanced by these eddies and contributes to the enrichment of the continental shelf (Brandini et al. 1989; Gaeta and Brandini 2006). The cyclonic eddies of the BC also favor the movement of the upwelling frontal zone toward the inshore area (Campos et al. 1995; 2000). The position of the bottom thermal front changes seasonally and is closer to the coast during summer and farther offshore during winter.

Meandering features of the Brazil Current provide an additional mechanism of upward motion of the cold and nutrient-rich South Atlantic Central Water, increasing phyto- and zooplankton biomass and production on mid- and outer shelves (Lopes et al. 2006). Reports indicate that larval fish abundance is lower at the center of this feature, suggesting that the eddy advected poorer offshore waters of tropical origin toward the inner shelf, concentrating the larvae around the eddy (Franco et al. 2006). The analyses of one of these eddies revealed that fish larvae of Myctophidae, Bregmacerotidae, Clupeidae, Synodontidae, and Engraulidae were most abundant on the inshore flow side of the meander. In the offshore flow region, the intrusion and mixing of coastal and subtropical waters were associated to abundant larvae of Engraulidae and Bregmacerotidae.

The continental slope presents a patchy distribution of ichthyoplankton, marked by high and low concentrations of fish eggs and larvae (Macedo-Soares et al. 2014). The outer shelf is marked by larvae of Scorpaenidae and larvae of pelagic fishes such as Myctophidae, Phosichthyidae, and Paralepididae. The slope registered high abundance of *Maurolicus muelleri* eggs and yolk sac larvae, indicating an important spawning area. High larval abundance at the outer shelf and slope may be related to the eddy-induced upwelling or due to larval retention by meanders or eddies (Franco et al. 2006; Macedo-Soares et al. 2014).

The meandering and frontal system of the BC and the extension of the CW and SACW influence the distribution of ichthyoplankton assemblages. During spring, the shelf assemblage is composed of larvae of small pelagic fishes, such as *S. brasiliensis*, *E. anchoita*, and *T. lathami*. In autumn, when hydrographic conditions are similar to the winter, the shelf assemblage is dominated by *Bregmaceros cantori* and *E. anchoita*. The transitional assemblage only occurred during spring and was composed of mesopelagic and coastal species. The oceanic assemblage was dominated by the mesopelagic families, Myctophidae, Sternoptychidae, and Phosichthyidae, and was similar in both seasons (Katsuragawa et al. 2014) (Table 1).

Eddy-meandering activity is also suggested as an important retention mechanism for fish larvae along the shelf-break. Mesopelagic species, especially *Maurolicus stehmanni*, spawn at continental slopes in onshore moving waters to maintain the larvae in a favorable retention area between the oceanic and coastal waters (Weiss et al. 1988). At the South Brazilian shelf, mesopelagic larvae associated with tropical waters (Franco and Muelbert 2003) and an increase in larval fish abundance around the eddy periphery (Franco et al. 2006; Katsuragawa et al. 2014) indicate this mechanism as a possibility for retention on more productive areas.

5.1.2 The Argentine Shelf-Break System

Near the continental shelf-break southward 36°S, the shelf waters meet the cooler and more saline waters of the Malvinas current, thus producing a thermohaline front (Martos and Piccolo 1988). This shelf-break front is a permanent feature that characterizes the border of the shelf. It may be followed from the Burdwood Bank along the shelf-break to the east, around Malvinas Islands, and northward up to the Brazil-Malvinas Confluence (Acha et al. 2004). This confluence, generated at the meeting of the major oceanic currents in the Southwestern Atlantic, produces an extended region of interrelated fronts extending offshore to the oceanic domain.

A few sampling stations near the Brazil-Malvinas Confluence show high species richness in a region where several water masses converge. Abundance is dominated by pelagic species such as *S. brasiliensis*, *E. anchoita*, and Myctophidae (*Hygophum* spp., *Lampanyctus* spp., *Lepdophanes guentheri*, *Myctophum* spp., *Notoscopelus* spp., and *Symbolophorus rufinus*). Other species or taxonomic categories recorded were *Hoplunnis tenuis*, *Bathylagus tenuis*, *Bathylagus gracilis*, *Argyropelecus aculeatus*, *Valencienellus tripunctulatus*, *Vinciguerria nimbaria*, *Idiacanthus* sp., *Lestidiops jayakari*, *Paralepis atlantica*, *Coelorhynchus* sp., *Urophycis* sp., *Melamphaes* sp., *Helicolenus dactylopterus*, and species belonging to Paralepididae, *Scopelosaurus* sp., Scopelarchidae, Macrouridae, and Astronesthidae (Table 1).

Southward the Confluence, the shelf-break is characterized by high concentrations of Chl-*a* in spring and summer (Romero et al. 2006; Carreto et al. 2016) and high concentrations of zooplankton, composed in part by small copepods in a size range that could be useful as prey for fish larvae (Sabatini and Alvarez Colombo 2001). The interaction of MC and the bottom topography produces a cross-shelf divergence of the current transport that leads to shelf-break upwelling, with the intensity proportional to the transport of the MC (Matano and Palma 2008). The upwelling along the SBF provides a large source of macronutrients and probably the dissolved iron needed to sustain the intense spring phytoplankton blooms (Carreto et al. 2016).

The ichthyoplankton of the southern Patagonian shelf and the slope has been poorly studied. Most of the sampling effort in the outer shelf and shelf-break was concentrated between 47°S and 55°S, focused on the reproductive concentrations of blue whiting *Micromesistius australis* around the Malvinas Islands (Ehrlich et al. 1999). Ichthyoplankton belong to species characteristic of the Magallanes province that inhabits also the oceanic realm. Eggs of *M. australis* and the grenadiers *Coelorinchus fasciatus* and *Macrourus whitsonia* were identified. More than 80% of the larvae found belong to the cosmopolitan family Myctophidae (*Krefftichthys anderssoni*, *Gymnoscopelus* spp., and *Protomyctophum* spp.). Other species or taxonomic categories recorded were *S. fuegensis*, *Bathylagus antarcticus*, *Notolepis coatsi*, *M. australis*, *Dissostichus eleginoides*, *Idiacanthus atlanticus*, Nototheniidae, and Zoarcidae.

6 Conclusions

The upwelling is highly variable and pulsating environments, mainly controlled by wind dynamics. Upwelling by winds generates the upward flow of nutrient-rich waters, promoting primary production. Small pelagics characterize the ichthyoplankton. The estuarine zone is dominated by coastal demersal species, notably by Sciaenidae. Estuaries are biologically productive due to the input of allochthonous nutrients and local processes. The meeting of salty and diluted waters generates density interfaces that could concentrate prey for fish larvae. Water dynamics is complex, characterized by 3D movements that could be employed by fish larvae to stay retained in the nursery grounds. The region is highly dynamic showing variability of high (days) and low frequency (interannual, ENSO cycle). The main forcing is continental discharge and winds.

Tidal energy generates in several productive Patagonian fronts related to specific geomorphological features. Those fronts show high abundance of Chl-*a* and zooplankton during spring and summer, and nutrient enrichment largely depends on the intensified vertical mixing in the homogeneous side of the front. Species of ecological and fishing relevance, such as *E. anchoita* and *M. hubbsi*, spawn in relation to these fronts. Retention seems to be mediated by larvae movements in a vertically structured flux. Although *E. anchoita* and *M. hubbsi* are dominant, other shelf species take advantage of those spawning and nursery grounds.

Southward of the Patagonian Current zone, in the highly advective Beagle Channel, the dominant reproductive strategy is parental care of eggs and larvae at the intertidal zone or hiders in live invertebrates. Large and well-developed larvae characteristic of such strategies remain near the kelp forests where water flow is perturbed by the mats increasing their chances to be retained. On the other side, the species with small and free planktonic eggs, like the Fuegan sprat, take advantage of a large anticlockwise circulation cell in which eggs, larvae, and juveniles are embedded. High abundances of Chl-*a* and zooplankton occur at this circulation cell, creating feeding opportunities for fish larvae.

In the north of the shelf-break zone, eddies and meanders characterize the highly energetic Brazil Current. Shelf-break upwelling of nutrient-rich waters is enhanced by these eddies and meanders, increasing phyto- and zooplankton biomass on mid- and outer shelves, playing also a role in larvae retention. Pelagic and mesopelagic fish assemblages dominate the ichthyoplankton. The Brazil-Malvinas Confluence is a highly complex and energetic environment. The few studies there show high species richness with a dominance of pelagic and mesopelagic species. Southward of the Confluence, the shelf-break is characterized by high concentrations of Chl-*a* in

Table 2 Main processes responsible for the Bakun's triads at the zones analyzed

Zone	Nutrient enrichment	Food particle concentration	Retention
South Brazil upwelling	Wind-driven upwelling	FCUS: wind-driven upwelling SMC: bottom turbulence results in subsurface Chl-*a* maximum	?
Temperate estuarine	Allochthonous nutrients supplied by river discharges	Haloclines	Wind-driven circulation
Patagonian tidal	Tidal mixing	Surface convergence; thermoclines	Vertically structured water dynamics (stratified side)
Patagonian current	Allochthonous nutrients supplied by continental runoff?	?	Wind-driven anticyclonic cell
Shelf-break	Eddies and meanders; interaction of currents and bottom topography	Thermoclines?	Eddies and meanders (Brazil current)

spring and summer, being the interaction of MC and bottom topography the upwelling mechanism that promotes primary production. Ichthyoplankton of this region is poorly studied. Mesopelagic species such as *M. australis*, grenadiers, and Myctophidae dominate this highly advective scenario.

The diversity of fronts and other oceanographic structures that characterize this region seems to play a key role in allowing the life cycle closure of a range of very different fishes, from tropical/subtropical to subantarctic, and from estuarine to oceanic species. Several forcing such as continental discharge, wind, tides, sea-atmosphere heat exchanges, and oceanic circulation interacts with bottom topography or coastline orientation to generate processes able to fulfill the requirements of Bakun's triad in the different zones (Table 2). In this way, a complex mosaic of opportunities for different fish spawning and larvae nursery was generated. *

References

Acha EM, Mianzan H, Guerrero R et al (2004) Marine fronts at the continental shelves of austral South America. Physical and ecological processes. J Mar Syst 44:83–105

Acha EM, Mianzan H, Guerrero RA et al (2008) An overview of physical and ecological processes in the Rio de la Plata Estuary. Cont Shelf Res 28(13):15759–11588

Acha EM, Orduna M, Rodrigues K et al (2012a) Caracterización de la zona de El Rincón (Provincia de Buenos Aires) como área de reproducción de peces costeros. Revista de Investigación y Desarrollo Pesquero, INIDEP 21:31–43

Acha EM, Simionato CG, Carozza CR et al (2012b) Climate–induced year classes' fluctuations of whitemouth croaker Micropogonias furnieri (Pisces, Sciaenidae) in the Río de la Plata estuary, Argentina–Uruguay. Fish Oceanogr 21(1):58–77

Acha EM, Piola AR, Iribarne O, Mianzan H (2015) Ecological processes at marine fronts: oases in the ocean. Springer briefs in environmental science. Springer, New York

Álvarez–Colombo G, Dato C, Macchi GJ et al (2011) Distribution and behavior of argentine hake larvae: evidence of a biophysical mechanism for self–recruitment in northern Patagonian shelf waters. Cs Mar 37(4B):633–657

Álvarez–Colombo GL, Dato CV, Machinandiarena L et al (2014) Daylight vertical segregation of young–of–the–year Argentine hake Merluccius hubbsi: advances in assessment of juvenile abundance with acoustic methods. Fish Res. https://doi.org/10.1016/j.fishres.2014.03.014

Auad G, Martos P (2012) Climate variability of the northern Argentinean shelf circulation: impact on *Engraulis anchoita*. Int J Ocean Clim Syst 3(1):17–43

Bakun A (1996) Patterns in the ocean: ocean processes and marine population dynamics. Sea Grant, La Jolla

Bakun A, Parrish RH (1991) Comparative studies of coastal pelagic fish reproductive habitats: the anchovy (Engraulis anchoita) of the southwestern Atlantic. ICES J Mar Sci 48:343–361

Balestrini CF, Manzella G, Lovrich GA (1998) Simulación de corrientes en el canal Beagle y Bahía Ushuaia mediante un modelo bidimensional, 98. Informe Técnico Servicio de Hidrografía Naval

Berasategui AD, Acha EM, Fernández Araoz NC (2004) Spatial patterns of ichthyoplankton assemblages in the Rio de la Plata estuary (Argentina–Uruguay). Estuar Coast Shelf Sci 60:599–610

Bisbal GA (1995) The Southeast South American shelf large marine ecosystem. Mar Policy 19(1):21–38

Brandini FP, Moraes CLB, Thamm CA (1989) Shelf break upwelling, subsurface maxima of chlorophyll and nitrite, and vertical distribution of a subtropical nano– and microplankton community off southeastern Brazil. In: Brandini FP (ed) Memórias do III Encontro Brasileiro de Plâncton. Editora UFPR, Curitiba, pp 47–56

Bruno DO, Acha EM (2015) Winds vs. tides: factors ruling the recruitment of larval and juvenile fishes into a microtidal and shallow choked lagoon (Argentina). Environ Biol Fish 98:1449–1458

Bruno DO, Delpiani SM, Cousseau MB et al (2014) Ocean–estuarine connection for ichthyoplankton through the inlet channel of a temperate choked coastal lagoon (Argentina). Mar Freshw Res 65:1116–1130

Bruno DO, Victorio MF, Acha EM et al (in press) Fish early life stages associated with giant kelp forests in sub–Antarctic coastal waters (Beagle Channel, Argentina). Polar Biol. https://doi.org/10.1007/s00300–017–2196–y

Burrage D, Wesson AJ, Martinez C et al (2008) Patos Lagoon outflow within the Río de la Plata plume using an airborne salinity mapper: observing an embedded plume. Cont Shelf Res 28:1625–1638

Busoli RO (2001) Transporte e retenção de ovos e larvas de *Engraulis anchoita* na Plataforma Continental Sul do Brasil. M.Sc. Dissertation. Universidade Federal do Rio Grande, Brazil, 141p.

Campos EJD, Gonçalves JE, Ikeda Y (1995) Water mass structure and geostrophic circulation in the South Brazil Bight – summer of 1991. J Geophys Res 100(C9):18537–18550

Campos EJD, Velhote D, da Silveira ICA (2000) Shelf break upwelling driven by Brazil current cyclonic meanders. Geophys Res Lett 27:751–754

Campos PC, Möller OO, Piola AR, Palma ED (2013) Seasonal variability and coastal upwelling near Cape Santa Marta (Brazil). J Geophys Res Oceans 118:1420–1433

Carreto JI, Benavides HR, Negri RM et al (1986) Toxic red–tide in the Argentine Sea. Phytoplankton distribution and survival of the toxic dinoflagellate *Gonyaulax excavata* in a frontal area. J Plankton Res 8:15–28

Carreto JI, Carignan MO, Montoya NG et al (2007) Ecología del fitoplancton en los sistemas frontales del Mar Argentino. In: El mar Argentino y sus recursos pesqueros, vol 5. Instituto Nacional de Investigación y Desarrollo Pesquero (INIDEP), Mar del Plata, pp 11–31

Carreto JI, Montoya NG, Carignan MO et al (2016) Environmental and biological factors controlling the spring phytoplankton bloom at the Patagonian shelf–break front – degraded fucoxanthin pigments and the importance of microzooplankton grazing. Prog Oceanogr 146:1–21. https://doi.org/10.1016/j.pocean.2016.05.002

Cassia MC, Booman CI (1985) Distribución del ictioplancton en el Mar Argentino en los años 1981–1982. Physis 43(105):91–111

Castelão RM, Barth JA (2006) Upwelling around Cabo Frio, Brazil: the importance of wind stress curl. Geophys Res Lett 33(L03602). https://doi.org/10.1029/2005GL025182
Castello JP (1997) Pelagic teleosts. In: Seeliger U, Odebrecht C, Castello JP (eds) Subtropical convergence environments. The coast and the sea in the Southwestern Atlantic. Springer, New York, pp 123–128
Castello JP, Vasconcellos MC (1995) Growth rate of anchovy (Engraulis anchoita) larvae caught off Cape Santa Marta Grande (Brazil). Arch Fish Mar Res 42(3):263–281
Ciechomski JD, Ehrlich MD, Lasta CA et al (1979) Distribución de huevos y larvas de peces en el mar argentino y evaluación de los efectivos desovantes de anchoíta y de merluza. Contrib Inst Nac Invest Des Pesq (INDEP) 383:59–79
Coelho–Souza SA, Soledad López M, Guimarães JRD et al (2012) Biophysical interactions in the Cabo Frio upwelling system, Southeastern Brazil. Braz J Oceanogr 60(3):353–365
Costa CSB, Seeliger U, Kinas P (1988) The effect of wind velocity and direction on the salinity regime in the lower Patos Lagoon estuary. Ciência e Cultura 40(9):909–912
Cousseau MB (1997) Peces, crustáceos y moluscos registrados en el sector del Atlántico Sudoccidental comprendido entre 34° y 55°S, con indicación de las especies de interés pesquero. INIDEP Informe Técnico 5:96 pp.
Cousseau MB, Perrotta RG (1998) Peces Marinos de Argentina. Biología, distribución, pesca. INIDEP, Mar del Plata
Cury JC, Araujo FV, eta l C–SAS (2011) Microbial diversity of a Brazilian coastal region influenced by an upwelling system and anthropogenic activity. PLoS One 6:e16553
Derisio C, Alemany D, Acha EM et al (2014a) Influence of a tidal front on zooplankton abundance, assemblages and life histories in Península Valdés, Argentina. J Mar Syst 139:475–485
Derisio C, Braverman M, Gaitán E et al (2014b) The turbidity front as a habitat for Acartia tonsa (Copepoda) in the Río de la Plata Estuary. J Sea Res 85:197–204
Dias JF, Petti MAV, Corbisier TN (2016) Trophic position and nutritional condition of the anchovy Engraulis anchoita larvae in the Cabo Frio region, Brazil. Vie et milieu – Life and Environment 66(3–4):275–285
Ehrlich MD (1998) Los primeros estadios de vida de la merluza Merluccius hubbsi Marini 1933, como aporte al conocimiento de su reclutamiento y estructura poblacional, Universidad de Buenos Aires
Ehrlich MD, Ciechomski JD (1994) Reseña sobre la distribución de huevos y larvas de merluza (*Merluccius hubbsi*) basada en veinte años de investigaciones. Frente Mar 15:37–50
Ehrlich MD, Sánchez RP, Ciechomski JD et al (1999) Ichthyoplankton composition, distribution and abundance on the southern patagonian shelf and adjacent waters. INIDEP Doc Cient 5:37–65
Fernandes EHL, Dyer KR, Möller OO et al (2002) The Patos Lagoon hydrodynamics during an El Nino event (1998). Cont Shelf Res 22:1699–1713
Framiñan MB, Brown OB (1996) Study of the Río de la Plata turbidity front, Part I: spatial and temporal distribution. Cont Shelf Res 16(10):1259–1282
Franco BC, Muelbert JH (2003) Distribuição e composição do ictioplâncton na quebra de plataforma do Sul do Brasil. Atlantica 25:75–86
Franco BC, Muelbert JH, Mata MM (2006) Mesoscale physical processes and the distribution and composition of ichthyoplankton on the southern Brazilian shelf break. Fish Oceanogr 15:37–43
Freitas DM, Muelbert JH (2004) Ichthyoplankton distribution and abundance off Southern Brazil. Braz Arch Biol Technol 47(4):601–612
Gaeta SA, Brandini FP (2006) Produção primária do fitoplâncton na região entre o Cabo de São Tomé (RJ) e o Chuí (RS). In: Rossi-Wongtschowski CLDB, Madureira LS (eds) O Ambiente oceanográfico da Plataforma Continental e do Talude na Região Sudeste–Sul do Brasil. Editora da USP, São Paulo, pp 219–264
Garcia A, Vieira JP, Winemuller K (2001) Dynamics of the shallow–water fish assemblage of the Patos Lagoon estuary (Brazil) during cold and warm ENSO episodes. J Fish Biol 59:1218–1238
Gonzalez Rodriguez E (1994) Yearly variation in primary productivity of marine phytoplankton from Cabo Frio (RJ, Brazil) region. Hydrobiologia 294:145–156

Guerrero RA, Piola AR (1997) Masas de agua en la plataforma continental. In: Boschi EE (ed) El Mar Argentino y sus recursos pesqueiros. Tomo 1. Antecedentes históricos de las exploraciones en el mar y las características ambientales. Instituto Nacional de Investigación y Desarrollo Pesquero, Mar del Plata, pp 107–118

Guerrero RA, Acha EM, Framiñan MB, Lasta CA (1997) Physical oceanography of the Río de la Plata Estuary, Argentina. Cont Shelf Res 17(7):727–742

Hansen JE, Martos P, Madirolas A (2001) Relationship between spatial distribution of the Patagonian stock of Argentine anchovy, Engraulis anchoita, and sea temperatures during late spring–early summer. Fish Oceanogr 10(2):193206

Hoffmeyer MS, Menéndez MC, Biancalana F et al (2009) Ichthyoplankton spatial pattern on the inner shelf off Bahía Blanca Estuary, SW Atlantic Ocean. Estuar Coast Shelf Sci 84:383–392

Jones CG, Lawton JH, Shachak M (1994) Organisms as ecosystem engineers. Oikos 69:373–386

Katsuragawa M, Matsuura Y (1992) Distribution and abundance of carangid larvae in the Southeastern Brazilian Bight, during 1975–1981. Bolm Inst Oceanogr Sao Paulo 40:55–78

Katsuragawa M, Dias JF, Harari J, Namiki C et al (2014) Patterns in larval fish assemblages under the influence of the Brazil current. Cont Shelf Res 89:103–117

Lopes RM, Katsuragawa M, Dias JF et al (2006) Zooplankton and ichthyoplankton distribution on the southern Brazilian shelf: an overview. Sci Mar 70(2):189–202

Lucas A, Guerrero RA, Mianzan HW et al (2005) Coastal oceanographic regimes of northern Argentina (34°–43°S). Cont Shelf Res 65:405–420

Lutz VA, Segura V, Dogliotti AI, Gagliardini D et al (2010) Primary production in the Argentine Sea during spring estimated by field and satellite models. J Plankton Res 32(2):181–195

Macchi GJ, Acha EM (1998) Aspectos reproductivos de las principales especies de peces muestreadas durante la campaña costera H–13/94. INIDEP Inf Téc 21:67–89

Macchi GJ, Martos P, Reta R, Dato C (2010) Offshore spawning of the Argentine hake (Merluccius hubbsi) Patagonian stock. Pan Am J Aquat Sci 5(1):22–35

Macedo–Soares LCP, Garcia CAE, Freire AS et al (2014) Large–scale Ichthyoplankton and water mass distribution along the South Brazil Shelf. PLoS One 9(3):e91241

Machado I, Conde D, Rodríguez–Graña L (2011) Composition and spatial distribution of ichthyoplankton in intermittently–open coastal lagoons of Uruguay. Pan Am J Aquat Sci 6(3):237–243

Machado I, Calliari D, Denicola A, Rodríguez–Graña L (2017) Coupling suitable prey field to in situ fish larval condition and abundance in a subtropical estuary. Estuar Coast Shelf Sci 187:31–42. https://doi.org/10.1016/j.ecss.2016.12.021

Marrari M, Viñas MD, Martos P, Hernández D (2004) Spatial patterns of mesozooplankton distribution in the Southwestern Atlantic Ocean (34°–41° S) during austral spring: relationship with the hydrographic conditions. ICES J Mar Sci 61:667–679

Marrari M, Signorini S, Mcclain C, Pájaro M, Martos P, Viñas MD, Hansen J, Dimauro R, Cepeda G, Buratti C (2013) Reproductive success of the Argentine anchovy, *Engraulis anchoita*, in relation to environmental variability at a mid–shelf front (Southwestern Atlantic Ocean). Fish Oceanogr 22(3):247–261

Marrari M, Piola AR, Valla D, Wilding JG (2016) Trends and variability in extended ocean color time series in the main reproductive area of the Argentine hake, *Merluccius hubbsi* (Southwestern Atlantic Ocean). Remote Sens Environ 177:1–12. https://doi.org/10.1016/j.rse.2016.02.011

Martos P, Piccolo MC (1988) Hydrography of the Argentine continental shelf between 38° and 42°S. Cont Shelf Res 8(9):1043–1056

Matano R, Palma ED (2008) On the upwelling of Downwelling currents. J Phys Oceanogr 38:2482–2500

Matsuura Y (1998) Brazilian sardine (*Sardinella brasiliensis*) spawning in the Southeast Brazilian Bight over the period 1976–1993. Rev Bras Oceanogr 46(1):33–43

Matsuura Y, Kitahara EM (1995) Horizontal and vertical distribution of anchovy Engraulis anchoita eggs and larvae off Cape Santa Marta Grande in southern Brazil. Arch Fish Mar Res 42:239–250

Matsuura Y, Spach HL, Katsuragawa M (1992) Comparison of spawning patterns of the Brazilian sardine (Sardinella brasiliensis) and anchoita (*Engraulis anchoita*) in Ubatuba region, southern Brazil during 1985 through 1988. Bolm Inst Oceanogr Sao Paulo 40:101–115
Matsuura Y, de Silva GAC, Katsuragawa M, Suzuki K (1993) Distribution and abundance of two species of codlet (Teleostei, Bregmacerotidae) larvae from the south-eastern Brazilian bight. Fish Oceanogr 2:82–90
Mazzini PLF, Barth JA (2013) A comparison of mechanisms generating vertical transport in the Brazilian coastal upwelling region. J Geophys Res Oceans 118:1–7
Miranda LB (1982) Análise de massas de água da plataforma continental e da região oceânica adjacente: Cabo de São Tomé (RJ) e Ilha de São Sebastião (SP). Universidade de São Paulo, São Paulo
Möller OO, Castaing P, Salomon J-C, Lazure P (2001) The influence of local and non local forcing effects on the subtidal circulation of Patos Lagoon. Estuaries 24(2):275–289
Moraes LES, Gherardi DFM, Katsuragawa M, Paes ET (2012) Brazilian sardine (Sardinella brasiliensis Steindachner, 1879) spawning and nursery habitats: spatial-scale partitioning and multiscale relationships with thermohaline descriptors. ICES J Mar Sci 69:939–952
Moser GAO, Gianesella-Galvão SMF (1997) Biological and oceanographic upwelling indicators at Cabo Frio (RJ). Rev Bras Oceanogr 45(1–2):11–23
Muelbert JH, Weiss G (1991) Abundance and distribution of fish larvae in the channel area of the Patos Lagoon Estuary, Brazil. NOAA Tech Rps NMFS 95 pp 43-54
Muelbert JH, Acha M, Mianzan H, Guerrero R, Reta R, Braga ES, Garcia VMT, Berasategui A, Gomez-Erache M, Ramírez F (2008) Biological, physical and chemical properties at the Subtropical Shelf Front Zone in the SW Atlantic Continental Shelf. Cont Shelf Res 28:1662–1673
Odebrecht C, Castello JP (2001) The convergence ecosystem in the Southwest Atlantic. In: Seeliger U, Kjerfve B (eds) Coastal marine ecosystems of Latin America, Ecological studies, vol 144. Springer, Berlin, pp 147–166
Odebrecht C, Djurfeldt L (1996) The role of nearshore mixing on phytoplankton size structure off Cape Santa Marta Grande, Southern Brazil (Spring 1989). Arch Fish Mar Res 43(3):217–230
Odebrecht C, Abreu PC, Bemvenuti CE, Coppertino M, Muelbert JH, Vieira JP, Seeliger U (2010) The Patos Lagoon Estuary: biotic responses to natural and anthropogenic impacts in the last decades (1979–2008). In: Kennisch M, Paerl H (eds) Coastal Lagoons: systems of natural and anthropogenic change. Taylor & Francis/CRC Press, Boca Raton, pp 437–459
Pájaro M (1998) El canibalismo como mecanismo denso-dependiente de mortalidad natural en la anchoíta argentina (Engraulis anchoita). Su relación con las estrategias reproductivas de la especie. Doctoral, Universidad Nacional de Mar del Plata, Mar del Plata
Pájaro M, Macchi GJ, Martos P (2005) Reproductive pattern of the Patagonian stock of Argentine hake (*Merluccius hubbsi*). Fish Res 72:97–108
Pájaro M, Macchi GJ, Leonarduzzi E et al (2009) Spawning biomass of Argentine anchovy (*Engraulis anchoita*) from 1996 to 2004 using the Daily Egg Production method. J Mar Biol Assoc UK 89(4):829–837
Palma ED, Matano RP, Piola AR et al (2004a) A comparison of the circulation patterns over the Southwestern Atlantic Shelf driven by different wind stress climatologies. Geophys Res Lett 31:1–5
Palma ED, Piola AR, Matano RP (2004b) A numerical study of the Southwestern Atlantic Shelf circulation: Barotropic response to tidal and wind forcing. J Geophys Res 109
Pereira GC, Coutinho R, Ebecken NFF (2008) Data mining for environmental analysis and diagnostic: a case of upwelling ecosystem of Arraial do Cabo. Braz J Oceanogr 56:1–12, pp 1-17
Piola AR, Rivas AL (1997) Corrientes en la plataforma continental. In: Boschi EE (ed) El Mar Argenrino y sus recursos pesqueros 1: Antecedentes históricos de las exploraciones en el mar y las características ambientales. Instituto Nacional de Investigación y Desarrollo Pesquero, Mar del Plata, pp 119–132
Piola AR, Campos EJD, Möller OO Jr et al (2000) The Subtropical Shelf Front off eastern South America. J Geophys Res 105(C3):6565–6578. https://doi.org/10.1029/1999JC000300

Piola AR, Matano RP, Palma ED et al (2005) The influence of the Plata River discharge on the western South Atlantic shelf. Geophys Res Lett 32:1603–1606

Resgalla C, Rocha CDL, Montú MA (2001) The influence of Ekman transport on zooplankton biomass variability off southern Brazil. J Plankton Res 23:641–650

Rodrigues RR, Lorenzzetti JA (2001) A numerical study of the effects of bottom topography and coastline geometry on the Southeast Brazilian coastal upwelling. Cont Shelf Res 21(4):371–394

Romero SI, Piola AR, Charo M et al (2006) Chlorophyll a variability off Patagonia based on SeaWiFS data. J Geophys Res 111:C05021

Sabatini ME, Alvarez Colombo G (2001) Seasonal pattern of zooplankton biomass in the Argentinian shelf off Southern Patagonia (45°–55°S). Sci Mar 65(1):21–31

Sabatini ME, Reta R, Lutz V et al (2016) Influence of oceanographic features on the spatial and seasonal patterns of mesozooplankton in the southern Patagonian shelf (Argentina, SW Atlantic). J Mar Syst 157:20–38. https://doi.org/10.1016/j.jmarsys.2015.12.006

Sánchez RP, Ciechomski JD (1995) Spawning and nursery grounds of pelagic fish species in the sea-shelf off Argentina and adjacent areas. Sci Mar 59(3–4):455–478

Sánchez RP, Remeslo AV, Madirolas A et al (1995) Distribution and abundance of post-larvae and juveniles of the Patagonian spratt, *Sprattus fuegensis,* and related hydrographic conditions. Fish Res 23:47–81

Sánchez RP, Madirolas A, Reta R et al (1997) The reproductive biology of the Patagonian spratt (*Sprattus fuegensis*): several facts and still some speculations. ICES CM1997/hh: 22 (pelagic fish committee), 24 pp

Secor HD (2015) Migration ecology of marine fishes. Johns Hopkins University Press, Baltimore

Sieg A (1998) A study on the histological classification of the in situ nutritional condition of larval South-West Atlantic anchovy, *Engraulis anchoita* Hubbs and Marini, 1935. Arch Fish Mar Res 46:19–36

Silveira ICA, Schmidt ACK, Campos EJD, Godoi SS, Ikeda Y (2000) A Corrente do Brasil ao largo da costa leste brasileira. Rev Bras Oceanogr 48:171–183

Simionato CG, Berasategui AD, Meccia V et al (2008) Short time-scale wind forced variability in the Río de la Plata Estuary and its role on ichthyoplankton retention. Estuar Coastal Shelf Sci 76:211–226

Sinque C, Muelbert JH (1997a) Environment and Biota of the Patos Lagoon Estuary. Ichthyoplankton. In: Subtropical convergence environments. The coast and sea in the Southwestern Atlantic. Springer, Berlin, pp 51–56

Sinque C, Muelbert JH (1997b) Ichthyoplankton. In: Seeliger U, Odebrecht C, Castello JP (eds) Subtropical convergence environments: the coastal and sea in the Southwestern Atlantic. Springer, Berlin, pp 120–123

Sinque C, Muelbert JH (1997c) Ichthyoplankton. In: Seeliger U, Odebrecht C, Castello JP (eds) Subtropical convergence environments: the coastal and sea in the Southwestern Atlantic. Springer, Berlin, pp 51–56

Temperoni B, Viñas MD (2013) Food and feeding of Argentine hake (*Merluccius hubbsi*) larvae in the Patagonian nursery ground. Fish Res 148:47–55

Torquato FO, Muelbert JH (2015) Spatial distribution of eggs and larvae of *Engraulis anchoita* (Hubbs & Marini, 1935) in the Southern Brazilian Shelf. Braz J Aquat Sci Technol 19(3):1–8

Valentin JL (1984) Analysis of hydrobiological parameters in the Cabo Frio (Brazil) upwelling. Mar Biol 82(3):259–276

Valentin JL (2001) The Cabo Frio upwelling system, Brazil. In: Seeliger U, Kjerfve B (eds) Coastal marine ecosystems of Latin America. Springer, Berlin, pp 97–105

Valentin JL, Andre DL, Jacob SA (1987a) Hydrobiology in the Cabo Frio (Brazil) upwelling two-dimensional structure and variability during a wind cycle. Cont Shelf Res 7:77–88

Valentin JL, Monteiro-Ribas WM, Mureb MA et al (1987b) Some abundant zooplankton in the Cabo-Frio upwelling (Brazil). J Plankton Res 9:1195–1216

Vaz AC, Möller OO, Almeida TL (2006) Análise quantitativa da descarga dos rios afluentes da Lagoa dos Patos. Atlantica 28(1):13–23

Viñas MD, Ramírez FC (1996) Gut analysis of first-feeding anchovy larvae from Patagonian spawning area in relation to food availability. Arch Fish Mar Res 43:231–256

Viñas MD, Santos B (2000) First-feeding of hake (*Merluccius hubbsi*) larvae and prey availability in the North Patagonian spawning area – comparison with anchovy. Arch Fish Mar Res 48:242–254

Viñas MD, Negri RM, Ramírez FC et al (2002) Zooplankton assemblages and hydrography in the spawning area of anchovy (*Engraulis anchoita*) off Río de la Plata estuary (Argentina-Uruguay). Mar Freshw Res 53:1031–1043

Vizziano D, Forni F, Saona G et al (2002) Reproduction of *Micropogonias furnieri* in a shallow temperate coastal lagoon in the southern Atlantic. J Fish Biol 61(A):196–206

Weiss G (1981) Ictioplancton del estuario de Lagoa dos Patos, Brasil. Ph.D. Dissertation, Universidad Nacional de La Plata, Facultad de Ciencias Naturales y Museo., La Plata

Weiss G, Hubold G, Bonecker ACT (1988) Eggs and larvae of *Maurolicus muelleri* (Cymelin, 1789) (Teleostei, Sternoptychidae) in the Southwest Atlantic. Meeresforsch 32:53–60

Wu Y, Hanna CG, O'Flaherty-Sproul M et al (2017) Representing kelp forests in a tidal circulation model. J Mar Syst 169:73–86

Part IV
Plankton of Coastal Systems

Flagellates Versus Diatoms: Phytoplankton Trends in Tropical and Subtropical Estuarine-Coastal Ecosystems

Clarisse Odebrecht, Maria C. Villac, Paulo C. Abreu, Lumi Haraguchi, Piter D. F. Gomes, and Denise Rivera Tenenbaum

Abstract Attempts to provide general patterns of phytoplankton and their regulating factors benefit from ecosystem comparisons, but these are strongly biased toward high-latitude environments of the northern hemisphere (> 20°N). In the present study, we compare the phytoplankton biomass and composition variability in two coastal environments in the southern hemisphere, the tropical Guanabara Bay, GB (23°S), and the subtropical Patos Lagoon Estuary, PLE (32°S), located on the South American southeast coast at the state of Rio de Janeiro and Rio Grande do Sul, respectively. These environments present contrasting features regarding the magnitude of anthropic impacts, the watershed size, geomorphology, and hydrology. Our goal was to identify the main factors that regulate the phytoplankton biomass and composition comparing data obtained at monthly intervals between the years 2011 and 2012 at a single station located in an area of significant water exchange in each environment. Surface water temperature, salinity, inorganic dissolved nutrients, chlorophyll *a*, phytoplankton biomass (carbon) and composition were analyzed. Phytoplankton biomass in the GB and PLE was dominated, respectively, by flagellates and diatoms, whereas cyanobacteria were more important in

C. Odebrecht (✉) · P. C. Abreu
Instituto de Oceanografia, Universidade Federal do Rio Grande (FURG),
Rio Grande, RS, Brazil
e-mail: doclar@furg.br; docpca@furg.br

M. C. Villac
Instituto de Biologia, Universidade Federal do Rio de Janeiro (UFRJ),
Rio de Janeiro, RJ, Brazil

Fish and Wildlife Research Institute, St. Petersburg, FL, USA

L. Haraguchi
Instituto de Oceanografia, Universidade Federal do Rio Grande (FURG),
Rio Grande, RS, Brazil

Institute of Bioscience, Aarhus University, Roskilde, Denmark

P. D. F. Gomes · D. R. Tenenbaum
Instituto de Biologia, Universidade Federal do Rio de Janeiro (UFRJ),
Rio de Janeiro, RJ, Brazil

M. S. Hoffmeyer et al. (eds.), *Plankton Ecology of the Southwestern Atlantic*,
https://doi.org/10.1007/978-3-319-77869-3_12

the former. Salinity was about twofold higher in the GB (mean 32.6 ± 1.5) than PLE (mean 15.4 ± 9.1) and, together with nutrient concentrations and their proportions, largely explained the observed different communities and much higher biomass in GB. GB presented strong eutrophication signals, with high ammonium and phosphate and lower, closer to limitation, silicate concentration. In contrast, high silicate concentration favored the predominance of diatoms in the PLE. Despite large environmental differences between both environments, the chlorophyll *a* presented a rather similar seasonal pattern, with maxima in austral summer/autumn and spring in both ecosystems. We suggest the seasonal pattern was associated to the incident light variation, but this hypothesis should be further explored.

Keywords Diatoms · Flagellates · Dinoflagellates · Cyanobacteria · Eutrophication · Brazil

1 Introduction

Phytoplankton in coastal ecosystems play a fundamental role due to its ecological importance as primary producer sustaining trophic interactions but also as indicator of environmental health, long-term variability, and climatic changes. The small size of the microalgae provides their fast growth rate, while the diverse taxonomic composition and species richness favor a wide range of physiological abilities allowing their growth, as a group, from the tropics to highest latitude, and from pristine to polluted environments. The main factors controlling phytoplankton species shifts, growth rate, and biomass accumulation have been studied in many coastal ecosystems, e.g., in estuaries, bays, and lagoons worldwide (Cloern and Jassby 2010; Cloern et al. 2016) and also in the Southwest Atlantic Ocean region (Abreu et al. 2010, 2016; Guinder et al. 2010; Odebrecht et al. 2010; Villac and Tenenbaum 2010; Haraguchi et al. 2015). Long-term projects are in progress in South America to distinguish phytoplankton changes due to natural variability from those due to external perturbations (anthropogenic effects) (e.g., ANTARES 2003; PELD 1999).

It is recognized that the physical factors, such as temperature, light availability, wind energy, and salinity variations, prompt short-term metabolic responses (hours-day), which will influence changes in the mesoscale (seasonal) and long-term (decades) species composition and abundance. However, the input of nutrients with the continental outflow is influenced by the cultural eutrophication in coastal regions and generally controls the potential maximum phytoplankton biomass, whereas nutrient recycling and changes in the relative proportion of the main elements (nitrogen, phosphorus, silicate) are crucial factors that determine species composition and abundance (Officer and Ryther 1980; Smayda 1990; Egge and Aksnes 1992). Long-term phytoplankton changes are still little understood, and factors relating to global warming and marine acidification must also be considered (Guinder and Molinero 2013).

Coastal ecosystems are very diverse considering the main factors influencing ecological functioning, i.e., biodiversity, productivity, trophic interactions, environment variability, resilience, and resistance to natural and anthropogenic perturbations. Attempts to provide general trends in phytoplankton chlorophyll and primary production based on comparison across ecosystems revealed that the large range of variability patterns is determined by several factors, from nutrient enrichment and anthropogenic impacts to climate shifts shaping seasonality and large event-scale variability (Cloern and Jassby 2010; Cloern et al. 2014, 2016). However, at any rate, these comparisons are strongly biased toward high-latitude environments of the northern hemisphere (> 20° N; Cloern and Jassby 2010, Cloern et al. 2014); hence, there is a call to make such comparisons across tropical and subtropical environments especially in the southern hemisphere.

In the present study, we compare phytoplankton (biomass and composition) variability over 2 years in two contrasting environments, located on the South American southeast coast at the Brazilian states of Rio de Janeiro and Rio Grande do Sul. The tropical Guanabara Bay and, further south, the subtropical Patos Lagoon Estuary contrast with regard to the magnitude of anthropic impacts, the size and geomorphology of their watersheds, and hydrography. Our goal was to recognize environmental factors that regulate the phytoplankton biomass and composition in these coastal ecosystems in the Southwest Atlantic Ocean and thus contribute to fill in gaps in our current understanding of the main patterns in phytoplankton ecology.

2 Study Area

Guanabara Bay (GB) in the state of Rio de Janeiro (GB: 22°15′–23°05′ S; 42°30′–43°30′ W) and Patos Lagoon (PL) in the state of Rio Grande do Sul (PL: 30°12′–32°12′ S; 50°40′–52°15′ W) (Fig. 1) are two important coastal ecosystems in Brazil. GB and PL are located, respectively, northward and southward of the Tropic of Capricorn and present distinct climate according to the classification of Köppen: GB at the transition between Aw (tropical with dry winter) and Cwa (subtropical, with dry winter and hot summer) and PL-Cfa (humid subtropical with hot summer) (Alvares et al. 2013).

The annual rainfall in the regions of GB (1000–1500 mm year^{-1}) and PL (1200–1500 mm) are comparable; however, the former presents a strong seasonal variation with highest values from December to March. On the other hand, the geomorphology, size of the watershed and embayment area, residence time, and oceanographic currents on the adjacent shelf differ largely between these environments (Table 1).

GB is a eutrophic, polluted marine ecosystem originated from a tectonic depression located at the foothill of the granitic and gneissic massifs of Serra do Mar mountains, lying northwest to northeastward. In the interior margin of the bay, mangrove wetlands parallel the southwest-northeast orientation of the basin and receive the main freshwater inflow (125 m^3 s^{-1}; Kjerfve et al. 2001) of the main rivers Macacu and Guapimirim. The coastal plain is occupied by the cities of Rio de

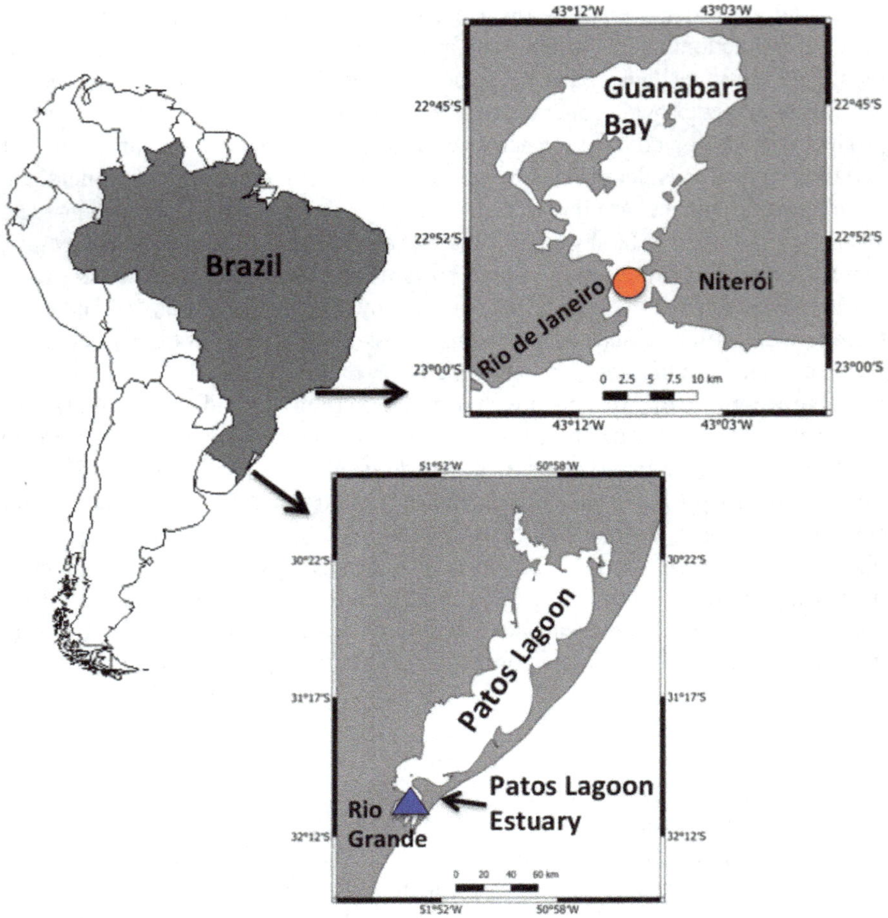

Fig. 1 Map of Guanabara Bay and Patos Lagoon with the sampling stations near the entrance

Janeiro and Duque de Caxias to the west and Niterói and São Gonçalo to the east. The area of GB (~380 km^2) is roughly 30 times smaller than that of the PL (~10,360 km^2); its maximum depth (58 m) occurs closer to the bay entrance, at the main N-S circulation channel (>20 m), but overall average depth is 5.7 m (Kjerfve et al. 1997). The hydrography and water quality of GB are very heterogeneous due to the interplay of pollution focuses along the margins and the action of semidiurnal tides (mean amplitude 0.7 m) that promote water exchange with the Atlantic Ocean through the N-S central channel. The NW region is the most exposed to human impact, whereas the southern channel area is the least impacted and best flushed in the system (Mayr et al. 1989). About 50% of the water volume in GB flushes every 11 days (Kjerfve et al. 2001). The adjacent continental shelf off GB is largely influenced by the southward-flowing nutrient-poor tropical water of the Brazil Current but also by cold water intrusions due to seasonal upwelling of nutrient-rich South

Table 1 Environmental characteristics of Guanabara Bay and Patos Lagoon

	Guanabara Bay	Patos Lagoon
Geographical position	22°15′–23°05′ S; 42°30′–43°30′W	30°12′–32°12′S; 50°40′–52°15′W
Evolutionary origin	Cenozoic, tectonic depression	Quaternary, multiple sand barrier complex due to sea-level changes
Classification	Costal bay	Choked lagoon
Watershed size	~4080 km^2	~200,000 km^2
Ecosystem size	~384 km^2 North: ~33 × 10 km Central: ~12 × 6–12 km Entrance Sector: ~8 × 1.6–8.5 km	~10,360 km^2 North+Central: ~200 × 25–60 km Southern Estuary: ~60 × 0.8–15 km Inlet width: 0.5 km
Depth	Maximum: 58 m at the entrance channel North/Central areas: ~8 m	Maximum: 15 m at the entrance channel North/Central areas: ~5 m Southern Estuary: 80% <1.5 m
Water volume	1.9 × 10^9 m^3 (Kjerfve et al. 2001)	50 × 10^9 m^3 (Möller 1996)
Annual rainfall	Mean 1173 mm (Denemet 1992)	1200–1500 mm
Annual evaporation	25 mm annual deficit Mean 1198 mm (Denemet 1992)	200–300 annual surplus
Coastal oceanography	Coastal water and tropical water of the Brazil Current prevail; upwelling of South Atlantic Central Water from September to March	Surface low-salinity plume in winter-spring is replaced by tropical water in summer. At subsurface, subtropical and subantarctic shelf waters prevail
Tides	Mixed, mainly semidiurnal Mean amplitude 0.7 m	Mixed, mainly diurnal Mean amplitude 0.47 m
Flushing time(half-life)	~11.4 days (Kjerfve et al. 2001)	North+Central: ~82 days Southern Estuary: ~3 days (Knoppers and Kjerfve 1999)
Population	12 millions	4.5 millions
Main economic activities	Industrial, harbor	Industrial, harbor, agriculture, and livestock

Atlantic Central Water at Cape Frio (23° S; 42° W), mainly in austral spring/summer from September to March (Valentin 2001; Coelho-Souza et al. 2012).

PL is a limnetic-oligohaline choked shallow lagoon (mean 5 m depth, except the inlet and navigation channels) extending for 250 km along the coastline of the southern Brazilian coastal plain. Two natural units, the inner (North + Central areas) and the southern estuary (PLE), are separated by sand banks. PL drains an approximately 50 times larger hydrological basin (200,000 km^2), compared to GB. Freshwater is discharged principally (85%) through the northern tributaries, followed by Camaquã River in the central area, and São Gonçalo channel in the south. River discharge (extremes of 12,000 $m^3\ s^{-1}$) is highest during El Niño periods (Möller et al. 1996; Fernandes et al. 2002; Marques 2012). Freshwater exchange with marine water is restricted to the southern funnel-shaped estuary of Patos Lagoon (PLE, approx. 1000 km^2, 80% <1.5 m depth). Due to relatively high freshwater discharge, the microtidal region (mean tide amplitude 0.47 m), and the narrow channel of the estuary (500 m wide at the jetties), PLE is a river-controlled environment (Möller et al. 1991). During flood periods, the seaward flow forced by river discharge can only be reversed by strong southwest winds. At low river discharge, the wind controls circulation, driving water in (southwest winds) or out (northeast winds) the PLE in accordance to the frequency of frontal systems (Möller et al. 1991). A flushing half-life time of 82 days was estimated for the Patos Lagoon and of 3 days for the southern PLE (Knoppers and Kjerfve 1999). However, extreme freshwater discharge alters significantly the residence time of Patos Lagoon (Fernandes et al. 2002). The adjacent continental shelf off PLE is influenced by continental runoff, the proximity of the La Plata River plume edging northward mainly in winter/spring, the Brazil-Malvinas confluence (Piola et al. 2008), and tropical waters especially in the summer season.

Both ecosystems provide important social and economic services for the local populations. The surroundings of GB are home for approximately 12 million people whose impact on the bay is severe. The industrial hub surrounding GB is the second largest in Brazil, and diverse economic activities result in acute and chronic environmental problems. Compared to GB, about one third of people live adjacent to PL. The human population is concentrated mainly in the cities located in the northern area (4 million inhabitants from Porto Alegre and vicinities) and less in the southern reaches (500,000 inhabitants in Pelotas, Rio Grande, and São José do Norte counties). Industrial development is more intense in the north, and agriculture of rice, soy, and afforestation prevail in the central and southwest area, whereas fisheries and harbor development are pronounced in the south. Here, the implementation of shipyards and the construction of oil platforms in recent years increased industrial and commercial activities.

3 Material and Methods

In order to compare the annual cycle of environmental and phytoplankton data between GB and the PLE, we selected one station in each system (Fig. 1) that represented areas of significant water exchange (GB: Mayr et al. 1989; PLE: Fujita and Odebrecht 2007). Here we report on surface samples collected monthly along 2011 and 2012. In situ water temperature was measured with a mercury thermometer (PLE) or electronic sensor (GB) and salinity with conductivity meter or refractometer in both ecosystems. Light penetration was measured as Secchi disk depth (Z_{SD}) and rainfall data were obtained from the Brazilian National Meteorological Institute (INMET 2017).

Water samples were filtered (GB and PLE) and stored frozen (PLE) for the analysis of dissolved inorganic nutrients, ammonium according to UNESCO (1983) and silicate, phosphate, nitrite, and nitrate according to Strickland and Parsons (1972) and Aminot and Chaussepied (1983). Chlorophyll *a* analyses started with gentle vacuum filtration (<25 cm of Hg) and pigments from cells collected on filters extracted overnight in 90% acetone at −20 °C. Chlorophyll *a* in GB samples (30–100 mL, cellulose membrane Millipore HAWP 0.45 μm filters) was measured with a Varian Cary Eclipse® spectrofluorometer, using a modified version of Neveux and Lantoine's (1993) method (Tenório et al. 2010). Chlorophyll *a* in PLE samples (25–50 mL, glass fiber Whatman GF/F) was measured using a calibrated fluorometer (Turner TD) following the recommendation of Welschmeyer (1994) and calibration with SIGMA chlorophyll *a*.

Phytoplankton samples were fixed with neutral Lugol's solution (1–2%), and cell identification and abundance were estimated using the inverted microscope technique (Utermöhl 1958) described by Hasle (1978). The whole chamber was inspected for the larger organisms (>20–50 μm), and transects or fields were analyzed to count the smaller cells. For each size fraction, at least 100 individuals (cells or colonies/trichomes) were counted to maintain the error $\leq 20\%$. Organisms were measured and identified to species level when possible, except for the flagellates that were counted by size and/or grouped in taxonomic categories (most often class). Cell volume was estimated by applying geometric formulae to each corresponding shape (Hillebrand et al. 1999; Sun and Liu 2003; Olenina et al. 2010), and cell carbon was determined using the equations proposed by Montagnes and Franklin (2001) for diatoms and Menden-Deuer and Lessard (2000) for other taxa except *Trichodesmium*, which was transformed using the factor proposed by Carpenter et al. (2005).

In order to evaluate the effects of abiotic parameters on the phytoplankton community of both localities, a redundancy analysis (RDA) was applied on log-transformed data of phytoplankton group biomass. Nine phytoplankton groups were defined based on class (chlorophytes, cryptophytes, cyanobacteria, dinoflagellates), class + origin (diatoms – freshwater, marine, and undefined), green flagellates (euglenophytes, prasinophytes), and Others (undetermined taxa). All the abiotic variables were tested for variance inflation factors (VIFs), and only the ones with $\sqrt{VIFs} < 2$ were kept. The abiotic variables considered in the analysis were salinity, temperature, $NO_2 + NO_3$, NH_4, PO_4, and SiO_2. A significance test (global and for individual axes) was performed by a permutation test with 1000 realizations and a significance level of 5% (Borcard et al. 2011). The analysis was performed using the package "vegan" (Oksanen et al. 2017) in R (v 3.3.2) (R Core Team 2017).

4 Results

Water temperature presented larger amplitude in the PLE (8–28 °C), with minima in July/August (austral winter) and maxima in the austral summer months January/February. In GB the temperature range was smaller (21–29 °C); however, an annual pattern was also observed with highest values in summer-autumn months mainly between February and May (Fig. 2).

Rainfall (Fig. 2) differed with regard to the magnitude and annual cycle, GB (mean 1883 mm year^{-1}) presenting higher values than PLE (mean 1208 mm year^{-1}). In both environments, however, 2012 was drier (GB 1866 mm y^{-1}; PL 1031 mm year^{-1}) in comparison to 2011 (GB 1900 mm year^{-1}; PLE 1385 mm year^{-1}). In GB

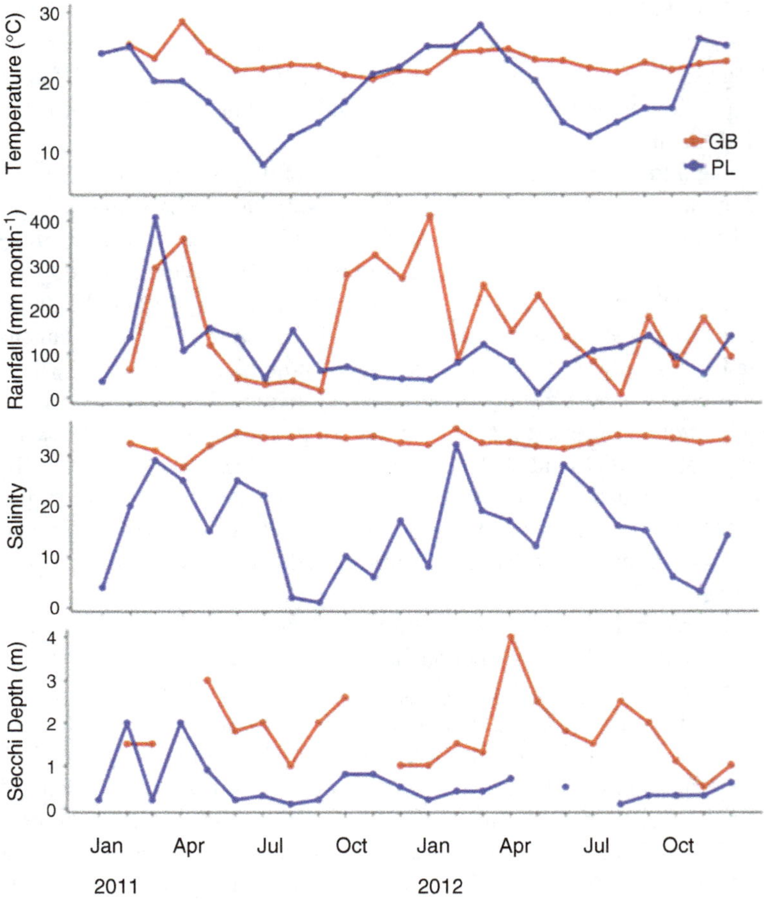

Fig. 2 Monthly values of water temperature, rainfall, salinity, and Secchi disk depth in Guanabara Bay and Patos Lagoon Estuary (2011–2012)

during the study period, maxima rainfall monthly values (>250 mm month^{-1}) were recorded in austral summer/autumn 2011 and spring/summer 2011/2012, except for February in both years. The lowest values (< 100 mm month^{-1}) were observed both years in the austral winter months (June to September 2011; July and August 2012). PLE presented relatively low values (<150 mm month^{-1}), except in March 2011. The spring/summer 2011/2012 period of extremely high rainfall at GB coincided with lowest rainfall (< 100 mm month^{-1}) at the PLE region.

Despite more rainfall, salinity was higher in GB (28–35) compared to PLE (1–32) (Fig. 2). In the PLE, lowest salinity values (<5) were observed in winter-spring months (August/September 2011; October/November 2012), when they were relatively high at GB (>33). A similar trend was observed for water transparency (Secchi disk depth), which was higher at GB (0.5–4.0 m) and extremely low at PLE (0.01–2.0 m).

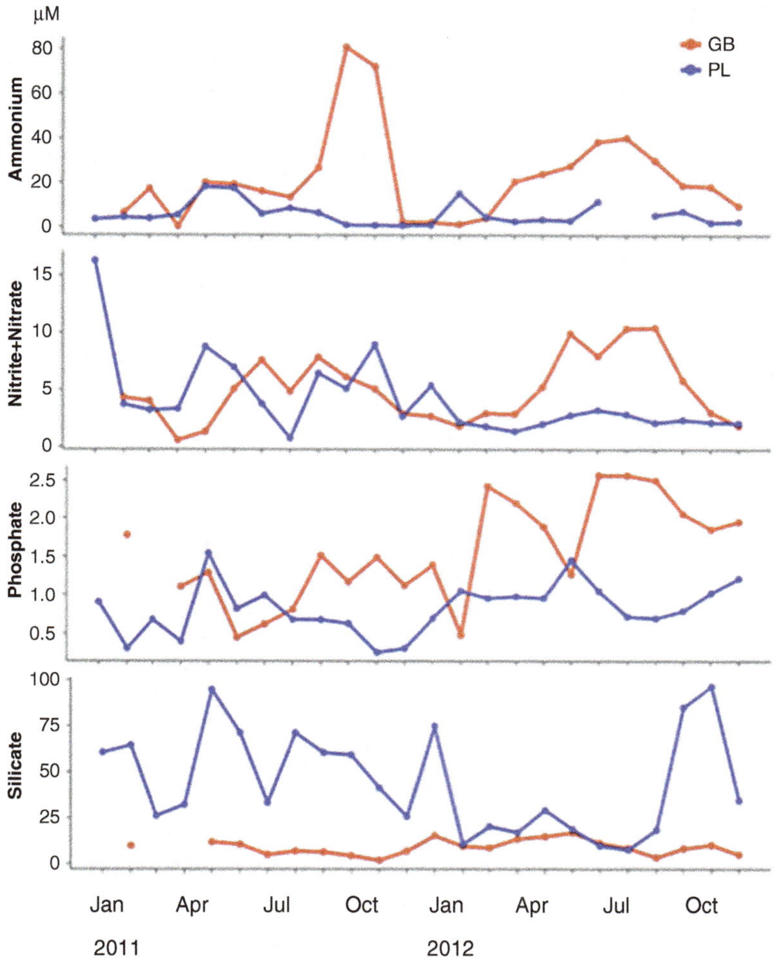

Fig. 3 Monthly values of dissolved inorganic nutrients: ammonium, nitrite+nitrate, phosphate, and silicate in Guanabara Bay and Patos Lagoon Estuary (2011–2012)

Each environment had particular features in terms of dissolved inorganic nutrients (Fig. 3): ammonium and phosphate concentrations were ca. three times higher in GB (mean 22.3 ± 20.6 μM and 1.6 ± 0.7 μM, respectively) than in the PLE (mean 6 ± 5.1 μM and 0.8 ± 0.3 μM, respectively), while silicate was five times higher in the PLE (mean 44.4 ± 28.0 μM; range 7.9–96.7 μM) than GB (mean 9.1 ± 4.0 μM; range 2.0–17.1 μM). Nitrite+nitrate concentrations were rather similar in both environments (mean 4.2–5.0 μM; range 0.6–10.5 μM), except the peak on January 2011 in the PLE (16.3 μM). Nitrogen nutrients had a pronounced seasonal variation in GB, ammonium increasing from January to October and decreasing afterward, and nitrite+nitrate with higher values in winter months (July to October). In the PLE, ammonium oscillated irregularly, and nitrite+nitrate values were higher in

Table 2 Environmental and phytoplankton parameters (mean ± standard deviation, minimum and maximum values) of Guanabara Bay and Patos Lagoon Estuary during the years 2011 and 2012: water temperature, annual rainfall, salinity, Secchi disk depth, dissolved inorganic ammonium, nitrite+nitrate, phosphate, silicate, atomic ratio N:P and Si:N, chlorophyll *a*, phytoplankton carbon, and species richness

	Guanabara Bay	Patos Lagoon Estuary
Water temperature (°C)	22.8 ± 1.8 (21.6–28.6)	19.0 ± 5.5 (8.0–28.0)
Annual rainfall (mm $year^{-1}$)	2011: 1900	2011: 1385
	2012: 1866	2012: 1031
Salinity	32.6 ± 1.5 (27.6–35.2)	15.4 ± 9.1 (1.0–32.0)
Secchi disk depth (m)	1.8 ± 0.8 (0.5–4.0)	0.5 ± 0.5 (0.01–2.3)
Ammonium (μM)	22.3 ± 20.6 (0.6–81.1)	6.0 ± 5.1 (0.8–18.2)
Nitrite+nitrate (μM)	5.0 ± 2.9 (0.6–10.5)	4.2 ± 3.4 (0.7–16.3)
Phosphate (μM)	1.6 ± 0.7 (0.4–2.6)	0.8 ± 0.3 (0.3–1.5)
Silicate (μM)	9.1 ± 4.0 (2.0–17.1)	44.4 ± 28.0 (7.9–96.7)
N:P (atomic ratio)	20.4 ± 19.2 (1.0–74.5)	14.3 ± 9.3 (4.1–39.7)
Si:N (atomic ratio)	0.71 ± 0.87 (0.03–3.15)	5.7 ± 4.5 (0.7–21.6)
Chlorophyll *a* (μg L^{-1})	35.3 ± 54.4 (2.6–266.5)	7.3 ± 4.7 (0.9–21.3)
Phytoplankton carbon (μg L^{-1})	1647.1 ± 2786.5 (111.2–11,829.1)	474.5 ± 1101.1 (142.0–5561.5)
Species richness (N)	29 ± 6 (18–46)	33 ± 9 (18–49)

2011 than 2012. Phosphate concentrations did not show a seasonal pattern in GB but presented higher values in 2012. In the PLE, higher values were observed both years in autumn/winter. Silicate values showed large interannual variations with higher (PLE) and lower (GB) values being observed in the year 2011 (Fig. 3).

The ratios N:P and Si:N were largely influenced by the concentrations of dissolved inorganic ammonium and phosphate (GB) and silicate (PLE) and clearly indicated highest anthropogenic nutrient enrichment in GB (Table 2). The N:P (atomic ratio) in GB (N:P mean 20.4 ± 19.2; range 1.0–74.5) varied largely in comparison to that found in the PLE (N:P mean 14.3 ± 9.3; range 4.0–39.7). The Si:N ratio was ca. 10 times higher in the PLE (Si:N mean 5.7 ± 4.5; range 0.6–21.6) than in GB (mean 0.7 ± 0.9; range 0.03–3.15).

Chlorophyll *a* monthly mean values were five times higher in GB than in the PLE (GB: 35.3 μg/L; PLE: 7.3 μg/L), and maxima monthly concentrations were up to ten times higher in GB (range 3–266 μg/L) than in the PLE (range 1–21 μg/L)

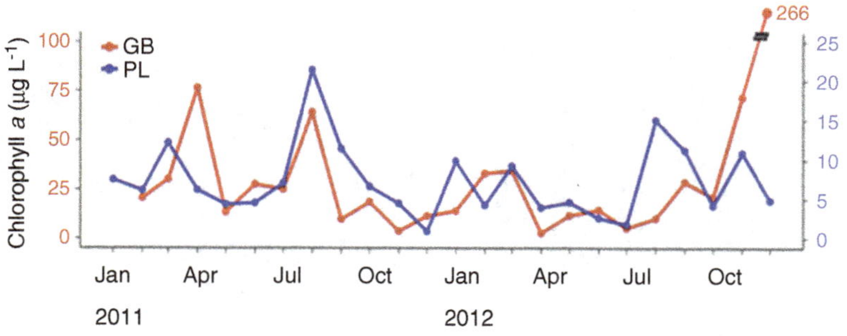

Fig. 4 Monthly values of phytoplankton chlorophyll *a* in Guanabara Bay and Patos Lagoon Estuary (2011–2012). Note the two axes: GB at the left and PLE on the right axis

(Table 2). Despite these great differences, the pattern of monthly variations was rather similar in both environments (Fig. 4, note different scales), with higher values in the austral fall (March to April) and in late winter (peak in August) in the year 2011 and in summer (January to March) and in late winter to spring (August to December) in 2012. Lowest values were observed from October to December 2011 and from April to July 2012 (Fig. 4).

Phytoplankton carbon (Fig. 5) roughly followed the pattern of chlorophyll *a*. Mean monthly values were three times higher in GB (GB: 1647 µg/L; PLE 475 µg/L), and maxima concentration in GB (range 111–11,829 µg/L) was twice the value observed in the PLE (range 42–5562 µg/L). On two occasions (September 2011, April 2012), the concentrations of chlorophyll *a* and phytoplankton carbon were larger in the PLE than GB.

The relative contribution of phytoplankton groups differed largely between the environments (Fig. 5). The main difference was the higher contribution to overall phytoplankton biomass (carbon) of flagellates and cyanobacteria in GB and of diatoms in the PLE. In GB, the green flagellates (Eugl. + Pras in Fig. 5), especially the prasinophytes (*Tetraselmis* spp., *Pyramimonas* spp.) and euglenophytes (*Eutreptia* spp., *Eutreptiella* spp.), prevailed most of the time, whereas the dinoflagellate *Levanderina fissa* comprised an important proportion of the biomass in November and December of 2012. Among diatoms, the most abundant taxa were *Leptocylindrus minimus* and *Cylindrotheca* (*Ceratoneis*) *closterium*. Cyanobacteria, comprised by thin filamentous species of the family *Leptolyngbyaceae*, appeared mainly in the spring and summer months of both years.

In the PLE, diatoms prevailed in both years mainly comprised by *Aulacoseira*, *Skeletonema* spp., *Ceratoneis* (*Cylindrotheca*) *closterium*, small *Chaetoceros* (*C. subtilis* and others), *Cerataulina bicornis*, and other centric and pennates. Dinoflagellates were mainly *Gymnodiniales* (< 20 µm) and *Prorocentrum cordatum*, and cyanobacteria were coccoid-forming colonies (*Aphanocapsa*, *Aphanothece*, and *Microcystis*) and filamentous *Nostocales* species with heterocytes (*Anabaena*, *Dolichospermum*, *Aphanizomenon*). Cryptophytes species were not identified, and the group Others included less abundant or sporadic species of chrysophytes,

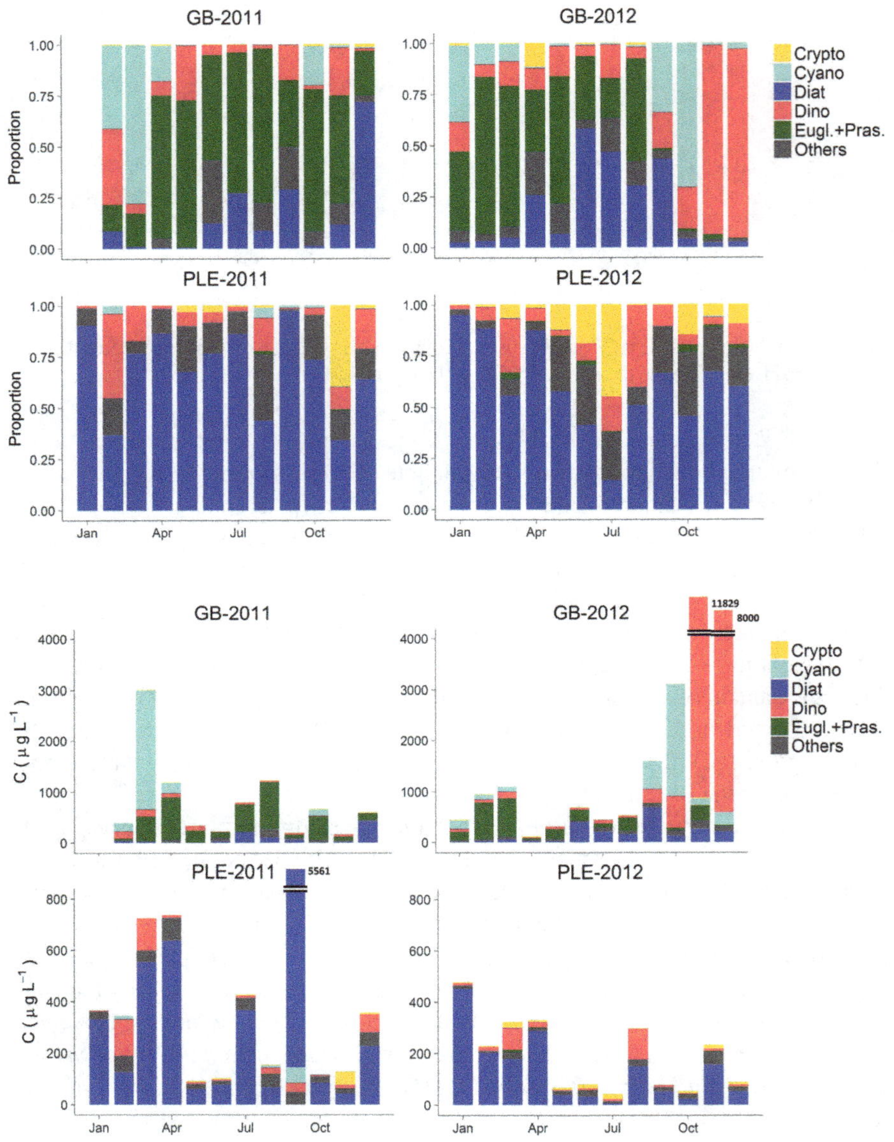

Fig. 5 Monthly values of phytoplankton group carbon as their relative contribution (upper pannel) and concentration (lower pannel)

chlorophytes, raphidophytes, or dictyochophytes. The main peak in September 2011 was associated with freshwater diatoms, mainly *Aulacoseira* spp.

Species richness (Table 2) was remarkably similar (mean GB 29 ± 6 species; PLE 33 ± 9 species) in both environments, despite the large differences in main taxonomic groups and phytoplankton biomass.

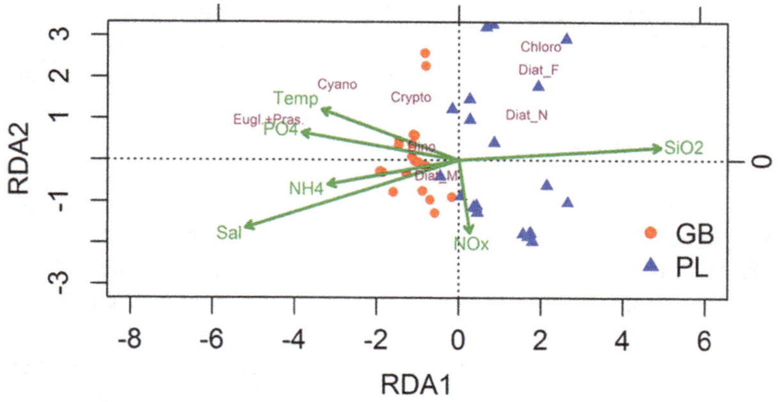

Fig. 6 Redundancy analysis (RDA) applied on log-transformed data of phytoplankton groups biomass: chlorophytes (Chloro), cryptophytes (Crypto), cyanobacteria (Cyano), dinoflagellates (Dino), diatoms of marine (Diat_M), freshwater (Diat_F) and undefined (Diat_N) origin, and euglenophytes and prasinophytes (Eugl. + Pras). Abiotic variables considered in the analysis were salinity (Sal), temperature (Temp), nitrite+nitrate (NOx), ammonium (NH_4), phosphate (PO_4), and silicate (SiO_2)

Multivariate redundancy analysis (RDA) provided the statistical comparison between the environments (Fig. 6). The RDA using six abiotic variables (salinity, temperature, dissolved inorganic nitrite+nitrate, ammonium, phosphate, and silicate) explained about 34% of the variance of the phytoplankton biomass data ($F < 0.001$); however, only the first axis was significant (F_{RDA1} 0.001). RDA1 explained about 72% of the constrained variance (about 24% of the total variance) and was mainly associated with salinity, temperature, silicate, ammonium, and phosphate. GB and PLE differences were best revealed by this axis, which represented the main environmental forcings in each environment: higher salinity, temperature, ammonium, and phosphate in GB in opposition to higher silicate in the PLE. Freshwater diatoms (Diat_F), non-identified diatoms (Diat_N), and freshwater chlorophytes (Chloro) were positively associated with silicate and negatively with salinity, temperature, and ammonium. Euglenophytes and prasinophytes (Eugl. + Pras.) were positively associated with salinity, temperature, ammonium, and phosphate, similarly to cyanobacteria (Cyano), which were more characteristic of BG.

5 Discussion

5.1 Phytoplankton Chlorophyll a *Seasonal Pattern*

In contrast to the PLE that has subtropical to temperate characteristics, tropical GB did not present a great variability in water temperature along the year. Temperature range between summer and winter was half the value in GB (**Δ** 10 °C) than in the

PLE (**Δ** 20 °C), but a seasonal pattern was evident in both ecosystems. The study site at GB was also relatively more stable in terms of salinity, that is, salinity variation was less than 5 in comparison to more than 30 in the PLE and seasonality was more evident in the PLE, despite the fact that minimum salinity values were trailing behind those for water temperature.

Considering the relative stability of these two major forcing functions in GB, the observed seasonal pattern of chlorophyll *a* in this ecosystem is noteworthy, with maxima in austral spring and late summer, a pattern very similar to that observed in the PLE, but with distinct phytoplankton communities. This similarity in the seasonal pattern of both coastal environments indicates that a large-scale forcing function is acting simultaneously on both systems.

The action of light is such a large-scale forcing function that could be claimed to justify the observed similarity in the seasonal variability of chlorophyll *a* in both systems (Kirk 2011). Data on incident light show analogous seasonal patterns in both places with similar maxima values (1200–1500 KJ m^{-2}), whereas small values (<100 KJ m^{-2}) were more frequent in the PLE (INMET 2017). It is also significant the fact that different phytoplankton groups generated similar seasonal chlorophyll *a* trends in both sites, suggesting that different species of diatoms in the PLE and green flagellates and dinoflagellates in GB presented similar light requirements to generate highest biomass in these ecosystems. However, further experiments in both ecosystems are necessary to support this hypothesis.

5.2 Phytoplankton Composition and Interannual Variability

In general, coastal and estuarine phytoplankton species tolerate salinity variations (euryhaline species) contrary to oceanic ones (Kirst 1989). Extreme variations like those observed in the PLE are, however, less well tolerated, and interspecific differences probably play an important role in structuring phytoplankton communities in systems alike (Islabão and Odebrecht 2015). Phytoplankton blooms in 86 estuarine-coastal sites with a large salinity spectrum (0–34) in temperate areas of the northern hemisphere were mostly composed of diatoms, seconded by dinoflagellates, whereas blooms of cyanobacteria were restricted to low-salinity waters (Carstensen et al. 2015). In the relatively more haline tropical GB, the large contribution of cyanobacteria composed of thin filamentous species (*Leptolyngbyaceae*) indicated the tolerance of this microalgae group to high salinity. Massive cyanobacteria abundances in GB tend to prevail, however, during the rainy season and in the inner reaches of the bay where freshwater contribution is more preponderant (Sevrin-Reyssac et al. 1979; Villac et al. 1991; Villac and Tenenbaum 2010).

The negative relationship between salinity and silicate confirms the fact that freshwater discharge was the main source of this nutrient. In fact, silicate input in aquatic ecosystems is in general not significantly associated with human activities (Officer and Ryther 1980), and the large hydrographic basin of Patos Lagoon provides extremely high silicate concentration (>200 μM) to the groundwater

(Niencheski et al. 2007) and water column of the Patos Lagoon (Odebrecht et al. 2005; Niencheski et al. 1999), whereas in the GB, silicate reached limiting concentration (±2.0 μM; Egge and Aksnes 1992). Thus, the five times higher silicate values in the PLE compared to the GB would justify the predominance of diatoms in the former. Dominant species of diatoms (*Skeletonema costatum* and *Skeletonema* spp., *Ceratoneis* (*Cylindrotheca*) *closterium*, *Chaetoceros subtilis*, and *Chaetoceros* spp.) were in general composed of small cells except for *Aulacoseira* spp. (intermediate size: ESD 40–45 μm). The highest biomass in September 2011 coincided with a bloom of the freshwater diatom *Aulacoseira*, which was transported southward from the inner Patos Lagoon to the estuarine region where it formed high biomass, as it has been previously observed (Odebrecht et al. 2005; Torgan et al. 2002). Indeed this bloom coincided with the lowest salinity of the studied period in the PLE (Fig. 2).

The positive relationship between salinity and ammonium and phosphate in GB indicates that other sources, besides freshwater input, are furnishing these nutrients to this ecosystem. Cultural eutrophication plays an important role in GB where in the last 100 years, a tenfold increase in the flux of organic matter was reported for the sediments (Carreira et al. 2002). Moreover, the low mean Si:N ratio (0.7 ± 0.9) observed in GB would justify the dominance of green flagellates in detriment of diatoms that need higher silicate concentrations to flourish. The shift from diatom-dominated communities, as exemplified by the PLE, to flagellate- and cyanobacteria-dominated communities, as exemplified by GB, is a pattern observed in long-term studies that report on the concomitant increase in eutrophication of aquatic ecosystems (Klais et al. 2011). Another aspect regarding cultural eutrophication in GB is the increase in phosphate concentration observed in 2012, when mean values almost doubled (2011: 1.1 ± 0.4 μM; 2012: 1.9 ± 0.6 μM). High chlorophyll *a* values measured in November and December 2012 in GB were mainly composed of dinoflagellates that appeared in high abundance during this period. In GB, a diverse community of flagellates, comprising the green line of prasinophytes (*Pyramimonas* spp., *Tetraselmis* spp.) and euglenophytes (*Eutreptia* spp., *Eutreptiella* spp.), prevailed most of the time, but the dominance in the flagellate community changed at the end of the study period with the occurrence of the dinoflagellate *Levanderina fissa* that formed high biomass especially in November and December 2012 (Fig. 5). This bloom of *L. fissa* followed high nitrogen (50 μM) and phosphate (2.5 μM) concentrations, well above the half-saturation nutrient concentration (K_S) reported for this species (K_S DIN = 8.5; DIP = 1.99; Wang et al. 2017).

Besides the availability of nutrients, it is recognized that the success of flagellates is also associated with the physical structure of the water column, thriving in less turbulent conditions usually leading to vertical stratification generated by calm weather (Margalef 1978). GB is less windy, with wind intensity smaller than 19 km/h, whereas in the region of the PLE at least 10 days in each month present wind velocities higher than 28 km/h (METEOBLUE 2006). Moreover, storm events are more frequent in the PLE than in GB. Thus, the stratification of the water column in the PLE is more transient than that observed in the main circulation channel in GB, which may be disrupted only during spring tide associated with strong winds

(Mayr et al. 1989; Villac et al. 1991). Although we did not estimate the degree of turbulence, the above mentioned conditions, in addition to the availability of nutrients, would favor the predominance of flagellates in the GB compared to the PLE.

Cyanobacteria formed high biomass more frequently in GB than PLE during the studied period. In a long-term analysis in the PLE, cyanobacteria presented large interannual variability with highest abundance in low-salinity waters associated with high rainfall in southern Brazil, characteristic of El Niño years (Haraguchi et al. 2015). In contrast, cyanobacteria in the GB formed by thin filamentous *Leptolyngbyaceae* were found in higher salinity (27.6–35.2). In the RDA, cyanobacteria were associated with phosphate concentration, indicating the importance of this nutrient for the cyanobacteria.

6 Concluding Remarks

Microalgae are indicators of water quality, and the observed phytoplankton biomass and composition in the GB and PLE clearly reflect conspicuous differences in these estuarine-costal environments. Besides the physical setting and the rainfall regime, the much higher anthropic impact leading to high nutrient inputs explains the higher phytoplankton biomass (chlorophyll *a* and carbon) in GB. Basically, as pointed out by the multivariate analysis, nutrients and salinity are the main driving forces that cause the strong differences in phytoplankton biomass and composition in these ecosystems. The predominance of green flagellates and dinoflagellates in GB was remarkable when compared to the community mainly comprised by diatoms in the PLE. However, despite large environmental differences, it is quite surprising that chlorophyll *a* presented a rather similar seasonal pattern, with maxima in late austral summer/autumn and spring. This finding is unexpected since a study with more than 100 estuaries around the globe did not show clear common patterns of chlorophyll *a* variability among temperate coastal environments (Cloern and Jassby 2010). Moreover, Abreu et al. (2016) observed difference in the seasonal variability of two sampling sites no more than 10 km apart in the PLE and in the adjacent coastal region. We suggest that the similarities in the chlorophyll *a* annual cycles observed in the PLE and GB are associated with incident light variation. Our results demonstrate a clear need for further comparative studies, especially in the tropical and subtropical regions of the southern hemisphere, in order to better understand the functioning of estuarine-coastal ecosystems.

Acknowledgments This study was funded by the Brazilian National Biodiversity Research Program (SISBIOTA-PELD Zonas Costeiras: CNPQ Proc. 563263/2015-5), the Brazilian Long Term Ecological Program (PELD CNPq Proc. 403809/2012-6, Proc. 403805/2012-0), and the research foundations of the states of Rio de Janeiro (FAPERJ Proc. E-26/110.114/2013) and Rio Grande do Sul (FAPERGS Proc. 12/3122-7). We would like to thank Ricardo Pollery for nutrient analysis (Laboratório de Biogeoquímica, UFRJ) and Márcio Tenório for chlorophyll analysis (Laboratório de Fitoplancton, UFRJ).

References

Abreu PC, Bergesch M, Proença LA et al (2010) Short- and long-term chlorophyll a variability in the shallow microtidal Patos Lagoon Estuary, Southern Brazil. Estuar Coasts 33:554–569

Abreu PC, Marangoni J, Odebrecht C (2016) So close, so far: differences in long-term chlorophyll a variability in three nearby estuarine-coastal stations. Mar Biol Res 13:1–13. https://doi.org/10.1080/17451000.2016.1189081

Alvares CA, Stape JL, Sentelhas PC (2013) Köppen's climate classification map for Brazil. Meteorol Z 22(6):711–728

Aminot A, Chaussepied M (1983) Manuel des analyses chimiques en milieu marin. Centre National pour l'Exploitation des Oceans, Brest

ANTARES (2003). Latin American network for the study of long term changes in coastal ecosystems. http://antares.ws. Accessed 10 Oct 2017

Borcard D, Gillet F, Legendre P (2011) Canonical ordination. In: Numerical ecology with R. Springer, New York, pp 153–225

Carpenter EJ, Subramaniam A, Capone DG (2005) Corrigendum to "biomass and primary productivity of the cyanobacterium *Trichodesmium* spp. in the tropical N Atlantic Ocean". Deep Sea Res Part 1 Oceanogr Res Pap 52:1787–1788

Carreira RS, Wagener ALR, Readman JW et al (2002) Changes in the sedimentary organic carbon pool of a fertilized tropical estuary, Guanabara Bay, Brazil: an elemental, isotopic and molecular marker approach. Mar Chem 79:207–227

Carstensen J, Klais R, Cloern JE (2015) Phytoplankton blooms in estuarine and coastal waters: seasonal patterns and key species. Estuar Coast Shelf Sci 162:98–109

Cloern JE, Jassby AD (2010) Patterns and scales of phytoplankton variability in estuarine-coastal ecosystems. Estuar Coasts 33:230–241

Cloern JE, Foster SK, Kleckner AE (2014) Phytoplankton primary production in the world's estuarine-coastal ecosystems. Biogeosciences 11:2477–2501

Cloern JE, Abreu PC, Carstensen J et al (2016) Human activities and climate variability drive fast-paced change across the world's estuarine–coastal ecosystems. Glob Chang Biol 22(2):513–529. https://doi.org/10.1111/gcb.13059

Coelho-Souza SA, López MS, Guimarães JRD et al (2012) Biophysical interactions in the Cabo Frio upwelling system, Southeastern Brazil. Braz J Oceanogr 60(3):353–365. https://doi.org/10.1590/S1679-87592012000300008

Egge JK, Aksnes DL (1992) Silicate as regulating nutrient in phytoplankton competition. Mar Ecol Prog Ser 83:281–289

Fernandes EHL, Dyer KR, Möller Jr OO et al (2002) The Patos Lagoon hydrodynamics during an El Niño event (1998). Cont Shelf Res 22:1699–1713

Fujita CC, Odebrecht C (2007) Short term variability of chlorophyll *a* and phytoplankton composition in a shallow area of the Patos Lagoon Estuary (Southern Brazil). Atlântica 29(2):93–106

Guinder VA, Molinero JC (2013) Climate changes effects on marine phytoplankton. In: Arias HA, Menendez AM (eds) Marine ecology in a changing world. CRC Press Taylor & Francis Group, Boca Raton, pp 68–90

Guinder VA, Popovich CA, Molinero JC et al (2010) Long-term changes in phytoplankton phenology and community structure in the Bahía Blanca Estuary, Argentina. Mar Biol 157:2703–2716

Haraguchi L, Carstensen J, Abreu PC et al (2015) Long-term changes of the phytoplankton community and biomass in the subtropical shallow Patos Lagoon Estuary, Brazil. Estuar Coast Shelf Sci 162:76–87

Hasle G (1978) Using the inverted microscope. In: Sournia A (ed) Phytoplankton manual, Monographs on oceanographic methodology, vol 6. UNESCO, Paris, pp 191–196

Hillebrand H, Dürselen CD, Pollingher U et al (1999) Biovolume calculation for pelagic and benthic microalgae. J Phycol 35:403–424

INMET (2017) Instituto Nacional de Meteorologia. http://www.inmet.gov.br/portal/

Islabão CA, Odebrecht C (2015) Influence of salinity on the growth of *Akashiwo sanguinea* and *Prorocentrum micans* (Dinophyta) under acclimated conditions and osmotic stress. Mar Biol Res 11:965–973

Kirk JTO (2011) Light and photosynthesis in aquatic ecosystems. Cambridge University Press, Cambridge

Kirst GO (1989) Salinity tolerance of eukaryotic marine algae. Annu Rev Plant Physiol Plant Mol Biol 40:21–53

Kjerfve B, Ribeiro CHA, Dias GMT et al (1997) Oceanographic characteristics of an impacted coastal bay: Baía de Guanabara, Rio de Janeiro, Brazil. Cont Shelf Res 17(13):1609–1643

Kjerfve B, Lacerda LD, Dias GMT (2001) Baía de Guanabara, Rio de Janeiro, Brazil. In: Seeliger U, Kjerfve B (eds) Coastal marine ecosystems of Latin America, Ecological studies, vol 144. Springer, Berlin, pp 107–117

Klais R, Tamminen T, Kremp A et al (2011) Decadal-scale changes of dinoflagellates and diatoms in the anomalous Baltic Sea spring bloom. PlosOne 6(6):e21567

Knoppers BA, Kjerfve B (1999) Coastal lagoons of Southeastern Brazil: physical and biogeochemical characteristics. In: Perillo G, Piccolo C, Pino-Quivira M (eds) Estuaries of South America. Springer, Berlin, pp 35–66

Margalef R (1978) Life forms of phytoplankton as survival alternatives in an unstable environment. Oceanol Acta 1(4):493–509

Marques WC (2012) The temporal variability of the freshwater discharge and water levels at the Patos Lagoon, Brazil. Int J Geosci 3:758–766. https://doi.org/10.4236/ijg.2012.34076

Mayr LM, Tenenbaum D, Villac MC et al (1989) Hydrobiological characterization of Guanabara Bay. In: Magoon O, Neves C (eds) Coastlines of Brazil. American Society of Civil Engineers, New York, pp 124–139

Menden-Deuer S, Lessard EJ (2000) Carbon to volume relationships for dinoflagellates, diatoms, and other protist plankton. Limnol Oceanogr 45(3):569–579. https://doi.org/10.4319/lo.2000.45.3.0569

METEOBLUE (2006) Previsões meteorológicas. https://www.meteoblue.com/pt/historyplus. Accessed 2 June 2017

Möller OO Jr, Paim PS, Loares ID (1991) Facteurs et mecanismes de la circulation des eaux dans l'estuaire de la lagune dos Patos. Bull lInst Geol Bass Aquitaine 49:15–21

Möller O, Lorenzetti JA, Stech J et al (1996) The Patos Lagoon summertime circulation and dynamics. Cont Shelf Res 16(3):335–351

Montagnes DJS, Franklin DJ (2001) Effect of temperature on diatom volume, growth rate, and carbon and nitrogen content: reconsidering some paradigms. Limnol Oceanogr 46(8):2008–2018

Neveux J, Lantoine F (1993) Spectrofluorometric assay of chlorophylls and pheopigments using the least squares approximation technique. Deep Sea Res Part 1 Oceanogr Res Pap 40(9):1747–1765. https://doi.org/10.1016/0967-0637(93)90030-7

Niencheski JF, Baumgarten MG, Fillmann G et al (1999) Nutrients and suspended matter behavior in the Patos Lagoon Estuary (Brazil). In: Perillo GME, Piccolo MC, Pino-Quivira M (eds) Estuaries of South America. Springer, Heidelberg, pp 67–81

Niencheski LF, Windom HL, Moore WS et al (2007) Submarine groundwater discharge of nutrients to the ocean along a coastal lagoon barrier, Southern Brazil. Mar Chem 106:546–561

Odebrecht C, Abreu PC, Möller OO et al (2005) Drought effects on pelagic properties in the shallow and turbid Patos Lagoon, Brazil. Estuaries 28(5):675–685

Odebrecht C, Bergesch M, Rörig LR et al (2010) Phytoplankton interannual variability at Cassino Beach, Southern Brazil (1992–2007), with emphasis on the surf zone diatom *Asterionellopsis glacialis*. Estuar Coasts 33:570–583

Officer CB, Ryther JH (1980) The possible importance of silicon in marine eutrophication. Mar Ecol Prog Ser 3:83–91

Oksanen J, Blanchet FG, Friendly M et al. (2017) vegan: Community Ecology Package

Olenina I, Hajdu S, Edler L et al (2010) Biovolumes and size-classes of phytoplankton in the Baltic Sea. HELCOM Balt Sea Environ Proc 106:144

PELD (1999) Programa de Pesquisa Ecológica de Longa Duração. http://cnpq.br/apresentacao-peld/. Accessed 10 Oct 2017
Piola A, Möller OO, Guerrero R et al (2008) Variability of the subtropical shelf front off eastern South America: winter 2003 and summer 2004. Cont Shelf Res 28:1639–1648
R Core Team (2017) R: A language and environment for statistical computing
Sevrin-Reyssac J, Machado MCS, Schutze MLM et al (1979) Biomasse et production du phytoplancton de la baie de Guanabara (État de Rio de Janeiro, Brésil) et du secteur océanique adjacent. Variations de mai à juillet 1978. Bul Mus Natn Hist Nat 4:329–354
Smayda TJ (1990) Novel and nuisance phytoplankton blooms in the sea: evidence for a global epidemic. In: Granéli E, Sundstrom B, Edler L, Anderson DM (eds) Toxic marine phytoplankton. Elsevier, New York, pp 29–40
Strickland JDH, Parsons TR (1972) A practical handbook of seawater analysis. J Fish Res Board Can, Ottawa
Sun J, Liu D (2003) Geometric models for calculating cell biovolume and surface area for phytoplankton. J Plankton Res 25(11):1331–1346. https://doi.org/10.1093/plankt/fbg096
Tenório MMB, Duarte R, Barrera-Alba JJ et al (2010) Plankton structure of shallow coastal zone at Admiralty Bay, King George Island, West Antarctic Peninsula (WAP): chlorophyll biomass and size-fractionated chlorophyll during austral summer 2009/2010. INCT–APA 1:115–120
Torgan LC, Tundisi JG, Niencheski LF (2002) Seasonal variation of planktonic diatoms in Patos Lagoon, Southern Brazil. In: John J (ed) Proceedings of 15th diatom symposium. Gantner Verlag, Lichtenstein, p 459470
UNESCO (1983) Chemical methods for use in marine environmental monitoring. Manual and Guides 12. IOC, Paris
Utermöhl H (1958) Zur Vervollkommnung der quantitativen Phytoplankton Methodik. Mitt Int Ver Limnol 9:1–38
Valentin JL (2001) The Cabo Frio upwelling system. In: Seeliger U, Kjerfve B (eds) Coastal marine ecosystems of Latin America, Ecological studies, vol 144. Springer, Berlin, pp 97–105
Villac MC, Mayr LM, Tenenbaum DR et al. (1991) Sampling strategies proposed to monitor Guanabara Bay, RJ, Brazil. In: Coastal zone 1991, Los Angeles, pp 1168–1182
Villac MC, Tenenbaum DR (2010) The phytoplankton of Guanabara Bay, Brazil. I. Historical account of its biodiversity. Biota Neotropica 10(2):271–293
Wang Z, Guo X, Qu L et al (2017) Effects of nitrogen and phosphorus on the growth of Levanderina fissa: how it blooms in Pearl River Estuary. J Ocean Univ China 16(1):114–120
Welschmeyer NA (1994) Fluorometric analysis of chlorophyll *a* in the presence of chlorophyll b and phaeopigments. Limnol Oceanogr 39:1985–1992

Phytoplankton Patterns and Processes in a Tropical-Subtropical Transition Region: Santa Catarina Coast, Southern Brazil

Leonardo R. Rörig, Marcio da Silva Tamanaha, Graziela da Rosa Persich, Carlos A. França Schettini, and Eliane C. Truccolo Schettini

Abstract The coast of Santa Catarina state is 561 km long and presents varied features and ecosystems. Bordered by the states of Paraná to the north and Rio Grande do Sul to the south, the coast of Santa Catarina shows a clear transition between these two neighboring states. In the coastal zone, this transition is marked by the limit of mangrove distribution (latitude 28°30′S) and the presence of the Cape of Santa Marta Grande, which induces upwelling and creates important oceanographic changes that affect the continental shelf. The Santa Catarina coast is one of the most important fishing regions of Brazil and contains 95% of the country's mariculture. Phytoplankton of the neritic zone presents considerable seasonality, as evidenced by moderate diatom blooms in the spring, reduction of phytoplankton biomass in the summer, and increased importance of dinoflagellates in the winter, patterns associated with peculiarities of local oceanography. Blooms of *Trichodesmium* spp. are also seasonal from early spring to late summer. Various estuaries impose strong influence over these seasonal patterns, forming estuarine or fluvial plumes where phytoplankton biomass increases at any time of the year. The high degree of pollution, most likely associated with climatic and oceanographic anomalies in some areas, has resulted in unusual blooms of diatoms, e.g., *Pseudo-nitzschia* spp. and *Amphitetras antediluviana*, and dinoflagellates, e.g., *Dinophysis*

L. R. Rörig (✉) · G. da Rosa Persich
Laboratório de Ficologia (LAFIC), Universidade Federal de Santa Catarina (UFSC), Florianópolis, SC, Brazil
e-mail: leonardo.rorig@ufsc.br

M. da Silva Tamanaha
Laboratório de Ficologia, Universidade do Vale do Itajaí (UNIVALI), Itajaí, SC, Brazil
e-mail: mt@univali.br

C. A. França Schettini · E. C. Truccolo Schettini
Laboratório de Hidrodinâmica Costeira, Centro de Tecnologia e Geociências, Departamento de Oceanografia, Universidade Federal de Pernambuco (UFPE), Recife, PE, Brazil

M. S. Hoffmeyer et al. (eds.), *Plankton Ecology of the Southwestern Atlantic*,
https://doi.org/10.1007/978-3-319-77869-3_13

spp., posing a risk to both local ecology and mariculture. On sandy beaches, surf diatom blooms are quite frequent, especially *Asterionellopsis glacialis* sensu lato. The surf diatom *Anaulus australis*, which occurs along the entire Brazilian coast, alternating with *A. glacialis*, has its southern limit of distribution in Santa Catarina. These and other peculiarities characterize the Santa Catarina coast as an area of abrupt ecological transition and, hence, a genuine hotspot for ecological studies based on phytoplankton. This chapter reviews published data, including theses and difficult-to-access publications, in an attempt to characterize these transitional features but also to highlight emerging patterns to allow a greater understanding of western South Atlantic pelagic ecology.

Keywords Brazil · Santa Catarina · HAB · Microalgae · Bloom

1 Study Area

1.1 Physiography

The coastline of Santa Catarina in Southern Brazil (from latitude 25°58′S to 29°19′S) is 531 km long and consists of varied features, including bays, coves, rocky shores, extensive sandy beaches, islands, estuaries, and coastal lagoons (Fig. 1). Bordered by the states of Paraná (PR) to the north and Rio Grande do Sul (RS) to the south, Santa Catarina (SC) presents a clear transition between the coastal landscapes typical of these two neighboring states. The north and central portions (from 25°58′S to 28°36'S) present a complex of uplands and cliffs that reach the sea, forming rocky headlands that alternate with sandy beaches. Babitonga Bay, which is located at the extreme north end, harbors one of the more extensive and rich mangroves of Brazil. From this sector to the mouth of the Imaruí Lagoon (Laguna City, 28°29′S), the coastline is characterized by relatively short sandy beaches, coves and rocky shores, outlets of medium and small rivers, and the Island of Santa Catarina, which forms two large shallow bays on its western side, facing the continent. The largest river in this region is the Itajaí-Açu River (26°54′S) whose plume generally drifts north-northeast and influences the sediment flow of the adjacent continental shelf. From the Imaruí Lagoon toward the south, the coastline becomes straight and is characterized by extensive and exposed sandy beaches, rivers with relatively small basins, and the presence of a broad coastal plain where a coastal lagoon system extends. The limit of distribution of mangroves along the southeast coast of South America is located in the estuarine part of Imaruí Lagoon (28°30′S). From this point southward, salt marshes begin to dominate the estuarine and lagoon fringes. The Cape of Santa Marta Grande (CSMG, 28°36′S) is located near this area. This geographical feature is also an important geographical accident that determined changes in the landscape of Santa Catarina, as well as the coastline, which shifted toward a more southeasterly direction. In addition, the position of the

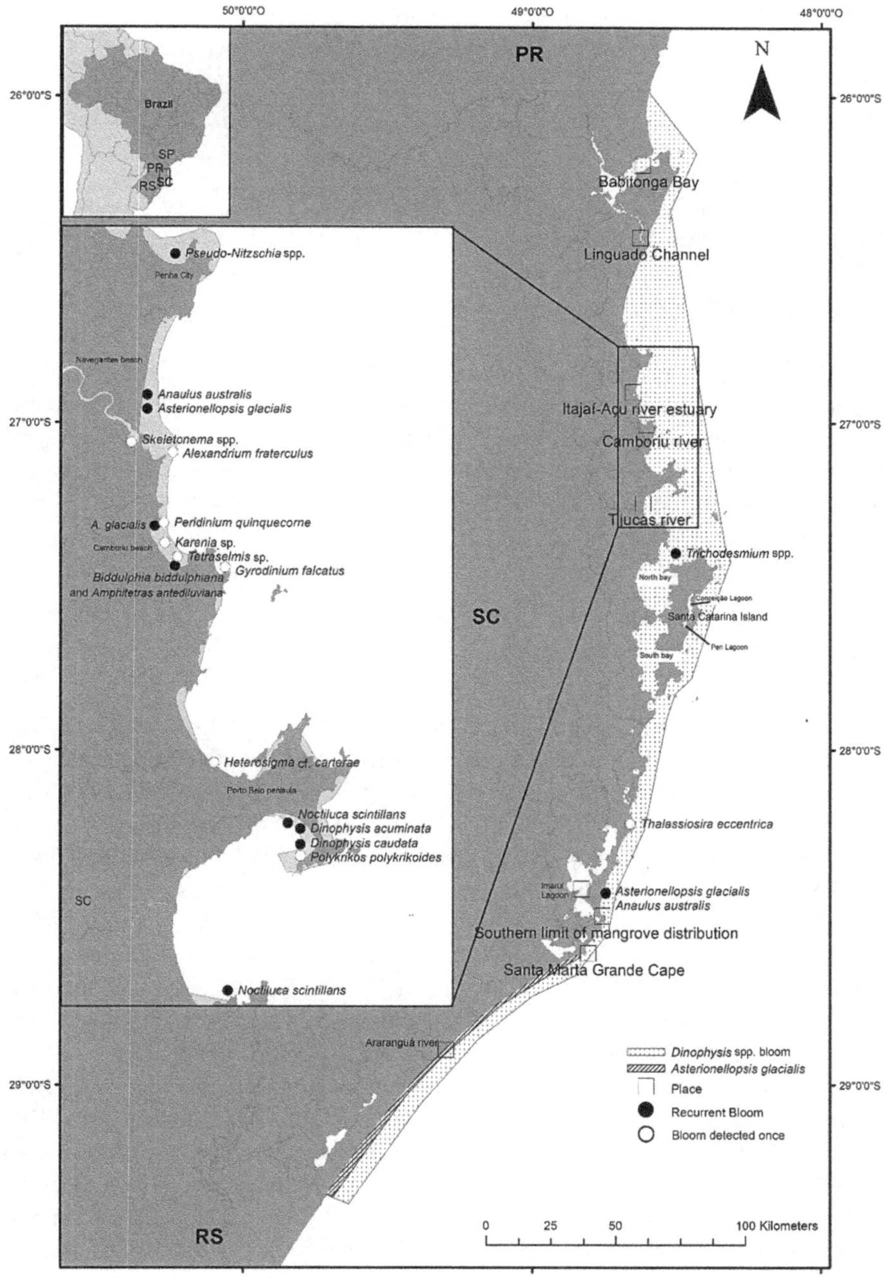

Fig. 1 Location of the coast of Santa Catarina in Southern Brazil with indications of the main occurrences of phytoplankton blooms

Table 1 Features of the southern states of Brazil coast

State coast	Latitude	Coast extension (km)	N° inlets/estuarine systems	N° inlets/100 km
SP	23°22′S–25°18′S	622	43	6.91
PR	25°18′S–25°58′S	98	7	7.14
SC	25°58′S–29°19′S	531	45	8.47
SC north	25°58′S–28°36′S	411	41	9.97
SC south	28°36′S–29°19′S	120	4	3.35
RS	29°19′S–33°44′S	623	4	0.67

SP São Paulo, *PR* Paraná, *SC* Santa Catarina, *RS* Rio Grande do Sul

cape determines the occurrence of coastal upwelling, strongly influencing the physical and ecological characteristics of coastal and shelf waters. South of CSMG, the decreasing number of inlets or estuaries of small- and medium-sized rivers presents topography more akin to that of extreme Southern Brazil, while the northern portion is more similar to the southeastern Brazilian coast by the greater number of inlets (Table 1).

1.2 Physical Oceanography

The southeast continental shelf of Brazil (SEBCS, from 21.5°S to 34.9°S) is divided into three areas according to physiographic features and major oceanographic processes: the southern Brazilian shelf (SBS) from Chuí (34.9°S) to CSMG (28.5°S), the southern Brazilian bight (SBB) from CSMG (28.5°S) to Cabo Frio (23°S), and the Abrolhos-Campos Region (ACR) from Cabo Frio (23°S) to Cape São Tomé (21.5°S) (Castro and Miranda 1998). The continental shelf of SC is covered by SBS and SBB. The shelf break along SBB and SBS occurs at a depth of about 180 m, ending at a depth of 2000–2900 m (Ito et al. 2016), and the shelf width ranges between 105 km off CSMG and 260 km off Babitonga Bay. Offshore circulation is dominated by the Brazilian Current (BC), which flows southward (limited to the upper 500 m), advecting tropical water (TW, salinity >36.4, temperature > 20 °C) and influencing the shelf water masses during summer (Castro and Miranda 1998). During winter, a northward coastal current advects the Plata Plume water (PPW, salinity <33.5, temperature > 19 °C). During summer, northerly winds favor an upwelling process and advection toward the shore of the South Atlantic Central Water (SACW, salinity 34.6–36.2, temperature 8.7–20 °C), which occupies the bottom layer and reaches the surface between CSMG and Santa Catarina Island. Mainly from Santa Catarina Island northward, the contribution of local freshwater produces coastal water (CW, salinity <35 and temperature varying seasonally). Figure 2 shows block diagrams that indicate the spatial occurrence of these water masses.

The SC coast receives significant input of nutrients from discharges emanating from the Río de la Plata and the Patos Lagoon (Ciotti et al. 1995), as well as small

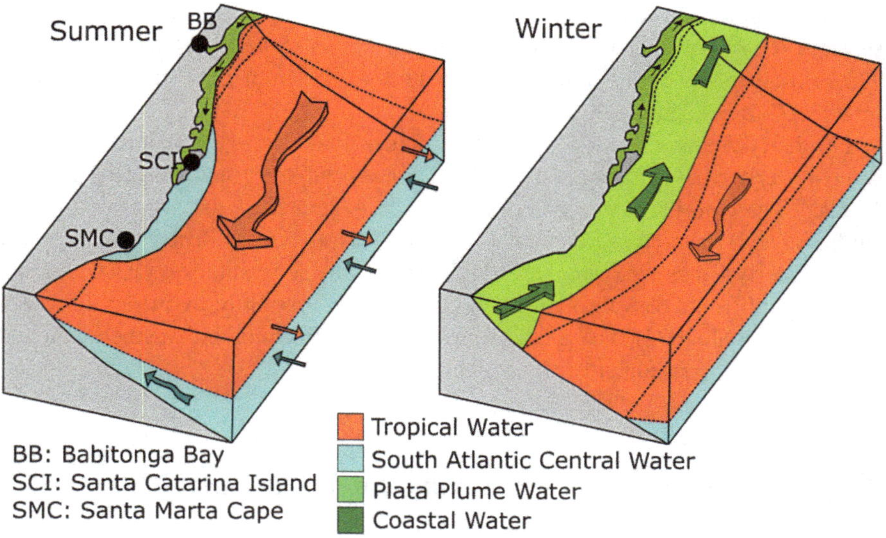

Fig. 2 Schematic block diagrams of the main distributions and currents of the water masses on the Santa Catarina continental shelf in summer and winter

contributions from local estuaries. Additionally, it is influenced by SACW upwelling directly in the coastal zone and from the external continental shelf. In winter, Campos et al. (1995) proposed that the mechanism responsible for pumping the SACW to the continental shelf involves the cyclonic meandering of the BC, inducing shelf break upwelling. During the summer, this process is combined with coastal upwelling from the northeast. The main coastal upwelling of SC occurs near the CSMG, but indications of small upwellings can be observed near coastal islands, especially in the case of both Santa Catarina and Arvoredo Islands (Rörig et al. 1997; Resgalla Jr. et al. 2004). These upwellings represent peculiarities of the SC coast, having the ability to induce special effects on local biological diversity and productivity (Aidar et al. 1993; Ciotti et al. 1995).

1.3 Regional Climatology

According to the Köppen Classification System, the climate of SC is humid subtropical, mesothermal, without a dry season (Cf), and with hot (Cfa) and temperate summers (Cfb) (Alvares et al. 2013; Pandolfo et al. 2002). Both temperature and precipitation present significant spatiotemporal variability, as determined by geographic position, relief characteristics, and atmospheric circulation systems acting in the region (Monteiro 2001; Grimm 2009a). Furthermore, significant variations occur on an interannual scale relative to El Niño/La Niña events (Grimm 2009b; Firpo et al. 2012).

The mean annual temperature in the region is 20 °C, with a winter (July) and summer (February) mean minimum of 10 and 31 °C, respectively (Pandolfo et al. 2002). Total annual precipitation is around 1100 to 1300 mm, but it can reach values up to 2300 mm (Grimm 2009a).

Northerly winds, mainly northeasterly, predominate along the year, followed by southwesterly winds, mainly in the fall and winter, with maximum velocities around 18 ms^{-1}. These northeasterly synoptic winds are generated by the Atlantic subtropical high (ASH), which predominates over the Atlantic Ocean. The components of southwesterly winds are strongly related to the passage of frontal systems across the region. Cold fronts pass through SC with an average monthly frequency of 3 to 4 (Rodrigues et al. 2004). These cold fronts strongly affect coastal wave climate and circulation and, consequently, the coastal ecology.

2 General Patterns of Coastal and Shelf Phytoplankton

Phytoplankton studies on the SC coast began in 1994, following the creation of the oceanography course at the Universidade do Vale do Itajaí (UNIVALI). Previously, only isolated studies reported on some estuaries and islands (e.g., Araujo et al. 1989; Odebrecht 1988; Cardoso et al. 1994). Some nationwide projects also conducted sampling at stations on the SC continental shelf and described general phytoplankton and primary production patterns, but without characterizing local specificities, owing to the degree of spatiotemporal spacing between samples (e.g., Brandini 1988; 1990a, b; Ciotti et al. 1995).

In 1994, a series of seasonal oceanographic cruises began in association with a project known as "Integrated Study of the Ecosystem of Itajaí-Açu River Estuary and Adjacent Coastal Zone" (FBB Project). It was developed by UNIVALI with funding from the Banco do Brasil Foundation (FBB) and support from the Brazilian Institute of the Environment and Renewable Natural Resources (IBAMA). Cruises aboard the Diadorim research vessel (IBAMA) occurred in the spring and summer of 1994 and in the fall and winter of 1995. These crises consisted of a spatiotemporal assessment of water mass characteristics, chemical characterization, primary production, and phytoplankton analysis. The sampling area covered the internal platform up to about 50 m depth, focusing on the influence of Itajaí-Açu River discharge, the main river flowing to the coast of SC, as well as other smaller rivers between the latitudes of 26°30′S and 27°17′S (Carvalho et al. 1998).

The physical oceanography results of this project showed seasonally different situations for the region. In spring and summer, the water column was stratified with a strong thermocline, where northerly winds, especially northeasterly, cause SACW upwelling and southerly winds cause subsidence of coastal waters with thermocline deepening. During autumn and winter, the water column is predominantly homogeneous as a result of both coastal subsidence, owing to the greater magnitude and persistence of southerly winds, and by the advection of subantarctic waters, which are influenced by the contribution of the Río de la Plata and Patos Lagoon (Carvalho

Table 2 Primary productivity data of Santa Catarina coast (Brazil) during a seasonal cycle survey between 1994 and 1995

Season	Average surface irradiance (μmol m^{-2} h^{-1})	Average surface chlorophyll *a* (mg m^{-3})	Average surface photosynthesis (mgC mgChl *a* h^{-1})	Irradiance of maximum primary productivity (μmol m^{-2} h^{-1})	Primary productivity at optimum irradiance (mgC mgChl a^{-1} h^{-1})	n
Spring	1569	2.26	3.50	1008	6.12	6
Summer	1757	1.27	2.69	645	4.19	4
Autumn	879	0.83	3.14	716	3.70	4
Winter	1101	1.72	5.42	1101	7.58	8

L.R. Rörig, previously unpublished data

et al. 1998). These results corroborate previously defined features found in larger-scale studies along SCSB and guided a series of regional studies in chemical and biological oceanography.

For the first time, data on the primary production of phytoplankton were generated, and the main species occurring in the coastal zone were detected, as well as typical blooms (see below). A considerable part of the information from the FBB Project has not been published in international journals, and the data are only found in technical reports, monographs, and authors' databases.

Phytoplankton photosynthesis data per unit of chlorophyll *a* (photosynthetic efficiency) were generated by the ^{14}C method (Richardson 1987) at stations selected from a grid of 63. Using an in situ simulated system, surface samples were incubated in a gradient of irradiances to obtain preliminary data on the optimal irradiance for the phytoplankton community during each season of the year. Table 2 shows a summary of these previously unpublished data. Since they are average values, we can clearly see the seasonal differences in irradiance, chlorophyll a, and photosynthesis. The highest chlorophyll a values were recorded in spring, followed by winter and summer, with lower values in fall. Surface photosynthesis values were highest in winter, followed by spring, which could be associated with higher nutrient concentration. Apparently, photosynthesis was photoinhibited by surface irradiance in all seasons, except winter, when the maximum surface irradiance occurred. Thus, although the mean value was higher in winter than in spring, an integration of the data along the water column would probably show maximum values in spring because of the lower irradiance and photoperiod in winter (Pellens 1997; Rörig et al. 1998b). These values were higher than those recorded by other authors in samples from the SC continental shelf and from northern states (PR and SP, Table 3). Such differences may be related to the sampling region where data obtained in the FBB Project were from more coastal stations, even though chlorophyll a values were similar.

Two types of diatom blooms were detected in 1994. The first was a relatively strong diatom spring bloom with subsurface biomass peaks often associated with the thermocline or in the top 10 m of the water column. This bloom was dominated

Table 3 Primary productivity data from southern and southeastern Brazilian continental shelf off São Paulo, Paraná, and Santa Catarina states

Water mass	Average value of chlorophyll a (mg m^{-3})	Average value of photosynthesis (mgC mgChl a^{-1} h^{-1})
Coastal water	2.9	4.24
Plata Plume water	0.46	2.0
Tropical water	0.21	3.59

Extracted from Gaeta and Brandini (2006)

by *Pseudo-nitzschia* spp., *Skeletonema tropicum*, *Chaetoceros* spp., *Bacteriastrum* sp., *Lauderia borealis*, and *Hemiaulus* spp., among others. The presence of dinoflagellates was more discrete, but *Prorocentrum* spp., *Protoperidinium* spp., and *Neoceratium* spp. were quite common. Another highlight was the considerable density of micro- and macrozooplanktonic organisms, even using phytoplankton nets (20 μm), indicating intense trophic flow.

The other bloom, detected simultaneously, but with patchy distribution, was composed of cyanobacteria *Trichodesmium* spp. The dominant species was identified at the time as *T. erythraeum* (Guimarães and Rörig 1997). Subsequent analyses suggested it to be *T. hildebrandtii* (Rörig et al. 1998a). *T. thiebautii*, a species already recognized as toxic at that time, was also detected, although rare. These blooms were easily visible on the surface of the water, forming in-line accumulations influenced by the wind (Langmuir circulation). White patches were revealed to be formed by senescent populations (Fig. 3c), and pale brown patches were formed by healthy populations. Frozen samples of the healthy populations developed a strong reddish color, suggesting the presence of phycoerythrin, an accessory pigment typical of marine cyanobacteria.

This positive buoyancy behavior in *Trichodesmium* spp. is a well-known phenomenon (Carpenter and Price 1977). Low levels of dissolved nitrogen in water cause trichomes to group into colonies and increase buoyancy by the action of intracellular gas vesicles. Once at the water-air interface, the colonies are susceptible to wind transport. Peripheral trichomes in the colonies remain photosynthetically active, while the internal ones, chlorophyll a is inactivated, and nitrogenase, responsible for N_2 fixation, is activated. Photosynthesis and nitrogen fixation are incompatible functions in the same cellular environment since oxygen produced by photosynthesis inactivates nitrogenase (Fay 1992). This ability of *Trichodesmium* spp. allows survival and proliferation in oligotrophic waters. Thus, this phenomenon results in a functional nitrogen link that fertilizes oligotrophic waters and promotes increases in biological productivity. In fact, in those blooms, biomass and diversity of other phytoplankton species and also zooplanktonic species were higher when compared to areas without *Trichodesmium* patches (Guimarães and Rörig 1997).

Previously, blooms of these cyanobacteria had only been recorded in the tropical northeastern region of Brazil (Satô et al. 1963) and on the coast of São Paulo (Gianesella-Galvão 2000). Despite this, it is likely that they always occurred in the

Fig. 3 Photographs of blooms (arrow) on the coast of Santa Catarina, Brazil, and the Towed Oceanographic Vehicle (VOR). (**a**) Bloom of the surf diatom *Asterionellopsis glacialis* on Rincão Beach. (**b**) Bloom of the surf diatom *Anaulus australis* on Navegantes Beach (kindly provided by Dr. Luis A. Oliveira Proença, IFSC, Florianópolis, Brazil). (**c**) Senescent bloom of *Trichodesmium* spp. near Arvoredo Island. (**d**) Bloom of invasive bryozoans and epibenthic diatoms on Camboriú Beach (kindly provided by Dr. Fernando Luiz Diehl – Acquaplan – Balneário Camboriú, Brazil). (**e**) VOR technical drawing. (**f**) VOR after operation; photo in detail at top right showing a *Neoceratium* sp. cell retained in the mesh

region and were never detected because of the absence of detailed studies. Evidence of this is revealed by the detection of such blooms every year since then between September and February in monitoring programs (CIDASC 2017). Although recurrent on the coast of SC, the origin of the *Trichodesmium* patches is in the most oceanic, oligotrophic waters associated with the Brazil Current. Their penetration toward the coast is associated with transport by wind and oceanic fronts. It is possible that *Trichodesmium* populations increase as they find phosphate-rich waters close to the coast related to small upwellings (Rörig et al. 1998a) or to continental inputs (Schettini et al. 2005), but they do not persist for long when entering less saline and more turbid coastal waters dominated by r-strategist species, such as diatoms. Even so, the inputs to biological productivity are abundant, as are the risks to mariculture and fish resources, as further studies have revealed toxicity associated with these patches (Proença et al. 2009; Detoni et al. 2016). Over the last years, in spring and summer, *Trichodesmium* patches were also recorded in the coastal region of Paraná (Siqueira et al. 2006) and Rio Grande do Sul (Silva et al. 2008).

FBB Project samplings also revealed general patterns for phytoplankton in summer, fall, and winter. In summer, a decrease in phytoplankton biomass was reported, especially in areas farthest from the coast. However, near the coast, the enrichment pulses generated by small upwellings and the contribution of fluvial discharge, owing to high regional rainfall, were able to sustain rich populations, especially diatoms. *Trichodesmium* patches also occurred, albeit less frequently (Rörig et al. 1998b).

Autumn showed the lowest chlorophyll a values (Table 2), probably from zooplankton grazing and strong water column stratification (Carvalho et al. 1998). It is important to highlight the occurrence of central diatoms of the genus *Coscinodiscus*, including *C. wailesii*, *C. gigas*, and *C. radiatus*. The large cell size of these species prevents grazing by most zooplankton, making them remnants of diatom populations that have developed since the spring of the previous year. It is possible that these populations are grazed by gelatinous organisms, such as siphonophores, salps, or cnidarians, which are common at this time in the region (Resgalla Jr. et al. 2005), and even fish (e.g., *Mugil* spp.). Another likely strategy is sinking so that they can act as food for benthic fauna or enter the debris chain.

The winter pattern was quite peculiar in that it showed an increase in the relative importance of dinoflagellates (Rörig et al. 1998b). A number of subsequent monitoring programs have shown that this greater importance of dinoflagellates in winter is a recurrent pattern (unpublished data from monitoring projects of UNIVALI and UFSC). It is likely that these populations originated from the greater influence of subantarctic waters in winter, which are richer in dinoflagellates (Islabão and Odebrecht 2011). Once on the coast of Santa Catarina, some dinoflagellate species probably find optimal conditions to form blooms, such as the presence of sites of lower turbulence and high nutrient concentration. It is precisely in this type of environment that mariculture, in particular mussels and oysters, was developed in the region, indicating that winter is a critical period for harmful algal bloom (HAB) monitoring.

These initial studies were fundamental in describing the general pattern of SC coastal phytoplankton. More systematic studies have confirmed these trends and support the idea that this region has heterogeneous and complex patterns in terms of abundance, diversity, and distribution of phytoplankton, as well as the occurrence of blooms. These data also supported the first major Brazilian HAB monitoring program established on the North Central Coast of SC (see below).

More recently, Becker (2014), in a comprehensive study involving phytoplankton and zooplankton on the SC continental shelf, highlighted the influence of estuarine discharges, coastal upwelling, and seasonal variations of water masses in structuring the planktonic community. It was evident from this study that the SC inner shelf undergoes dramatic annual variability with strong thermocline during summer and almost no stratification during winter when the influence of PPW is greater (Fig. 2). The highest abundances of diatoms and copepods were associated with the maximum chlorophyll *a* layers from SACW intrusions near the surface (Fig. 2). The intermittent upwelling and river plume events bring nutrients to surface layers. This surfacing of nutrients results in variation in phytoplankton size structure such that larger, or chain-forming, cells have an advantage that ends up influencing the size structure of copepods. If these enrichment processes do not occur, or are less frequent, either as a matter of time or place, the predominance of smaller phytoplankton (pico- and nanophytoplankton) determines longer trophic chains typical of tropical conditions. Becker (2014) also highlighted nutrient enrichment caused by the Itajaí-Açu River (latitude 26°54′S) and Babitonga Bay (latitude 26°10′S) and their influence on the biomass and composition of phytoplankton and zooplankton, as also pointed out in previous studies (Schettini et al. 1998; Resgalla Jr. 2009; Rörig et al. 2003). These data indicate that the combined influence of the Itajaí-Açu River plume and CSMG represents the major regional factor affecting the structure of plankton assemblage, often overlapping with the patterns defined by the seasonal influence of water masses.

3 Surf Zone Diatoms

The occurrence of dense diatom blooms or accumulations in the surf zone is a typical phenomenon of many exposed, intermediate to dissipative, sandy beaches (Talbot et al. 1990). The accumulations appear as neustonic, flocculent, and brownish patches at the inner surf zone and are also deposited on the beach face, representing intense autochthonous contributions of organic matter to the beach ecosystem with extremely high chlorophyll *a* concentration (Talbot and Bate 1988). These accumulations are mainly formed by a combination of resuspended epibenthic stocks deposited beyond the surf zone and special adaptations of the species involved (Talbot and Bate 1988). The increase of wave energy, as determined mainly by the passage of frontal systems, resuspends the stocks. Suspended cells release mucus allowing them to stick to wave foam and form patches. After a few days, the mucus properties change, and particles of suspended material adhere to cell masses, favoring sedimentation, and return to the original epibenthic position, as wave energy dissipates. In Brazil, surf diatoms accumulations are more common in the southernmost part of the country (RS coast) and also in the southern portion of SC state (Rörig et al. 1997; Odebrecht et al. 2010) where extensive exposed sandy beaches dominate the coastal landscape. However, these phenomena are also recorded along the entire Brazilian coast (Odebrecht et al. 2014). In SC, a 6 km stretch between headlands (Balneário Camboriú, 26°59′S) was the shortest beach where these accumulations were observed. Of the seven species that form this phenomenon around the world (Odebrecht et al. 2014), at least two occur in Brazilian beaches, and *Asterionellopsis glacialis* and *Anaulus australis* are part of it. Recently, molecular analyses have shown that the species traditionally identified as *A. glacialis* represents a complex of species (Kaczmarska et al. 2014). In this chapter, which reports works made before this discovery, the possible different species are grouped as *Asterionellopsis glacialis* sensu lato.

On the SC coast, accumulations of *A. glacialis* s.l. and *Anaulus australis* show different distribution patterns (Fig. 3a, b). *A. glacialis* s.l. occurs as an exclusive species on the beaches south of CSMG and co-occurs with *A. australis* on northern beaches, with monospecific or bispecific accumulations having different degrees of dominance among species (Rörig et al. 1997). This pattern of mixed accumulation has also been verified on beaches in the state of Paraná (Rezende and Brandini 1987) and in the northeastern part of the country (Tedesco et al. 2017). What draws attention to these data is the apparent limit of distribution of *A. australis* in CSMG. This species has neither been recorded on the coast of RS nor on the beaches of Uruguay and Argentina where only *A. glacialis* s.l. and *Attheya armata* have been recorded (Odebrecht et al. 2014). It is still not possible to assert what determines this limit of distribution, whether the physical barrier (lower temperature), as represented by the coastal upwelling south of the CSMG, or competition that would favor *A. glacialis* s.l. by its adaptation to the morphodynamic patterns of extensive beaches which lie south of this cape. In any case, this aspect reinforces the ecological transition that takes place on the southern Brazilian coast, as represented by the CSMG.

The beaches that are active for this phenomenon are usually close to river mouths or estuaries. In these areas, changes in sedimentation patterns, especially where dredging activities take place, can affect the cycle of surf diatoms, reducing, or even blocking, their occurrence. This is likely to occur, for example, at Cassino Beach near the Patos Lagoon estuary (RS) and at Navegantes Beach (SC), suggesting the wisdom of monitoring actions, given the central trophic importance of these blooms for regional fisheries, not only quantitatively but also qualitatively, as sources of essential fatty acids (Rörig et al. 2017a).

4 HAB Monitoring and Research on the Santa Catarina Coast

In the early 1990s, mariculture was established on the coast of SC, as an option to complement the income of artisanal fishermen, and it gradually became an important activity, changing the economic profile of the region. In this context, in 1995, UNIVALI began research at Armação do Itapocorói, in the city of Penha, to identify and monitor harmful algal blooms (Proença and Rörig 1995; Rörig et al. 1998b; Proença et al. 1998). In 1997, the first monitoring program of harmful algae and phycotoxins in Brazil began, and they consisted of identifying microalgae and analyzing phycotoxins with bioassays and high-performance liquid chromatography (HPLC). The program followed the model of Galicia, Spain, where UNIVALI researchers were trained under the guidance of the Spanish Oceanographic Institute (IEO), Vigo. After a few years, the methods were standardized for routine analyses of saxitoxins (paralytic shellfish poisoning – PSP), okadaic acid (diarrheic shellfish poisoning – DSP), and domoic acid (amnesic shellfish poisoning – ASP) (Hallegraeff et al. 1995). Subsequently, the monitoring was expanded to other mariculture-producing regions of SC with the participation of EPAGRI (Agricultural Research and Rural Extension Company of Santa Catarina), aiming to guarantee the sanitary quality of the shellfish.

The first effective results of this program came in 2001 when the microalga *Gymnodinium catenatum*, a PSP producer (Proença et al. 2001), was identified for the first time and isolated. In the following years, four potentially harmful taxa, including *Peridinium quinquecorne*, *Karenia* sp., *Tetraselmis* sp., and *Gyrodinium falcatus*, were recorded in Balneário Camboriú, SC (M. S. Tamanaha, unpublished data) (Fig. 1). In 2007, high biomasses of *Pseudo-nitzschia* spp. (> 40,000 cells L^{-1}, Tamanaha et al. (2008)) and *Alexandrium* cf. *tamarense* (Miotto and Tamanaha 2012) were recorded for the first time in SC. In the same year, the SC coast was affected by a bloom of *Dinophysis acuminata*, a potential DSP producer (unpublished data from UNIVALI Monitoring Program). According to official data (ANVISA), more than one hundred people were poisoned by contaminated mussels. At that time, the harvesting and consumption of shellfish were prohibited. It would be the first of many official ordinances recommending discontinuation of shellfish consumption, attesting to the program's effectiveness in regional food security. Both HAB and phycotoxin monitoring became part of the National Program for Sanitary Control of Bivalve Molluscs (PNCMB). This program defined maximum permitted limits for phycotoxins in shellfish meat at 0.8 mg 100 g^{-1} for saxitoxin, 0.16 mg100 g^{-1} for okadaic acid, 20 mg 100 g^{-1} for domoic acid, 1 mg 100 g^{-1} for yessotoxins, and 0.16 mg 100 g^{-1} for azaspiracid (Instrução Normativa Interministerial MPA/MAPA n ° 07/2012). These interinstitutional efforts culminated in December 2011 with the establishment of a website administered by the state inspection service. This provided a database where reports of phycotoxin analyses of almost all SC cultivation areas could be published (CIDASC 2017).

In August 2011, another bloom of *D. acuminata* was detected in Governador Celso Ramos (Latitude 27°18′S). It was a timely alert, preventing people from consuming this toxin. Recurrent blooms of *D. acuminata* and *D. caudata* are currently reported throughout the SC coast, especially in winter. In 2016, the bloom was particularly large, extending to the coastline of São Paulo. A very interesting phytoplankton succession was recorded after a few days, with the presence of *Noctiluca scintillans* and *Polykrikos polykrikoides*, as predators, owing to the high biomass of *Dinophysis* spp. This phenomenon has intensified and occurred at least every 2 years since 2007.

According to EPAGRI, SC's mariculture counts on the participation of 572 producers and generates a production of 20438 tons per year (oysters and mussels). This history of research and monitoring of HABs in SC showed a very important evolution in parallel with the growth of shellfish production. As also pointed out in classical studies (Hallegraeff 1993; Wells et al. 2015), an apparent increase in the occurrence of HABs was detected. This increase may be related to the increase in aquaculture operations that can act as "bioassay systems" for harmful algae, making previously unknown problem organisms detectable. Such increase could also be caused by anthropogenic eutrophication whereby coastal nutrient enrichment from wastewater stimulates the growth of some harmful species (Hallegraeff et al. 1995). Both possibilities are feasible in the case of the SC coast where mariculture has increased, as well as pollution, as a consequence of the high population growth rate in the region, which is one of the largest in Brazil. It is certain that winter is the period of high dinoflagellate incidence, which is clearly related to the influence of cold water masses that come from the south, carrying dinoflagellate-rich populations. The diversity and heterogeneity of microenvironments on the SC coast, in association with increased nutrients, can result in the greater proliferation of these algae when compared to the southernmost regions (RS state) or northernmost regions (PR and SP states). In other words, southern water masses act as inocula, while local conditions favor the development of some harmful microalgae species.

Regarding *Pseudo-nitzschia* spp., the occurrence of the potentially toxic species *P. calliantha* and *P. multiseries* has been confirmed on the SC coast. However, the production of domoic acid was first verified in Brazilian waters in 2009, and it caused the closing of two shellfish farms for 25 days in the region of Penha (Fernandes and Brandini 2010). Some authors have verified that *P. calliantha* and *P. multiseries* present higher growth rates and increase in domoic acid production in response to high concentrations of ammonium (Quijano-Scheggia et al. 2008; Ljubesic et al. 2011). Since ammonium is a nitrogenous nutrient often associated with organic pollution, it is possible that the large population increase on the SC coast, combined with insufficient wastewater treatment, could have contributed to the increased occurrence of *Pseudo-nitzschia* spp. blooms, as has been recorded in estuaries and bays of Santa Catarina Island (Rörig et al. in prep.).

The epibenthic diatoms *Amphitetras antediluviana* and *Biddulphia biddulphiana*, which occur in association with large masses of invasive bryozoans on the beach of Balneário Camboriú (Fig. 3d), were reported to cause an important non-toxic bloom, albeit still resulting in aesthetic and economic damage (Rörig et al. 2017b). These two filamentous species grow as epibionts on bryozoan masses, which are deposited on the beach after increased wave energy. Most deposited biomass can be attributed to bryozoans, but the relative importance of diatoms increases in winter and spring. This is another effect related to intense urbanization in the absence of adequate waste management.

5 New Challenges and Advances: The Towed Oceanographic Vehicle (VOR)

The Towed Oceanographic Vehicle (VOR) (Fig. 3e–f) was developed by Faccin et al. (2014), as a modified version of the CPR (Continuous Plankton Recorder; www.sahfos.ac.uk), for use in fishing vessels operating along the southeastern and southern Brazilian coast. The VOR can obtain samples of the planktonic community at low cost in ships of opportunity, but it can also be implemented in areas of greater ecological importance and economic interest, such as fishing areas. The first cruises with the VOR were carried out in May and June of 2013, covering the southern Brazilian continental shelf off the states of SP, PR, SC, and RS in areas with depths between 50 and 100 m. The mesh analysis resulted in the identification of 75 phytoplankton taxa. The main differences were recorded between transects off SC state and RS state, with high density of the cyanobacteria *Trichodesmium* spp. and high densities of *Thalassiosira* spp. and *Octotactis octonaria*, respectively (Tamanaha et al. 2016). The high density of diatoms and silicoflagellates was related to the higher concentration of silicates and other dissolved nutrients off RS state, especially during the winter, when the influence of the Río de la Plata and Patos Lagoon is more pronounced (Piola et al. 2000; Rigual-Hernandez et al. 2010).

Tamanaha et al. (2016) also reported that mesh size is a limitation for VOR samplings, essentially because pico- and nanoplankton were rarely collected, thus overestimating the presence of microphytoplankton, e.g., dinoflagellates, chain-forming diatoms, and filamentous cyanobacteria. Even so, some nanoplankton species were observed in high abundance, as in the case of the cyanobacteria *Johannesbaptistia* sp., which was very common in samples from transects of SC. VOR sampling efficiency was not evaluated, but CPR has been compared to traditional samplers in different studies, and it has proven to be robust (Batten et al. 2003). Tamanaha et al. (2016) concluded that VOR is a new and practical tool to help understand plankton trophic relationships and long-term community changes.

6 Conclusions and Suggestions

Historical data and recent research support the hypothesis that the coast of SC, Brazil, has features in common with the nearby regions to the north and south but also emerging and exclusive features with respect to the ecology of phytoplankton. The coastal transition is marked by the CSMG where coastal orientation and landscapes change. To the north of the CSMG, the coastline is more indented, with greater diversity of habitats, more creeks and inlets, and to the south, the coastline is more homogeneous, rectilinear, with fewer inlets. The relatively frequent occurrence of coastal upwelling in the CSMG also imposes a physical barrier such that the lower mean water temperature may interfere with the distribution, dispersion, and even survival of planktonic, benthic, and nektonic species. This set of environmental heterogeneities certainly influences the distribution, composition, and biomass of phytoplankton, especially when compared to areas to the north and to the south of SC.

This ecotone condition warrants more detailed studies in the region because in relatively small areas, several factors can affect pelagic ecology, which is very dynamic and variable both at seasonal and interannual scales. The importance of continuous and high-frequency monitoring to understand these processes is clear, as evidenced after the start of HAB monitoring when many previously unknown phenomena were detected and explained, leading to sound guidance for management activities and future studies.

Several topics can be suggested as pertinent for studies on phytoplankton on the coast of SC based on previous information, including the following:

- Verify the influence of nutrient enrichment by the pollution that flows to the coast, but also its probable toxic effect, and its potential to cause anoxia, owing to features of the regional watersheds with high level of contamination and toxic loads (Rörig 2005; Pereira-Filho and Rörig 2016).
- Carry out studies directed to areas of high fishing productivity in the region to understand trophic relationships.
- Evaluate the effects of highly acidic waters of the Araranguá River (latitude 28°53′S), which acts as drainage for an extensive coal mining region (Couceiro and Schettini 2010). The area can be considered a large-scale bioassay to assess the effects of acidification in the oceans.
- Understand the causes of the increased incidence of potentially harmful species in the winter, and improve the predictability of possible HABs for the management of mariculture.
- Evaluate the apparent increase in *Pseudo-nitzschia* spp. populations in the coastal zone, as an effect of ammonium excess from organic pollution.
- Determine the relationship between mullet populations (*Mugil* sp.), the main fishing resource on the SC coast, and surf diatom blooms, probably the main food source for this fish.
- Evaluate the possible disturbance of the surf diatom cycle at Navegantes Beach by the sediments dredged from the Itajaí-Açu River estuary.
- Verify the possible toxic effect of *Cylindrospermopsis raciborskii*, a saxitoxin producer (Miotto et al. 2017) from the Peri Lagoon (Santa Catarina Island) on shellfish from the coastal region where its waters flow.
- Understand the causes of *A. antediluviana* and *B. biddulphiana* blooms associated with bryozoans at Camboriú Beach in order to control the process, as well as anticipate or avoid its proliferation to other beaches.

References

Aidar E, Gaeta SA, Gianesella-Galvão SMF, Kutner MB, Teixeira C (1993) Ecossistema costeiro subtropical: nutrientes dissolvidos, fitoplâncton e clorofila-a e suas relações com as condições oceanográficas na região de Ubatuba, SP. Publ Esp Inst Oceanogr 10:9–43

Alvares CA, Stape JL, Sentelhas PC, Gonçalves JLM, Sparovek G (2013) Köppen's climate classification map for Brazil. Meteor Zeitschrift 22(6):711–728

Araujo N, Souza-Mosimann RM, Fernandes LF (1989) Contribuição ao estudo das diatomáceas (Bacillariophyceae) da Foz do Rio Araranguá – SC – Brasil. Ínsula 19:305–334

Batten SD, Flinkman R, Clark J, Hays G, John E, John AWG, Jonas T, Lindley JA, Stevens DP, Walne A (2003) CPR sampling: the technical background, materials and methods, consistency and comparability. Prog Oceanogr 58:193–215

Brandini FP (1988) Hydrography, phytoplankton biomass and photosynthesis in shelf and oceanic waters off Southeastern Brazil during autumn (May/June 1983). Braz J Oceanogr 36:63–72

Brandini FP (1990a) Producão primária e características fotossintéticas do fitoplâncton na Região Sueste do Brasil. Braz J Oceanogr 38:147–159

Brandini FP (1990b) Hydrography and characteristics of the phytoplankton in shelf and oceanic waters off Southeastern Brazil during winter (July/August 1982) and summer (February/March 1984). Hydrobiologia 196:111–148

Becker EC (2014) Comunidade planctônica da plataforma continental de Santa Catarina (26-29°S):distribuição e estrutura de tamanho de fitoplâncton e zooplâncton, com ênfase em copépodes. Dissertação deMestrado, Universidade Federal de Santa Catarina

Campos EJ, Goncalves J, Ikeda Y (1995) Water mass characteristics and geostrophic circulation in the South Brazil Bight: summer of 1991. J Geophys Res 100:18537–18550

Cardoso LS, Baptista LRM, Jebram DHA (1994) Dinoflagelados da ilha do Arvoredo e da praia de Ponta das Canas-Santa Catarina, Brasil (Setembro de 1991 a Fevereiro de 1992): condições hidrológicas e lista de Espécies. Iheringia 45:143–155

Carpenter EJ, Price CC (1977) Nitrogen fixation, distribution, and production of Oscillatoria (Trichodesmium) spp. in the western Sargasso and Caribbean seas. Limnol Oceanogr 22:60–72

Carvalho JLB, Schettini CAF, Ribas TM (1998) Estrutura termohalina do litoral centro-norte catarinense. Notas Técnicas da FACIMAR 2:181–197

Castro BM, Miranda LB (1998) Physical oceanography of Western Atlantic continental shelf located between 4°N and 34°S. In: Robinson AR, Brink KH (eds) The sea, vol 2. Wiley, New York, pp 209–251

CIDASC - Companhia Integrada de Desenvolvimento Agrícola de Santa Catarina (2017) Florianópolis. Santa Catarina, Brazil. http://www.cidasc.sc.gov.br/. Accessed 22 Sept 2017

Ciotti AM, Odebrecht C, Fillmann G, Moller O Jr (1995) Freshwater outflow and Subtropical Convergence influence on phytoplankton biomass on the southern Brazilian continental shelf. Cont Shelf Res 15:1737–1756

Couceiro MAA, Schettini CAF (2010) Estudo da dinâmica dos sedimentos em suspensão do estuário do rio Araranguá (SC): possíveis efeitos da drenagem ácida da atividade de mineração de carvão. Geociências 29:251–266

Detoni AMS, Costa LF, Abrão LP, Sarkis JY (2016) Toxic *Trichodesmium* bloom occurrence in the southwestern South Atlantic Ocean. Toxicon 110:51–55

Faccin J, Cunha DMC, Barddal R, Resgalla C Jr (2014) Development of an oceanographic towing vehicle adapted for fishing craft: prototype and protocol for use. Methods Oceanogr 9:61–74

Fernandes LF, Brandini FP (2010) The potentially toxic diatom *Pseudo-nitzschia* H. Peragallo in the Paraná and Santa Catarina States, Southern Brazil. Iheringia 65:47–62

Firpo MAF, Sansigolo CA, Assis SV (2012) Climatologia e variabilidade sazonal do número de ondas de calor e de frio no Rio Grande do Sul associadas ao ENOS. Rev Bras Meteorologia 27(1):95–106

Fay P (1992) Oxygen relations of nitrogen fixation in cyanobacteria. Microbiol Rev. 56(2): 340–373

Gaeta AS, Brandini FP (2006) Produção primária do fitoplâncton na região entre o Cabo de São Tomé (RJ) e o Chuí (RS). In: Rossi-Wongtschowski CL, Madureira LS (eds) O Ambiente Oceanográfico da Plataforma Continental e do Talude na Região Sudeste-Sul do Brasil. EDUSP, São Paulo, pp 219–264

Gianesella-Galvão SMF (2000) Variabilidade da camada de clorofila máxima na região de quebra de plataforma continental sudeste brasileira. Dissertation, Instituto Oceanográfico da Universidade de São Paulo

Grimm AM (2009a) Clima da Região Sul do Brasil. In: Cavalcanti IFA et al. (orgs) Tempo e Clima no Brasil. Oficina de Textos, São Paulo

Grimm AM (2009b) Variabilidade Interanual do Clima no Brasil. In: Cavalcanti IFA et al.(orgs) Tempo e Clima no Brasil. Oficina de Textos, São Paulo

Guimarães SCP, Rörig LR (1997) Microrganismos associados a florações de *Trichodesmium hildebrandtii* na costa de Santa Catarina - Brasil. Notas Técnicas da FACIMAR 1:43–47

Hallegraeff GM (1993) A review of harmful algal blooms and their apparent global increase. Phycologia 32(2):79–99

Hallegraeff GM, Anderson DM, Cembella AD (1995) Manual on harmful marine microalgae. UNESCO, Paris

Islabão CA, Odebrecht C (2011) Dinoflagelados (Peridiniales, Prorocentrales) do microplâncton na plataforma continental e talude do extremo sul do Brasil (inverno 2005, verão 2007). Biota Neotrop 11(3):153–166

Ito RG, Garcia CAE, Tavano VM (2016) Net sea-air CO2 fluxes and modelled pCO2 in the southwestern subtropical Atlantic continental shelf during spring 2010 and summer 2011. Cont Shel Res 119:68–84

Kaczmarska I, Mather L, Luddington IA, Muise F, Ehrman JM (2014) Cryptic diversity in a cosmopolitan diatom known as Asterionellopsis glacialis (Fragilariaceae): implications for ecology, biogeography, and taxonomy. Am J Bot 101(2):267–286

Ljubesic Z, Bosak S, Vilicic D et al (2011) Ecology and taxonomy of potentially toxic *Pseudo-nitzschia* species in Lim Bay (north-eastern Adriatic Sea). Harmful Algae 10:713–722

Miotto MC, Tamanaha MS (2012) Ocorrência de dinoflagelados tecados potencialmente tóxicos em cultivo de moluscos situados no município de Penha, SC. Braz J Aquat Sci Technol 16:53–67

Miotto MC, Costa LF, Bretano DM, Nader C et al (2017) Ecophysiological characterization and toxin profile of two strains of *Cylindrospermopsis raciborskii* isolated from a subtropical lagoon in Southern Brazil. Hydrobiologia 802(1):97–113

Monteiro MA (2001) Caracterização climática do estado de Santa Catarina: uma abordagem dos principais sistemas atmosféricos que atuam durante o ano. Geosul 16(31):69–78

Odebrecht C (1988) Variações espaciais e temporais do plâncton (fitoplâncton, protozooplâncton e metazooplâncton) na Lagoa da Conceição, Ilha de Santa Catarina, Brasil. Atlântica 10:21–40

Odebrecht C, Bergesch M, Rörig LR et al (2010) Phytoplankton interannual variability at Cassino Beach, Southern Brazil (1992:2007), with emphasis on the surf zone diatom *Asterionellopsis glacialis*. Estuar Coasts 33:570–583

Odebrecht C, Du preez DR, Abreu PCOV et al (2014) Surf zone diatoms: a review of the drivers, patterns and role in sandy beaches food chains. Estuar Coast Shelf Sci 150:24–35

Pandolfo C, Braga HJ, Silva VP Jr et al (2002) Atlas climatológico do Estado de Santa Catarina. Epagri, Florianópolis. CD-ROM

Pellens IC (1997) Dinâmica de Nutrientes Inorgânicos no Estuário do Rio Itajaí-Açu / SC. Trabalho de Conclusão de Curso, Universidade do Vale do Itajaí

Pereira-Filho J, Rörig LR (2016) Balanço de massa para nutrientes inorgânicos (N e P) no estuário do rio Itajaí-Açu, SC. Rev Bras de Recur Hídr 21:168–187

Piola A, Campos ED, Moller OO Jr et al (2000) Subtropical shelf front off eastern South America. J Geophys Res 105:6565–6578

Proença LAO, Rörig LR (1995) Mussel production and toxic algal blooms in Santa Catarina state, southern Brazil. Harmful Algae News 12/13:5

Proença LAO, Schmit F, Costa TF et al (1998) Evidences of diarrhetic shellfish poisoning in Santa Catarina - Brazil. Ciência e Cultura (SBPC) 50(6):458–462

Proença LAO, Tamanaha MS, Souza NP (2001) The toxic dinoflagellate *Gymnodinium catenatum* in Southern Brazilian Waters: occurrence, pigments and toxins. Atlântica 3:59–65

Proença LAO, Tamanaha MS, Fonseca RS (2009) Screening the toxicity and toxin content of a bloom of the cyanobacterium *Trichodesmium erythraeum* (Ehrenberg) in northeast Brazil. J Venomous Anim Toxins Incl Trop Dis. 15:204–215

Quijano-Scheggia S, Garcés E, Sampedro N et al (2008) Identification and characterisation of the dominant Pseudo-nitzschia species (Bacillariophyceae) along the NE Spanish coast (Catalonia, NW Mediterranean). Sci Mar b 72:343–359

Resgalla Jr C (2009) Zooplâncton do estuário do rio Itajaí-açu e zona costeira adjacente. In: Branco JOB, Lunardon-Branco MJ, Bellotto VR (Org). Estuário do Rio Itajaí-açu, Santa Catarina: caracterização ambiental e alterações antrópicas. 1ed. Editora Univali, Itajaí SC (1): 171–179

Resgalla Jr C, Carvalho JLB, Pereira-Filho J et al. (2004) Migração vertical e taxas fisiológicas de *Thalia democratica* (Salpidae: Thaliacea) na Reserva Marinha do Arvoredo, Santa Catarina. Notas Técnicas da FACIMAR (8): 45–54

Resgalla C Jr, Souza VGC, Klein AHF (2005) The occurrence of jellyfish stings on the Santa Catarina coast, southern Brazil. Ver Bras Oceanogr 53(3/4):183–186

Rezende KRV, Brandini FP (1987) Variacão sazonal do fitoplâncton na zona de arrebentação na Praia de Pontal do Sul (Pontal do Paraná, PR). Nerítica 11:12–19

Richardson K (1987) Primary production: guidelines for measurement by 14C incorporation. The Danish Institute for Fisheries and Marine Research, Charlottenlund. 21p

Rigual-Hernandez AS, Barcena MA, Sierro FJ et al (2010) Seasonal to inter-annual variability and geographic distribution of the silicoflagellate fluxes in the Western Mediterranean. Mar Micropal 77:46–57

Rodrigues MLG, Franco D, Sugahara S (2004) Climatologia de frentes frias no litoral de Santa Catarina. Rev Bras Geof 22(2):135–151

Rörig LR (2005) Usos múltiplos e qualidade das águas da bacia do baixo Itajaí-Açu-SC: elementos para um gerenciamento integrado. Tese de Doutorado, Universidade Federal de São Carlos

Rörig LR et al (1997) Variação sazonal do fitoplâncton na área da Reserva Biológica Marinha do Arvoredo (Santa Catarina - Brasil). In: Resumos do VII COLACMAR - Congresso Latino-Americano de Ciências del Mar, Universidade de São Paulo, 1997

Rörig LR, Yunes, JS, Kuroshima KN et al. (1998a). Studies on the ecology and toxicity of *Trichodesmium* spp blooms in Southern Brazilian Coast. In: Harmful Algae. Ed. Pontevedra, Paris: Xunta de Galicia, IOC – UNESCO, pp 22–25

Rörig LR, Guimarães SCP, Lugli DO et al (1998b) Monitorização de microalgas planctônicas potencialmente tóxicas na área de maricultura da Enseada de Armação do Itapocorói - Penha - SC. Notas Técnicas da FACIMAR 2:71–79

Rörig LR, Resgalla C Jr, Schettini CAF (2003) Estrutura da assembleia planctônica através do estuário e da pluma do Rio Itajaí-açu. Rev Est Amb 5(1):76–94

Rörig LR, Honorato M, Itokazu AG et al. (2017a) Ecophysiological and biochemical variation of the surf zone diatom Asterionellopsis glacialis sensu lato from Santa Catarina, Southern Brazil. Braz J Oceanogr. 65(4):702–715

Rörig LR, Ottonelli M, Itokazu, AG et al. (2017b) Blooms of bryozoans and epibenthic diatoms in an urbanized sandy beach (Balneário Camboriú - SC - Brazil): dynamics, possible causes and biomass characterization. Braz J Oceanogr. 65(4):685–701

Satô S, Paranaguá MN, Eskinazi E (1963) On the mechanism of red tide of *Trichodesmium* in Recife north eastern Brazil, with some considerations of the relation to the human disease Tamandaré Fever. Trab Inst Oceanogr (Univ Recife) 5(6):7–49

Schettini CAF, Kuroshima KN, Pereira-Filho J et al (1998) Oceanographic and ecological aspects on the Itajaí-açu river plume during a high discharge period. Anais Acad Bras de Ciênc 70(2):335–351

Schettini CAF, Resgalla C Jr, Pereira-Filho J et al (2005) Variabilidade temporal das características oceanográficas e ecológicas da região de influência fluvial do rio Itajaí-Açu. Braz J Aquat Sci Tech 9(2):93–102

Silva LM, Matthiensen A, Yunes JS (2008) Florações de *Trichodesmium* na Costa Brasileira. In: Baptista Neto JA, Kersanach MW, Patchineelam SM (Org.). Poluição Marinha. 1ed. Intersciencia, Rio de Janeiro, 03–19 p

Siqueira A, Kolm HE, Brandini FP (2006) Offshore distribution patterns of the cyanobacterium *Trichodesmium erythraeum* Ehrenberg and associated phyto- and bacterioplankton in the southern Atlantic coast (Paraná, Brazil). Braz Arch Biol Technol 49:323–337

Talbot MM, Bate GC (1988) The use of false buoyancies by the surf diatom *Anaulus birostratus* in the formation and decay of cell patches. Estuar Coast Shelf Sci 26:155–167

Talbot MM, Bate GC, Campbell EE (1990) A review of the ecology of surf zone diatoms, with special reference to *Anaulus australis*. Oceanogr Mar Biol Annu Rev 28:155–175

Tamanaha MS, Proença LAO, Schramm MA et al. (2008) Gênero *Pseudo-nitzschia* H. Peragallo 1899 no parque de maricultura do município de Penha, SC: taxonomia e prodição de toxina amnésica durante uma floração. In: Resumos do XII Congresso Brasileiro de Ficologia, 2008, Brasilia

Tamanaha MS, Cunha DMC, Resgalla C Jr (2016) The first continuous plankton sampling by VOR (Towed Oceanographic Vehicle) in southeastern and southern Brazil waters. Lat Am J Aquat Res 44(5):935–946

Tedesco EC, Ribeiro SMMS, Pompeu M, Gaeta AS, Cavalcante KP (2017) Low-latitude accumulation of the surf-zone diatoms *Anaulus australis* Drebes & Schulz and *Asterionellopsis glacialis* (Castracane) Round species complex in the eastern coast of Brazil. Braz J Oceanogr 65(2):324–331

Wells ML, Trainer VL, Smayda TJ, Karlson BSO, Trick CG, Kudela RM, Ishikawa A, Bernard S, Wulff A, Anderson DM (2015) Harmful algal blooms and climate change: learning from the past and present to forecast the future. Harmful Algae 49:68–93

Near-Surface Biogeochemistry and Phytoplankton Carbon Assimilation in the Rio de la Plata Estuary

Danilo L. Calliari, Mónica Gómez-Erache, Denise Vizziano Cantonnet, and Cecilia Alonso

Abstract The Rio de la Plata estuary (RPE) is considered a highly productive ecosystem, but knowledge of its functioning is sparse, particularly at basal trophic levels. Direct measurements of primary production are scarce, and the mechanisms that drive biological production and biogeochemistry of nutrients and other key compounds are largely unknown. This review summarizes the current information available in published literature in standard journals and other sources and explores driving mechanisms for photosynthetic carbon assimilation (PCA). A database was compiled which includes photosynthetic rates and ancillary environmental variables, i.e. salinity; chlorophyll a (as surrogate for phytoplankton biomass); dissolved inorganic macronutrients N, P and Si; suspended particulate matter (SPM); and underwater light environment. Information gathered covers an extended time period but clustered into an early (1980–1987) and a more recent one (1999–2009). Data was unequally distributed between both periods; for example, PCA data exist only for the most recent period. Data indicate prevalence of high photosynthetic rates in the RPE (mean of 29.25 ± 22.61 mg C m^{-3} h^{-1}). The concentration of nutrients, SPM and indicators of underwater light regime covered wide ranges of variability with remarkable differences between both time periods. All these variables were strongly affected by salinity, showing either a decreasing pattern of concentration from freshwaters to marine waters (e.g. nutrients) or from marine waters to

D. L. Calliari (✉)
Oceanografía y Ecología Marina, Facultad de Ciencias, Universidad de la República, Montevideo, Uruguay

Centro Universitario Regional del Este (CURE), Universidad de la República, Rocha, Uruguay
e-mail: dcalliar@fcien.edu.uy

M. Gómez-Erache · D. V. Cantonnet
Oceanografía y Ecología Marina, Facultad de Ciencias, Universidad de la República, Montevideo, Uruguay
e-mail: mge@fcien.edu.uy

C. Alonso
Centro Universitario Regional del Este (CURE), Universidad de la República, Rocha, Uruguay

M. S. Hoffmeyer et al. (eds.), *Plankton Ecology of the Southwestern Atlantic*,
https://doi.org/10.1007/978-3-319-77869-3_14

freshwaters (i.e. better light conditions in marine-influenced waters). These results were valid irrespective of the period and of the light environment indicator considered (Kd, turbidity). PCA was highest at intermediate surface salinities (10–20), decreasing both towards fresh and towards marine areas. Observed variability patterns of nutrients, SPM, light regime and carbon assimilation along the salinity gradient were consistent with mechanisms predicted by the theory developed for turbid estuaries, which propose that PCA is regulated by inverse horizontal gradients of light and nutrients.

Keywords Phytoplankton · Chlorophyll · Photosynthetic carbon assimilation · Nutrients · Turbidity

1 Introduction

Primary production is the synthesis of organic molecules from inorganic compounds and constitutes the fundamental process that sustains the metabolism of ecosystems on Earth (Libes 2009). Oxygenic photosynthesis is the dominant form of primary production and implies the transformation of light into chemical energy with the concomitant carbon fixation and synthesis of simple sugars. Currently, oceanic photosynthesis provides nearly 50% of global primary production (Falkowski et al. 1997; Westberry et al. 2008) and constitutes the main energetic support for marine trophic webs. Within ocean basins, those waters overlying continental shelves, coastal areas and estuaries constitute hotspots for biological production (Mann and Lazier 2006; Day et al. 2013; Valiela 2015). High biological production results from complex biophysical mechanisms that include allochthonous nutrient sources, turbulent mixing of shallow water columns and hydrographic processes that facilitate retention of water masses within illuminated portions of the water column (Cloern 1999; Lucas et al. 1999).

The Río de la Plata estuary (RPE) is a large, shallow and turbid estuary on the Southwest Atlantic at 35°S–57°W (Fig. 1). Main tributaries are Paraná and Uruguay rivers which together drain a basin >3 million km^2 spread across the territories of Argentina, Bolivia, Brazil, Paraguay and Uruguay. With an annual average flow of 22,000–25,000 m^3 s^{-1} (Guerrero et al. 1997), RPE represents the main point source of freshwater into the South Atlantic Ocean. It is a partially mixed, vertically structured estuary where marine waters enter as a salt wedge through the bottom layer (Guerrero et al. 1997; Nagy et al. 1998). The estuarine plume may extend >1000 km (mostly northwards) and has a strong impact on shelf and open ocean ecosystems as it affects the physical structure of the water column and delivers significant amounts of particulate matter, chemical and biological species to the adjacent sea (Nagy et al. 1998; Piola et al. 2005, 2008). A zonation of RPE based on morphology and mean salinity and turbidity fields (Nagy et al. 1998) proposes an upper region comprised by the Paraná delta and the freshwater tidal reaches of the

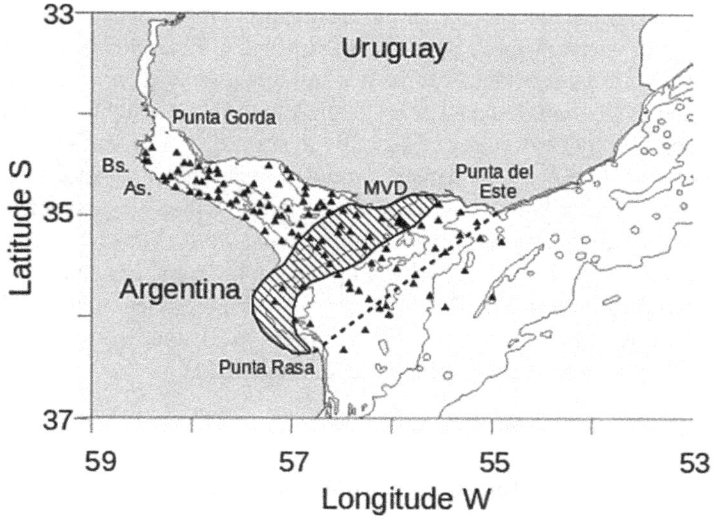

Fig. 1 Map of the Rio de la Plata estuary. Upper limit is marked by the entrance of the Uruguay River at Punta Gorda. The outer limit is marked with a dashed line between Punta del Este (Uruguay) and Punta Rasa (Argentina). Dashed grey area provides a schematic representation of the average position of the turbidity front, adapted from Framiñan and Brown (1996) and Derisio et al. (2014). Triangles indicate sampling stations. Isobaths correspond to 5, 10, 20 and 50 m

estuary; an intermediate region – the main mixing zone – from the maximum reach of the salt intrusion to a point where the section of the estuary abruptly broadens and features a large depositional shoal (Barra del Indio); and the outer estuary, where marine influence is strongest.

The RPE is a plankton-dominated system (Nagy et al. 2002) and constitutes a reproduction and nursery area for several fish species (Vizziano 2001; Acuña and Viana 2001; Acha et al. 2008, 2012). It hosts abundant large-sized consumers including molluscs, crustaceans, reptiles, birds and mammals, besides fishes (Masello and Menafra 1998). That fauna includes highly charismatic species (e.g. the estuarine dolphin *Pontoporia blainvillei*) and populations of economic significance exploited by Argentinian and Uruguayan artisanal- and industrial-scale fleets (Lercari et al. 2014).

In marine systems, production of large consumers may result from local primary production, from a detritus food web fuelled by allochthonous organic matter or from a combination of both (Mann and Lazier 2006). Based on fisheries yield, RPE is usually considered a highly productive ecosystem (Acha et al. 2008). Important efforts have been devoted to characterize populations of commercial interest, their standing stocks and variability in time or space. However, knowledge on ecosystem functioning and particularly on the ecological processes affecting basal trophic levels (i.e. those supporting exploited populations) has lagged behind. Paucity of our knowledge extends to the typical rates of biological production, their variability patterns in space and time, the biogeochemistry of key elements and dissolved organic matter and the mechanisms involved in the modulation of biological production.

Within the last decade, direct measurements of photosynthetic carbon assimilation (PCA) rates started to become available for the RPE which allowed a first glimpse to the primary productivity of this important ecosystem. We discriminate here between PCAs, which result from single-point measurements of organic carbon incorporation by phytoplankton (usually in surface or near-surface waters) and the ecosystem-level process of primary production, i.e. the depth-integrated flux of organic matter, its magnitude derived from multiple discrete assimilation rates measured within the photic layer. The scope of this work is to review information available to date regarding PCA and phytoplankton biomass. In addition, it aims to discuss such information together with ancillary environmental data within the context of relevant theoretical frameworks in order to advance the understanding of regulatory processes for biological production in the RPE.

2 Theoretical Frameworks and Working Hypothesis

In order to explore regulation mechanisms of PCA, we turn to theory on production in turbid estuaries and plumes of large river systems (i.e. Cloern 1987, 1999, 2001; Cole and Cloern 1987; May et al. 2003). The salinity gradient between riverine and marine areas is the fundamental structuring factor that modulates forcing variables for PCA, i.e. nutrient concentrations, seston load and the light environment. The concentration of dissolved nutrients and the suspended particles load are usually much higher in the continental runoff compared to marine waters. Nutrient- and seston-rich freshwaters are progressively diluted along the river-sea axis by mixing with clearer and more oligotrophic marine waters, generating inverse horizontal gradients of light and nutrients. Phytoplankton growth in the turbid freshwaters is presumed to be light-limited (Alpine and Cloern 1988). Spatial variability in the light environment along the main longitudinal estuarine axis results from physical and chemical processes (the estuarine geochemical filter, sensu Sharp et al. 1984). In the freshwater reaches, the behaviour of clay particles is dominated by negative electrostatic charges on the mineral lattice which create repulsive electrochemical forces and prevent flocculation, contributing to keeping particles in suspension. But as salinity increases (and so does the concentration of cations Ca^{+}, Mg^{2+}, Na^{+}), repulsive forces are destabilized, van der Waals forces become important, and flocculation proceeds to facilitate the sedimentation of seston (Bianchi 2007).

The transition from highly turbid to more clear waters is usually sharp, defining a turbidity front. From the above, it follows that this turbidity front is mechanistically linked with a salinity front, and it is normally found in the oligohaline zones between salinities of 2 and 10. Improved light penetration on the marine side of the turbidity front boosts assimilation of dissolved nutrients by phytoplankton (the estuarine biochemical filter, sensu Sharp et al. 1984). At higher salinities, after dissolved nutrients have been incorporated into biomass, nutrient levels drop markedly and – rather than light – become limiting for primary production. The general expected pattern for PCA resulting from the above mechanism is thus one of light limitation in the freshwater end (estuarine head), nutrient limitation in the marine end (estuarine mouth) and maximum assimilation rates and production on the marine side of the turbidity front, i.e. at intermediate salinities.

Control of PCA by concomitant gradients of light and nutrients had been earlier suggested based on indirect evidence, i.e. distribution of nutrients and the light regime (Nagy et al. 1987, 2002; Calliari et al. 2005), and by a physical-biogeochemical model (Huret et al. 2005). A qualitative and robust test of that mechanism would be to compare observed patterns of nutrient concentrations, SPM, light availability and photosynthesis along the salinity gradient vs. those expected under the proposed hypothesis. If that mechanism is valid for the estuary, the actual data on nutrients, SPM and light along the salinity gradient should show the inverse horizontal gradients, and PCA should be maximum at an intermediate salinity region.

3 General Approach

An extensive search was performed on studies reporting original results on photosynthetic rates and on phytoplankton biomass. The search also included ancillary environmental information expected to contribute to the understanding of the regulation of PCA and the biogeochemistry of key elements in the study area. Those variables included the concentrations of macronutrients such as total inorganic nitrogen (TIN, i.e. nitrate + nitrite + ammonia), phosphate-P and silicate-Si, as well as suspended particulate material (SPM), and the underwater light environment as represented by the vertical diffuse light extinction coefficient (Kd, m^{-1}) obtained by discrete measurements of irradiance along the water column, and an indirect estimator of water turbidity (as NTU) measured in surface waters.

The search included papers in peer-reviewed journals as well as reports of more restricted circulation. Given the rationale background, salinity is a core variable in the data analysis. It follows that salinity values tied to either SPM, Kd, turbidity, nutrients, PCA or chlorophyll *a* were a requisite for the inclusion of data sets into current analysis. Data were also filtered in order to retain only measurements obtained in open waters, i.e. excluding sites in nearshore locations or within small embayments (e.g. Montevideo bay or harbour), in order to exclude bias due to small scale local processes. Due to scope and space constrains, those matters related to biodiversity and structure of phytoplankton communities (i.e. taxonomic structure, size structure) are not covered in this review (but see Ferrari and Pérez 2002; Gómez et al. 2004; Kruk et al. 2014). Also, due to known issues regarding the estimation of phytoplankton biomass and production from satellite-borne sensors in highly turbid waters (Martínez et al. 2005), such type of information was not taken into account. Table 1 summarizes sources of information effectively considered for the present review.

Data analysis was pattern-oriented, i.e. intended to qualitatively and quantitatively assess the distribution of variables along the salinity gradient. In all cases,

Table 1 Sources of information on phytoplankton carbon assimilation, phytoplankton biomass, nutrient concentrations and light environment in the Rio de la Plata estuary (RPE)

Period	Spatial coverage	Data type	St #	References
1980–1985	HM	Turbidity	17	Nagy et al. (1987)
1980–1985	HM	Nut, SPM	45[a]	Carp (1989)
1999	SFR	PCA, nut, Kd	6	Gómez-Erache et al. (2001)
2001	HM	Nut, Kd, Chl *a*, SPM	31	Calliari et al. (2005)
52,003	OM (shelf)	PCA, Chl *a*, Kd	10	Calliari et al. (2009a)
1999–2003	HM	Chl *a*, SPM	15	Derisio et al. (2014), Carreto and Hozbor unpublished
2009	OM (shelf)	PCA, nut, Kd, Chl *a*	9	Kruk et al. (2014)

References for spatial coverage: *HM* head to mouth, *SFR* salinity front region, *RPE-OH-Shelf* from oligohaline waters to estuarine mouth (includes salinity front). *PCA* phytoplankton carbon assimilation rate, *nut* nutrient concentration, Chl *a* chlorophyll *a* concentration, *Kd* PAR attenuation coefficient

[a]Data are average values from 11 to 15 measurements according to station between 1980 and 1985

response variables were grouped according to salinity ranges. Quantitative statistical analyses also included the evaluation of significant differences in nutrient concentrations between data subsets corresponding to an earlier and a more recent period (see below). Effects were evaluated for each variable through use of generalized linear models; in each case, competitive model formulations were constructed using alternative error family distributions (i.e. Gaussian, Gamma) and link functions (inverse, identity) and selected based on Akaike information criterion (AIC, Burnham and Anderson 2002). All statistical analyses were performed using R (R Core team 2015) and the RKWard interface (Friedrichsmeier et al. 2015).

4 Rates, State Variables and Potential Drivers of Primary Production in RPE

A vast majority of data was represented by measurements performed on surface waters. Only three studies measured PCA in the RPE, the first during summer and spring 1999 (Gómez-Erache et al. 2001), the second in spring 2003 (Calliari et al. 2009a) and the third in autumn 2009 (Kruk et al. 2014). All three studies measured rates in the surface mixing layer over a wide salinity range and followed similar methodological approaches, i.e. measurement of photosynthesis as ^{14}C incorporation during short on-board incubations. In total, they sum 21 independent PCA measurements obtained during 4 cruises. Core results on PCA and complementary variables are summarized in Tables 2 and 3 and in Figs. 2–4. The three studies evidenced moderate to high photosynthetic rates as compared to other marine systems (Behrenfeld and Falkowski 1997), ranging between 4.32 and 81.93 mg C m^{-3} h^{-1} (mean of 29.25 ± 22.61 mg C m^{-3} h^{-1}). Such estimates fall within the upper range reported for other estuaries using comparable approaches (Abreu et al. 1994; Conde et al. 2000; Kocum et al. 2002; Montes-Hugo et al. 2004). This small data set on

Table 2 Data sets considered in this study

Variable	1980–1985	1980–1985	1999	2001	2003	1999–2003	2009
Salinity	12 (5.5–25) 17	0.42 (0.14–25.01) 45	22 (2.5–25.9) 6	24.3 (0.07–26.31) 31	17.69 (0.1–28.72) 10	24.81 (2.33–31.66) 15	20.41 (4.27–28.61) 9
TIN; µM		18.95 (1.04–32.45) 45	3.6 (1.72–11.14) 6	9.74 (2.63–62.68) 30			1.94 (0.65–22.09) 9
Phosphate-P; µM		1.16 (0.55–2.12) 45	1.06 (0.35–1.69) 6	1.94 (0.65–6.46) 30			0.93 (0.5–2.26) 9
Silica-Si; µM		152 (39.5–180) 45	36.74 (4.2–131.21) 6	203 (122–228) 30			27.38 (13.31–72.1) 9
SPM; mg L^{-1}		103.5 (74–169) 28		62.34 (26.52–118.3) 30	22.67 (0.77–190.2) 10	13.49 (0–92.8) 15	
TURB/KD; NTU/m^{-1}	6.94 (3.01–78) 17		1.38 (0.58–2.66) 6	1.76 (0.39–10.09) 28	1.75 (0.15–6.23) 10		0.33 (0.13–2.07) 9
PCA; mgC m^{-3} h^{-1}			10.9 (4.3–27.2) 6		13.8 (4.4–64.2) 5		40.6 (13.2–40.6) 9
Chl *a*; mg m^{-3}			2 (1.4–8.0) 6	4.17 (1.00–7.01) 31	4.9 (0.5–22.7) 8	1.19 (0.99–1.37) 8	4.42 (0.84–18.14) 9

Mean, range (between brackets) and number of observations considered in current analyses are shown for each variable and dataset
TURB was reported in the 1980–1985 dataset; KD was reported in 1999, 2001, 2003 and 2009 datasets.

Table 3 Effect of salinity range and data set period on response variables

Response	Model specification	Parameters sign
TIN	$B_0 + B_1$ SALCLASS+B_2 PERIOD+B_3 SAL*PERIOD + $\varepsilon \sim G(\alpha, \beta)$, *link* = inverse	B_0^{**} B_1^{**} B_2^{*} B_3 = ns
Phosphate	$B_0 + B_1$ SALCLASS +B_2 PERIOD+B_3 SAL*PERIOD + $\varepsilon \sim G(\alpha, \beta)$, *link* = identity	B_0^{**} B_1^{**} B_2^{**} B_3 = ns
Silicate[a]	$B_0 + B_1$ SALCLASS +B_2 PERIOD+B_3SAL*PERIOD +$\varepsilon \sim N(\mu, \sigma^2)$, *link* = identity	B_0^{**} B_1^{**} B_2 = ns, B_3*
SPM	$B_0 + B_1$ SALCLASS + $\varepsilon \sim G(\alpha, \beta)$, *link* = identity	B_0^{**} B_1^{**}
Kd	$B_0 + B_1$ SALCLASS + $\varepsilon \sim G(\alpha, \beta)$, *link* = inverse	B_0^{**} B_1^{**}
TURB	$B_0 + B_1$ SALCLASS + $\varepsilon \sim G(\alpha, \beta)$, *link* = identity	B_0^{**} B_1^{**}
Chl *a* HLO[b]	$B_0 + B_1$ SALCLASS + $\varepsilon \sim N(\mu, \sigma^2)$, *link* = identity	B_0^{**} B_1^{*}
PCA[c]	$B_0 + B_1$ SALCLASS + $\varepsilon \sim N(\mu, \sigma^2)$, *link* = identity	B_0^{**} B_1^{**}

Total inorganic nitrogen (TIN), phosphate-P, silicate-Si, suspended particulate matter (SPM), vertical light diffuse attenuation coefficient (Kd), turbidity (TURB), Chl *a* and phytoplankton carbon assimilation (PCA)

Model specifications include parameters assigned to each factor, family distribution for the stochastic component (N = Gaussian, G = Gamma) and link function

Significance codes: $*p < 0.05$; $**p < 0.01$

[a]Silicate as 1/Ln(Silicate)

[b]Chl *a* as Ln(Chl *a*)

[c]PCA as Ln(PCA)

PCA supports previous contentions about high potential of the estuary for carbon assimilation.

Information on phytoplankton biomass (as chlorophyll *a*), SPM, dissolved inorganic nutrients and the light environment was more extensively available (Table 1). All these variables evidenced very wide ranges of variability, e.g. between 0.5 and 27.3 mg Chl *a* m^{-3}, 1.04 and 62.7 μM for TIN and 0.35 and 6.46 μM for phosphate-P (Table 2, Figs. 2–3). It is noteworthy that independently of the amount of information available or the length of the time period covered, the structure of the resultant database precluded an analysis of variability patterns at the seasonal time scale, particularly in the case of PCA, but also for the other variables. In the case of TIN, phosphate-P and silicate-Si, information was available for the periods 1980–1985 and 1999–2009. An exploratory comparison indicated that, besides clear differences in TIN and phosphate-P along the salinity gradient, remarkable differences existed also between survey periods. In particular, TIN decreased, while phosphate-P increased (Fig. 2, Table 3).

System-scale distribution of macronutrients evidenced a nonlinear pattern of high concentrations in freshwaters and oligohaline waters (<10) and a sharp non-conservative decline in a salinity range between 10 and 20. That general pattern was valid irrespective of the period considered (Fig. 2). Nutrient levels at high salinities (>25) were low, i.e. 1.6 μM TIN and 0.7 μM phosphate-P (but silicate-Si levels remained high, 25.9 μM), strongly suggesting RPE outflow may not be a significant N and P source for shelf waters. That is consistent with earlier findings of a minimum in nutrient concentration in the transition between the estuary and the inner continental shelf at salinities between ca. 30 and 33 (Carreto et al. 1986). SPM and

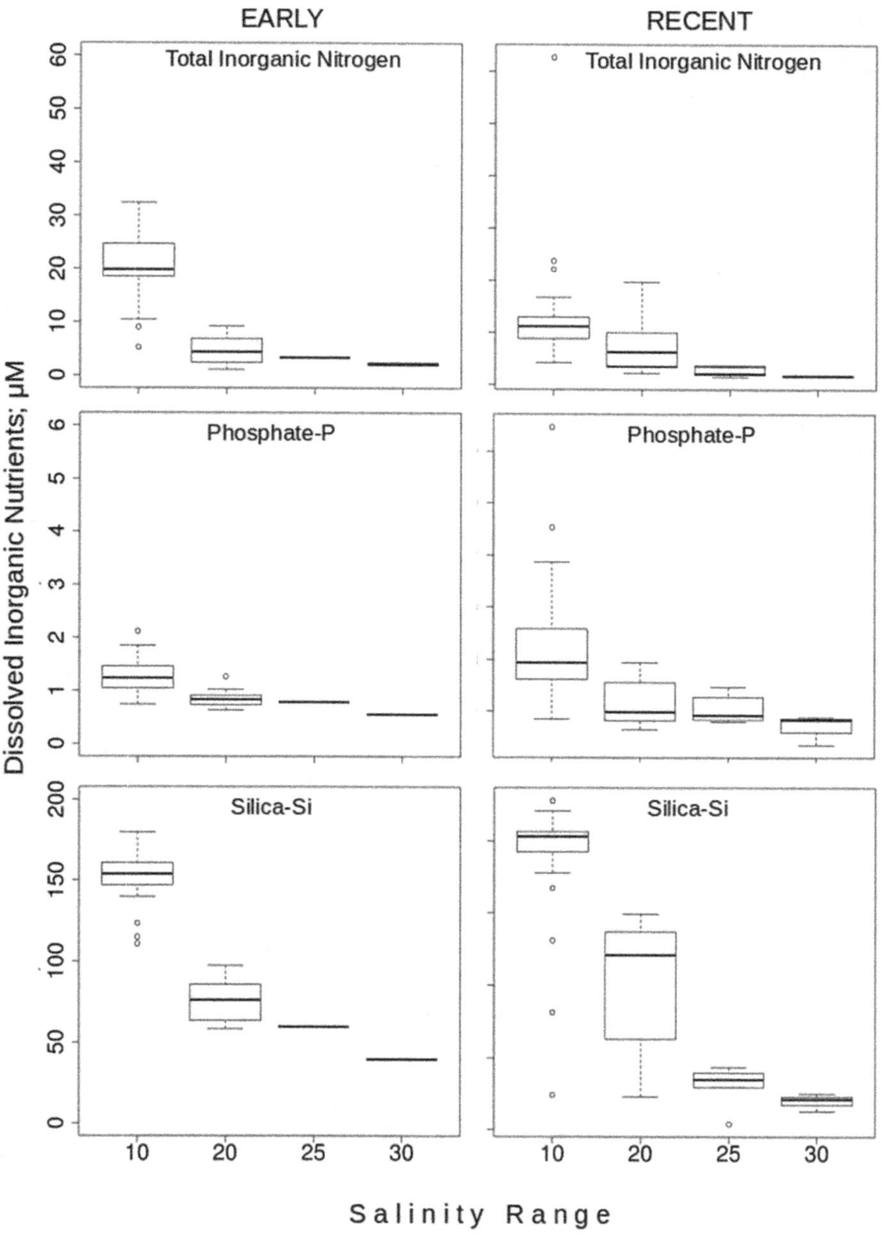

Fig. 2 Range of total inorganic nitrogen (TIN: $NO_3 + NO_2 + NH_4$), phosphate-P and silica-Si along the salinity gradient between 1980–1987 (left panel) and 1999–2009 (right panel). Salinity data are grouped into classes representing different geographic sections of the estuary (abscissa labels are upper limits of the corresponding class)

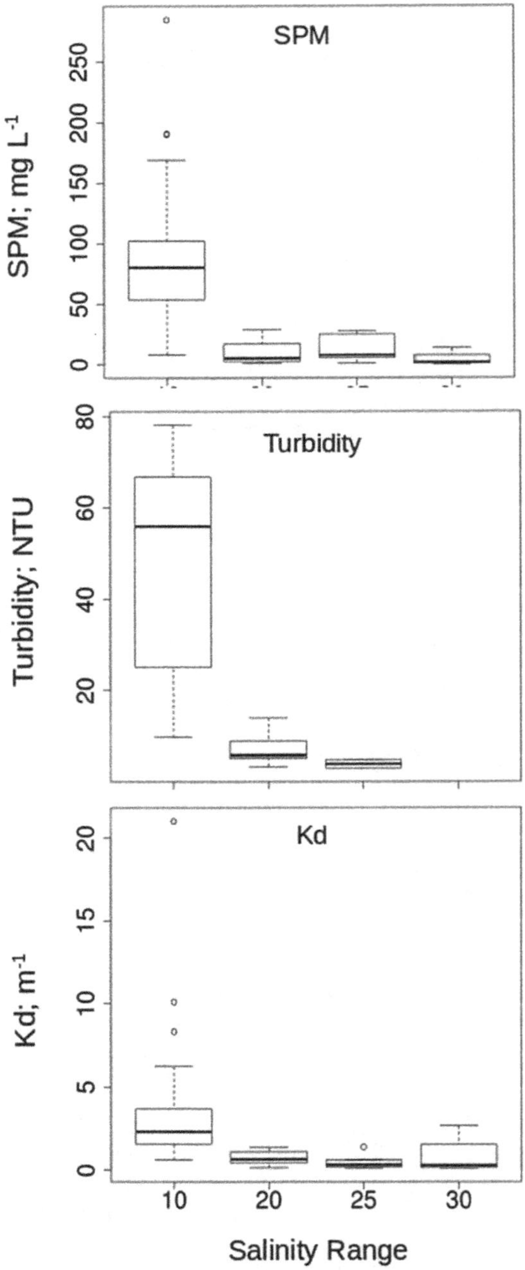

Fig. 3 Underwater light environment proxies along the salinity gradient in the RPE. Light environment is represented by turbidity (NTU) for the early period (1980–1987, above) and by the diffuse vertical attenuation coefficient (Kd, m^{-1}) for the recent period (1999–2009, below). Salinity data are grouped into classes representing different geographic sections of the estuary (abscissa labels are upper limits of the corresponding class)

light regime followed a similar pattern to that of nutrients: maximum values occurred in all cases at the lowest salinities, and a strong decline was evident in the range 10–20. For the light regime, that pattern was true both for surface water turbidity and for the vertical diffuse light extinction coefficient (Fig. 4). However, a difference between patterns observed for nutrients vs. SPM and light regime was that

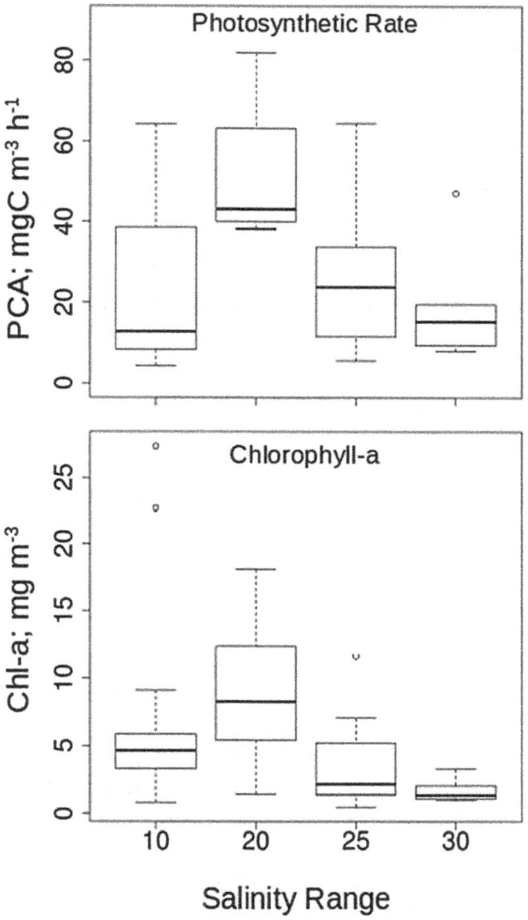

Fig. 4 Distribution of phytoplankton carbon assimilation rates and phytoplankton biomass (as chlorophyll *a*) along the salinity gradient in the RPE. Salinity data are grouped into classes representing different geographic sections of the estuary (abscissa labels are upper limits of the corresponding class)

variability in the latter levelled off at salinities 20 and higher, while nutrient concentrations monotonically decreased along the whole salinity gradient.

In summary, freshwater reaches of the estuary were characterized by a poor light regime, a high suspended particle load and elevated nutrient concentrations, typically >10 μM TIN, >2 phosphate-P and > 150 μM silica-Si (Figs. 2 and 3), while the outer reaches and inner shelf present better light penetration and lower SPM and nutrient values, i.e. <5 μM TIN, <2 phosphate-P and < 50 μM silica-Si. That scenario closely resembles expected conditions in turbid estuaries where the light regime responds to SPM variability driven by the mixing of freshwaters and marine waters.

Stoichiometry of N and P suggests TIN as the potential limiting nutrient of phytoplankton growth as TIN:phosphate-P ratio in the pooled database was systematically below the Redfield ratio of 16:1, decreasing along the salinity gradient from 6.2 (±3.65 sd) at salinities <10 to 1.92 (±0.10 sd) at salinities between 25 and 30. Due to increased phosphate-P and decreased TIN between the 1980s and 2000s, limitation by TIN seaward of the turbidity front is probably exacerbated in recent times compared to the earlier period. However, variability patterns of TIN, phosphate-P and light along the salinity gradient were consistent with the scenario expected in turbid estuaries irrespective of the time period or the light *proxy* considered.

Referred nutrient concentrations in the freshwater and oligohaline areas (salinities <10) may be considered high. They lay in the same range than those reported for other estuaries in the same region (e.g. Laguna de Rocha, Calliari et al. 2009b; Solís Grande, Rodriguez-Graña 2010; Patos Lagoon, Niencheski and Baumgarten 1997) but are below those in other large estuaries elsewhere (i.e. for DIN: Delaware, Sharp et al. 1984; Mississippi, Cloern 2001; San Francisco, Alpine and Cloern 1992; Colne estuary, Kocum et al. 2002).

PCA also differed among salinity classes (Table 3, Fig. 4) but followed a non-monotonic pattern with a maximum in the 10–20 salinity range (average 54.21 ± 20.34 mg C m^{-3} h^{-1}, Fig. 4). Improved light regime driven by decreased amounts of SPM at low salinities induced an up-regulation of phytoplankton growth rates. Nutrient losses observed within the same salinity range (Fig. 2) likely resulted from higher phytoplankton assimilation. Such responses fit closely to the variability pattern predicted by theory for turbid estuaries (e.g. Cloern 1987; Cole and Cloern 1987; May et al. 2003; Bianchi 2007). Thus, overall results observed for SPM, nutrients, light regime and PCA variability along the salinity gradient are consistent with the hypothesis of the double and inverse horizontal gradients in light and nutrients as a core mechanism driving PCA in the RPE.

Chlorophyll *a* concentration ranged between 1.98 and 9.16 mg Chl *a* m^{-3} (Fig. 4) and mirrored photosynthetic rates, also with maximum values in the intermediate 10–20 salinity range. That is a close match with a spatially explicit map of SPM and Chl *a* distribution in the estuary provided by Derisio et al. (2014, their Fig. 2) which neatly shows the spatial succession from highly turbid to clearer waters and the chlorophyll peak immediately seaward of the turbidity front. Independent biomass estimates based on phytoplankton biovolume (Gómez et al. 2004) or simply cell counts (Ferrari 2008) also support the general biomass distribution pattern described here based on Chl *a*.

A relevant issue to elucidate is whether higher biomass in the salinity range 10–20 strictly results from in situ growth or if it is also favoured by hydrographically driven phytoplankton accumulation in the frontal area. If active growth was the unique – or most important – process responsible for the biomass distribution pattern, biomass-specific assimilation rates (mg C mg Chl a^{-1} h^{-1}) should also be higher at the 10–20 salinity range. Mean biomass-specific PCA, as well as P^{B}_{max} (the maximum chlorophyll-normalized photosynthesis, derived from P vs. E functional response models fitted to data; see Gómez-Erache et al. 2001 and Calliari et al. 2009a) tended to be higher in the salinity range from 10 to 20, but differences were non-significant (data not shown). Results on that issue are thus not conclusive, and both growth and accumulation likely contribute to high biomass at intermediate salinities.

A physical model of the RPE coupled to a five-compartment biogeochemical NPZ-type formulation produced results which are also to a large extent consistent with present findings (Huret et al. 2005). Within the estuarine domain, that model reproduced a permanent zone of maximum chlorophyll concentration seaward of the turbidity front, very much like the distribution reported in Derisio et al. (2014). Modelled spatial pattern of primary production was broadly similar to that of chlorophyll *a* and also consistent with current results. Instead, model results suggested a more complex behaviour of limiting factors than here considered, indicating strong and permanent light limitation in the inner estuary, and a switch from light limitation at the frontal zone during early spring to nutrient limitation in summer.

5 Restrictions to Current Knowledge and Perspectives

The current discussion of PCA patterns and environmental drivers within the RPE made use of a dataset gathered in an extensive search and represents knowledge based on the best evidence available today. Still, the amount of data is low and critically low for PCA. Also, available data is "patchy", meaning that in most cases they were derived from snapshot observations, with little or no time replication which hinders analysis of time variability. In spite of that, clear trends were detected for most variables, which likely reflect the strength and robustness of variability patterns along the salinity gradient.

Data limitations identified here clearly show the need for further efforts to advance our knowledge on the regulation of biological production in the estuary. For that, planning of future field investigations should consider issues beyond the general "amount of data" problem. For example, results discussed here on phytoplankton biomass and carbon assimilation correspond to surface observations expressed on a volumetric basis. That constitutes a valid approximation to the problem, but future research should also consider the vertical distribution of Chl *a* and production. In strongly stratified systems – like the RPE (Guerrero et al. 1997; Acha et al. 2008), Chl *a* and photosynthesis are likely to be vertically structured. In order to determine whether primary production patterns per unit area (i.e. mg C m^{-2} d^{-1}) are consistent with those here reported, explicit consideration of the vertical dimension is in order.

Another issue is related to the temporal scale of the analysis. Pooling of data collected at different times of the year during an extended period results in strongly time-averaged patterns. Again, that was the strategy followed due to stringent restrictions in data availability. It is well documented for forcing factors of PCA to fluctuate at different time scales. For instance, prevailing winds and freshwater runoff control key ecosystem processes, e.g. residence time of water and plankton, the mixing-stratification cycle and geometry and position of salinity and turbidity fronts (Framiñan and Brown 1996; Guerrero et al. 1997; Framiñan et al. 1999; Nagy et al. 2002; Gómez-Erache et al. 2001; Acha et al. 2008). Those forcings fluctuate at scales that range from hours to interannual coupled to ENSO cycles (Mechoso and

Pérez-Iribarren 1992; Depetris et al. 1996; Pasquini et al. 2006; Depetris and Pasquini 2008). Also, seasonal variability of incoming irradiance may imply an exacerbation of light limitation during winter and a relaxation during summer months, as suggested by Huret et al. (2005). Whether such factors modify the broad picture depicted here remains an open question.

Finally, the destruction of organic matter – its catabolic use by autotrophs and heterotrophs – can be thought of as the other side of the coin in relation to photosynthesis. That is generally measured as the community respiration (R). PCA data dealt within the present paper were obtained using the ^{14}C incorporation method, which has limitations in terms of measuring the real amount of carbon incorporated into organic matter. The ultimate primary production values estimated with this method approximately equal gross primary production (GPP). The precise determination of the relative amount of net carbon assimilation and respiration by phytoplankton in estuarine environments can contribute to a finer understanding of the biogeochemistry, energy and matter fluxes within an ecosystem (del Giorgio and Williams 2005; Libes 2009). The GPP:R ratio is a powerful descriptor of the autotrophic vs. heterotrophic behaviour of an ecosystem (del Giorgio and Duarte 2002; Duarte and Regaudie-de-Gioux 2009), with implications on its role as sink vs. source of inorganic carbon to the atmosphere. There is growing consensus regarding the global ocean as a net heterotrophic system, but different sub-systems behave differently, i.e. the oligotrophic open ocean is systematically heterotrophic, but more productive coastal seas may alternate autotrophic and heterotrophic phases (del Giorgio and Duarte 2002; Duarte et al. 2004). There are yet no clear patterns established for P:R ratios in estuaries, as they can range from net heterotrophic to net autotrophic on an annual basis (Caffrey 2004). Elucidating the behaviour of a large-scale estuarine system like the RPE is highly relevant in a regional and global perspective.

A deeper understanding of spatial and temporal trends in the biogeochemistry of the RPE in relation to physical environmental forcing calls for systematic observations at relevant frequencies along time (i.e. a time series approach) combining broad spatial coverage, an explicit consideration of the vertical dimension and a wide array of field and experimental methods. Development of research programs designed to tackle such issues will be highly demanding in terms of costs and effort, even with the help of automated observation technologies available today. That should be viewed as an opportunity for the implementation of collaborative approaches among institutions and countries.

6 Conclusions

Best evidence currently available for the Rio de la Plata estuary indicates moderate to high PCA with a maximum in the salinity range between 10 and 20. That is consistent with regulation by light penetration at low salinities (<10), and nutrient regulation at salinities >20, according to a mechanism mediated by physicochemical forcing of SPM distribution (Sharp et al. 1984; Cole and Cloern 1987; May et al. 2003). The theory of production regulation in turbid estuaries thus seems to apply to this large-scale system. That statement should be taken cautiously as it is based on a still limited amount of data. Future research will further support or challenge and rectify this conclusion.

Acknowledgements Results presented in this paper were obtained during research programs and projects funded by diverse agencies: ECOPLATA program (IDRC-Canadá, UNESCO-PNUD, Ministry of Housing and Environment of Uruguay), FREPLATA program (UNDP-GEF, Argentina-Uruguay), Comisión Administradora del Río de la Plata (Argentina-Uruguay), FEMCIDI-OAS and Max Planck Partner Group projects. Data provided by INIDEP (Argentina) were produced by Dr. Constanza Hozbor (SPM, Laboratory of Molecular Biology and Microbiology) and Dr. José I. Carreto (chlorophyll *a*, Program for the Marine Environment and Red Tides) and are deeply acknowledged. Dr. Carla Derisio helped with making INIDEP data accessible for the present paper and contributed with discussions of SPM and chlorophyll data distribution. Comments by reviewers and editors – particularly Dr. F. P. Brandini – contributed to improve this paper and are greatly appreciated.

References

Abreu P, Odebrecht C, Gonzalez A (1994) Particulate and dissolved phytoplankton production of the Patos lagoon estuary, southern Brazil: comparison of methods and influencing factors. J Plankton Res 16:737–735

Acha EM, Mianzan H, Guerrero R et al (2008) An overview of physical and ecological processes in the Rio de la Plata estuary. Cont Shelf Res 28:1579–1588

Acha EM, Simionato C, Carozza C, Mianzan H (2012) Climate-induced year-class fluctuations of whitemouth croaker Micropogonias furnieri (Pisces, Sciaenidae) in the Río de la Plata estuary, Argentina–Uruguay. Fish Oceanogr 21:58–77

Acuña A, Viana F (2001) Ciclo reproductivo y características ambientales del área de desove de la pescadilla de red (*Macrodon ancylodon*) y la pescadilla de calada (Cynoscion guatucupa) en la costa uruguaya. In: Vizziano D et al (eds) El Río de la Plata. Investigación para la Gestión del Ambiente, los Recursos Pesqueros y la Pesquería en el Frente Salino. Programa Ecoplata, Montevideo, Uruguay, pp 71–84

Alpine A, Cloern J (1992) Trophic interactions and direct physical effects control phytoplankton biomass and production in an estuary. Limnol Oceanogr 37:946–955

Alpine A, Cloern J (1988) Phytoplankton growth rates in a light limited environment, San Francisco Bay. Mar Ecol Prog Ser 44:167–173

Behrenfeld M, Falkowski P (1997) Photosynthetic rates derived from satellite-based chlorophyll concentration. Limnol Oceanogr 42:1–20

Bianchi T (2007) Biogeochemistry of estuaries. Oxford University Press, New York

Burnham K, Anderson D (2002) Model selection and multimodel inference. A practical information-theoretic approach, 2nd edn. Springer, New York

Caffrey J (2004) Factors controlling net ecosystem metabolism in U.S. estuaries. Estuaries 27:90–101

Calliari D, Gómez-Erache M, Gómez N (2005) Biomass and composition of the phytoplankton in the Río de la Plata: large-scale distribution and relationship with environmental variables during a spring cruise. Cont Shelf Res 25:197–210

Calliari D, Brugnolli E, Ferrari G et al (2009a) Phytoplankton distribution and production along a wide environmental gradient in the South-West Atlantic off Uruguay. Hydrobiol 620:47–61

Calliari D, Britos A, Conde D (2009b) Testing the relationship between primary production and *Acartia tonsa* grazing pressure in an estuarine lagoon. J Plankton Res 31:1045–1058
Carp (1989) Estudio para la evaluación de la contaminación en el Rio de la Plata. Informe de avance. SHN–SOHMA.
Carreto J, Negri R, Benavides H (1986) Algunas caracteristicas del florecimiento del fitoplancton en el frente del Río de la Plata. I: los sistemas nutritivos. Rev Inv Des Pesq 5:7–29
Cloern J (1987) Turbidity as a control on phytoplankton biomass and productivity in estuaries. Cont Shelf Res 7:1367–1381
Cloern J (1999) The relative importance of light and nutrient limitation of phytoplankton growth: a simple index of coastal ecosystem sensitivity to nutrient enrichment. Aquat Ecol 33:3–15
Cloern J (2001) Our evolving conceptual model of the coastal eutrophication problem. Mar Ecol Prog Ser 210:223–253
Cole B, Cloern J (1987) An empirical model for estimating phytoplankton productivity in estuaries. Mar Ecol Prog Ser 36:299–305
Conde D, Aubriot L, Sommaruga R (2000) Changes in UV penetration associated with marine intrusions and freshwater discharge in a shallow coastal lagoon of the Southern Atlantic Ocean. Mar Ecol Prog Ser 207:19–31
Day J, Kemp W, Yáñez-Arancibia A, Crump B (eds) (2013) Estuarine ecology, 2nd edn. Wiley, Hoboken
Depetris P, Pasquini A (2008) Riverine flow and lake level variability in Southern South America. EOS Trans Am Geophys Union 89:254–255
Depetris P, Kempe S, Latif M et al (1996) ENSO-controlled flooding in the Paraná River (1904–1991). Naturwissenschaften 83:127–129
Derisio C, Braverman M, Gaitan E et al (2014) The turbidity front as a habitat for Acartia tonsa (Copepoda) in the Río de la Plata, Argentina-Uruguay. J Sea Res 85:197–204
Duarte C, Regaudie-de-Gioux A (2009) Thresholds of gross primary production for the metabolic balance of marine planktonic communities. Limnol Oceanogr 54:1015–1022
Duarte C, Agustí S, Vaqué D (2004) Controls on planktonic metabolism in the Bay of Blanes, northwestern Mediterranean littoral. Limnol Oceanogr 49:2162–2170
Falkowski P, Barber R, Smetacek V (1997) Biogeochemical controls and feedbacks on ocean primary production. Science 281:200–206
Ferrari G (2008) Caracterización de las masas de agua presentes en el Río de la Plata y su frente oceánico a través del fitoplancton, Tesis de Maestría, PEDECIBA – Biologia, Universidad de la República, Uruguay.
Ferrari G, Pérez M d C (2002) Fitoplancton de la costa platense y atlántica del Uruguay (1993–1994). Iheringia Ser Bot 57:263–278
Framiñan M, Brown O (1996) Study of the Río de la Plata turbidity front, part I: spatial and temporal distribution. Cont Shelf Res 16:1259–1282
Framiñan M et al (1999) Physical characteristics and processes of the Rio de la Plata estuary. In: Perillo G, Piccolo C, Pino-Quivira M (eds) Estuaries of South America: their geomorphology and dynamics. Springer, Berlin/New York, pp 161–194
Friedrichsmeier T, P Ecochard, Roediger S et al (2015) Interfaz gráfica para el lenguaje estadístico R. http://rkward.kde.org Licencia Publica General de GNU
del Giorgio P, Duarte C (2002) Respiration in the open ocean. Nature 420:379–384
del Giorgio P, le B Williams P (eds) (2005) Respiration in aquatic ecosystems. Oxford University Press, Oxford, 315p
Gómez N, Hualde PR, Licursi M et al (2004) Spring phytoplankton of Río de la Plata: a temperate estuary of South America. Est Coast Shelf Sci 61:301–309
Gómez-Erache M, Lagomarsino JJ, Nuñez K et al (2001) Producción fitoplanctónica en la región frontal del Río de la Plata. In: Vizziano D et al (eds.) El Río de la Plata. Investigación para la gestión del ambiente, los recursos pesqueros y la pesquería en el frente salino. Programa Ecoplata, pp 33–45

Guerrero R, Acha M, Framiñan MB et al (1997) Physical oceanography of the Río de la Plata estuary. Cont Shelf Res 17:727–742
Huret M, Dadou I, Dumas F et al (2005) Coupling physical and biogeochemical processes in the Río de la Plata plume. Cont Shelf Res 25:629–653
Kocum E, Underwood G, Nedwell D (2002) Simultaneous measurement of phytoplanktonic primary production, nutrient and light availability along a turbid, eutrophic UK east coast estuary (the Colne estuary). Mar Ecol Prog Ser 231:1–12
Kruk C, Martinez A, Nogueira L et al (2014) Morphological traits variability reflects light limitation of phytoplankton production in a highly productive subtropical estuary (Río de la Plata, South America). Mar Biol. https://doi.org/10.1007/s00227-014-2568-6
Lercari D, Horta S, Martinez G et al (2014) A food web analysis of the Río de la Plata estuary and adjacent shelf ecosystem: trophic structure, biomass flows, and the role of fisheries. Hydrobiol. https://doi.org/10.1007/s10750-014-1964-8
Libes (2009) Introduction to marine biogeochemistry, 2nd edn. Academic, Burlington
Lucas LV, Koseff JR, Cloern JE et al (1999) Processes governing phytoplankton blooms in estuaries. II: the role of horizontal transport. Mar Ecol Prog Ser 187:17–30
Mann K, Lazier J (2006) Dynamics of marine ecosystems: biological-physical interactions in the oceans, 3rd edn. Blackwell Science Publishing, Malden
Martínez G, Brugnoli E, Hernández J et al (2005) How valid is the SeaWiFS estimation of chlorophyll-a at the Río de la Plata estuary and its area of influence?. In: Frouin R, Kawamura H, Pan D (eds) Active and Passive Remote Sensing of the Oceans. Proc of SPIE 5656. https://doi.org/10.1117/12.582665.
Masello A, Menafra R (1998) Comunidades macrobentónicas de la zona costera uruguaya y areas adyacentes. In: Wells P, Daborn G (eds). El Río de la Plata. Una Revisión Ambiental. Un informe de Antecedentes del Proyecto EcoPlata. Dalhousie University, Halifax, Nova Scotia, Canada, pp 142–193.
May C, Koseff JR, Lucas LV et al (2003) Effects of spatial and temporal variability of turbidity on phytoplankton blooms. Mar Ecol Prog Ser 254:111–128
Mechoso C, Pérez-Iribarren G (1992) Streamflow in Southeastern South America and the Southern Oscillation. J Clim 5:1535–1539
Montes-Hugo MA, Alvarez-Borrego S, Gaxiola-Castro G (2004) Annual phytoplankton production in a coastal lagoon of the southern California Current System. Mar Ecol Prog Ser 277:51–60
Nagy G, López-Laborde J, Anastasía L (1987) Caracterización de ambientes en el Río de la Plata Exterior (salinidad y turbiedad óptica). Inv Oceanológicas 1:31–56
Nagy G, Martínez CM, Caffera RM, et al (1998) Marco hidrológico y climático del Río de la Plata. In: Wells P, Daborn G (eds). El Río de la Plata. Una Revisión Ambiental. Un informe de Antecedentes del Proyecto EcoPlata. Dalhousie University, Halifax, Nova Scotia, Canada, p 17–70
Nagy G, Gómez-Erache M, López CH, Perdomo AC (2002) Distribution patterns of nutrients and symptoms of eutrophication in the Rio de la Plata River Estuary System. Hydrobiol 475/476:125–139
Nienchesky L, Baumgarten M (1997) Environment and biota of the Patos Lagoon estuary: environmental chemistry. In: Seeliger U, Odebrecht C, Castello J (eds) Subtropical convergence environments. the coast and sea in the Southwestern Atlantic. Springer, Berlin, pp 20–23
Pasquini A, Lecomte KL, Piovano EL et al (2006) Recent rainfall and runoff variability in Central Argentina. Quaternary Internacional 158:127–139
Piola A, Matano RP, Palma E et al (2005) The influence of the Plata River discharge on the western South Atlantic shelf. Geophys Res Lett 32:1–4
Piola A, Romero S, Zajaczkovski U (2008) Space-time variability of the Plata plume inferred from ocean color. Cont Shelf Res 28:1556–1567
R Core team (2015) R: A Language and Environment for Statistical Computing. R Foundation for Statistical Computing. Vienna, Austria, https://www.R-project.org. Accessed 3 July 2017.

Rodriguez-Graña L (2010) Retinal development in flatfish larvae: the effect of dietary essential fatty acids. Final activity and management report. Marie Curie Mobility Actions – EU, 8pp

Sharp J, Pennock JR, Church TM et al (1984) The estuarine interaction of nutrients, organics, and metals: a case study in the Delaware Estuary. In: Kennedy V (ed) The estuary as a filter. Academic, San Diego, pp 241–258

Valiela I (2015) Marine ecological processes. Springer, New York

Vizziano D (2001) Determinación del ciclo reproductivo de la corvina *Micropogonias furnieri* (Pisces: Scianidae) y los factores que inciden en su estacionalidad en la zona frontal del Río de la Plata. In: Vizziano D et al (eds.) El Río de la Plata. Investigación para la gestión del ambiente, los recursos pesqueros y la pesquería en el frente salino. Programa Ecoplata, pp 105–114.

Westberry T, Behrenfeld MJ, Siegel DA, Boss E (2008) Carbon-based primary productivity modeling with vertically resolved photoacclimation. Global Biogeochem Cycles. https://doi.org/10.1029/2007GB003078

Satellite-Measured Phytoplankton and Environmental Factors in North Patagonian Gulfs

Gabriela N. Williams, Miriam E. Solís, and José L. Esteves

Abstract An extensive series of high-resolution satellite images from the Sea-Viewing Wide Field-of-View Sensor (SeaWiFS, 2000–2006) was used in the characterization of the phytoplankton biomass seasonal cycle of the north Patagonian gulfs (NPG). The NPG system is formed by the San Matías, San José, and Nuevo gulfs (between 40°47′and 43°00′S and 63°00′ and 65°1.2′W) and is an area of ecological importance and of great significance for marine conservation in the Patagonian Argentinean Shelf. The spatio temporal variability of phytoplankton biomass in each of these environments was characterized by chlorophyll *a* data from satellite images (Chl*a*-sat). The observed seasonal variability was explained by factors influencing the growth of phytoplankton: photosynthetically available radiation (PAR from the SeaWiFS sensor) and sea surface temperature (SST from the Advanced Very High-Resolution Radiometer, AVHRR). In situ temperature, nutrient, and chlorophyll *a* concentration data from oceanographic cruises carried out in Nuevo (four cruises: 1982–1983), San José (four cruises: 1984–1985), and San Matías (four cruises: 1986–1994) gulfs were also used to explain the observed patterns. Cycles of phytoplankton and SST over the NPG are typical of temperate waters. However, Chl*a*-sat cycles were different among gulfs. At the same time, Chl*a*-sat cycles over the gulfs were different from that over the adjacent middle continental shelf. SMG was characterized by a bimodal cycle, although in winter mean values were higher than the mean concentration for the whole area. SJG was characterized by a unimodal cycle with relatively high values of chlorophyll *a* concentration in spring-summer. NG showed a bimodal cycle with maximum values in autumn and spring and minimum values in winter and summer. Particularities of each gulf are discussed in relation to the seasonal hydrographic characteristics of the water column (temperature and nutrients) and in the context of the Patagonian shelf ecosystem.

G. N. Williams (✉) · M. E. Solís · J. L. Esteves
Laboratorio de Oceanografía Química y Contaminación de Aguas (LOQyCA), Centro para el Estudio de Sistemas Marinos (CESIMAR), Consejo Nacional de Investigaciones Científicas y Técnicas (CONICET), Puerto Madryn, Chubut, Argentina
e-mail: williams@cenpat-conicet.gob.ar; miriam@cenpat-conicet.gob.ar

M. S. Hoffmeyer et al. (eds.), *Plankton Ecology of the Southwestern Atlantic*,
https://doi.org/10.1007/978-3-319-77869-3_15

Keywords Chlorophyll *a* · Nutrients · Remote sensing · Vertical profiles · North Patagonian gulfs

1 Introduction

The ecosystem of the north Patagonian gulfs (NPG) in the Southwestern Atlantic Ocean (40°30′–43°00′S and 65°18′–62°30′W) is comprised by San Matías (SMG), San José (SJG), and Nuevo gulfs (NG) (Fig. 1). SJG and NG are part of the "Península Valdés" Natural Protected Area, which was declared a World Natural Heritage site by UNESCO in 1999. Also several productive activities of economic relevance are developed within these gulfs: industrial (Ocampo Reinaldo et al. 2013; Romero et al. 2013) and artisanal (Orensanz et al. 2006; Amoroso et al. 2011) fisheries, wildlife watching, and maritime transport (Argüelles et al. 2016; Chalcobsky et al. 2017).

The SMG covers an area of approximately 20,000 km^2, being the second largest gulf of Argentina. Around 55% of its total area is deeper than 100 m, with a maximum depth of 180 m in the center. This gulf communicates with the Patagonian shelf (PS) through a sill 100 km long and 70 m deep. The SJG has a subelliptic shape, and it is the shallowest (mean depth 30 m, maximum 80 m) and smallest (817 km^2) of the NPG. It does not connect directly with the PS, but it opens to the SMG through a narrow sill 6.9 km wide and 30 m average depth (Rivas 1990). The NG covers approximately 2400 km^2, and more than one third of its area is deeper than 120 m. It is connected to the PS through a sill 16 km wide and 44 m deep on average (Rivas and Ripa 1989). The maximum depths of SMG and NG, around 180 m, are greater than those of the adjacent PS (~70 m). Because of the physiographical features above, the NPG can be regarded as deep, semi-open, coastal ecosystems.

The area where these gulfs are located is characterized by low rainfall (annual average is 180 mm), and there are no direct river discharges on their coasts. Circulation is influenced by high tidal amplitudes (Glorioso and Flather 1997; Tonini and Palma 2017). Tidal energy at the mouths of SMG and NG is among the highest around the world (Tonini et al. 2013).

The surrounding PS integrates a large marine ecosystem, which is home to a rich and diverse community of species which are supported by high concentrations of plankton. The PS is characterized by the presence of several oceanographic fronts associated with high chlorophyll *a* concentrations relative to the deep adjacent ocean (Fig. 1a, Romero et al. 2006, Carreto et al. 2016). These frontal systems are marine structures which support feeding, reproduction, and migratory patterns of local populations of fish, birds, and mammals (Ocampo Reinaldo et al. 2013; Svendsen et al. 2015). They are associated with relatively high rates of primary (Garcia et al. 2008; Lutz et al. 2010) and secondary productivity (Acha et al. 2004). In this case, the Valdés front is the one closest to the NPG area (Carreto et al. 1986; Acha et al. 2004) and has certain influence over the mouth of the SMG (Romero et al. 2006).

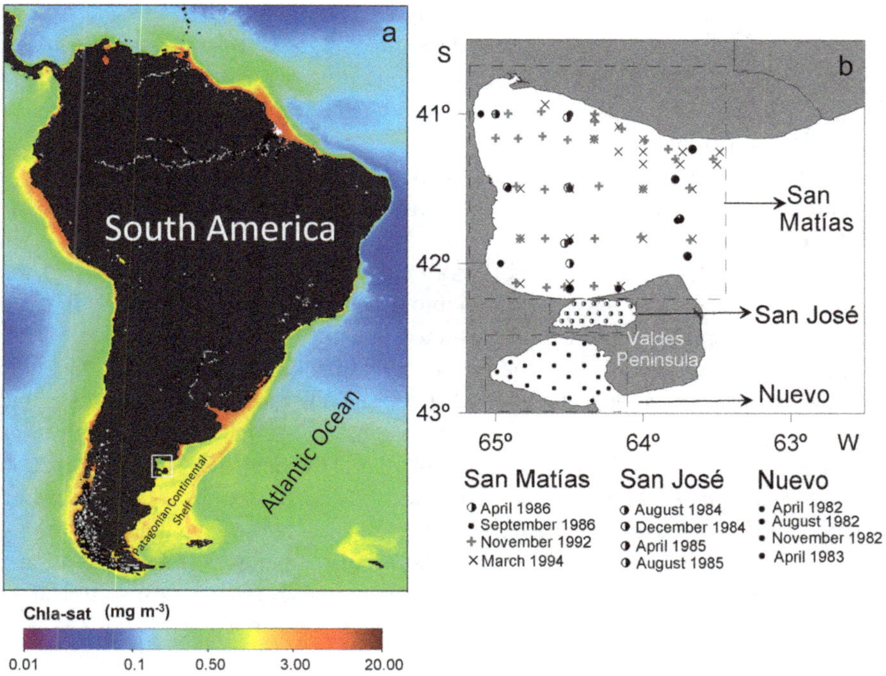

Fig. 1 (**a**) Location of the study area (white line) in the Southwest Atlantic Ocean (composite of chlorophyll *a* concentration from the SeaWiFS entire mission), (**b**) Sampling during oceanographic research cruises carried out in the north Patagonian gulf

At the latitude of the NPG, phytoplankton blooms over the Patagonian middle shelf initiate in early spring (chlorophyll *a* concentration > 3.5 mg.m^{-3}), decay at the end of this season, reach the minimum in summer, and have a second maximum at the beginning of winter (Romero et al. 2006). The seasonal and interannual variability of the chlorophyll *a* cycle in the PS has been linked to the variability of physical and chemical forcing (Romero et al. 2006; Garcia et al. 2008; Carreto et al. 2016). However, this type of approach has not been addressed in the NPG, and it is not known whether the chlorophyll *a* cycle matches that of the nearby Patagonian shelf. Studies on phytoplankton ecology over the NPG have aimed at identifying the typical phytoplankton species of each season, particularly in the SMG (Sastre et al. 2001), or from environments subject to anthropic disturbance (Esteves et al. 1996; Santinelli 2008). Toxic dinoflagellate blooms associated with an upwelling process in NG have been also studied (Esteves et al. 1992). Other works have mainly focused on the physical dynamics of the water column in the gulf (Rivas and Ripa 1989; Rivas 1990; Rivas and Beier 1990). Recent studies based on both past and recent ocean color data suggest that since 2004 there has been a large increase in the magnitude of the spring phytoplankton bloom in the Nuevo and San José gulfs (Wilson et al. 2016). However, the environmental mechanisms that drive and sustain the chlorophyll *a* cycle have not been fully studied.

Chlorophyll *a* concentration estimated from remote sensing ocean color images is useful to determine the magnitude and spatio temporal changes of phytoplankton blooms (IOCCG 2008). From an economic point of view, phytoplankton is the main support of the food webs that sustain world fisheries. Also, the occurrence of phytoplankton blooms, particularly of harmful species, has direct impacts on human health and well-being, mainly through their effects on coastal ecosystem services (fisheries, tourism, and recreation) and on some marine organisms and environments (IOCCG 2008).

The overall objective of this work is to contribute to the understanding of the climatological cycle of phytoplankton biomass in north Patagonian semi-open coastal ecosystems. Specific aims were focused on the following questions: (i) What is the intra-annual (seasonal) variability of phytoplankton in the NPG? Does this differ from that in the adjacent Patagonian shelf? (ii) What are the major hydrographic characteristics that drive phytoplankton blooms in the NPG? The study was performed on the basis of images of chlorophyll *a* of the SeaWiFS sensor from 2000 to 2006 and other environmental characteristics estimated by remote sensing (photosynthetically available radiation, PAR, and sea surface temperature, SST). Historical data of surface and vertical patterns of temperature, nutrients (nitrate, phosphate, silicic acid), and chlorophyll *a* from oceanographic research cruises carried out over the NPG between 1982 and 1994 were also used.

2 Material and Methods

2.1 Remote Sensing Data: Study Area and Image Processing

All images were provided by Argentina's National Commission for Spatial Activities (CONAE) and cover the NPG and the adjacent continental shelf, between 40°47′S and 43°00′S and 63°00′W and 65°1.2′W. They include:

- Chlorophyll *a* images (Chl*a*-sat). A time series of 7 years (2000–2006) of high-resolution (1.1 km) images of local area coverage (LAC) from the SeaWiFS sensor was examined. Chl*a*-sat concentration was derived from the OC4 ocean color algorithm (O'Reilly et al. 2000) using SeaDAS (SeaWiFS Data Analysis System version 4, www.seadas.gsfc.nasa.gov).
- SST images. A time series of 9 years (2000–2008) of high-resolution (1.1 km) images from the AVHRR sensor was analyzed. SST was calculated through a method based on the "split-window technique" algorithm (McClain et al. 1985). Clouds were removed using a combination of threshold values from channels 2 and 4 (Kelly 1985). SST calculation, further calibration, georeferencing, and image analysis were performed using ERDAS Imagine 8.7.

All images were mapped to a WGS84 reference system (datum WGS84, ellipsoid WGS84) and co-registered using a calibrated coastline. Afterward they were subset in order to circumscribe them to the study area. A total of 360 5-day maximum composites of SST and 510 chlorophyll *a* daily images were available from January 2000 to December 2008 and to 2006, respectively.

Monthly mean PAR images with 9 km spatial resolution (SeaWiFS_L3m_PAR v2014) were obtained from http://giovanni.gsfc.nasa.gov/. Because PAR data were not significantly different among gulfs for the period 2000–2006 (data not shown), a single monthly climatological mean was calculated for the whole study area over that period.

Monthly mean images of Chl*a*-sat and SST were produced by arithmetically averaging all scenes available for each month, on a pixel-by-pixel basis, in order to obtain a series of 12 climatological images for each variable. Land and cloud pixels were flagged to zero and were not considered when calculating the means. Calculations were performed using ERDAS Imagine 8.7. The resulting monthly mean images kept the same spatial resolution of the input images from both sensors (1.1 km).

Monthly climatological data of Chl*a*-sat for the whole area and for each gulf were adjusted to an annual plus semiannual cycle (Eq. 1, Espinosa-Carreon et al. 2004). Monthly climatological data of PAR and SST were adjusted to an annual cycle (Eq. 2, Rivas 2010):

$$\text{Chl}a\text{-sat} = \text{Chl}a\text{-sat}_0 + T_1 \cos\left(w\left(t - t_0\right)\right) + T_2 \cos 2\left(w\left(t - t_{00}\right)\right) \quad (1)$$

$$\text{PAR/SST} = \text{PAR}_0/\text{SST}_0 + T_1 \cos\left(w\left(t - t_0\right)\right) \quad (2)$$

where PAR_0, SST_0, and Chl*a*-sat$_0$ are the mean PAR, SST, and Chl*a*-sat temporal values, T_1 is the annual harmonic amplitude, T_2 is the semiannual harmonic amplitude, w is the frequency ($w = 2\pi/12$), t_0 is the annual harmonic phase, and t_{00} is the semiannual harmonic phase.

Seasonal maps of SST and Chl*a*-sat were produced by arithmetically averaging all monthly scenes available for each season in order to obtain four maps: summer (January to March), autumn (April to June), winter (July to September), and spring (October to December).

2.2 *Field Data from Oceanographic Research Cruises*

Seasonal vertical profiles of mean temperature and mean nutrient concentration (nitrate, phosphate, and silicic acid for SJG and SMG and nitrate and phosphate for NG) were obtained from surface down to 100 m depth in NG and from surface down to 75 and 50 m in SJG and SMG, respectively. Data were obtained during 12 oceanographic cruises carried out on board de R/V "El Austral" in NG (April, August, and November 1982 and April 1983), SJG (August and December 1984 and April and August 1985), and SMG (April and September 1986, November 1992, and March 1994) (Fig. 1b).

Water was sampled with Nansen bottles in 11–34 points in SMG, 18 in SJG, and 21 in NG. Temperature was recorded using reversing thermometers Kahlsico, and nutrient samples were frozen to −20 °C until further analysis. Nitrate and phosphate samples of NG cruises were analyzed by colorimetric techniques and with a UV-VIS Hitachi-110A Spectrophotometer according to Strickland and Parsons (1972). In samples from the SJG and SMG, nitrate and silicic acid were measured using a Technicon AutoAnalyzer II, and phosphate was measured with a UV-VIS Hitachi-110A Spectrophotometer (Strickland and Parsons 1972). Mean temperature and mean chlorophyll *a* from surface down to 10 m depth were also calculated for each cruise using the same data set.

Vertical profiles of nitrate for January 1983 (NG), December 1984 (SJG), and November 1992 (SMG) were estimated by the Kriging interpolation method. Data were obtained at stations located along transects oriented southeast-northeast in NG, west-east in SJG, and north-south in SMG. Profiles in SMG and SJG were calculated along areas of maximum SST gradients based on previous studies (Solís 1998; Amoroso and Gagliardini 2010; Rivas and Pisoni 2010). Chlorophyll *a* samples were obtained on board by filtering 1 L water samples through a 0.5 μm Millipore cellulose membrane. Filters were stored at −20 °C until further analysis at the lab. Chlorophyll *a* was extracted with 90% acetone and measured with a Turner Designs fluorometer following Strickland and Parsons (1972).

3 Results and Discussion

3.1 *Space-Time Variability of Chl*a*-sat and SST*

In autumn and winter, the spatial distribution of seasonal climatological Chl*a*-sat and SST was relatively uniform inside the gulfs in comparison with the adjacent continental shelf. SST values were around 16.5 °C and 12.0–12.5 °C, respectively, and Chl*a*-sat around 1.2 and 1.0 mg m^{-3}, respectively (Fig. 2a, b).

In spring and summer (Fig. 2c, d), the spatial distribution of these variables showed certain heterogeneity: the northern area of SMG displayed lower Chl*a*-sat values (~ 0.5 mg m^{-3}) and higher SST (13.5–19 °C) as compared to the southern area of the SMG and the nearby continental shelf (Chl*a*-sat >1 mg m^{-3}; SST between 12.5 and 17 °C). For these seasons, the observed spatial pattern is explained by the development of a tidal front which has been identified north of the mouth and toward the south in the inner part of the gulf (Piola and Scasso 1988; Gagliardini et al. 2004; Rivas and Pisoni 2010; Pisoni et al. 2015). In the northern portion of the gulf, this front defines a stratified area with higher temperature and salinity and a circulation dominated by a cyclonic subgyre about 70 km in diameter (Piola and Scasso 1988; Tonini et al. 2013); the southern region is dominated by a smaller cyclonic subgyre and is influenced by intrusions of water from the shelf, which result in high nitrate concentrations and less stratification (Carreto et al. 1974; Scasso and Piola 1988; Rivas and Beier 1990; Gagliardini and Rivas 2004; Williams 2011).

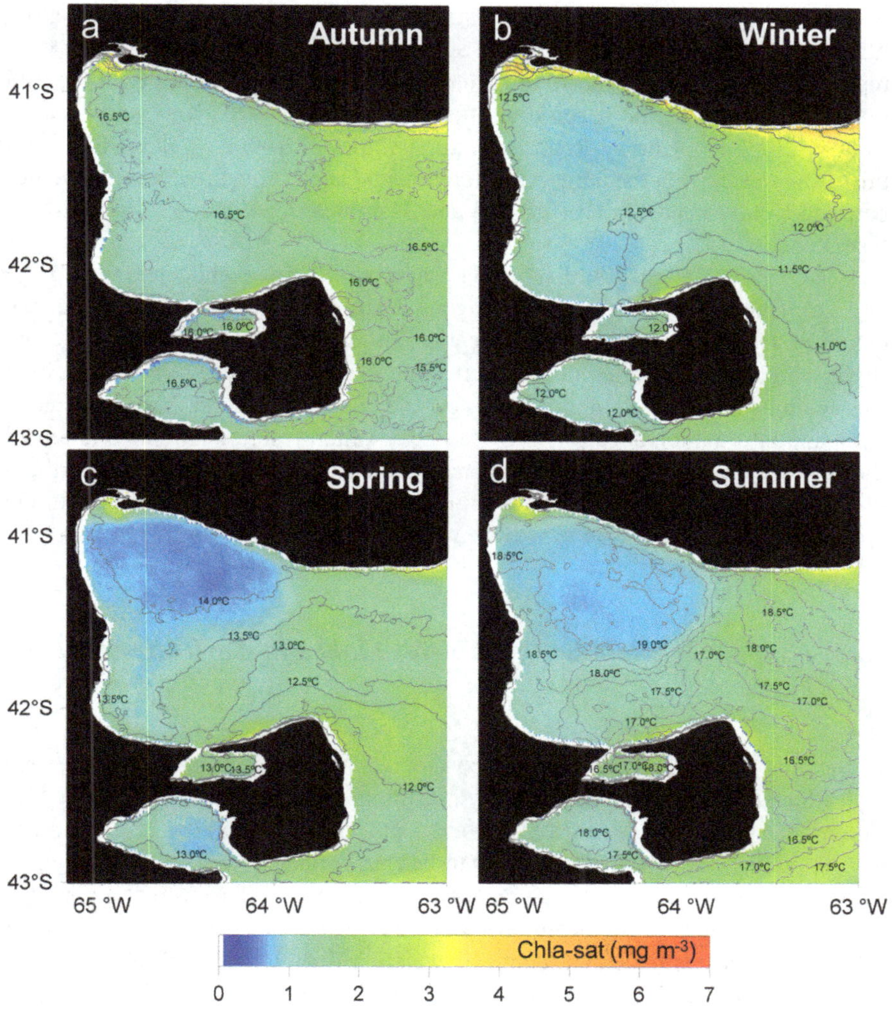

Fig. 2 Climatological seasonal Chl*a*-sat (mg m^{-3}) and SST contours (°C) over the study area

In the SJG, the spatial Chl*a*-sat distribution was relatively homogeneous and similar to the adjacent continental shelf (~2 mg m^{-3}) in both seasons, though mainly in summer. SST was higher in the eastern sector (18 °C) than in the western sector (16.5–17 °C). Analyses of high-resolution remote-sensed data (Amoroso and Gagliardini 2010) have shown that this gulf is consistently divided into two domains, east and west of a thermal and turbidity front that develops in spring and summer; turbulence and eddies dominate the circulation in the west domain. On the other hand, the flows between the SJG and the adjacent SMG involve a complex and recurrent pattern of jets and eddy, presumably resulting from the interaction of tidal currents with bottom topography and basin shape (Gagliardini et al. 2004; Amoroso and Gagliardini 2010). Numerical models suggest that the mean circulation in SJG is dominated year-round by a pair of strong counter-rotating eddies produced by tidal rectification (Tonini et al. 2013). Studies conducted at a coastal station have reported that the relatively high nutrient concentration and productivity of the SJG may be attributed to the penetration of nutrient-rich water (Charpy and Charpy-Roubaud 1980a, b; Charpy-Roubaud et al. 1982, 1983) from the Valdés Front. This tidal front develops in the neighboring continental shelf, offshore Península Valdés between October and March (Sabatini and Martos 2002; Rivas and Pisoni 2010; Pisoni et al. 2015).

In NG, the spatial distribution of SST during spring was homogeneous (~12 °C, Fig. 2c). Chl*a*-sat (~2 mg m^{-3}) was higher in the southern coastal zone near the mouth of the gulf than over any other area. In summer, the SST pattern showed slightly higher values in the middle of the gulf (18 °C), and Chl*a*-sat (~2 mg m^{-3}) was again higher in the southern coast near the mouth but also in the western area. Available information about the NG is scarce. A high-resolution numerical model showed only one gulf-wide cyclonic gyre (Tonini et al. 2013). Same as the northern subgyre in SMG, the gyre in NG is explained by the interaction of tides with the evolving stratification driven by surface heat (Tonini et al. 2013).

3.2 Temporal Variability of Chla-sat Linked to Environmental Factors

The annual cycle of mean Chl*a*-sat, considering the three gulfs as a whole area (black line, Fig. 3a), is characterized by maximum concentrations (~1.5 mg m^{-3}) in autumn (April) and spring (October) and lower relative concentrations (~1.00 mg m^{-3}) in summer (December to February) and winter (July and August). Highest values in April agree with records by Rivas et al. (2006) who observed that the autumn bloom seems to begin earlier in the coastal zone than in the middle shelf (May). This cycle is typical of temperate waters, although the spring bloom would be expected to be higher than the autumn bloom (Mann and Lazier 2006). The fit of the data to a mean plus annual and semiannual harmonics model explains more than 90% of the variance ($r^2_{annual + semiannual}$, Table 1). The contribution of each harmonic (annual and semiannual) to the cycle was determined by fitting Chl*a*-sat data to each harmonic separately (first and second term of Eq. 1, respectively). As a result, the whole study area shows a bimodal pattern explained mainly by the semiannual harmonic ($r^2_{semiannual} = 0.91$, Table 1).

Monthly mean values of incident PAR vary throughout the year (Fig. 3a) between a minimum of 10.78 einsteins m^{-2} day^{-1} at the end of autumn (May) and a maximum of 59.84 einsteins m^{-2} day^{-1} in spring (November). The PAR cycle shows a marked annual signal evidenced by the high percentage of the explained variance by the annual harmonic ($r^2 > 0.99$, Eq. 2).

The monthly climatological mean SST (Fig. 3b) for the whole area ranged between 11.48 °C at the end of winter (August) and 19.12 °C at the end of summer (February). In this case, in agreement with Rivas (2010), the marked annual signal is evidenced by the high percentage of variance that is explained by the annual har-

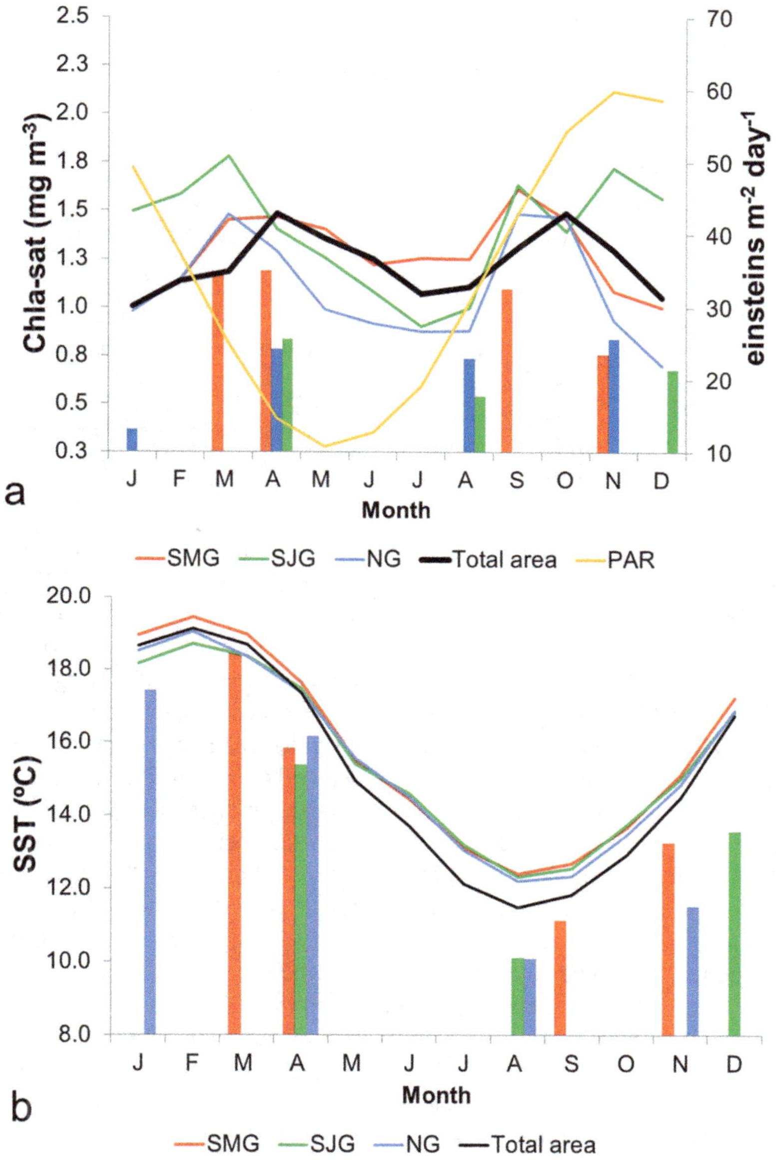

Fig. 3 (**a**) Monthly mean values at each gulf of Chl*a*-sat and PAR (SeaWiFS) for the period 2000–2006 (lines) and in situ mean chlorophyll *a* from surface down to 10 m depth during cruises 1982–1994 (bars), (**b**) SST (AVHRR) monthly mean values for the period 2000–2008 (lines) and in situ mean temperature 0–10 m during cruises 1982–1994 (bars)

monic ($r^2 > 0.99$, Table 2; coefficients of the fitted model are shown as supplementary material). The SST cycle is explained mainly by an annual harmonic signal, as the surface heat flux is also dominated by this frequency (Rivas 1990, 2010). The SST annual cycle showed a similar trend to that of PAR, with a 3-month lag between their minimum and maximum values.

Table 1 Descriptive statistics of climatological Chl*a*-sat (mg m^{-3}) over each gulf based on monthly values

Area	Mean	Max1	Max2	Min	SD	Month max1	Month max2	Month min	$r^2_{annual + semiannual}$	r^2_{annual}	$r^2_{semiannual}$
Total	1.23	1.49	1.48	1.00	0.16	O	A	J-F	0.94	0.24	0.91
SMG	1.28	1.61	1.47	1.00	0.20	S	A	D-J	0.94	0.49	0.79
SJG	1.40	1.78	1.72	0.90	0.29	M	N	J	0.90	0.79	0.53
NG	1.10	1.49	1.48	0.70	0.27	S	M	D	0.90	0.07	0.89

Table 2 Descriptive statistics of climatological SST (°C) over each gulf based on monthly values

Area	Mean	Max	Min	SD	Month max	Month min	r^2_{annual}
Total	15.18	19.12	11.48	2.84	F	Ag	0.99
SMG	15.76	19.44	12.40	2.59	F	Ag	0.99
SJG	15.54	18.72	12.33	2.33	F	Ag	0.99
NG	15.51	19.04	12.20	2.48	F	Ag	0.99

The climatological annual cycle of satellite-measured chlorophyll *a* in NPG agrees with that observed in the northern middle shelf (Romero et al. 2006; Rivas et al. 2006) and is typical of temperate waters (Mann and Lazier 2006), although the amplitude is lower (0.5 mg m^{-3}) than the amplitude in the middle shelf (6.46 mg m^{-3}, Romero et al. 2006). Also, unlike in the northern middle shelf, spring values of Chl*a*-sat in the NPG area were similar to those recorded in autumn.

Although in situ data used in this work were collected prior to the satellite images used here, in situ mean values of chlorophyll *a* and temperature within the first 10 meters of the water column match relatively well with the annual cycle of satellite climatological data (Fig. 3). Despite that differences between satellite and in situ data have been reported in the Patagonian shelf (Dogliotti et al. 2009), the analysis of Chl*a*-sat seasonal variability over the shelf (Rivas et al. 2006; Romero et al. 2006), and particularly over the SMG, shows that it reproduces quite well the patterns obtained using field data (Williams et al. 2013) .

The overall annual Chl*a*-sat cycle for the entire NPG area may be explained considering the seasonal progress of hydrographic characteristics. The decrease in water temperature that starts in February (Fig. 3b) produces an increase in the density of the upper layer of the water column with the resulting sinking of water masses. This generates a mixing process with vertical homogenization (Rivas and Beier 1990; Mann and Lazier 2006). A signal of this event can be seen in the average profiles of temperature and nutrients of the gulfs in autumn, within the upper 50 m layer of the water column (Fig. 4a–d). The beginning of the mixing process, which can also be favored by local winds, generates the breakdown of the thermocline that in turn facilitates the ascent of nutrients toward the photic zone (Mann and Lazier 2006). These conditions, plus the facts that daytime is still relatively long (13:03 h, Servicio de Hidrografía Naval 2017) and predators' metabolism may be lower due to decreasing temperature (Hoffmeyer 1994; Mann and Lazier 2006), would allow the development of the autumn phytoplankton bloom, even when PAR (Fig. 3a) reaches only 25% of the annual maximum value for the region.

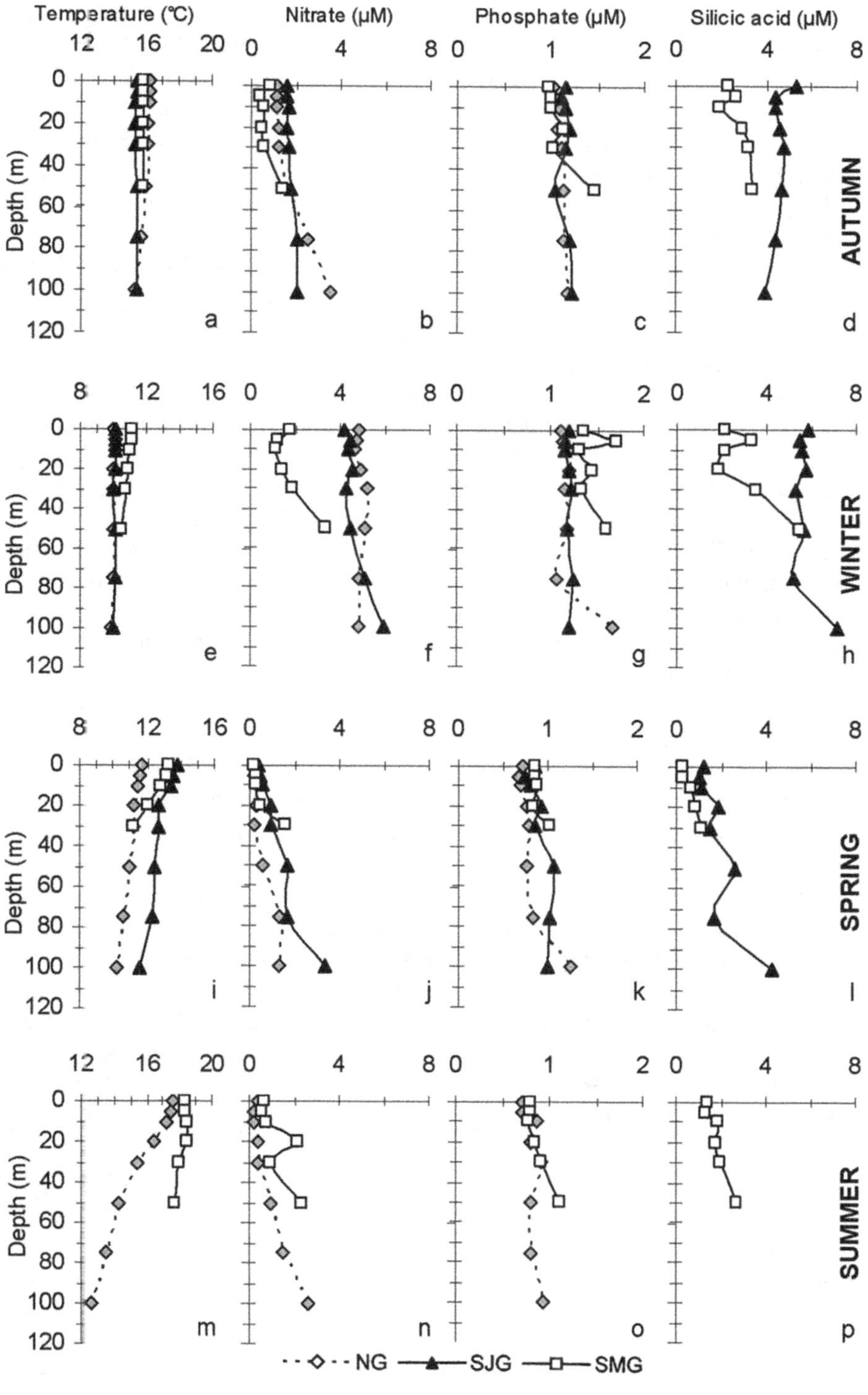

Fig. 4 Vertical profiles of in situ mean temperature, nitrate, phosphate, and silicic acid recorded during cruises 1982–1994

In winter (July) Chl*a*-sat is minimum (Fig. 3a). Average vertical profiles of SST and nutrient concentration in SJG and NG (Fig. 4e–h) are relatively homogeneous. Nutrient concentration is higher than in other seasons, particularly for nitrate and silicic acid (Fig. 4b–d, f–h, j–l, n–p). The hydrographic conditions and PAR intensity described above are similar to those found during the autumn bloom. However, in winter the mixed layer is deeper and daytime hours are minimum (~ 9 h, Servicio de Hidrografía Naval 2017), and thus conditions are not favorable for phytoplankton growth (Sverdrup 1953).

Differences observed between the vertical profiles of SMG and those of SJG and NG could be due to the sampling date. In SMG, data were collected during a cruise carried out at the end of winter (September), while data from SJG and NG were collected in midwinter (August).

In May, PAR shows the annual minimum (Fig. 3a) and begins to increase, reaching 75% of the maximum value in October. PAR and daytime hours increase simultaneously. Consequently, a rise in SST (Fig. 3b), the resulting formation of the thermocline, and a shallowing of the mixed layer are expected to occur from around September (Sverdrup 1953; Mann and Lazier 2006). In NG, the analysis of seasonal SST profiles from winter (Fig. 4f), spring (Fig. 4i), and summer (Fig. 4m) suggests that thermal stratification begins in spring, and it is well established in summer. In SMG the profiles evidence stronger stratification in spring (Fig. 4i, November 1992) and a typical season ending in summer (Fig. 4m, March 1994). In temperate zones, the spring bloom is associated with the stabilization caused by thermal stratification, which traps phytoplankton cells within the euphotic zone when nutrient concentration is high (Mann and Lazier 2006). In this sense, examination of the gulfs' mean vertical profiles from winter (Fig. 4–h) through spring (Fig. 4j–l) shows that there is an actual decrease in nutrient concentration, which indicates the consumption associated with the spring bloom, in turn evidenced by high Chl*a*-sat concentrations recorded during this period (Fig. 3a).

After the spring bloom, Chl*a*-sat decreases and reaches its minimum in summer (December and January) (Fig. 3a). At this time growth and cell division of phytoplankton would be limited by the low replenishment of nutrients in the mixed layer due to the thermocline occurrence and zooplankton grazing (Mann and Lazier 2006).

Field data available for summer in NG show a thermocline above 20 m depth (Fig. 4m), depleted nitrate (Fig. 4n), and phosphate concentrations (Fig. 4o) slightly lower than spring values (Fig. 4j–k). The relatively homogeneous temperature profile in SMG is probably due to data having been collected at the end of summer (March).

The evolution of the annual cycle and vertical distribution of temperature have already been described by Rivas and Beier (1990) analyzing field data from April and September 1986, which were also used in this work. More details on the analysis of the seasonal thermohaline structure of these gulfs may be found in Rivas and Ripa (1989), Rivas (1990), and Rivas and Beier (1990). Raw data from these analyses were used to obtain the temperature vertical profiles shown in Fig. 4a, e, i, m. These previous works focused on the analysis of the physical conditions of the water column that are necessary for nutrients to reach the well-lit upper layers, so that the bimodal cycle of phytoplankton biomass can occur as discussed above.

3.3 Particularities of the Annual Chla-sat Cycle for each of the North Patagonian Gulfs

The Chl*a*-sat cycle showed some particular differences between each gulf and the pattern observed over the whole study area. Thus, the SMG had minimum Chl*a*-sat concentrations at the end of spring and during summer (November to February, ~1.00 mg m^{-3}) with relatively higher values in March and September (1.50 mg m^{-3}) and mean concentrations in winter (May to August, ~1.20 mg m^{-3}) (Fig. 3a). The Chl*a*-sat annual cycle in this gulf was bimodal (explained mainly by the semiannual harmonic, $r^2 = 0.79$, Table 1). In winter particularly, mean Chl*a*-sat values were higher than those from SJG and NG. This could be due to the presence of species adapted to lower light intensities and temperature (Popovich and Marcovecchio 2008), in an environment with high nutrient concentration (Fig. 4f–h) and favored by a low grazing rate (Hoffmeyer 1994). The relatively low concentrations of Chl*a*-sat observed during summer months could be explained by a lower nutrient availability as a consequence of the increasing water column stratification. This is evidenced in the spring vertical profile from historical data of the SMG showing stratification in the northern area, with low nitrate concentrations (0.5 μM) above 40 m and between 1.0 and 6.0 μM at greater depths, while relatively uniform concentrations (1.0–1.5 μM) are recorded in the south (Fig. 5a). These results agree with previous studies in the area (Carreto et al. 1974; Rivas and Beier 1990; Williams 2011). A high rate of zooplankton grazing may also explain the low concentrations of chlorophyll *a* in this gulf (Ramírez 1996; Mann and Lazier 2006).

The SJG showed minimum concentrations of Chl*a*-sat in winter (May to August, ~1.0 mg m^{-3}) and relatively higher concentrations from October to March (~1.6 mg m^{-3}) (Fig. 2a). The cycle was unimodal, and its variability is explained mainly by the annual harmonic ($r^2_{\text{annual}} = 0.79$, Table 1). In this case, the high Chl*a*-sat concentrations during summer could be explained by the interaction of tidal currents with the morphology and bottom topography of the SJG, which has a narrow mouth 30 m deep that communicates with the SMG. This generates turbulence and gyres, which hinder the stratification of the water column (Amoroso and Gagliardini 2010; Amoroso et al. 2011) and force the upwelling of nutrients (Mann and Lazier 2006) mainly in the western sector of the gulf. Thus, a nitrate profile from early summer showed concentrations relatively higher in the western area (0.5–2.0 μM) than in the eastern sector (0.5 μM) (Fig. 5b).

The NG showed minimum Chl*a*-sat concentrations in winter (June to August, ~ 0.90 mg m^{-3}) and summer (December and January, ~0.80 mg m^{-3}), a peak in autumn (March), and another one in spring (September to October, ~1.5 mg m^{-3}). The NG was characterized by a bimodal cycle explained mainly by the semiannual harmonic ($r^2_{\text{semiannual}} = 0.89$, Table 1), and the annual harmonic explains only 7% of the variability (Table 1). This cycle would be related to the topography of the gulf, which is deep and has a mouth that because of its characteristics limits the exchange of water with the adjacent continental shelf. The limiting factors for phytoplankton growth in this gulf would be radiation in winter and water column stratification

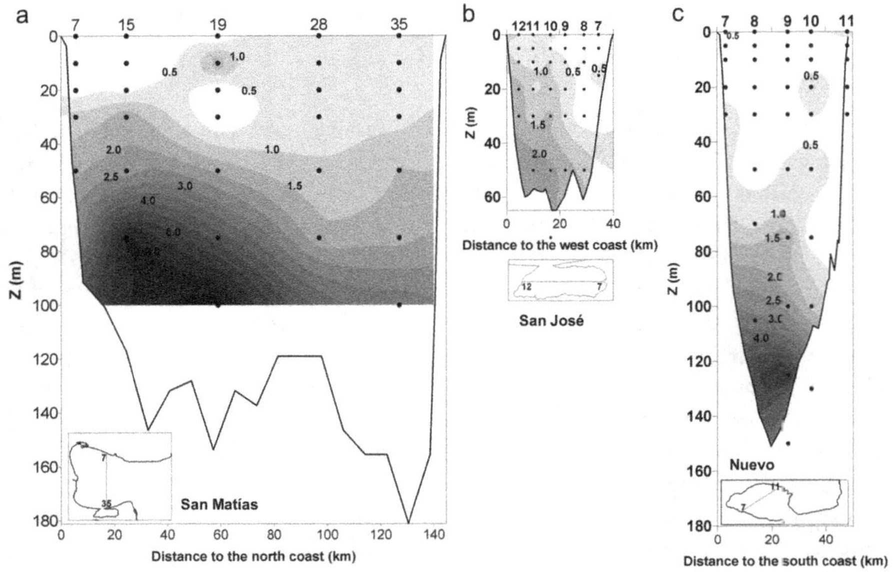

Fig. 5 (**a**) Water column distribution of nitrate (μM) in SMG (November 1992), (**b**) Water column distribution of nitrate (μM) in SJG (December 1984), (**c**) Water column distribution of nitrate (μM) in NG (January 1983). Location of vertical sections at each gulf is shown in the respective insets. Labels at the top of each profile indicate sampling station numbers

together with nutrients in summer. This environmental condition is demonstrated by the summer vertical profile of nitrate from historical data of NG (Fig. 5c) showing concentrations <1 μM above 75 m in the southern area and above 100 m depth to the north of the gulf.

Finally, the thermal cycles in the study area showed different SST ranges across gulfs between August and April, being SST in the SMG higher than in SJG and NG. Between April and June, the average SST was similar in the three gulfs (Fig. 3b, Table 2). Overall, the SST amplitude in the NPG ranged 3.14–3.49 °C mainly due to atmospheric influence (Krepper and Bianchi 1982; Scasso and Piola 1988; Rivas and Beier 1990).

4 Conclusions

This work has shown that the chlorophyll *a* cycle over the NPG area is similar to that over the middle continental shelf. However, when each gulf is considered separately, the NG and SJG show some differences with the middle shelf due to their particular hydrographic characteristics.

The general relationship between the annual cycle of phytoplankton biomass (Chl*a*-sat) and the environmental conditions considered in this study is synthesized in the conceptual scheme illustrated in Fig. 6. Different vertical and horizontal

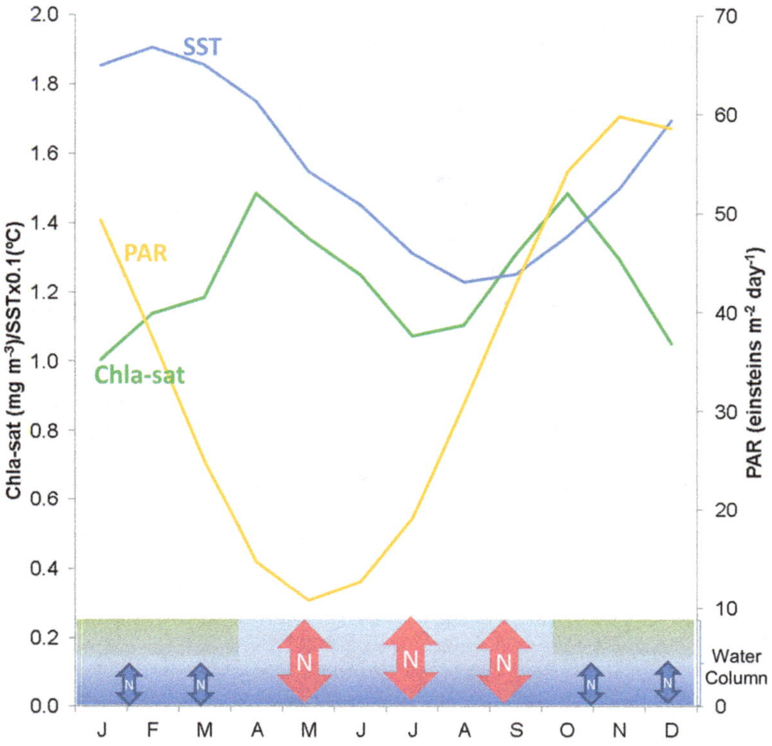

Fig. 6 Conceptual scheme showing seasonal changes of phytoplankton biomass (Chl*a*-sat), sea surface temperature (SST), nitrate concentration (arrows), and water column stratification (blue-green tones)

patterns were observed along the three gulfs. In the vertical dimension, winter is represented by a homogenous distribution of variables, summer by stratified conditions, and autumn and spring by transitional stages toward vertical homogenization and development of a thermocline, respectively. The result is a bimodal chlorophyll *a* cycle (Rivas and Beier 1990; Solís 1998; Williams 2011). In areas with high tidal dissipation (Tonini et al. 2013; Palma et al. 2004) and therefore characterized by a mixed water column, the chlorophyll *a* cycle is unimodal.

Future research in the NPG should examine the cycle of phytoplankton biomass taking into account the particular characteristics of each gulf, such as level of stratification, attenuation of light in the water column, presence of tidal fronts, and the effect of winds, among other variables.

The high temporal and spatial resolution of the remote sensing data used in this study enabled the characterization of the annual cycle of phytoplankton biomass from the top layer of the sea, while field data provided environmental information for understanding, at least partially, the forcing of the observed cycles. There are still many questions to answer: Does the spring bloom occur after or before the stratification and stabilization of the water column? Have the autumn and spring blooms always a similar magnitude? Is there any interannual variability? What is the influence that winds have on the phytoplankton dynamics in the NPG region? This last variable, which was not considered in this study, is known as a key factor of seasonal variability of phytoplankton blooms in temperate waters. In this regard, Pisoni et al. (2014), among others, have demonstrated the occurrence of upwelling events in the SMG which are associated with northwest winds.

The results of this work provide an overview of the annual chlorophyll *a* cycles and their relation to the physical environment of these gulfs. This information may prove to be useful for future investigations addressing the annual and interannual variability of phytoplankton biomass over the NPG and the related biogeochemical processes.

Acknowledgments We dedicate this chapter to Dr. Domingo Antonio Gagliardini who was a pioneer in the field of marine remote sensing in Argentina. The authors thank the Comisión Nacional de Actividades Espaciales (CONAE, Argentina) for the provided images and the Ocean Biology Processing Group (Code 614.2) at the GSFC, Greenbelt, MD20 771, for the distribution of the ocean color data. We also thank M. Sapoznik, N. Pérez de la Torre, and M.R. Marin for their assistance in processing the satellite images and N. Glembocki for language improvement. An anonymous reviewer helped to improve the manuscript. The work was supported by funding from the Consejo Nacional de Investigaciones Científicas y Técnicas (CONICET, Argentina) and the Agencia Nacional de Promoción Científica y Tecnológica (ANPCyT, Argentina) through projects PICT 2003 N° 15221, 2006 N° 1575, 2006 N° 649, and 2013 N°0687.

References

Acha EM, Mianzan HW, Guerrero RA et al (2004) Marine fronts at the continental shelves of austral South America. Physical and ecological processes. J Mar Syst 44:83–105

Amoroso RO, Gagliardini DA (2010) Inferring complex hydrographic processes using remote-sensed images: turbulent fluxes in the Patagonian gulfs and implications for scallop metapopulation dynamics. J Coast Res 262:320–332. https://doi.org/10.2112/08–1095.1

Amoroso RO, Parma AM, Orensanz JM, Gagliardini DA (2011) Zooming the macroscope: medium-resolution remote sensing as a framework for the assessment of a small-scale fishery. ICES J Mar Sci 68:696–706. https://doi.org/10.1093/icesjms/fsq162

Argüelles MB, Fazio A, Fiorito C et al (2016) Diving behavior of southern right whales (*Eubalaena australis*) in a maritime traffic area in Patagonia, Argentina. Aquat Mamm 42:104–108. https://doi.org/10.1578/AM.42.1.2016.104

Carreto JI, Verona CA, Casal A, Laborde MA (1974) Fitoplancton, pigmentos y condiciones ecológicas del Golfo San Matías: Marzo 1971 (I), Mayo 1971 (II) y Noviembre de 197I (III). Anal Inf Com Inv Cient La Plata, Argentina, pp 1–76

Carreto JI, Benavides HR, Negri RM et al (1986) Toxic red-tide in the Argentine Sea. Phytoplankton distribution and survival of the toxic dinoflagellate *Gonyaulax excavata* in a frontal area. J Plankton Res 8:15–28

Carreto JI, Montoya NG, Carignan MO et al (2016) Environmental and biological factors controlling the spring phytoplankton bloom at the Patagonian shelf-break front – degraded fucoxanthin pigments and the importance of microzooplankton grazing. Prog Oceanogr 146:1–21. https://doi.org/10.1016/j.pocean.2016.05.002

Chalcobsky BA, Crespo EA, Coscarella MA (2017) Whale-watching in Patagonia: what regulation scheme should be implemented when the socio-ecological system is changing? Mar Policy 75:165–173. https://doi.org/10.1016/j.marpol.2016.11.010

Charpy LJ, Charpy-Roubaud CJ (1980a) La production primaire des eaux du golfe San José (Peninsula Valdés, Argentina): populations phytoplanctoniques et composition du seston. Hydrobiologia 75:215–224

Charpy LJ, Charpy-Roubaud CJ (1980b) La production primaire des eaux du golfe San José (Peninsula Valdés, Argentina): estimation de la production phytoplanctonique annuelle. Hydrobiologia 75:225–229

Charpy-Roubaud CJ, Charpy LJ, Maestrini SY (1982) Fertilité des eaux cotières nord-patagoniques: facteurs limitant la production du phytoplancton e potentialités d'exploitation mytilicole. Oceanol Acta 2(5):188–197

Charpy-Roubaud CJ, Charpy LJ, Maestrini SY (1983) Nutrient enrichments of waters of "Golfo de San José" (Argentina, 42°S), growth and species selection of phytoplankton. Mar Ecol 4(1):1–18

Dogliotti AI, Schloss IR, Almandoz GO, Gagliardini DA (2009) Evaluation of SeaWiFS and MODIS chlorophyll *a* products in the Argentinean Patagonian Continental Shelf (38°S–55°S). Int J Remote Sens 30:251–273. https://doi.org/10.1080/01431160802311133

Espinosa-Carreon L, Beier E., Ocampo Torres F. et al (2004) Seasonal and interannual variability of satellite-derived chlorophyll pigment, surface height, and temperature off Baja California. J Geophys Res 109:1–20. https://doi.org/10.1029/2003JC002105

Esteves JL, Santinelli N, Sastre V, Díaz R, Rivas O (1992) A toxic dinoflagellate bloom and PSP production associated with upwelling in Golfo Nuevo, Patagonia, Argentina. Hydrobiologia 242:115–222

Esteves JL, Solís ME, Sastre V, Santinelli N, Gil M et al (1996) Evaluación de la contaminación urbana de la Bahía de San Antonio. Informes técnicos del Plan de Manejo Integrado de la Zona Costera Patagónica. Fundación Patagonia Natural, Puerto Madryn

Gagliardini DA, Rivas AL (2004) Environmental characteristics of San Matías Gulf obtained from Landsat-TM and ETM+ Data. Gayana (Concepción) 68:1–10. https://doi.org/10.4067/S0717-65382004000200034

Gagliardini DA, Amoroso RO, Dell' Arciprete OP, Yorio P, Orensanz JM (2004) Detection of small-scale coastal oceanographic processes through Landsat-TM/ETM+ images: implications for the study of biological processes along the Patagonian Coasts of Argentina. Gayana (Concepción) 68:194–200. https://doi.org/10.4067/S0717-65382004000200035

Garcia V, Garcia C, Mata M et al (2008) Environmental factors controlling the phytoplankton blooms at the Patagonia shelf-break in spring. Deep Sea Res Part I Oceanogr Res Pap 55:1150–1166. https://doi.org/10.1016/j.dsr.2008.04.011

Glorioso PD, Flather RA (1997) The Patagonian Shelf tides. Prog Oceanogr 40:263–283. https://doi.org/10.1016/S0079-6611(98)00004-4

Hoffmeyer MS (1994) Seasonal succession of Copepoda in the Bahía Blanca estuary. Hydrobiologia 292(293):303–308

IOCCG (2008) Why ocean colour? The societal benefits of ocean-colour technology. In Platt T, Hoepffner N, Stuart V, Brown C (eds) Reports of the International Ocean-Colour Coordinating Group, No. 7, IOCCG, Dartmouth, Canada

Kelly KA (1985) Separating clouds from ocean in infrared images. Remote Sens Environ 17:67–83

Krepper C, Bianchi AA (1982) Balance calórico del Mar Epicontinental Argentino. Acta Oceanogr Arg 3(1):119–133

Lutz VA, Segura V, Dogliotti AI et al (2010) Primary production in the Argentine Sea during spring estimated by field and satellite models. J Plankton Res 32:181–195. https://doi.org/10.1093/plankt/fbp117

Mann KH, Lazier JR (2006) Dynamics of marine ecosystems. Biological-physical interactions in the oceans, 3rd edn. Blackwell Science Publications, Cambridge, MA

McClain EP, Pichel WG, Walton CC (1985) Comparative performance of AVHRR-based multi-channel sea surface temperature. J Geophys Res 90:11587–11601

Ocampo Reinaldo M, González R, Williams G et al (2013) Spatial patterns of the Argentine hake *Merluccius hubbsi* and oceanographic processes in a semi-enclosed Patagonian ecosystem. Mar Biol Res 9:394–406. https://doi.org/10.1080/17451000.2012.739700

O'Reilly JEO, Maritorena S, Siegel DA et al (2000) Ocean color chlorophyll-a algorithms for SeaWiFS, OC2 and OC4: version 4. SeaWiFS Postlaunch Technical Report Series Volume 11. SeaWiFS Postlaunch Calibration and Validation Analyses, Part 3. S. B. Hooker and E. R. Firestone

Orensanz JM, Parma AM, Turk T, Valero J (2006) Dynamics, assessment and management of exploited natural populations. In: Shumway S, Parsons GJ (eds) Scallops: biology, ecology and aquaculture, 2nd edn. Elsevier, Amsterdam, pp 765–868

Palma ED, Matano RP, Piola AR (2004) A numerical study of the Southwestern Atlantic Shelf circulation: Barotropic response to tidal and wind forcing. J Geophys Res 109:C08014. https://doi.org/10.1029/2004JC002315

Piola AR, Scasso L (1988) Circulación en el golfo San Matías. Geoacta 15:33–51

Pisoni JP, Rivas AL, Piola AR (2014) Satellite remote sensing reveals coastal upwelling events in the San Matías Gulf? Northern Patagonia. Remote Sens Environ 152:270–278. https://doi.org/10.1016/j.rse.2014.06.019

Pisoni JP, Rivas AL, Piola AR (2015) On the variability of tidal fronts on a macrotidal continental shelf, Northern Patagonia, Argentina. Deep Res Part II Top Stud Oceanogr 119:61–68. https://doi.org/10.1016/j.dsr2.2014.01.019

Popovich CA, Marcovecchio JE (2008) Spatial and temporal variability of phytoplankton and environmental factors in a temperate estuary of South America (Atlantic coast, Argentina). Cont Shelf Res 28:236–244. https://doi.org/10.1016/j.csr.2007.08.001

Ramírez FC (1996) Composición, abundancia y variación estacional del zooplancton de red del Golfo San Matías. Frente Marítimo (Sec A) 16:157–167

Rivas AL (1990) Análisis estacional de la estructura termohalina en el Golfo San José, Argentina. Geoacta 17(1):37–48

Rivas AL (2010) Spatial and temporal variability of satellite-derived sea surface temperature in the southwestern Atlantic Ocean. Cont Shelf Res 30:752–760

Rivas A, Beier E (1990) Temperature and salinity fields in the Northpatagonic Gulfs. Oceanol Acta 13:15–20

Rivas AL, Pisoni JP (2010) Identification, characteristics and seasonal evolution of surface thermal fronts in the Argentinean Continental Shelf. J Mar Syst 79:134–143. https://doi.org/10.1016/j.jmarsys.2009.07.008

Rivas A, Ripa P (1989) Variación estacional de la estructura termohalina de Golfo Nuevo, Argentina. Geofis Int 28:3–24

Rivas AL, Dogliotti AI, Gagliardini DA (2006) Seasonal variability in satellite-measured surface chlorophyll in the Patagonian Shelf. Cont Shelf Res 26:703–720. https://doi.org/10.1016/j.csr.2006.01.013

Romero SI, Piola AR, Charo M, Garcia CAE (2006) Chlorophyll-a variability off Patagonia based on SeaWiFS data. J Geophys Res C05021. doi:https://doi.org/10.1029/2005JC003244

Romero MA, Reinaldo MO, Williams G et al (2013) Understanding the dynamics of an enclosed trawl demersal fishery in Patagonia (Argentina): a holistic approach combining multiple data sources. Fish Res 140:73–82. https://doi.org/10.1016/j.fishres.2012.12.002

Sabatini M, Martos P (2002) Mesozooplankton features in a frontal area off northern Patagonia (Argentina) during spring 1995 and 1998. Sci Mar 66:215–232

Santinelli NH (2008) Fitoplancton de un ambiente costero sometido a perturbación antrópica: Bahía Nueva, Provincia de Chubut. Tesis Doctoral, Universidad Nacional de la Patagonia San Juan Bosco, Trelew, Argentina

Sastre AV, Santinelli NH, Esteves JL, Ferrario ME (2001) Aspectos ecológicos de especies de *Pseudo-nitzschia* en aguas costeras patagónicas (Argentina). In: Alveal K, Antezana T (eds) Sustentabilidad de la biodiversidad. Universidad de Concepción, Concepción, pp 217–235

Scasso ML, Piola AR (1988) Intercambio neto de agua entre el mar y la atmósfera en el Golfo San Matías. Geoacta 15(1):13–31

Servicio de Hidrografía Naval (2017). Salida y puesta del sol http://www.hidro.gov.ar. Accessed 22 Jun 2017

Solís ME (1998) Monitoring in Nuevo Gulf (Argentina): Analysis of oceanographic data by geographic information systems (GIS). M. Sc. Thesis D.E.W. 021. Unesco-IHE Delft. The Netherlands

Strickland JDH, Parsons TR (1972) A practical handbook of the seawater analysis. 2nd edn. Bull J Fish Res Bd Can 167: 311 pp.

Svendsen GM, Romero MA, Williams GN et al (2015) Environmental niche overlap between common and dusky dolphins in North Patagonia, Argentina. PLoS One 10:1–20. https://doi.org/10.1371/journal.pone.0126182

Sverdrup HU (1953) On conditions for the vernal blooming of phytoplankton. J Cons Perm Int Exp Mer 18:287–295

Tonini MH, Palma ED (2017) Tidal dynamics on the North Patagonian Argentinean Gulfs. Estuar Coast Shelf Sci 189:115–130. https://doi.org/10.1016/j.ecss.2017.02.026

Tonini MH, Palma ED, Piola AR (2013) A numerical study of gyres, thermal fronts and seasonal circulation in austral semi-enclosed gulfs. Cont Shelf Res 65:97–110. https://doi.org/10.1016/j.csr.2013.06.011

Williams GN (2011) Caracterización ambiental del golfo San Matías mediante sensores remotos y parámetros océanográficos. Relación con la distribución y abundancia de los recursosbiológicos de interés pesquero. Tesis Doctoral, Universidad Nacional del Comahue, San Carlos de Bariloche, Argentina

Williams GN, Dogliotti AI, Zaidman P et al (2013) Assessment of remotely-sensed sea-surface temperature and chlorophyll-a concentration in San Matías Gulf (Patagonia, Argentina). Cont Shelf Res 52:159–171. https://doi.org/10.1016/j.csr.2012.08.014

Wilson C, Sastre AV, Hoffmeyer MS et al (2016) Southern right whale (*Eubalaena australis*) calf mortality at Península Valdés, Argentina: are harmful algal blooms to blame? Mar Mamm Sci 32:423–451. https://doi.org/10.1111/mms.12263

Mesozooplankton Structure and Seasonal Dynamics in Three Coastal Systems of Argentina: Bahía Blanca Estuary, Pirámide Bay, and Ushuaia Bay

Anabela A. Berasategui, María C. López Abbate, Valeria C. D'Agostino, María L. Presta, Román Uibrig, Tami M. García, Eugenia Nahuelhual, Carlo J. Chazarreta, María S. Dutto, Maximiliano Garcia, Fabiana Capitanio, and Mónica S. Hoffmeyer

Abstract Mesozooplankton communities in coastal ecosystems have successfully adapted to a wide range of environments. However, the current rate of coastal modification is challenging the survival of resident species. In this chapter, we describe the structure and annual dynamics of the mesozooplankton community in recent years for three coastal systems in Argentina that are subject to human disturbance: (1) Bahía Blanca Estuary, (2) Pirámide Bay in Nuevo Gulf and (3) Ushuaia Bay in

A. A. Berasategui (✉) · M. C. López Abbate · R. Uibrig · T. M. García · E. Nahuelhual
C. J. Chazarreta · M. S. Dutto · M. Garcia
Instituto Argentino de Oceanografía (IADO), Consejo Nacional de Investigaciones científicas y Técnicas–Universidad Nacional del Sur (CONICET–UNS), Bahía Blanca, Buenos Aires, Argentina
e-mail: mclabbate@iado-conicet.gob.ar

V. C. D'Agostino
Centro para el Estudio de Sistemas Marinos (CESIMAR), Consejo Nacional de Investigaciones científicas y Técnicas (CONICET), Puerto Madryn, Chubut, Argentina

M. L. Presta · F. Capitanio
Departamento de Biodiversidad y Biologia Experimental (DBBE FCEN UBA), Instituto de Biodiversidad y Biología Experimental y Aplicada (IBBEA CONICET UBA), Buenos Aires, Argentina
e-mail: capitani@ege.fcen.uba.ar

M. S. Hoffmeyer
Facultad Regional Bahía Blanca, Universidad Tecnológica Nacional, Bahía Blanca, Buenos Aires, Argentina

Instituto Argentino de Oceanografía (IADO), Consejo Nacional de Investigaciones científicas y Técnicas–Universidad Nacional del Sur (CONICET–UNS), Bahía Blanca, Buenos Aires, Argentina

M. S. Hoffmeyer et al. (eds.), *Plankton Ecology of the Southwestern Atlantic*,
https://doi.org/10.1007/978-3-319-77869-3_16

the Beagle Channel. The seasonal dynamics of mesozooplankton abundance in Bahía Blanca Estuary during 2009–2010 exhibited a unimodal pattern with an increase during the warm seasons, while organism abundance increased linearly with both salinity and turbidity. In Nuevo Gulf, research on the seasonal dynamics of mesozooplankton conducted during 2014–2015 exhibited a bimodal abundance pattern with peaks in late summer and spring. No significant relationships between mesozooplankton and the environmental variables included in Nuevo Gulf were found. The seasonal dynamics of mesozooplankton in Ushuaia Bay in the years 2006–2008 exhibited a bimodal abundance pattern with peaks in summer and early autumn; this pattern was mainly driven by the concentration of chlorophyll *a*. In Bahía Blanca Estuary, a gradual decrease in species richness was observed throughout the last four decades, while changes in species composition were also observed, suggesting that some species have the ability to acclimate to higher salinity and turbidity as well as to pollution. In Nuevo Gulf, the mesozooplankton community showed no significant modifications over time, which may be related to the low anthropogenic pressure. Although no profound shifts in mesozooplankton were observed in Ushuaia Bay, eutrophication may have an impact in the future through its effect on primary producers.

Keywords Copepods · Mero-holoplankton · Anthropogenic impact · Bay · Estuary

1 Introduction

Mesozooplankton in aquatic ecosystems are the functional link between primary producers and higher trophic levels (Turner 2004). These organisms rapidly respond to changes in the environment and are therefore considered valuable bioindicators (Silva et al. 2004; Chang et al. 2009; Uriarte and Villate 2004). Even though estuarine and coastal mesozooplankton communities are adapted to a wide range of environments, they are also vulnerable to sudden natural or anthropogenic changes and may thus undergo profound structural and phenological changes (Biancalana et al. 2012a; Berasategui et al. 2017).

Mesozooplankton is an important size fraction within planktonic heterotrophs in estuaries and coastal areas of Argentina (Hoffmeyer 2004; Fernández Severini and Hoffmeyer 2005; Menendez et al. 2011; Dutto et al. 2012). These organisms play a relevant role as agents in carbon cycling (Hoffmeyer et al. 2008; Diodato and Hoffmeyer 2008; Biancalana et al. 2012a; Dutto et al. 2014) and represent a fundamental energetic resource for fish (Capitanio et al. 2005; Lopez-Cazorla et al. 2011; Sato et al. 2011).

In this chapter, we describe the structure and annual dynamics of mesozooplankton community in recent years within three coastal systems of Argentina; these are located at different latitudes and subject to different human disturbance: (1) Bahía

Blanca Estuary in the margins between Pampas and Patagonia; (2) Pirámide Bay in Nuevo Gulf, on the northern Patagonian coast; and (3) Ushuaia Bay in the Southern margins of the Patagonian shelf, at the Beagle Channel. The ecological significance of these highly vulnerable areas has encouraged management actions and has therefore led authorities to designate protected areas either within the area or closest to these ecosystems. Given the relevance of the mesozooplankton as a bioindicator of environmental change (Gadzała-Kopciuch et al. 2004; Biancalana et al. 2012b; Dutto et al. 2012; Diodato 2013; Berasategui et al. 2017) and of its role in coastal food webs (Sommer et al. 2002; Fernández Severini and Hoffmeyer 2005; Hoffmeyer et al. 2008; Hoffmeyer et al. 2010a, b; D'Agostino 2016), we herein provide a synthesis of results from published literature as well as recent knowledge of new research, both of which will contribute as a basis to set the framework for future management actions.

2 Bahía Blanca Estuary

2.1 Main Features

This estuary is located at the northern margin of the Patagonian Shelf (38°42′–39°25'S and 61°50′–62°22'W), in a temperate region (Fig. 1a). It is the second largest estuary on the Argentinean coast and is characterised by an extent of wide tidal flats and several tidal channels (Perillo et al. 2007). This feature makes it a key spot for feeding and breeding for crustaceans, demersal fish, and migratory bird species (Garcia-Borboroglu and Yorio 2007; Hoffmeyer and Cervellini 2007; López-Cazorla 2007; Petracci et al. 2008; Simonetti et al. 2013). This ecosystem is part of the multipurpose natural reserve "Bahía Blanca, Bahía Falsa and Bahía Verde", founded in 1998 with the aim of preserving resident biological communities (Fidalgo 2007).

This mesotidal estuary has a tidal amplitude that varies between 3.5 m at the head to 2.2 m at the mouth on a semidiurnal cycle (Perillo et al. 2007). The inner zone is eutrophic and turbid (annual mean of suspended particulate matter, SPM of 77.6 mg L^{-1}) (Cuadrado et al. 2007; Perillo et al. 2007; Marcovecchio et al. 2008). Dissolved oxygen concentration is generally near saturation conditions (Marcovecchio et al. 2008; Guinder et al. 2009). Mean sea surface temperature (SST) is approximately 13 °C, with fluctuations between 8.5 °C in winter and 21.6 °C in summer (Perillo et al. 2007). Salinity ranges between 41 and 16, with a mean of approximately 33 in the inner zone (Freije et al. 2008). The estuarine salinity gradient is sometimes inverted (a negative estuary) as a result of the limited water circulation and freshwater inputs from the Sauce Chico River and Napostá Grande Stream (1.9 and 0.8 m^3 s^{-1}, respectively) (Perillo et al. 2007).

This estuary is subject to progressive deterioration due to the growing development of urban, industrial and port activities near the watershed (Marcovecchio et al. 2008; Arias et al. 2010; Grecco et al. 2011; Fernández Severini et al. 2011; Dutto

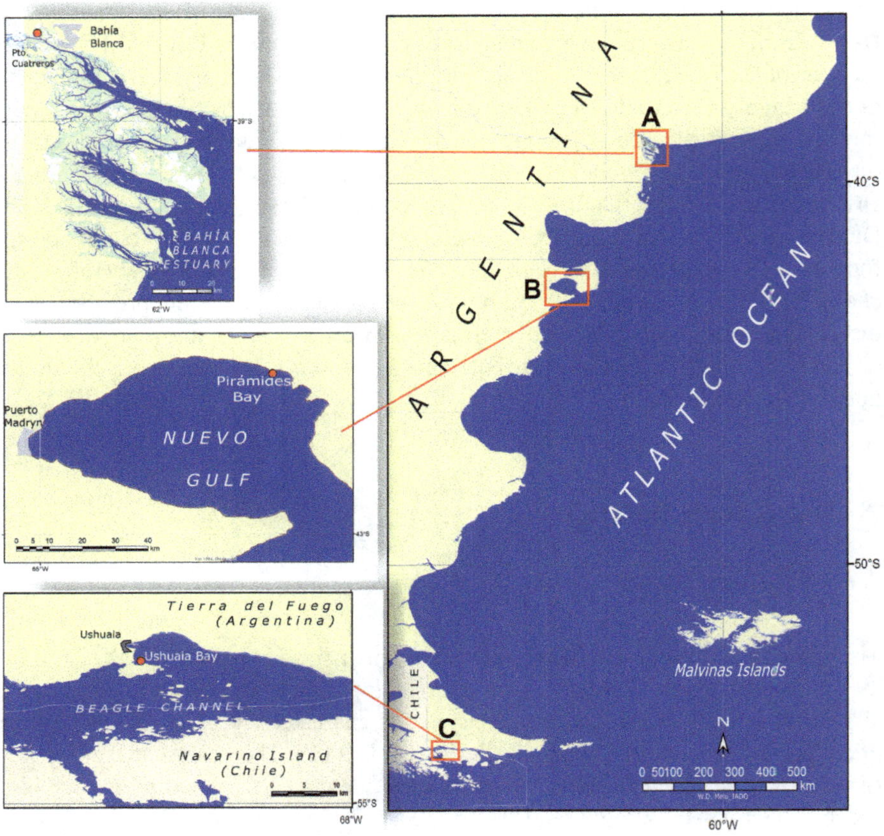

Fig. 1 Map showing the location of the sampling sites in coastal systems studied on the Argentine Coast

et al. 2012; López Abbate et al. 2012; Fernández Severini et al. 2013). For instance, the inner area of the estuary receives raw urban sewage from nearby cities (350,000 inhabitants) and untreated industrial effluents, the latter containing toxic substances such as heavy metals and hydrocarbons (Marcovecchio et al. 2008). It also receives runoff water containing organochlorines coming from its catchment area (approximately 4200 km^2), which consists almost entirely of cultivated land (Marcovecchio et al. 2008; La Colla et al. 2015; Oliva et al. 2015; Spetter et al. 2015). The main navigation channel has been periodically dredged since 1958; increasing maritime traffic has also propitiated the introduction of invasive species, probably through the hulls and ballast water from ships (Hoffmeyer 2004; Hoffmeyer et al. 2009; Dos Santos and Fiori 2010; Fiori et al. 2016).

2.2 Mesozooplankton Structure and Seasonal Dynamics: Seasonal Dynamics of Physical Environment

Mesozooplankton composition and dynamics have been studied in this estuary since more than four decades (Hoffmeyer et al. 2008 and references therein). Here, we analysed data (Hoffmeyer et al. 2010a, b, 2011) obtained during a monitoring programme from 2007 to 2010 for the Bahía Blanca Local Government. Samples used in this analysis were collected at the Cuatreros Port, located in the inner zone of estuary at 3–4-week intervals.

A total of 23 mesozooplankton samples were collected with a 200 μm mesh, 0.30 m open-mouth net containing a mechanical flowmeter. Mesozooplankton sampling was conducted by using horizontal-oblique tows in the surface layer (< 3 m), and samples were preserved in 4% formalin. Each tow was carried out on board a small boat at a speed of 2 knots, for 5 min (15–30 m^3, volume of water filtered by the net), during daytime and ebb tides (only at BBE). The type of net and general methodology used for samples collection was equivalent between the periods compared. In situ temperature, salinity and turbidity were recorded at the surface using a HORIBA multiparameter probe. Surface seawater samples were collected with a Van Dorn bottle to analyse chlorophyll *a*.

The concentration of chlorophyll *a* was estimated by spectrophotometry according to Lorenzen (1967). Mesozooplankton samples were qualitatively analysed under a Wild M5 stereomicroscope; taxon identification was primarily from descriptions of taxa and identification keys in Bradford Grieve et al. (1999), particular chapters in Boltovskoy (1981a, b, 1999) and other taxon-specific literature. Quantitative analyses of samples were carried out by means of sample aliquots in most of cases, following Boltovskoy (1981a, b).

The seasonal dynamics of mesozooplankton during the years 2009 and 2010 was analysed with generalised additive models (GAMs), using the monthly total mesozooplankton abundance records in individuals per cubic metre (TMA, ind m^{-3}) as a function of time (months). The same analyses were applied to evaluate the annual seasonal patterns of environmental variables (SST, salinity, chlorophyll *a* and turbidity).

Additionally, we modelled the response of TMA to seawater surface temperature (SST) and chlorophyll *a* concentration. However, given the known importance of salinity and turbidity in the spatio-temporal dynamics of estuarine mesozooplankton species (e.g. Roman et al. 2001), these variables were also included in the analysis to consider the particular behaviour of these two variables in the estuary. The effects of environmental variables on the TMA were quantified by generalised linear models (GLMs).

The holoplanktonic fraction (nine taxa) was mainly represented by copepod species (*Acartia tonsa*, *Paracalanus parvus*, *Eurytemora americana*, *Euterpina acutifrons*, *Halicyclops* aff. *crassicornis*, *Oithona nana*, *Labidocera fluviatilis* and *Calanoides carinatus*), while *A. tonsa* was the dominant key species in the warm months. Chaetognaths, cnidarians and cladocerans were also important within the

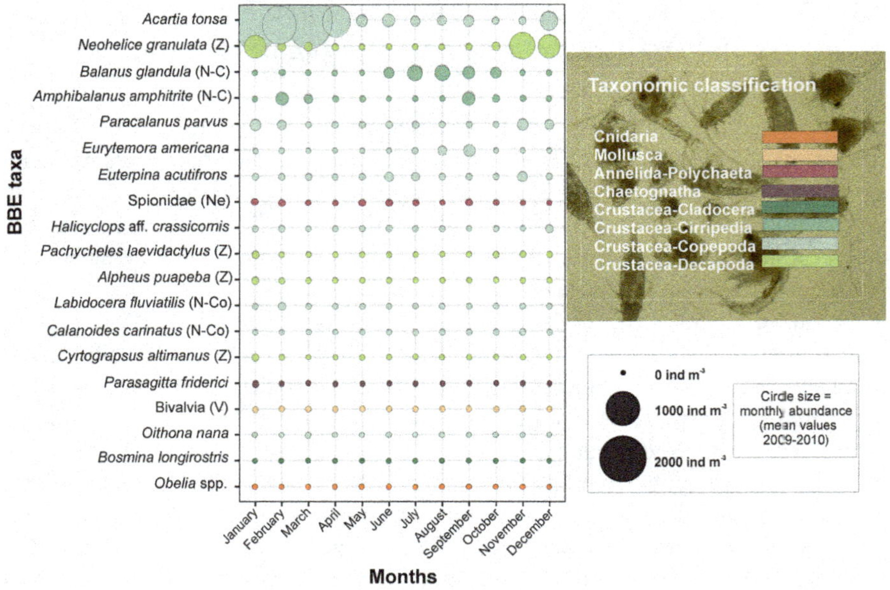

Fig. 2 Annual pattern of mesozooplankton taxa abundance (ind m^{-3}) in the Bahía Blanca Estuary. Abundance of each taxon is represented by circle size in decreasing order. Taxa without abbreviations between brackets include adults and the rest of developmental stages. Abbreviations of names of early metazoan stages are Z, zoea (Decapoda); N and Co, nauplius and copepodite stages (Copepoda); N and C, nauplius and cypris (Cirripedia); Ne, nectochaet (Polychaeta); V, veliger (Bivalvia, Gastropoda). Taxonomic categories are denoted by the colour scale\

holoplankton although each group was represented by only one taxa. Meroplankton was represented by ten taxa from the groups of cirripeds, polychaetes, decapods and molluscs. Within these groups, larvae of *Neohelice granulata*, which is an emblematic crab species of Bahía Blanca Estuary, showed the highest abundance in the warm months and the cirriped *Balanus glandula* larvae dominated the mesozooplankton abundance in the cold months (Fig. 2). TMA ranged between 0 and 2000 ind m^{-3} (Fig. 2); 19 mesozooplankton taxa were found, belonging to 8 taxonomic groups. The analysis revealed a unimodal pattern with an increase in the organism abundance (TMA) towards the warm months, while minimum values occurred during the cold months (Fig. 3).

Mean SST during the period analysed was 15.1 °C (range, 6–25 °C). Mean salinity was 33.1 (range, 23.2–42.9). Mean turbidity was 93.5 NTU (range, 23.4–245.0 NTU). The concentration of chlorophyll *a* ranged between 1.5 and 17.3 μg l^{-1}, while its mean concentration was 6.3 μg l^{-1}. Annual seasonal patterns were not synchronised among these environmental variables (Fig. 3). GLM models revealed significant effects for salinity and turbidity on the TMA in Bahía Blanca Estuary (Table 1). The partial response of mesozooplankton abundance to explanatory variables in the GLM (Fig. 4a, b) showed that TMA increased linearly with increasing salinity and turbidity.

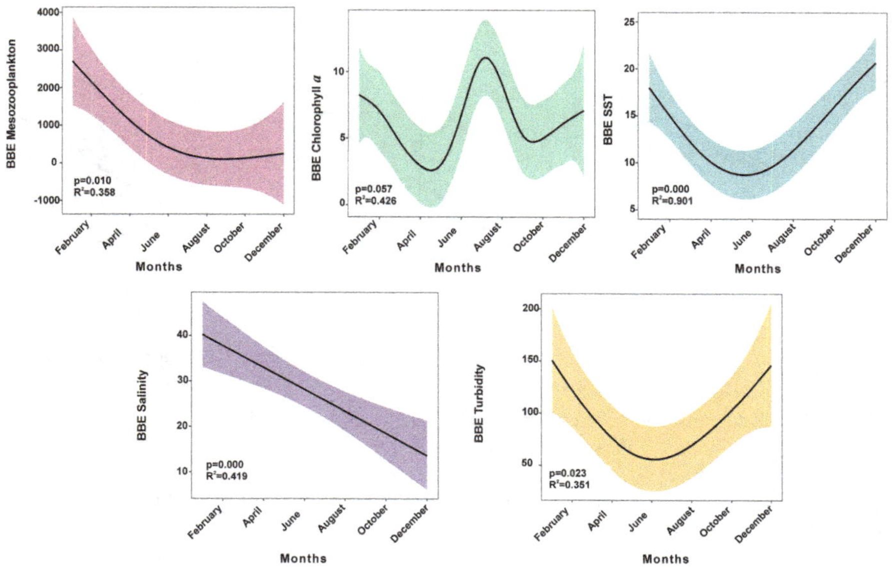

Fig. 3 Plots represent generalised additive models (GAMs) of total mesozooplankton abundance (TMA) and measured environmental variables in Bahía Blanca Estuary (BBE) as a function of time. Significant p-values denote that the temporal variability of the response variable (y) fits the pattern illustrated by the plot, while the adjusted R^2 denotes the proportion of variance explained by the model. Environmental variables were temperature, salinity, turbidity and chlorophyll *a* concentration in BBE. Monthly records of mesozooplankton abundance and environmental variables were fitted with a logit link function using the R-package mgcv

Table 1 Results of the generalised linear model (GLM) showing the standardised effect size (SES) of environmental variables on the abundance of mesozooplankton in Bahía Blanca Estuary (BBE) and in Ushuaia Bay (UB)

Explanatory variable	Site	SES	SE	p-value	Model R^2
Salinity	BBE	0.30	0.12	0.02	0.37
Turbidity	BBE	0.46	0.12	0.00	
Chlorophyll *a*	UB	0.82	0.34	0.03	0.22

The standard error and p-value of each SES and the variance explained by the models in BBE and UB (model R^2) are also shown

The seasonal dynamics of the mesozooplankton community shifted from a typical bimodal curve of abundance (during the 1990–1991 period), with maximum values at the end of winter-spring and late summer (Hoffmeyer 2004), to a unimodal curve with maximal abundances during the warm season during the 2009–2010 period. The bimodal pattern documented in the 1990s and early 2000s mainly was due to the effects of temperature and phytoplankton availability (Sabatini 1989; Gayoso 1999; Hoffmeyer et al. 2008). In recent years, the total mesozooplankton abundance showed a linear relationship with both salinity and turbidity.

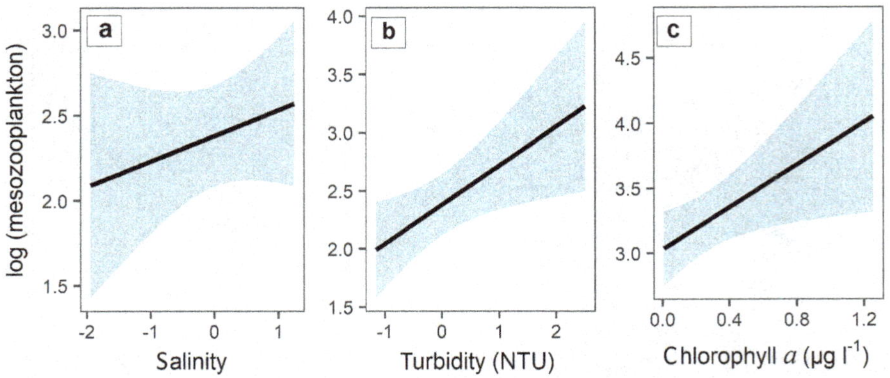

Fig. 4 Partial response of log-transformed mesozooplankton abundance to the environmental variables that produced significant ($p < 0.05$) effects according to GLM. In Bahía Blanca Estuary, the response of mesozooplankton was due to the joint effect of salinity (**a**) and turbidity (**b**), while in Ushuaia Bay, the concentration of chlorophyll *a* was the main variable that explained the response of mesozooplankton abundance (**c**)

The positive relation found between mesozooplankton abundance and salinity suggests that mesozooplankton are well acclimated to high salinity (Fig. 4a). These findings agree with other studies, which highlight the importance of salinity in determining the spatio-temporal distribution of mesozooplankton in estuaries. For instance, Mouny and Dauvin (2002) analysed the spatial variability of mesozooplankton in the Seine Estuary and demonstrated a strong effect of salinity on abundance, composition and distribution of metazoan organisms within the estuary. The most abundant species in our study have also been found in estuaries from the northern hemisphere; in general, their populations thrive at lower salinity values (e.g. Mouny and Dauvin 2002; Lawrence et al. 2004; David et al. 2005). Thus, a presumption is that the Bahía Blanca Estuary populations, probably belonging to a different lineage (Drillet et al. 2008), may have gained tolerance to higher salinity levels over generations. A positive relationship between TMA and turbidity was also found in our study (Fig. 4b). Water turbidity normally increases during the warm season in the inner zone of the Bahía Blanca Estuary because of the complex interaction among tide and wind patterns (Guinder et al. 2013). Nevertheless, the mechanisms underlying the relationship between the abundance of mesozooplankton and turbidity are still not clear.

Previous research on mesozooplankton composition focusing on those species with abundances greater than 0.25 ind m^{-3} (Hoffmeyer 2004) reported approximately 30 species in the 1980s and 22 species in the 1990s; our study (2009–2010) found 19 species, following the same selection criteria. In addition to this negative trend in biodiversity over time, a notable difference was observed in the mesozooplankton composition. The composition of the mesozooplankton community in 1990–1991 (Hoffmeyer 2004) was mainly copepods. The dominant species were, in decreasing order of mean annual abundance, *A. tonsa*, *P. parvus*, *Labidocera fluviatilis*, *C. carinatus*, *E. americana*, *O. nana*, *E. acutifrons* and *Tisbe varians*. The meroplankton was scarce compared to copepod abundance, mainly represented by larvae of Ceriantharia, Grapsidae spp. and *B. glandula*. Low-abundance values (<1 ind m^{-3}) of larval stages of Spionidae spp., Bryozoa and Mollusca were also reported (Hoffmeyer 2004).

Our results on annual mean abundance show that the mesozooplankton community in the years 2009–2010 was composed mainly of *A. tonsa* (695.48 ind m^{-3}), *N. granulata* (81.97 ind m^{-3}), *B. glandula* (28.89 ind m^{-3}), *A. amphitrite* (13.35 ind m^{-3}), *P. parvus* (9.63 ind m^{-3}), *E. americana* (3.91 ind m^{-3}), *E. acutifrons* (3.62 ind m^{-3}) and Spionidae spp. (1.15 ind m^{-3}). These results highlight a restructuring of mesozooplankton community towards the dominance of typical estuarine species. A decrease in mean abundances occurred compared to those recorded in 1990–1991. These species appear to be more adapted to coexist in a polluted and eutrophic environment than organisms inhabiting the surrounding shelf area (Dutto et al. 2012; Biancalana et al. 2012b). Organotin compounds and heavy metals were found in zooplankton in the inner reach of the estuary, which is the area most affected by human activities (Marcovecchio et al. 2008; Fernández Severini et al. 2009, 2011, 2013). Our results showed a decrease in the abundance of some copepod species typical of the adjacent shelf area, such as *L. fluviatilis* and *C. carinatus* (Boltovskoy 1981a, b; Bradford Grieve et al. 1999), presumably because of their low tolerance to anthropogenic pollutants. The small copepod *O. nana* (adults size between 440 and 720 μm, Bradford Grieve et al. 1999) found in the inner zone of BBE is common as well in the shelf waters adjacent to this estuary (Ramirez 1981). The apparent decrease of its abundance in the inner zone of EBB during recent years would be linked to a low efficiency of capture of the 200 μm mesh-size plankton net used in these studies (Di Mauro et al. 2009). This small copepod, usually registered in the innermost areas of European estuaries, has been mentioned as a biological indicator of those perturbed systems (Richard and Jamet 2001; Williams and Muxagata 2006).

Another change observed in the mesozooplankton community was a shift in the distribution of larvae of the decapod species *Pachycheles laevidactylus*, *Alpheus puapeba* and *Cyrtograpsus altimanus*. These species have retreated inward the estuary in the last few decades (Cervellini 2001; Hoffmeyer 2004), possibly benefiting from dredging and the tidal transport of larvae and subsequent retention in the inner zone (Chazarreta et al. 2015). The recent appearance of larval stages of the invasive oyster *Crassostrea gigas* (Chazarreta 2015) occurred in concurrence with this phenomena and led to the settling and permanence of oyster beds in the inner zone of the estuary (Dos Santos and Fiori 2010; Fiori et al. 2016).

3 Pirámide Bay in Nuevo Gulf

3.1 Main Features

Nuevo Gulf is located in the north centre of the Patagonian shelf, south of Península Valdés, Argentina (42° 30′–42° 56' S and 65° 02′–64° 08' W), in a semiarid, temperate climate region (Fig. 1b) (Rivas and Beier 1990). It is connected to the Atlantic shelf by a 16 km-wide mouth and reaches maximal depths of approximately 170 m in its central zone (Rivas and Ripa 1989). Mean surface temperature is 13 °C, with a 20 °C maximum in summer and an 8 °C minimum in winter (Rivas and Beier 1990). It is characterised by high water transparency, saturated oxygen conditions and homogeneous concentrations of nutrients during the annual cycle (Solís 1998). During winter, the water column is fully mixed, whereas vertical stratification occurs during summer (Rivas and Ripa 1989). It receives no freshwater runoff from rivers, resulting in nearly uniform salinity varying from 33.5 to 33.9 (Rivas and Beier 1990). Salinity within the gulf is higher than in the surrounding shelf area during most of the year (Rivas and Ripa 1989). Tidal regimes are semidiurnal, with an amplitude ranging from 3.84 m to 5.55 m (Mazio et al. 2004).

Nuevo Gulf is part of the protected area "Península Valdés", a natural reserve designated as a World Heritage Site by UNESCO because of its importance for marine conservation. This gulf is internationally known as one of the most important calving grounds for the southern right whale (*Eubalaena australis*) as well as an important feeding ground for the dusky dolphin (*Lagenorhynchus obscurus*) and South American tern (*Sterna hirundinacea*) (Degrati et al. 2008; Gatto 2009). Close to the port area, a high percentage of species found are exotic, such as the cirriped *Balanus glandula*, the ascidiacean *Ciona intestinalis* and *Ascidiella aspersa* and the alga *Undaria pinnatifida* (Boltovskoy et al. 2011; Irigoyen et al. 2011; Schwindt et al. 2014) as a consequence of the high maritime traffic. This gulf is impacted by both intensive tourist activities and effluents containing organic matter and heavy metals from the urban settlements in Puerto Madryn (81,995 inhabitants) (Esteves et al. 1996; Ferrando et al. 2010; Gil et al. 2014) and Puerto Pirámides (565 inhabitants). Although the coastal area of Puerto Pirámides does not receive organic and toxic effluents, it is subject to intense ship traffic due to whale-watching practices (Rivarola et al. 2001; Hoffmeyer pers. obs.).

3.2 Mesozooplankton Structure and Seasonal Dynamics: Seasonal Dynamics of the Physical Environment

At present, there is limited knowledge on the mesozooplankton composition and abundance in this gulf (Wilson et al. 2015), with no systematic sampling yet conducted. For this study, we analysed TMA from two adjacent sampling stations in the coastal zone of Pirámide Bay that were sampled during the year 2015 (D'Agostino, unpubl. results). Mesozooplankton sampling and the measurement of chlorophyll *a* and SST were performed following the same methodology applied in the Bahía Blanca Estuary, except for the use of a coarser (335 μm) mesh net. Generalised additive models (GAM) were applied to evaluate the behaviour of total mesozooplankton abundance records (TMA, ind m^{-3}) and their link to the environmental variables. The effects of environmental variables on the TMA were quantified by generalised linear models (GLMs).

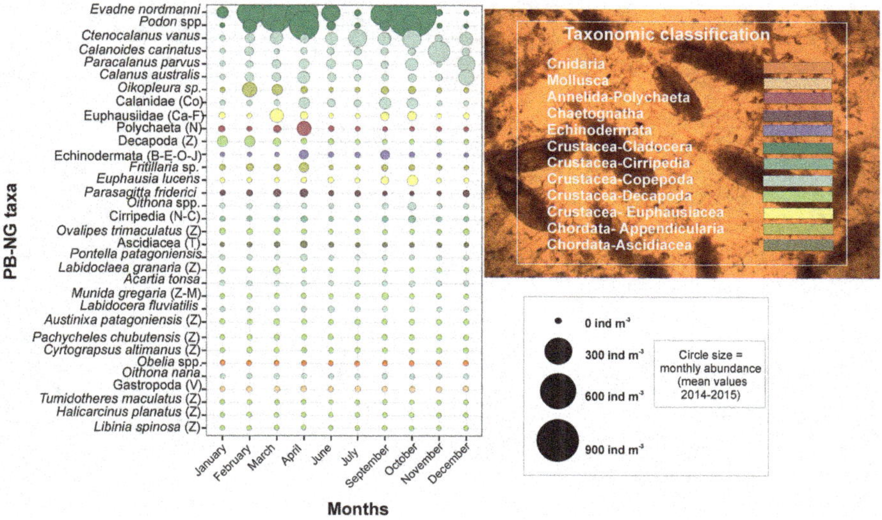

Fig. 5 Annual pattern of mesozooplankton taxa abundance (ind m^{-3}) in Pirámide Bay, Nuevo Gulf. Abundance of each taxon is represented by circle size in decreasing order. Taxa without abbreviations between brackets include adults and the rest of developmental stages. Abbreviations of names of early metazoan stages are as in Fig. 2, plus Ca and F, caliptopes and furcilia (Euphausiidae); B, bipinnaria; E, echinopluteus; O, ophiopluteus, and J, juvenile (Echinodermata); T, tadpole (Ascidiacea); M, megalopa (*M. gregaria*); V, veliger (Gastropoda). Taxonomic categories are denoted by the colour scale

Mesozooplankton was composed of 33 taxa (Fig. 5) belonging to 12 taxonomic groups. Holoplankton was represented by 18 taxa; the most abundant during seasonal peaks were the cladocerans *Evadne nordmanni* and *Podon* spp. and the copepods *Ctenocalanus vanus*, *P. parvus*, *Calanus australis* and *Calanoides carinatus*. Within meroplankton (15 taxa), decapods, polychaetes, bryozoans and ichthyoplankton larvae were the most abundant taxa. Compared with copepods and cladocerans, these latter groups, together with euphausiids and appendicularians, had a lower abundance throughout the year. The seasonal dynamics of mean monthly TMA was bimodal, with two peaks in abundance in late summer and spring and the minimum abundance in winter (Fig. 6a). Regarding environmental patterns, mean SST during the period analysed was 15.7 °C (range, 11–19 °C). The concentration of chlorophyll *a* ranged between 0.4 and 4.4 μg l^{-1} with a mean of 1.0 μg l^{-1}. No significant relationships between TMA, SST and chlorophyll *a* were found.

The seasonal dynamics of mesozooplankton were similar to those previously reported for Pirámide Bay and other areas from Nuevo gulf (Esteves et al. 1996; Hoffmeyer et al. 2010a, b; Menéndez et al. 2011). Seasonal mesozooplankton dynamics for the San José, Nuevo and San Matías gulfs also exhibited a typical bimodal pattern in Wilson et al. (2015), similar to that reported herein. In our analysis, both seasonal peaks were typically dominated by the copepods *P. parvus*, *C.*

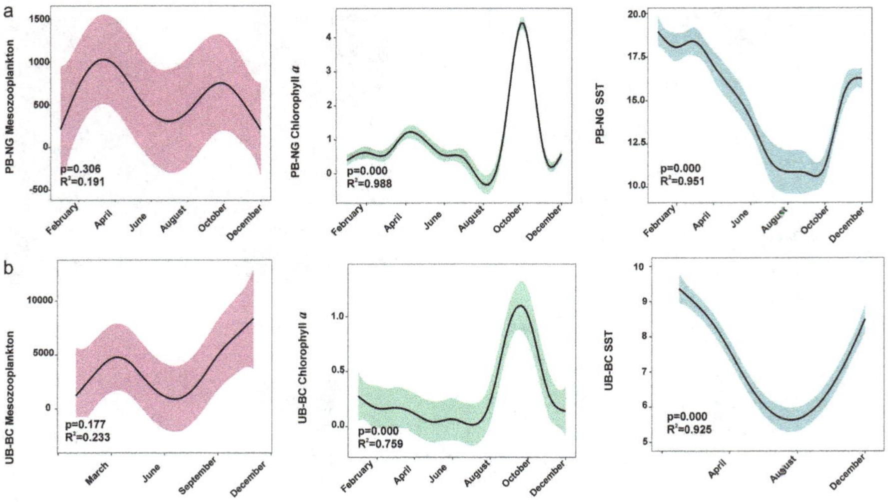

Fig. 6 (**a**) Plots represent generalised additive models (GAM) of total mesozooplankton abundance (TMA) and measured environmental variables as a function of time in Pirámices Bay and (**b**) Ushuaia Bay. Significant p-values denote that the temporal variability of the response variable (y) fits the pattern illustrated by the plot, while the adjusted R^2 denotes the proportion of variance explained by the model. Environmental variables were chlorophyll *a* concentration and temperature in both PB-NG (Pirámide Bay-Nuevo Gulf) and UB-BC (Ushuaia Bay-Beagle Channel). Monthly records of mesozooplankton abundance and environmental variables were fitted with a logit link function using the R-package mgcv

vanus, *C. australis* and *C. carinatus* and the cladocerans *Evadne nordmanni* (D'Agostino 2016; D'Agostino et al. 2016). After statistical analyses of the 2015 data set using GAM, no significant linear relationships between TMA, SST and chlorophyll *a* were found. Instead, we found that the first seasonal peak of mesozooplankton in late summer-autumn co-occurred with the highest annual SST record, while the second seasonal peak in spring was concurrent with.

the annual peak of chlorophyll *a* (Fig. 6a), revealing nonlinear relationships among variables. Overall, the results indicate that this community has not undergone appreciable changes over time during the last few decades, which could be related to the low anthropogenic pressure in this environment.

4 Ushuaia Bay

4.1 Main Features

Ushuaia Bay is located at the southern margin of the Patagonian shelf, in the inner region of the Beagle Channel (54°79'S–68°22'W), in a wet temperate, subpolar climate region (Fig. 1c) (Torres et al. 2009). This cold subantarctic bay is an important zone for the development of commercially important decapod species, such as the king crab *Lithodes santolla* and the stone crab *Paralomis granulosa* (Lovrich et al. 2003; Thatje et al. 2005), as well as ecologically relevant fish species, such as the sprat *Sprattus fuegensis* (Hansen 1999). Ushuaia Bay and the Bridges Islands, which are located in the outer region of the bay, not only exhibit the highest seabird diversity in the channel but also offer shelter habitat for imperial cormorants (*Phalacrocorax atriplex*) and southern giant petrels (*Macronectes giganteus*), which are considered flagship species (Raya Rey and Schiavini 2000). However, the region encompassing Ushuaia Bay and the Beagle Channel has been subject to a great industrial expansion during the last five decades (Amin et al. 1996; Amin et al. 2011; Gil et al. 2011). In Ushuaia City (60,000 residents), the urban and industrial areas have shown fast growth and development in the last few decades. Ushuaia Bay receives raw sewage and industrial effluent discharges as well as hydrocarbon spills associated with maritime traffic and port operations (Esteves et al. 2006; Torres et al. 2009; Gil et al. 2011). The northwestern coast of Ushuaia Bay, close to the urban and industrial centres, is the most affected by cultural eutrophication (Torres et al. 2009; Amin et al. 2011; Gil et al. 2011; Diodato 2013) and industrial toxic inputs containing heavy metals (Amin et al. 1996) and hydrocarbons (Esteves et al. 2006; Commendatore et al. 2012).

Net water circulation in the bay is counterclockwise; current velocities vary from 5.5 to 16.3 cm s^{-1} (Balestrini et al. 1998). SST ranges between 3 and 12.8 °C (Isla et al. 1999). The bay is characterised by high freshwater input coming mostly from the Olivia River and Grande Stream, providing 6.8 and 2.6 m^3 s^{-1}, respectively (Gil et al. 2011). Salinity ranged between 32 in winter and 25 in spring-summer. Minimum values are related to increases in the rate of ice and snow melting (Isla et al. 1999). Seawater is always over-oxygenated because of the mixing effect of local winds (Amin et al. 2011).

4.2 Mesozooplankton Structure and Seasonal Dynamics: Seasonal Dynamics of the Physical Environment

The data used in the present study was derived from samples obtained in Ushuaia Bay during the years 2006 and 2007, sampled monthly. Sampling was performed under the framework of Subproject B-CB-05 (GEF/BIRF N° 28.385). Mesozooplankton sampling followed the same methodology as in Bahía Blanca Estuary, explained above. Mesozooplankton abundance and environmental variables were also analysed for relationships using GAMs and GLMs.

Thirty-five mesozooplankton taxa were recorded (Fig. 7), belonging to 13 taxonomic groups.

The holoplankton fraction of the mesozooplankton community was mainly characterised by the presence of the copepods *Drepanopus forcipatus*, *Oithona similis* sensu *lato* (i.e. identification by optical microscopy, according to Mazzochi et al. 1995 description), *Ctenocalanus citer* and *Clausocalanus brevipes* as well as

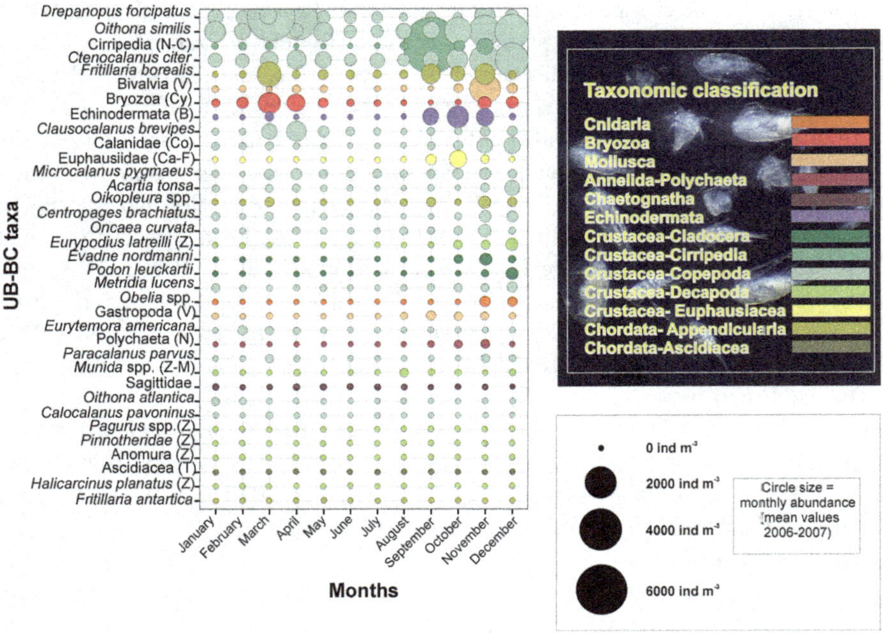

Fig. 7 Annual pattern of mesozooplankton taxa abundance (ind m^{-3}) in Ushuaia Bay, Beagle Channel. Abundance of each taxon is represented by circle size in decreasing order. Taxa without abbreviations between brackets include adults and the rest of developmental stages. Abbreviations of names of early metazoan stages are as in Fig. 2, plus Cy, cyphonautes (Bryozoa). Taxonomic categories are denoted by the colour scale

the appendicularian *Fritillaria borealis*, with maximum abundance during late summer and spring. The meroplankton was dominated by larval stages of cirripedes, bryozoans, bivalves and echinoderms in those same periods of the year.

The seasonal dynamics of TMA during the period analysed (2006–2007) had bimodal abundance peaks, in summer and early autumn; the minimum abundance occurred during winter (Fig. 6b). The increase in TMA paralleled an increase in the chlorophyll *a* concentration (Fig. 6b).

Mean SST during the years 2006–2007 was 7.2 °C (range, 4.6–9.9 °C). The concentration of chlorophyll *a* ranged between 0.0 and 1.2 μg l^{-1}, while its mean concentration was 0.3 μg l^{-1}. Our study used GLM analysis to test the linkage between TMA and environmental variables. This revealed a significant linear relationship between total mesozooplankton abundance and chlorophyll *a*, while no significant relationships with SST were found (Fig. 4c). High abundances of meroplanktonic larvae were observed during late winter-spring and late summer, concurrent with high chlorophyll *a* concentration, which demonstrated the existence of strong benthopelagic coupling. Fast-growing zooplankters, such as the appendicularian species *F. borealis* and *Oikopleura* spp., also reached maximum abundances during these periods. The availability of phytoplankton in this bay also triggers more pronounced changes in the mesozooplankton dynamics than those produced by physical factors such as SST (Presta et al. unpubl. results). These findings emphasise the relevance of phytoplankton in shaping the functional response of mesozooplankton in Ushuaia Bay. This suggests vulnerability of mesozooplankton that are subject to future coastal eutrophication, when phytoplankton dynamics may be modified.

The first studies on mesozooplankton in Ushuaia Bay were performed in 2001–2004 (Fernández Severini and Hoffmeyer 2005; Biancalana et al. 2012a; Biancalana et al. 2014). These investigations revealed differences in abundance and composition between the inner and outer zones of the bay, which were related with the strong influence of human activities on the coastal area. Anthropogenic impact in the coastal area is mainly perceived as the runoff of nutrient loads from land (Biancalana et al. 2014). Mesozooplankton in this area was characterised by the dominance of *Eurytemora americana*, *Acartia tonsa* and the cladoceran *Podon leuckarti*, while the outer bay was represented by copepods, such as *D. forcipatus*, *C. citer* and *C. brevipes* (Biancalana et al. 2012a). Our study, conducted in outer Ushuaia Bay, partially agrees with previous reports in that mesozooplankton community was characterised by the dominance of Clausocalanidae and Oithonidae copepods. Although highest abundances of *D. forcipatus*, *C. citer*, *C. brevipes* and *O. similis s.l.* were observed in later summer and spring, they were also well represented during winter (chlorophyll $a < 0.1$ μg l^{-1}). This could indicate the use of alternative food resources (Presta et al. unpubl. results). Opportunistic trophic behaviour is well known in this small cyclopoid (Turner 2004), *D. forcipatus* populations from adjacent Patagonian shelf waters (Antacli et al. 2014) and antarctic populations of *C. citer* (Niehoff et al. 2002).

5 Concluding Remarks and Future Prospects

The coastal systems included in this study are among the most important of Argentina in terms of biological productivity (Biancalana et al. 2012a; Hoffmeyer et al. 2008; Menendez et al. 2011). However, shoreline human populations, industrial and port development and exploitation of coastal fish resources are fostering pronounced changes in their functioning (Esteves et al. 2006; Marcovecchio et al. 2008; Ferrando et al. 2010; Amin et al. 2011; Gil et al. 2011; Gil et al. 2014; Torres et al. 2009; Biancalana et al. 2014; López Abbate et al. 2012; Dutto et al. 2012, 2014; Fernandez Severini et al. 2011). Although the mesozooplankton data presented here are limited and do not cover an adequate period of time for retrospective analyses, our recent data analysis revealed a strong dependence of the mesozooplankton composition and abundance on the seasonal cycle and on the degree of habitat deterioration. In the Bahía Blanca Estuary, habitat deterioration was perceived as modifications in mesozooplankton assemblages and phenology with occurrence of true estuarine and invasive species but lower in numbers and diversity in the last few decades. Ushuaia Bay is suffering from chronic organic pollution due to persistent nutrient loading, which visibly appears to affect the plankton communities through changes in composition and dominance patterns. Finally, no significant changes were observed in mesozooplankton composition and seasonal succession in Pirámide Bay, in Nuevo Gulf, which is exposed to slight anthropogenic disturbance. Multiple environmental drivers acting at different scales modulate the structure and function of populations and move ecosystems towards alternative regimes of variability (Scheffer and Carpenter 2003; Mieszkowska et al. 2014). Both the prevention of habitat degradation and potential restoration of these systems depend on their resilience capacity, in turn dependent on local biodiversity. Within this context, it is mandatory to outline conservation priorities that take into account both the present and future rate of environmental change (Mori et al. 2013).

Acknowledgements This research was funded by a grant (PICT 2014-3091, ANPCYT Argentina) to M.S.H. and a grant (PIP0630, CONICET) to A.A.B. We thank CONICET (National Council of Scientific and Technical Research) and ANPCYT (National Agency for Scientific and Technical Promotion), the Ministry for Science, Technology and Productive Innovation (Argentina) fellowships for some co-authors of the chapter. We would also like to thank to staff of the Austral Centre of Scientific Research (CADIC)-CONICET in Ushuaia, the Centre of Marine Systems (CESIMAR)-CONICET in Puerto Madryn and the Argentine Institute of Oceanography (IADO)-CONICET in Bahía Blanca, for their help during sampling and nutrient determination. Special thanks are due to W. Melo from IADO for his help in drawing figures.

References

Amin OA, Ferrer L, Marcovecchio J (1996) Heavy metal concentrations in littorals sediments from the Beagle Channel, Tierra del Fuego, Argentina. Environ Monit Assess 41:219–231

Amin O, Comoglio L, Spetter C et al (2011) Assessment of land influence on a high-latitude marine coastal system: Tierra del Fuego, southernmost Argentina. Environ Monit Assess 175:63–73

Antacli JC, Sabatini ME, Silva RI et al (2014) Feeding and reproductive activity of the copepods *Drepanopus forcipatus* and *Calanus australis* during late summer on the southern Patagonian shelf (Argentina, 47°-55° S). Braz J Oceanogr 62:295–314

Arias AH, Vazquez-Botello A, Tombesi N et al (2010) Presence, distribution, and origins of polycyclic aromatic hydrocarbons (PAHs) in sediments from Bahía Blanca estuary, Argentina. Environ Monit Assess 160:301–314

Balestrini C, Manzella G, Lovrich GA (1998) Simulación de corrientes en el Canal Beagle y Bahía Ushuaia, mediante un modelo bidimensional. Servicio de Hidrografía Naval, Departamento de Oceanografía, Technical Report 98:1–58

Berasategui AA, Biancalana F, Fricke A et al (2017) The impact of sewage effluents on the fecundity and survival of *Eurytemora americana* in a eutrophic estuary of Argentina. Estuar Coast Shel Sci. https://doi.org/10.1016/j.ecss.2017.08.034

Biancalana F, Diodato SL, Hoffmeyer MS (2012a) Seasonal and spatial variation of mesozooplankton biomass in Ushuaia and Golondrina bays (Beagle Channel, Argentina). Braz J Oceanogr 60:99–106

Biancalana F, Menéndez MC, Berasategui AA et al (2012b) Sewage pollution effects on mesozooplankton structure in a shallow temperate estuary. Environ Monit Assess 184:3901–3913

Biancalana F, Dutto MS, Berasategui AA et al (2014) Mesozooplankton assemblages and their relationship with environmental variables: a study case in a disturbed bay (Beagle Channel, Argentina). Environ Monit Assess 186:8629–8647

Boltovskoy D (1981a) Atlas del zooplancton del Atlántico Sudoccidental y métodos de trabajo con el zooplancton marino. Publicación especial del INIDEP, Mar del Plata, Argentina

Boltovskoy D (1981b) Recuentos y análisis de los datos. Manipulación del material. In: Boltovskoy D (ed) Atlas del zooplancton del Atlántico Sudoccidental y métodos de trabajo con el zooplancton marino. Publicación especial del INIDEP, Mar del Plata, pp 7–107

Boltovskoy D (ed) (1999) South Atlantic zooplankton, vol 1–2. Backhuys Publishers, Leiden

Boltovskoy D, Almada P, Correa N (2011) Biological invasions: assessment of threat from ballast-water discharge in Patagonian (Argentina) ports. Environ Sci Pol 14:578–583

Bradford Grieve JM, Markoseva EL, Rocha CEF et al (1999) Copepoda. In: Boltovskoy D (ed) South Atlantic zooplankton. Backhuys Publishers, Leiden, pp 869–1098

Capitanio F, Pájaro M, Esnal GB (2005) Appendicularians: an important food supply for the Argentine anchovy *Engraulis anchoita* in coastal waters. J Appl Ichthyol 21:414–419

Cervellini P (2001) Variabilidad en la abundancia y retención de larvas de crustáceos decápodos en el estuario de Bahía Blanca, Provincia de Buenos Aires, Argentina. Investig Mar 29(2):25–33

Chang KH, Doi H, Nishibe Y et al (2009) Spatial and temporal distribution of zooplankton communities of coastal marine waters receiving different human activities (fish and pearl oyster farmings). The Open Mar Biol J. https://doi.org/10.2174/1874450809030010083

Chazarreta CJ (2015) Dinámica del transporte de zooplancton en el estuario de Bahía Blanca. PhD dissertation, Universidad Nacional del Sur. Bahía Blanca. Argentina

Chazarreta CJ, Hoffmeyer MS, Cuadrado DG (2015) Tidal effects on short-term mesozooplankton distribution in small channels of a temperate-turbid estuary, Southwestern Atlantic. Braz J Oceanogr 63(2):83–92

Commendatore MG, Nievas ML, Amin O et al (2012) Sources and distribution of aliphatic and polyaromatic hydrocarbons in coastal sediments from the Ushuaia Bay (Tierra del Fuego, Patagonia, Argentina). Mar Environ Res 74:20–31

Cuadrado DG, Ginsberg SS, Gomez EA (2007) Geomorfología. In: Piccolo MC, Hoffmeyer MS (eds) Ecosistema del Estuario de Bahía Blanca, 3rd edn. EdiUNS, Bahía Blanca, pp 29–38

D'Agostino VC (2016) Dinámica planctónica y transferencia de biotoxinas marinas a la Ballena Franca Austral (*Eubalaena australis*) a través de su alimentación en los golfos Nuevo y San José, Península Valdés, Chubut. PhD dissertation, Universidad Nacional del Sur, Bahía Blanca

D'Agostino VC, Hoffmeyer MS, Degrati M (2016) Faecal analysis of southern right whales (*Eubalaena australis*) in Península Valdés calving ground, Argentina: *Calanus australis*, a key prey species. J Mar Biol Assoc UK 96(4):859–868

David V, Sautour B, Chardy P et al (2005) Long-term changes of the zooplankton variability in a turbid environment: the Gironde estuary (France). Estuar Coast Shelf Sci 64(2–3):171–184

Degrati M, Dans SL, Pedraza SN et al (2008) Diurnal behaviour of dusky dolphins, *Lagenorhynchus obscurus*, in Golfo Nuevo. J Mammal 89(5):1241–1247

Di Mauro R, Capitanio F, Viñas MD (2009) Capture efficiency for small dominant mesozooplankters (Copepoda, Appendicularia) off Buenos Aires Province (348S–418S), Argentine Sea, using two plankton mesh sizes. Braz J Oceanogr 57:205–214

Diodato S (2013) Respuesta de especies locales a procesos de eutrofización en la zona costera próxima a la ciudad de Ushuaia, Tierra del Fuego. PhD dissertation, Universidad Nacional del Sur, Bahía Blanca

Diodato S, Hoffmeyer MS (2008) Contribution of planktonic and detritic fractions to the natural diet of mesozooplankton in Bahía Blanca Estuary. Hydrobiologia 614:86–90

Dos Santos EP, Fiori SM (2010) Primer registro sobre la presencia de *Crassostrea gigas* en el estuario de Bahía Blanca (Argentina). Com Soc Malacol Urug 9:245–252

Drillet G, Goetze E, Jepsen PM et al (2008) Strain-specific vital rates in four *Acartia tonsa* cultures I: strain origin, genetic differentiation and egg survivorship. Aquaculture 280:109–116

Dutto MS, López Abbate MC, Biancalana F et al (2012) The impact of sewage on environmental quality and the mesozooplankton community in a highly eutrophic estuary in Argentina. ICES J Mar Sci 69(3):399–409

Dutto MS, Kopprio GA, Hoffmeyer MS et al (2014) Planktonic trophic interactions in a human impacted estuary of Argentina: a fatty acid marker approach. J Plankton Res. https://doi.org/10.1093/plankt/fbu012

Esteves JL, Solís M, Gil M et al (1996) Dos Bahías Patagónicas: Bahía Nueva-Bahía Engaño. In: Abstracts of the international conference on pollution processes in coastal environments, Mar del Plata December 1996

Esteves JL, Commendatore MG, Nievas ML et al (2006) Hydrocarbon pollution in coastal sediments of Tierra del Fuego Islands, Patagonia Argentina. Mar Pollut Bull 52:582–590

Fernandez Severini MD, Hoffmeyer MS (2005) Mesozooplankton assemblages in two bays in the Beagle Channel (Argentina) during January 2001. Sci Mar 69:27–37

Fernández Severini MD, Botté SE, Hoffmeyer MS et al (2009) Spatial and temporal distribution of cadmium and copper in water and zooplankton in the Bahia Blanca estuary, Argentina. Estuar Coast Shelf Sci 85:57–66

Fernández Severini MD, Botté ES, Hoffmeyer MS et al (2011) Lead concentrations in zooplankton, water, and particulate matter of a southwestern Atlantic temperate estuary (Argentina). Arch Environ Contam Toxicol 61(2):243–260

Fernández Severini MF, Hoffmeyer MS, Marcovecchio J et al (2013) Heavy metals concentrations in zooplankton and suspended particulate matter in a southwestern Atlantic temperate estuary (Argentina). Environ Monit Assess 185(2):1495–1513

Ferrando A, Esteves JL, Elías R et al (2010) Intertidal macrozoobenthos in sandy beaches of Bahía Nueva (Patagonia, Argentina) and their use as bioindicators of environmental impact. Sci Mar 74(2):345–352

Fidalgo LG (2007) Mamíferos Marinos. In: Piccolo MC, Hoffmeyer MS (eds) Ecosistema del Estuario de Bahía Blanca, 3rd edn. EdiUNS, Bahía Blanca, pp 221–222

Fiori SM, Pratolongo P, Zalba SM et al (2016) Spatially explicit risk assessment for coastal invaders under different management scenarios. Mar Biol 163:245. https://doi.org/10.1007/s00227-016-3017-5

Freije RH, Spetter CV, Marcovecchio JE et al (2008) Water chemistry and nutrients of the Bahía Blanca estuary. In: Neves R, Baretta J, Mateus M (eds) Perspectives on integrated coastal zone Management in South America. IST Press, Lisboa, pp 241–254

Gadzała-Kopciuch R, Berecka B, Bartoszewicz J et al (2004) Some considerations about Bioindicators in environmental monitoring. Pol J Environ Stud 13(5):453–462

Garcia-Borboroglu P, Yorio P (2007) Breeding habitat requirements and selection by the threatened Olrog's Gull *Larus atlanticus*. Auk 124:1201–1212

Gatto AJ (2009) Ecología trófica de tres especies simpátricas de gaviotines en la costa norte de Patagonia. PhD dissertation, Universidad Nacional de Buenos Aires, Buenos Aires

Gayoso AM (1999) Seasonal succession patterns of phytoplankton in the Bahía Blanca Estuary (Argentina). Bot Mar 42:367–375

Gil MN, Torres AI, Amin O et al (2011) Assessment of recent sediment influence in an urban polluted subantarctic coastal ecosystem. Beagle Channel (Southern Argentina). Mar Pollut Bull 62:201–207

Gil MN, Torres AI, Marinho CH et al (2014) Caracterización de Sedimentos Costeros y Flujos Bentónicos en una Bahía Patagónica de Argentina: Antes y Después de la Eliminación del Efluente Urbano. In: Marcovecchio J, Botté SE, Freije RH (eds) Procesos Geoquímicos de la Superficie en América Latina. SIFyQA, Bahía Blanca, pp 22–37

Grecco, L.E., Gómez, E.A., Botté et al (2011) Natural and anthropogenic heavy metals in estuarine cohesive sediments: geochemistry and bioavailability. Ocean Dyn 61, 285–293

Guinder VA, Popovich CA, Perillo GME (2009) Particulate suspended matter concentrations in the Bahía Blanca Estuary, Argentina: implication for the development of phytoplankton. Estuar Coast Shelf Sci 85:157–165

Guinder VA, Popovich CA, Molinero JC et al (2013) Phytoplankton summer bloom dynamics in the Bahía Blanca Estuary in relation to changing environmental conditions. Cont Shelf Res 52:150–158

Hansen JE (1999) Estimación de parámetros poblacionales del efectivo de sardina fueguina (*Sprattus fuegensis*) de la costa continental Argentina. INIDEP Informe Técnico:27–28
Hoffmeyer MS (2004) Decadal change in zooplankton seasonal succession in the Bahía Blanca estuary, Argentina, following introduction of two zooplankton species. J Plankton Res 26:181–189
Hoffmeyer MS, Cervellini PM (2007) Zooplancton temporario. In: Piccolo MC, Hoffmeyer MS (eds) Ecosistema del Estuario de Bahía Blanca, 3rd edn. EdiUNS, Bahía Blanca, pp 153–161
Hoffmeyer MS, Fernandez-Severini MD, Menendez MC et al (2008) Composition and dynamics of mesozooplankton assemblages in the Bahía Blanca estuary. In: Neves R, Baretta J, Mateus M (eds) Perspectives on integrated coastal zone Management in South America. IST Press, Lisboa, pp 271–276
Hoffmeyer MS, Berasategui AA, Beigt D et al (2009) Environmental regulation of the estuarine copepods *Acartia tonsa* and *Eurytemora americana* during coexistence period. J Mar Biol Assoc UK 89:355–361
Hoffmeyer MS, Lindner MS, Carribero A et al (2010a) Planktonic food and foraging of *Eubalaena australis*, on Península Valdés (Argentina) nursery ground. Rev Biol Mar Oceanogr 45:131–139
Hoffmeyer MS, Barría MS, Pettigrosso RE et al (2010b) Informe año 2009. Programa de monitoreo de la calidad ambiental de la zona interior del estuario de Bahía Blanca. Anexo III Estudio del fitoplancton, micro- y mesozooplancton. Convenio MBB-UNS, 51 p
Hoffmeyer MS, Barría MS, Pettigrosso RE et al (2011). Informe año 2010. Programa de monitoreo de la calidad ambiental de la zona interior del estuario de Bahía Blanca. Anexo III Estudio del fitoplancton, micro- y mesozooplancton. Convenio MBB-UNS, 50 p
Irigoyen AJ, Eyras C, Parma A (2011) Alien algae *Undaria pinnatifida* causes habitat loss for rocky reef fishes in North Patagonia. Biol Invasions 13:17–24
Isla F, Bujalesky G, Coronato A (1999) Procesos estuarinos en el Canal Beagle, Tierra del Fuego. Rev Asoc Geol Argent 54:307–318
La Colla NS, Negrin VL, Marcovecchio JE et al (2015) Dissolved and particulate metals dynamics in a human impacted estuary from the SW Atlantic. Estuar Coast Shelf Sci 166:45–55
Lawrence D, Valiela I, Tomasky G (2004) Estuarine calanoid copepod abundance in relation to season, salinity, and land-derived nitrogen loading, Waquoit Bay, MA. Estuar Coast Shelf Sci 61:547–557
López Abbate MC, Molinero JC, Guinder VA et al (2012) Microplankton dynamics under heavily anthropogenic pressure. The case of the Bahía Blanca estuary, southwestern Atlantic Ocean. Mar Pollut Bull 95:305–314
Lopez-Cazorla AC (2007) Peces. In: Piccolo MC, Hoffmeyer MS (eds) Ecosistema del Estuario de Bahía Blanca, 3rd edn. EdiUNS, Bahía Blanca, pp 191–201
Lopez-Cazorla AC, Pettigrosso RE, Tejera L et al (2011) Diet and food selection by *Ramnogaster arcuata* (Osteichthyes, Clupeidae). J Fish Biol 78(7):2052–2066
Lorenzen CJ (1967) Vertical distribution of chlorophyll and phaeopigments: Baja California. Deep Sea Res 14:735–745
Lovrich G, Thatje S, Calcagno JA et al (2003) Changes in biomass and chemical composition during lecithotrophic larval development of the southern king crab *Lithodes santolla* (Molina). J Exp Mar Biol Ecol 288:65–79
Marcovecchio JE, Botté SE, Delucchi F et al (2008) Pollution processes in Bahía Blanca estuarine environment. In: Neves R, Baretta J, Mateus M (eds) Perspectives on integrated coastal zone Management in South America. IST Press, Lisboa, pp 301–314
Mazio CA, Dragani WC, Caviglia FJ et al (2004) Tidal hydrodynamics in Golfo Nuevo, Argentina, and the adjacent continental shelf. J Coast Res 20(4):1000–1011
Mazzochi MG, Zagami G, Ianora A et al (1995) Copepods. In: Guglielmo L, Ianora A (eds) Atlas of Marine Zooplankton Straits of Magellan. Springer Verlag, Berlin, pp 219–222
Menéndez MC, Biancalana F, Berasategui AA et al (2011) Mesozooplankton composition and spatial distribution, Nuevo Gulf, Patagonia, Argentina. Check List 7:101–107

Mieszkowska N, Sugden H, Firth LB et al (2014) The role of sustained observations in tracking impacts of environmental change on marine biodiversity and ecosystems. Phil Trans R Soc A 372(2025):1–13

Mori AS, Furukawa T, Sasaki T (2013) Response diversity determines the resilience of ecosystems to environmental change. Biol Rev 88:349–364

Mouny P, Dauvin JC (2002) Environmental control of mesozooplankton community structure in the Seine estuary (English Channel). Influence des paramètres environnementaux sur la répartition spatio-temporelle du mésozooplancton de l'estuaire de la Seine. Oceanol Acta 25:13–22

Niehoff B, Schnack-Schiel S, Cornils A et al (2002) Reproductive activity of two dominant Antarctic copepod species, *Metridia gerlachei* and *Ctenocalanus citer*, in late autumn in the eastern Bellingshausen Sea. Polar Biol 25:583–590

Oliva AL, Quintas PY, La Colla NS et al (2015) Distribution, sources, and potential ecotoxicological risk of polycyclic aromatic hydrocarbons in surface sediments from Bahía Blanca estuary, Argentina. Arch Environ Contam Toxicol 69(2):163–172

Perillo GME, Piccolo MC, Palma ED et al (2007) Oceanografía física. In: Piccolo MC, Hoffmeyer MS (eds) Ecosistema del Estuario de Bahía Blanca, 3rd edn. EdiUNS, Bahía Blanca, pp 61–67

Petracci P, Sotelo MR, Díaz LI (2008) A new colony of Olrog's Gull (*Larus atlanticus*) in the Bahía Blanca, Bahía Falsa and Bahía Verde Natural Reserve, Buenos Aires, Argentina. Hornero 23:37–40

Ramírez FC (1981) Zooplancton y producción secundaria. Parte I: distribución y variación estacional de los copépodos. Campañas de investigación pesquera realizadas en el Mar Argentino por los B/I "Shinkai Maru"y "Walter Herwig" y el B/P "Marburg", años 1978 y 1979. Resultados de la parte argentina. Contrib. INIDEP 383:202–212

Raya Rey A, Schiavini ACM (2000) Distribution, abundance and associations of seabirds in the Beagle Channel, Tierra del Fuego, Argentina. Polar Biol 23(5):338–345

Richard S, Jamet JL (2001) An unusual distribution of *Oithona nana* Giesbrecht (1892) (Crustacea: Cyclopoida) in a Bay: The Case of Toulon Bay (France, Mediterranean Sea). J Coast Res 17(4):957–963

Rivarola M, Campagna C, Tagliorette A (2001) Demand-driven comercial whale-watching in Peninsula Valdés (Patagonia): conservation implications for right whales. J Cet Res Manage 2 (Special Issue): 145–151

Rivas AL, Beier EJ (1990) Temperature and salinity fields in the north Patagonian gulfs. Oceanol Acta 13(1):15–20

Rivas A, Ripa P (1989) Variación estacional de la estructura termohalina de Golfo Nuevo, Argentina. Geofis Int 28:3–24

Roman MR, Holliday DV, Sanford LP (2001) Temporal and spatial patterns of zooplankton in the Chesapeake Bay turbidity maximum. Mar Ecol Prog Ser 213:215–227

Sabatini ME (1989) Ciclo anual del copépodo *Acartia tonsa* Dana, 1849 en la zona interna de la Bahía Blanca (Pcia. de Buenos Aires, Argentina). Sci Mar 53:847–856

Sato NE, Hernández D, Viñas MD (2011) Feeding habits of *Engraulis anchoita* (Hubbs & Marini, 1935) larvae in coastal waters off Buenos Aires Province, Argentina. Lat Am J Aquat Res 39:6–24

Scheffer M, Carpenter SR (2003) Catastrophic regime shifts in ecosystems: linking theory to observation. Trends Ecol Evol 18:648–656

Schwindt E, Gappa JL, Raffo MP et al (2014) Marine fouling invasions in ports of Patagonia (Argentina) with implications for legislation and monitoring programs. Mar Environ Res 99:60–68

Silva AP, Neumann-Leitão S, Schwamborn R et al (2004) Mesozooplankton of an impacted bay in North Eastern Brazil. Braz Arch Biol Techol 47(3):485–493

Simonetti P, Fiori SM, Botté SE et al (2013) Nidificación del ostrero común (*Haematopus palliatus*) en el estuario de Bahía Blanca, Buenos Aires, Argentina. Hornero 28(2):51–58

Solís M (1998) Monitoring in Nuevo Gulf (Argentina): Analysis of oceanographic data by Geographic Information Systems (GIS). Dissertation, International Institute for Infrastructural, Hydraulic and Environmental Engineering

Sommer U, Stibor H, Katechakis A et al (2002) Pelagic food web configurations at different levels of nutrient richness and their implications for the ratio fish production: primary production. Hydrobiologia 484:11–20

Spetter CV, Popovich CA, Arias A et al (2015) Role of nutrients in phytoplankton development during a winter diatom bloom in a eutrophic south American estuary (Bahía Blanca, Argentina). J Coast Res 31:76–87

Thatje S, Anger K, Calcagno JA et al (2005) Challenging the cold: crabs reconquer the Antarctic. Ecology 86:619–625

Torres AI, Gil MN, Amin OA et al (2009) Environmental characterization of a eutrophicated semi-enclosed system: nutrient budget (Encerrada Bay, Tierra del Fuego Island, Patagonia, Argentina). Water Air Soil Pollut 204:259–270

Turner JT (2004) The importance of small planktonic copepods and their roles in pelagic marine food webs. Zool Stud 43:255–266

Uriarte I, Villate F (2004) Effects of pollution on zooplankton abundance and distribution in two estuaries of the Basque coast (Bay of Biscay). Mar Pollut Bull 49:220–228

Williams JA, Muxagata E (2006) The seasonal abundance and production of *Oithona nana* (Copepoda:Cyclopoida) in Southampton Water. J Plankton Res 28(II):1055–1065

Wilson C, Sastre AV, Hoffmeyer MS et al (2015) Southern right whale (*Eubalaena australis*) calf mortality at Península Valdés, Argentina: are harmful algal blooms to blame? Mar Mamm Sci 32(2):423–451

Trophic Ecology of the White Croaker (*Micropogonias furnieri* Desmarest, 1823) and Rough Scad (*Trachurus lathami* Nichols, 1920) Larvae in the Río de la Plata Estuary

Laura Rodríguez-Graña, Mario Vera, Guillermo Cervetto, and Danilo L. Calliari

Abstract This study describes the feeding ecology of white croaker (*Micropogonias furnieri*) and rough scad (*Trachurus lathami*) larvae in the Río de la Plata Estuary (RPE) during a breeding season. The working hypotheses were that the coexistence of the species under study is favored by spatial segregation and divergence in mouth gape size, which mirrors their diet composition and trophic niche breadth (TNB_s), based on prey size. *Micropogonias furnieri* was more abundant at the innermost stations, while *T. lathami* was most abundant at outermost stations of the estuary. However, their abundances were uncorrelated to environmental salinity and temperature. *Micropogonias furnieri* and *T. lathami* larvae showed differences in their gape growth patterns and in the type of preys ingested, which resulted in low dietary

---

Electronic supplementary material: The online version of this chapter (https://doi.org/10.1007/978-3-319-77869-3_17) contains supplementary material, which is available to authorized users.

L. Rodríguez-Graña (✉)
Ecología Funcional de Sistemas Acuáticos, Centro Universitario Regional del Este, Universidad de la República, Rocha, Uruguay
e-mail: laurod@fcien.edu.uy

M. Vera
Oceanografía y Ecología Marina, Facultad de Ciencias, Universidad de la República, Montevideo, Uruguay

G. Cervetto
Dirección Nacional de Medio Ambiente. Ministerio de Vivienda, Ordenamiento Territorial y Medio Ambiente, Montevideo, Uruguay
e-mail: guillermo.cervetto@mvotma.gub.uy

D. L. Calliari
Ecología Funcional de Sistemas Acuáticos, Centro Universitario Regional del Este, Universidad de la República, Rocha, Uruguay

Oceanografía y Ecología Marina, Facultad de Ciencias, Universidad de la República, Uruguay
e-mail: dcalliar@fcien.edu.uy

M. S. Hoffmeyer et al. (eds.), *Plankton Ecology of the Southwestern Atlantic*,
https://doi.org/10.1007/978-3-319-77869-3_17

overlap. This did lead to differences in the TNB_S and supported the prediction of the trophic differentiation hypothesis and could be interpreted as a mechanism to minimize food competition in the RPE ecosystem.

Keywords Niche breadth · Larval morphometry · Trophic differentiation · *Micropogonias furnieri* · *Trachurus lathami*

1 Introduction

White croaker *Micropogonias furnieri* and rough scad *Trachurus lathami* (Perciformes) are widespread coastal species in the South West Atlantic found between 20° N and 41° S (Saccardo and Katsuragawa 1995; Cosseau and Perrota 2000). They are abundant in estuaries that serve as nurseries, and they are important components of the fish community of the Río de la Plata Estuary (RPE) and adjacent shelf waters (Saccardo and Katsuragawa 1995; Acha et al. 1999). *Micropogonias furnieri* supports coastal artisanal- and industrial-scale fisheries in Brazil, Uruguay, and Argentina and is one of the main commercially exploited resources in the RPE (Norbis 1995; Lasta and Acha 1996). Less is known about *T. lathami*, the only species of the genus described for the western Atlantic (Berry and Cohen 1974). This species is considered a potential food resource (Saccardo and Katsuragawa 1995) and is captured as bycatch of other species like *Engraulis anchoita* and *Scomber japonicus* (Pacheco 1989). For both *M. furnieri* and *T. lathami*, most of the attention has focused on their adult life history (e.g., Jaureguizar et al. 2003 and references therein, Saccardo and Katsuragawa 1995; de Carvalho and Hashimoto 2006). Few studies addressed their early stages and mostly focused on spatial and temporal distribution of larvae (e.g., Katsuragawa and Matsuura 1992; Berasategui et al. 2004; Sampaio de Souza and Mafalda Jr. 2008; Braverman et al. 2009; Katsuragawa et al. 2014). Current knowledge on the ecological interactions within the ichthyoplankton community in the RPE is sparse; in particular, information on larval trophic ecology is scarce for this and other estuaries of the region (Braverman 2011; Vera 2011; Machado et al. 2017). Such knowledge is critical for the design of ecosystem-based fisheries management and particularly for conservation-oriented management policies for overexploited species – as is the case of *M. furnieri* (Pin and Defeo 2000) – that take into account recruitment variability and its effect on fish population dynamics.

At RPE, *M. furnieri* and *T. lathami* often co-occur with a partial overlap in their distribution range (García et al. 2010) and exhibit similar reproductive strategies: (i) they are pelagic egg spawners (Balon 1975), (ii) they spawn inside the estuary (Berasategui et al. 2004 and references therein), and (iii) they reproduce between spring and summer, with a main peak around November (Saccardo and Katsuragawa 1995; Macchi and Acha 2000). As a consequence, peak abundances of *M. furnieri* and *T. lathami* larvae may overlap seasonally within the estuary.

The fact that both larvae are pelagic, abundant (Berasategui et al. 2004), and zooplanktivorous (Pedreira 1997; Machado et al. 2017) may increase the potential for inter specific competition.

Coexistence of species in natural communities can result from access to unlimited resources or from niche specialization (i.e., species evolve to occupy different feeding niches; Moser 1984; Sabatés and Saiz 2000; Rodríguez-Graña et al. 2005). Alternatively, larvae could exhibit interspecific spatial segregation based on differential environmental preferences in concordance with changes in their physiological and metabolic requirements along ontogeny (Pörtner et al. 2010).

Since fish larvae usually swallow prey whole, mouth size is a restrictive factor for food ingestion, and prey width is considered the limiting dimension for prey selectivity (Hunter 1984; Houde 1997). During ontogeny, mouth size and searching capabilities of larvae increase and so does the size of ingested prey (Hunter 1984; Houde 1997). However, trophic niche breadth tends to remain relatively constant throughout growth (Pearre 1986), which could been seen as a way to reduce intraspecific competition. But, constancy in trophic niche breadth may not to be a universal pattern (Pepin and Penney 1997; Gonzáles-Quiroz and Anadón 2001).

The present study was designed to explore ontogenetic differences in mouth morphology in *M. furnieri* and *T. lathami*, to evaluate their use of trophic resources, and to discuss whether morphological differences contribute to resource partitioning within these two components of the fish larval community at RPE ecosystem. In order to address such issues, feeding habits of larval stages of *M. furnieri* and *T. lathami* at RPE were analyzed in two periods of a breeding season and compared for differences in diet composition, feeding periodicity and incidence, and prey-size selection. Breadth of the prey-size niche and an overlap index were used to explore potential competition between and within species. In addition, morphological measurements were made in an attempt to discern if either species has a competitive advantage at some particular stage due to differential development. With this procedure, we were able to test the hypothesis that the coexistence of the species under study is favored by spatial segregation and divergence in the ontogeny of mouth morphology, as determinant of diet composition and trophic niche.

2 Material and Methods

2.1 Study Site and Sample Collection

The RPE is a funnel-shaped, salt wedge, microtidal estuary (34°10′–36°10’ S, 55°00′–58°10’ W) between Uruguay and Argentina (Fig. 1a). It is nearly 38,000 km^2, drains the second largest basin of South America, and provides the main source of freshwater runoff into the southwest Atlantic Ocean (annual average

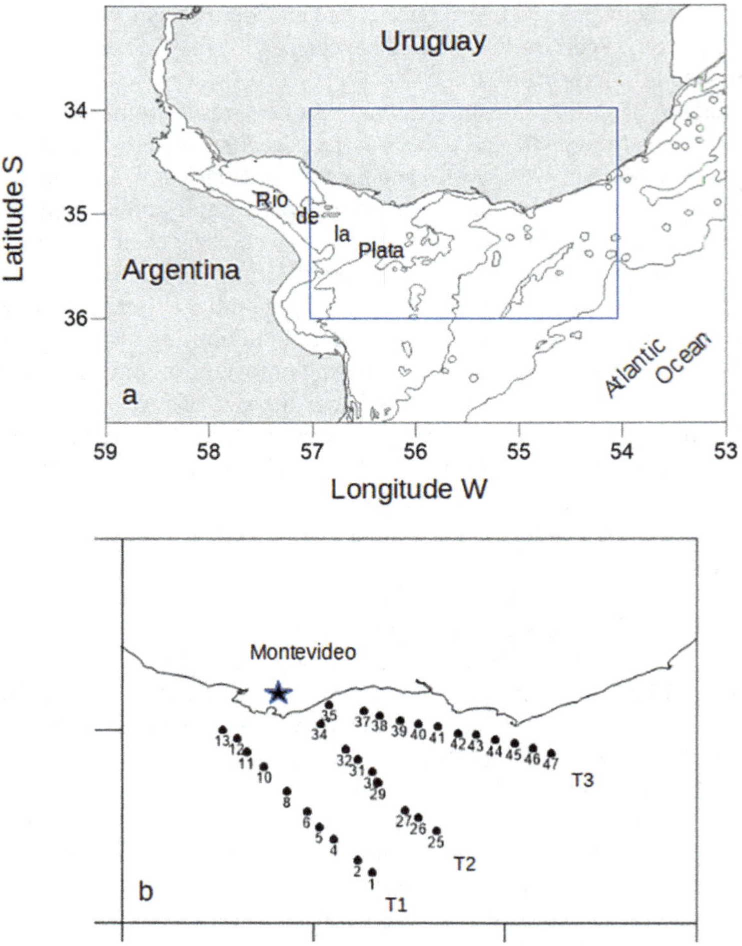

Fig. 1 (**a**) Synoptic map of the study area showing location of Río de la Plata. (**b**) sampling stations. Full circles with numbers correspond to hydrographic and plankton stations; T transects

ca. 22,000 $m^3 s^{-1}$; Nagy et al. 1997). Average depth is less than 10 m over >50% of its surface with a maximum of approx. 25 m at the mouth. Salinity varies from <1 in the upper estuary to 33 in the lower estuary strongly influenced by Atlantic coastal waters (a combination of Brazil and Malvinas Currents) and by freshwaters from continental runoff (Guerrero et al. 1997). Upper and lower regions are separated by salinity and turbidity fronts that almost overlap and show complex and variable geometries (Guerrero et al. 1997). These frontal structures influence diverse processes including phytoplankton and zooplankton distribution (Calliari et al. 2005; Berasategui et al. 2006; Kruk et al. 2015), primary production (Calliari et al. 2009), and fish spawning (Acha et al. 1999; Acha and Macchi 2000). Zooplankton diversity in the estuary is dominated by small copepods (Berasategui et al. 2006), and gelatinous plankton (ctenophores; Mianzán and Guerrero 2000; Alvarez et al. 2003), chaetognaths (Milstein and Juanicó 1985), and mysids (Schiariti et al. 2006; Calliari et al. 2007; Cervetto et al. 2016) are important components, among others. The RPE and inner shelf are important spawning and nursery areas for several fish species including *Micropogonias furnieri* (Macchi 1997; Acuña et al. 2000) where the saline frontal structure and wind patterns modulate larval retention within the estuary (Simionato et al. 2008; Braverman et al. 2009).

Environmental data and plankton samples were collected on board the R./V. Aldebarán in austral spring (October 2006) and summer (March 2007), over a salinity gradient from ca. 4 to >30 in a total of 38 stations arranged in three transects (T1, T2, T3) (Fig. 1b). Vertical profiles of salinity and water temperature were recorded with a SeaBird CTD profiler. Fish larvae were collected with a Bongo net (50 cm mouth diameter, 500 μm mesh). The availability of zooplankton as potential food was estimated at the same stations and collected with a mini-Bongo net (40 cm mouth diameter, 63 μm mesh). Both nets were fitted with General Oceanic flow meters to estimate larval and zooplankton abundances. All tows were performed obliquely from near the bottom to the surface (bottom depth ranged from 6 to 36 m) and mostly at daylight hours from 6:30 to 18:30 local time. Only one station was sampled during night in autumn (21:00 h). All samples were preserved in 4% marine formaldehyde buffered with borax.

2.2 Laboratory and Data Analysis

2.2.1 Habitat Preference and Spatial Overlap

To explore salinity and temperature preferences, the approach proposed by Perry and Smith (1994) was applied, simplified for a single-stratum case. Briefly, the method compares the empirical cumulative distribution function (cdf) of an environmental variable (surface and bottom salinity and surface and bottom temperature in our case) and of the species abundance using a Kolmogorov-Smirnov test based on maximum vertical distance between estimated cdf's (d_{max}); d_{max} is compared with tabulated critical values $d_{P,k,N}$ for probability level p, with *k* (number of classes) and N (number of observations) degrees of freedom under the null hypothesis of random association between organism distribution and habitat condition (Perry and Smith 1994; Zar 1996).

2.2.2 Spatial Overlap

Intra- and inter specific spatial overlap were determined by Morisita-Horn (MH) index (Horn 1966):

$$\mathrm{MH} = \frac{\left[\sum_{i}^{n} 2a_i^* b_i\right]}{\left[\sum_{i}^{n} a_i^2 + \sum_{i}^{n} b_i^2\right]}, \quad \text{station } i \ldots n$$

where a_i is the proportion of species "*a*" at station *i* and b_i is the proportion of species "*b*" at station *i*. For intraspecific comparison (preflexion and postflexion stages), a_i is the proportion of "stage a" at station *i*, and b_i is the proportion of "stage b" at station *i*. This index varies according to the proportion and coincidence of each species (stages, species) in sampled stations: 0 if the two species do not co-occur at any station, and 1 if both species co-occur in the same proportions in all stations. A significant overlap is commonly assumed for MH > 0.6 (Zaret and Rand 1971). In this study, three overlap categories were considered: low overlap (0.0–0.3), medium overlap (0.31–0.6), and high overlap (> 0.6).

2.2.3 Morphometric and Diet Analyses

Fish larvae were sorted out from whole plankton samples under a stereoscopic microscope, identified, counted, and abundance expressed as individuals 10 m^3. Prior to dissection for gut content analysis, body length (BL), upper jaw length (JL), and mouth width (MW) were measured. Body length was measured as notochord length (NL) in preflexion (PF) larvae and as standard length (SL) in postflexion (PsF) larvae to the nearest 0.1 mm using a stereoscopic microscope. JL was measured from the tip of the snout to the posterior end of the maxilla. MW was determined in ventral view as the width between the posterior edges of the maxillae. For gut content analysis, the entire digestive tract from each larva was dissected under high-magnification microscope. Prey items in the gut were identified, counted, and measured to the nearest 0.01 mm using an inverted microscope. Prey size was expressed as its maximum width (Wmax; μm) according to Busch (1996). Additionally, the zooplankton community – as natural food offer for fish larvae – was described in terms of taxa and abundance (ind m^{-3}); mesozooplankton aliquots were prepared for identification and counting under a stereoscopic microscope.

Feeding incidence (% FI) was estimated as the percentage of individuals with prey in their guts out of the total number of larvae examined and was considered a measure of feeding success in the field. Composition of the overall diet of each species was summarized using an index of relative importance (IRI; Govoni et al. 1983) for each prey type, calculated as the product of the percent number (N %) and frequency of occurrence (F %) of prey items, and expressed as percentage (% IRI) for standardization purposes.

Diversity of prey in the diet of both species and stages was estimated using equitability index E_H (Begon et al. 1996) and expressed in percentage.

$$E_H = \frac{H}{H\max} = \frac{H}{\ln S}$$

where H is the Shannon-Wiener index, Hmax is the maximum prey diversity possible, and S total number of preys (richness). H assumes a value between 0 and 1 with 1 being complete evenness.

The diets of *M. furnieri* and *T. lathami* larvae were compared based on dietary niche overlap estimated as the amount of food items shared using the similarity index D (Schoener 1968):

$$D = 1 - 0.5 * \sum \left| \left(p_i - q_i \right) \right|$$

where p_i is the proportion of item i in the diet of species p and q_i is the proportion of item i in the diet of species q. D ranges between 0 (diets do not overlap) and 1 (diets completely overlap) with values >0.6 considered to be biological significant overlap (Wallace 1981). Index D was estimated for prey items classified according to two alternative criteria: prey type (taxonomic affiliation) and size.

The standard deviation of log10-transformed prey size was taken as an estimator of the breadth of the trophic niche based on prey size (TNB_S; Sabatés and Saiz 2000). Larvae of each species were pooled in 250 μm mouth width size classes. To explore interspecific differences, the average of the standard deviations on log10 prey size was compared between species as a function of their corresponding mouth: body-size ratios (Sabatés and Saiz 2000).

The relationships between jaw length and body length and between mouth width and body length were explored for each species using generalized linear models (GLM) and regressions fitted by maximum likelihood. Alternative models were explored using Gamma and Gaussian families with inverse and identity link functions. Models were selected based on Akaike information criterion (Burnham and Anderson 2002). The range of prey sizes ingested during larval development was explored using nonlinear squares. All the statistical analyses were performed using the software R (R Core team 2015).

3 Results

3.1 *Abundance, Environmental Preferences, and Spatial Overlap*

Temperature ranged from 10.7 to 27.9 and followed expected seasonal differences. In spring, bottom temperature varied along the estuary in a broader range than in the surface, where in autumn variability was much less pronounced (Fig. 2a, d). Salinity ranged from 3.6 to 34.7 and evidenced a clear horizontal

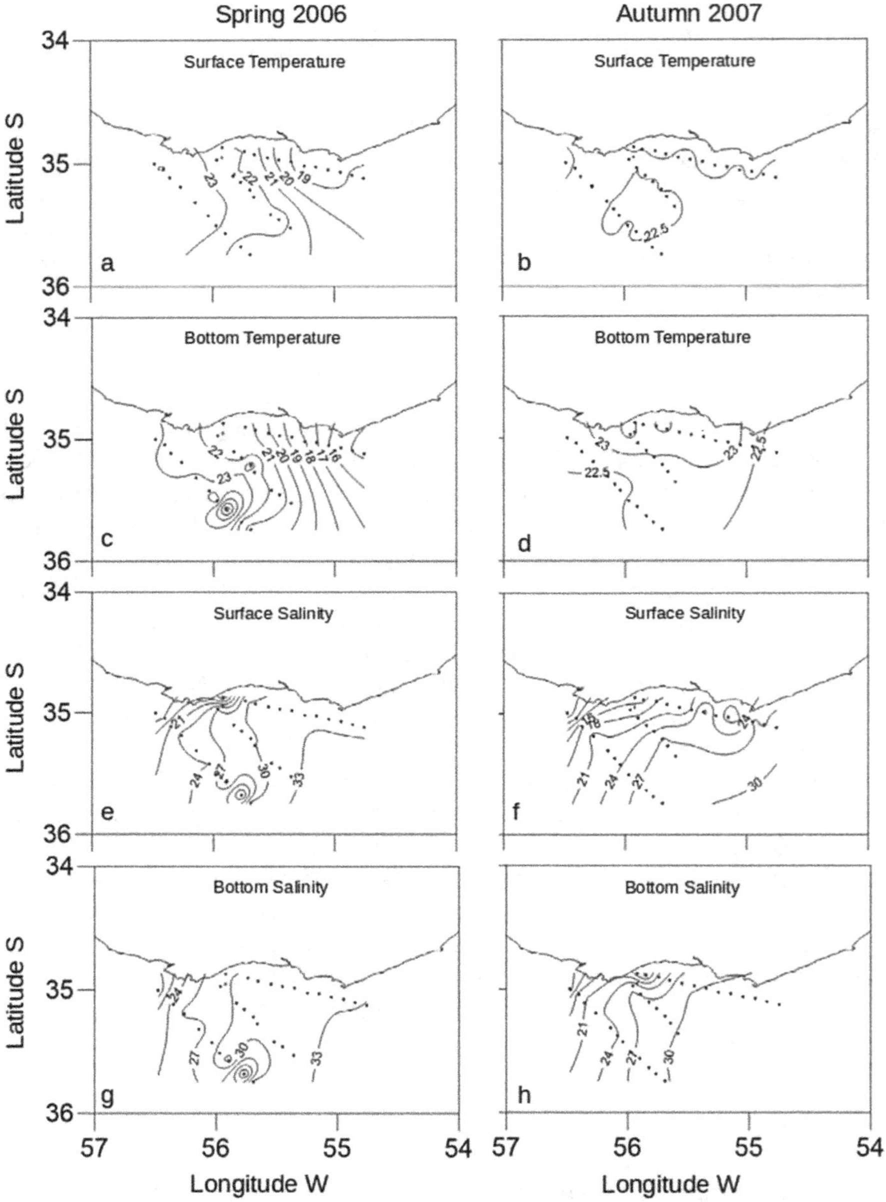

Fig. 2 Horizontal section at surface and bottom of temperature °C (**a** and **d**) and salinity (**e** and **h**) at Río de la Plata Estuary during spring 2006 and autumn 2007. Full circles correspond to sampled stations

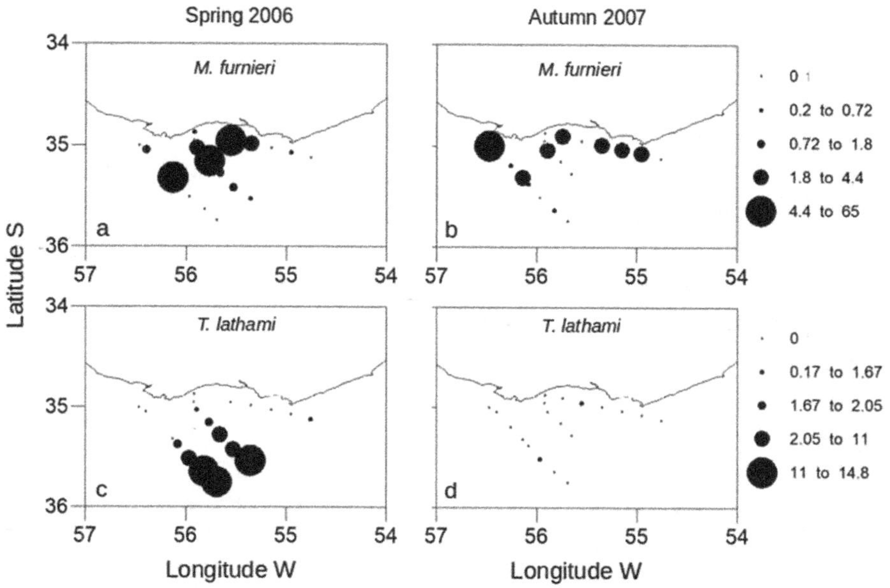

Fig. 3 Spatial distribution of (**a** and **b**) *Micropogonias furnieri* and (**c** and **d**) *Trachurus lathami* larvae at the Río de la Plata Estuary during spring 2006 (left) and autumn 2007 (right). Abundances are expressed as individuals 10 m^{-3}

gradient along the main axis of the estuary, both at surface and near the bottom, in both periods (>20 units) (Fig. 2e, h).

Micropogonias furnieri was more abundant in spring (mean ± SD; 5.3 ± 15.2 ind. 10 m^{-3}; range 0.2 to 64.9 ind. 10 m^{-3}) than in autumn (2.7 ± 7 ind. 10 m^{-3}; range 0.5 to 30.5 ind 10 m^{-3}) (Fig. 3a, b). In both periods, most of the larvae were at preflexion stage. Larval size ranged between 1.7 and 18.8 mm (3.17 ± 2.09 mm) in spring and 1.74 and 18.20 mm (5.35 ± 3.34 mm) in autumn. Larval abundance did not evidence an association with temperature or salinity (surface or bottom), for neither of the two species (cumulative frequency analysis, example is shown for bottom salinity; Fig. 4). Spatial overlap among stages was low in both periods (MH, 0.15 spring and 0.01 autumn).

Trachurus lathami was more abundant in spring (2.9 ± 4.7 ind. 10 m^{-3}; range 0.2 to 14.8) than in autumn when only two preflexion larvae were found (0.1 ± 0.2 ind. 10 m^{-3}) (Fig. 3c, d). For that reason, results for *T. lathami* were based only on the analysis of the spring data set. Larval size ranged between 1.72 and 24.7 mm (3.27 ± 2.8) in spring. Spatial overlap among stages was medium (MH, 0.51).

In spring, *M. furnieri* and *T. lathami* co-occurred in 33% of stations and interspecific spatial overlap as measured by Morisita-Horn index was low (MH, 0.02). At those stations where both species co-occurred, *M. furnieri* was more abundant than *T. lathami*, and *M. furnieri* tended to predominate in the inner estuary, while *T. lathami* did so in the outer estuary (Fig. 3a, d).

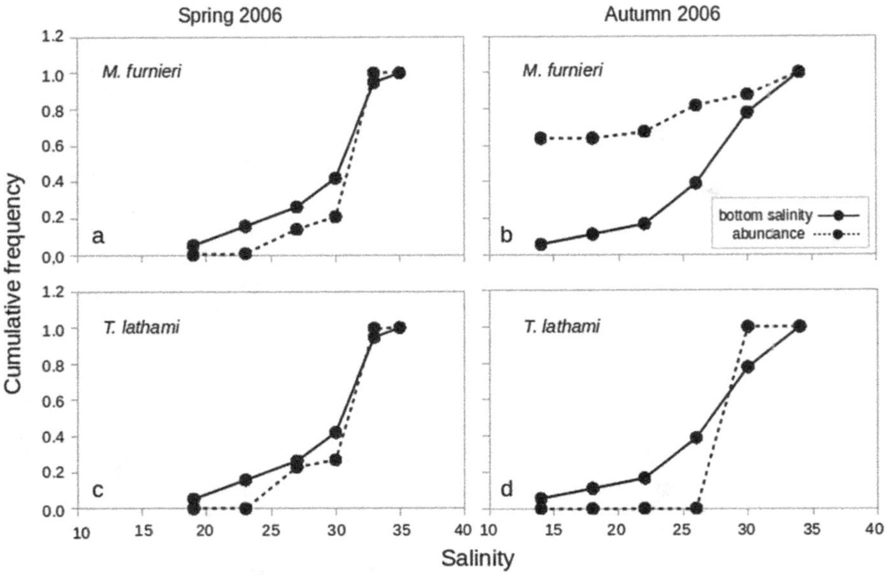

Fig. 4 Estimated cumulative distribution functions for bottom salinity data (continuous line) and abundance (dashed line) in (**a** and **b**) *Micropogonias furnieri* and (**c** and **d**) *Trachurus lathami* at the Río de la Plata Estuary during spring 2006 and autumn 2007. $p > 0.05$ in all cases. Cumulative frequencies are expressed as fractions in the 0–1 range

3.2 Larval Morphology

Micropogonias furnieri and *T. lathami* differed in the patterns of morphological development. *M. furnieri* showed positive allometric growth for both mouth width and jaw length with body length (Fig. 5a, c). The increase in gape size (as mouth width and jaw length) during development (slope of MW-BL and JL-BL relationship) was similar in both cases and significantly different from 1 ($b > 1$, $p << 0.01$) (Fig. 5a, c). The range of prey sizes ingested was rather wide for larvae along the whole larval size range. Prey size (Wmax) ranged between 13 and 175.7 μm (63.6 ± 32.5 μm). The size of the main items ingested ranged from 27 to 69 μm for tintinnids (57.1 ± 6.5 μm), 47 to 65 μm for copepod nauplii (55.8 ± 7.1 μm) and 80 to 160 μm for copepods (118.9 ± 28.1 μm). Prey size correlated to mouth width. But there was a threshold at about 0.6 mm in mouth width where larvae switched to prey sizes >100 um wide (Fig. 6a).

Trachurus lathami exhibited an isometric growth for both mouth width and jaw length with body length, and for both mouth size descriptors, the increment in size with body length was similar and not different from 1 ($b = 1$, $p = <<0.01$) (Fig. 5b, d). Also for *T. lathami*, the prey size correlated to mouth width, but a higher scatter in prey sizes at a given MW was observed, compared to *M. furnieri* (Fig. 6b). At a mouth width threshold of ca. 0.3 mm larvae started to ingest

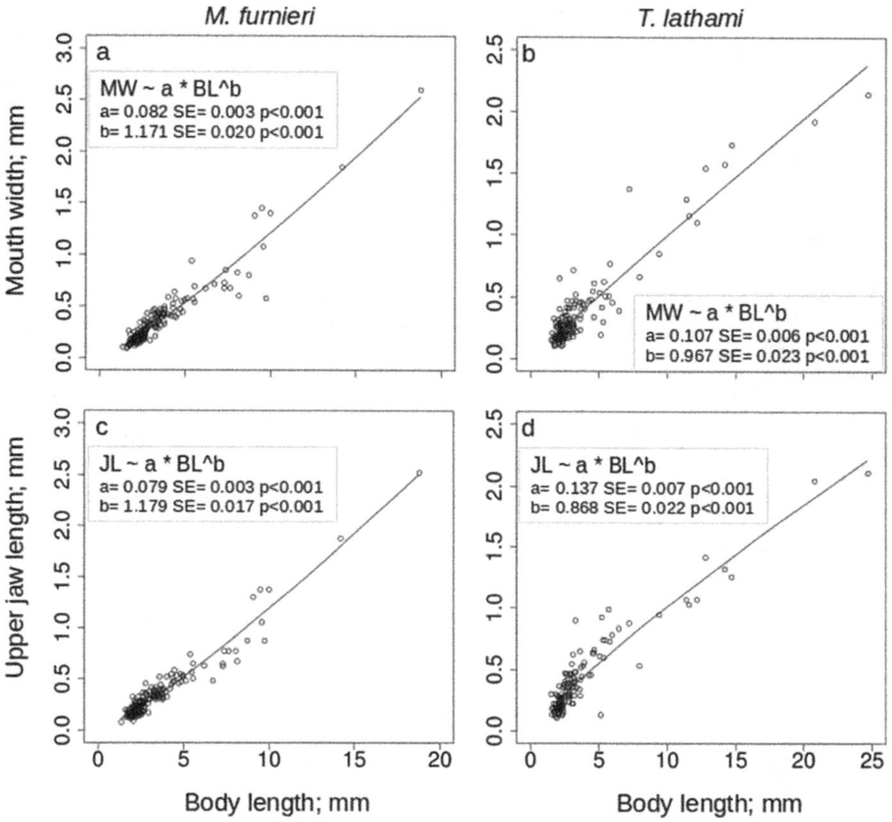

Fig. 5 Relationship between body size (mm) and mouth size (mm) (upper panels) and body length and upper jaw size (mm) (lower panels) of (**a** and **c**) *Micropogonias furnieri* ($n = 204$ in both cases) and (**b** and **d**) *Trachurus lathami* (b, $n = 157$, and d, $n = 145$). Regression analyses are shown in each panel. SE, standard deviation. n, number of cases analyzed

relatively large prey (200 μm and wider), but that did not preclude ingestion of smaller items, even at the largest mouth width classes.

3.3 Zooplankton Community

A total of 22 groups were identified in the zooplankton community as potential preys for *M. furnieri* and *T. lathami* larvae from those stations where species co-occurred. During spring, 18 potential prey groups were observed where copepods – mainly *Paracalanus* sp. and *Oithona* sp. and their early stages – were the most abundant, followed by cladocerans (*Podon* sp. and *Penilia* sp.) and larvaceans (*Oikopleura* sp.) (See Supplementary Data Table 1). Autumn presented 21

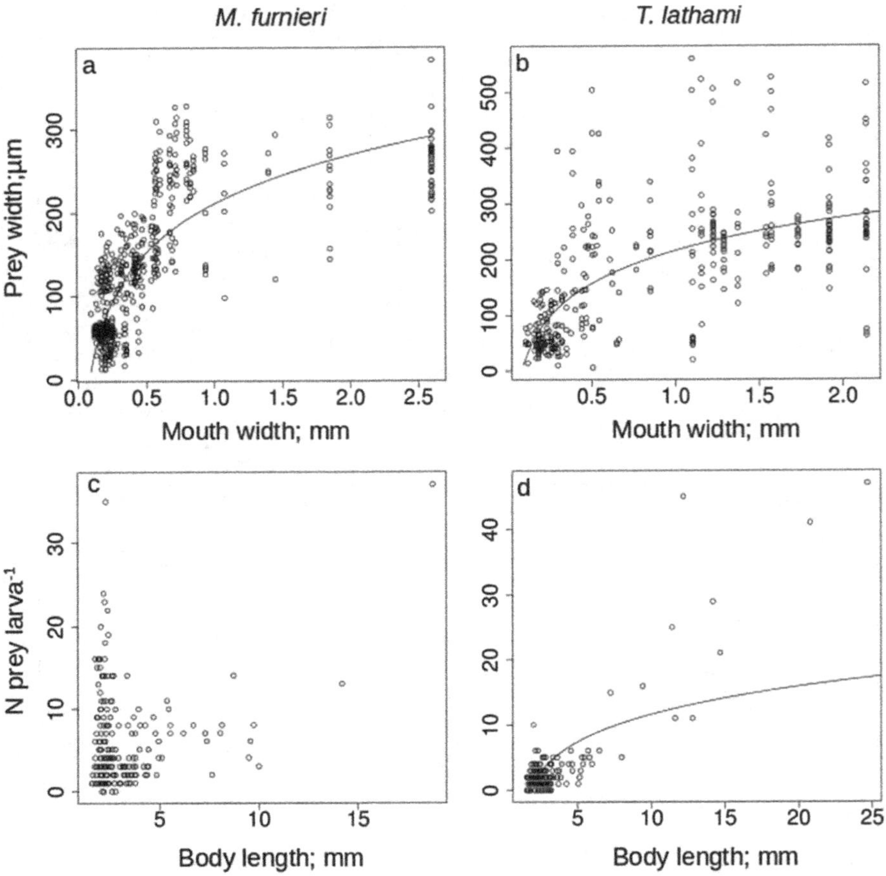

Fig. 6 Relationship between mouth width (MW; mm) and prey size (maxW; µm) (upper panels) and body length (BL; mm) and number of prey ingested (lower panels) of (**a** and **c**) *Micropogonias furnieri* and (**b** and **d**) *Trachurus lathami*. Regression in A: max W = a + b * log (MW); a = 210.93 (SD 2.55); b = 84.036 (SD 1.73); n = 932. Regression in B: max W = a + b * log (MW); a = 216.147 (SD 5.15); b = 85.556 (SD 5.07); n = 367. Regression in D: Nprey = a + b * BL; a = 0.433 (SD 0.05); b = 0.161 (SD 0.00); n = 189. $p < 0.01$ in all cases. n number of cases analyzed

potential prey groups where cladocerans (*Penilia* sp.), copepod nauplii, and small copepods (e.g., *Paracalanus* sp. and *Oithona* sp.) dominated the zooplankton community (See Supplementary Data Table 2).

3.4 Diet Composition and Prey Diversity

A total of 220 larvae of *M. furnieri* and 242 of *T. lathami* were dissected for diet analysis. Preflexion and postflexion stages of *M. furnieri* exhibited a high feeding incidence; however, postflexion larvae showed a slightly higher FI (92.3%) than preflexion larvae (90.5%). This species always exceeded 50% of larvae with prey in their guts at all hours and stations where it occurred (See Supplementary Data Fig. 1a). The number of prey in the guts (6.9 ± 5.9, range from 1 to 37 preys) did not show a clear trend when plotted vs. body size (Fig. 6c).

Diet of *M. furnieri* was composed by a wide spectrum of prey items (32) and showed an equitability over 50%, and H_E was higher in PsF larvae (85.7%) than in PF larvae (76.3%). Tintinnids (IRI 71%), copepod nauplii (IRI 9%), and a small unidentified copepod (IRI 6.4%) constituted the bulk of the diet. However, the IRI index showed differences between stages and periods (Table 1). In spring, PF larvae fed mainly on tintinnids (IRI 84.9%), an unidentified copepod (IRI 4.8%), and copepod nauplii (IRI 4%). PsF larvae fed on unidentified copepods (IRI 38.1%), *Paracalanus* sp. (IRI 23.8%), and *Oithona* sp. (IRI 14.3%). In autumn, PF larvae fed mainly on copepod *Oncaea* sp. (IRI 41%), copepod nauplii (IRI 31%), and cladoceran *Penilia* sp. (IRI 13.4%), while PsF larvae fed almost exclusively on copepod *Acartia tonsa* (IRI 92.8%).

Larvae of *T. lathami* exhibited a high feeding incidence (83.4%); all postflexion larvae had preys in their stomachs (FI 100%), while preflexion larvae showed FI = 81.3%. Similar to *M. furnieri* larvae, this species always exceeded 50% of larvae with prey in their guts at all hours and stations where it occurred (See Supplementary Data Fig. 1b). The number of prey in the guts (5.3 ± 8.7, ranged from 1 to 47 preys) increased with larval body size (Fig. 6d). For *T. lathami* since only two larvae were captured in autumn, all estimations were based on the cruise performed in spring. *Trachurus lathami* diet was composed by 26 items and also showed an equitability over 50%, and H_E was higher in PsF larvae (76.2%) than PF larvae (66.4%). Copepod nauplii (IRI 48.3%), a small unidentified copepod (IRI 27.3%), and *Penilia* sp. (IRI 9.90%) constituted the bulk of the diet. However, IRI showed differences between stages (Table 2). Preflexion larvae fed mainly on copepod nauplii (IRI 72.5%) and an unidentified copepod (IRI 16.6%). In turn, PsF larvae fed on *Penilia* sp. (IRI (40.8%), *Paracalanus* sp. (29.4%), and an unidentified copepod (IRI 27.2%).

3.5 *Diet Overlap and Trophic Niche Breadth*

Interspecific diet overlap based on the taxonomic criteria was low in both preflexion ($D = 0.37$) and postflexion ($D = 0.17$) larvae. Based on prey size, both species and stages showed a high dietary overlap (>0.60).

Prey-size niche breadth varied among species, where *T. lathami* showed a wider TNB_s (0.23 ± 0.11) than *M. furnieri* (0.11 ± 0.1) (Fig. 7a). Each species exhibited different trends in the TNB_S regarding morphological changes with development. In *Micropogonias furnieri*, TNB_S ranged from 0.05 to 0.20 and decreased with development by increasing the mouth width: body length ratio (Fig. 7b). In *Trachurus lathami*, TNB_S ranged from 0.06 to 0.40 and did not exhibit a clear pattern with development (Fig. 7c).

Table 1 Prey composition of *Micropogonias furnieri* of preflexion and postflexion larvae collected at Río de la Plata Estuary in Spring 2006 and Autumn 2007. Values were expressed as percent frequency of occurrence (F%) in larval guts, percent of the total number (N%) of items in the diet, and the index of relative dietary importance F% × N%; those preys with highest IRI are highlighted in bold. n = number of larvae examined, *n/i* = not identified

Micropogonias furnieri															
				Spring 2006						Autumn 2007					
	Pooled data (n = 198)			Preflexion (n = 152)			Postflexion (n = 3)			Preflexion (n = 21)			Postflexion (n = 22)		
Food items	F%	N%	IRI%	F%	N%	IRI%	F%	N%	IRI%	F%	N%	IRI%	F%	N%	IRI%
Phytoplankton															
Actinocyclus sp.	4.32	0.94	0.16							4.76	2.06	0.34	31.8	5.06	2.83
Diatoms n/i	0.54	0.09	<0.01	0.65	0.11	<0.01									
Ceratium sp.	0.54	0.09	<0.01	0.65	0.11	<0.01									
Protoperidinium sp.	7.57	2.92	0.84	9.09	3.67	1.0									
Dinoflagellate n/i	16.8	3.43	2.19	19.5	4.48	2.5				4.76	1.03	0.17			
Phytoplankton n/i	2.16	0.43	0.04	2.60	0.57	0.04									
Cladocera															
Bosmina sp.	1.08	0.17	0.01										9.09	1.12	0.18
Evadne sp.	0.54	0.09	<0.01							4.76	1.03	0.17			
Penilia sp.	4.32	1.37	0.23							28.6	13.4	**13.4**	9.09	1.69	0.27
Cladocerans n/i	3.24	0.60	0.07	1.95	0.34	0.02				9.52	3.09	1.03	4.55	0.56	0.04
Copepoda															
Acartia tonsa	12.4	11.8	5.62	3.90	0.92	0.1				4.76	1.03	0.17	72.7	72.5	**92.8**
Centropages sp.	0.54	0.17	<0.01										4.55	1.12	0.09
Copillia sp.	0.54	0.09	<0.01	0.65	0.11	<0.01									
Corycaeus sp.	2.16	0.34	0.03	0.65	0.11	0.00				4.76	1.03	0.17	9.09	1.12	0,18
Euterpina sp.	1.08	0.26	0.01				33.3	11.8	9.52	4.76	1.03	0.17			
Lubbockkia sp.	0.54	0.17	<0.01										4.55	1.12	0.09

Microsetella sp.	1.08	0.17	0.01	0.65	0.11	<0.01				4.76	1.03	0.17			
Oithona sp.	10.8	2.75	1.13	7.79	2.07	0.5	33.3	17.6	14.3	14.3	4.12	2.07	18.2	3.93	1.26
Oncea sp.	5.41	3.95	0.81							33.3	35.1	**41.0**	13.6	6.74	1.62
Paracalanus sp	3.24	1.55	0.19	1.95	1.03	0.06	33.3	29.4	**23.8**	4.76	2.06	0.34	4.55	1.12	0.09
Scolecithricella sp.	1.08	0.34	0.01	0.65	0.11	<0.01	33.3	17.6	14.3						
Calanoid n/i	1.62	0.26	0.02							9.52	2.06	0.69	4.55	0.56	0.04
Copepod n/i	28.1	7.21	**6.41**	25.3	7.81	**4.8**	66.6	23.53	**38.1**	38.1	9.28	7.93			
Copepods egg	0.54	1.12	0.02	0.65	1.49	0.03									
Copepod nauplii	29.7	7.98	**9.05**	16.2	8.6	**4.0**				47.6	18.6	**31.0**			
Flagellates															
Flagellated n/i	0.54	0.09	<0.01	0.65	0.11	<0.01									
Mollusks															
Bivalve larvae	10.2	1.72	0.67	10.4	1.95	0.58				9.52	2.06	0.69	4.55	0.56	0.04
Mysida															
Mysids n/i	1.08	0.17	0.01	0.65	0.11	<0.01							4.55	0.56	0.04
Ostracoda															
Ostracods n/i	0.54	0.09	<0.01										4.55	0.56	0.04
Ciliates															
Tintinnids	41.1	45.4	**71.1**	48.7	60.6	**84.9**				4.76	1.03	0.17			
Others															
Crustacea n/i	10.3	2.75	1.08	11.7	3.56	1.2				4.76	1.03	0.17			
Invertebrates eggs	4.86	1.37	0.25	5.19	1.84	0.27									

Table 2 Prey composition of *Trachurus lathami* of preflexion and postflexion larvae collected at Río de la Plata Estuary in Spring 2006 and Autumn 2007. Values were expressed as percent frequency of occurrence (F%) in larval guts, percent of the total number (N%) of items in the diet, and the index of relative dietary importance F% × N%; those preys with highest IRI are highlighted in bold. n = number of larvae examined, *n/i* = not identified

Trachurus lathami									
Spring 2006									
	Pooled data (n = 171)			Preflexion (n = 152)			Postflexion (n = 18)		
Food items	F%	N%	IRI%	F%	N%	IRI%	F%	N%	IRI%
Phytoplankton									
Peridinium sp.	0.55	0.16	<0.01	0.65	0.33	0.01			
Prorocentrum sp	1.65	0.49	0.04	1.94	0.98	0.07			
Dinoflagellate n/i	2.75	0.97	0.14	3.23	1.96	0.24			
Phytoplankton n/i	1.10	0.32	0.02	1.29	0.65	0.03			
Cirripedia									
Cirripeds nauplii	0.55	0.16	<0.01	0.65	0.33	0.01			
Cladocera									
Penilia sp.	9.34	20.7	**9.90**	2.58	3.27	0.32	68.4	38.7	**40.8**
Cladocerans n/i	0.55	0.16	<0.01	0.65	0.33	0.01			
Copepoda									
Acartia tonsa	2.75	1.94	0.27	2.58	1.96	0.19	10.53	1.97	0.32
Corycaeus sp.	1.10	0.49	0.03	0.65	0.33	0.01	5.26	0.66	0.05
Euterpina sp.	1.10	0.49	0.03				10.5	0.98	0.16
Microsetella sp.	0.55	0.16	<0.01	0.65	0.33	0.01			
Oithona sp.	3.85	1.94	0.38	0.65	0.33	0.01	26.32	3.61	1.46
Paracalanus sp.	9.34	13.9	6.65	1.94	2.29	0.17	73.7	25.9	**29.4**
Temora sp.	0.55	0.16	<0.01				5.26	0.33	0.03
Calanoid n/i	2.75	1.29	0.18	2.58	1.63	0.16	5.26	0.98	0.08
Copepod n/i	28.0	19.6	**27.3**	23.9	19.0	**16.6**	84.2	21.0	**27.2**
Copepods egg	1.10	0.32	0.02	1.29	0.65	0.03			
Copepod nauplii	42.9	22.0	**48.3**	45.2	41.8	**72.5**	10.5	0.66	0.11
Flagellates									
Flagellated n/i	0.55	0.16	<0.01						
Larvacea									
Oikopleura sp.	0.55	0.16	<0.01	0.65	0.33	0.01			
Larvaceans n/i	0.55	0.32	0.01				5.26	0.66	0.05
Mollusks									
Bivalve larvae	7.69	3.07	1.21	9.03	6.21	2.15			
Mysida									
Mysids n/i	0.55	0.16	<0.01				5.26	0.33	0.03
Ciliates									
Tintinnids	10.4	5.02	2.68	11.6	6.21	2.77	5.26	3.93	0.32
Others									
Crustacea n/i	11.5	4.21	2.48	13.5	7.84	**4.08**	5.26	0.33	0.03
Invertebrates eggs	4.40	1.62	0.36	5.16	3.27	0.65			

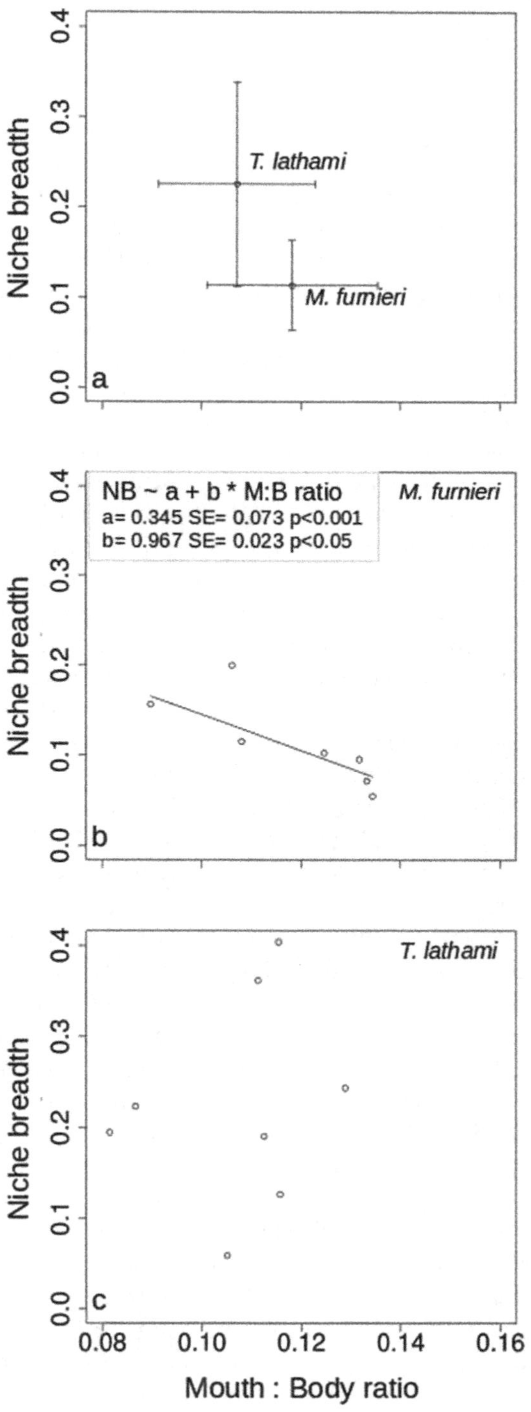

Fig. 7 Relationship between mouth width-body length ratio and prey-size niche breadth (average of SD of log10 prey for each mouth size class) for (**a**) *Micropogonias furnieri* and *Trachurus lathami* (averaged data), (**b**) *Micropogonias furnieri*, and (**c**) *Trachurus lathami*. Error bars, 1 SE. Data are means and range for each size classes

4 Discussion

This study linked the spatial distribution and trophic interactions of white croaker and rough scad larvae that coexist at RPE. Both species spawn and breed in spring and summer within the estuary (Jaureguizar et al. 2003; García et al. 2010), and trophic overlap and competition is a potential scenario. It was hypothesized that coexistence was favored by spatial segregation between larvae of *M. furnieri* and *T. lathami* within the estuary. Due to environmental characteristics of the RPE – i.e., spatial gradients – spatial segregation implies that both species would be subjected to different environmental conditions. During the 2006–2007 breeding season, the estuary exhibited the typical salt wedge configuration (Guerrero et al. 1997) with freshwater advected offshore over denser shelf waters that intruded into the estuary. That intrusion was more remarkable during spring and resulted in strong stratification, horizontal salinity gradients, and high saline waters at the estuarine mouth. Spatial distribution of both species was partially overlapped: *M. furnieri* was most abundant in the upper estuarine section of the studied area, whereas *T. lathami* was mainly present in the outermost region, as shown in previous studies (Berasategui et al. 2004). However, larvae co-occurred in 33% of sampled stations in spring.

Known habitat preferences for both species could explain such spatial segregation. *M. furnieri* larvae tend to be associated with the bottom salinity front of the RPE (Berasategui et al. 2004) at temperatures ranging from 20 °C to 25 °C and salinities from 4 to 30. *Trachurus lathami* larvae have been recorded in high abundances on the continental shelf off Rio Grande do Sul (Brazil) and Uruguay and outside the RPE at temperatures from 11 °C to 27 °C and salinities from 22 to >36 (Katsuragawa and Matsuura 1992; Berasategui et al. 2004). In the present study, the distribution of both species was not clearly associated to specific salinity and temperature conditions. Instead, both species were present over rather wide salinity and temperature ranges.

The distribution patterns of fish larvae relate to the reproductive activity of adult populations and to local topography and hydrodynamics that modulate larval dispersal. Adult *M. furnieri* at RPE spawn preferentially in the vicinity of the salinity front (Acha et al. 1999) taking advantage of retention processes (Braverman et al. 2009). That area offers shelter, aggregation of zooplankton, and higher temperatures which would favor larval survival (Berasategui et al. 2004). *Trachurus lathami* present on the Uruguayan continental shelf move inshore during spring and presumably spawn in the outer RPE between spring and summer (Saccardo 1987; Saccardo and Katsuragawa 1995). Off the Brazilian coast, *T. lathami* larvae have wide vertical distribution from 16 m to 200 m depth, but their preferred depth range is between 40 and 100 m (Katsuragawa and Matsuura 1992). The very low abundances for *T. lathami* found in autumn were consistent with the general pattern of adult distribution off Uruguayan coast (Saccardo 1987; Saccardo and Katsuragawa 1995). The dominance of preflexion stages in both species and the small sizes registered during all the breeding season reinforce the role of RPE as a spawning area for estuarine and marine species (Berasategui et al. 2004).

The second part of the hypothesis focused on how both species use food resources and established that *M. furnieri* and *T. lathami* differentiate their trophic niches to diminish any potential competence. In terms of food offer, RPE did not seem to be a restrictive habitat for larval feeding since it offered a wide spectrum of potential preys during the breeding season but of variable abundance according to the period studied. Among them, copepods, nauplii, and cladocerans were well represented coinciding with previous studies (Calliari et al. 2004; Colombo et al. 2003; Berasategui et al. 2006). In addition, the size range of preys available and the nutritional value of some of those (e.g., calanoid copepods *Acartia tonsa*, *Paracalanus* sp.; Machado et al. 2017) represented good scenario for larval survival.

Development of mouth throughout larval growth influence the size and type of prey ingested (Houde 1997; Sabatés and Saiz 2000; Rodríguez-Graña et al. 2005). *Micropogonias furnieri* and *T. lathami* belong to Perciformes and have looped gut type with high retention of gut content and long digestion times (Sánchez-Velazco et al. 1999). That contributes to explain the high feeding incidence of both species for all stages and daytime hours. But *M. furnieri* and *T. lathami* differed in mouth development patterns. The former showed a positive allometric growth of the mouth, i.e., mouth grew faster than body length, while for the latter, mouth and body length grew at similar rates. That would impinge on different energy/biomass investment strategies along larval development. From gap-limitation theory, that could represent a potential advantage for *M. furnieri*, which would have the chance for a faster increase in the size of consumed prey during development, compared to *T. lathami*. Nonetheless, *T. lathami* actually fed on larger prey than *M. furnieri* along the whole body length range here considered (Fig. 6b). That result suggests that mouth gape was not the key determinant for size/type of prey ingested. In fact, in most cases, prey sizes consumed by *M. furnieri* represented less than 50% of the mouth width, despite the wide range of prey sizes available at stations where this species occurred.

That finding is consistent with the fact that at preflexion stage *M. furnieri* fed intensively on tintinnids and nauplii, while *T. lathami* fed – but to a lesser extent – on nauplii and focused on juvenile and adult copepods and cladocerans. Thus, at early stages, *M. furnieri* larvae tended to compensate size of prey with quantity; i.e., they ingested many small items like tintinnids, while *T. lathami* consumed fewer prey of larger size (compare Fig. 6c, d). Both strategies are common and relate with energetic requirements of larvae during development and to the nutritionally quality of their preys (Machado et al. 2017). Tintinnids and nauplii tend to be more abundant and have more limited mobility than larger prey, so they are easy targets with low capture costs for predators (energy, time). On the contrary, positive selection toward adult copepods would imply a higher energetic reward since adult stages tend to present high fatty acid content (Evjemo et al. 2003; Kattner and Hagen 2009), a strategy followed mainly by *T. lathami* and older stages of *M. furnieri.*

Dietary overlap between both species was high when the corresponding index was estimated according to a prey-size criterion. But when taxonomic affiliation of prey was taken into account, overlap among the two species was low both in preflexion (D = 0.37) and particularly at postflexion stages (D = 0.17). Higher dietary overlap at preflexion stage could result from stronger morphological similarities for early larvae and thus in their capability to detect, capture, and ingest prey. That is a known pattern for early larval stages (Hunter 1984; Østergaard et al. 2005). However, even for early stages overlap was well below 50%, and thus it can conclude that at RPE *M. furnieri* and *T. lathami* fed on different prey taxa of similar size ranges.

Breadth of the prey-size niche and its relationship with changes in morphology along larval development (e.g., mouth size-body size ratio) showed different patterns in both species. *M. furnieri* exhibited a negative relationship between TNB_S and MW/BL, suggesting a gradual change in feeding strategy: from consumption of a wide range of prey sizes with dominance of small preys at PF stages to a focus on large-sized prey at PsF stage. The strategy of *M. furnieri* to boost energetic gains along development seems to be the change in targeted sizes with a steady consumption rate (i.e., number of prey ingested). A similar pattern was reported for blue whiting *Micromesistius poutassou* (Gonzales-Quiroz and Anadón 2001), and it has been related with morphological and physiological changes during the flexion of the notochord and with the formation of the hypural complex. Such development increases larval motility and swimming speed and thus the capture efficiency of motile and larger preys (Hunter 1984; Sabatés and Saiz 2000). In *T. lathami*, TNB_S and the WM-BL ratio did not seem to change along development. This species fed on a wide spectrum of preys along the whole larval stage, with PsF incorporating prey of increasing size while also keeping small preys (<100 μm) as part of their diet. The mean size of prey increased along development, but following a weaker and much more variable pattern than *M. furnieri* (Fig. 6b). The strategy of *T. lathami* to increase energetic gains along development was based on enhancing consumption rates, with minor increases in prey size as larvae grow (Fig. 6d). This strategy probably provides advantages in environments with moderate food availability like neritic zones and open marine waters, where this species tends to be most abundant.

In synthesis, *M. furnieri* and *T. lathami* tended to be spatially segregated within the estuary, but differences in environmental preferences were not straightforward. Differences in the development of mouth during ontogeny resulted in different feeding strategies, which in turn were reflected in the trophic niche of both species. Prey-size niche breadth was higher for *T. lathami*, and decreased along development for *M. furnieri* consistent with previous studies for other species and ecosystems (Pearre 1986; Munk 1992; Sabatés and Saiz 2000).

Acknowledgments Thanks to the working team at Dirección Nacional de Recursos Acuáticos (DINARA) G. Mantero, C. Mesones, A. Martínez, and L. Ortega for their help with identification of fish larvae and for providing environmental data and access to plankton samples. Special thanks to I. Machado, D. Cambón, and R. Castiglioni (Facultad de Ciencias, Universidad de la República) for their collaboration during field work. This study was supported by contract PDT S/C/OP/36/10 (DINACYT, Uruguay). M Vera was partially funded by Programa para el Desarrollo de las Ciencias Básicas (PEDECIBA) and by Marie Curie Grant (EU- IIF – FP 6, MIF1-CT-2006-021529).

References

Acha EM, Macchi GJ (2000) Spawning of Brazilian menhaden, *Brevoortia aurea*, in the Río de la Plata estuary off Argentina and Uruguay. Fish Bull 98:227–235

Acha EM, Mianzán H, Lasta C et al (1999) Estuarine spawning of the withe mouth croaker *Micropogonias furnieri* (Pisces: Scienidae) in the Río de la Plata. Argentina Mar Freshwater Res 50:57–65

Acuña A, Viana F, Vizziano D, Danulat E (2000) Reproductive cycle of female Brazilian codling *Urophycis brasiliensis* (Kaup 1858) caught off the Uruguayan coast. J Appl Ichthyol 16:48–55

Alvarez Colombo G, Mianzán H, Madirolas A (2003) Acoustic characterization of gelatinous-plankton aggregations: four case studies from the Argentine continental shelf. ICES J Mar Sci 60:650–657

Balon EK (1975) Reproductive guilds of fishes: a proposal and definition. J Fish Res Board Can 32:821–864

Begon M, Harper JL, Townsend CR (1996) Ecology: individuals, populations, and communities, 3rd edn. Blackwell Science Ltd., Cambridge, MA

Berasategui AD, Acha EM, Araoz NCF (2004) Spatial patterns of icthyoplankton assemblages in the Río de la Plata Estuary (Argentina–Uruguay). Est Coast Shelf S 60:599–610

Berasategui AD, Menu Marque S, Gómez-Erache M et al (2006) Copepod assemblages in a highly complex hydrographic region. Estuar Coast Shelf Sci 66:483–492

Berry FH, Cohen L (1974) Synopsis of the species of *Trachurus* (Pisces: Carangidae). Quart J Fla Acad Sci 35(4):177–211

Braverman MS (2011) Historia de vida temprana de la corvina rubia (*Micropogonias furnieri*, Sciaenidae) en el estuario del Río de la Plata. Dissertation Universidad de Buenos Aires

Braverman MS, Acha M, Gagliardini DA et al (2009) Distribution of whitemouth croaker (*Micropogonias furnieri*. Desmarest 1823) larvae in the Río de la Plata estuarine front. Estuar Coast Shelf Sci 82:557–565

Burnham K, Anderson D (2002) Model selection and multimodel inference. A practical information-theoretic approach, 2nd edn. Springer, New York

Busch A (1996) Transition from endogenous to exogenous nutrition:larval size parameters determining the start of external feeding and size of prey ingested by Ruegen spring herring *Clupea harengus*. Mar Ecol Prog Ser 130:39–46

Calliari D, Cervetto G, Castiglioni R (2004) Summertime herbivory and egg production by *Acartia tonsa* at the Montevideo coast – Río de la Plata. Ophelia 58:115–128

Calliari D, Gómez M, Gómez N (2005) Biomass and composition of the phytoplankton in the Río de la Plata: large-scale distribution and relationship with environmental variables during a spring cruise. Cont Shelf Res 25:197–210

Calliari D, Cervetto G, Castiglioni R et al (2007) Salinity preferences and habitat partitioning between dominant mysids at the Río de la Plata estuary (Uruguay). J Mar Biol Assoc UK 87:1–6

Calliari D, Brugnoli E, Ferrari G, Vizziano D (2009) Phytoplankton distribution and production along a wide environmental gradient in the South-West Atlantic off Uruguay. Hydrobiologia 620:47–61

Cervetto G, Castiglioni R, Rodríguez-Graña L et al (2016) Mysids from the Río de la Plata estuary, with new record of *Mysidopsis rionegrensis*, Hoffmeyer 1993, and *Promysis atlantica* Tattersall, 1923 (Mysida, Mysidae). Panamjas 11(3):179–187

Cosseau MB, Perrota RG (2000) Peces marinos de Argentina: biología, distribución, pesca. INIDEP, Mar del Plata

de Carvalho MR, Hashimoto Soares LS (2006) Diel feeding pattern and diet of rough scad *Trachurus lathami* Nichols, 1920 (Carangidae) from the Southwestern Atlantic. Neotrop Ichthyol 4(4):419–426

Evjemo JO, Kjell IR, Olsen Y (2003) Copepods as live food organisms in the larval rearing of halibut larvae (*Hippoglossus hippoglossus* L.) with special emphasis on the nutritional value. Aquaculture 227:191–210

García ML, Jaureguizar AJ, Protogino LC (2010) From fresh water to the slope: fish community ecology in the Río de la Plata and the sea beyond. Lat Am J Aquat Res 38(1):81–94
Gonzáles-Quiroz R, Anadón R (2001) Diet breadth variability in larval blue withing as a response to plankton size structure. J Fish Biol 59:1111–1125
Govoni JJ, Hoss DE, Chester AJ (1983) Comparative feeding of three species of larval fishes in the northern Gulf of Mexico: *Brevoortia patronus*, *Leiostomus xanthurus*, and *Micropogonias undulatus*. Mar Ecol Prog Ser 13:189–199
Guerrero RA, Acha M, Framiñan MB et al (1997) Physical oceanography of the Río de la Plata Estuary, Argentina. Cont Shelf Res 17(7):727–742
Horn H (1966) Measurement of "overlap" in comparative ecological studies. Am Nat 100:419/424
Houde ED (1997) Patterns and consequences of selective processes in teleosts early life histories. In: Chambers RC, Trippel EA (eds) Early life history and recruitment in fish populations. Chapman & Hall, London, pp 173–196
Hunter JR (1984) Feeding ecology and predation of marine fish larvae. In: Lasker R (ed) Marine fish larvae: morphology, ecology and relation to fisheries. Washington Sea Grant Program, Seattle, pp 34–77
Jaureguizar A, Bava J, Carozza CR et al (2003) Distribution of whitemouth croaker *Micropogonias furnieri* in relation to environmental factors at the Río de la Plata estuary, South America. Mar Ecol Prog Ser 255:271–282
Katsuragawa M, Matsuura Y (1992) Distribution and abundance of carangid larvae in the southeastern Brazilian Bight during 1975–1981. Bol Inst Oceanogr S Paulo 40(1/2):55–78
Katsuragawa M, Dias JF, Harari J et al (2014) Patterns in larval fish assemblages under the influence of the Brazil current. Cont Shelf Res 89:103–117
Kattner G, Hagen W (2009) Lipids in marine copepods: latitudinal characteristics and perspective to global warming. In: Arts MT, Brett MT, Kainz M (eds) Lipids in aquatic ecosystem. Springer, New York, pp 257–280
Kruk C, Martínez A, Nogueira L et al (2015) Morphological traits variability reflects light limitation of phytoplankton production in a highly productive subtropical estuary (Río de la Plata, South America). Mar Biol 162(2):331–341
Lasta CA, Acha EM (1996) Cabo San Antonio: su importancia en el patrón reproductivo de peces marinos. Publ Com Tec Mix Frente Marit 16:39–46
Macchi GJ (1997) Reproducción de la corvina rubia (*Micropogonias furnieri*) del sector rioplatense Su relación con los gradientes horizontales de salinidad. Rev Invest Des Pesq (Argentina) 11:73–94
Macchi GJ, Acha EM (2000) Spawning frequency and batch fecundity in the Brazilian menhaden, *Brevoortia aurea*, in the Río de la Plata estuary off Argentina and Uruguay. Fish Bull 98:283–289
Machado I, Calliari D, Denicola A et al (2017) Coupling suitable prey field to in situ fish larval condition and abundance in a subtropical estuary. Estuar Coast Shelf Sci 187:31–42
Mianzan H, Guerrero R (2000) Environmental patterns and biomass distribution of gelatinous macrozooplankton. Three study cases in the Southwestern Atlantic. Sci Mar 64(1):215–224
Milstein A, Juanicó M (1985) Zooplankton dynamics in Maldonado Bay (Uruguay). Hydrobiologia 126:155–164
Moser HG (1984) Morphological and functional aspects of marine fish larvae. In: Lasker R (ed) Marine fish larvae: morphology, ecology and relation to fisheries. Washington Sea Grant Program, Seattle, pp 90–131
Munk P (1992) Foraging behaviour and prey size spectra of larval herring *Clupea harengus*. Mar Ecol Prog Ser 80:149–158
Nagy G, Martínez C, Caffera RM et al (1997) The hydrological and climatic setting of the Río de la Plata. In: Well PG, Daborn GR (eds) The Río de la Plata, an environmental overview. Dalhousie University, Halifax, pp 17–68
Norbis W (1995) Influence of wind, behaviour and characteristic of the croaker (*Micropogonias furnieri*) artisanal fishery in the Rio de la Plata (Uruguay). Fish Res 22:43–58

Østergaard P, Munk P, Janekarn V (2005) Contrasting feeding patterns among species of fish larvae from the tropical Andaman Sea. Mar Biol 146:595–606

Pacheco Tak RL (1989) Contribución al conocimiento de la biología pesquera del Surel *Trachurus picturatus australis* Nani, 1950 (Piscis: Carangidae) con algunas consideraciones ecológicas. Dissertation Universidad Nacional de La Plata

Pearre S (1986) Ratio-based trophic niche breadths of fish, the Sheldon spectrum, and the size efficiency hypothesis. Mar Ecol Prog Ser 29:299–314

Pedreira MM (1997) Alimentação e hábito alimentar de larvas de *Trachurus lathami* (família Carangidae), na região de Ubatuba, Estado de São Paulo. Dissertation Universidade de São Paulo

Pepin P, Penney RW (1997) Patterns of prey size and taxonomic composition in larval fish: are there general size dependent models? J Fish Biol 51(A):84–100

Perry RI, Smith SJ (1994) Identifying habitat associations of marine fishes using survey data: an application to the Northwest Atlantic. Can J Fish Aquat Sci 51:589–602

Pin O, Defeo O (2000) Modelos de producción captura -mortalidad para la pesquería de corvina *Micropogonias furnieri* (Desmarest, 1823) en el Río de la Plata y zona común de pesca Argentino- Uruguaya (1975–1986). In: Rey M, Arena G (eds) Modelos de producción excedente aplicado a los recursos corvina y pescadilla. Proyecto URU/92/003 INAPE-PNUD, Montevideo, pp 31–65

Pörtner HO, Schulte PM, Wood CM et al (2010) Niche dimensions in fishes: an integrative view. Physiol Biochem Zool 83(5):808–826

R Core team (2015) R: A Language and Environment for Statistical Computing. R Foundation for Statistical Computing. Vienna, https://www.R-project.org. Accessed 3 July 2017

Rodríguez- Graña L, Castro L, Loureiro M et al (2005) Feeding ecology of dominant larval myctophids in an upwelling area of Humbolt current. Mar Ecol Prog Ser 290:119–134

Sabatés A, Saiz E (2000) Intra- and interspecific variability in prey size and Niche breadth of mictophyform fish larvae. Mar Ecol Prog Ser 201:261–271

Saccardo SA (1987) Morfologia, distribuição e abundância de Trachurus /atham; Nichols, 1920 (Teleostei: Carangidae) na região sudeste-sul do Brasil. Bolm Inst Oceanogr S Paulo 35(1):65–95

Saccardo SA, Katsuragawa M (1995) Biology of rough scad *Trachurus lathami*, on the southeastern coast of Brazil. Sci Mar 59:265–277

Sampaio de Souza C, Mafalda P Jr (2008) Distribution and abundance of Carangidae (Teleostei, Perciformes) associated with oceanographic factors along the northeast Brazilian exclusive economic zone. Braz Arch Biol Technol 51(6):1267–1278

Sánchez-Velasco L, Contreras-Arredondo I, Esqueda-Escárcega G (1999) Diet composition of *Euthynnus lineatus* and *Auxis* sp. larvae (Pisces: Scombridae) in the Gulf of California. Bull Mar Sci 65(3):687–698

Schiariti A, Berasategui AD, Giberto DA et al (2006) Living in the front: *Neomysis americana* (Mysidacea) in the Río de la Plata estuary, Argentina-Uruguay. Mar Biol 149:483–489

Schoener T (1968) The Anolis lizards of Bimini: resource partitioning in a complex fauna. Ecology 49:704–726

Simionato CG, Berasategui A, Meccia VL et al (2008) On the short time-scale wind forced variability in the Río de la Plata Estuary and its role on ichthyoplankton retention. Estuar Coast Shelf Sci 76:211–226

Vera M (2011) Distribución y ecología trófica en larvas de corvina (*Micropogonias furnieri*) y surel (*Trachurus lathami*) en el Río de la Plata. Dissertation, Universidad de la República

Wallace RK Jr (1981) An assessment of diet-overlap indexes. Trans Am Fish Soc 110:72–76

Zar JH (1996) Biostatistical analysis, 3rd edn. Prentice-Hall Inc. Upper Saddle River

Zaret TM, Rand AS (1971) Competition in tropical stream fishes: support for the competitive exclusion principle. Ecology 52:336–342

Part V
Gelatinous Zooplankton

Diversity, Species Composition and Assemblage Dynamics of Estuarine Gelatinous and Semi-Gelatinous Zooplankton from Brazil

Miodeli Nogueira Jr., Lorena Silva do Nascimento, Pedro Vieira Maciel, Sebastião Tilbert, and Lívia Dias Oliveira

Abstract The literature on gelatinous (Cnidaria, Ctenophora, Tunicata) and semi-gelatinous (Chaetognatha) zooplankton from 32 Brazilian estuaries is reviewed. Altogether 104 species have been recorded, 70 cnidarians, 2 ctenophores, 14 chaetognaths, 13 appendicularians and 5 thaliaceans. All groups are understudied with only few detailed data available. Most of these gelatinous and semi-gelatinous taxa are typically oceanic, and thus the low diversity in estuarine systems was expected and is not probable to increase much with increasing sampling effort. Contrary, the meroplanktonic hydromedusae from the orders Anthoathecata and Leptothecata are diversified in estuarine ecosystems and the respectively 29 and 19 species currently reported from Brazilian estuaries are an underestimation and certainly will increase considerably with increased sampling effort. The best-studied taxa are the chaetognaths, followed by appendicularians, and very little is known about the other taxa, particularly concerning ctenophores and thaliaceans. We summarize the main tendencies of each taxon regarding diversity, abundance, temporal and spatial variations, species composition, and the most abundant species. In addition, we also provide general guidelines for future research on gelatinous and semi-gelatinous zooplankton taxa from Brazilian estuaries.

Keywords Cnidaria · Ctenophora · Chaetognatha · Tunicata · Brazilian estuaries

M. Nogueira Jr. (✉) · P. V. Maciel · L. D. Oliveira
Departamento de Sistemática e Ecologia,
Universidade Federal da Paraíba, João Pessoa, PB, Brazil

L. S. do Nascimento
Centro de Estudos do Mar, Universidade Federal do Paraná, Paraná, Brazil

S. Tilbert
Departamento de Sistemática e Ecologia,
Universidade Federal da Paraíba, João Pessoa, PB, Brazil

Laboratório de Meiofauna, Centro de Educação e Saúde,
Universidade Federal de Campina Grande, Cuité, PB, Brazil

M. S. Hoffmeyer et al. (eds.), *Plankton Ecology of the Southwestern Atlantic*,
https://doi.org/10.1007/978-3-319-77869-3_18

1 Introduction

Zooplankton is highly diversified and a paramount component of aquatic ecosystems mainly due to their trophic role as link between the primary producers and higher trophic levels. Thus, understanding the factors influencing its abundance, biomass, assemblage composition and dynamics is fundamental to the general comprehension of the functioning of these ecosystems (e.g., Boltovskoy 1999). Many zooplankton taxa, not directly related phylogenetically, have aqueous tissues in different degrees, such as cnidarians, ctenophores, polychaetes, chaetognaths and tunicates, in addition to the larvae of other groups such as echinoderms (Hamner et al. 1975; Larson 1986). These watery tissues are an evolutive convergence with advantages such as:

(i). Higher sizes may be achieved with relatively little carbon investments and consequently the metabolism can be relatively low (Larson 1986; Acuña et al. 2011).
(ii). Rapid individual and population growth and production (Paffenhöfer and Lee 1987; Hopcroft and Roff 1998; Acuña et al. 2011).
(iii). Larger sizes increase the possibility of encounters as well as the size range of a possible prey and, in addition, reduce the chances of predation by smaller organisms (Larson 1986; Arai 1997; Kremer 2002).
(iv). Greater corporal plasticity, allowing them to grow, shrink, and grow again depending on the environmental conditions and food availability (Arai 1997).
(v). They are frequently transparent and invisible to visual predators which is a way to hide in the tridimensional and relatively homogeneous pelagic habitat (Hamner et al. 1975; Johnsen 2001).
(vi). Large amount of water in the tissues results in a better buoyancy (Larson 1986).

Thus, among the many functional zooplankton classifications, such as according to the life-cycle strategy or to the body size, the relative amount of water in the tissues can be used to classify zooplankton taxa as gelatinous, semi-gelatinous, or non-gelatinous organisms (Larson 1986).

These watery organisms are often understudied, in part due to their body fragility that frequently hinders their sampling and preservation in conditions to allow accurate taxonomic classification and sometimes even to enumerate them (Hamner et al. 1975; Esnal and Daponte 1999a, b; Mianzan 1999). However, they represent an important fraction of the marine zooplankton biodiversity and may reach high densities, typically having an important ecological role in the pelagic realm and energy flow (Boltovskoy 1999). Both carnivores (mainly cnidarians, ctenophores, and chaetognaths) and herbivores (mainly tunicates) commonly have high feeding rates, with crucial role as consumers in the pelagic domain. It is not uncommon for gelatinous and semi-gelatinous zooplankton to negatively affect populations of other organisms, including those with commercial value, either through direct predation or competition for food (Mills 1995; Kremer

2002). In this way, it can usually be expected that these organisms are relevant to the diversity, dynamics, and community structuring of the pelagic ecosystems and also affect several human activities at the sea such as tourism and fisheries (e.g., Nagata et al. 2009).

Estuaries can generally be defined as coastal water bodies, usually semi-enclosed, which are permanently or intermittently connected to the sea and where marine and freshwater from continental runoff interact (Potter et al. 2010; Whitfield and Elliot 2011). These ecosystems have high ecological, economic and cultural importance. Their biological productivity is typically very high, harboring an abundant, diversified and characteristic biota which includes many important fishing resources (Daily et al. 1997; Lana et al. 2001; Cremer et al. 2006). Additionally, human occupation for habitation commercial and recreational purposes is historic and massive nearby estuaries worldwide, posing a great threat to these environments.

Since estuaries are transitional ecosystems, spatial gradients are typical particularly regarding the salinity (Potter et al. 2010; Whitfield and Elliot 2011). Physical conditions of an estuary are vital in determining its biological components such as primary production, biomass, diversity, trophic pathways and species composition. This may be particularly true for zooplankton assemblages which are under direct influence of the hydrography and are commonly sensitive to changes in the physical structure of the water column. Features such as salinity, tides, temperature, rainfall and continental runoff are typically the most important ones. These parameters change horizontally and/or vertically mainly influenced by the mixing, circulation and stratification processes of the estuarine systems (Daily et al. 1997), as well as seasonally, when it is mostly controlled by local rainfall regime and general latitudinal climatic patterns.

The Brazilian coast is enormous with over 8,000 km spanning from the tropics (~2°N) to almost temperate areas (~33.7°S) and embraces countless estuarine/brackish-water ecosystems, most of them with high ecological, economic, cultural importance and under high human pressure (e.g., Lana et al. 2001; Cremer et al. 2006; Mahiques et al. 2013). Gelatinous and semi-gelatinous zooplankton taxa are primarily marine. Yet, representatives of all major groups can be found inside estuaries where they may reach high abundances, biomass, and/or production rates and have a significant role in the trophic chains (Dagg et al. 1996; Dagg and Brown 2005; Sato et al. 2008; Spinelli et al. 2009; Marques et al. 2017). Most of the (few) information available on these estuarine gelatinous and semi-gelatinous zooplankton taxa from Brazil is superficial and scattered through general zooplankton publications. In this study, we made an effort to summarize all the available data on gelatinous (cnidarians, ctenophores, and tunicates) and semi-gelatinous (chaetognaths) zooplankton from Brazilian estuaries, focusing on species composition, diversity, abundance and assemblage structure trends. We were able to gather data for at least one major gelatinous/semi-gelatinous taxon from 32 estuaries (Fig. 1, Table 1), spanning a wide latitudinal range (0–32°S). Not all estuaries and/or taxa are known at the same depth, but we tried to include all information available and summarize the general tendencies of each taxon.

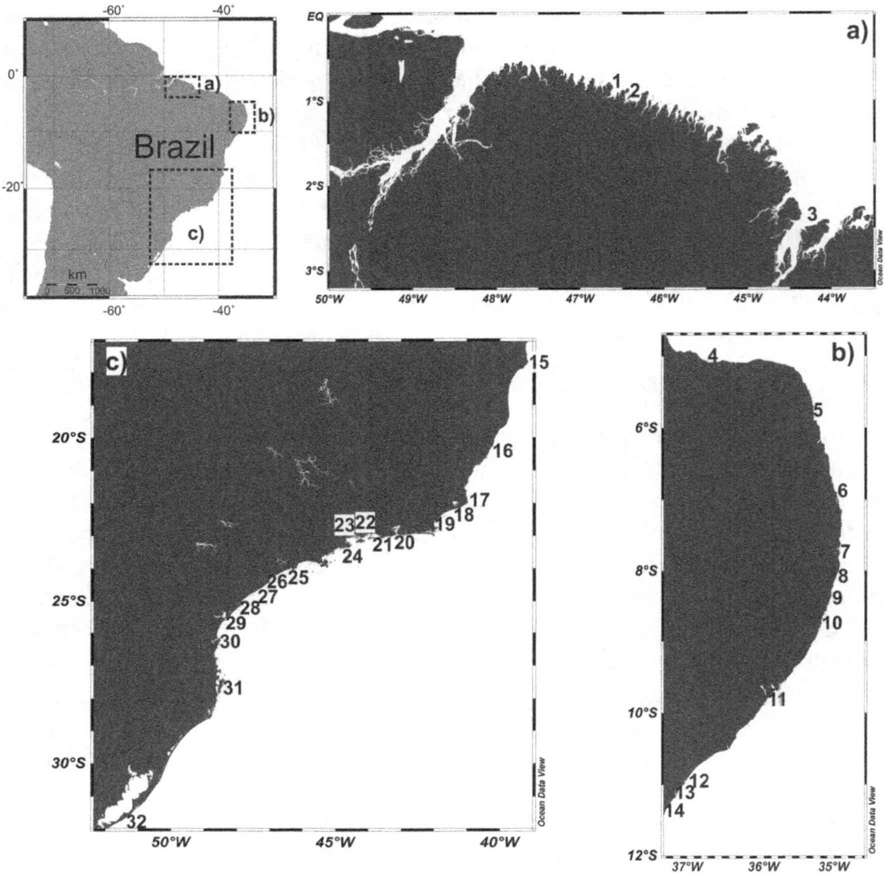

Fig. 1 Map of the Brazilian coast showing the location of the estuaries considered in the present review (1–32). For codes of the numbers and data sources, please see Table 1

2 Cnidaria

There are quite few studies dealing with estuarine planktonic cnidarians from Brazil (e.g., Navas-Pereira 1980; Montú and Cordeiro 1988; Mesquita et al. 2006). A total of 70 species have been recorded (Table 2); most of them are meroplanktonic hydromedusae, particularly those from the orders Anthoathecata (29 spp.) and Leptothecata (19 spp.) which commonly are highly diversified in shallow coastal waters. This number is probably underestimated considering that most Brazilian estuaries have not been studied adequately or at all, along with the inability to properly identify down to species level the medusae of some genera such as *Clytia* and *Obelia*; at least three different medusa species of the former have been reported from Brazilian estuaries (Mesquita et al. 2006; Nogueira Júnior 2012, Table 2); however we pulled all together as *Clytia* spp. since we cannot exactly determine which species each

Table 1 Summary of the Brazilian estuaries included in this study showing their minimum and maximum recorded temperature (Temp, in °C) and salinity (Sal), data availability for each taxa (Cnid, Cnidaria; Cten, Ctenophora; Chae, Chaetognatha; Tun, Tunicata) along with the source of the information. N°, number of the estuary in the Fig. 1; nda, no data available

Estuary	N°	Cnid	Cten	Chae	Tun	Temp	Sal	Data source
Taperaçu River (0°45'S/46°44'W)	1			X	X	25.9–30.1	9.1–40	Costa et al. (2008)
Caeté River (0°55'S/45°36'W)	2	X		X	X	25.7–28.3	0–37	Krumme and Liang (2004), Mesquita et al. (2006)
São Marcos Bay (2°32'S/44°36'W)	3			X	X	22.7–28.9	16.4–24.1	Bonecker and Dias (2009)
Conceição and Casqueira Rivers (5°06'S/36°35'W)	4			X	X	27.5–31	31–43.5	Sankarankutty et al. (1997)
Potengi River (5°45'S/35°12'W)	5			X	X	25.5–30	26.8–37.5	Sankarankutty et al. (1979, 1995, 1996), Esnal et al. (1985)
Paraíba River (6°58'S/34°51'W)	6	X	X	X	X	25–30	2.5–36.8	Singarajah (1978), MNJ unpublished
Santa Cruz Channel and Itamaracá estuarine system (7°42'S/34°54'W)	7			X	X	25–31	17.4–35.5	Porto Neto et al. (1999), Silva et al. (2003), Freire et al. (2014)
Capibaribe and Jiquiá Rivers (8°05'S/34°51'W)	8	X				26.8–29.5	2.2–36	Freire et al. (2014)
Barra das Jangadas estuarine system (8°14'S/34°56'W)	9	X		X	X	25.2–29.6	0.3–33.4	Paranaguá (1963), Cavalcanti et al. (2008)
Suape Bay (8°22'S/35°05'W)	10	X		X	X	23.4–33	0.05–36	Neumann-Leitão and Matsumura-Tundisi (1998), Silva et al. (2004)
Mundaú/Manguaba estuarine system (9°35'S/35°42'W)	11			X		23–29	0.13–37	Magalhães et al. (1996)
Sergipe River (10°8'S/37°2'W)	12	X		X	X	23–29.9	8.6–32.6	Araújo et al. (2008), Freire et al. (2014)

(continued)

Table 1 (continued)

Estuary	N°	Cnid	Cten	Chae	Tun	Temp	Sal	Data source
Tabatinga River (11°32'S/37°29'W)	13	X		X		24.5–30	5.9–26.3	Marcolin (2008), Marcolin et al. (2010)
Itapicuru River (11°47'S/37°30'W)	14	X		X	X	25–31	8.6–36.3	Marcolin (2008), Marcolin et al. (2010)
Caravelas River (17°44'S/39°04'W)	15				X	nda	nda	Carvalho and Bonecker (2010)
Vitória Bay estuarine system (20°18'S/40°20'W)	16			X	X	21–28	12–38	Loureiro Fernandes et al. (2005), Sterza and Loureiro Fernandes (2006)
Macaé River (22°22'S/41°46'W)	17				X	23–24.5	18.5–22.5	Carvalho et al. (2016)
Imboassica Lagoon (22°24'S/44°42'W)	18				X	18.1–31.1	0–34	Branco et al. (2007)
São João River (22°36'S/41°59'W)	19				X	22.5–26	21–23.5	Carvalho et al. (2016)
Guanabara Bay (22°45'S/43°08'W)	20			X		22–26	12–31.2	Marazzo and Nogueira (1996), Mendes et al. (2012)
Sepetiba Bay (22°55'S/43°40'W)	21	X		X	X	20–29.9	0.003–33.9	Navas-Pereira (1980), Coelho-Botelho et al. (1999)
Bracuí River (22°57'S/44°23'W)	22				X	21–29	20–28.5	Carvalho et al. (2016)
Ribeira Bay (22°59'S/44°25'W)	23			X	X	21–36.4	29–36	Dias and Bonecker (2008)
Perequê-Açu River (23°12'S/44°42'W)	24				X	21.5–30	17.5–22	Carvalho et al. (2016)
Santos estuarine system (23°50'S/46°25'W)	25	X		X		19.2–32.5	4.2–35.8	Pereira (2011), Nogueira Júnior and Silva Nascimento (2018), MNJ unpublished

Una do Prelado River (24°S/47°03'W)	26	X		X	X	20–31	0–34.4	Lansac-Tôha and Lima (1993)
Guaraú River (24°22'S/7°06'W)	27			X	X	17–29	0.5–35.4	Lopes (1994)
Cananéia estuarine system (25°S/47°46'W)	28	X	X	X	X	18.6–32.6	4.5–33	Vannucci (1954, 1956, 1957), Teixeira et al. (1965), Morandini (2003), Liang et al. (2003), Bardi and Marques (2009), Bardi (2011), Martinez et al. (2015), Nogueira Júnior and Silva Nascimento (2018), MNJ unpublished
Paranaguá estuarine system (25.5°S/48°17'W)	29	X	X	X	X	16.9–30.5	0–34	Montú and Cordeiro (1988), Lopes et al. (1998), Nogueira Júnior and Oliveira (2006), Bardi and Marques (2009), Bardi (2011), Miyashita et al. (2012), Haddad et al. (2014), Nascimento (2016), Nogueira Júnior and Silva Nascimento (2018), L.K. Miyashita, personal communication, MNJ unpublished
Babitonga estuarine system (26.2°S/48.6°W)	30	X	X	X	X	19.2–27	10.1–33.6	Bardi and Marques (2009), Oliveira (2007), Bardi (2011), Nogueira Júnior (2012), Nogueira Júnior et al. (2015b,), Nogueira Júnior and Silva Nascimento (2018), MNJ unpublished
North and South Bays (27°30'S/48°32'W)	31	X	X	X	X	17.5–28	29.3–37.8	Resgalla (2001), Nogueira Júnior et al. (2010), MNJ unpublished
Patos lagoon (31°50'S/52°2'W)	32	X		X		8.4–31	0–35	Montú (1980), Teixeira-Amaral et al. (2017)

Table 2 List of gelatinous and semi-gelatinous estuarine zooplankton species recorded from Brazil. N = 104 spp. See Table 1 and Fig. 1 to see data source, estuary codes and locations. To avoid artificial inflation of the number of species due to taxonomic problems, only unique morphotypes were considered in the total species count. The interrogation indicates doubtful taxa, probably misidentifications which were disregarded from the total species count (see text)

Taxa	Estuary	Taxa	Estuary
Phylum Cnidaria		**Order Narcomedusae**	
Class Hydrozoa		**Family Cuninidae**	
Actinula larvae	25, 28–30	*Cunina octonaria* McCrady, 1859	21, 25, 28–30
Subclass Hydroidolina		**Family Solmarisidae**	
Order Anthoathecata		*Solmaris* sp.	28, 29
Suborder Filifera		**Order Trachymedusae**	
Family Bougainvilliidae		**Family Geryoniidae**	
Bougainvillia muscus Allman, 1863	12–14, 21, 25, 28, 29	*Liriope tetraphylla* Chamysso and Eyesenhardt, 1821	1, 12–14, 21, 25, 28–32
Bougainvillia pagesi Nogueira et al. 2013	30	*Geryonia proboscidalis* Forskål, 1775	29
Bougainvillia frondosa Mayer, 1900	29	**Family Rhopalonematidae**	
Bougainvillia carolinensis McCady, 1859	25, 29–30	*Aglaura hemistoma* Péron and Lesueur, 1810	28, 29
Family Hydractiniidae		?*Aglantha* sp.?	29
Cnidostoma fallax Vanhöffen, 1911	25, 28–30, 32	*Crossota* sp.	1
Hydractinia sp.	30	*Rhopalonema velatum* Gegenbaur, 1857	21
Podocoryna loyola Haddad, Bettim and Miglieta, 2014	25, 28–30	**Class Scyphozoa**	
Podocoryna sp.	26	**Subclass Discomedusae**	
Family Oceaniidae		**Order Semaeostomeae**	
Turritopsis nutricula McCrady, 1857	25, 28–30	**Family Pelagiidae**	
Family Pandeidae		*Chrysaora lactea* Eschscholtz, 1829	6, 25, 28–31
Amphinema australis Mayer, 1900	21	**Family Ulmaridae**	
Amphinema dinema Perón and Lesueur, 1810	28–30	*Aurelia* sp.	28, 31
Amphinema sp.	1, 25	**Order Rhizostomeae**	
Family Proboscidactylidae		**Family Lychnorhizidae**	
Proboscidactyla ornata McCrady, 1859	28–30	*Lychnorhiza lucerna* Haeckel, 1880	6, 25, 28–31
Family Protiaridae		**Family Rhizostomatidae**	
Halitiara formosa Fewkes, 1882	30	?*Rhizostoma pulmo* (Macri, 1778)?	6
Protiara sp.	30	**Family Mastigiidae**	

(continued)

Table 2 (continued)

Taxa	Estuary	Taxa	Estuary
Family Rathkeidae		*Phyllorhiza punctata* Lendenfeld, 1884	28–31
Podocorynoides minima Trinci, 1903	21, 28, 30	**Family Stomolophidae**	
Lizzia blondina Forbes, 1848	28, 29	*Stomolophus meleagris* Agassiz, 1862	28, 29
Suborder Capitata		**Class Cubozoa**	
Family Calycopsidae		**Order Carybdeida**	
Heterotiara sp.	1	**Family Tamoyidae**	
Family Corymorphidae		*Tamoya haplonema* Müller, 1859	28, 31
Corymorpha gracilis Brooks, 1883	21, 28–30	**Order Chirodropida**	
Corymorpha forbesii Mayer, 1894	21, 25, 28–30	**Family Chiropsalmidae**	
Corymorpha januarii Steenstrup, 1854	25, 29, 30	*Chiropsalmus quadrumanus* Müller, 1859	28–31
Family Corynidae		**Phylum Ctenophora**	
Stauridiosarsia reesi Vannucci, 1956	25, 28–30	**Class Nuda**	
Coryne eximia Allman, 1859	25, 28, 29	**Order Beroida**	
Sarsia sp.1	1	**Family Beroidae**	
Sarsia sp.2	1	*Beroe ovata* Bruguière, 1789	6, 25, 28–31
Family Halimedusidae		**Class Tentaculata**	
Tiaricodon sp.**	28	**Order Cydippida**	
Family Moerisiidae		**Family Pleurobrachiidae**	
Moerisia inkermanica Paltschikowa-Ostroumova, 1925	9, 28, 29, 30	?*Pleurobrachia* sp.?	6
Family Tubulariidae		**Order Lobata**	
Ectopleura dumortieri van Beneden, 1844	21, 25, 28–30	**Family Bolinopsidae**	
Capitata incertae sedis		*Mnemiopsis leidyi* A. Agassiz, 1865	6, 25, 28–31
Paulinum sp.	30	**Phylum Chaetognatha**	
Family Zancleopsidae		**Class Sagittoidea**	
Zancleopsis dichotoma Mayer, 1900	21	**Order Aphragmophora**	
Zanclea costata Gegenbaur, 1857	29	**Family Sagittidae**	
Order Leptothecata		*Caecosagitta macrocephala* Fowler, 1904	6, 23
Family Aequoreidae		*Decipisagitta decipiens* Fowler, 1905	16, 23

(continued)

Table 2 (continued)

Taxa	Estuary	Taxa	Estuary
Rhacostoma atlanticum L. Agassiz, 1851	30, 31	*Ferosagitta hispida* Conant, 1895	3, 16, 20, 21, 23, 25, 28, 31
Aequorea sp.	31	*Flaccisagitta enflata* Grassi, 1881	5, 6, 11–14, 16, 21, 23, 25, 28–32
Family Blackfordiidae		*Flaccisagitta hexaptera* d'Orbigny, 1836	23
Blackfordia virginica Mayer, 1910	7–9, 25, 28–30	*Mesosagitta minima* Grassi, 1881	16, 23
Family Cirrholovenidae		*Parasagitta friderici* Ritter–Záhony, 1911	1, 2, 7, 9, 16, 20, 21, 23, 25–32
Cirrholovenia tetranema Kramp, 1959	28–30	*Parasagitta tenuis* Conant, 1896	3–5, 7, 9, 10, 12–14, 21, 28–31
Family Eirenidae		*Sagitta helenae* Ritter–Záhony, 1911	5
Eirene sp.	28, 30	*Sagitta bipunctata* Quoy and Gaimard, 1827	6, 23
Eutima mira McCrady, 1859	25, 28–30	*Serratosagitta serratodentata* Krohn, 1853	6, 23
Eutima sp.	12	**Family Krohnittidae**	
Helgicirrha sp.	25, 28–30	*Krohnitta pacifica* Aida, 1897	23, 28
Family Laodiceidae		*Krohnitta subtilis* Grassi, 1881	6, 23
Laodicea minuscula Vannucci, 1957	25, 28–30	**Family Pterosagittidae**	
Family Lovenellidae		*Pterosagitta draco* Krohn, 1853	23
Eucheilota duodecimalis A. Agassiz, 1862	21, 25, 28–30	**Phylum Chordata**	
Eucheilota maculata Hartlaub, 1894	25, 28–30	**Subphylum Tunicata**	
Eucheilota paradoxica Mayer, 1900	21, 25, 28–30	**Class Appendicularia**	
Family Malagazziidae		**Order Copelata**	
Malagazzia carolinae Mayer, 1900	21, 25, 28–30	**Family Oikopleuridae**	
Malagazzia sp.	1	*Oikopleura (Coecaria) cornutogastra* Aida, 1907	23
Octophialucium haeckeli Vannucci and Moreira, 1966	30	*Oikopleura (Coecaria) fusiformis* Fol, 1872	15, 16, 21, 22
?*Octophialucium funerarium* (Quoy and Gaimard, 1827)?	6	*Oikopleura (Coecaria) gracilis* Lohmann, 1896	21, 23
Family Tiaropsidae	1	*Oikopleura (Coecaria) intermedia* Lohmann, 1896	21

(continued)

Table 2 (continued)

Taxa	Estuary	Taxa	Estuary
Tiaropsidium sp.	1	*Oikopleura (Coecaria) longicauda* Vogt, 1854	7, 9, 10, 12, 15–19, 22, 23, 25, 28–30
Suborder Proboscoida		*Oikopleura (Vexillaria) albicans* Leuckart, 1853	21
Family Campanulariidae		*Oikopleura (Vexillaria) cophocerca* Gegenbaur, 1855	15, 23
Obelia spp.	10, 21, 25, 26, 28–30	*Oikopleura (Vexillaria) dioica* Fol, 1872	1–4, 7, 10, 12, 13, 15–17, 19, 21–31
Clytia hemisphaerica Linnaeus, 1767	21	*Oikopleura (Vexillaria) parva* Lohmann, 1896	31
Clytia spp. (1–3 spp. on each location)	1, 21, 25, 28–30	*Oikopleura (Vexillaria) rufescens* Fol, 1872	3, 7, 10, 15, 16, 23
Order Siphonophora		*Oikopleura* spp.	13, 14, 31
Suborder Calycophora		**Family Fritillariidae**	
Family Abylidae		*Appendicularia sicula* Fol, 1874	28
Abylopsis tetragona Otto, 1823	29, 30	*Fritillaria haplostoma* Fol, 1872	17, 23
Family Diphyidae		*Fritillaria tenella* Lohmman, 1896	23
Diphyes bojani Eschscholtz, 1825	30	*Fritillaria* spp.	10, 25, 29
Lensia sp.	10	**Class Thaliacea**	
Muggiaea kochii Will, 1844	28–30	**Order Doliolida**	
Suborder Physonectae		**Family Doliolidae**	
Family Agalmatidae		*Dolioletta gegenbauri* Uljanin, 1884	23, 28
Nanomia bijuga delle Chiaje, 1844	25, 28–30	*Doliolina* sp.	6
Subclass Trachylina		*Doliolum nationalis* Borgert, 1893	3, 23, 28, 30
Order Limnomedusae		*Doliolum* sp.	12
Family Olindiasidae		**Order Salpida**	
Gossea brachymera Bigelow, 1909	29, 30	**Family Salpidae**	
Family Olindiidae		*Salpa* sp.	6
Aglauropsis kawari Moreira and Yamashita, 1972	25, 29, 30	*Thalia democratica* Forskål, 1775	12, 23, 25, 28–31
Olindias sambaquiensis Müller, 1861	25, 28–31		

record refers to; three species of *Obelia* polyps have been recorded from coastal Brazilian waters (Oliveira et al. 2016) and probably medusae of all of them can be found inside estuaries. Also, the distribution of these studies is highly uneven through the Brazilian coast. Except for Paranaguá (1963) and Mesquita et al. (2006), and from general zooplankton studies which mention a few cnidarian species (e.g., Singarajah 1978; Araújo et al. 2008; Marcolin 2008; Marcolin et al. 2010), studies are concentrated at South Brazilian Bight (~23–28°S) subtropical estuaries (e.g., Montú and Cordeiro 1988; Navas-Pereira 1980; see Table 1), reflecting the proximity of historically consolidated research groups studying cnidarians (Haddad and Marques 2009).

The records of *Octophialucium funerarium* (as *Octocanna funeraria*) and *Rhizostoma pulmo* from the Paraíba River estuary (Singarajah 1978) were considered doubtful since the former species has no other record from the South Atlantic (Bouillon 1999) and *R. pulmo* has only been found at the African coast (Mianzan and Cornelius 1999). We suspect that the record of *R. pulmo* is a misidentification with *Lychnorhiza lucerna*, which is relatively similar and known to commonly occur throughout the Brazilian coast (Oliveira et al. 2016) including inside the Paraíba estuary (MNJ personal observations). The record of *Aglantha* sp. from Paranaguá (Montú and Cordeiro 1988) also is considered doubtful, probably a misidentification with *Aglaura hemistoma* (see Nagata et al. 2014a). These doubtful records were retained in the species list (marked with an interrogation) but were not included in the total species count. The records of *Dipurena* sp. from Santos (Pereira 2011), Paranaguá and Babitonga (Bardi 2011) were considered as *Stauridiosarsia reesi*. The genus *Dipurena* is no longer accepted with the species being moved either to *Slabberia* or *Stauridiosarsia* (Schuchert 2011). *S. reesi* is common and the unique species of these genera reported (Table 2) from these fairly well-studied subtropical estuaries (Vannucci 1956; Nogueira Júnior 2012; Martinez et al. 2015; MNJ unpublished).

Among the few relatively well-known Brazilian estuaries regarding species composition, 36 hydromedusae species have been recorded at Babitonga (Nogueira Júnior 2012) and 34 at Paranaguá (Nascimento 2016) and Cananéia (Martinez et al. 2015). These values are typically higher than the adjacent shelf (16–27 spp., Vannucci 1957, 1963; Nagata et al. 2014a), highlighting the high estuarine hydromedusae diversity and the necessity of thorough sampling and analysis of other Brazilian estuaries, particularly those from tropical latitudes which are less studied and potentially harbor high diversity. Lower number of hydromedusae species recorded from some estuaries (e.g., Navas-Pereira 1980; Mesquita et al. 2006; Marcolin et al. 2010; Bardi 2011, Table 2) is not conclusive due to the small sampling and/or analysis effort (only aliquots analyzed and/or from a reduced number of samples), particularly considering that most species are not common. For instance, in tens of thousands of individuals analyzed from hundreds of samples, 17 and 18 species were represented by <10 individuals at Paranaguá and Babitonga bays, respectively (Nogueira Júnior 2012; Nascimento 2016). This number of rare species alone is larger than the 12–15 total hydromedusae species recorded by Navas-Pereira (1980) and Mesquita et al. (2006), highlighting the need of comprehensive efforts to fully depict the estuarine planktonic cnidarian diversity.

Holoplanktonic cnidarians are considerably less diversified on Brazilian estuaries and 12 species have been reported thus far, 5 trachymedusae, 5 siphonophores, and 2 narcomedusae (Table 2). This relatively low species richness is a regular feature, since these holoplanktonic cnidarians are typically oceanic and more common, abundant and diversified in offshore open waters (Bouillon 1999; Pugh 1999; Nogueira Júnior et al. 2014). This may be particularly true for the South Brazilian Bight estuaries since the continental shelf in this region is large, reaching up to 200 km wide, which may be difficult for these oceanic species to reach coastal waters. Yet, species such as *Abylopsis tetragona* and *Diphyes bojani*, more typical from mid- to outer shelf on this area (Nogueira Júnior et al. 2014, 2015a), may occasionally enter estuaries (Table 2) in low abundances. The short shelf from the Northeastern Brazil may allow species with more oceanic affinities reach the coast and even penetrate estuaries occasionally, such as the occurrence of the siphonophore *Lensia* sp. at Suape (Neumann-Leitão and Matsumura-Tundisi 1998).

There are few quantitative data available. Higher hydromedusae densities (Fig. 2a) were reported from Sepetiba and Paranaguá (early 2010s) and considerably lower at Cananéia (<15 ind. m^{-3}, Bardi 2011). The seasonality and succession of planktonic cnidarian assemblages from Brazilian estuaries apparently are less predictable (Nagata et al. 2014b) than in high-latitude estuaries (e.g., Zamponi and Genzano 1994; Ballard and Myers 2000; Primo et al. 2012) and temporal variations seem to differ between years (Fig. 2a, b), species and populations (see below). At Sepetiba, high hydromedusae densities (80–90 ind. m^{-3}) occur throughout most of the year (Fig. 2a), while at Paranaguá peaks were observed during spring or summer, with high abundance differences between the early 1980s and early 2010s (Fig. 2a, b). Spring/summer peaks, associated with the dry season and higher salinities, have also been observed for the tropical Caeté estuary (Mesquita et al. 2006). Abundance peaks at Babitonga and Potengi were reported during spring, averaging ~15 ind. m^{-3} on both, while at Cananéia hydromedusae typically peak during winter (Fig. 2b), but in 2007 summer peaks were recorded (Bardi 2011). Such longer temporal variations may be associated with large-scale climatic events in some cases, as already observed for these organisms in other estuaries worldwide (e.g., Purcell 2005; Primo et al. 2012); however it is not known their effects in the community on Brazilian estuaries.

The holoplanktonic *Liriope tetraphylla* is ubiquitous, occurring in nearly all the few estuaries with available data, typically being the most common and abundant species from subtropical Brazilian estuaries reaching up to 117 ind. m^{-3} (Teixeira et al. 1965; Navas-Pereira 1980; Nogueira Júnior et al. 2015b). Its direct holoplanktonic development, lacking a hydroid stage (Russell 1953), associated with its high adaptability to environmental variables and ability to take advantage of food available, results in fast population responses (Purcell 2005; Yilmaz 2015) and is probably the reason why this species is so ubiquitous in these estuaries. These medusae may have a variable seasonal pattern; at Babitonga and Paranaguá, higher abundances have been reported during spring, when young medusae predominated followed by the dominance of adults during summer (Montú and Cordeiro 1988; Pukanski 2011; Nogueira Júnior et al. 2015b). However, winter

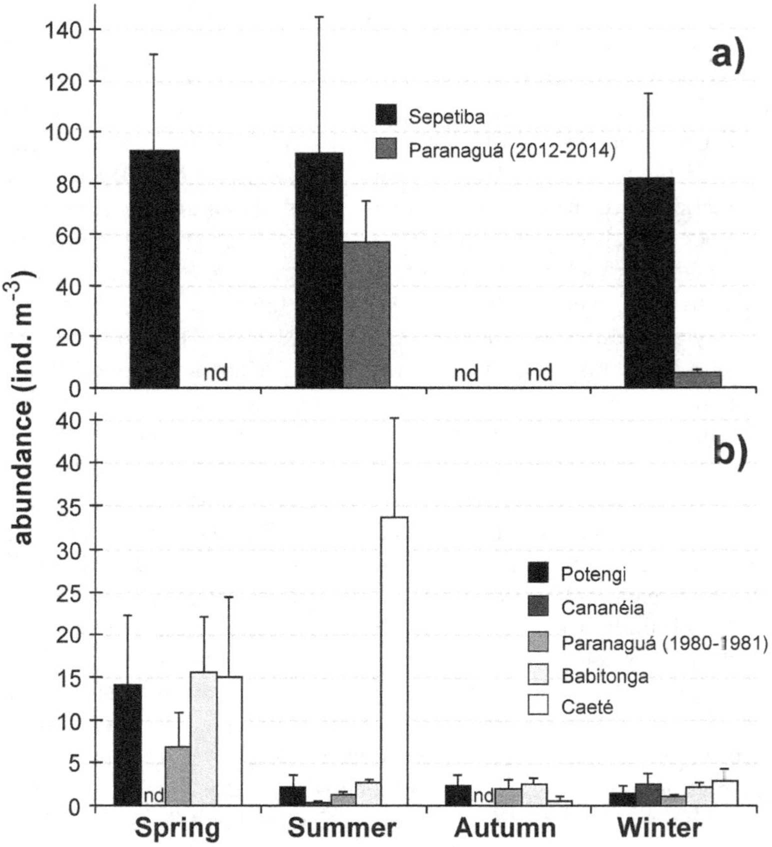

Fig. 2 Mean seasonal variation of hydromedusae abundance (ind. m^{-3}) from different Brazilian estuaries. (**a**) Sepetiba (data from Navas–Pereira 1980) and Paranaguá (2012–2014, data from Nascimento 2016), (**b**) Potengi (data from Sankarankutty et al. 1995), Cananéia (data from Bardi 2011), Paranaguá (1980–1981, data from Montú and Cordeiro 1988), Babitonga (data from Nogueira Júnior et al. 2015b), and Caeté (data from Mesquita et al. 2006) estuaries. The bars indicate the standard error. nd, no available data

peaks from these estuaries have also been reported (Bardi 2011), similar to the observed for Sepetiba (Navas-Pereira 1980). In the tropical Caeté estuary (Mesquita et al. 2006), *L. tetraphylla* peaked during summer but reached considerably lower densities (<4 ind. m^{-3}) and ranked in fifth, with hydromedusae abundance being dominated by *Clytia* spp. (up to ~32 ind. m^{-3}) and *Sarsia* spp. (up to ~15 ind. m^{-3}).

Other abundant and/or frequent hydromedusae from Brazilian estuaries are the meroplanktonic *Clytia* spp., *Obelia* spp., *Bougainvillia* spp., *Corymorpha gracilis*, *Blackfordia virginica*, *Malagazzia carolinae*, *Podocoryna loyola* and *Cnidostoma fallax* (Navas-Pereira 1980; Mesquita et al. 2006; Bardi 2011; Nogueira Júnior 2012; Martinez et al. 2015; Nascimento 2016; Teixeira-Amaral et al. 2017; Nogueira

Júnior and Silva Nascimento 2018). Polyps from *Clytia*, *Obelia*, *P. loyola*, and *B. muscus* are frequently captured in these estuaries (Bardi 2011; Haddad et al. 2014; Oliveira et al. 2016; Bettim and Haddad 2017); however polyps from *C. gracilis*, *B. virginica*, *M. carolinae* and *C. fallax* have never been found. *Blackfordia virginica* polyps are probably very difficult to find in the field, probably due to its short benthic life and very small size (Moore 1987; Mills and Rees 2000), while polyps of the other three species have not been described yet and perhaps are also small and/or cryptic or short-lived. The presence and abundance of these meroplanktonic medusae is largely controlled by factors affecting polyp growth and medusae production (e.g., Boero et al. 2008) and may occur in seasonal pulses, as the summer peak of *C. fallax* (Nascimento 2016; Teixeira-Amaral et al. 2017), *P. loyola*, *T. nutricula* and *B. virginica* or late autumn peak of *Corymorpha gracilis* and *Proboscidactyla ornata* (Nogueira Júnior et al. 2015b). The factors controlling such peaks have not been evaluated, and differences between the estuaries and different years are expected (see below).

With a few exceptions, only scant punctual nonquantitative records are available for Scyphomedusae. The most common and abundant species are *Chrysaora lactea* and *Lychnorhiza lucerna* which are known to typically occur year-round on subtropical estuaries, reaching densities of up to ~60 and ~40 ind. ha^{-}, respectively, at Babitonga bay (MNJ, unpublished). Both medusa species usually are less abundant during summer; *L. lucerna* abundance typically peaks during spring, reaching higher biomass during winter when most medusae are large, while *C. lactea* may peak during spring or autumn (Morandini 2003; Nogueira Júnior et al. 2010; Nogueira Júnior and Souza-Conceição 2010).

Phyllorhiza punctata is a typical estuarine species (e.g., Rippingale and Kelly 1995) widely distributed on Brazilian coastal and estuarine waters (Oliveira et al. 2016). This medusa has intermittent periods of abundant occurrence, such as those observed between the mid-1950s and early 1960s and between the 1990s and early 2000s, followed by disappearance few years later on both periods (Silveira and Cornelius 2000; Haddad and Nogueira Júnior 2006; Nagata et al. 2013). Since *P. punctata* has a well-documented history of invasions (e.g., González-Duarte et al. 2016), it is not clear whether these fluctuations are (ir)regular populational variations or recurrent invasions (Nagata et al. 2013). When present, *P. punctata* medusa has a well-marked seasonal cycle on subtropical estuaries, growing from late spring/early summer reaching larger sizes in late summer or early autumn, later becoming senescent and disappearing from the water column (Haddad and Nogueira Júnior 2006). Populations of *Stomolophus meleagris* may be in decline since large quantities of this medusa have been recorded in the 1950s from Cananéia, Paranaguá, and Guaratuba estuaries (Vannucci 1954; Prof. Dr. Jayme de Loyola-e-Silva 2005 personal communication), while in more recent evaluations only a few specimens have rarely been found (Morandini 2003; Nogueira Júnior and Haddad 2006).

There is no detailed information regarding the Cubomedusae, but both species are not common, and only a few individuals have been occasionally recorded inside subtropical estuaries (Morandini 2003; Nogueira Júnior et al. 2010; Nogueira Júnior 2012). These species are typical from the open shallow waters where *Chiropsalmus quadrumanus* may reach high concentrations and *Tamoya haplonema*, though not abundant, is commonly found (Nogueira Júnior and Haddad 2008). Both are known stingers that often cause accidents in fishermen and bathers.

Muggiaea kochii is typically the most common and abundant siphonophore, reaching densities of up to 5 col. m^{-3} and found in salinities down to 21 (Nogueira Júnior et al. 2015b; Nascimento 2016). This is considered one of the few typical neritic siphonophores (Mackie et al. 1987; Pugh 1999), being common and abundant over the shallow shelf mostly under influence of the Coastal Water (Nogueira Júnior et al. 2014; Nogueira Júnior and Brandini, this volume). Although not particularly abundant, the frequent presence of the physonect *Nanomia bijuga* is noteworthy both inside estuaries such as Babitonga, Paranaguá, Cananéia, and Santos Bays (Table 2) in salinities down to 11.5 (MNJ unpublished observations) and over the shelf more than a 100 km offshore in salinities >36 (Nogueira Júnior et al. 2014, 2015a), suggesting a large habitat plasticity of this species.

Exotic species in Brazilian estuaries are all meroplanktonic; between them, *P. punctata* (see above) and *Blackfordia virginica* are the best known. The latter is known to have established populations in some estuaries of tropical (Freire et al. 2014) and subtropical Brazil (Nogueira Júnior and Oliveira 2006; Bardi and Marques 2009; Nogueira Júnior 2012). Although recurrently found, it is not abundant being mainly found during summer, reaching densities up to 1.2 ind. m^{-3} at Babitonga and commonly lower (<1 ind. m^{-3}) in other estuaries (Nogueira Júnior and Oliveira 2006; Bardi and Marques 2009; Nascimento 2016). *Blackfordia virginica* can tolerate a wide salinity range (2–35, Moore 1987), being more frequent and abundant in inner and intermediate estuarine portions (Pukanski 2011; Nascimento 2016) in brackish water with intermediate salinities (~20–30) both on Brazilian (e.g., Bardi 2011) estuaries and elsewhere (e.g., Genzano et al. 2006; Chícharo et al. 2009; Marques et al. 2017). Moreover, polyps are known to recruit in salinities between 15 and 22 (Wintzer et al. 2011) and medusae production seems to be stimulated under low salinity conditions (Moore 1987).

The hydromedusae *Moerisia inkermanica* (Paranaguá 1963; Nogueira Júnior and Oliveira 2006), *Podocoryna loyola* (Haddad et al. 2014), and more recently *Cnidostoma fallax* (Nascimento 2016) have also been appointed as probably nonindigenous species. In spite of local studies since the 1980s, *P. loyola* hydroids were detected in 2007 growing on artificial substrata, and subsequently both polyps and medusae have been recurrently found on Brazilian subtropical estuaries (Nogueira Júnior 2012; Haddad et al. 2014; Nogueira Júnior et al. 2015b; Bettim and Haddad 2017), suggesting it has been introduced. *Cnidostoma fallax* has only recently been detected in Brazilian estuaries, but it seems widespread between ~24 and 34°S, occasionally dominating estuarine assemblages with recorded abundances up to 11,369 and 3,542 ind. m^{-3} at Patos Lagoon (Teixeira-Amaral et al. 2017) and Paranaguá Bay (Nascimento 2016), respectively. Considering the absence of *C. fallax* in the previous studies on these estuaries (Montú 1980; Montú and Cordeiro 1988; Lopes et al. 1998; Bardi 2011), it seems that this medusa appeared recently.

Although the species richness tends to be highest in portions of high and more stable salinity, most of these estuarine medusae species are euryhaline, occurring in a wide salinity range (Navas-Pereira 1980; Mesquita et al. 2006; Bardi 2011; Nogueira Júnior 2012; Nascimento 2016), with few species being considered exclusively estuarine or oceanic. Exceptions are *Halitiara formosa* and *Moerisia inkermanica* which occurred only in salinities <20 and are typically brackish-freshwater species, the latter commonly occurring in salinities down to 0 (Nogueira Júnior and Oliveira 2006; Bardi 2011). On the other hand, *Bougainvillia frondosa*, *Amphinema australis*, *Protiara* sp., *Rhopalonema velatum*, *Clytia hemisphaerica*, *Aglauropsis kawari*, and *Aglaura hemistoma* were only captured in salinities >30, and most of them are more common in shelf waters (Vannucci 1957, 1963; Nagata et al. 2014b), corroborating the hypothesis that they are nonresident species that can enter the estuary under specific conditions. *Rhopalonema velatum* and *Aglaura hemistoma*, for instance, are typical from the Tropical Water (Vannucci 1957, 1963, Navas-Pereira 1973; Nogueira Júnior and Brandini this volume) and reported inside estuaries only during winter (Navas-Pereira 1980; Bardi 2011; Nascimento 2016). The typical conditions for this season at these subtropical latitudes, such as low precipitation rates, high salinities, and frequent incidence of south winds favor their higher abundance in the inner shelf (Nagata et al. 2014b) and occasional entries into estuaries.

Some frequent euryhaline species are *Obelia* spp., *P. ornata*, and *L. tetraphylla*, the three occurring in lower salinities down to 10 and higher up to 37 for the latter species and to 34 for the two others (Bardi 2011; Nogueira Júnior 2012). *Obelia* spp. and *P. ornata* were very tolerant to diluted seawater in laboratory (Moreira 1978), showing decrease of the swimming movement only at salinity 20; at 15 they moved only after mechanical stimulation, and at salinity 10 the movements of the medusae stopped completely, but some of them were able to recover when placed again in appropriated conditions (Moreira 1978). In contrast, *L. tetraphylla* did not tolerate very low salinities, surviving well only in salinities above 20 (Moreira 1978). *Obelia* spp. have established populations in Brazilian estuaries, commonly occurring in all estuarine portions but often being more abundant in salinities >20 (Nascimento 2016). In the adjacent shelf, medusae from *Obelia* are also common and abundant, mostly associated with the Coastal Water (Nagata et al. 2014b; Nogueira Júnior et al. 2014). *Proboscydactila ornata* is a shelf water mass indicator (Vannucci 1957), occurring in the shelf more than 150 km far from the coast (Nogueira Júnior et al. 2014), although it is also found in coastal waters after strong winds or vertical mixing (Vannucci 1963), only occurring in outer portions of the estuary (Nascimento 2016). Similarly, *L. tetraphylla* is typically most abundant in intermediate and outer sectors with higher marine influences, being less abundant in inner portions (Pukanski 2011; Nascimento 2016).

3 Ctenophora

The study of the ctenophores is challenging mostly due to their large fragility and consequent difficulties for identification and quantification (Mianzan 1999). Nondestructive specific sampling techniques are highly recommended, but not always possible to perform. Consequently, ctenophores are commonly understudied worldwide, which is particularly true for Brazilian waters where less than ten studies dealing with the phylum have been published (Oliveira et al. 2007, 2016). Regarding estuarine ecosystems, there are few scattered records of *Mnemiopsis leidyi* and *Beroe ovata* from tropical (~7°S, Singarajah 1978) and subtropical latitudes (~23–26°S, e.g., Montú and Cordeiro 1988, see Tables 1 and 2). Yet, these two species are expected to dwell in the majority, if not all, Brazilian estuarine environments, considering both their wide distribution on southwestern Atlantic (0–>40°S) and the wide environmental conditions they can withstand (Mianzan 1999; Costello et al. 2012). *Pleurobrachia* sp. has been recorded exclusively in the tropical Paraíba river estuary (Singarajah 1978), what may be a misidentification since in the southwestern Atlantic this genus is only known to occur in temperate Argentina (37–47°S, Mianzan 1999; Oliveira et al. 2016). Moreover, larvae of lobate ctenophores (e.g., *M. leidyi*) are morphologically similar to the adults of the cydippids as *Pleurobrachia* sp. (Oliveira et al. 2007), particularly for non-experts on the phylum.

Distributional and quantitative ecological data are almost completely nonexistent for Brazilian ctenophores whether they are estuarine or not. Summer densities between <1 and 14 ind. 10 m^{-3} were reported for *M. leidyi* at different subtropical estuaries (25–26°S, Oliveira 2007). Between November 2007 and August 2008 at Babitonga, *M. leidyi* peaked during early summer (January) and winter (August), with average densities in the main channel around 1.1–1.2 ind. 10 m^{-3} in both seasons; its biomass (wet weight), on the other hand, was considerably higher during winter (mean ~11 g 10 m^{-3}) than summer (mean ~0.4 g 10 m^{-3}) due to differences in the population structure, dominated by small juveniles during summer and larger individuals during winter (MNJ unpublished data). Detailed studies on Brazilian estuarine ctenophores such as population dynamics and biological parameters such as growth, reproduction and predation rates, including their relationship with the environment and other organisms, are lacking and strongly encouraged.

4 Chaetognatha

Only a few studies provide detailed quantitative accounts on Brazilian estuarine chaetognaths (Magalhães et al. 1996; Marazzo and Nogueira 1996; Liang et al. 2003; Loureiro Fernandes et al. 2005; Mendes et al. 2012), apart from studies that dealt with the entire zooplankton community and presented brief additional information on the group (e.g., Krumme and Liang 2004; Araújo et al. 2008). Altogether, 14 chaetognath species have been recorded from Brazilian estuaries (Table 2), which represents 56% of the 25 species recorded from the whole Brazilian coast (Bonecker et al. 2017). This is a relatively high value, considering that most chaetognaths are regarded as typically oceanic and not expected to dwell in brackish waters (see below).

Species richness did not have any latitudinal tendency (Fig. 3a). This is not surprising considering chaetognath diversity remains relatively constant between 40°N

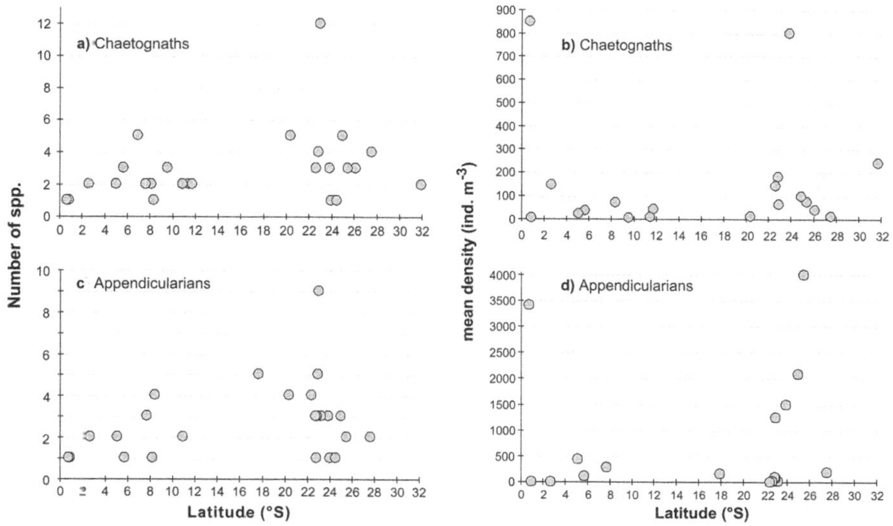

Fig. 3 Latitudinal trends of Brazilian estuarine chaetognath (**a**, **b**) and appendicularian (**c**, **d**) species richness (**a**, **c**) and mean density (**b**, **d**). See Table 1 for data sources

and 40°S in the Atlantic (Macpherson 2002). Moreover, even though up to 12 species have been recorded in a single estuary (Dias and Bonecker 2008), 79% of the estuaries harbor ≤3 species (Fig. 3a). However, it is important to emphasize that most, if not all, of the Brazilian estuaries, with a single chaetognath species recorded, have not been comprehensively studied and the records are from studies that focused mostly on crustaceans and analyzed only small aliquots of the samples (Lansac-Tôha and Lima 1993; Lopes 1994; Neumann-Leitão and Matsumura-Tundisi 1998; Cavalcanti et al. 2008; Costa et al. 2008). Hence, these studies may have missed scarce species. Furthermore, taxonomic confusion with the two *Parasagitta* species may also have occurred (see below). In any case, this low species richness is in general accordance with what is normally found at different tropical or subtropical brackish-water ecosystems worldwide, such as those from India (Srinivasan 1971, 1980), Pacific side of Costa Rica (Hossfeld 1996), Mexican Caribbean (Hernández et al. 2005), or subtropical China (Du et al. 2011; Liu et al. 2013). This reflects both the low diversity and the typical oceanic affinities of the phylum, with only a few species able to tolerate lower and variable salinities of estuaries (Srinivasan 1971, 1980; Boltovskoy 1975; Pierrot-Bults and Nair 1991; Hossfeld 1996; Hernandéz et al. 2005). The relatively high diversity reported from the Ribeira Bay (12 spp., Dias and Bonecker 2008, Table 2, Fig. 3a) is an exception, probably related to the high openness and sea influence on this ecosystem where salinities are always >29 (Dias and Bonecker 2008). Most of these species are typical from outer shelf and oceanic waters (Liang 2002; Liang and Vega-Pérez 2001, 2002; Souza et al. 2014) and thus are not expected in most of the other less saline estuaries.

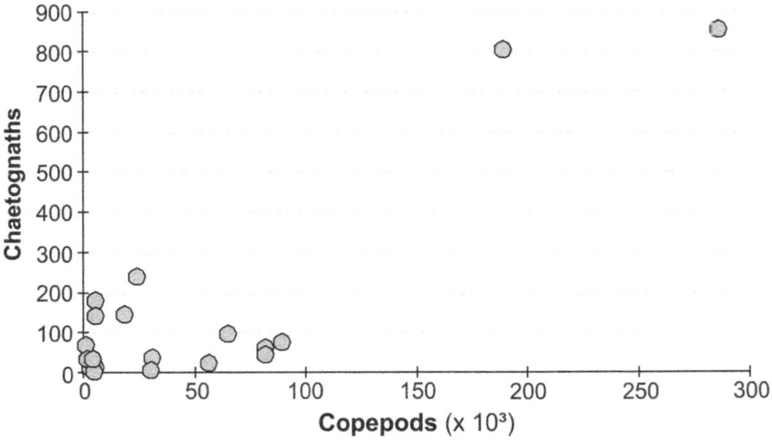

Fig. 4 Scatterplot between mean chaetognath and maximum copepod abundances (ind. m^{-3}) from 19 Brazilian estuaries (see Table 1 for data source). Whenever possible, data was taken from the same study or year for both chaetognaths and copepods. Both variables have a significant positive correlation (Pearson, $r = 0.85$, $n = 19$, $p < 0.05$)

Chaetognath densities from different Brazilian estuaries vary a lot, without any clear latitudinal trend (Fig. 3b). In most ecosystems, maximum and mean densities are lower than 800 and 200 ind. m^{-3} respectively (Fig. 3b). The lowest densities are from Mundaú-Manguaba, with chaetognath triennial mean of 0.75 ind. m^{-3} and maximum values of ~5 ind. m^{-3} (Magalhães et al. 1996). Highest abundances were reported from Taperaçu and Santos estuaries with maximum chaetognath densities between 5,200 and 5,750 ind. m^{-3} and mean densities between 800 and 850 ind. m^{-3} (Fig. 3b, Costa et al. 2008; Pereira 2011). Due to data scarcity, it is difficult to determine the factors controlling such huge density differences, in some cases in ecosystems close to each other (see Fig. 3b), but food availability certainly plays an important role.

Copepods, typically the main prey item of chaetognaths (Vega-Pérez and Liang 1992; Liang and Vega-Pérez 1995; Sato et al. 2011), have very high abundances (maximum densities >190,000 ind. m^{-3}, Costa et al. 2008; Pereira 2011) in the two estuarine systems with higher chaetognath abundances (Taperaçu and Santos). On the other hand, in most of the estuaries with lower chaetognath abundances such as Caeté, Potengi, Itamaracá, Suape, Vitória and Sepetiba, copepod densities are also considerably lower, rarely exceeding 5,000 ind. m^{-3} (Sankarankutty et al. 1979, 1995; Navas-Pereira 1980; Neuman-Leitão and Matsumura-Tundisi 1998; Silva et al. 2003, 2004, Krumme and Liang 2004; Sterza and Loureiro Fernandes 2006). Indeed, mean and maximum chaetognath densities have a relatively high positive ($r = 0.85$ and 0.83, respectively) and significant ($p < 0.05$) correlation to maximum copepod abundance (Fig. 4). Yet, this trend must be seen with caution considering that the two ecosystems with very high abundances of both chaetognaths and copepods may bias the analysis and the tendency is not always valid. For instance, Tabatinga and Casqueira and Conceição estuaries sustain mid-to-high copepod densities (up to 30,000 and 56,000 ind. m^{-3}, respectively) and relatively low chaetognath abundance levels (Sankarankutty et al. 1997; Marcolin 2008; Marcolin et al. 2010), suggesting that other factors are also important. Geomorphology of the estuary (e.g., degree of openness, depth), physical and biological characteristics (salinity variations, tidal circulation, temperature, quality of food, parasites, abundance of predators, etc.), and the intricate interactions between all of these factors have been shown to be important (Srinivasan 1971, 1980; Mulkana and McIlwain 1973; Grant 1977; Nair and Sankarankutty 1988; Hossfeld 1996; Tse et al. 2008), but data is mostly missing from Brazilian estuaries.

Parasagitta friderici, *Flaccisagitta enflata* and *Parasagitta tenuis* are the most common and widespread chaetognaths, reported between 60 and 64% of the 25 Brazilian estuaries with data available (Table 2, Fig. 5a). In addition, these species are also the most abundant ones; most sites are dominated by *P. friderici* and/or *P. tenuis* and a few by *F. enflata* (Fig. 6a). The two *Parasagitta* species have a wide distribution (Srinivasan 1971, 1980; Mclelland 1980, 1989; Pierrot-Bults and Nair 1991) on coastal tropical to mid-temperate east Pacific and east and west Atlantic coasts, including the Mediterranean (*P. friderici*) or Indian estuaries and coastal open waters (*P. tenuis*). They are among the few chaetognath species restricted to and abundant at coastal waters (Srinivasan 1971, 1980; Boltovskoy 1975; Pierrot-Bults and Nair 1991; Hossfeld 1996; Casanova 1999), including brackish-water ecosystems. Indeed, they have been commonly found in salinities <15 (Montú and Cordeiro 1988; Magalhães et al. 1996; Marazzo and Nogueira 1996; Loureiro Fernandes et al. 2005) and even in salinities <5 for *P. friderici* (Montú 1980; Liang et al. 2003), supporting the hypothesis that they are well-adapted to the brackish-water condition of estuaries maintaining permanent populations in many of them.

It is noteworthy that most estuaries in latitudes <11.5°S are dominated by *P. tenuis*, while most of those in latitudes >22°S are dominated by *P. friderici* (Fig. 6a). Exceptions are Paranaguá and Taperaçu and Caeté estuaries; in the former, both species are nearly equivalent in abundance (Miyashita et al. 2012; L.K. Miyashita personal communication), but *P. friderici* may dominate in some years (Lopes et al. 1998). Concerning the two latter estuaries, only *P. friderici* has been recorded (Fig. 6a, Table 2, Krumme and Liang 2004; Costa et al. 2008). It is difficult to determine the factors leading to such latitudinal shift, but a taxonomic caution is necessary here. These two species are morphologically very similar and their validity has been greatly discussed even with the proposition for their synonymization in the past (e.g., Pierce 1951; Furnestin 1957; Almeida-Prado 1961; Boltovskoy 1981). Currently, there is a relatively long-lasting consensus that both are valid separated species (e.g., Mclelland 1980, 1989; Casanova 1999; Thuesen 2009). Yet, in some of the Brazilian studies, both species were explicitly considered synonyms (Coelho-Botelho et al. 1999; Resgalla 2010), probably following the identification guide of Boltovskoy (1981). This may have been the case of other studies that did not disclose this particular piece of information. Therefore, this latitudinal shift may be an artifact of taxonomic problems. Detailed taxonomic studies, including morphological and molecular analyses from as many estuaries as possible, are recommended to test the validity of this latitudinal dominance shift from *P. tenuis* in tropical (~2–12°S) to *P. friderici* in subtropical (22–32°S) Brazilian estuaries (Fig. 6a).

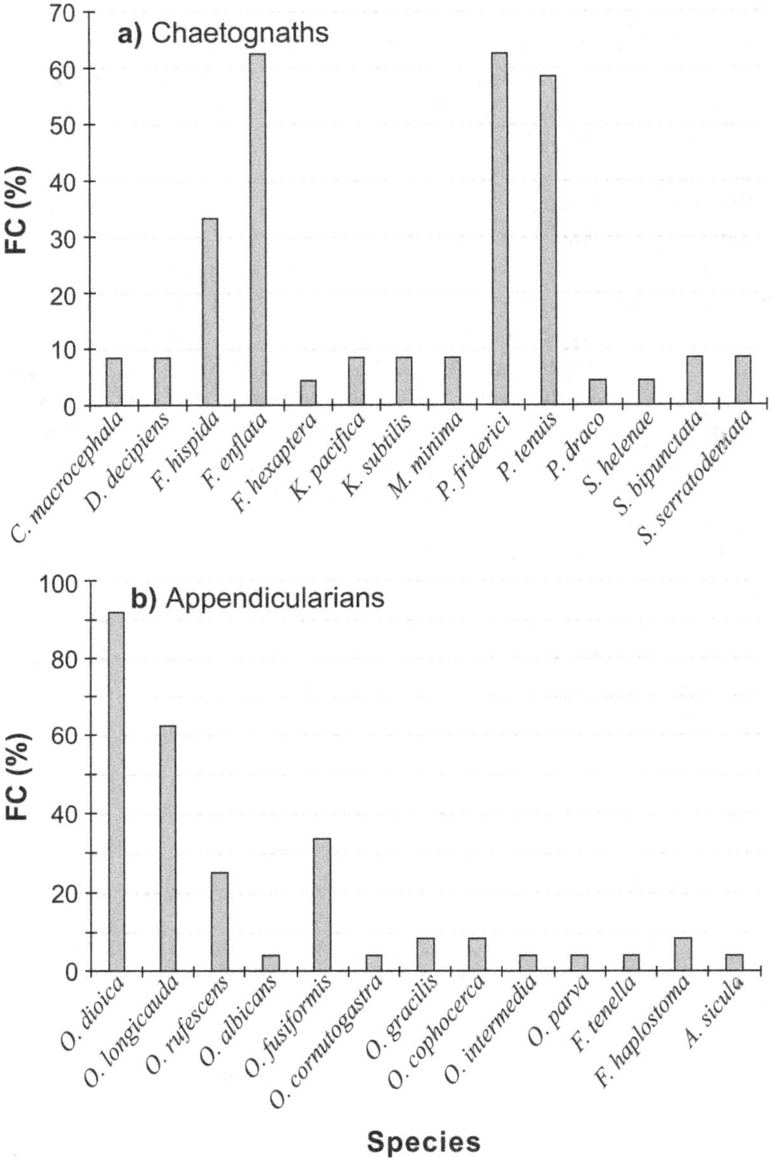

Fig. 5 Relative (%) number of Brazilian estuaries each chaetognath (**a**) and appendicularian (**b**) species have been recorded. 100% = 25 estuaries for chaetognaths and 24 for appendicularians

Flaccisagitta enflata is distributed worldwide between ca. 40°N and 40°S including both coastal and oceanic waters. It usually is considered as a typical oceanic epipelagic species (Pierrot-Bults and Nair 1991; Casanova 1999), more associated with shelf and offshore warm Tropical Water off Brazil (Souza et al. 2014;

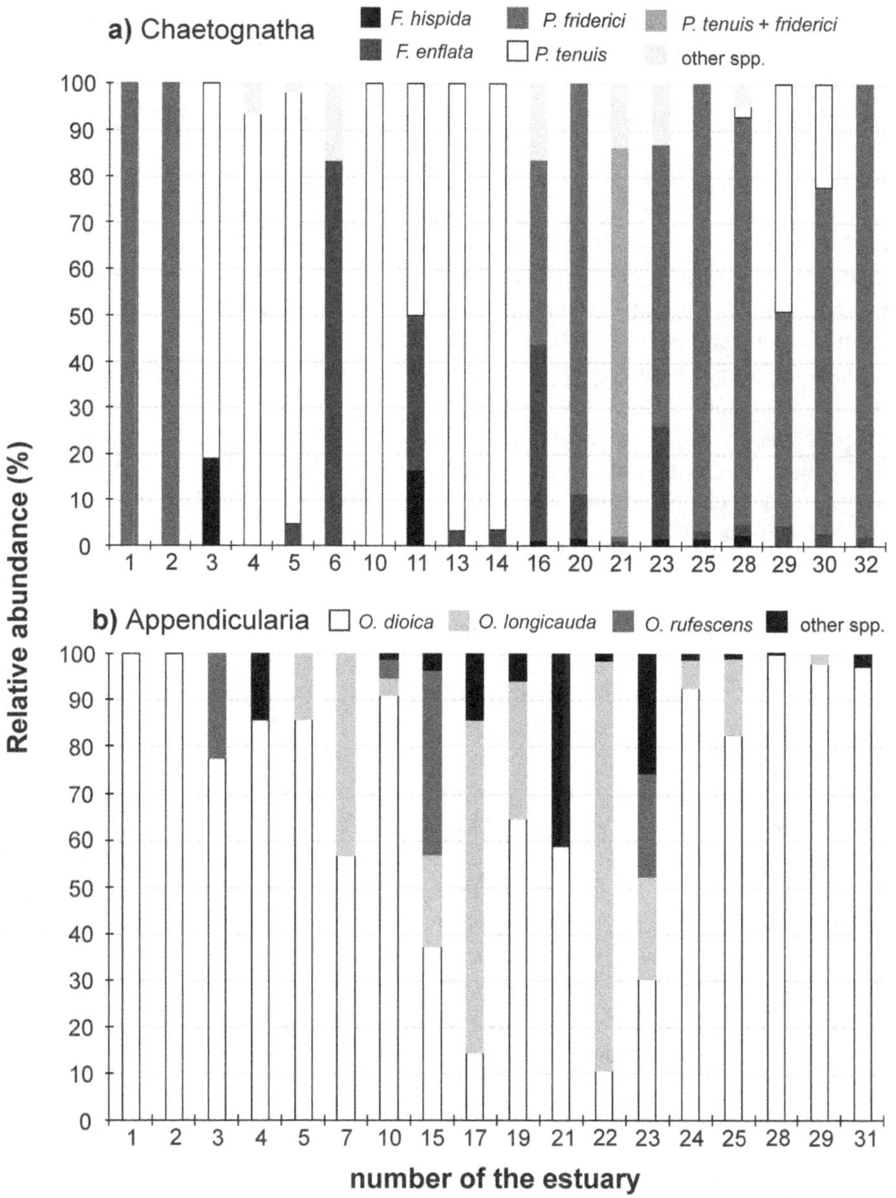

Fig. 6 Relative abundance (% of ind. m^{-3}) of the dominant chaetognath (**a**) and appendicularian (**b**) species from different Brazilian estuaries. Numbers of the estuaries correspond to the numbers of Fig. 1 and Table 1. See the latter for data source. Coelho-Botelho et al. (1999) considered *P. friderici* as a junior synonym of *P. tenuis*, and thus it is not possible to know the contribution of each species

Nogueira Júnior and Brandini this volume). Nevertheless, *F. enflata* has frequently been reported from Brazilian brackish waters (Figs. 5a, 6a) and also from many brackish-water ecosystems worldwide (Srinivasan 1971, 1980; Mulkana and McIlwain 1973; Grant 1977; Hernandez et al. 2005; Du et al. 2011; Liu et al. 2013). It is the dominant species from Paraíba and Vitória estuaries, typically associated with the outer sectors and higher salinities (Singarajah 1978; Loureiro Fernandes et al. 2005).

Beyond these three abovementioned species, *Ferosagitta hispida* has been reported in ~36% of the Brazilian estuaries (Fig. 5a), usually in low densities (Fig. 6a). It is probably not resident from most of these estuaries but rather transported from open marine waters (Coelho-Botelho et al. 1999; Liang et al. 2003; Loureiro Fernandes et al. 2005; Pereira 2011). Yet, at São Marcos Bay (2.7°S) *F. hispida* reaches relatively high densities (up to ~89 ind. m^{-3}, averaging 22 ± 34 ind. m^{-3}) and can be found throughout the estuary (Bonecker and Dias 2009), representing ca. 20% of total chaetognath abundance (Fig. 6a) which seems in accordance with its tropical affinities (Casanova 1999). It is endemic from brackish, neritic, and oceanic Atlantic warm waters (Reeve 1970; Pierrot-Bults and Nair 1991; Casanova 1999).

All the other 10 chaetognath species have been reported only from 1–2 sites each (<9%, Fig. 5a, Table 2), sporadically found near the mouth of the estuary in low abundances (<1 ind. m^{-3}, e.g., Singarajah 1978; Coelho-Botelho et al. 1999; Liang et al. 2003; Loureiro Fernandes et al. 2005). These data indicate that they probably do not sustain permanent estuarine populations. Indeed, all these species typically have oceanic affinities (Pierrot-Bults and Nair 1991; Liang and Vega-Pérez 2001, 2002; Souza et al. 2014) and their occasional occurrence inside estuaries is probably related to tidal- and/or wind-driven transport from adjacent shelf (Nair and Sankarankutty 1988; Loureiro Fernandes et al. 2005).

Data on temporal dynamics of estuarine Brazilian chaetognaths is lacking for most ecosystems. The few available ones suggest that there is no clear general seasonal trend neither for total chaetognaths nor for dominant species, greatly varying between different environments and/or species. Seasonal variations were small or absent in some of the low-latitude (<7°S) estuaries, such as *P. friderici* from Taperaçu (Costa et al. 2008), *F. enflata* from Paraíba river (Singarajah 1978), and *P. tenuis* from Casqueira and Conceição rivers (Sankarankutty et al. 1997). Although, a slight peak in December was noticeable at the innermost stations but not in the others in the latter case (Sankarankutty et al. 1997). *Parasagitta tenuis* populations from other tropical estuaries may vary seasonally such as winter and/or summer peaks at Potengi estuary (Nair and Sankarankutty 1988; Sankaranutty et al. 1995, 1996), higher abundance during rainy (August) and dry (December) seasons at Tabatinga and Itapicurú rivers, respectively (Marcolin 2008), and October/November peaks at Mundaú-Manguaba (Magalhães et al. 1996). This latter ecosystem was sampled through three consecutive years and in spite of great interannual abundance variations, the seasonal trend was analogous (Magalhães et al. 1996).

All information from higher latitudes (20–32°S) indicate seasonal variations, more commonly with densities peaking between spring and/or summer such as reported for Patos Lagoon (Montú 1980), Paranaguá (Montú and Cordeiro 1988; Miyashita et al. 2012; Salvador and Bersano 2017), Cananéia (Liang et al. 2003) and Santos bays (Pereira 2011). Some minor variations have also been reported, such as a winter smaller peak in Paranaguá (Montú and Cordeiro 1988; Miyashita et al. 2012). While abundance peaks of both *P. friderici* and *F. enflata* are common during spring and/or summer in the outer sector of Guanabara Bay, in the innermost areas, they peak during winter because of the low salinities during summer associated with higher rainfall (Marazzo and Nogueira 1996; Mendes et al. 2012). At Vitória Bay, differently, higher abundances for both *P. friderici* and *F. enflata* have been reported during winter and spring, with considerably lower densities during summer (Loureiro Fernandes et al. 2005).

Biomass and production of Brazilian estuarine chaetognaths have not been evaluated in details, with the exception of *P. friderici* populations from Cananéia, where mean annual biomass is ca. 30 mg DW m^{-3}, reaching daily means of up to 57 mg DW m^{-3} during summer, and mean annual production is of ~1 mg C m^{-3} d^{-} reaching up to ~4.5 mg C m^{-3} d^{-} (Liang et al. 2003). The few available data on reproductive dynamics show that all *P. friderici* developmental stages occur through the year at Guanabara and Cananéia, with dominance of adults and increased reproduction during spring followed by a recruitment peak mostly in summer (Mendes et al. 2012) or juvenile dominance throughout the year with three apparent annual cohorts (Liang et al. 2003) respectively. Three annual spawning peaks (May, September and December) have also been recorded for *P. tenuis* in the Potengi estuary, where juveniles and developing individuals can be found through the year and mature ones occur sporadically, mostly during winter (Nair and Sankarankutty 1988).

Apart from seasonal dynamics, small-scale temporal variations (e.g., lunar, nictemeral, tidal cycles) may be large (Sankarankutty et al. 1979; Liang et al. 2003; Mendes et al. 2012) and particularly important considering the high level of these short-scale variations in the estuarine physical environment, mainly controlled by the tidal patterns and ultimately by the salinity. In general, higher species richness and densities of chaetognaths have been found more typically during high tide (Sankarankutty et al. 1979; Nair and Sankarankutty 1988; Loureiro-Fernandes et al. 2005; Marcolin 2008) tending to positively correlate with tidal height (Liang et al. 2003). However, greater densities of *P. friderici* have been reported during ebb at Itapicurú estuary (Marcolin 2008).

Neither vertical nor horizontal distribution of Brazilian estuarine chaetognaths has been evaluated in details thus far, but the few data available indicate that abundance tend to be higher closer to the estuary mouth (Montú 1980; Pekala 1980; Montú and Cordeiro 1988; Marazzo and Nogueira 1996; Sankarankutty et al. 1997; Loureiro Fernandes et al. 2005), a pattern similar to other estuaries elsewhere (e.g., Grant 1977; Srinivasan 1980), probably associated with higher sea influence. Still, *P. friderici* is highly resistant to low salinities (see above) and commonly found in the inner portions, being probably a permanent resident in estuaries (Montú and Cordeiro 1988; Loureiro Fernandes et al. 2005). In addition, wind patterns (e.g., Hernández et al. 2005) along with tidal cycles can be particularly important in the horizontal distribution of estuarine chaetognaths probably increasing their extent (as well as abundance; see above) during flood and decreasing in the ebb, which obviously will also be influenced by the geomorphology of the estuary and the flow of the tidal current. Interactions between spatial and seasonal factors in the abundances may occur (Marazzo and Nogueira 1996; Sankarankutty et al. 1997; Miyashita et al. 2012), resulting in different seasonal cycles for different parts of the estuary; for instance, as commented above, at Guanabara outer sector chaetognaths peak during summer and/or winter, but in the innermost areas, peaks have been recorded only during winter, apparently constrained by low salinities during summer (Marazzo and Nogueira 1996; Mendes et al. 2012).

5 Appendicularia

Similar to the previous taxa, information on Brazilian estuarine appendicularians is derived mostly from general zooplankton studies which provide variable and usually superficial information about the group (e.g., Bonecker and Dias 2009; Miyashita et al. 2012; Salvador and Bersano 2017). A total of 13 appendicularian species have been recorded from Brazilian estuaries (Table 2), which represents 37% of the 35 species recorded from Brazil (Rocha et al. 2017). Similar to the chaetognaths, no latitudinal trend in the appendicularian species richness can be noticed (Fig. 3c), which is in accordance with the relatively constant species richness of the group in the Atlantic Ocean between ~40°N and 40°S (Macpherson 2002).

Appendicularian species richness from Brazilian brackish-water environments is usually low (Fig. 3c). Up to 9 species have been recorded from a single estuary, but 75% of them harbor 1–3 species (Fig. 3c), which is in general accordance to other estuaries worldwide (Brunetti et al. 1990; Spinelli et al. 2009; Du et al. 2011; Liu et al. 2013). Although deep temperate estuaries such as Toyama Bay, Japan, may harbor considerably higher number of species including the presence of many oceanic epipelagic and deepwater species (Tomita et al. 2003). Similar to the chaetognaths, most of the Brazilian estuaries with a single species recorded have not been comprehensively studied, and the records are derived from studies focusing on crustaceans that analyzed only small aliquots of the samples (e.g., Lopes 1994; Cavalcanti et al. 2008; Costa et al. 2008) and that may have missed rarer species. In any case, the class Appendicularia is not particularly diversified (Esnal 1999; Rocha et al. 2017), and, in parallel to the observed for chaetognaths, highest species richness at Ribeira Bay probably reflects the higher openness and sea influence on this ecosystem (see above).

No latitudinal pattern has been observed for appendicularian densities (Fig. 3d). In most estuaries from Brazil, maximum and mean densities are lower than 1,000 and 280 ind. m^{-3}, respectively. The lowest densities are from São Marcos Bay, never exceeding 1 ind. m^{-3}, yet these estimates were made from a single sampling campaign (Bonecker and Dias 2009) and probably have missed the full extent of local abundance variability. Low appendicularian densities have also been reported from Caeté (mean of 0.65 and up to 4.5 ind. m^{-3}, Krumme and Liang 2004) and from Perequê-Açu, Sepetiba, Macaé, São João, and Itapicuru (mean densities between 5 and 20 ind. m^{-3}, Fig. 3d). Highest densities have been reported from Cananéia, Santos, Ribeira and Taperaçu, with mean densities between 1190 and 3400 ind. m^{-3} (Fig. 3d) and maximum between ~13,000 and 31,550 ind. m^{-3}.

Oikopleura dioica and *Oikopleura longicauda* are the most common species, respectively, reported from 91.7 and 66.7% of the 24 estuaries considered for the group (Fig. 5b, Table 2). These two species also are by far the most abundant appendicularian, particularly the former which represent ≥70% of total appendicularian abundance in ~60% of the estuaries (Fig. 6b). The latter species represented around 80 and 90% of all appendicularians at Macaé and Bracuí rivers, respectively, and 43% at Itamaracá (Fig. 6b). This is similar to the observed elsewhere in the world where *O. dioica* and/or *O. longicauda* are common and typically the dominant appendicularian species in brackish-water ecosystems from temperate (Brunetti et al. 1990 Spinelli et al. 2009), subtropical (Du et al. 2011; Liu et al. 2013) and tropical areas (Hopcroft and Roff 1998; Hoover et al. 2006). Both *O. dioica* and *O. longicauda* are abundant and widely distributed throughout South Atlantic up to ~55 and 60°S, respectively (Esnal 1999). Both species have coastal affinities, but while the latter is neritic and oceanic (Esnal and Castro 1977), commonly being the dominant species at the inner shelf off Brazil (Tundisi 1970; Miyashita and Lopes 2011), *O. dioica* is more neritic/estuarine being usually the most abundant appendicularian in brackish-water ecosystems (Figs. 5b, 6b). Along Brazil, *O. dioica* is commonly found in salinities <20 (e.g., Lopes et al. 1998; Miyashita et al. 2012; Bonecker and Dias 2009) and occasionally even in salinities down to nearly zero (Coelho-Botelho et al. 1999), being rarely found in offshore waters under oceanic influence (Tundisi 1970; Esnal and Castro 1977).

Apart from these two abovementioned appendicularians, *Oikopleura fusiformis* has been found in 33% of the estuaries and *Oikopleura rufescens* in 25% (Fig. 5b). The last one contributed up to ca. 40% (Fig. 6b) to the total appendicularian abundance at Caravelas (mean of 26 and up to 242 ind. m^{-3}, Carvalho and Bonecker 2010) and around 20% at São Marcos and Ribeira Bays, reaching densities up to 1,990 ind. m^{-3} in the latter (Dias and Bonecker 2008; Bonecker and Dias 2009, Fig. 6b). *Oikopleura fusiformis* also can reach relatively high densities at Ribeira Bay (up to 849 ind. m^{-3}, Dias and Bonecker 2008); however its relative contribution to total appendicularian abundances is typically low (<5%), and low densities (mean < 1 ind. m^{-3}) have been reported from all other estuaries this species has been recorded (Coelho-Botelho et al. 1999; Carvalho and Bonecker 2010; Carvalho et al. 2016). All the other nine appendicularian species have been reported only from 1–2 estuaries (<9%, Fig. 5b, Table 2), in lower abundances in the outer sectors, probably not maintaining permanent populations. These species typically have more oceanic affinities and are more common and abundant on salinities >35 (Tundisi 1970; Miyashita and Lopes 2011; Carvalho et al. 2016). Their eventual presence in estuaries probably is resultant of advective processes transporting them from the adjacent shelf.

With a few exceptions (e.g., Esnal et al. 1985), temporal dynamics of appendicularians from Brazilian estuaries have not been evaluated in details. The few data available suggest different patterns for different ecosystems and/or species. At Taperaçu, the unique available data very close to the Equator, *O. dioica* remain in similarly high densities (~3,000–3,800 ind. m^{-3}) both in the dry and rainy seasons (Costa et al. 2008). Southward (~5–11°S), at Casqueira and Conceição (Sankarankutty et al. 1997), Potengi (Esnal et al. 1985), and Tabatinga (Marcolin 2008), appendicularians, mostly *O. dioica*, tend to be more abundant between September and March, corresponding to the summer-dry season. Yet, variations also occur such as (i) a smaller winter peak of *O. dioica* at Casqueira and Conceição (Sankarankutty et al. 1997) and Potengi (Esnal et al. 1985), depending on the tide and/or the location within the estuary, (ii) *O. longicauda* peaking in the rainy season and no clear seasonal pattern for *O. dioica* at Itamaracá (Porto Neto et al. 1999), and (iii) *Oikopleura* spp. more abundant in the beginning of the rainy season at Itapicuru (Marcolin 2008). Further south at Caravelas, the dominant species *O. dioica* and *O. rufescens* also tended to peak during the dry season, which in this region correspond to the months between May and October; however the very high standard deviations observed (Carvalho and Bonecker 2010) clearly indicate a high variability between different sampling campaigns. Data from subtropical latitudes such as Santos (Pereira 2011), Guaraú (Lopes 1994), Paranaguá (Montú and Cordeiro 1988; Miyashita et al. 2012; Salvador and Bersano 2017) and Babitonga (MNJ unpublished) suggest higher appendicularian abundances typically during spring and/or summer, occasionally with a second smaller peak during winter.

Larger scales of temporal variability have not been addressed in details, but data from Salvador and Bersano (2017) suggest abundance variations of 1.5 to 2 times between 2 consecutive years at Paranaguá. Small-scale temporal variations have also not been evaluated in details, except for the Potengi estuary, where *O. dioica* reaches highest abundances during low tide, particularly at night, and *O. longicauda* only appear during high tide (Esnal et al. 1985), supporting the estuarine affinities of the former species.

Biomass and secondary production of Brazilian estuarine appendicularians have not been evaluated at all but probably are significative considering the high abundances they may attain in some of these ecosystems (Fig. 3d) and the typical high production rates of appendicularians from warm waters (Hopcroft and Roff 1998). For instance, estimates from subtropical Brazilian shallow shelf indicate appendicularian secondary production represents 77% that of the copepods (Miyashita and Lopes 2011). Regarding population dynamics, the only data available is from Potengi, with the presence of both mature and juvenile *O. dioica* throughout the year, indicating a continuous reproductive period, but with mature individuals attaining smaller sizes during winter (Esnal et al. 1985). It is interesting to note a consistent nictemeral pattern throughout the year, with mature individuals predominating at night, being nearly absent during day, and juveniles dominating during day and representing only around 10% of the nighttime population (Esnal et al. 1985).

6 Thaliacea

Studies reporting thaliaceans from Brazilian estuaries are quite few and five species have been recorded thus far (Table 2), apart from some unidentified records (e.g., Sterza and Loureiro Fernandes 2006; Araújo et al. 2008). Although local studies on the group are quite scant, this low diversity is not expected to increase considerably, since thaliaceans are not particularly species-rich. There are only 81 species currently recognized as valid worldwide (Purushothaman et al. 2017) and most of them are typical from outer shelf and oceanic areas, not tolerating brackish estuarine waters (Esnal and Daponte 1999a, b). *Doliolum nationalis*, *Dolioletta gegenbauri*, and *Thalia democratica* are among the few thaliaceans that thrive in coastal waters (e.g., Paffenhöfer and Lee 1987; Paffenhöfer et al. 1995; Menard et al. 1997; Nakamura 1998) and also are the most commonly reported from Brazilian estuaries. Aside from salinity variations, coastal and estuarine waters are particularly problematic for salps considering their filters clog under high concentrations of particulate organic matter (Bone 1998; Esnal and Daponte 1999a) which is common of estuarine high productive waters.

The salp *T. democratica* and the doliolids *D. gegenbauri* and *D. nationalis* are common and abundant in open coastal and offshore waters throughout most of the Brazilian coast (Tavares 1967; Esnal and Daponte 1999a, b; Nogueira Júnior and Brandini this volume) and also in many regions of the world (Purushothaman et al. 2017), mostly in open waters. Their presence inside Brazilian estuaries is quite occasional (Pereira 2011; Nogueira Júnior 2012), suggesting they do not maintain resident populations but are rather transported from adjacent shelf waters. Truly, they are mostly associated with the outer sectors of the estuaries where salinities are generally higher, but *T. democratica* and *D. nationalis* have occasionally been found in salinities down to ~21 and 26, respectively (Nogueira Júnior 2012; MNJ unpublished).

While both doliolid species have always been found in very low numbers, *T. democratica*, though not frequent, usually found in <15% of the samples, may represent a considerable fraction of the gelatinous zooplankton assemblage (Nogueira Júnior 2012; Martinez et al. 2015), reaching densities of up to 600 and 3,330 ind. 10 m^{-3} at Babitonga and Cananéia estuaries respectively (MNJ unpublished). Such high concentrations are probably related with high rates of asexual reproduction (aggregate zooids asexually produced normally dominate these estuarine aggregations), along with its capacity to tolerate broad changes in environmental conditions (Bone 1998; Gibbons 1997; Esnal and Daponte 1999a). Their spatial and temporal dynamics have not been explored but are probably related to local wind and tide patterns along with geomorphological aspects, such as the degree of openness of the estuary. The fate of these estuarine aggregations justifies detailed investigation, to test whether these populations do not persist inside the estuaries because they perish (and then sink and enrich the estuarine benthos) or because they are transported back to open waters during ebb.

7 Conclusions

A total of 104 gelatinous and semi-gelatinous zooplankton species have been recorded from Brazilian estuaries, 70 cnidarians, 2 ctenophores, 14 chaetognaths, 13 appendicularians and 5 thaliaceans (Table 2). All major taxa are understudied, and detailed taxonomic and/or quantitative information is not available for most of the numerous Brazilian estuarine ecosystems. Despite this, the number of siphonophores, ctenophores, chaetognaths, appendicularians, and thaliaceans species from Brazilian estuaries is not expected to increase considerably with increasing sampling effort since most of these taxa are not particularly species-rich and typically have oceanic affinities, with only a few species occurring in estuarine ecosystems. The same is not true for meroplanktonic hydrozoans, particularly those from the orders Anthoathecata and Leptothecata which are diversified in coastal environments with several species tolerating lower and variable salinities of estuaries, representing ~28 and 18%, respectively, of all species recorded here, and whose diversity is probably underestimated due to the lack of data and thus probably will increase considerably with increasing sampling effort, particularly from the less studied tropical estuaries.

For all higher taxa, higher abundances and diversity tend to occur in the outer sectors of the estuaries, because of the higher marine influence and sporadic occurrence of nonresident species, transported by tidal and/or wind processes from the adjacent shelf to inside the estuaries. Density peaks are more common between spring and summer, especially for estuaries from subtropical latitudes (20–32°S), probably due to the nutrient enrichment associated with higher rainfall in these seasons at those latitudes. However, a great variation has been reported from different years, locations, and/or species, which seems to be more related to specific environmental preferences, tolerances, and favorable conditions for their occurrence within estuaries. In lower latitudes (0 to ~11°S), patterns are even more variable, with absence of seasonal variations or peaks in the rainy or dry seasons, but they were less studied. Larger-scale temporal variations, such as interannual and interdecadal changes, have not been explored in details yet, although preliminary data (Nascimento 2016; Salvador and Bersano 2017; Teixeira-Amaral et al. 2017) suggest they may be large. Similarly, small-scale temporal analyses also have been only superficially evaluated. Most typically, chaetognaths tend to increase abundance with the high tide, while the unique available data for appendicularians indicate the opposite. There is no such data for the other gelatinous taxa.

Among cnidarians, the most abundant and ubiquitous species are the holoplanktonic *L. tetraphylla* and the meroplanktonic *Clytia* spp. and *Obelia* spp. along with *P. loyola* and *C. fallax* on subtropical estuaries which are considerably best studied. The Scyphomedusae *C. lactea* and *L. lucerna* also are common throughout Brazilian estuaries and may attain high biomass, but quantitative studies are lacking. Nearly nothing is known on ctenophores, but both *M. leidyi* and *B. ovata* are probably widespread and resident from most, if not all, Brazilian estuaries. No latitudinal trend was found for species richness or abundance neither for chaetognaths nor appendicularians, with most estuaries harboring between one and three species of each taxa. *Parasagitta friderici* and *Parasagitta tenuis* are the most widespread and abundant chaetognaths, and an apparent latitudinal shift in their dominance must be carefully tested since it may be resultant from taxonomic problems. *Oikopleura dioica* and *Oikopleura longicauda* are the most ubiquitous appendicularians, particularly the former species reported from >90% of the estuaries and most commonly being the dominant species. Thaliaceans have only a few scattered records and probably do not maintain permanent populations inside these estuaries, being transported from the adjacent shelf. Yet, *Thalia democratica* may reach dense aggregations apparently associated with high rates of asexual reproduction.

Basically, any future research on estuarine gelatinous and semi-gelatinous zooplankton from Brazil would be welcome. Species composition is the best-known aspect, but efforts are still necessary, particularly regarding the cnidarians, which are more species-rich, and tropical estuaries which have been less studied and potentially are highly diversified. Abundance data and its temporal variations in many different scales (e.g., tidal, nictemeral, seasonal, interannual, etc.), along with environmental data and prey and/or predator levels, are mostly available for a few subtropical estuaries. Long-term monitoring programs may be particularly interesting considering the presence of exotic species, some of them apparently well-established, in order to track possible changes in abundances and/or species composition and predict their effects on the ecosystem. Studies on any biological parameter (i.e., reproduction, population dynamics, life cycle, secondary production, feeding rates, trophic interactions, etc.) with rare exceptions are missing and should also be emphasized in the future, particularly associating field data with laboratory experiments.

Acknowledgments We are deeply indebted to Dras. Graciela Esnal and Tulia Aguilar Martinez for kindly providing important references and to Dr. Leonardo Kenji Miyashita and Dra. Catarina Marcolin for generously providing their original data abundances from different estuaries. Dra Elena Guerrero read the text and provided useful suggestions. Dra. Monica Susana Hoffmeyer and Dr. Frederico Pereira Brandini are acknowledged for their invitation and patience during the development of this study. MNJ also acknowledges Dr. Renato Mitsuo Nagata for his friendship and encouragement.

References

Acuña JL, López-Urrutia Á, Colin S (2011) Faking giants: the evolution of high prey clearance rates in jellyfishes. Science 333:1627–1629

Almeida-Prado MS (1961) Distribuição dos Chaetognatha no Atlântico Sul Ocidental. Bol Inst Ocean 12:15–49

Arai MN (1997) A functional biology of Scyphozoa. Chapman and Hall, London. 316 pp

Araújo HMP, Nascimento-Vieira DA, Neumann-Leitão S et al (2008) Zooplankton community dynamics in relation to the seasonal cycle and nutrient inputs in an urban tropical estuary in Brazil. Braz J Biol 68:751–762

Ballard L, Myers A (2000) Observations on the seasonal occurrence and abundance of gelatinous zooplankton in Lough Hyne, Co. Cork, South-West Ireland. Proc R Ir Acad 100:75–83

Bardi J (2011) Comunidades de hidrozoários (Cnidaria) estuarinos do sudeste e sul do Brasil. PhD thesis, Universidade de São Paulo

Bardi J, Marques AC (2009) The invasive hydromedusae *Blackfordia virginica* Mayer, 1910 (Cnidaria: Blackfordiidae) in southern Brazil, with comments on taxonomy and distribution of the genus *Blackfordia*. Zootaxa 2198:41–50

Bettim AL, Haddad MA (2017) Seasonal recruitment of the hydroid *Podocoryna loyola* (Hydractiniidae) in the Paranaguá Bay, South of Brazil. Mar Biol Res 13(5.) https://doi.org/10.1080/17451000.2017.1307990

Boero F, Bouillon J, Gravili C et al (2008) Gelatinous plankton: irregularities rule the world (sometimes). Mar Ecol Prog Ser 356:299–310

Boltovskoy D (1975) Some biometrical, ecological, morphological and distributional aspects of Chaetognatha. Hydrobiologia 46:515–534

Boltovskoy D (1981) Chaetognatha. In: Boltovskoy D (ed) Atlas del zooplancton del Atlántico Sudoccidental y métodos de trabajo con zooplancton marino. INIDEP, Mar del Plata, pp 759–791

Boltovskoy D (1999) South Atlantic zooplankton. Backhuys Publishers, Leiden

Bone Q (1998) The biology of pelagic tunicates. Oxford University Press, New York

Bonecker SLC, Dias CO (2009) Zooplankton composition in an estuarine area of State of Maranhão, Northeastern Brasil. Animal Biol J 1:69–79

Bonecker SLC, Oliveira OMP, Carvalho PF (2017) Chaetognatha in Catálogo Taxonômico da Fauna do Brasil. PNUD. Available in: http://fauna.jbrj.gov.br/fauna/faunadobrasil/140164. Accessed 11 Sept 2017

Bouillon J (1999) Hydromedusae. In: Boltovsky D (ed) South Atlantic zooplankton. Backhuys Publishers, Leiden, pp 385–465

Branco CWC, Kozlowsky-Suzuki, B, Esteves FA (2007) Environmental changes and zooplankton temporal and spatial variation in a disturbed Brazilian coastal lagoon. Braz J Biol 67:251–262

Brunetti R, Baiocchi L, Bressan M (1990) Seasonal distribution of *Oikopleura* (Larvacea) in the lagoon of Venice. Boll Zool 57:89–94

Carvalho PF, Bonecker LC (2010) Seasonal and spatial variability of Appendicularian density and taxonomic composition in the Caravelas estuary (Northeastern Brazil) and adjacent coastal area. Braz Arch Biol Technol 53:161–169

Carvalho PF, Bonecker SLC, Nassar CAG (2016) Analysis of the Appendicularia class (subphylum Urochordata) as a possible tool for biomonitoring four estuaries of the tropical region. Environ Monit Assess. https://doi.org/10.1007/s10661–016–5616–5

Casanova JP (1999) Chaetognatha. In: Boltovskoy D (ed) South Atlantic zooplankton. Backhuys Publishers, Leiden, pp 1353–1374

Cavalcanti EAH, Neumann-Leitão A, DAN V (2008) Mesozooplâncton do sistema estuarino de Barra das Jangadas, Pernambuco, Brasil. Rev Brasil Zool 25:436–444

Chícharo A, Leitão T, Range P (2009) Alien species in the Guadiana Estuary (SEPortugal/SW–Spain): *Blackfordia virginica* (Cnidaria, Hydrozoa) and *Palaemon macrodactylus* (Crustacea, Decapoda): potential impacts and mitigation measures. Aquat Invasions 4:501–506

Coelho-Botelho MJ, JBN M, Dias CO et al (1999) Aspectos do zooplâncton da Baía de Sepetiba (RJ, Brasil). Oecologia Brasiliensis 7:1–33

Costa KG, Pereira LCC, Costa RN (2008) Short and long–term temporal variation of the zooplankton in a tropical estuary (Amazon region, Brazil). Bol Mus Para Emílio Goeldi sér Ciências Naturais 3:127–141

Costello JH, Bayha KM, Mianzan HW (2012) Transitions of *Mnemiopsis leidyi* (Ctenophora: Lobata) from a native to an exotic species: a review. Hydrobiologia 690:21–46

Cremer MJ, Morales PRD, Oliveira TMN (2006) Diagnóstico Ambiental da baía da Babitonga. Editora da Univille, Joinville

Dagg MJ, Brown SL (2005) The potential contribution of fecal pellets from the larvacean *Oikopleura dioica* to vertical flux of carbon in a river dominated coastal margin. In: Gorsky G, Youngbluth MJ, Deibel D (ed) Response of marine ecosystems to global change Ecological impact of Appendicularians, pp 293–397

Dagg MJ, Green EP, Mckee BA et al (1996) Biological removal of fine–grained lithogenic particles from a large river plume. J Mar Res 54:149–160

Daily GCS, Alexander PR, Ehrlich L et al (1997) Ecosystem services: benefits supplied to human societies by natural ecosystems. Issues in Ecology 2:1–16

Dias CO, Bonecker SLC (2008) Long–term study of zooplankton in the estuarine system of Ribeira Bay, near a power plant (Rio de Janeiro, Brazil). Hydrobiologia 614:65–81

Du P, Xu X, Liu J, Zeng J, Chen Q, Shou L, Liao Y, Zhou Q (2011) Distribution of zooplankton in the Jiaojiang Estuary and its response to environment factors in spring and autumn. Chin J Appl Environ Biol 17:486–494. (in Chinese with English abstract)

Esnal GB (1999) Appendicularian. In: Boltovsakoy D (ed) South Atlantic zooplankton. Backhuys Publishers, Leiden, pp 1375–1399
Esnal GB, Castro RJ (1977) Distributional and biometrical study of appendicularians from the west South Atlantic Ocean. Hydrobiologia 56:241–246
Esnal GB, Daponte MC (1999a) Doliolida. In: Boltovsakoy D (ed) South Atlantic zooplankton. Backhuys Publishers, Leiden, pp 1409–1421
Esnal GB, Daponte MC (1999b) Salpida. In: Boltovskoy D (ed) South Atlantic zooplankton. Backhuys Publishers, Leiden, pp 1423–1444
Esnal GB, Sankarankutty C, Castro RJ (1985) Diurnal and seasonal fluctuations of *Oikopleura dioica* fol 1872 (Tunicata, Appendicularia) in the mouth of the River Potengi (North Brazil). Physis, Secc A 43:65–71
Freire M, Genzano GN, Neumann-Leitão S (2014) The nonindigenous medusa *Blackfordia virginica* (Hydrozoa, Leptothecata) in tropical Brazil: 50 years of unnoticed presence. Biol Invasions 16:1–5
Furnestin ML (1957) Chaetognathes et zooplancton du secteur Atlantique Marocain. Revue des Travaux de l'Institut des Pêches Maritimes 21:1–356
Genzano GN, Mianzan H, Acha EM (2006) First record of the invasive medusa *Blackfordia virginica* (Hydrozoa: Leptomedusae) in the Río de la Plata estuary, Argentina–Uruguay. Rev Chil Hist Nat 79:257–261
Gibbons MJ (1997) Vertical distribution and feeding of *Thalia democratica* on the Agulhas Bank during march 1994. J Mar Biol Assoc UK 77:493–505
González-Duarte MM, Megina C, López-González PJ (2016) Cnidarian alien species in expansion. In: Goffredo S, Dubinsky Z (eds) The Cnidaria, past, present and future. Springer, Basel, pp 139–160
Grant GC (1977) Seasonal distribution and abundance of the Chaetognatha in the lower Chesapeake Bay. Estuar Coast Shelf Sci 5:809–824
Haddad MA, Marques AC (2009) Cnidaria. In: Rocha RM, Boeger WA (eds) Zoologia. Estado da arte e perspectivas. Editora da UFPR, Curitiba, pp 29–48
Haddad MA, Nogueira Júnior M (2006) Reappearance and seasonality of *Phyllorhiza punctata* von Lendenfeld (Cnidaria, Scyphozoa, Rhizostomeae) medusae in southern Brazil. Rev Brasil Zool 23:824–831
Haddad MA, Bettim AL, Miglietta MP (2014) *Podocoryna loyola*, n. sp. (Hydrozoa, Hydractiniidae): a probably introduced species on artificial substrate from southern Brazil. Zootaxa 3796:494–506
Hamner WM, Madin LP, Alldredge AL et al (1975) Underwater observations of gelatinous zooplankton: sampling problems, feeding biology, and behavior. Limnol Oceanogr 20:907–917
Hernández RM, Suarez-Morales E, Gasca R (2005) Seasonal distribution of Chaetognatha in a Mexican neotropical bay during a year cycle. J Coast Res 21:186–192
Hoover RS, Hoover D, Miller M et al (2006) Zooplankton response to storm runoff in a tropical estuary: bottom–up and top–down controls. Mar Ecol Prog Ser 318:187–201
Hopcroft RR, Roff JC (1998) Production of tropical larvaceans in Kingston Harbour, Jamaica: are we ignoring an important secondary producer? J Plankton Res 17:205–220
Hossfeld B (1996) Distribution and biomass of arrow worms (Chaetognatha) in Golfo de Nicoya and Golfo Dulce, Costa Rica. Rev Biol Trop 44:157–117
Johnsen S (2001) Hidden in plain sight: the ecology and physiology of organismal transparency. Biol Bull 201:301–318
Kremer P (2002) Towards understanding of salp swarm dynamics. CM 12:1–15
Krumme U, Liang TH (2004) Tidal–induced changes in a copepod–dominated zooplankton community in a macrotidal mangrove channel in Northern Brazil. Zool Stud 43:404–414
Lana PC, Marone E, Lopes RM et al (2001) The subtropical estuarine complex of Paranaguá Bay, Brazil. In: Seeliger U, Lacerda LD, Kjerfve B (eds) Coastal Marine ecosystems of Latin America. Springer Verlag, Basel, pp 131–145
Lansac Tôha FA, Lima AF (1993) Ecologia do zooplâncton do Estuário do Rio Una do Prelado (São Paulo, Brasil). Acta Limnol Bras 6:82–95

Larson RJ (1986) Water content, organic content, and carbon and nitrogen composition of medusae from the Northeast Pacific. J Exp Mar Biol Ecol 99:107–120
Liang TH (2002) Chaetognath seasonality and distribution at surface layers off Ubatuba, South Brazil Bight coastal waters. Oceánides 17:103–111
Liang TH, Vega-Pérez LA (1995) Studies on chaetognaths of Ubatuba region, Brazil. II. Feeding habits. Bol Inst Oceanogr 43:27–40
Liang TH, Vega-Pérez LA (2001) Diversity, abundance and biomass of epiplanktonic chaetognath off South Atlantic Western sector, from Cabo Frio (23°S, 42°W) to São Pedro and São Paulo rocks (01°N, 29°W). Oceánides 16:34–48
Liang TH, Vega-Pérez LA (2002) Distribution, abundance and biomass of Chaetognaths off São Sebastião region, Brazil in February 1994. Rev Bras Oceanogr 50:1–12
Liang TH, Ara K, Miranda LB, Bérgamo AL, Bernardes M (2003) On the variability of the chaetognath *Sagitta friderici* Ritter–Z'ahony at Cananéia Lagoon Estuarine system, São Paulo, Brazil. Hydrobiologia 510:91–102
Liu H, Li K, Huang H et al (2013) Seasonal community structure of mesozooplankton in the Daya Bay, South China Sea. J Ocean U China 12:452–458
Lopes RM (1994) Zooplankton distribution in the Guaraú River Estuary (South–Eastern Brazil). Estuar Coast Shelf Sci 39:287–302
Lopes RM, Vale R, Brandini FP (1998) Composição, abundância e distribuição espacial do zooplâncton no complexo estuarino de Paranaguá durante o inverno de 1993 e o verão de 1994. Rev Bras Oceanogr 46:195–211
Loureiro Fernandes L, Sterza JM, Neves KO (2005) Seasonal chaetognath abundance and distribution in a tropical estuary (Southeastern, Brazil). Braz J Oceanogr 53:47–53
Mackie GO, Pugh PR, Purcell JE (1987) Siphonophore biology. Adv Mar Biol 24:97–262
Macpherson E (2002) Large–scale species–richness gradients in the Atlantic Ocean. Proc Biol Sci 269:1715–1720
Magalhães EMM, PO M–J, Silva JJ et al (1996) Chaetognatha planctônicos do complexo estuarino lagunar Mundaú/Manguaba, AL. Boletim de Estudos de Ciências do Mar 9:63–87
Mahiques MM, Figueira RCL, Salaroli AB et al (2013) 150 years of anthropogenic metal input in a biosphere reserve: the case study of the Cananéia–Iguape coastal system, Southeastern Brazil. Environ Earth Sci 68:1073–1087
Marazzo A, Nogueira CSR (1996) Composition, spatial and temporal variations of Chaetognatha in Guanabara Bay, Brazil. J Plankton Res 18:2367–2376
Marcolin CR (2008) Estrutura da comunidade zooplanctônica estuarina em relação ao impacto da atividade de carcinicultura no litoral norte da Bahia, Brasil. Masthers' dissertation, Universidade Federal da Bahia
Marcolin CR, Conceição BL, Nogueira MM et al (2010) Mesozooplankton and ichthyoplankton composition in two tropical estuaries of Bahia, Brazil. Check List 6(2):210–216
Marques F, Angélico MM, Costa JL (2017) Ecological aspects and potential impacts of the non–native hydromedusa *Blackfordia virginica* in a temperate estuary. Estuar Coast Shelf Sci 197:69–79
Martinez TIM, Costa BSP, Miyashita LK et al (2015) Diversity of gelatinous zooplankton (Cnidaria, Ctenophora, Chaetognatha and Tunicata) from the Cananéia–Iguape Estuarine System (São Paulo, Brazil). Proceedings of the XVI Colacmar and XVI Senalmar, Santa Marta, Colombia, p 35
McLelland JA (1980) Notes on the northern Gulf of Mexico occurrence of *Sagitta friderici* Ritter-Zahony (Chaetognatha). Gulf Res Rep 6:343–348
McLelland JA (1989) An illustrated key to the Chaetognatha of the Northern Gulf of Mexico with notes on their distribution. Gulf Res Rep 8:145–172
Menard F, Fromentin JM, Goy J et al (1997) Temporal fluctuations of doliolid abundance in the bay of Villefranche–sur–Mer (Northwestern Mediterranean Sea) from 1967 to 1990. Oceanol Acta 20:733–742
Mendes F, Figueiredo GM, Valentin JL (2012) Reproduction and structure of the population of the Chaetognath *Parasagitta friderici* in Guanabara Bay (Brazil) based on short term sampling. An Acad Brasil Ciênc 84:103–111

Mesquita SDSA, Costa RM, Pereira LCC (2006) Composição, ocorrência e distribuição das hidromedusas no estuário do rio Caeté, litoral do estado do Pará. Bol Mus Para Emílio Goeldi Ciênc Nat 1:113–119
Mianzan HW (1999) Ctenophora. In: Boltovskoy D (ed) South Atlantic zooplankton, vol 1. Backhuys Publishers, Leiden, pp 51–573
Mianzan HW, Cornelius PFS (1999) Cubomedusae and Scyphomedusae. In: Boltovskoy D (ed) South Atlantic zooplankton, vol 1. Backhuys Publishers, Leiden, pp 513–559
Mills CE (1995) Medusae, siphonophores, and ctenophores as planktivorous predators in changing global ecosystems. ICES J Mar Sci 52:575–581
Mills C, Rees JT (2000) New observations and corrections concerning the trio of invasive hydromedusae *Maeotias marginata* (= *M. inexpectata*), *Blackfordia virginica*, and *Moreisia* sp. in the San Francisco Estuary. Sci Mar 64:151–155
Miyashita LK, Lopes RM (2011) Larvacean (Chordata, Tunicata) abundance and inferred secondary production off southeastern Brazil. Estuar Coast Shelf Sci 92:367–375
Miyashita LK, Brandini FP, JE M–F et al (2012) Comparison of zooplankton community structure between impacted and non–impacted areas of Paranaguá Bay Estuarine Complex, South Brazil. J Nat Hist 46:1557–1571
Montú MA (1980) Zooplâncton do estuário da Lagoa dos Patos I – Estrutura e variações temporais e espaciais da comunidade. Atlantica 4:53–72
Montú MA, Cordeiro TA (1988) Zooplancton del complejo estuarial de la Bahía de Paranaguá. I. Composición, dinámica de las especies, ritmos reproductivos y acción de los factores ambientales sobre la comunidade. Neritica 3:61–83
Moore SJ (1987) Redescription of the leptomedusan *Blackfordia virginica*. J Mar Biol Assoc UK 67:287–291
Morandini AC (2003) Estrutura populacional de *Chrysaora lactea* e *Lychnorhiza lucerna* (Cnidaria, Scyphozoa) em amostras de plâncton, com a redescrição das espécies. PhD thesis, Universidade de São Paulo
Moreira GS (1978) A preliminary laboratory study on the salinity and temperature tolerances of some medusae from the São Paulo coast, Brazil. Bol Inst Oceanogr 27:45–55
Mulkana MS, McIlwain TD (1973) The seasonal occurrence and abundance of Chaetognatha in Mississippi Sound. Gulf Res Rep 4:264–261
Nagata RM, Haddad MA, Nogueira Jr M (2009) The nuisance of jellyfish to shrimp trawls in central part of Southern Brazilian Bight, from the perspective of artisanal fishermen. Panm J Aquat Sci 4:312–325
Nagata RM, Santos L, Stampar SN et al (2013) *Phyllorhiza punctata* in Brazilian coast: population oscillations or recurrent invasions? Proceedings of the fourth international Jellyfish blooms symposium, Hiroshima, p 128
Nagata RM, Nogueira-Junior M, Brandini FP (2014a) Spatial and temporal variation of planktonic cnidarians density in subtropical waters of the Southern Brazilian Bight. J Mar Biol Assoc UK 94:1387–1400
Nagata RM, Nogueira-Junior M, Haddad MA (2014b) Faunistic survey of Hydromedusae (Cnidaria, Medusozoa) from the coast of Paraná State, Southern Brazil. Zootaxa 3768:291–326
Nair VR, Sankarankutty C (1988) Chaetognaths of the Potengi estuary (Natal, Northeast Brazil). Atlantica 10:5–20
Nakamura Y (1998) Blooms of tunicates *Oikopleura* spp. and *Dolioletta gegenbauri* in the Seto Inland Sea, Japan, during summer. Hydrobiologia 385:183–192
Nascimento LS (2016) Hidrozoários planctônicos do complexo estuarino de Paranaguá–Brasil. Masthers' dissertation, Universidade Federal do Paraná
Navas-Pereira D (1973) Relatório sobre a segunda pesquisa oceanográfica e pesqueira do Atlântico Sul entre Torres e Maldonado (Lat. 29 S–35 S). Análise qualitativa e quantitativa do zooplâncton. Publ Espuar Inst Oceanogr Sci Paulo 3:29–92
Navas-Pereira D (1980) Hydromedusae of the Bay of Sepetiba (Rio de Janeiro, Brazil). Rev Bras Biol 40:817–824
Neumann-Leitão S, Matsumura-Tundisi T (1998) Dynamics of a perturbed estuarine zooplanktonic community: Port of Suape, PE, Brazil. Verh Internat Verein Limnol 26:1981–1988

Nogueira Júnior M (2012) Gelatinous zooplankton fauna (Cnidaria, Ctenophora and Thaliacea) from Baía da Babitonga (southern Brazil). Zootaxa 3398:1–21

Nogueira Júnior M, Brandini FP (2018) Community structure and spatial–temporal dynamics of the zooplankton in the South Brazilian Bight – a review. In: Hoffmeyer MS, Sabatini ME, Brandini FP, Calliari D, Santinelli N (eds) Plankton ecology of the Southwestern Atlantic – from the subtropical to the subantarctic realm, Springer, chapter 8, this volume

Nogueira Júnior M, Haddad MA (2006) Macromedusae (Cnidaria) from the Paraná Coast, Southern Brazil. J Coast Res SI 39:1161–1164

Nogueira Júnior M, Haddad MA (2008) The diet of Cubomedusae (Cnidaria, Cubozoa) in southern Brazil. Braz J Oceanogr 56(3):157–164

Nogueira Júnior M, Oliveira JS (2006) *Moerisia inkermanica* Paltschikowa–Ostroumova (Hydrozoa, Moerisidae) e *Blackfordia virginica* Mayer (Hydrozoa, Blackfordiidae) na Baía de Antonina, Paraná, Brasil. Panam J Aquat Sci 1:35–42

Nogueira Júnior M, Souza-Conceição JMS (2010) Seasonal distribution and size class composition of the Scyphomedusae *Chrysaora lactea* (Semaeostomeae) and *Lychnorhiza lucerna* (Rhizostomeae) in a subtropical estuary in Southern Brazil. Proceedings of the third international Jellyfish blooms symposium, Mar del Plata, p 82

Nogueira Júnior M, Nagata RM, Haddad MA (2010) Seasonal variation of macromedusae (Cnidaria) at North Bay, Florianópolis, southern Brazil. Fortschr Zool 27:377–386

Nogueira Júnior M, Brandini FP, Codina JC (2014) Distribution of planktonic cnidarians in response to South Atlantic central water intrusion in the South Brazilian Bight. Cont Shelf Res 89:93–102

Nogueira Júnior M, Brandini FP, Codina J (2015a) Diel vertical dynamics of gelatinous zooplankton (Cnidaria, Ctenophora and Thaliacea) in a subtropical stratified ecosystem (South Brazilian Bight). PLoS One 10:0144161–0144128

Nogueira Júnior M, Pukanski LEDM, Conceição JMS (2015b) Mesh size effects on assessments of planktonic hydrozoan abundance and assemblage structure. J Mar Syst 144:117–126

Nogueira Júnior M, Silva Nascimento L (2018) The ecology and developmental changes of meristic characters of the medusa *Malagazzia carolinae* (Hydrozoa: Leptothecata) from subtropical Southwestern Atlantic estuaries. Zoologischer Anzeiger 274:34–45

Oliveira MP (2007) Ctenóforos da costa brasileira: considerações taxonômicas e biológicas. PhD thesis, Universidade de São Paulo

Oliveira OMP, Mianzan H, Migotto AE et al (2007) Chave de identificação dos Ctenophora da costa brasileira. Biol Neotrop 7:341–350

Oliveira OM, Miranda TP, Araujo EM et al (2016) Census of Cnidaria (Medusozoa) and Ctenophora from south American marine waters. Zootaxa 4194:1–256

Paffenhöfer GA, Lee TN (1987) Development and persistence of patches of Thaliacea. In: Payne, A.I.L., Gulland, J.A. and Brink, K.H. (Eds.). The Benguela and comparable ecosystems. Afr J Mar Sci 5:305–318

Paffenhöfer GA, Sherman BK, Lee TN (1987) Abundance, distribution and patch formation of zooplankton. Prog Oceanogr 9:403–436

Paffenhöfer GA, Atkinson LP, Lee TN et al (1995) Distribution and abundance of thaliaceans and copepods off the southeastern U.S.A. during winter. Cont Shelf Res 15:255–280

Pekala GA (1980) Nota preliminar sobre a composição do zooplâncton do estuário do Rio Paraíba do Norte, PB, Brasil. Bol Inst Oceanogr 29:291–296

Pereira JB (2011) Composição, distribuição, biomassa e produção secundária do zooplâncton do sistema estuarino de Santos, São Paulo, Brasil. PhD thesis, Universidade de São Paulo

Pierce EL (1951) The Chaetognatha of the west coast of Florida. Biol Bull 100:206–228

Pierrot-Bults AC, Nair VR (1991) Distribution patterns in Chaetognatha. In: Bone Q, Kapp H, Pierrot-Bults AC (eds) The biology of Chaetognaths. Oxford University, Oxford, pp 86–116

Porto Neto FF, Neumann-Leitão S, Gusmão LMO, Nascimento Vieira DA, Silva AP, Silva TA (1999) Variação sazonal e nictemeral do zooplâncton no Canal de Santa Cruz, Itamaracá, PE, Brasil. Trab Oceanog Univ Fed PE 27):43–58

Potter IC, Chuwen BM, Hoeksema SD et al (2010) The concept of an estuary: a definition that incorporates systems which can become closed to the ocean and hypersaline. Estuar Coast Shelf Sci 87:497–500

Primo AL, Marques SC, Falcão J (2012) Environmental forcing on jellyfish communities in a small temperate estuary. Mar Environ Res 79:152–159

Pugh PR (1999) Siphonophorae. In: Boltovsky D (ed) South Atlantic zooplankton. Backhuys Publishers, Leide, pp 467–511

Pukanski LEDM (2011) Dinâmica populacional das hidromedusas *Liriope tetraphylla* (Trachymedusae) e *Blackfordia virginica* (Leptothecata) na Baía da Babitonga. Masthers' dissertation, Universidade Federal do Paraná

Purcell JE (2005) Climate effects on formation of jellyfish and ctenophore blooms: a review. J Mar Biol Assoc UK 85:461–476

Purushothaman J, Hansda S, Dey J et al (2017) An annotated checklist of thaliaceans (Chordata: tunicates). Mar Biodivers:1–28

Reeve MR (1970) The biology of Chaetognatha I. Quantitative aspects of growth and egg production in *Sagitta hispida*. In: Steele JH (ed) Marine food chains. University of California Press, Berkeley, pp 1–550

Resgalla C (2001) Estudo de impacto ambiental sobre a comunidade do zooplâncton na enseada so Saco dos Limões, Baía Sul da ilha de Santa Catarina, Brasil. Atlântica 23:5–16

Resgalla C (2010) Parâmetros populacionais de *Sagitta friderici* Ritter–Záhony (Chaetognatha) na zona de arrebentação da praia do Cassino, sul do Brasil. Atlantica 32:141–149

Rippingale RJ, Kelly SJ (1995) Reproduction and survival of *Phyllorhiza punctata* (Cnidaria: Rhizostomeae) in a seasonally fluctuating salinity regime in Western Australia. Mar Freshw Res 46:1145–1151

Rocha RM, Bonecker SLC, Carvalho PF (2017) Appendicularia in: Catálogo Taxonômico da Fauna do Brasil. http://fauna.jbrj.gov.br/fauna/faunadobrasil/156717 Accessed on 21 Sept 2017

Russell FS (1953) The Medusae of the British Isles. Anthomedusae, Leptomedusae, Limnomedusae, Trachymedusae, and Narcomedusae. Cambridge University Press, Cambridge, p 529

Salvador B, Bersano JGF (2017) Zooplankton variability in the subtropical estuarine system of Paranaguá Bay, Brazil, in 2012 and 2013. Estuar Coast Shelf Sci. https://doi.org/10.1016/j.ecss.2017.09.019

Sankarankutty C, Medeiros GF, Santos NQ (1979) On diurnal variations of zooplankton in a tidal estuary. J Mar Biol Assoc India 21(1):187–190

Sankarankutty C, Medeiros GF, Oliveira JE (1995) Diurnal and seasonal fluctuations of estuarine zooplankton of Potengi estuary, Natal, Northeast of Brazil. J Mar Biol Assoc India 37:201–211

Sankarankutty C, Mendonça KMA, Ferreira AC et al (1997) On zooplankton of a mangrove ecosystem close to Macau, Rio Grande do Norte, Brazil. Trab Oceanog Univ Fed PE 25:47–59

Sankarankuty C, Medeiros GF, Silva ACC et al (1996) On an unusual concentration of *Sagitta tenuis* conant in the estuary of Potengi, Natal, Brazil. J Mar Biol Assoc India 38:153–155

Sato R, Ishibashi Y, Ishimaru T, Dagg MJ (2008) Productivity and grazing impact of Oikopleura dioica (Tunicata, Appendicularia) in Tokyo Bay. J Plankton Res 30:299–309

Sato NE, Hernández D, Viñas MD (2011) Hábitos alimentarios de *Sagitta friderici* Ritter–Zahony em las águas costeras de la Provincia de Buenos Aires, Argentina. Bol Invest Mar Cost 40:59–74

Schuchert P (2011) *Dipurena* McCrady, 1859. In: Schuchert, P. (2017). World Hydrozoa database. Accessed through: World Register of Marine Species at http://marinespecies.org/aphia.php/aphia.php?p=taxdetailsandid=117069. Accessed on 6 Oct 2017

Silva TA, Neumann-Leitão S, Schwamborn R et al (2003) Diel and seasonal changes in the macrozooplankton community of a tropical estuary in Northeastern Brazil. Rev Bras Zootec 20:439–446

Silva AP, Neumann-Leitão S, Schwamborn R et al (2004) Mesozooplankton of an impacted Bay in North Eastern Brazil. Braz Arch Biol Technol 47:485–493

Silveira FL, Cornelius PFS (2000) New observations on medusae (Cnidaria, Scyphozoa, Rhizostomae) from the northeast and South Brazil. Acta Biol Leopold 22:9–18
Singarajah KV (1978) Hydrographic conditions, composition and distribution of plankton in relation to potential resources of Paraíba River estuary. Revta Nordest Biol 1:125–144
Souza CS, Luz JAG, Mafalda-Júnior P (2014) Relationship between spatial distribution of chaetognaths and hydrographic conditions around seamounts and islands of the tropical southwestern Atlantic. An Acad Brasil Ciênc 86:1151–1165
Spinelli ML, Martos P, Esnal GB (2009) Appendicularian assemblages and their space–time variability off the la Plata River, SW Atlantic Ocean. Estuar Coast Shelf Sci 85:97–106
Srinivasan M (1971) Biology of chaetognaths of the estuarine waters of India. J Mar Biol Assoc India 13:173–181
Srinivasan M (1980) Life cycle and seasonal fluctuation of Chaetognatha in Ennore estuary, Madras. Bull Zoo Surv India 3:55–61
Sterza JM, Loureiro Fernandes L (2006) Zooplankton community of the Vitória Bay Estuarine system (Southeastern Brazil). Characterization during a three–year study. Braz J Oceanogr 54:95–105
Tavares DQ (1967) Occurrence of doliolids and salps during 1958, 1959, and 1960 off the São Paulo coast. Bolm Inst oceanogr 16:87–97
Teixeira C, Tundisi J, Kutner MB (1965) Plankton studies in a mangrove environment II. The standing stock and some ecological factors. Bol Inst Oceanogr 14:13–41
Teixeira-Amaral P, WJA A, de Ortiz DO (2017) The mesozooplankton of the Patos Lagoon Estuary, Brazil: trends in community structure and secondary production. Mar Biol Res 13:48–61
Thuesen EV (2009) *Parasagitta* Tokioka, 1965. Accessed through: World Register of Marine Species at http://www.marinespecies.org/aphia.php?p=taxdetailsandid=105408 Accessed on 19 Sept 2017
Tomita M, Shiga N, Ikeda T (2003) Seasonal occurrence and vertical distribution of appendicularians in Toyama Bay, southern Japan Sea. J Plankton Res 25:579–589
Tse P, Souissi S, Hwang JS et al (2008) Spatial and seasonal variations in chaetognath assemblages in two subtropical marine inlets with different hydrographical characteristics. Zool Stud 47:258–267
Tundisi TM (1970) On the seasonal occurrence of appendicularians in waters off the coast of São Paulo State. Bol Inst Oceanogr 19:131–144
Vannucci M (1954) Hydrozoa e Scyphozoa existentes no Instituto Oceanográfico. II Bol Inst Oceanogr 5:95–149
Vannucci M (1956) Biological notes and description of a new species of *Dipurena* (Hydrozoa, Corynidae). J Zool 127:479–487
Vannucci M (1957) On Brazilian Hydromedusae and their distribution in relation to different water masses. Bol Inst Oceanogr 8:23–109
Vannucci M (1963) On the ecology of Brazilian medusae at 25° lat. S. Bol Inst Oceanogr 13:143–184
Vega-Pérez LA, Liang TH (1992) Feeding of a pelagic chaetognath, *Sagitta friderici* Ritter-Záhony off Ubatuba region (São Paulo, Brazil). Bol Inst Oceanogr 40:93–100
Whitfield A, Elliot M (2011) Ecosystem and biotic classifications of estuaries and coasts. In: Wolanski E, Mclusky DS (ed). Estuar Coast Shelf Sci 1:99–124
Wintzer AP, Meek MH, Moyle PB (2011) Ecological insights into the polyp stage of non-native hydrozoans in the San Francisco Estuary. Aquatic Ecol 45:151–161
Yilmaz IN (2015) Collapse of zooplankton stocks during *Liriope tetraphylla* (Hydromedusa) blooms and dense mucilaginous aggregations in a thermohaline stratified basin. Mar Ecol 36:595–610
Zamponi MO, Genzano G (1994) Seasonal distribution of hydromedusae from Samborombon Bay (Buenos Aires, Argentina). Plankton. Newsletter 19:51–56

An Overview of the Medusozoa from the Southwestern Atlantic

Agustín Schiariti, María S. Dutto, André Carrara Morandini, Renato M. Nagata, Daiana Y. Pereyra, Francisco A. Puente Tapia, Luciana Díaz Briz, and Gabriel Genzano

Abstract Medusozoans are critical components of coastal and marine ecosystems. They are ubiquitous, living from the surface to the bottom layers of the world's oceans and tolerating a wide range of environmental conditions. They modulate food webs not only by consuming large quantities of ichthyoplankton and other zooplankton and acting as predators and competitors of varied pelagic organisms including fish but also by being consumed by other predators. Population outbreaks of these gelatinous animals commonly occur over a variety of spatiotemporal scales. These population explosion events have implications for the ecosystem and, usually,

A. Schiariti (✉)
Instituto Nacional de Investigación y Desarrollo Pesquero (INIDEP), Mar del Plata, Buenos Aires, Argentina

Instituto de Investigaciones Marinas y Costeras (IIMyC), CONICET, Universidad Nacional de Mar del Plata, Mar del Plata, Buenos Aires, Argentina
e-mail: agustin@inidep.edu.ar

M. S. Dutto
Instituto Argentino de Oceanografía (IADO), Centro Científico Tecnológico Bahía Blanca, CONICET – UNS, Bahía Blanca, Buenos Aires, Argentina
e-mail: msdutto@cribs.edu.ar

A. C. Morandini
Departamento de Zoologia, Instituto de Biociências, Universidade de São Paulo (USP), São Paulo, Brazil
e-mail: acmorand@ib.usp.br

R. M. Nagata
Instituto de Oceanografia Universidade Federal do Rio Grande (FURG) Rio Grande, Rio Grande, RS, Brazil

D. Y. Pereyra
Instituto Nacional de Investigación y Desarrollo Pesquero (INIDEP), Mar del Plata, Buenos Aires, Argentina

F. A. P. Tapia · L. D. Briz · G. Genzano
Departamento de Ciencias Marinas, Facultad de Ciencias Exactas y Naturales, Universidad Nacional de Mar del Plata (UNMdP), Mar del Plata, Buenos Aires, Argentina

M. S. Hoffmeyer et al. (eds.), *Plankton Ecology of the Southwestern Atlantic*,
https://doi.org/10.1007/978-3-319-77869-3_19

for human enterprise. Despite their ecological and socioeconomical importance, there is as yet no attempt to compile existing information on the medusozoan species of the Southwestern Atlantic (SWA). We provide here an overview of the information available regarding several aspects of Medusozoa in the SWA: the characteristics of their life cycles, life histories and "blooms," and the ecological implications for SWA ecosystems. Guidelines for future research and perspectives on the field are also provided.

Keywords Jellyfish · Gelatinous zooplankton · Medusae · Ecological roles · Societal impacts

1 Introduction

Medusozoans (i.e., non-anthozoan cnidarians) comprise roughly 3700 species worldwide and are ordinarily divided into four classes: Scyphozoa (true jellyfish), Cubozoa (box jellies), Hydrozoa (hydroids, hydromedusae, siphonophores), and Staurozoa (stalked jellyfish) (Marques and Collins 2004; Daly et al. 2007). The majority of Scyphozoa and Hydrozoa, and all Cubozoa, include pelagic stages (i.e., medusae) in their metagenetic life cycles (Fig. 1). Free-swimming medusae have transparent, soft bodies consisting of ca. 95% of water, conferring a gelatinous consistency typical of all species commonly grouped as "jellyfish." This common term is also used for other gelatinous zooplankton groups, and it has the advantage of emphasizing the convergent features of transparency, fragility, and planktonic existence that unite these disparate creatures, without the complex taxonomic terminology associated with other terms (Haddock 2004). Species belonging to four different phyla are included under this term. Therefore, "jellyfish" is meant to have no taxonomic implication but rather to evoke these diverse groups of non-crustacean zooplankton that are too fragile to be sampled with conventional net-based systems (Haddock 2004). In the context of this chapter, the word "jellyfish" refers only to the planktonic/pelagic stages of medusozoan species, i.e., scypho-, cubo-, and hydromedusae as well as siphonophores. We deliberately exclude the staurozoans due to a lack of any pelagic/planktonic stages.

Medusozoans, as do all cnidarians, have highly specialized cells, i.e., the cnidocytes, that contain specialized intracellular structures, the cnidae, unique within the animal kingdom (Morandini et al. 2016a). Nematocysts, a type of cnidae that are probably the most complex secretion products of single animal cells, have enabled the group to achieve enormous success as predators with little investment in the elaboration of sensory and morphological specializations that characterize most predators. In a sense, they have been considered as "little more than a gut with tentacles" (Kass-Simon and Scappaticci 2002). With very few exceptions, cnidarian jellyfish are carnivores that use their cnidae to kill their prey, which include, depending on the species, other jellyfish, crustaceans, veliger larvae, fish eggs, and

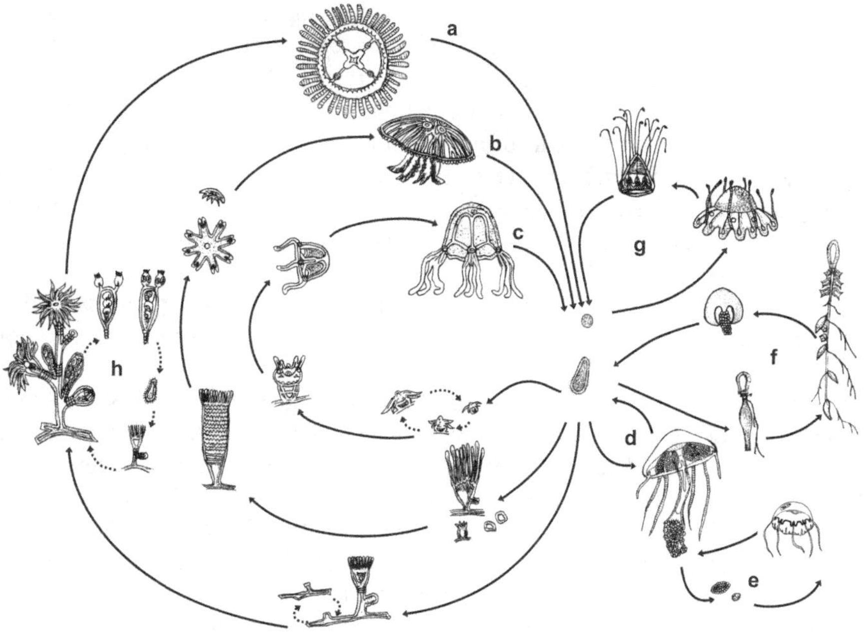

Fig. 1 Schematic representation of the medusozoan life cycles. The metagenetic pattern is represented for Hydrozoa (**a**), Scyphozoa (**b**), and Cubozoa (**c**). Holoplanktonic life cycles in Hydrozoa, Trachymedusae (**d**), Narcomedusae (**e**), siphonophore colonies (**f**), and some Scyphozoa (**g**). Holobenthic life cycles, with fixed abortive medusoid stages in Hydrozoa (**h**). See details in text (Section 3)

larvae (Genzano et al. 2014). Some jellyfish, however, are microphagous or even contain symbiotic zooxanthellae. Jellyfish are ubiquitous, inhabiting all marine ecosystems of the world over more than 500 million years. They can range in size from a few mm to more than 1 m in bell diameter, thus being parts of the meso-, macro-, and megazooplankton. They are ecologically and evolutionarily important and relevant to human affairs. They present interesting conceptual problems, not only because of the key position they occupy in the evolutionary tree but also due to distinctive aspects of their biology and development (Mackie 2002; Morandini et al. 2014; Technau et al. 2015).

Research on zooplankton has been historically oriented toward the non-gelatinous species (mostly copepods) that provide the bulk of the food for commercially important fish stocks (Haddock 2004). These organisms present size, body structures, and spatial and temporal distributions that can be reliably represented with conventional plankton net samples. However, different devices are needed to establish patterns of abundance, distribution, behavior, and trophic ecology of soft-bodied gelatinous zooplankters, particularly when they occur in mass aggregations. The watery bodies of medusozoans are badly damaged by traditional plankton tows, often posing unanswerable questions regarding taxonomy and phylogeny and preventing accurate abundance and biomass estimates. Therefore, pelagic cnidarians usually have been omitted from routine oceanographic sampling. Since the mid-twentieth century, the development and improvement of collection and observation techniques (e.g., submersibles, remotely operated vehicles, SCUBA, acoustics, video and optical devices) have painted a dramatically different picture of medusozoan diversity from that provided by typical net tows. These sampling techniques provide access to specimens which can be used in natural experiments and not just poorly enumerated. This is essential to study their natural history, physiology, and ecology and is perhaps the most significant impact of modern collection techniques (Haddock 2004).

Research takes place within the larger context of human society, and as with any investigation, one must consider potential benefits that justify support using public funds.

Within this framework, interest in medusozoans has expanded in response to the problems generated for human enterprises by their population outbreaks (e.g., Purcell et al. 2007). Meanwhile, the economic exploitation of different scypho- and hydromedusae species has increased, which is the other side of the same coin (e.g., Purcell et al. 2013; Lange et al. 2016; Brotz et al. 2017). Because of these negative and positive interactions between "blooming species and business," the factors regulating jellyfish abundance have emerged as one of the big questions in jellyfish research. Today jellyfish are widely acknowledged as keystone species shaping pelagic communities in marine ecosystems (Pauly et al. 2009; Boero 2013). The worldwide knowledge about jellyfishes is patchy, as it is the distribution of the animals; this means that some geographic areas are better known than others. In the Southwestern Atlantic (SWA), jellyfish research is still in its infancy. Our main goal here is to review the state of the art in jellyfish research and additionally drawing attention to the need for directed research programs and international cooperation. Thus, our contribution is organized in the following topics: species composition, biological patterns of life cycles and blooms, ecological roles of jellyfishes, interaction with man, and future perspectives.

2 Jellyfish Species Composition in the SWA

In the SWA region, 689 medusozoan species or morphotypes have been recognized so far (see Oliveira et al. 2016). Of these, 335 (49%) have a pelagic medusa-like stage (Table 1), although a few species have "unusual" habits and can live on the bottom or on other substrates like algae (e.g., *Cassiopea*, *Cladonema*, *Staurocladia*). In addition, the pelagic phase can be comprised of polypoid forms that can live in the planktonic or neustonic environment (e.g., *Porpita*, *Velella*). The 335 species/morphotypes with a pelagic medusa stage can be classified into the following classes:

- Cubozoa: 5 species are reported for the SWA, from a total of ~50 species known worldwide (Kingsford and Mooney 2014): *Alatina alata*, *Chiropsalmus quadrumanus*, *Chiropsalmus zygonema*, *Tamoya haplonema*, and *Tripedalia cystophora*.

Table 1 Medusozoan species composition reported from the Southwestern Atlantic waters. Only taxa determined to species level having planktonic/pelagic stages were considered, comprising 268 hydrozoan, 20 scyphozoan, and 5 cubozoan species or morphotypes (according to Oliveira et al. 2016). (*) species with described life cycles. For Hydrozoa species, see Section 3.3

Podocoryna loyola (Haddad et al., 2014) (*)	*Dimophyes arctica* (Chun, 1897) (*)
Podocoryna tenuis (Browne, 1902) (*)	*Diphyes bojani* (Eschscholtz, 1829) (*)
Podocoryna uniformis (Stampar et al., 2006) (*)	*Diphyes chamissonis* (Huxley, 1959) (*)
Stylactaria hooperii (Sigerfoos, 1899) (*)	*Diphyes dispar* (Chamisso & Eysenhardt, 1821) (*)
Niobia dendrotentaculata (Mayer, 1900)	*Eudoxoides mitra* (Huxley, 1859) (*)
Oceania armata (Kölliker, 1853)	*Eudoxoides spiralis* (Bigelow, 1911) (*)
Turritopsis nutricula (McCrady, 1857b) (*)	*Gilia reticulata* (Totton, 1954) (*)
Amphinema australis (Mayer, 1900) (*)	*Lensia achilles* (Totton, 1941) (*)
Amphinema dinema (Perón & Lesueur, 1809) (*)	*Lensia ajax* (Totton, 1941) (*)
Annatiara affinis (Hartlaub, 1914)	*Lensia campanella* (Moser, 1917) (*)
Cirrhitiara superba (Mayer, 1900)	*Lensia challengeri* (Totton, 1954) (*)
Halitholus intermedius (Browne, 1902) (*)	*Lensia conoidea* (Keferstein & Ehlers, 1860) (*)
Leuckartiara octona (Fleming, 1823) (*)	*Lensia cossack* (Totton, 1941) (*)
Leuckartiara zacae (Bigelow, 1940) (*)	*Lensia fowleri* (Bigelow, 1911) (*)
Merga tergestina (Neppi & Stiasny, 1912) (*)	*Lensia grimaldii* (Leloup, 1933) (*)
Neoturris pileata (Forskal, 1775) (*)	*Lensia hardy* (Totton, 1941) (*)
Pandea conica (Quoy & Gaimard, 1827) (*)	*Lensia havock* (Totton, 1941) (*)
Stomotoca atra (L. Agassiz, 1862) (*)	*Lensia hostile* (Totton, 1941) (*)
Proboscidactyla mutabilis (Browne, 1902) (*)	*Lensia hotspur* (Totton, 1941) (*)
Proboscidactyla ornata (McCrady, 1859) (*)	*Lensia hunter* (Totton, 1941) (*)
Halitiara Formosa (Fewkes, 1882) (*)	*Lensia leloupi* (Totton, 1954) (*)
Lizzia blondina (Forbes, 1848)	*Lensia lelouveteau* (Totton, 1941) (*)
Podocorynoides minima (Trinci, 1903)	*Lensia meteori* (Leloup, 1934) (*)
Rathkea formosissima (Browne, 1902)	*Lensia multicristata* (Moser, 1925) (*)
Rathkea octopunctata (M. Sars, 1835)	*Lensia subtilis* (Chun, 1886) (*)
Order *Incertae sedis*	*Lensia subtiloides* (Lens & van Riemsdijk, 1908) (*)
Dipleurosoma collapsum (Mayer, 1900)	*Lensia* cf. *tottoni* (Daniel & Daniel, 1963) (*)
Hebella furax (Millard, 1957) (*)	*Muggiaea atlantica* (Cunningham, 1892) (*)
Hebella scandens (Bale, 1888) (*)	*Muggia eakochii* (Will, 1844) (*)
Order *Laodicea*	*Sulculeolaria biloba* (Sara, 1846) (*)
Laodicea indica (Browne, 1905)	*Sulculeolaria chuni* (Lens & van Riemsdijk, 1908) (*)
Laodicea minuscula (Vannucci, 1957)	*Sulculeolaria monoica* (Chun, 1888) (*)
Laodicea ocellata (Babnik, 1948)	*Sulculeolaria quadrivalvis* (Blainville, 1834) (*)
Laodicea pulchra (Browne, 1902)	*Sulculeolaria turgida* (Gegenbaur, 1853) (*)

(continued)

Table 1 (continued)

Laodicea undulata (Forbes & Goodsir, 1851)	*Hippopodius hippopus* (Forskål, 1776)
Staurophora mertensii (Brandt, 1835)	*Vogtia glabra* (Bigelow, 1918)
Modeeria rotunda (Quoy & Gaimard, 1827) (*)	*Vogtia pentacantha* (Kölliker, 1853)
Order Statocysta	*Vogtia serrata* (Moser, 1925)
Tetracanna octonema (Goy, 1979)	*Vogtia spinosa* (Keferstein & Ehlers, 1861)
Cirrholovenia tetranema (Kramp, 1959) (*)	*Amphicaryon acaule* (Chun, 1888) (*)
Cosmetirella davisi (Browne, 1902)	*Amphicaryon ernesti* (Totton, 1954) (*)
Cosmetira pilosella (Forbes, 1848)	*Amphicaryon peltifera* **(Haeckel, 1888)** (*)
Halopsis ocellata (A. Agassiz, 1865)	**Phylum Cnidaria**
Mitrocomella brownei (Kramp, 1930)	**Subphylum Medusozoa**
Mitrocomella frigida (Browne, 1910)	**Class Hydrozoa (cont.)**
Mitrocomella polydiademata (Romanes, 1876)	**Order Calycophorae(cont.)**
Phialella falklandica (Browne, 1902) (*)	*Lilyopsis rosea* (Chun, 1885)
Aequorea coerulescens (Brandt, 1838) (*)	*Maresearsia praeclara* (Totton, 1954) (*)
Aequorea forskalea (Péron & Lesueur, 1809) (*)	*Nectadamas diomedeae* (Bigelow, 1911)
Aequorea globosa (Eschscholtz, 1829) (*)	*Nectopyramis natans* (Bigelow, 1911) (*)
Aequorea macrodactyla (Brandt, 1835) (*)	*Nectopyramis thetis* (Bigelow, 1911) (*)
Rhacostoma atlanticum (L. Agassiz, 1851)	*Praya dubia* (Quoy & Gaimard, 1833) (*)
Zygocanna vagans (Bigelow, 1912)	*Rosacea cymbiformis* (Delle Chiaje, 1822)
Blackfordia virginica (Mayer, 1910) (*)	*Rosacea plicata*sensu (Bigelow, 1911)
Eirene lactea (Mayer, 1900) (*)	*Sphaeronectes fragilis* (Carré, 1967) (*)
Eirene viridula (Péron & Lesueur, 1809)	*Sphaeronectes koellikeri* (Huxley 1859) (*)
Eutima coerulea (L. Agassiz, 1862) (*)	**Order Limnomedusae**
Eutima cf. *gegenbauri* (Haeckel, 1874) (*)	*Aglauropsis agassizii* (F. Müller, 1865) nomen dubium
Eutima gentiana (Haeckel, 1879)	*Aglauropsis conantii* (Browne, 1902)
?*Eutima gracilis* (Forbes & Goodsir, 1851)	*Aglauropsis kawari* (Moreira & Yamashita, 1972)
Eutima mira (McCrady, 1859) (*)	*Cubaia aphrodite* (Mayer, 1894)
Phylum Cnidaria	*Gonionemus vertens* (A. Agassiz, 1862) (*)
Subphylum Medusozoa	*Gossea brachymera* (Bigelow, 1909)
Class Hydrozoa (cont.)	*Olindias sambaquiensis* (F. Müller, 1861)
Order Statocysta (cont.)	*Vallentinia falklandica* (Browne, 1902)
Eutima sapinhoa (Narchi & Hebling, 1975) (*)	*Vallentinia gabriellae* (Vannucci Mendes, 1948)
Eutonina scintillans (Bigelow, 1909) (*)	**Order Narcomedusae**
Irenium teuscheri (Haeckel, 1879)	*Aegina citrea* (Eschscholtz, 1829) (*)
Phialopsis diegensis (Torrey, 1909)	*Aeginura grimaldii* (Maas, 1904) (*)
Mitrocomium cirratum (Haeckel, 1879) (*)	*Solmundella bitentaculata* (Quoy & Gaimard, 1833) (*)
Malagazzia carolinae (Mayer, 1900)	*Cunina duplicata* (Maas, 1893) (*)

(continued)

Table 1 (continued)

Octophialucium haeckeli (Vannucci & Moreira, 1966) (*)	*Cunina frugifera* (Kramp, 1948) (*)
Octophialucium bigelowi (Kramp, 1955) (*)	*Cunina globosa* (Eschscholtz, 1829) (*)
Eucheilota diademata (Kramp, 1959b) (*)	*Cunina octonaria* (McCrady, 1859) (*)
Eucheilota duodecimalis (A. Agassiz, 1862) (*)	*Cunina peregrina* (Bigelow, 1909) (*)
Eucheilota foresti (Goy, 1979) (*)	*Solmissus atlantica* (Zamponi, 1983) nomen dubium (*)
Eucheilota maculata (Hartlaub, 1894) (*)	*Solmissus faberi* (Haeckel, 1879) (*)
Eucheilota menoni (A. Agassiz, 1862) (*)	*Solmissus marshalli* (Agassiz & Mayer, 1902) (*)
Eucheilota paradoxica (Mayer, 1900) (*)	*Pegantha clara* (Bigelow, 1909) (*)
Eucheilota ventricularis (McCrady, 1859) (*)	*Pegantha laevis* (Bigelow, 1909) (*)
Clytia brunescens (Bigelow, 1904) (*)	*Pegantha martagon* (Haeckel, 1879) (*)
Clytia discoida (Mayer, 1900) (*)	*Pegantha rubiginosa* (Kölliker, 1853) (*)
Clytia elsaeoswaldae (Stechow, 1914) (*)	*Pegantha triloba* (Haeckel, 1879) (*)
Clytia gracilis (M. Sars, 1850) (*)	*Solmaris corona* (Keferstein & Ehlers, 1861) (*)
Clytia hemisphaerica (Linnaeus, 1767) (*)	*Solmaris flavescens* (Kölliker, 1853) (*)
Clytia linearis (Thornely, 1900) (*)	*Tetraplatia volitans* (Busch, 1851) (*)
Clytia lomae (Torrey, 1909) (*)	**Order Trachymedusae**
Clytia noliformis (McCrady, 1859) (*)	*Geryonia proboscidalis* (Forskål, 1775) (*)
Clytia simplex (Browne, 1902) (*)	*Liriope tetraphylla* (Chamisso & Eysenhardt, 1821) (*)
Gastroblasta ovale (Mayer, 1900) (*)	*Botrynema brucei* (Browne, 1908) (*)
Obelia bidentata (Clark, 1875) (*)	*Halicreas minimum* (Fewkes, 1882) (*)
Obelia dichotoma (Linnaeus, 1758) (*)	*Halitrephes maasi* (Bigelow, 1909a) (*)
Obelia geniculata (Linnaeus, 1758) (*)	*Aglantha digitale* (F. Müller, 1776) (*)
Obelia longissima (Pallas, 1766) (*)	*Aglantha elata* (Haeckel, 1879) (*)
Order Cystonectae	*Aglaura hemistoma* Péron & Lesueur, 1809 (*)
Physalia physalis (Linnaeus, 1758) (*)	*Amphogona apicata* (Kramp, 1957) (*)
Rhizophysa filiformis (Forskål, 1775) (*)	*Amphogona apsteini* (Vanhöffen, 1903) (*)
Order Physonectae	*Colobonema sericeum* (Vanhöffen, 1903) (*)
Agalma elegans (Sars, 1846)	*Crossota brunnea* (Vanhöffen, 1903) (*)
Agalma okeni (Eschscholtz, 1825) (*)	*Homoeonema platygonon* (Browne, 1903) (*)
Athorybia rosacea (Forskål, 1775)	*Pantachogon haeckeli* (Maas, 1893) (*)
Halistemma rubrum (Vogt, 1852)	*Persa incolorata* (McCrady, 1859) (*)
Melophysa melo (Quoy & Gaimard, 1827)	*Rhopalonema velatum* (Gegenbaur, 1856) (*)
Erenna richardi (Bédot, 1904)	*Sminthea eurygaster* (Gegenbaur, 1856) (*)
Forskalia contorta (Milne Edwards, 1841)	**Class Scyphozoa**
Forskalia edwardsi (Kölliker, 1853)	**Order Coronatae**
Physophora hydrostatica (Forskål, 1775) (*)	*Atolla chuni* (Vanhöffen, 1902)
Bargmannia elongata (Totton, 1954) (*)	*Atolla wyvillei* (Haeckel, 1880)
Pyrostephos vanhoeffeni (Moser, 1925) (*)	*Linuche unguiculata* (Swartz, 1788) (*)

(continued)

Table 1 (continued)

Rhodalia rotunda (Haeckel, 1888) (*)	*Nausithoe atlantica* (Broch, 1914)
Cordagalma ordinatum (Haeckel, 1888) (*)	*Nausithoe aurea* (Silveira & Morandini, 1997) (*)
Lychnagalma utricularia (Claus, 1879) (*)	*Nausithoe punctata* (Kölliker, 1853) (*)
Marrus antarcticus (Totton, 1954)	*Periphylla periphylla* (Péron & Lesueur, 1810) (*)
Marrus cf. *orthocanna* (Kramp, 1942)	**Order Rhizostomeae**
Halistemma striata (Totton, 1965)	*Cassiopea andromeda* (Forskål, 1775) (*)
Melophysa melo (Quoy & Gaimard, 1827)	*Lychnorhiza lucerna* (Haeckel, 1880) (*)
Nanomia bijuga (Delle Chiaje, 1841) (*)	*Phyllorhiza punctata* (von Lendenfeld, 1884) (*)
Apolemia uvaria (Lesueur, 1811)	*Stomolophus meleagris* (L. Agassiz, 1860) (*)
Order Calycophorae	**Order Semaeostomeae**
Abyla bicarinata (Moser, 1925)	*Desmonema comatum* (Larson, 1986)
Abyla haeckeli (Lens & van Riemsdijk, 1908)	*Desmonema gaudichaudi* (Lesson, 1832)
Abyla trigona (Quoy & Gaimard, 1827)	*Drymonema gorgo* (Müller, 1883)
Abylopsis eschscholtzii (Huxley, 1859)	*Chrysaora lactea* (Eschscholtz, 1829) (*)
Abylopsis tetragona (Otto, 1823)	*Chrysaora plocamia* (Lesson, 1830) (*)
Bassia bassensis (Quoy & Gaimard, 1833) (*)	*Pelagia noctiluca* (Forskål, 1775) (*)
Ceratocymba dentata (Bigelow, 1918)	*Phacellophora camtschatica* (Haeckel, 1880) (*)
Ceratocymba leuckarti (Huxley, 1859)	*Aurelia* sp. (*)
Ceratocymba sagittata (Quoy & Gaimard, 1827)	*Stygiomedusa gigantea* (Browne, 1910) *Poralia rufescens* (Vanhöffen, 1902)
Enneagonum hyalinum (Quoy & Gaimard, 1827)	**Class Cubozoa**
Chuniphyes moserae (Totton, 1954)	**Order Carybdeidae**
Chuniphyes multidentata (Lens & van Riemsdijk, 1908)	*Alatina alata* (Reynaud, 1830) (*)
Crystallophyes amygdalina (Moser, 1925)	*Tamoya haplonema* (F. Müller, 1859)
Heteropyramis crystallina (Moser, 1925)	*Tripedalia cystophora* (Conant, 1897) (*)
Heteropyramis maculata (Moser, 1925)	**Order Chirodropida**
Chelophyes appendiculata (Eschscholtz, 1829) (*)	*Chiropsalmus quadrumanus* (F. Müller, 1859)
	Chiropsalmus zygonema **(Haeckel, 18 = 80)**

- Scyphozoa: there are 21 species (20 confirmed species + *Aurelia* s.l.) registered for the SWA, out of ~200 known worldwide (Daly et al. 2007): *Atolla chuni*, *Atolla wyvillei*, *Nausithoe atlantica*, *Nausithoe aurea*, *Nausithoe punctata*, *Linuche unguiculata*, *Periphylla periphylla*, *Cassiopea andromeda*, *Lychnorhiza lucerna*, *Phyllorhiza punctata*, *Stomolophus meleagris*, *Desmonema comatum*, *Desmonema gaudichaudi*, *Chrysaora lactea*, *Chrysaora plocamia*, *Pelagia noctiluca*, *Drymonema gorgo*, *Phacellophora camtschatica*, *Poralia rufescens*, *Aurelia* sp., and *Stygiomedusa gigantea*.

- Hydrozoa: there are 278 species (222 confirmed species + 56 identified only to genus) recorded for the SWA area, from ~3500 species known worldwide (Daly et al. 2007): 53 Trachylinae (21 Trachymedusae, 23 Narcomedusae, 9 Limnomedusae), 92 Siphonophorae (2 Cystonectae, 22 Physonectae, 68 Calycophorae), 71 Anthoathecata (13 Aplanulata, 16 Capitata, 42 Filifera), and 62 Leptothecata (3 *Incertae sedis*, 7 Laodiceida, 52 Statocysta).

3 Life Cycles, Life Histories, and Jellyfish Blooms

3.1 Medusozoan Life Cycles: General Patterns

Life cycle can be defined as the continuous sequence of changes undergone by an organism from one primary form, as a gamete, to the development of the same form again (Stearns 1992). In the general medusozoan life cycle, fertilization of gametes results in a planula larva that settles onto the substrate and metamorphoses into a sessile polyp. The polyps reproduce only asexually by producing more polyps. When certain environmental conditions are met, polyps begin producing ephyrae (i.e., young medusae), which are released into the water column and become adult medusae (Fig. 1). This basic scheme has been described as a *metagenetic* life cycle, *alternation (or succession) of generations* (*or stages*) (see Morandini et al. 2016b).

From this general life cycle, an unparalleled diversity and plasticity have been reported, perhaps because the anatomical and physiological simplicity of cnidarians makes them evolutionarily plastic (Boero et al. 1997; Fautin 2002; Jarms 2010). Yet, an acceptable approach can be derived from the general pattern described above: the life cycle of the typical medusozoan comprises a polyp stage, which reproduces asexually in the benthos, and a medusoid stage, which reproduces sexually in the plankton, with the gametes and the planula larvae as links in between them (Fig. 1).

This general pattern is present in the majority of scyphozoans, most hydrozoans and all cubozoans, but differs from staurozoan life cycles in which the pelagic stage has been reduced to a creeping benthic planula (Miranda et al. 2010). While there is great variation in medusozoan life cycles, there is congruence between the variations and the origins of major medusozoan taxa (Collins 2002). The way in which the medusa stage is produced is a good example: the Hydrozoa produce medusae by budding, the Scyphozoa by strobilation, and the Cubozoa by complete metamorphosis of a polyp into a medusa (Fig. 1).

Scyphozoa is the taxon with the highest proportion of species among medusozoans with the typical metagenetic life cycle. Most scyphomedusae are produced asexually by individual polyps (i.e., scyphistomae) through the process of *strobilation* (i.e., transverse fission followed by metamorphosis) (Fig. 1) (Jarms 2010). During strobilation, polyps lose their tentacles and mouth and reduce their size. After ephyrae are released into the water column, polyps can reattain their normal size, shape, and function within a few days. Polyps can be then capable of producing new polyps and ephyrae again, if suitable environmental conditions are met.

Within the general metagenetic life cycle, hydrozoans exhibit the greatest number of variations. In the typical case, planula larvae metamorphose into primary polyps that reproduce asexually in the benthos. However, the polyps of the majority of species bud off additional polyps that remain connected, producing a branching colony. The colony, in turn, can produce new buds that either enlarge the mother colony or detach and form another colony. Hydromedusae are produced by *budding* instead of by strobilation (Boero et al. 1997) (Fig. 1).

The life cycle of cubozoans differs from that of scyphozoans and hydrozoans in the way medusae are produced. Cubomedusae also develop asexually from benthic polyps; however, the whole polyp (in most species) transforms into a medusa by metamorphosis, leaving no polyp remnant behind (Fig. 1). In Cubozoa, each fertilized ovum becomes a planula larva, which in turn forms a single polyp that becomes a single medusa. In contrast, the Scyphozoa and Hydrozoa life cycles involve what has been referred to as "larval" *amplification* (Boero 2013), when a single polyp (or a colony) can produce asexually thousands of polyps and medusae by repeating the processes strobilation and budding several times within a single annual cycle (Bouillon et al. 2006; Morandini et al. 2016a).

Although with some variations, the majority of Scyphozoa and Hydrozoa, and all Cubozoa, present the described general life cycle characterized by these alternations of body forms (polypoid/medusoid), types of reproduction (asexual/sexual), and environments (benthos/plankton). Hereafter, deviations from this pattern that we comment upon are based on examples among the species present in the SWA. More detailed revisions of medusozoan life cycles can be consulted in Collins (2002), Jarms (2010), and Toshino et al. (2015).

3.2 Medusozoan Life Cycles: Deviations from the Pattern

Several different variations of the general metagenetic life cycle can be found among medusozoans groups, consisting of reduction of the polypoid stage or the medusa stage and concomitant simplifications (or increased complexity) within each form (Fig. 1). The simplest examples can be found at both the extremes of the pattern; there are species that lack "one half" of the general scheme. Several species exhibit holopelagic (=holoplanktonic) life cycles, with medusae reproducing sexually in the plankton with no benthic, polypoid, or asexually reproducing stages. The examples from the SWA are the scyphozoans *Pelagia noctiluca* and *Periphylla periphylla*. Also, species of Trachymedusae (Hydrozoa) are holoplanktonic, with a medusa stage that reproduces sexually in the plankton and a completely absent benthic polypoid stage (Table 1). But even within holopelagic life cycles, several variations can be found with representatives in the SWA. For example, Narcomedusae species have holoplanktonic life cycles that include a polypoid stage parasitic on other medusae. Despite having an asexually reproducing polypoid stage, these species occur only in the pelagic realm. Holoplanktonic life cycles are also present in siphonophores that have a polymorphic organization with polypoid and medusoid forms included in the same colony (Fig. 1) (Carré 1969; Kirkpatrick and Pugh 1984; Carré and Carré 1991).

A special case that perhaps constitutes the best example of the extraordinary diversity of medusozoan life cycles is the so-called immortal jellyfish *Turritopsis nutricula* (Hydrozoa) (see Piraino et al. 1996) that is also present in the SWA. Life cycles generally include a progression of developmental stages leading to sexually mature adults (e.g., the medusa stage). Although in a few hydrozoans ontogeny reversal is possible (i.e., a change the sequence of stages), this never occurs after the onset of sexual reproduction. Normally, medusae have a limited life span, with a growth phase leading to sexual maturity and spawning, followed by cell disintegration and death. The onset of sexual reproduction has been hypothesized to be a point of no return in the ontogenetic sequence of any living organism (Stearns 1992). However, Piraino et al. (1996) have found in *T. nutricula* one of the most surprising cases in the entire animal kingdom: all its stages of development – from newly liberated to fully mature individuals – can transform back into colonial hydroids. Because of this extraordinary ability, this species has been considered as "immortal." This case does not deviate from the described general pattern, just adds the most astonishing variation.

Briefly, whereas the basic scheme of medusozoan life cycles can be thought of as alternation between a pelagic, sexually reproducing medusa stage and a benthic, asexually reproducing polyp (or colony) stage, it is possible to find many of the imaginable deviations from this general pattern. Thus, some species are pelagic as polypoid forms, and in some others the benthic stage is a medusa (or some medusoid form), and both types of reproduction (sexual and asexual) can be conducted in different species by medusae or by polyps, and either can be found in both the planktonic and benthic environments. As stated by Fautin (2002), cnidarian reproduction is more variable than previously thought, and examples of this unparalleled diversity of life cycles and reproductive strategies are found within the region considered in this study, including one species that "seems to cheat death."

3.3 Jellyfish Blooms

Medusozoans regularly show natural events of massive proliferation that are triggered by the cyclical occurrence of favorable environmental conditions. As an intrinsic feature of cnidarians, pelagic stages can be extremely abundant during certain periods; even if rarely found for years, they may massively (and oddly) reappear (Boero 2013). It is likely that at least three conditions must be favorable simultaneously to allow jellyfish population explosions (see Box 1): (1) optimum values of physical factors, such as temperature, (2) suitable food in terms of both quality and quantity for the various life stages, and (3) relatively low mortality rates from predators, parasites, and diseases compared to the growth rate of the bloom former (Kremer 2001). These simultaneous environmental conditions may enable these pelagic pulses to develop over a relatively short period (i.e., days to weeks). In general, blooming species have several specific properties: wide food spectrum, high ingestion rates, rapid population growth rate, and life stages competitive with those of other species (Dawson and Hamner 2009).

Box 1 How to Define a Medusozoan Population?

In ecology, the term *population* has been defined in different manners according to the authors and their research goals. Whereas some authors consider a population simply as a group of organisms of the same species that coexist in a given area, others add to this concept the condition of self-sustainability (Sinclair 1988). In other words, while the members of a population are able to exchange genes, those which belong to different populations are not able to do so (Jummars 1993).

In population ecology, the emerging characteristics of the populations are studied (e.g., abundance, spatial distribution, sex ratio, mortality and birth rates, size structure) (Begon et al. 1988). In turn, population dynamic focuses on the variations through time of the abovementioned features, the factors which cause these variations, and the mechanisms by which they occur (Ricklefs 1979). Thus, the first thing a researcher should do when studying the population ecology of a particular species is to clearly define which the population under study is and which area it occupies. This decision, although sometimes fairly trivial, presents certain complications in the study of medusozoan populations which may reflect that these ecological concepts have been thought to species (mostly terrestrial) with rather different life cycles.

The first complication emerges from the metagenetic life cycle typical of most medusozoans. The alternation between sexually reproducing planktonic stages (medusae) and asexually reproducing benthic stages (polyps) poses the following questions: What is a jellyfish population? Does it include the medusa and polyp stages? Is there a medusa population and a separate polyp population?. Considering that a given population has to be able to self-sustain, we can state the existence of a population of polyps, which can perpetuate themselves through asexual reproduction. Conversely, we cannot consider only the medusae as a population since they need the polyps to perpetuate themselves. If genetic interchange must exist within a population, we should include both medusae and polyps in the concept of population, with medusae reproducing sexually and polyps propagating asexually, ensuring the perpetuation of the species.

These events have been classified according to the observed patterns and their likely causes categorized as accumulations, mass occurrences, aggregations, swarms, population outbreaks, and blooms (true or apparent) (see Box 2.2 in Lucas and Dawson 2014). Unfortunately, in most cases they have been referred to in the literature simply as "jellyfish blooms" for reasons of simplicity, ignorance, lack of proper classification, and lack of historical records over long-term time series.

Considering their individual size and the frequency and magnitude of their "blooms," it can be said that medusozoans are the most conspicuous components of the planktonic community. For phylogenetic, ecological, and socioeconomic reasons, these phenomena have gained increased attention during the last two or three decades, with funds available for study of the causes and mechanisms of jellyfish blooms and for the development of management and adaptation strategies (e.g., Purcell 2009; Lucas et al. 2014).

There are two main ways in which medusozoans form their "blooms": through qualitative adjustments of their life cycle or through quantitative fluctuations in the life history (Giangrande et al. 1994; Boero et al. 2008). Whereas life cycle adjustments are related to the biology and reproductive strategies (e.g., how many stages are involved?, free-living or sessile?, benthic or planktonic?, sexual or asexual?), life history adjustments deal with ecological aspects of growth and reproduction. For example, at what age and size should reproduction start? Should it be once or more than once? Should it be continuous or seasonal? How much energy and time should be allocated to reproduction? (Stearns 1992). Therefore, the evolution of life cycles and life history traits determines the population dynamics of species including the timing and magnitude of jellyfish blooms. Hence, the abundance of the medusa stage will be determined by the "success" of each part of the life cycle defined, in turn, by their interaction with the environment, including both biotic and abiotic factors. Therefore, in the metagenetic life cycle, the potential number of medusae forming a bloom will depend on fertilization rates, the ability of the planulae to settle and metamorphose, the polyp survival, and their capacity to produce medusae and medusa growth, reproduction, and survival. Jellyfish species persist locally at different times during the different life cycle stages that inhabit different environments (e.g., benthic polyps and pelagic medusae). Life cycle adjustments are, therefore, the outcome of the evolution of life cycle stages; life history adjustments are, instead, the outcome of the population dynamics (Boero 2013). In this case, species may undergo seasonal or irregular peaks of rarity and abundance in their populations through growth and reproduction of their stages (Boero et al. 2008).

In several species the presence of resting stages (i.e., cysts) adds another source of variability in later medusa abundance. Under particular (species-dependent) environmental conditions, polyps (or planulae) form (or transform into) cysts that are thought to withstand unfavorable environmental conditions (see Schiariti et al. 2014, 2015). Particular environmental stimuli (also species-dependent and poorly studied) trigger encystment, later originating new batches of polyps that, in turn, grow and reproduce leading to population peaks of the medusa stage. Those can be more intense than usual and are frequently considered to be jellyfish blooms (Arai 2009; Kawahara et al. 2013; Schiariti et al. 2014).

Trying to understand how the environment defines the timing and magnitude of jellyfish blooms and how these environmental changes potentially affect marine ecosystem functioning is impossible without a basic knowledge of the life cycles and life history of the species in question. In 2001, Mills commented that "Knowledge about the ecology of both the medusa and the polyp phases of each life cycle is necessary if we aim to understand the true causes of these increases and decreases, but in most cases where changes in medusa populations have been recognized, we know nothing about the field ecology of the polyps." Over the past two decades, this knowledge gap has started to be addressed. The percentage of described life cycles at present reaches 40% for Cubozoa (2 described life cycles for 5 reported species) and 60% for Scyphozoa (12 life cycles for 20 reported species/morphotypes) (Table 1). In Hydrozoa, descriptions of entire life cycles remain unknown for most species. However, for many of them, there are morphological descriptions of both phases of the cycle, gonophores in the polyp and the mature medusa or description of cormidia (i.e., a cluster of zooids usually consisting of a helmet-shaped bract, a gastrozooid, and one or more gonophores) in calycophoran colonies. Consequently, it is possible to provide a general scheme of the type of life cycle for these species. Besides, if the life cycle of a certain species is known, it could be assumed that cogeneric species have a similar life cycle (e.g., *Podocoryna*, *Clytia*, *Bougainvillia*). Reproductive modes seem to be similar for species belonging to the order Trachymedusae (holoplanktonic, with larva stages directly developing into medusa in water column) or Narcomedusae (holoplanktonic, with larva stage parasitizing other medusa species; see comments in Section 2). Taking this into account, we can infer the reproductive way of most Hydrozoa species (7186%, 191 life cycles for 222 confirmed species) in the SWA (Table 1). However, studies about life history traits and population dynamics of key species in the SWA are still very few and urgently needed.

4 Ecological Roles of Medusozoans in Marine Ecosystems

Historically kept on the dark side of the marine ecology and ignored during traditional oceanographic cruises, medusozoans are now acknowledged as crucially modulating the dynamics of marine ecosystems. They can not only act in trophic webs as predators and prey, but they play important roles related to nutrient cycling and also establish a variety of interspecific associations that cannot be neglected. In this section, a brief review of the different ecological roles that medusozoans can play is given, focusing on the examples reported for the SWA.

4.1 Trophic Interactions

4.1.1 Medusozoans as Predators

The role of jellyfish as predators is probably the best documented among their potentially important ecological roles (Arai 2005 and references therein). Medusae and siphonophores have been tagged as "deadly creatures" by journalists and researchers (Mackie 2002; Doyle et al. 2014) and categorized as the most important predators of the sea (Pauly et al. 2009). This trend for attention on medusozoans as predators is owed to their capacity to prey on early life stages (eggs, larvae, and juveniles) of valuable fishing resources and/or to compete with them for food resources (e.g., Brodeur et al. 2002; Lynam et al. 2005).

Medusozoans display a vast diversity of feeding mechanisms and body sizes, allowing them to feed on a large range of prey types and sizes, from micro- to macroplankton (e.g., Arai 2005; Boero 2013). Most medusozoans are generalist carnivorous and prey on a variety of zooplankton, including crustaceans (small adults and larvae), fish eggs and larvae, other jellyfish (including ctenophores), and chaetognaths, among others. However, there are also specialists, some reported from the SWA, like the siphonophore *Hippopodius hippopus* that feeds exclusively on ostracods (Purcell 1981). Their considerable functional diversity requires more detailed information for better understanding of the roles of medusozoan in pelagic trophic webs.

Jellyfish feeding strategies (e.g., ambush, cruising predator) and characteristics of their cnidomes are related to capture of distinct prey (e.g., Purcell 1997; Costello et al. 2008). Among the hydromedusae, feeding habits are diverse, with a few species feeding on bacterioplankton and protozoans (Colin et al. 2005; Boero et al. 2007), while others feed preferentially on gelatinous prey (Larson et al. 1989; Purcell 1997), crustaceans, or fish (Zamponi and Mianzan 1985). Scyphomedusae have broader diets, including crustaceans (e.g., copepods, cladocerans, ostracods), soft-bodied animals (e.g., eggs, fish larvae, ctenophores, small hydromedusae), and meroplanktonic larvae (bivalve veligers) (Suchman et al. 2008; Riascos et al. 2014). There are also some scyphozoans (especially in the order Rhizostomeae) that can consume both micro- and mesozooplankton (Larson 1987; Fancett 1988; Nagata 2015), which are retained along their complex oral arms (Nagata et al. 2016). In cubomedusae, age-specific diets are probably related to the maturation of their toxins and to an ontogenetic shift in their cnidome. These jellyfish generally feed on crustaceans at smaller stages, while adults can consume fish (Carrette et al. 2002; Nogueira Jr and Haddad 2008; Kingsford and Mooney 2014).

When jellyfish occur at high densities, they can remove a considerable fraction of zooplankton standing stock (ca. 20–60%) (e.g., Behrends and Schneider 1995; Purcell 2001; Uye and Shimauchi 2005). High predation impacts have been reported for scyphomedusae (*Aurelia* and *Chrysaora*: Hayet al. 1990; Purcell et al. 1994; Mills 1995; Purcell 2003; Hansson et al. 2005) and some large hydromedusae (*Aequorea*: Purcell and Grover 1990; Purcell 2003). Smaller species (<2 cm), such as *Liriope tetraphylla* and *Muggiaea atlantica*, may also consume large fractions of zooplankton when they occur in extremely high densities (>500 org m^{-3}) (Greve 1994; Yilmaz 2014).

Studies about diet and feeding strategies of medusozoans from the SWA are mostly recent and restricted to laboratory observations and lists of prey items found in their gastric cavities. The prey of several medusozoans have been described in the SWA, including dinoflagellates (*Noctiluca* sp.), chaetognaths, barnacle larvae, cladocerans, copepods, appendicularians, other jellyfish, fish eggs, and larvae and other medusozoan species (Zamponi 1985; Zamponi and Mianzan 1985; Nagata

2015; Carrizo et al. 2016; Díaz Briz unpublished data, Dutto pers. obs.). A further and more detailed examination of gut contents, as well as diet analyses using stable isotopes, is needed to assess the impacts of medusozoans on local food web dynamics within the SWA.

4.1.2 Medusozoans as Prey

Whereas the significance of jellyfish as predators is widely acknowledged, their role as prey has been less studied. Jellyfish have been historically considered as trophic "dead ends" in marine food webs, because of their absence from gut-content analyses and their supposedly low nutritional value. They were assumed to be insufficiently nutritious to supply the energetic demands of the vertebrates preying on them (Verity and Smetacek 1996; Richardson et al. 2009). However, there is considerable evidence now indicating that a wide range of taxa consume exclusively, or opportunistically, different medusan species. There are also some reports demonstrating that polyps are consumed by a variety of predators.

Predation upon jellyfish has been reviewed on several occasions (e.g., Arai 1988, 2005; Ates 1988; Purcell 1997; Acuña et al. 2011). In these reviews, consumption of different medusan species by other jellyfish, mollusks, arthropods, fish, reptiles, and birds has been documented. Observation of intraguild predation in the SWA includes only a laboratory observation of *Aurelia* and *Chrysaora* ephyrae being eaten by the rhizostome *Lychnorhiza lucerna* (Carrizo et al. 2016). Although in situ local observations are lacking from the SWA, there are reports in the literature from other regions that involve species occurring in our study area, such as *Aequorea* feeding on several smaller hydromedusan species (Purcell 1991).

While reports of intraguild predation are few worldwide, examples of other taxa consuming jellyfish are much more common. Jellyfish comprise the diet of more than 120 fish species, 39 of them reported for the SWA (Arai 2005; Pauly et al. 2009; Díaz Briz et al. 2017), several seabirds (Harrison 1984; McInnes et al. 2017; Phillips et al. 2017; Thiebot et al. 2017), and some marine turtles (Houghton et al. 2006; Gonzalez Carman et al. 2013). Examples of medusae as prey in the SWA exist for Argentina and Uruguay, which are recorded in the historical database built by the National Institute for Fishery Research and Development of Argentina (INIDEP). This database contains results from more than 100 fish gut-content analyses (see Díaz Briz 2014).

The leatherback turtle *Dermochelys coriacea* and the green sea turtle *Chelonia mydas* represent other vertebrates that feed on jellyfish in the SWA. These two species of sea turtles are present each summer in the Río de la Plata estuary (Argentina-Uruguay), coinciding with blooms of *Lychnorhiza lucerna*, *Chrysaora lactea*, and *Liriope tetraphylla* (Estrades et al. 2007; Gonzalez Carman et al. 2013). During recent years, the miniaturized animal-borne video data loggers have enabled feeding events to be monitored from a predator's perspective. Video recordings obtained from cameras placed on four penguin species revealed that Magellanic penguins (*Spheniscus magellanicus*) consume jellyfish (likely *Chrysaora plocamia* and

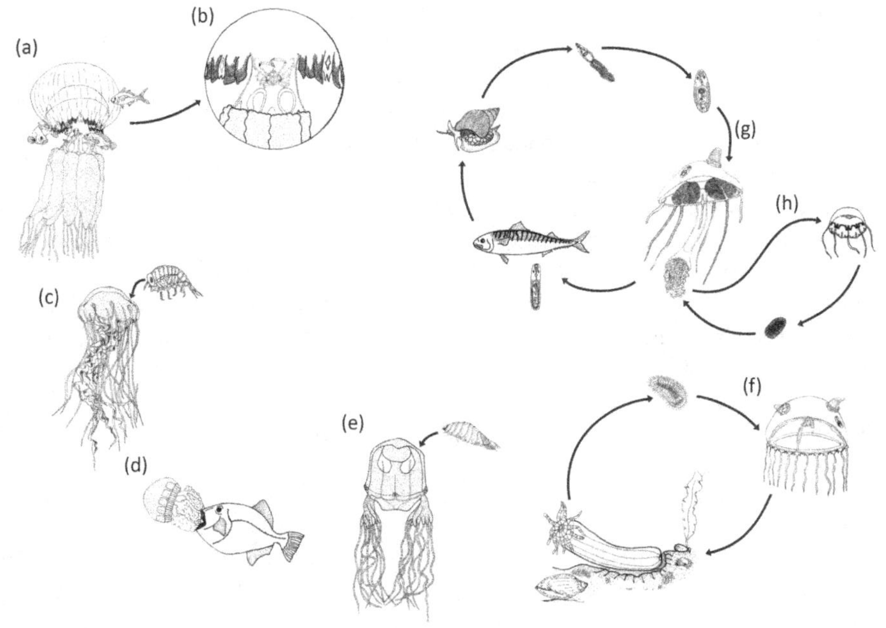

Fig. 2 Schematic representation of the different interspecific relationships among medusozoan and other species. Scyphomedusae with associated fishes (**a**), with symbiont crabs (**b**), with hyperiid parasites (**c**), and predation by fishes (**d**). Cubomedusae with isopod parasites (**e**). Hydromedusae with sea anemone parasites (**f**), secondary hosts in the life cycle of digenean (**g**), and with Narcomedusae parasites (**h**)

Aequorea forskalea), particularly along the Patagonian shelf (Thiebot et al. 2017). Contributions of jellyfish to the diets of different marine vertebrate species are common, likely more so than previously thought, supporting a potentially important role of medusozoans in transferring energy within marine trophic webs.

4.2 Nontrophic Interactions

Pelagic stages of medusozoans have a broad range of potentially important ecological roles in addition to their trophic interactions (Fig. 2). These include diverse interspecific associations that have been classified under terms like phoresy, parasitism, parasitoidism, kleptoparasitism, ectocommenalism, endocommensalism, amensalism, mutualism, micropredation, and epizoism (see Towanda and Thuesen 2006; Ohtsuka et al. 2009; Chiaverano et al. 2015). Although these categorizations have mostly been vaguely defined (see Sal Moyano et al. 2012), we briefly describe these interactions that suggest keystone status in marine ecosystems, particularly from the SWA.

Medusae can provide structures in the water column that may be used as shelter or as focus for aggregation (Ohtsuka et al. 2009; Sal Moyano et al. 2012). A simple example is the widely documented presence of schools of small fish around the conspicuous oral arms of scyphomedusae and siphonophore colonies (Fig. 2) (Purcell and Arai 2001). The nature of these associations may change as the fish grow, but in all reported cases, large adult medusae were involved. It seems likely the fish near a medusa gain protection from predators (Mansueti 1963; Purcell and Arai 2001). However, as they grow they may consume parts of it and may steal its food. Fish survive contact with these "toxic" hosts due to a variety of special characteristics (reviewed in Arai 1988). In the SWA (particularly in Brazil), different fish species (Carangidae and Serranidae) have been observed in association with the semaeostome *Chrysaora lactea* and the rhizostomes *Lychnorhiza lucerna* and *Phyllorhiza punctata* (Morandini 2003; Bonaldo et al. 2004; Sobolewski et al. 2017). Several unidentified fish species have been documented in association with *C. lactea* and *L. lucerna* in the Río de la Plata estuary (Argentina-Uruguay) and with *Chrysaora plocamia* off Patagonia (Argentina) (Mianzan et al. 2014; Schiariti pers. obs.).

Other species, including crabs and shrimp, apparently benefit from association with medusae (Fig. 2). Brachyuran crabs riding scyphomedusae have been widely documented in coastal and estuarine environments from Brazil, Uruguay, and Argentina. In general, crabs are thought to benefit from shelter, from enhanced mobility, from access to food, and in some cases by preying on medusae (see Sal Moyano et al. 2012 and references therein). Larval, juvenile, and adult crabs all occur with medusae, suggesting varied benefits to each of the life stages. Benefits of the associations have rarely been noted for medusae, but neither is there conspicuous damage produced by crabs. The benefits obtained by crabs may be diverse, but in most cases they are only based on speculations. However, findings like that of recently molted megalopae of *Libinia* crabs (*L. spinosa* and *L. ferreirae*) associated with *L. lucerna* indicate that the crabs gain protection from their host during their most vulnerable periods. Several other associations between crabs, shrimp, and scyphomedusae have been documented in the SWA, always suggesting benefits for the symbionts with neither discernible benefits nor damages to the medusan host (Moreira 1961; Chace 1969; Mianzan et al. 1988; Nogueira Jr and Haddad 2005; Martinelli Filho et al. 2008; Santos et al. 2008; Schiariti et al. 2012; Gonçalves et al. 2016).

A different kind of association is found between medusae and hyperiid amphipods (Fig. 2). These associations reported worldwide have been considered clearly parasitic, because they are nearly always detrimental to the host, which is devoured when the hyperiid reaches adulthood (Laval 1980; Gasca et al. 2015). Adult hyperiid females may also deposit their offspring directly into the tissue of their host, which then feed on prey caught by the host or directly on the host, consuming it partially or totally (Laval 1980; Sullivan and Kremer 2011). Thus, medusae provide not only food but a reproductive habitat for some hyperiid species (Dittrich 1988). Energy from medusae may be channeled into fishes that feed on these amphipod parasites (Riascos et al. 2012). In the SWA, some cases have been recorded encompassing *Hyperia galba* and the scyphozoan *Desmonema gaudichaudi* (as *D. chierchianum*) (Mianzan 1986), between *Brachyscelus rapacoides* and *C. lactea*, or the hydromedusa *Olindias sambaquiensis* (Puente Tapia et al. submitted).

Platyhelminthes also form parasitic associations with medusozoan pelagic stages. Digenean worms, such as *Monascus* and *Opechona*, have medusae and ctenophores as secondary hosts (Fig. 2). Their transmission to fish, in which they complete their life cycles, occurs when fish consume the gelatinous zooplankton, and several examples are reported from the SWA (see Díaz Briz et al. 2012, 2015 and references therein). Four taxa of digenean metacercariae (*Monascus filiformis*, *Opechona* sp., *Bacciger* sp., and species of Hemiuridae) have been found parasitizing more than 20 hydromedusae and 4 scyphomedusae species (Morandini et al. 2005; Díaz Briz et al. 2012; Nogueira Jr et al. 2013, 2015).

A few studies have shown parasitic associations between medusae and isopods, and some examples have been found in the region. Isopods can affect the reproductive performance and growth of their hosts by feeding on them (Fig. 2). The isopods have been found on different parts of medusae, including the sub- and exumbrella, manubrium, oral arms, and subgenital cavities (Nogueira Jr and Silva 2005). They may use medusae as food, protection, and probably transportation (Saito et al. 2002; Nogueira Jr and Silva 2005). Several examples of these associations have been reported for the SWA, including between *L. lucerna* and *Synidotea marplatensis*, *C. lactea* and the isopods *Cymothoa catarinensis* and *S. marplatensis* (Nogueira Jr and Silva 2005), *P. punctata* and an identified Cymothoidae species (Moreira 1961), the cubozoan *Chiropsalmus quadrumanus* and the isopods *Nerocila fluviatilis* and *Ancinus brasiliensis* and the hydromedusa *Olindias sambaquiensis* parasitized by the isopod *S. marplatensis* (Nogueira Jr and Silva 2005).

Medusozoans can also act as parasites; for example, the actinula larva of two genera of narcomedusae (*Cunina* and *Pegantha*) settles onto hydro- and scyphomedusae establishing parasitic associations (Fig. 2). These larval stages develop into a polypoid stage using their hosts as a substrate in the pelagic realm (Bouillon 1987; Osborn 2000). In the SWA, the polypoid phase of *Cunina octonaria* was observed parasitizing the Trachymedusae *L. tetraphylla* (Puente Tapia pers. obs.), but further studies are needed for a proper characterization of this association. Another type of parasitic relationship involving non-medusozoan cnidarians involves the larvae of anemones (Spaulding 1972; Sullivan and Kremer 2011). In our region, larvae of *Peachia* sp. have been observed attached to the scyphomedusae *C. lactea* and *C. plocamia* (identified as *C. hysoscella*) (Mianzan 1986) and the hydromedusae *L. tetraphylla* and *Eucheilota ventricularis* (Puente Tapia unpubl. data.).

4.3 Nutrient Cycling

Medusozoans may contribute to support primary production (Pitt et al. 2009). The products generated by medusan excretion (C, N, and P), mucus production, and "sloppy feeding" can be significant (Pitt et al. 2005; West et al. 2009). In a similar way, regenerated products released by medusae become available to bacteria and can create the so-called jelly loop of carbon cycling between jellyfish, bacteria, heterotrophic nanoflagellates, and ciliates (Condon et al. 2011). Jellyfish may contribute to nutrient recycling through the transport of nutrients and other dissolved matter across physicochemical boundaries. Considering the abundance of some medusozoans and the scale of their diel vertical migrations (e.g., *Periphylla periphylla*, siphonophores), such mixing can impact ecosystem function (Doyle et al. 2014). The input of organic material to benthic and pelagic trophic webs when medusae die constitutes another nutrient flow in coastal marine environments.

Depending on their sinking speeds and water depth, jellyfish bodies may decompose within the water column or near the bottom. The decay of jellyfish may involve both leaching of dissolved organic carbon (DOC) from the medusae and mineralization by bacteria. The release of DOC during decomposition may support bacterioplankton production (Pitt et al. 2009). Decomposition of jellyfish is also likely to affect oxygen dynamics. The complete oxidation of jellyfish tissues would require considerable oxygen, resulting in local hypoxia when the decomposing biomass is large enough. The consumption of oxygen may be more severe in the benthos leading to potential anoxia. This would be particularly intense in closed areas or with slow mixing of the water. The declines and decomposition of jellyfish blooms could induce "boom and bust" dynamics in a given ecosystem. Decay of bloom abundances can even annihilate benthic fauna (e.g., Pitt et al. 2009).

5 Medusozoans and *Homo sapiens*

While jellyfish are some of the most ancient multicellular organisms on Earth, man only started to take notice of their impact on human activities from about the 1960s. For the general public, jellyfish are largely synonymous with stinging. However, when abundant, jellyfish can negatively affect human enterprise in a number of ways beyond the economic losses caused to the tourism industry (Lucas et al. 2014). In some regions of the world, jellyfish blooms can have substantial impacts on human activities, including the clogging of fishing nets and cooling water intakes in power plants, and damage to aquaculture systems (Purcell et al. 2007; Boero 2013). Given these sometimes dramatic consequences, the number of reports has increased, and the public has acquired an overwhelmingly negative perception of these creatures (e.g., Vandendriessche et al. 2013; Graham et al. 2014). However, as research and knowledge increase, jellyfish are also being portrayed in a more positive light. From the study of marine ecosystem functioning and biodiversity to their importance as ecosystem services providers and through the discovery of their potential economic value as food or as source of biochemical compounds, it is now clear that there is far more to jellyfish than bad news (Doyle et al. 2014). In the following section, we briefly review the variety of negative and positive interactions between medusozoans and *Homo sapiens*, focusing on species present in the SWA.

5.1 Medusozoans as "Troublesome" Species

5.1.1 Public Health and Tourism

Problems for tourism generated by stinging species are perhaps the most "attractive" to public media, as suggested by frequent headlines around the world. Globally, nearly 100 species have been recognized as threats to human health, a few of those being fatal (Burnett 1991). In the SWA, several medusozoans do give nasty stings, but none of those have caused fatalities so far. In general, the most serious effect is panic; clinical records are scarce, and species have not been clearly identified because few people are able to identify them with certainty. The widely distributed "Portuguese man-of-war" (*Physalia physalis*) is probably the most dangerous species in the region. Fortunately, although serious, injuries from it (which typically include ulceration, local muscle contractures, and tissue necrosis) are not frequent (Freitas et al. 1995; Haddad Jr et al. 2002; Failla Siquier pers. com.).

The hydrozoan *Olindias sambaquiensis* is one of the most frequently reported species as causing health problems (malaise, vomiting, dyspnea, and tachycardia) for swimmers off Brazil, Uruguay, and Argentina (Kokelj et al. 1993; Chiaverano et al. 2004; Haddad Jr 2008; Resgalla Jr et al. 2011; Mosovich and Young 2012). In coastal environments, *Liriope tetraphylla*, locally known as "tapioca," is another hydromedusa that can be troublesome during summer because of their intense blooms (Mianzan et al. 2000; Dutto et al. 2017).

Some frequent and abundant scyphozoans can become problematic when blooming off different regions of Brazil, Uruguay, and Argentina. In southern Brazil, the larva of *Linuche unguiculata* has been responsible for the "sea bather's eruption," a pruritic erythematous papular eruption that develops in areas covered by swimsuits (Haddad Jr et al. 2001, 2010; Rossetto et al. 2015). From other regions of the SWA, erythematous lesions due to the sting of *Chrysaora lactea* have also been reported (Marques et al. 2014).

Box jellyfish (Cubozoa) are among the most toxic marine animals because their venom contains hemolytic, neurotoxic, and cardiotoxic elements. Along the tropical coasts of Brazil, two species have been responsible for a few reported cases: *Tamoya haplonema* and *Chiropsalmus quadrumanus*. These medusae cause an intense pain and long linear plaques (Haddad Jr 2003; Haddad Jr et al. 2002, 2009). Although it is present also in Uruguayan coasts, no accidents related to this species have been documented (Leoni et al. 2016).

Literature on the treatment of jellyfish envenomation is abundant but usually controversial and lacking scientific support. No universal therapeutic remedy exists. Thus, development species- or genus-specific therapies are needed, since the nature of the venom is organism-dependent (see Montgomery et al. 2016). Many anecdotal treatments are available (gas oil, onion, and pee are among the most surprising), but species-specific first aid response is essential for an effective treatment. The removal of tentacles followed by treatment of the stung area can be crucial to avoid further release of nematocysts. Freshwater should never be used as a treatment for jellyfish stings, because changes in osmotic concentration can trigger nematocyst release. Vinegar and seawater have been established to be an effective painkiller, and baking soda slurry can be used, both as an immediate therapy and to wash off tentacles (Mianzan et al. 2001). The training of lifeguards and education on jellyfish envenomation are strongly recommended.

5.1.2 Fisheries

Medusozoans and fisheries interact in a number of ways: (1) jellyfish feed on the eggs and larvae of the species we commercially exploit (e.g., fish, crustaceans, mollusks); (2) they are competitors with zooplanktivorous fish for food resources; (3) they can transmit parasites and bacterial pathogens to fish (Purcell and Arai 2001; Delannoy et al. 2011). Although these interactions have been documented, the majority of the cases are just speculations based on logical assumptions or literature records of similar species from other regions. The economic costs associated with these ecological interactions are very difficult to evaluate. Particularly for the SWA, we are not in a position to assess the extent of impacts that jellyfish may be having on the fishing industry (directly or indirectly). There is no reliable evidence of these interactions in the SWA, and research on medusan feeding strategies, diet, and impacts on zooplankton, ichthyoplankton, and fish communities remain relatively unknown (but see Schiariti et al. 2012, 2015; Nagata 2015; Macchi and Schiariti 2016; Nagata et al. 2016). In addition, characterization of the zooplankton community in spawning areas of the commercially important species is still not adequate to test hypotheses regarding jellyfish impacts in the region.

In addition to the potential negative effects of jellyfish on fish stocks (through their impact on recruitment rates), mass occurrences of medusae can directly affect net-based fisheries through clogging and bursting of nets, decreasing fish catch, killing and spoiling the targeted species, costing time and effort during the removal of jellyfish bycatch, and even causing fishing boats to capsize (Lucas et al. 2014 and references therein). These problems have been reported in several regions of the world but have seldom been quantified. The impact of large jellyfish blooms, such as those of *Nemopilema nomurai* in Japan and *Phyllorhiza punctata* in the Gulf of Mexico, is among the best studied cases with economic losses of several million dollars (Graham et al. 2003; Uye 2008). In South America, a few specific cases have been reported from Brazil and Argentina that were caused by year-round blooms of the large (20–30 cm bell diameter) medusa *Lychnorhiza lucerna*, which reduce total fish captures and catch quality, damage nets, and prevent fishermen from operating (Schiariti 2008; Schiariti et al. 2008; Nagata et al. 2009). Although blooms of *C. lactea* are also frequent and intense in the same regions, no impacts of this species on fishing operations have yet been documented. To a lesser extent, large catches of *Desmonema gaudichaudi* have interfered fishing operations off Southern Argentina (Schiariti unpubl. data).

5.1.3 Aquaculture, Power Plants and Ship Operations

The problems caused by jellyfish blooms for aquaculture, ship operations, and power plants have been reviewed by Purcell et al. (2013) and Lucas et al. (2014). In general, damage occurs for aquaculture operations when massive numbers of medusae are transported by tidal currents and accumulate around the fish cages (Doyle et al. 2008; Mianzan et al. 2014). Damage to fish may be indirect, through hypoxia, or direct by stinging as medusae or pieces passing through the mesh of the cages (Mitchell et al. 2012; Mianzan et al. 2014). Also, damaged gills may become infected by fish pathogens (Delannoy et al. 2011).

Power stations and desalination plants are located in coastal regions worldwide, because of the large amounts of cooling water needed for condensers and of seawater for desalination. Large quantities of medusae can block the screened intakes, preventing the water inflow (Purcell et al. 2007). Provision of power and desalinated water to customers can be reduced or temporarily halted altogether. A study by the Association of Nuclear Operators reported in 2006 that 44 power outages and load reductions have occurred during medusa blooms at nuclear plants (Lucas et al. 2014), but none of those were in the SWA region. Similarly to power stations, many ship operations are affected by the accumulation of jellyfish in on their cooling water uptake screens. There again, the reports of this problem in the SWA are poorly documented.

5.2 *Medusozoans as "Beneficial" Species*

The benefits that can be obtained from jellyfish have been obscured. However, several wealthy industries are based on particular medusan species, including their utilization as human food, as partial feedstocks for a variety of animals (e.g., fish, farmed chickens, and pigs), and as sources of biochemical compounds utilized in pharmacology and medical research (Hsieh and Rudloe 1994; Kingsford et al. 2000; Brotz et al. 2017). However, as clearly stated by Doyle et al. (2014), "the benefits of particular species for society are often more cryptic and emerge from research rather than commerce." Paradoxically, research (number of papers and specific funding) has been growing during the last two decades, much of it in response to the problems generated by jellyfish blooms. Therefore, favored by the pressure from "bad press," our knowledge about gelatinous species among which medusozoans are included has remarkably improved. Consequently, general acknowledgment of the ecological importance of jellyfish in marine ecosystems has developed, not only within the small jellyfish scientific community but also among fishery biologists, modelers, policymakers, businessmen, and the general public.

Medusozoan fisheries (primarily for scyphomedusae) have a long history in Asia, where jellyfish have been caught and processed as food for centuries. More recently, jellyfish fisheries have expanded to the Western Hemisphere, often driven by demand from Asian buyers and by collapse of more traditional local fish stocks.

As many as 35 species of jellyfish have reportedly been consumed by humans, with the majority of commercial jellyfish fisheries focusing on species from the scyphozoan order Rhizostomeae (see Table 2 in Brotz et al. 2017). The Chinese savor jellyfish as cuisine to be served regularly, as well as for holidays, weddings, and celebrations (Hsieh and Rudloe 1994). Consumption of jellyfish is popular in other Asian countries, including Japan, Malaysia, Korea, Taiwan, and Singapore, sustaining strong market demand (Kingsford et al. 2000; Hsieh et al. 2001; Omori and Nakano 2001). Comprehensive reviews of jellyfish fisheries have been published recently describing fishing and processing techniques, diversity of targeted species, the edible products, and other uses of jellyfish (Kingsford et al. 2000; Hsieh et al. 2001; Omori and Nakano 2001; Brotz and Pauly 2017; Brotz et al. 2017). Therefore, we only describe here the main features of this industry focusing on the few ongoing local experiences.

Chinese emigrants likely first introduced jellyfish fisheries to Southeast Asia, initiating them to several countries (Brotz et al. 2017). To keep up with demand, jellyfish fisheries have spread to the Western Hemisphere, often preceded by local collapses of more traditional finfish and shrimps resources. While development of jellyfish fisheries has been explored in more than 20 Western Hemisphere countries, the degree to which they have successfully established varies (see Brotz and Pauly 2017). Most consumption continues to be in Asia, with the majority of the traded product being exported to China, Japan, and South Korea (Huang 1986, 1988; Hsieh and Rudloe 1994; Omori and Nakano 2001; Kitamura and Omori 2010). At present, catches of jellyfish as food for humans are significant, with global landings only recently exceeding 1 million tonnes (Brotz and Pauly 2017).

Jellyfish fisheries in the SWA are currently under consideration only in Argentina, targeting the rhizostome *Lychnorhiza lucerna* (Fig. 3) (Schiariti and Mianzan 2013; Brotz et al. 2017). To date, the life history and population dynamics of this species have been studied, as well as the development of the processed product and evaluation of the potential markets (Schiariti 2008; Schiariti and Mianzan 2013; Schiariti et al. 2015). Processing of jellyfish has been performed by fisheries researchers under instruction from potential buyers, and initial responses from Chinese and Malaysian importers have been positive. However, a major hurdle to the establishment of a permanent jellyfish fishery in Argentina is uncertainty regarding how much jellyfish can be produced from the region on a consistent basis, as buyers require a minimum availability to remain involved (Brotz et al. 2017). Significant investment is required to undertake proper biomass assessment, to investigate the costs involved, and to acquire a better understanding of jellyfish population dynamics in the region. Policymakers in the area continue to approach a potential jellyfish fishery with incredulity and are dismissive about jellyfish providing significant economic value. Conversely, fishermen in the region are motivated and have been working directly with fisheries researchers and potential buyers for several years. Until the economic and ecologic knowledge gaps can be filled, a jellyfish fishery in Argentina will remain undeveloped. *Lychnorhiza lucerna* also occurs along the neighboring coasts of southern Brazil and Uruguay, as shown by bycatch records and scientific studies

Fig. 3 Development of a jellyfish fishery in Argentina targeting the rhizostome *Lychnorhiza lucerna*. Two different phenotypes of *L. lucerna* medusae with whitish or purplish-blue Margilan lobs (**a**, **b**). Captures of *L. lucerna* medusae in the Buenos Aires Province, Argentina (**c**, **d**). Processing (salting) stages (**e**, **f**). Processed medusae exhibited in the Chinese market in Buenos Aires (**g**)

(Schiariti 2008; Nagata et al. 2009; Schroeder et al. 2014), suggesting that the area of potential exploitation for this species can be expanded.

Jellyfish may be targeted for a number of reasons other than as food for humans. Jellyfish have been used successfully as partial feedstock for a variety of animal's foods in traditional and aquaculture farms (Hsieh and Rudloe 1994; Gopakumar et al. 2008; Miyajima et al. 2011; Wakabayashi et al. 2012). Jellyfish may be used as bait, as is done in Japan where parts of the giant jellyfish *Nemopilema nomurai* are used for sea bream fishing (Omori and Kitamura 2004). Historically, fishermen in Peru used large blooms of *C. plocamia* to locate leatherback sea turtles (*Dermochelys coriacea*), which were hunted for their meat during the 1960s, 1970s, and 1980s (Brotz et al. 2017). On the other hand, different medusa species have also been studied for their potential utilization in medical, biomedical, and pharmacological research. Among the most remarkable examples are two Nobel Prizes: one in 1913 for the discovery of anaphylaxis and another in 2008 for the discovery and development of green fluorescent protein (GFP). The processing of *L. lucerna* to extract collagen in Brazil is among the ongoing local examples; the extraction of collagen from other rhizostomes, to be utilized in cosmetics and pharmaceuticals, is being studied (Addad et al. 2011). One company based in France (www.javenech.com, accessed 26 June 2015) processes several tonnes of *Rhizostoma pulmo* caught in the Atlantic Ocean for collagen each year.

There are other uses of medusae. In design engineering their biomechanics are often mimicked due to their simple and efficient designs (Gemmell et al. 2013). Among other industrial applications, jellyfish have been successfully added to cement in Russia, which increased the mechanical strength of traditional cement by 50%, although the details are unfortunately vague (see Brotz et al. 2017). Experiments have also demonstrated that jellyfish can successfully be used as fertilizer for a variety of plants, trees, and crops (see Brotz et al. 2017 and references therein). There are even recent reports that a company in Israel has developed an absorbent and biodegradable material from jellyfish that could be used in products such as diapers and paper towels (Shamah 2014). Most of the technologies that propose to use jellyfish in medical and industrial applications are in their infancy, and it will likely be sometime before there is significant demand for jellyfish other than as food. Nonetheless, it is conceivable that jellyfish could be used in a variety of future applications, some of them under current consideration in the SWA region.

6 Concluding Remarks and Guidelines for the Future

In the SWA, jellyfish can be found across and along the shelf, as well as in the oceanic environment to seaward. As a group, they are abundant all year-round, but particular species bloom seasonally. Whereas some species can be troublesome from time to time, others have shown potential to become valuable economical fishing resources. Furthermore, beyond their positive and negative impacts on different industries, the abundances some species can reach during their blooms give them important ecological roles in the marine ecosystems. Yet, although empirical data for the region have been increasing recently, too much of our understanding remains speculative because of the scarcity of specific studies on the resident species.

The SWA extends along about 10,000 km of marine coastline from northern Brazil (ca. 4°N) to southern Argentina (ca. 56°S). Opportunities and possibilities of scientific studies vary from one subregion to another in this vast area as a result of differences in funding, availability of qualified personnel, equipment, and several intricacies of the countries. The distribution of funds and research effort (hence, advances of knowledge) are heterogeneous according to the distribution of wealth. Brazilian states such as São Paulo, Rio de Janeiro, Paraná and Santa Catarina, the Río de la Plata estuarine zone (Argentina-Uruguay), and Buenos Aires Province coast (Argentina) are the regions where general knowledge about Medusozoa is relatively better. In contrast, the coasts of northern and northeast Brazil and southern Patagonia in Argentina are still poorly studied because of limited funds, lack of specialists, and the scarcity of collections from such regions. Moreover, the majority of the available records come from neritic environments and the upper layers of the oceanic realm. Therefore, the fauna from oceanic deep waters of the SWA remains underestimated or unevenly unknown.

There are reasons for this state of the art in jellyfish research in the SWA. In general, jellyfish research has been triggered worldwide in response to the socioeconomic problems for human enterprises caused by their blooms (Purcell et al. 2007). In the SWA, blooms of different medusa species (e.g., *Lychnorhiza lucerna*, *Chrysaora lactea*, *Liriope tetraphylla*, *Olindias sambaquiensis*) have been frequently documented at different locations; however, there have been no serious troubles documented so far (see Section 5.1). Therefore, the pressure from affected sectors and public media, which have produced funds and human resources dedicated to study of jellyfish elsewhere (e.g., Mediterranean and North Sea), is still absent in our region. A similar situation occurs in respect to knowledge needed for economic exploitation of different medusae. Research about the potential utilization of medusozoans as foods, or sources of collagen, are ongoing in the region (Section 5.2) but still at a small scale because of the lack of research funding from private or public sources. Consequently, research on medusozoans is recent and scarce and has been provided budgets too limited for field sampling or to build long-term databases. The available grants have been used to collect medusozoan specimens in particular areas for isolated periods and to improve the laboratory equipment of some institutions in Brazil, Uruguay, and Argentina. However, they have been insufficient to answer most of the questions concerning larger spatial and temporal scales and the deep details of species biology.

One of the main limitations for study of jellyfish is the limits on logistical resources (mainly ship time) available to jellyfish researchers. Therefore, alternatives must be found, and indirect sources of information from surveys not specifically for study of jellyfish, although far from ideal, can be very useful. In this context, the integration of goals and methodologies between jellyfish researchers and fishery biologists has become an interesting approach (see Pauly et al. 2009). Only when interdisciplinary research starts across in a number of SWA areas we will be in a better position to address the roles of jellyfish in marine ecosystems and their potential impacts on human activities. Some countries, like Argentina and Uruguay, organize regular fisheries resource surveys across extensive national and international waters, with many of these incidentally involving a bycatch of large jellyfish species. Although the fishing gear used is rarely suitable for accurate quantitative sampling of jellyfish (either to collect undamaged specimens for taxonomy or live specimens for experimentation), the equipment and methods are reasonably controlled, and records of such bycatch events could provide good relative indices of the distributions and abundance of large scypho- and hydromedusae (Schiariti 2008; Purcell 2009; Bastian et al. 2010; Schiariti et al. 2013; Rodriguez et al. 2017). Furthermore, over several years the data collected during these standardized surveys forms interesting time series, useful for investigating interannual variations of relative jellyfish abundances (Brodeur et al. 1999; Lynam et al. 2005; Schiariti 2008; Bastian et al. 2010). In Argentina, the National Institute for Fishery Research and Development (INIDEP) has performed routine fishery research cruises since the 1980s, cruises being utilized to build a gelatinous zooplankton database covering most of the Argentinean continental shelf (ca. 1 million km^2). Demersal (mostly) and pelagic fishing trawls and a variety of plankton devices have been utilized with different aims, and several macromedusa species have been caught as bycatch (>5 cm bell diameter) (Schiariti et al. in press). Important information about the occurrence and spatial distribution of these species has been also obtained with acoustic devices (Álvarez Colombo et al. 2003; Cabreira et al. 2006).

However, the opportunities provided by fishery research surveys are not completely appropriate because their methods and sampling designs that are inadequate for study of jellyfish. The majority of the specimens are severely damaged and useless for experimentation or even taxonomic identification. Besides, jellyfish blooms of any kind might impair plankton nets or typical demersal trawling in next to no time or, if the specimens are sparse, might not be evaluated in the right way. In addition, historically jellyfish bycatch has not even been recorded on a regular basis. Most of the time, the jellyfish catch has not been recorded at all, and they were simply discarded. Therefore, this information is reliable when jellyfish blooms were recorded, but the reverse is not true. Therefore, the way that medusozoans are being studied in the SWA needs to be reevaluated, because it is mostly linked to episodic observations that can be considered as almost anecdotal. Therefore, we consider that jellyfish research in the SWA is still in its infancy.

Funding is not the only trouble for the study of jellyfish in the SWA. Key obstacles to a more widespread study of medusozoans are still the limited time to collect samples and data onboard due to the demanding workload of the fishery surveys as well as the insufficient human resources. Currently, the number of specialists in the region is too low to cover such a vast and diverse region. Indeed, few than 20 researchers from the 3 countries are dedicated to medusozoans, and the number of specialists on particular taxa is even lower, making this field potentially attractive for masters and PhD students. Solutions for these problems demand organization and financial support. We believe that training in the systematics and ecology of medusozoans and stretching the gap between researchers, fishermen, journalist, and policymakers should be a priority. Comprehensive monographs on specific groups should be undertaken, because they generate thorough and qualified results over a relatively short period. Improvement in the quality and number of collections is of utmost importance for continuity of studies, preservation of data, and availability of specimens for future comparisons (Marques et al. 2003). Given that recognition of the important ecological interactions between finfish and jellyfish populations has grown wider, we hope that it may become easier to initiate fruitful collaborations and to leverage appropriate resources.

Better understanding of medusozoan taxonomy should be the first step and will facilitate other research, such as on pharmacology, ecology, and phylogeny. The detailed description of more life cycles and the study of the life histories and population dynamics of medusozoans are also of paramount importance for understanding the ecological roles these species play in marine ecosystems and the regulation of their blooms by environmental factors. Finally, integration among jellyfish researchers, oceanographers, and fishery biologists is essential to obtaining broadscale datasets useful not only to jellyfish research but to everyone with an interest in developing a better understanding of the effects of climate and other environmental and biological factors on oceanic ecosystems. Such integration will be a critical element in development of an ecosystem-based approach to fisheries management. In addition, research outputs generated by jellyfish datasets are usually welcomed by the scientists in charge of fisheries surveys, who value them to help them justify the surveys and secure future funding. Medusozoans have become too cogently significant to be neglected.

Acknowledgments This paper was supported by INIDEP, CONICET PIP 2013-00615, FONCyT PICT 2013-1773, and FONCYT PICT 2015-1151. ACM was supported by grants 2010/50174–7, 2011/50242–5, and 2015/21007-9 São Paulo Research Foundation (FAPESP) and by CNPq (301039/2013–5 and 304961/2016-7). This is a contribution of NP-BioMar, USP. This is INIDEP contribution N° 2119.

References

Acuña JL, López-Urrutia A, Colin S (2011) Faking giants: the evolution of high prey clearance rates in jellyfishes. Science 333:1627–1629

Addad S, Exposito JY, Faye C et al (2011) Isolation, characterization and biological evaluation of jellyfish collagen for use in biomedical applications. Mar Drugs 9(6):967–983

Arai MN (1988) Interactions of fish and pelagic coelenterates. Can J Zool 66:1913–1927

Arai MN (2005) Predation on pelagic coelenterates: a review. J Mar Biol Assoc UK 85:523–536

Arai MN (2009) The potential importance of podocyst to the formation of scyphozoan blooms: a review. Hydrobiologia 616:241–246

Ates RML (1988) Medusivorous fishes, a review. Der Zool Meded 62:29–42

Álvarez Colombo G, Mianzan H, Madirolas A (2003) Acoustic characterization of gelatinous plankton aggregations: four case studies from the Argentine continental shelf. ICES J Mar Sci 60(3):650–657

Bastian T, Lilley MKS, Stokes D (2010) How fish surveys provide a 'backbone' for jellyfish research. ICES J Mar Sci 06:1–6

Begon M, Harper JL, Townsend CR (1988) Ecología: Individuos, poblaciones y comunidades. Ediciones Omega SA, Barcelona

Behrends G, Schneider G (1995) Impact of *Aurelia aurita* medusae (Cnidaria, Scyphozoa) on the standing stock and community composition of mesozooplankton in the Kiel Bight (western Baltic Sea). Mar Ecol Prog Ser 127:39–45

Boero F (2013) Review of jellyfish blooms in the Mediterranean and Black Sea. Studies and reviews. General fisheries Commission for the Mediterranean, vol 92. Rome, FAO, p 53

Boero F, Bouillon J, Piraino S et al (1997) Diversity of hydroidomedusan life cycles: ecological implications and evolutionary patterns. In: den Hartog JC (ed) Proc 6th international conference on Coelent Biology, pp 53–62

Boero F, Bucci C, Colucci AMR et al (2007) *Obelia* (Cnidaria, Hydrozoa, Campanulariidae): a microphagous, filter–feeding medusa. Mar Ecol 28:178–183
Boero F, Bouillon J, Gravili C et al (2008) Gelatinous plankton: irregularities rule the world (sometimes). Mar Ecol Prog Ser 356:299–310
Bonaldo RB, Krajewski JP, Sazima I (2004) Does the association of young fishes with jellyfishes protect from predation? A report on a failure due to damage to the jellyfish. Neotrop Ichthyol 2:103–105
Bouillon J (1987) Considerations sur le développment des Narcoméduses et sur leur position phylogénétique. Indo–Malayan Zool 4:189–278
Bouillon J, Gravili C, Pagès F et al (2006) An introduction to Hydrozoa. Mém Mus Natl Hist Nat 194:1–591
Brodeur RD, Mills CE, Overland JE et al (1999) Evidence for a substantial increase in gelatinous zooplankton in the Bering Sea, with possible links to climate change. Fish Oceanogr 8:296–306
Brodeur RD, Sugisaki H, Hunt Jr GL (2002) Increases in jellyfish biomass in the Bering Sea: implications for the ecosystem. Mar Ecol Prog Ser 233:89–103
Brotz L, Pauly D (2017) Studying jellyfish fisheries: toward accurate national catch reports and appropriate methods for stock assessments. In: Mariottini GL (ed) Jellyfish: ecology, distribution patterns and human interactions. Nova Publishers, Hauppauge, pp 313–329
Brotz L, Schiariti A, López-Martínez J et al (2017) Jellyfish fisheries in the Americas: origin, state of the art, and perspectives on new fishing grounds. Rev Fish Biol Fish 27:1–29
Burnett JW (1991) Jellyfish envenomation syndromes worldwide. MAP Tech Rep Ser, vol 47. UNEP, Athens, pp 227–235
Cabreira AG, Madirolas A, Álvarez-Colombo G et al (2006) Acoustic study of the Río de la Plata estuarine front. ICES J Mar Sci 63:1718–1725
Carré D (1969) Étude histologique du développement de *Nanomia bijuga* (Chiaje, 1841), siphonophore physonecte, Agalmidae. Cah Biol Mar 10:325–341
Carré C, Carré D (1991) A complete life cycle of the calycophoran siphonophore *Muggiaea kochi* (will) in the laboratory, under different temperature conditions: ecological implications. Philos Trans R Soc Lond B 334:27–32
Carrette T, Alderslade P, Seymour J (2002) Nematocyst ratio and prey in two Australian cubomedusans, *Chironex fleckeri* and *Chiropsalmus* sp. Toxicon 40(11):1547–1551
Carrizo SS, Schiariti A, Nagata RM et al (2016) Preliminary observations on ephyrae predation by *Lychnorhiza lucerna* medusa (Scyphozoa; Rhizostomeae). Der Zool Gart 85(1):74–83
Chace FA (1969) A new genus and five new species of shrimps (Decapoda: Palaemonidae, Pontoniinae) from the Western Atlantic. Crustaceana (Leiden) 16(3):251–272
Chiaverano L, Mianzan H, Ramírez F (2004) Gonad development and somatic growth patterns of *Olindias sambaquiensis* (Limnomedusae, Olindiidae). Hydrobiologia 530(531):373–381
Chiaverano LM, Graham WM, Costello JH (2015) Parasites alter behavior, reproductive output, and growth patterns of *Aurelia* medusae in a marine lake. Mar Ecol Prog Ser 540:87–98
Colin SP, Costello JH, Graham WM (2005) Omnivory by the small cosmopolitan hydromedusa *Aglaura hemistoma*. Limnol Oceanogr 50(4):1264–1268
Collins AG (2002) Phylogeny of Medusozoa and the evolution of cnidarian life cycles. J Evol Biol 15:418–432
Condon RH, Steinberg DK, del Giorgio PA (2011) Jellyfish blooms result in a major microbial respiratory sink of carbon in marine systems. Proc Natl Acad Sci 108(25):10225–10230
Costello J, Colin SP, Dabiri JO (2008) Medusan morphospace: phylogenetic constraints, biomechanical solutions, and ecological consequences. Invertebr Biol 127(3):265–290
Daly M, Brugler M, Cartwright P et al (2007) The phylum Cnidaria: a review of phylogenetic patterns and diversity 300 years after Linnaeus. Zootaxa 1668:127–182
Dawson MN, Hamner WM (2009) A character–based analysis of the evolution of jellyfish blooms: adaptation and exaptation. Hydrobiologia 616(1):193–215
Delannoy CMJ, Houghton JDR, Fleming NEC et al (2011) Mauve Stingers (*Pelagia noctiluca*) as carriers of the bacterial fish pathogen *Tenacibaculum maritimum*. Aquaculture 311:255–257

Díaz Briz LM (2014) Rol del Plancton Gelatinoso en el Mar Argentino: ¿Cuán importante es en la trama trófica marina? Doctoral Thesis. Universidad Nacional de Mar del Plata, Mar del Plata: Departamento de Ciencias Marinas. 150 pp

Díaz Briz LMD, Martorelli SR, Genzano GN et al (2012) Parasitism (Trematoda, Digenea) in medusae from the southwestern Atlantic Ocean: medus hosts, parasites prevalences, and ecological implications. Hydrobiologia 690(1):215–226

Díaz Briz LM, Martorelli SR, Genzano G (2015) The parasite *Monascus filiformis* (Trematoda, Digenea, Fellodistomidae) on *Stromateus brasiliensis* (Pisces, Perciformes, Stromateidae): possible routes of transmission involving jellyfish. J Mar Biol Assoc UK 96(7):1483–1489

Díaz Briz LM, Sánchez F, Mari N (2017) Gelatinous zooplankton (ctenophores, salps and medusae): an important food resource of fishes in the temperate SW Atlantic Ocean. Mar Biol Res 13(6):630–644

Dittrich B (1988) Studies on the life cycle and reproduction of the parasitic amphipod *Hyperia galba* in the North Sea. Helgoländer Wiss Meeresunters 42:79–98

Doyle TK, De Haas H, Cotton D et al (2008) Widespread occurrence of the jellyfish *Pelagia noctiluca* in Irish coastal and shelf waters. J Plankton Res 30:963–968

Doyle TK, Hays GC, Harrod C et al (2014) Ecological and societal benefits of jellyfish. In: Pitt KA, Lucas CH (eds) Jellyfish blooms. Springer, Dordrecht, pp 105–127

Dutto MS, Genzano GN, Schiariti A et al (2017) Medusae and ctenophores from the Bahía Blanca Estuary and neighboring inner shelf (Southwest Atlantic Ocean, Argentina). Mar Biodivers Rec 10(1):14

Estrades A, López-Mendilaharsu M, Fallabrino A (2007) *Dermochelys coriacea* (Leatherback Sea turtle) diet. Herpetolo Rev 38:330

Fancett MS (1988) Diet and prey selectivity of scyphomedusae from Port Philip Bay, Australia. Mar Biol 98:503–509

Fautin DG (2002) Reproduction of Cnidaria. Can J Zool 80:1735–1754

Freitas JC, Schiozer WA, Malpezzi ELA (1995) A case of envenoming by Portuguese man–of–war from the Brazilian coast. Toxicon 33:859–861

Gasca R, Hoover R, Haddock SHD (2015) New symbiotic associations of hyperiid amphipods (Peracarida) with gelatinous zooplankton in deep waters off California. J Mar Biol Assoc UK 95:503–511

Gemmell BJ, Costello JH, Colin SP et al (2013) Passive energy recapture in jellyfish contributes to propulsive advantage over other metazoans. Proc Natl Acad Sci U S A 110(44):17904–17909

Genzano GN, Schiariti A, Mianzan HW (2014) Cnidaria. In: Calcagno JA (ed) Los Invertebrados Marinos. Fundación Félix de Azara, Buenos Aires, pp 67–85

Giangrande A, Geraci S, Belmonte G (1994) Life–cycle and life–history diversity in marine invertebrates and the implications in community dynamics. Oceanogr Mar Biol Annu Rev 32:305–333

Gonçalves GRL, Wolf MR, da Costa RC et al (2016) Decapod crustacean associations with scyphozoan jellyfish (Rhizostomeae: Pelagiidae) in the Southeastern Brazilian coast. Symbiosis 69(3):193–198

Gonzalez Carman V, Botto F, Gaitán E et al (2013) A jellyfish diet for the herbivorous green turtle *Chelonia mydas* in the temperate SW Atlantic. Mar Biol 161(2):339–349

Gopakumar G, Madhu K, Jayashankar R et al (2008) Live feed research for larviculture of marine finfish and shellfish. Mar Fisher Infor Serv T&R Ser 197:1–6

Graham WM, Martin DL, Felder DL et al (2003) Ecological and economic implications of a tropical jellyfish invader in the Gulf of Mexico. Biol Invasions 5:53–69

Graham WM, Gelcich S, Robinson KL et al (2014) Linking human well–being and jellyfish: ecosystem services, impacts, and societal responses. Front Ecol Environ 12(9):515–523

Greve W (1994) The 1989 German Bight invasion of *Muggiaea atlantica*. ICES J Mar Sci 51:355–358

Haddad V Jr (2003) Aquatic animals of medical importance in Brazil. Rev Soc Bras Med Trop 36(5):591–597

Haddad V Jr (2008) Animais aquáticos potencialmente perigosos do Brasil; guia médico e biológico. Editora Roca, São Paulo, p 145
Haddad V Jr, JLC C, da Silveira FL (2001) Seabather's eruption: report of five cases in Southeast region of Brazil. Rev Inst Med Trop São Paulo 43:171–172
Haddad Jr V, Silveira FL, Cardoso JLC et al (2002) A report of 49 cases of cnidarian envenoming from southeastern Brazilian coastal waters. Toxicon 40:1445–1450
Haddad Jr V, Lupi O, Lonza JP et al (2009) Tropical dermatology: marine and aquatic dermatology. J Am Acad Dermatol 61(5):733–750
Haddad Jr V, Silveira FL, Migotto AE (2010) Skin lesions in envenoming by cnidarians (Portuguese man of war and jellyfish): etiology and severity of accidents on the Brazilian coast. Rev Inst Med Trop São Paulo 52:47–50
Haddock SHD (2004) A golden age of gelata: past and future research on planktonic ctenophores and cnidarians. Hydrobiol 530/531:549–556. In: Fautin DG, Westfall JA, Cartwright P et al (eds) Coelenterate biology: trends in research on Cnidaria and Ctenophora. Kluwer Academic Publishers, Dordrecht, pp 549–556
Hansson LJ, Moeslund O, Kiørboe T et al (2005) Clearance rates of jellyfish and their potential predation impact on zooplankton and fish larvae in a neritic ecosystem (Limfjorden, Denmark). Mar Ecol Prog Ser 304:117–131
Harrison NM (1984) Predation on jellyfish and their associates by seabirds. Limnol Oceanogr 29(6):1335–1337
Hay SJ, Hislop JRG, Shanks AM (1990) North Sea Scyphomedusae; summer distribution, estimated biomass and significance particularly for 0–group gadoid fish. Neth J Sea Res (1/2):113–130
Houghton JD, Doyle TK, Wilson MW et al (2006) Jellyfish aggregations and leatherback turtle foraging patterns in a temperate coastal environment. Ecology 87(8):1967–1972
Hsieh Y–HP, Rudloe J (1994) Potential of utilizing jellyfish as food in western countries. Trends Food Sci Technol 5:225–229
Hsieh Y–HP, Leong F–M, Rudloe J (2001) Jellyfish as food. Hydrobiologia 451:11–17
Huang Y–W (1986) The processing of Cannonball jellyfish (*Stomolophus meleagris*) and its utilization. Proceedings of the 11th annual conference of the tropical and subtropical fisheries technological society of de Americas, Tampa, Florida, 13–16 Jan 1986, pp 141–146
Huang Y–W (1988) Cannonball jellyfish (*Stomolophus meleagris*) as food resource. J Food Sci 53(2):341–343
Jarms G (2010) The early life history of Scyphozoa with emphasis on Coronatae. Verh Naturwiss Ver Hamburg 45:17–31
Kass-Simon G, Scappaticci AA Jr (2002) The behavioral and developmental physiology of nematocysts. Can J Zool 80:1772–1794
Kawahara M, Ohtsu K, Uye SI (2013) Bloom or non–bloom in the giant jellyfish *Nemopilema nomurai* (Scyphozoa: Rhizostomeae): roles of dormant podocysts. J Plankton Res 35(1):213–217
Kingsford MJ, Mooney CJ (2014) The ecology of box jellyfishes (Cubozoa). In: Pitt KA, Lucas CH (eds) Jellyfish blooms. Springer, Dordrecht, pp 267–302
Kingsford MJ, Pitt KA, Gillanders BM (2000) Management of jellyfish fisheries, with special reference to the order Rhizostomeae. Oceanogr Mar Biol Annu Rev 38:85–156
Kirkpatrick PA, Pugh PR (1984) Siphonophores and Velellids: keys and notes for the identification of the species. In: Brill EJ, Backhuys W (eds) Synopses of the British Fauna (new series), vol 29. The Linnean Society of London and The Estuarine and Brackish–Water Sciences Association, Leiden, p 154
Kitamura M, Omori M (2010) Synopsis of edible jellyfishes collected from Southeast Asia, with notes on jellyfish fisheries. Plankon Benthos Res 5(3):106–118
Kokelj F, Mianzan H, Avian M et al (1993) Dermatitis due to *Olindias sambaquiensis*: a case report. Cutis 51(5):339–342
Kremer BP (2001) Opportunistic lifestyles of the gelatinous and abundant: what gives a species "the right stuff"? In: Briand F (ed) CIESM Workshop Series "Gelatinous zooplankton outburst: theory and practice", Naples, 29 August–1 September, 14, pp 87–92

Lange J, Tai M, Kaiser R (2016) Husbandry of jellyfish, from the beginning until today. Der Zool Gart 85:52–63

Larson RJ (1987) Trophic ecology of planktonic gelatinous predators in Saanich Inlet, British Columbia: diets and prey selection. J Plankton Res 9(5):811–820

Larson RJ, Mills CE, Harbison GR (1989) In situ foraging and feeding behaviour of narcomedusae (Cnidaria: Hydrozoa). J Mar Biol Assoc UK 69(4):785–794

Laval P (1980) Hyperiid amphipods as crustacean parasitoids associated with gelatinous zooplankton. Oceanogr Mar Biol Annu Rev 18:11–56

Leoni V, González S, Ortega L et al (2016) *Tamoya haplonema* (Cnidaria: Cubozoa) from Uruguayan and adjacent waters: oceanographic context of new and historical findings. Mar Biodivers Rec 9(1):92

Lucas CH, Dawson MN (2014) What are jellyfishes and thaliaceans and why do they bloom? In: Pitt KA, Lucas CH (eds) Jellyfish blooms. Springer, Dordrecht, pp 9–44

Lucas CH, Gelcich S, Uye SI (2014) Living with jellyfish: management and adaptations strategies. In: Pitt KA, Lucas CH (eds) Jellyfish blooms. Springer, Dordrecht, pp 129–150

Lynam CP, Hay SJ, Brierley AS (2005) Jellyfish abundance and climate variation: contrasting responses in oceanographically distinct regions of the North Sea, and possible implications for fisheries. J Mar Biol Assoc UK 85:435–450

Macchi GJ, Schiariti A (2016). Taller sobre estudios de zooplancton, en el marco del Proyecto Reclutamiento del Efectivo Patagónico de Merluza (REC). Informe de Gestión N°13. INIDEP. Dirección: Pesquerías Pelágicas y Ambiente Marino; Programa: Ambiente Marino y Ecosistemas Pesqueros, p 6

Mackie GO (2002) What's new in cnidarian biology? Can J Zool 80:1649–1653

Mansueti R (1963) Symbiotic behavior between small fishes and jellyfishes, with new data on that between the stromateid, *Peprilus alepidotus*, and the scyphomedusae, *Chrysaora quinquecirrha*. Copeia 1:40–80

Marques AC, Collins AG (2004) Cladistic analysis of Medusozoa and cnidarian evolution. Invertebr Biol 123(1):23–42

Marques AC, Morandini AC, Migotto AE (2003) Synopsis of knowledge on cnidaria medusozoa from Brazil. Biota Neotrop 3(2):1–18

Marques AC, Haddad Jr V, Rodrigo L et al (2014) Jellyfish (*Chrysaora lactea*, Cnidaria, Semaeostomeae) aggregations in southern Brazil and consequences of stings in humans. Lat Am J Aquat Res 42(5):1194–1199

Martinelli Filho JE, Stampar SN, Morandini AC et al (2008) Cleaner shrimp (Caridea: Palaemonidae) associated with scyphozoan jellyfish. Vie Milieu 58:133–140

McInnes JC, Alderman R, Lea MA et al (2017) High occurrence of jellyfish predation by black-browed and Campbell albatross identified by DNA metabarcoding. Mol Ecol 26(18):4831–4845

Mianzan HW (1986) Estudio sistemático y bioecológico de algunas medusas Scyphozoa de la región subantártica. PhD thesis, Universidad Nacional de La Plata, Argentina, pp 196

Mianzan HW, Olague G, Montero R (1988) Scyphomedusae de las aguas uruguayas. Spheniscus 6:1–9

Mianzan H, Sorarrain D, Burnett JW et al (2000) Mucocutaneous junctional and flexural paresthesias caused by the holoplanktonic trachymedusa *Liriope tetraphylla*. Dermatology 201(1):46–48

Mianzan HW, Fenner PJ, Cornelius PFS et al (2001) Vinegar as a disarming agent to prevent further discharge of the nematocysts of the stinging hydromedusa *Olindias sambaquiensis*. Cutis 68:45–48

Mianzan H, Quiñones J, Palma S et al (2014) *Chrysaora plocamia*: A poorly understood jellyfish from South American waters. In: Pitt KA, Lucas CH (eds) Jellyfish blooms. Springer, Dordrecht, pp 219–236

Mills CE (1995) Meduase, shiphonophores and ctenophores as planktivorous predators in changing global ecosystems. ICES J Mar Sci 52:575–581

Mills CE (2001) Jellyfish blooms: are populations increasing globally in response to changing ocean conditions? Hydrobiologia 451:55–68

Miranda LS, Collins AG, Marques AC (2010) Molecules clarify a cnidarian life cycle – the "hydrozoan" *Microhydrulalimopsicola* is an early life stage of the staurozoan *Haliclystus antarcticus*. PLoS One 5. https://doi.org/10.1371/journal.pone.0010182

Mitchell SO, Baxter EJ, Holland C et al (2012) Development of a novel histopathological gill scoring protocol for assessment of gill health during a longitudinal study in marine–farmed Atlantic salmon (*Salmo salar*). Aquac Int 20(5):813–825

Miyajima Y, Masuda R, Kurihara A et al (2011) Juveniles of threadsail file–fish, *Stephanolepis cirrhifer*, can survive and grow by feeding on moon jellyfish *Aurelia aurita*. Fish Sci 77(1):41–48

Montgomery L, Seys J, Mees J (2016) To pee, or not to pee: a review on envenomation and treatment in European jellyfish species. Mar Drugs 14(7):127

Morandini AC (2003) Estrutura populacional de *Chrysaora lactea* e *Lychnorhiza lucerna* (Cnidaria; Scyphozoa) em amostras de plâncton, com a redescrição das espécies. PhD thesis. Universidade de São Paulo, São Pablo, p 115

Morandini AC, Martorelli SR, Marques AC et al (2005) *Digenean metacercaria* (Trematoda, Digenea, Lepocreadiidae) parasitizing "coelenterates" (Cnidaria, Scyphozoa and Ctenophora) from southeastern Brazil. Braz J Oceanogr 53:39–45

Morandini AC, Medina M, Collins AG (2014) Eumetazoans. The emergence of tissues and guts. In: Vargas P, Zardoya R (eds) The tree of life. Sinauer Associates Inc Publishers, Sunderland, pp 207–216

Morandini AC, Custódio MR, Marques AC (2016a) Phylum Porifera and Cnidaria. In: Gopalakrishnakone P, Haddad Jr V, Kem WR, Tubaro A, Kim E (eds) Marine and freshwater toxins. Springer, Dordrecht, pp 287–316

Morandini AC, Schiariti A, Stampar SN et al (2016b) Succession of generations is still the general paradigm for scyphozoan life cycles. Bull Mar Sci 92(3):343–351

Moreira MGBS (1961) Sobre *Mastigias scintillae* sp. nov. (Scyphomedusae, Rhizostomeae) das costas do Brasil. Bol Inst Oceanogr USP 11:5–30

Mosovich JH, Young P (2012) Picadura de medusa *Olindias sambaquiensis*: Análisis de 49 casos. Medicina 72(5):380–388

Nagata RM (2015) Bases morfo–funcionais da alimentação e o papel trófico de *Lychnorhiza lucerna* (Scyphozoa, Rhizostomeae). PhD thesis. Universidade de São Paulo, São Paulo, p 172

Nagata RM, Haddad MA, Nogueira Jr M (2009) The nuisance of medusae (Cnidaria, Medusozoa) to shrimp trawls in central part of southern Brazilian Bight, from the perspective of artisanal fishermen. Pan–Am. J Aquat Sci 4:312–325

Nagata RM, Morandini AC, Colin SP et al (2016) Transitions in morphologies, fluid regimes, and feeding mechanisms during development of the medusa *Lychnorhiza lucerna*. Mar Ecol Prog Ser 557:145–159

Nogueira Jr M, Haddad MA (2005) *Lychnorhiza lucerna* Haeckel (Scyphozoa, Rhizostomeae) and *Libinia ferreirae* Brito Capello (Decapoda, Majidae) association in southern Brazil. Rev Brasil Zool 22:908–912

Nogueira Jr M, Haddad MA (2008) The diet of Cubomedusae (Cnidaria, Cubozoa) in southern Brazil. Braz J Oceanogr 56(3):157–164

Nogueira Jr M, Silva JL (2005) Associações entre medusas (Cnidaria) e isópodos (Crustacea) nos litorais do Paraná e Santa Catarina, Brasil. Acta Biol Par 34:127–138

Nogueira Jr M, Díaz Briz L, Haddad MA (2013) New records of *Opechona* sp. metacercariae (Digenea: Trematoda) on hydromedusae from South Brazil. Mar Biodivers Rec 6:1–4

Nogueira Jr M, DíazBriz LM, Haddad MA (2015) Monthly and inter–annual variations of *Opechona* sp. (Digenea: Lepocreadiidae) parasitizing scyphomedusae off southern Brazil. Mar Biol 162:391–400

Ohtsuka S, Koike K, Lindsay D et al (2009) Symbionts of marine medusae and ctenophores. Plank Benth Res 4:1–13

Oliveira O, Araújo E, Ayón P et al (2016) Census of Cnidaria (Medusozoa) and Ctenophora from South American marine waters. Zootaxa 4194:1–256

Omori M, Kitamura M (2004) Taxonomic review of three Japanese species of edible jellyfish (Scyphozoa: Rhizostomeae). Plankton Biol Ecol 51:36–51

Omori M, Nakano E (2001) Jellyfish fisheries in Southeast Asia. Hydrobiologia 451:19–26

Osborn DA (2000) Cnidarian "Parasites" on *Solmissus incisa*, a Narcomedusa. Sci Mar 64:157–163

Pauly D, Graham W, Libralato S et al (2009) Jellyfish in ecosystems, online databases and ecosystem models. Hydrobiologia 616:67–85

Phillips N, Eagling L, Harrod C et al (2017) Quacks snack on smacks: mallard ducks (*Anas platyrhynchos*) observed feeding on hydrozoans (*Velella velella*). Plankon Benthos Res 12(2):143–144

Piraino S, Boero F, Aeschbach B et al (1996) Reversing the life cycle: Medusae transforming into polyps and cell transdifferentiation in *Turritopsis nutricula* (Cnidaria, Hydrozoa). Biol Bull 190:302–312

Pitt KA, Koop K, Rissik D (2005) Contrasting contributions to inorganic nutrient recycling by the co–occurring jellyfishes, *Catostylus mosaicus* and *Phyllorhiza punctata* (Scyphozoa, Rhizostomeae). J Exp Mar Biol Ecol 315:71–86

Pitt KA, Welsh DT, Condon RH (2009) Influence of jellyfish blooms on carbon, nitrogen and phosphorus cycling and plankton production. Hydrobiologia 616:133–149

Puente Tapia FA, Gasca R, Genzano G et al New records of association between *Brachyscelus rapacoides* (Arthropoda: Amphipoda) and medusae (Cnidaria: Scyphozoa and Hydrozoa) from São Sebastião Channel, SE Brazil. Mar Biodivers (submitted)

Purcell JE (1981) Dietary composition and diel feeding patterns of epipelagic siphonophores. Mar Biol 65(1):83–90

Purcell JE (1991) A review of cnidarians and ctenophores feeding on competitors in the plankton. Hydrobiologia 216(217):335–342

Purcell JE (1997) Pelagic cnidarians and ctenophores as predators: selective predation, feeding rates, and effects on prey populations. Ann l'Inst Océanogr 73(2):125–137

Purcell JE (2001) Interactions of pelagic cnidarians and ctenophores with fish: a review. Hydrobiologia 451:27–44

Purcell JE (2003) Predation on zooplankton by large jellyfish *Aurelia labiata*, *Cyanea capillata* and *Aequorea aequorea*, in Prince William Sound, Alaska. Mar Ecol Prog Ser 246:137–152

Purcell JE (2009) Extension of methods for jellyfish and ctenophore trophic ecology to large–scale research. Hydrobiologia 616:23–50

Purcell JE, Arai MN (2001) Interactions of pelagic cnidarians and ctenophores with fish: a review. Hydrobiologia 451:27–44

Purcell JE, Grover JJ (1990) Predation and food limitation as causes of mortality in larval herring at a spawning ground in British Columbia. Mar Ecol Prog Ser 59:55–61

Purcell JE, White JR, Roman MR (1994) Predation by gelatinous zooplankton and resource limitation as potential controls of *Acartia tonsa* copepod populations in Chesapeake Bay. Limnol Oceanogr 39:263–278

Purcell JE, Uye S–i, Lo W–T (2007) Anthropogenic causes of jellyfish blooms and their direct consequences for humans: a review. Mar Ecol Prog Ser 350:153–174

Purcell JE, Baxter EJ, Fuentes V (2013) Jellyfish as products and problems for aquaculture. In: Allan G, Burnell G (eds) Advances in aquaculture hatchery technology, Woodhead publishing series in food science, technology and nutrition, vol 242. Woodhead Publishing, Cambridge, pp 404–430

Resgalla C Jr, Rosseto AL, Haddad V Jr (2011) Report of an outbreak of stings caused by *Olindias sambaquiensis* in Southern Brazil. Braz J Oceanogr 59:391–396

Riascos JM, Villegas V, Cáceres I et al (2012) Patterns of a novel association between the scyphomedusa *Chrysaora plocamia* and the parasitic anemone *Peachia chilensis*. J Mar Biol Assoc UK 93:919–923

Riascos JM, Villegas V, Pacheco AS (2014) Diet composition of the large scyphozoan jellyfish *Chrysaora plocamia* in a highly productive upwelling Centre off northern Chile. Mar Biol Res 10(8):791–798

Richardson AJ, Bakun A, Hays GC et al (2009) The jellyfish joyride: causes, consequences and management responses to a more gelatinous future. Trends Ecol Evol 24:312–322

Ricklefs RE (1979) Ecology. Chiron Press Inc., New York

Rodriguez CS, Marques AC, Mianzan HW et al (2017) Environment and life cycles influence distribution patterns of hydromedusae from austral South America. Mar Biol Res 13(6):659–670

Rossetto AL, Silveira FL, Morandini AC et al (2015) Seabather's eruption: report of fourteen cases. An Acad Brasil Ciênc 8:431–436

Saito N, Kurata Y, Moku M (2002) Note on a meso–bathypelagic isopodean genus *Anuropus* (Crustacea: Isopoda: Anuropidae) collected in the western North Pacific off northern Honshu, Japan. Bull Plankton Soc Japan 49:88–94. (in Japanese with English abstract)

Sal Moyano MP, Schiariti A, Giberto DA et al (2012) The symbiotic relationship between *Lychnorhiza lucerna* (Scyphozoa, Rhizostomeae) and *Libinia spinosa* (Decapoda, Epialtidae) in the Río de la Plata (Argentina–Uruguay). Mar Biol 159:1933–1941

Santos LA, Feres SJC, Lopes ATL (2008) Primeiro registro da associação entre o caranguejo *Libinia ferreirae* (Crustacea; Decapoda; Majidae) e a medusa *Lychnorhiza lucerna*(Cnidaria; Scyphozoa; Rhizostomeae) para o Golfão Maranhense. Bol Lab Hidrobiol 21(1):99–102

Schiariti A (2008) Historia de vida y dinámica de poblaciones de *Lychnorhiza lucerna* (Scyphozoa) ¿Un recurso pesquero alternativo? PhD thesis. Universidad de Buenos Aires, Buenos Aires, p 209

Schiariti A, Mianzan HW (2013) La medusa *Lychnorhiza lucerna* (Cnidaria: Scyphozoa) como recurso pesquero alternativo en Argentina. Informe de Investigación N° 36 INIDEP. Dirección: Pesquerías Pelágicas y Ambiente Marino. Proyecto: Ecología Pesquera, Argentina, p 9

Schiariti A, Kawahara M, Uye S–I et al (2008) Life cycle of the jellyfish *Lychnorhiza lucerna* (Scyphozoa: Rhizostomeae). Mar Biol 156:1–12

Schiariti A, Sal Moyano MP, Giberto DA et al (2012) First record of the association between *Lychnorhiza lucerna* (Scyphozoa, Rhizostomeae) and *Cyrtograpsus affinis* (Decapoda, Varunidae). Lat Am J Aquat Res 40(4):1090–1093

Schiariti A, Gaitán E, Mianzan HW (2013) Cartilla de identificación y protocolo básico de muestreo para el monitoreo de medusas del Mar Argentino. Informe de Asesoría y Transferencia N°37 INIDEP. Dirección: Pesquerías Pelágicas y Ambiente Marino; Programa: Ambiente Marino y Ecosistemas Pesqueros, p 7

Schiariti A, Morandini AC, Jarms G et al (2014) Asexual reproduction strategies and blooming potential in Scyphozoa. Mar Ecol Prog Ser 510:241–253

Schiariti A, Melica V, Kogovšek T et al (2015) Density–dependent effects control the reproductive strategy and population growth of *Aurelia aurita* s.l. scyphistomae. Mar Biol 162(8):1665–1672

Schroeder R, Branco JO, Freitas Jr F et al (2014) Preliminary assessment of the jellyfish bycatch captured off southern and southeastern Brazil. Lat Am J Aquat Res 42:289–300

Shamah D (2014) Israeli tech turns jellyfish into paper towels. Times of Israel, edition of April 8, Jerusalem

Sinclair M (1988) Marine populations: an assay on population regulation and speciation. Washington Sea Grant Program, Seattle

Sobolewski M, Rebuli KB, Haddad MA et al (2017) Ocorrência da associação entre *Phyllorhiza puntacta* Von Lendenfeld, 1884 (Cnidaria: Scyphozoa) e peixes na Baía de Paranaguá, Estado do Paraná. Arq Ciên Mar 37(1–2):29–34

Spaulding JG (1972) The life cycle of *Peachia quinquecapitata*, an anemone parasitic on medusae during its larval development. Biol Bull 143:440–453

Stearns SC (1992) The evolution of life histories. Oxford University Press, Oxford, p 250

Suchman CL, Daly EA, Keister JE et al (2008) Feeding patterns and predation potential of scyphomedusae in a highly productive upwelling region. Mar Ecol Prog Ser 358:161–172

Sullivan LJ, Kremer P (2011) Gelatinous zooplankton and their trophic roles. In: Wolanski E, McLusky D (eds) Treatise on estuarine and coastal science. Elsevier, New York, pp 127–171

Schiariti A, Dutto MS, Pereyra DY, Failla Siquier G, Genzano GN (in press) Scyphozoa and Cubozoa from southwestern Atlantic and Subantarctic region (32-60°S; 34-70°W): species composition, spatial distribution and life history traits. Latin American Journal of Aquatic Research

Technau U, Genikhovich G, Kraus JEM (2015) Cnidaria. In: Wanninger A (ed) Evolutionary developmental biology of invertebrates 1: introduction, Non–Bilateria, Acoelomorpha, Xenoturbellida, Chaetognatha. Springer, Vienna, pp 115–163

Thiebot J–B, Arnould J, Gómez Laich A et al (2017) Jellyfish and other gelata as food for four penguin species–insights from predator–borne videos. Front Ecol Environ 15(8):437–441

Toshino S, Miyake H, Ohtsuka S et al (2015) Monodisc strobilation in Japanese giant box jellyfish*Morbakka virulenta* (Kishinouye, 1910): a strong implication of phylogenetic similarity between Cubozoa and Scyphozoa. Evol Dev 17:231–239

Towanda T, Thuesen EV (2006) Ectosymbiotic behavior of *Cancer gracilis* and its trophic relationships with its host *Phacellophora camtschatica* and the parasitoid *Hyperia medusarum*. Mar Ecol Prog Ser 315:221–236

Uye S–i (2008) Blooms of the giant jellyfish *Nemopilema nomurai*: a threat to the fisheries sustainability of the East Asian Marginal Seas. Plankon Benthos Res 3:125–131

Uye S, Shimauchi H (2005) Population biomass, feeding, respiration and growth rates, and carbon budget of the scyphomedusa *Aurelia aurita* in the Inland Sea of Japan. J Plankton Res 27(3):237–248

Vandendriessche S, Vansteenbrugge L, Hostens K et al (2013) Jellyfish, jellypress and jellyperception. ILVO Mededeling 142:21

Verity PG, Smetacek V (1996) Organism life cycles, predation, and the structure of marine pelagic ecosystems. Mar Ecol Prog Ser 130:277–293

Wakabayashi K, Sato R, Ishii H et al (2012) Culture of phyllosomas of *Ibacus novemdentatus* in a closed recirculating system using jellyfish as food. Aquaculture 330:162–166

West EJ, Welsh DT, Pitt KA (2009) Influence of decomposing jellyfish on the sedimentoxygen demand and nutrient dynamics. Hydrobiologia 616:151–160

Yilmaz IN (2014) Collapse of zooplankton stocks during *Liriope tetraphylla* (Hydromedusa) blooms and dense mucilaginous aggregations in a thermohaline stratified basin. Mar Ecol 36(3):595–610

Zamponi MO (1985) La alimentación de algunas especies de Hydromedusae. Neotropica 31(86):155–162

Zamponi MO, Mianzan H (1985) La mecánica de captura y alimentación de *Olindias sambaquiensis* Müller, 1961 (Limnomedusae) en el medio natural y en condiciones experimentales. Hist Nat 5:269–278

Part VI
Harmful Algae and Their Impacts

Alexandrium tamarense/catenella Blooms in the Southwestern Atlantic: Paralytic Shellfish Toxin Production and Its Trophic Transference

Nora G. Montoya, Mario O. Carignan, and José I. Carreto

Abstract Some species of the dinoflagellate genus *Alexandrium* (Halim) Balech produce potent neurotoxins called saxitoxins (STXs), which can be accumulated in several invertebrates whose ingestion can cause the human intoxication named paralytic shellfish poisoning (PSP). Many of these STXs are transferred through the food chain and thus can also affect a variety of wildlife. The species *A. tamarense/catenella* were mainly responsible for PSP in Southwestern Atlantic. However, unexpected patterns of diversity in genetic, reproductive compatibility, toxicity, and physiological response were found in several studies. The present review seeks to provide a broad overview of the species *A. tamarense/catenella* distribution, their toxicity and toxin composition in clonal isolates, and natural populations, using historical plankton, cyst, and toxin records. Also, the metabolic changes that occur in the toxic composition of primary and secondary consumers and the impact produced by their transference through the food chain are evaluated.

Keywords *Alexandrium tamarense/catenella* blooms · Southwestern Atlantic · Toxin composition · Trophic transference · Impacts

1 Introduction

Harmful dinoflagellate blooms are a worldwide phenomenon that have apparently been increasing in frequency, duration, and magnitude in recent years (Anderson et al. 2012 and references therein). The impacts of these phenomena include mass mortalities of wild and farmed fish and shellfish, human intoxications, and alterations of marine ecosystem structure through adverse effects on larvae and other life history stages of marine organisms. Due to their negative economic and health

N. G. Montoya (✉) · M. O. Carignan · J. I. Carreto
Instituto Nacional de Investigación y Desarrollo Pesquero (INIDEP), Mar del Plata, Argentina
e-mail: nmontoya@inidep.edu.ar; marioc@inidep.edu.ar; jcarreto@inidep.edu.ar

M. S. Hoffmeyer et al. (eds.), *Plankton Ecology of the Southwestern Atlantic*,
https://doi.org/10.1007/978-3-319-77869-3_20

impacts, harmful algal blooms (HABs) are often carefully monitored, through preventive controls and mitigation measures.

HABs cause ecological damage due to their high biomass and because some species produce potent toxins which can kill mice within minutes (GEOHAB 2001 and references therein). Among the few algal species that can synthesize toxins affecting humans are some species of the dinoflagellate genus *Alexandrium* (Halim) Balech. These toxic species produce potent neurotoxins called saxitoxin (STX) analogs (about 57 known STX derivatives that differ in structure and toxicity) (Wiese et al. 2010), which can be accumulated and metabolized in several invertebrates without lethal effect to them (Etheridge 2010 and references therein), but their ingestion can cause the human intoxication named paralytic shellfish poisoning (PSP). Beyond the well-established effects that these toxins have on humans via shellfish poisoning events (Anderson et al. 2012 and references therein), many of these are transferred through the food chain and thus can affect a variety of wildlife, including fish, marine mammals, seabirds (Landsberg et al. 2014 and references therein), and sea turtles (Hattenrath-Lehmann et al. 2017 and references therein).

Alexandrium catenella (Whedon et Kof.) Balech, *A. tamarense* (M. Lebour) Balech, and *A. fundyense* Balech comprise the *A. tamarense* complex, dinoflagellates responsible for PSP worldwide. Although recently recognized as *Alexandrium catenella* (Prud'homme and Willem 2017), this study retains the names *A. tamarense* and *A. catenella* because of the large body of previously published literature. In South America, blooms of several toxic species of the genus *Alexandrium* (*A. tamarense*, *A. catenella*, *A. ostenfeldii*, *A. minutum*, *and A. tamiyavanichii*) have been well documented (Carreto et al. 1986; 2001; Odebrecht et al. 1997; Guzmán et al. 2002; Menezes et al. 2008; Varela et al. 2012; Almandoz et al. 2014; Fabro et al. 2017). However, thus far, the increased reports on PSP human intoxication and associated economic impacts had been caused by the increased of intensity and geographic spread of *A. tamarense/catenella* blooms (Carreto et al. 1986, 2001; Guzmán et al. 2002; Persich et al. 2006; Varela et al. 2012). However, in spite of the importance of *Alexandrium* species as a source of PSTs in the southwestern coast of South America, a comprehensive view of their biogeography, ecology, and toxicology is lacking.

The present review seeks to provide a broad overview of the *A. tamarense/catenella* distribution in Southwestern Atlantic, using historical plankton, cyst, and toxicity records. Another key issue is the toxin content and composition in both, cultured and natural populations of these species. Finally, the metabolic changes that occur in the toxic composition of primary and secondary consumers and the impact produced by their transference through the food chain are evaluated.

2 Biogeography

In the Western Atlantic, the first PSP outbreak associated with a bloom of *A. tamarense* (= *Gonyaulax excavatum*) was detected in 1980 in the area near the tidal front of Valdés Peninsula (Carreto et al. 1981, 1986). Since then, blooms of this species were observed in almost the entire coastal ecosystem of Argentina (Carreto et al. 1998a), Uruguay (Méndez et al. 2001), and Brazil (Odebrecht et al. 1997) demonstrating a clear northward advance in its distribution. In the southern tip of South America (Magellan Strait), the first PSP outbreak was detected in 1972 and was associated with a bloom of *A. catenella* (Guzmán et al. 1975). This species also showed a clear northward expansion in Chile (Guzmán et al. 1975; Lembeye 2004). However, the earliest documentation of a PSP outbreak in this austral region is the report from Segers (1908). He described death and mass poisoning of the aboriginal inhabitants near Ushuaia (Argentina) following ingestion of mussels in year 1886. Therefore, it is possible that this species might then have been a common – and occasionally prominent – component of the phytoplankton community from the region, before than dispersal mechanisms were initiated.

The species *A. tamarense* and *A. catenella* were defined originally based on morphological characteristics. Among these, the absence of a ventral pore on the first apical plate and the formation of large chains were thought to be the most reliable characters to distinguish *A. catenella* from *A. tamarense* (Fukuyo 1985; Balech 1995). On this base, most of the studied cultured strains and natural populations from the eastern coast of South America were identified as *A. tamarense* (Benavides et al. 1983; Balech 1995; Krock et al. 2015). Nevertheless, Gayoso and Fulco (2006) reported that the presence of the ventral pore and the chain-forming ability in *A. tamarense* cells collected from field populations of Nuevo Gulf (Argentina), as well as between isolate clones and even in the same clone grown under different temperature and irradiance, varies considerably. For instance, during spring *A. tamarense* was observed mainly as solitary cells, whereas the autumn populations had the distinctive and unusual ability to form long, fast-moving chains of up to eight cells (Carreto et al. 2004; Gayoso and Fulco 2006). Morphotypes of *A. catenella* and *A. tamarense* have a sympatric distribution in several geographical areas, including the Beagle Channel (Almandoz et al. 2011) and the Magellan Plume (Benavides et al. 1995; Montoya et al. 2010). The compatible mating of these morphospecies has also been observed (Sako et al. 1990; Mackenzie et al. 2004; Mardones et al. 2016), and therefore, hybridization between them may be hypothesized. By contrast, *A. catenella* seems to be the only morphospecies of the *A. tamarense* complex recorded in the southern coast of Chile (Varela et al. 2012; Fraga et al. 2015; Mardones et al. 2016). Analysis of partial sequences of LSU rDNA showed that strains of *A. tamarense* complex, isolated from Southern Atlantic, clustered within the North American clade (Group I) (Persich et al. 2006; Lilly et al. 2007; Penna et al. 2008; Aguilera-Belmonte et al. 2011; Mardones et al. 2016). Nevertheless, when several strains have been studied, unexpected patterns of diversity in genetic, reproductive compatibility, toxicity, and physiological response were found (Córdova and Müller 2002; Montoya et al. 2010; Aguilera-Belmonte et al. 2011; Varela et al. 2012; Mardones et al. 2016). Surprisingly, one strain of *A. tamarense* isolated from subantarctic waters near Malvinas Islands present the ribotype of the Group IV, being closely related to the *A. tamarense* strains isolates from the China Sea (Ho, et al. 2003). Confirmation of the presence or absence of alternative ribotypes of this species complex in the Southern Atlantic might be more comprehensively investigated by additional screenings of planktonic cells by various molecular probe-based methods (John et al. 2014).

3 Geographic Distribution

The majority of records of *A. tamarense* vegetative cells are included in a continuous band ranging from the southern tip of the Beagle Channel to 34°S, and between the coast and the 100 m isobath (Carreto et al. 2007), which coincides with the area of influence of the Patagonian Current (Fig. 1). These waters of subantarctic origin, diluted by the continental drainage from the Beagle Channel and the Magellan Strait, flow close to the shore up to 47°S separating toward the central continental shelf until the Río de la Plata area. Exceptionally, *A. tamarense* populations can be transported up to southern Brazil (~32°S) as appear to be occurred during late spring of 1996 (Odebrecht et al. 1997) (see Méndez and Carreto, chapter "Harmful Algal Blooms in the Río de la Plata Region"). *A. tamarense* was only exceptionally observed in low concentrations in the more coastal waters north of 41°S (Carreto et al. 1998a, Gayoso 2001; Sunesen et al. 2014) and in the northern sector of San Matias Gulf (Fig. 1) (Carreto et al. 1981, 1998a), where the Patagonian Current waters are separated from the coast by a water mass of higher salinity created by local processes. By contrast many records of *A. tamarense* cells were obtained in the south sector of San Matias Gulf (Carreto et al. 1981; Sastre et al. 1997; Santinelli et al. 2002) and San José Gulf (Santinelli et al. 2002), as these gulfs are influenced by the intrusion of the non-stratified cold waters originated near the tidal front of Valdés Peninsula (Carreto et al. 1974, 1981; Williams et al. 2013). Nuevo Gulf is a semi-enclosed ecosystem in which there is only limited water exchange with the open-shelf waters (Mouzo et al. 1978; Rivas and Beier 1990). However, and probably due to human-assisted dispersal mechanisms, motile cells and temporary cysts of *A. tamarense* were also recorded in this ecosystem since 1987 (Esteves et al. 1992; Gayoso 2001; Santinelli et al. 2002).

In the San Jorge Gulf, the highest concentrations of *A. tamarense* were associated with the stratified region of the tidal frontal system (Carreto et al. 1985, 2007; Akselman 1996), but its distribution includes areas very close to the coast (Pérez et al. 2012). Although the abundance distribution of *A. tamarense* is highly heterogeneous, it is evident that the highest abundances (up to 1.8×10^6 cells L^{-1}) have been reported in the tidal front of Valdés Peninsula (Carreto et al. 1986). In a decreasing order of abundance are the bloom records of Nuevo Gulf (Santinelli et al. 2002) and those observed in the Buenos Aires coastal front, which extend

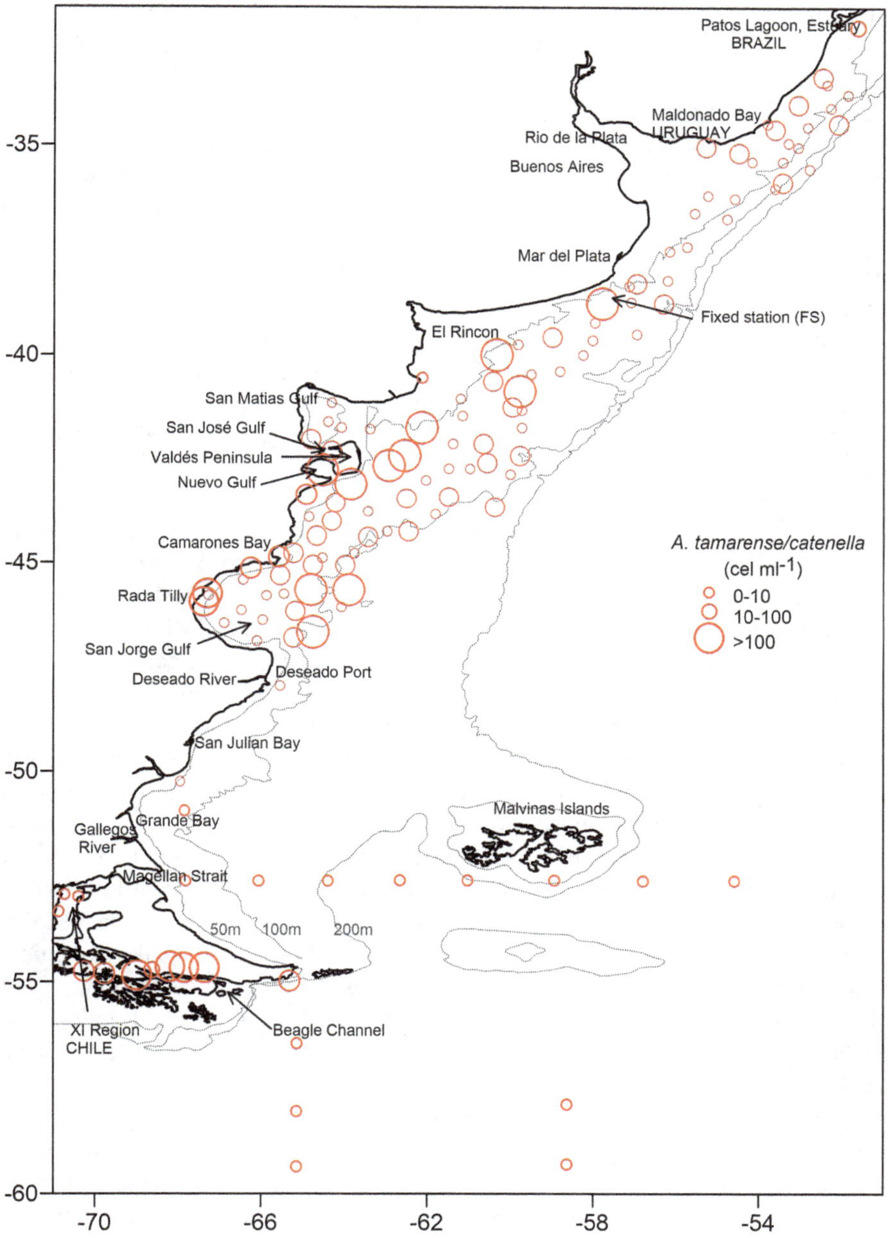

Fig. 1 Distribution area and recorded cellular abundances of *A. tamarense/catenella* in the Southwestern region of South America. (Modified from Carreto et al. 2007)

along the 50 m isobath (Montoya et al. 1996; Carreto et al. 1998b) and finally in the San Matias, San José, and San Jorge Gulfs, where the records were generally two orders of magnitude below those registered in the Patagonian tidal fronts (Gayoso 2001; Santinelli et al. 2002) with the exception of a toxic event in the year 2010 in Rada Tilly (Santinelli 2013). South of 47°S, the records are very scarce and seem to be restricted to Deseado Port and San Julián Bay (Carreto et al. 2007; Benavides unpubl. data). In very low concentrations (<20 cells L^{-1}), *A. tamarense* was also recorded in the outer-shelf waters (subantarctic shelf waters), between the southern end of South America and the Malvinas Islands and in the northern side of the polar front (Ho et al. 2003).

Although *A. tamarense* and *A. catenella* were recorded in the Beagle Channel (Almandoz et al. 2011), the Magellan Plume, and the coast of Malvinas Islands (Ho et al. 2003; Uhart et al. 2004), the PST outbreaks seem to be associated with summer blooms of *A. catenella* (Benavides et al. 1995; Uhart et al. 2004; Goya and Maldonado 2014; Pizarro et al. 2015).

Vegetative cells of *A. tamarense/catenella* produce, under certain conditions, gametes that fuse to form a motile zygotic cell (planozygote) which will eventually transform into a resting cysts (e.g., Dale 1983; Benavides et al. 1983) of nearly identical morphologies that reflect their close phylogenetic relationship.

The massive formation of these cysts in nature was associated with the last stage of the bloom (Carreto et al. 1986). Since the germination of cysts can lead directly to the initiation of blooms of vegetative cells, the knowledge of its distribution and abundance is essential to understand the ecology and dynamics of *Alexandrium* blooms. In addition, cysts act as a dispersing agent, as a source of toxicity and as a means for genetic recombination through sexuality. In spite of this, the studies related to their distribution and abundance in the surface sediments of the Southwest Atlantic are scarce. Carreto et al. (1981) described for the first time an important concentration of *A. tamarense* cysts in the sediments of the tidal front localized at the Valdés Peninsula. However, the more important study was developed in the sediments of the Patagonian shelf (42°–47°S) where both, the horizontal and vertical distribution of *A. tamarense* resting cysts were studied (Orozco and Carreto 1989) (Fig. 2). Other records are available for the surface sediments of a fixed station (FS) in the coastal front near Mar del Plata, along a seasonal cycle (Carreto et al. 1998b), in the fine sediments of a deep region of the Nuevo Gulf (Gayoso 2001), and, recently, in sediments near the tidal front of San Jorge Gulf (Krock et al. 2015) (Fig. 2). These studies show that the abundance of *A. tamarense* resting cysts in the Patagonian shelf (42–47°S) is very high. The higher concentrations occur in the first 2–3 centimeters and reach values of 9000 cysts cm^{-3} wet sediment (Fig. 2). However, the spatial distribution is highly heterogeneous, and concentration values were very different between stations (Orozco and Carreto 1989). Cysts were also very abundant in the FS (up to 2770 cysts cm^{-3}), being present in sediments throughout the year (Carreto et al. 1998b). In contrast, the *A. tamarense* cysts concentrations observed in Uruguayan (Méndez et al. 2001) and Brazilian (Persich et al. 2006) shores are substantially lower than in the FS (Carreto et al. 1998b). Cysts of *A.*

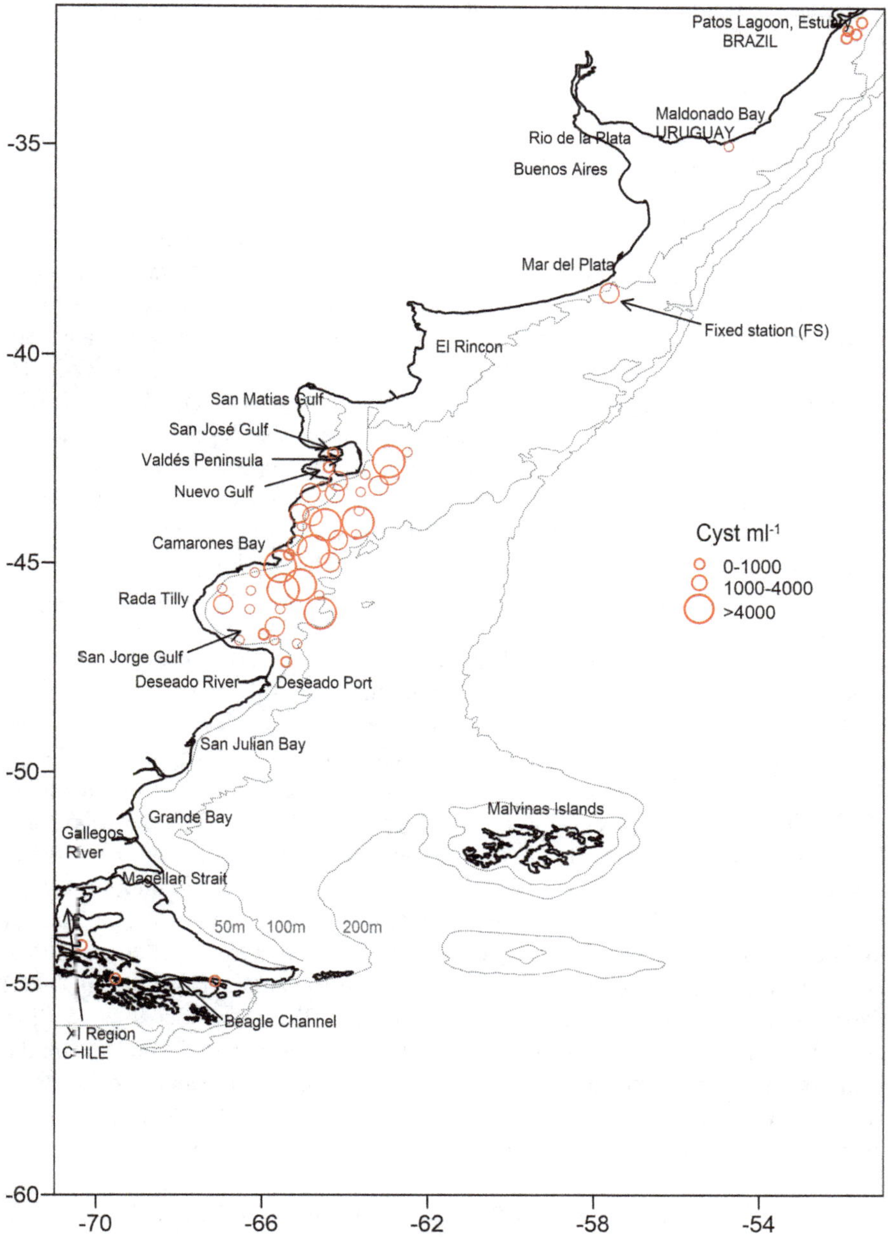

Fig. 2 Distribution area and recorded abundances of *A. tamarense/catenella* cysts, in the Southwestern region of South America. (Modified from Carreto et al. 2007)

tamarense/catenella only were detected in low concentration in the Beagle Channel (Lembeye 2004).

4 Toxin Content and Composition

Several reports indicated that *A. tamarense/catenella* can be either toxic or non-toxic. Results obtained are in agreement with these findings as all strains and populations from southern region of South America tested for toxicity were toxic. However, a 12-fold difference was found in the toxin contents among strains isolates from Argentina (Carreto et al. 1996, 2001; Montoya et al. 2010; Ho et al. 2003; Krock et al. 2015), Uruguay (Méndez et al. 2001), and Brazil (Persich et al. 2006). The toxic composition and genotypic characters were also highly diverse among isolates coming from the same geographic population (Persich et al. 2006; Alpermann et al. 2010; Montoya et al. 2010). These results support the hypothesis that it is advantageous to maintain a great variety of coexisting genotypes that would be available for selection under different environmental conditions.

Cell toxicity ranges (from 1.81 to 65.92 pg STX eq. cell^{-1} in cultured strains and 9.7–92.9 in field population) are similar to those described for strains from North America (Anderson et al. 1994) and Russia (Orlova et al. 2007). The dominance of N-sulfocarbamoil toxins (C1/2) and gonyautoxins (GTX1/4), minor proportions of GTX2/3 and neosaxitoxin (neoSTX), and the absence or trace of both STX and decarbamoil-saxitoxin (dcSTX) are the main characteristics of the toxin profile for *A. tamarense* cultured strains from Argentina (Carreto et al. 1996, 2001; Ho et al. 2003; Montoya et al. 2010; Krock et al. 2015), Brazil (Persich et al. 2006), and Uruguay (Méndez et al. 2001) (Table 1). They clustered together, but in different non-geographic subgroups, indicating the existence of a diverse continuous *A. tamarense* population (Montoya et al. 2010). Considering the highest percentages of toxin concentrations in each profile, at least two distinctive toxin patterns could be distinguished in cultured strains of *A. tamarense/catenella* (Carreto et al. 2001; Persich et al. 2006; Montoya et al. 2010; Varela et al. 2012; Krock et al. 2015). The dominant pattern is constituted by those less toxic strains, where C1 and C2 are the main toxins (ranging from 47% to 78.4%) (Carreto et al. 2001; Méndez et al. 2001; Ho et al. 2003; Persich et al. 2006; Montoya et al. 2010) and the other (Table 1) by the more toxic strains where GTX1/4 reach the higher concentrations (ranging from 54% to 87%).

It has been pointed out that most of the natural spring time populations of *A. tamarense* from Argentine Sea were several times more toxic (68.2–92.9 pg STX eq. cell^{-1}) than those observed in the majority of studied isolates (1.81–10.3 pg STX eq. cell^{-1}) during nutrient replete conditions in laboratory cultures (Montoya et al. 2010). This was due, in part, to the higher cell toxin content (163.9–261.4 fmol cell^{-1}) observed in almost all of the natural studied populations, but principally because they exhibited a high relative content of the potent carbamate derivatives GTX1/4 (69.1–93.6%) (Montoya et al. 2010). These concentrations, reached in the

Table 1 Toxin content, toxicity and composition of *A. tamarense/catenella* field populations and cultures strains

Code		Major toxins (mol %)								Toxicity	Toxin content
		GTX1/4	GTX2/3	C1/2	GTX5	GTX6	STX	neoSTX	dcSTX	(pg STX eq. $cell^{-1}$)	(fmol $Cell^{-1}$)
Field populations											
[a]ONPMDQ00	Mar del Plata, Argentina	7.74	88.52	3.74						9.77	45.53
[a]SNPMDQ00	Mar del Plata, Argentina	69.13	22.79	8.51						68.21	163.98
[a]SNPMDQ96	Mar del Plata, Argentina	89.45	4.66	4.34			0.20	1.18		–	–
[a]SNPPV05	Valdés peninsula, Argentina	87.82	12.58	0.00						80.08	225.52
[a]SNPC05	Camarones Bay, Argentina	93.64	6.36	0.00						86.77	228.68
[a]SNPBG05	Grande Bay, Argentina	88.46	11.54	0.00						92.96	261.4
[b]SJG	San Jorge gulf, Argentina		99				Trz				
[c]SNPPSJG	San José gulf, Argentina	89.20	7.60	3.20							
[c]SNPBSJG	San José gulf, Argentina	89.00	6.90	4.30							
[c]SNPLSJG	San José gulf, Argentina	75.80	20.80	3.40							
[c]SNPPNG	Nuevo gulf, Argentina	59.20	36.10	4.60							

(continued)

Table 1 (continued)

Code		Major toxins (mol %)								Toxicity	Toxin content
		GTX1/4	GTX2/3	C1/2	GTX5	GTX6	STX	neoSTX	dcSTX	(pg STX eq. cell^{-1})	(fmol Cell^{-1})
Cultured strains											
[a]PV1005	Valdés peninsula, Argentina	42.40	5.14	37.60			1.19	13.67		10.27	51.33
[d]MDQ1096	Mar del Plata, Argentina	25.7	1.5	69.3			0.2	3.3	0.1	9.18	50.50
[c]AM1	Mar del Plata, Argentina	25.6	2.1	57.8			0.1	11.7	0.1		
[a]at2c	Nuevo gulf, Argentina	40.28	2.32	57.40						3.00	16.30
[a]at2b	Nuevo gulf, Argentina	40.94	3.12	55.94						6.47	35.42
[a]at2b50	Nuevo gulf, Argentina	41.46	1.72	56.82						4.83	26.77
[a]atj2	Nuevo gulf, Argentina	7.45	18.22	72.83			1.5			2.19	17.46
[a]atab2	Nuevo gulf, Argentina	27.80	2.61	68.02				1.56		3.91	26.43
[a]atab1	Nuevo gulf, Argentina	26.63	2.15	71.2						1.81	13.30
[a]sja3	San José gulf, Argentina	19.9	19.35	58.61				2.14		3.43	20.12
[a]sj3a	San José gulf, Argentina	24.22	22.59	52.06			0.08	1.05		5.09	26.93

Code		Major toxins (mol %)								Toxicity	Toxin content
		GTX1/4	GTX2/3	C1/2	GTX5	GTX6	STX	neoSTX	dcSTX	(pg STX eq. $cell^{-1}$)	(fmol $Cell^{-1}$)
[a]sj1d	San José gulf, Argentina	20.02	4.54	69.66				5.81		2.51	17.19
[b]H5	San Jorge gulf, Argentina	56		41				3			74.0 (pg $cell^{-1}$)
[b]H7	San Jorge gulf, Argentina	55		41				3			60.0(pg $cell^{-1}$)
[f]Kci	Malvinas Island, Argentina			>90							
[g]ATUR 01A, 01B, 02	Maldonado Bay, Uruguay	2–46	3–24	48–67			1–4	1–11	0–1	10–26	
[h]2a, 2b, 2c, 2d, 2e, 2f, 2 g, 3a, 3c, 3d, 1a7, 1a10, C6	Patos lagoon, estuary, Brazil	7–55	0.3–29	30–85			0–4	1–24	0–1	7.07–65.92	
[i]ACC02, CHI	Region XI, Chile	24.00	3.00	57.00	13.00	1		2.00			
[d]CC08A	Region XI, Chile	5.1	8.0	76.1	8	0.9		2		4.65	50

References: [a]Montoya et al. (2010), [b]Krock et al. (2015), [c]Reyero et al. (1998), [d]Carreto et al. (2001), [e]Carreto et al. (1996), [f]Ho et al. (2003), [g]Méndez et al. (2001), [h]Persich et al. (2006), [i]Krock et al. (2007)

field populations, are only comparable to that showed by three different strains: one of them isolated from Brazil (Persich et al. 2006) and the other two from Argentina (Krock et al. 2015). Outside this scheme, a less toxic (45.5 fmol cell^{-1}) and morphologically distinctive autumn population containing GTX2/3 as the quasi unique (88.5%) toxin derivative, clustered separately (Montoya et al. 2010). Recently, autumn populations from San Jorge Gulf, whose toxins profiles consisted only of GTX2/3 were also recorded (Krock et al. 2015). It is interesting to note that whereas in culture strains the 11-hydroxysulfate toxins exist almost exclusively as β-epimers, some natural populations showed the net predominance of α-epimers, indicating that at least some degradation of 11-hydroxysulfate toxins occurs (Montoya et al. 2010).

Environmental parameters such as irradiance, temperature, salinity, or inorganic nutrients have been shown to affect toxin content and composition for several different *Alexandrium* strains. Therefore, a potential reason for differences in regional toxicity could be due to strain-specific responses to environmental conditions. Montoya et al. (2010) have shown that the variability in toxin content and composition of *A. tamarense* field populations from Argentine Sea were well correlated with in situ temperature and nitrate concentration. Whereas toxin cell content and GTX1/4 (mol %) increased following saturation functions, GTX2/3 (mol %) decrease exponentially with the increase of in situ nitrate concentration. The maximum value of PTS content obtained by Montoya et al. (2010) (261fmol cell^{-1}) was similar to that indicated by Persich et al. (2006) (199 fmol cell^{-1}), Varela et al. (2012) (295 fmol cell^{-1}), and Aguilera-Belmonte et al. (2011) (239 fmol cell^{-1}) for the *A. tamarense/catenella* strains from Argentina, Brazil, and Chile, respectively.

5 Paralytic Shellfish Toxicity (PST) in Bivalve Molluscs

All aquatic organisms are vulnerable to saxitoxin and other HAB poisons, but bivalve organisms (mussels, clams, oysters) are the primary transvector up to higher trophic levels. Being sessile, bivalves obtain nutrients via seawater filtration, accumulating toxins in their digestive organs and soft tissues without lethality to the organism. Therefore, as has been discussed from *A. tamarense/catenella* distribution, the majority of shellfish toxicity records are included in a continuous band ranging from the southern tip of the Beagle Channel to 34°S and between the coast and the 50 m isobaths (Carreto et al. 2007), which coincides with the area of influence of the Patagonian Current (Fig. 3). However, *A. tamarense* abundance and the bivalve toxicity records are not directly related, due to the variability observed in toxin content and composition of *A. tamarense* field populations (Montoya et al. 2010) and because bivalve species differ markedly in their ability to accumulate and eliminate these toxins.

Toxicity records (Fig. 3) indicate that in Region I (32–40°S), populations of intertidal bivalves (*Mesodesma mactroides, Donax hanleyanus, Brachidontes rodriguezii*) have been free from toxicity associated to *A. tamarense*, and only exception-

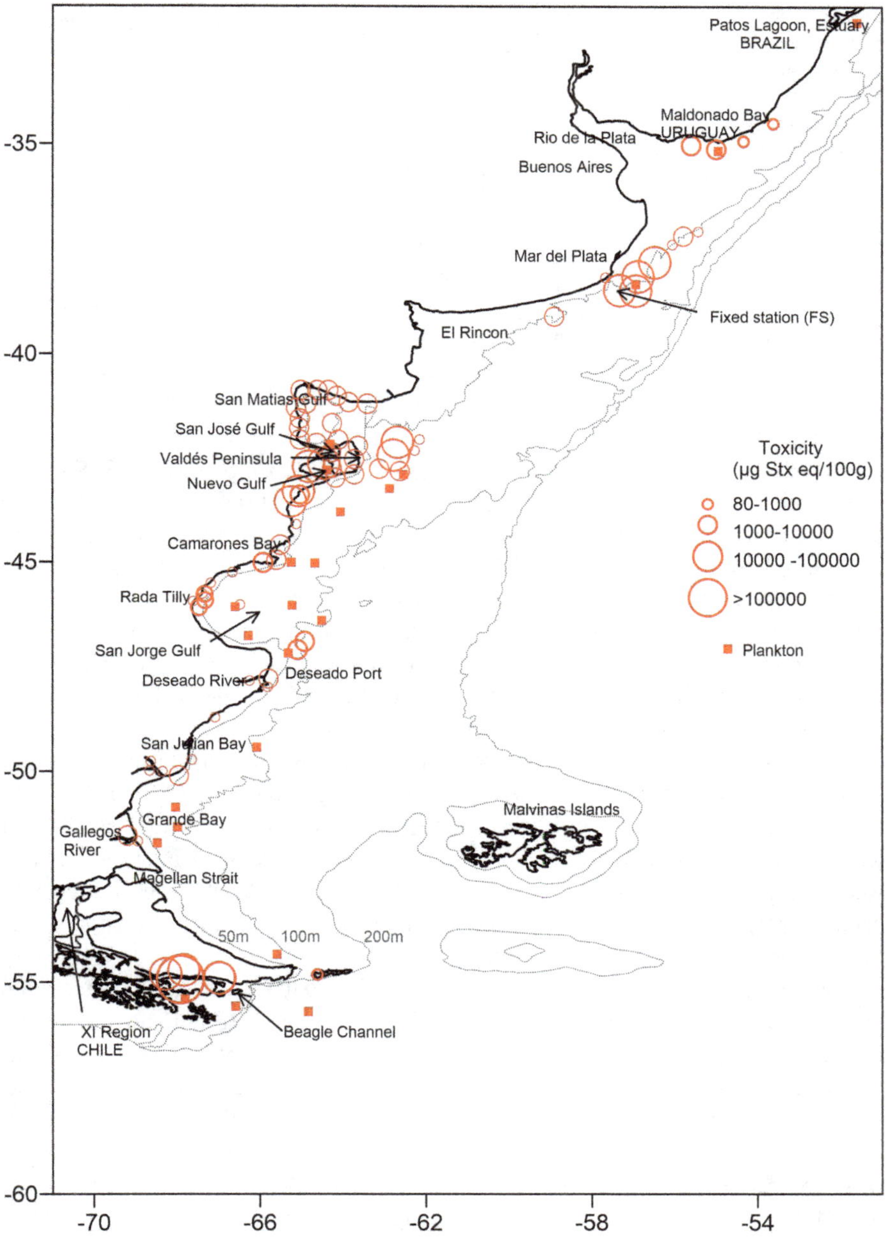

Fig. 3 Distribution area and recorded levels of paralytic shellfish toxins in bivalve molluscs from the Southwestern region of South America (O), showing their presence in plankton samples (■). (Modified from Carreto et al. 2007)

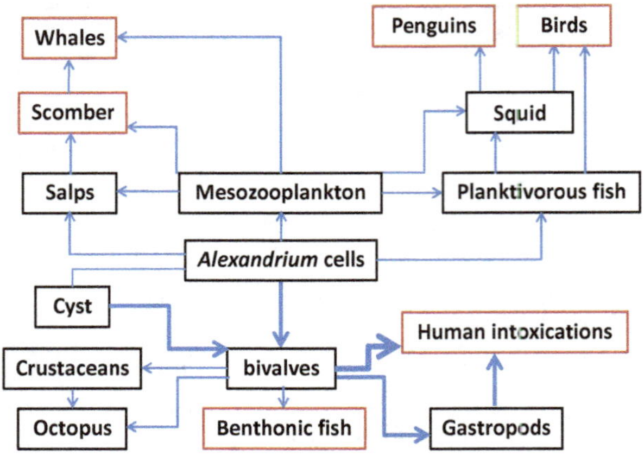

Fig. 4 Exposure and transference pathways of paralytic shellfish toxins through marine food webs. Their impact on humans and some marine organisms (fatal in *red*)

ally have recorded values higher to the security limit (Carreto et al. 2008; Sunesen et al. 2014). This contrasts with the high values (up to 30,000 μg STX eq. 100 g^{-1}) detected in the mussel banks (*Mytilus edulis*) located offshore and along the 50 m isobaths (Carreto et al. 1993). On the contrary, in Region II (40–48°S), both inshore and offshore bivalve populations have presented high toxicity values, with exceptional records (up to 50,000 μg STX eq. 100 g^{-1}) for mussel beds (*Mytilus edulis*) collected in the frontal region of Valdés Peninsula (Carreto et al. 1981). The northern region of the San Matías Gulf has presented an intermediate behavior as these coasts were free from toxicity until year 1990, when the toxicity outbreak was extraordinary (El Busto et al. 1992), not only in extension but also in severity, especially in the frontal region of Valdés Peninsula (up to 32,000 μg STX eq. 100 g^{-1}). This fact and the minor toxicity detected in populations of intertidal bivalves of Region I also suggest that *A. tamarense* populations developed in frontal waters were transported up to the coastal line. A similar transport mechanism has recently been suggested to explain toxicity phenomena recorded in Uruguayan shores (See Méndez and Carreto, in this volume).

Few toxicity records exist for the San Jorge Gulf and the more Austral Patagonian region, but they show that both coastal bivalve molluscs (Santinelli 2013) and frontal-related populations (Carreto et al. 1998a) have presented high toxicity values. Recently in a coastal locality of the San Jorge Gulf (Rada Tilly), a toxic outbreak reached high toxicity levels (9800 μg STX eq. 100 g^{-1}) in the mussel *Aulacomya ater* (Santinelli 2013). In the more Austral Patagonian region, toxicity was recorded in the mussels from the estuaries of the Deseado and Gallegos rivers (Fig. 3) and in the San Julian Bay. The highest global toxicity value (up to 127,000 μg STX eq. 100 g^{-1}) was recorded in the Beagle Channel (Ushuaia) during an intense bloom of *A. catenella* (Benavides et al. 1995). Since this exceptional outbreak, occurred during the summer of 1992, only moderate PSP toxin levels have been recorded in the mussels from the Beagle Channel (*Mytilus edulis chilensis* and *Aulacomya ater*). Unexpectedly, high PSP toxin levels (5600 μg STX eq.100 g^{-1})

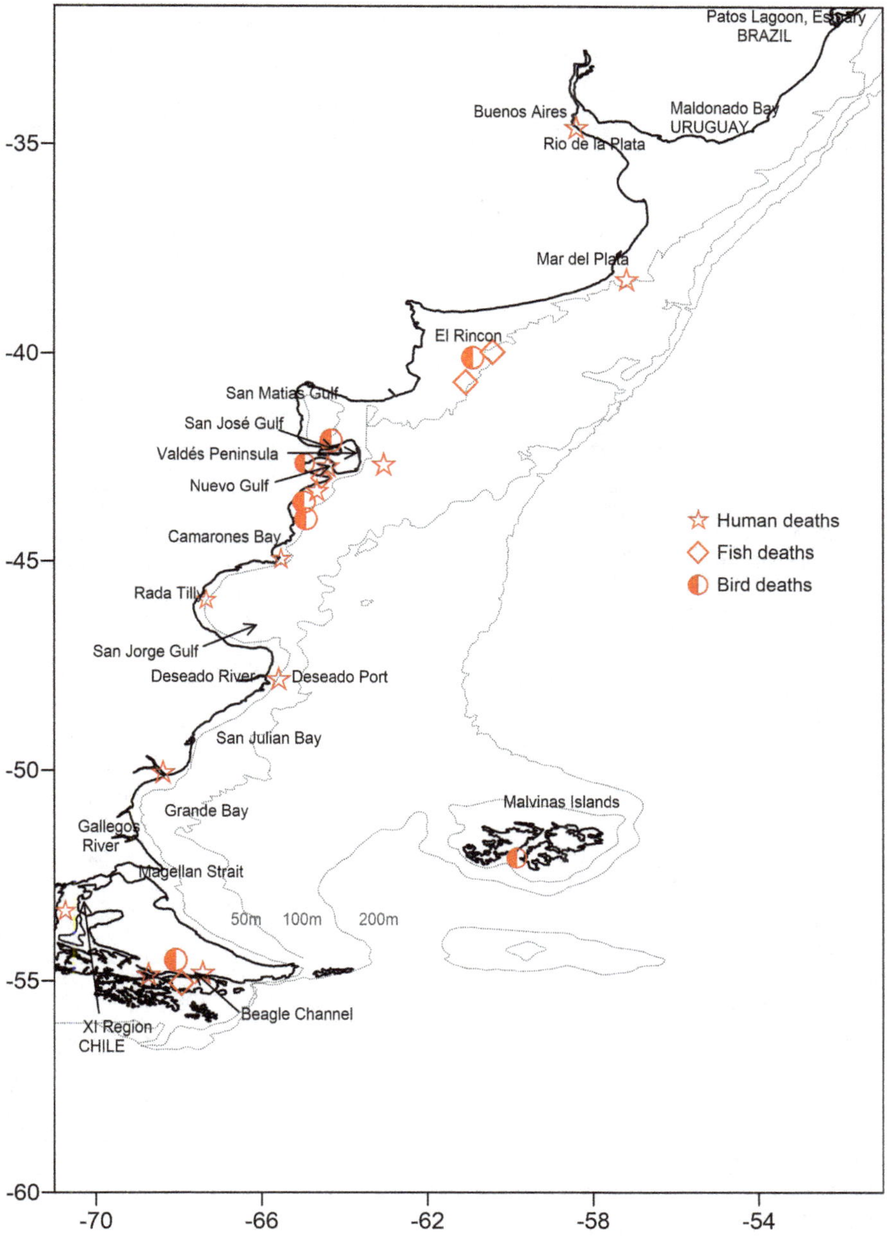

Fig. 5 Geographic distribution of fatal human intoxications (☆), fish mortality (◇), and seabird mortality (◐), reported in association with *A. tamarense/catenella* blooms in the Southwestern region of South America

Table 2 PSP composition reported in marine organisms

	Major toxins (mol %)						Toxicity	Toxin content
	GTX 1/4	GTX 2/3	C1/2	STX	neoSTX	dcSTX	(μg STX eq.100 g^{-1})	
[a]*Z. Angulata* Viscera	0.2	3.3		87.1	8.4	0.9		
Foot		0.6		95.2	3.6	0.6		
Mucus		0.8		99.3				
[a]*A. Brasiliana* Foot		0.7		97.9		1.2		
[b]*Mytilus edulis* Autumn whole	41	43	4	10	1			88.23 nmol g^{-1}
[b] Spring whole	85	10	3	1	1			98.7 nmol g^{-1}
[c]*Aulacomya ater*	67	29	4					
[c]*Aequipecten tehuelchus*	49.5	50.5						
[d]*Scomber japonicus* Stomach contents	67.7	24.5		7.8	Trace		2800	13,760 μg 100 g^{-1}
Liver		+		+	+		500	
Intestine							361	
Gills							72	
[e]*Engraulis anchoita* Viscera	73	29	Trace		Trace		101	
[f]*Eubalaena australis* Fecal sample								172–800 ng g^{-1}
[g]*Larus dominicus* Intestine	100							39 nmol g^{-1}
[g]*Sterna maxima* Intestine	100						26	37 nmol g^{-1}
Kidney	100				Traces		20	
[h]*Pygoscelis Papua* Intestine	+				+			
[h]*Spheniscus magellanicus* Intestine	+				+			
[i]mesozooplankton		100					43	
[i]*Pseudechinus magellanicus*		58.5		34.4	6.7		834.7	
[i]*Rochinia gracilipes*		100					2.4	
[i]*Peltarion spinosulum*		100					5.0	

(continued)

Table 2 (continued)

	Major toxins (mol %)						Toxicity	Toxin content
	GTX 1/4	GTX 2/3	C1/2	STX	neoSTX	dcSTX	(μg STX eq.100 g^{-1})	
[i]*Zygochlamys patagonica*		88.5		11.5			39.5	
[i]*Porifera indet.*		89.9		10.1			63.3	
[i]*Arbacia dufresnii*		100					9.6	
[i]*Cycethra verrucosa*		100					26.9	

References: [a]Carreto et al. (1996), [b]Carreto et al. (2004), [c]Reyero et al. (1998), [d]Montoya et al. (1996), [e]Montoya et al. (1998), [f]Wilson et al. (2016), [g]Montoya and Carreto (2007), [h]Uhart et al. (2004), [i]Montoya and Carignan (2011)

were recorded during summer of 2011, (Goya and Maldonado 2014) indicating a high interannual variability in the intensity of *A. catenella* blooms in this area.

6 Trophic Transference of PSP and Its Impacts

The impacts of PSP outbreaks in the Southwestern South America include human intoxications and death from contaminated shellfish, loss of seafood resources, and death of marine mammals, fish, and seabirds. However, the food web transfers (Fig. 4) from toxic phytoplankton to these organisms has not been fully elucidated.

In Argentina, the main cause of human poisonings and death by saxitoxin-contaminated shellfish was due to mussels' ingestion. Although probably not all cases have been reported, it is evident that in the period from 1980 to present, several human fatal intoxications have occurred along almost 5000 km of the coast, from the Beagle Channel to the Río de la Plata estuary (Fig. 5).

A bimodal PSP cycle was observed in a study conducted in FS. The mussel (*Mytilus edulis platensis*) toxin content in autumn was low in comparison with the spring values (88.23 nmol g^{-1} and 98.7 nmol g^{-1}, respectively) (Table 2) (Carreto et al. 2004). In both blooms the toxin profile was dominated by the gonyautoxins and in lower extension by C1/2, STX, and neo-STX (Carreto et al. 2004; Montoya et al. 2010). However, during spring, the toxic profile showed GTX1/4 as the main toxins whereas in autumn in agreement with the profile of *A. tamarense* associated to that event were GTX2/3 (Carreto et al. 2004; Montoya et al. 2010). Although the highest toxicity values have been detected in bivalve molluscs (*M. edulis platensis*, *Aulacomya ater*, *M. edulis chilensis*, *Zygochlamys patagonica*, *Aequipecten tehuelchus*), several species of gastropods of great commercial importance (e.g., *Zidona dufresnei*) also present toxicity, especially in their viscera (Carreto et al. 1996), and their consumption was implicated in several human intoxications, one of them fatal (Turner et al. 2014). In relation to the viscera, the foot, and the mucous secretion of *Z. dufresnei* exhibited lowest toxicity. Observed results suggest that these organisms obtain the toxins by consumption of toxic mussels and that several metabolic processes (interconversions, preferential retention, etc.) during trophic transfer, cause changes in their toxin composition. These changes (Table 2) take place in both the primary consumers (mussels), where the main PSP toxins are the gonyautoxins, and the secondary consumers (marine snails), where STX is the dominating toxin (Carreto et al. 1996; Reyero et al. 1998; Turner et al. 2014).

Although bivalve organisms can accumulate toxins without apparent lethality, during the exceptional outbreak occurred in the Beagle Channel in 1992 (Benavides et al. 1995), there were large and unusual mortalities of clams (*Eurhomalea exalbida*, *Ameghinomya antiqua*, *Mulinia edulis*, and *Tawera gayi*) and, in a lesser extent, mussels (*Mytilus edulis chilensis*). Among the sessile fauna was also observed the unusual death of large sponges (Vinuesa 1993).

Among the fish, the impact of an *A. tamarense* bloom (3.45×10^5 cells L^{-1}) was first recorded during a large outbreak of toxicity, when a high mortality of mackerel containing PSP toxins was detected in a frontal area of the Buenos Aires shelf (Fig. 5) (Montoya et al. 1996). Up to 2800, 500, 361, and 72 μg STX eq. 100 g^{-1} tissues were found in stomach contents, intestine, liver, and gills, respectively. Toxin profiles of the stomach content showed a large proportion of GTX1/4, whereas liver extracts showed that GTX2/3 were the main components, reflecting elimination and/or biotransformation of toxins during the accumulation processes (Montoya et al. 1996). In this event the gelatinous zooplankton was the principal vector organism of the *A. tamarense* toxins (Montoya et al. 1996). The anchovy (*Engraulis anchoita*) from the Buenos Aires coast also has presented toxicity in its viscera (101 μg STX eq. 100 g^{-1} wet tissue), but toxins did not accumulate to detectable levels in the muscle. The major toxin detected was GTX1 and in decreasing order GTX2 and GTX3, with neo STX and C1/4 only at trace levels (Montoya et al. 1998). Although the intoxication process is still unknown, the feeding behavior of the anchovy and the detection of PST (GTX2/3) in zooplankton samples of North Patagonia Gulfs (D'Agostino 2017) lead us to presume that the herbivorous zooplankton was the main vector of toxin transference. On the other hand, during a PSP outbreak at Valdés Peninsula (1980), the grouper (*Acanthistius brasilianus*) and sea salmon (*Pinguipes somnambula*) were subjected to a feeding experiment. The results of these tests showed that these fish were severely affected by the PST, resulting in death (See Carreto et al. 1981).

Large bird deaths of South American terns (*Sterna hirundinacea*), imperial cormorants (*Phalacrocorax albiventer*), great grebes (*Podiceps major*), the duck (*Tachyeres pteneres*), seagulls (*Larus dominicanus, Sterna maxima*), and penguins (*Spheniscus magellanicus*) have been several times observed along the Buenos Aires, Patagonian, and Beagle Channel coasts (Carreto et al. 1981; Vinuesa 1993; Uhart et al. 2004; Montoya and Carreto 2007). The toxin analysis of tissue samples from agonizing seagulls (*Larus dominicanus*) showed that GTX4 was present in all studied tissues (intestine, stomach, liver, and kidney) (Uhart et al. 2004; Montoya and Carreto 2007). Because the stomach of analyzed birds was empty, it was not possible to determine the vector of intoxication, but small filter-feeding fish such as anchovy and invertebrates like squid are an important part of the diet of many seabirds (Uhart et al. 2004).

Mortality of a large number of penguins (*Pygoscelis papua, Eudyptes chrysocome*, and *Spheniscus magellanicus*) has been also recorded in the Malvinas Islands. Between November 2002 and February 2003, many thousand penguins were poisoned (along with albatrosses, petrels, and prions) during a PSP outbreak event. Analysis of tissue and stomach samples from penguins and analysis of water samples showed that the toxins which caused these deaths (GTX4 and neoSXT) were originated by the dinoflagellates *A. tamarense* and *A. catenella* (Uhart et al. 2004) present in the plankton. Although the organism that transferred the toxins was unknown, cephalopods (octopods, cuttlefishes, and squids) have been reported to accumulate saxitoxin and its derivatives and, therefore, can act as PST vectors in marine food webs.

Since 2005, right whale (*E. australis*) mortality has increased at Valdés Peninsula, with most of the deaths (~90%) being calves <3 month old. Wilson et al. (2016) estimated that they may have been exposed to toxins in the uterus, while others were less than one month old at the time of death, suggesting that they may have been exposed to biotoxins through their mother's milk. Evidence of the whales' exposure to toxins includes trace levels of paralytic shellfish toxins (PSTs) in tissues of some dead whales (Wilson et al. 2016; D'Agostino 2017). Additionally, whales are present at Valdés Peninsula during both closures of the shellfish industry (due to high levels of PSTs) and periods with high concentrations of *A. tamarense* cells in the plankton (Wilson et al. 2016). Another study has been carried out on capacity to accumulate PSP in several organisms that inhabit the frontal region of Valdés Peninsula (Montoya and Carignan 2011). The only toxin found in the plankton was GTX2/3 (18–108 pmol L^{-1}), whereas STX were detected at trace levels. This result agrees with the low concentrations of *A. tamarense* cells found in this study which indicated that the toxic bloom was in the final stage of development. The mesozooplankton reached 43 μg STX eq. 100 g^{-1} wet tissue, and the toxin profile was similar to that of phytoplankton, with high percentages of toxins GTX3/2 (~100%) and traces of neoSTX and STX. In an indeterminate species of *Porifera* a similar result was found, with a toxin concentration of 66.3 μg STX eq. 100 g^{-1} tissue. The highest levels of toxicity were observed in the noncommercial sea urchin *Pseudechinus magellanicus*, which accumulated 834 μg STX eq. 100 g^{-1} in its soft tissue, and a higher percentage of STX (58.5% GTX2/3, 34.6% STX, and 6.7% neoSTX). The rest of the studied organisms, the crustaceans' decapods *Rochinia gracilipes, Peltarion spinulosum*, and the sea star *Cycethra verrucosa* presented low toxicity (2.5–26 μg STX eq./100 g) and a similar toxic profile. The mechanisms of accumulation/detoxification are not known, but it is likely that the difference in the toxic level found could be due to the different physiology of the organisms studied.

7 Conclusions

Paralytic shellfish poisoning (PSP), due to STX and related compounds, typically results from the consumption of filter-feeding molluscan that concentrates toxins from marine dinoflagellates. STXs are transferred and bioaccumulate throughout marine food webs and can be vectored to biota, including humans. The impacts of PSP outbreaks of *A. tamarense/catenella* in the Southwestern Atlantic include human intoxications and death from contaminated shellfish, loss of seafood resources, and death of marine mammals, fish, and seabirds. A high diversity in toxin content and composition of Brazilian, Uruguayan, and Argentine cultured *A. tamarense/catenella* strains was observed. They clustered together, but in different non-geographic subgroups, indicating the existence of a diverse continuous population, while the field populations are more closely related to one another than to the cultured strains.

References

Aguilera-Belmonte A, Inostroza I, Franco JM et al (2011) The growth, toxicity and genetic characterization of seven strains of Alexandrium catenella (Whedon and Kofoid) Balech 1985 (Dinophyceae) isolated during the 2009 summer outbreak in southern Chile. Harmful Algae 12:105–112

Akselman R (1996) Estudios ecológicos en el Golfo San Jorge y adyacencias (Atlántico Sudoccidental) Distribución, abundancia y variación estacional del fitoplancton en relación a factores físico-químicos y la dinámica hidrológica (Tesis Doctoral) Facultad de Ciencias Exactas y Naturales, Universidad de Buenos Aires, 234 pp

Almandoz GO, Hernando MP, Ferreyra GA et al (2011) Seasonal phytoplankton dynamics in extreme southern South America (Beagle Channel, Argentina). J Sea Res 66(2):47–57

Almandoz GO, Montoya N, Hernando MP et al (2014) Toxic strains of the *Alexandrium ostenfeldii* complex in South America (Beagle Channel, Argentina). Harmful Algae 37:100–109

Alpermann TJ, Tillmann U, Beszteri B et al (2010) Phenotypic variation and genotypic diversity in a planktonic population of the toxigenic marine dinoflagellate *Alexandrium tamarense* (Dinophyceae). J Phycol 46:18–32

Anderson DM, Kulis DM, Doucette GJ et al (1994) Biogeography of toxic dinoflagellates in the genus *Alexandrium* from the Northeast United States and Canada. Mar Biol 120:467–478

Anderson DM, Cembella AD, Hallegraeff GM (2012) Progress in understanding harmful algal blooms: paradigm shifts and new Technologies for Research, monitoring, and management. Annu Rev Mar Sci 4:143–176

Balech E (1995) The genus *Alexandrium Halim (Dinoflagellata)*. Marine Station, Sherkin Island Co Cork, Sherkin Island

Benavides HR, Negri RM, Carreto J (1983) Investigaciones sobre el ciclo de vida del dinoflagelado tóxico *Gonylaulax excavata* (Braarud) Balech (Dinophycea). Physis Secc A 41:135–142

Benavides H, Prado S, Diaz S et al (1995) An exceptional bloom of *Alexandrium catenella* in the Beagle Channel, Argentina. In: P Lassus, G Arzul, E Erard, P Gentien, C Marcaillou (eds) Harmful marine algal blooms. Proceedings of the Sixth International Conference on Toxic Marine Phytoplankton, October 1993 Nantes, France, pp 113–119

Carreto JI, Casal AB, Hinojal A et al (1974) Fitoplancton, pigmentos y condicones ecologicas del Golfo San Matias. Inform N°10 CIC, Buenos Aires, pp 76

Carreto JI, Lasta M, Negri RM et al (1981) Los fenómenos de marea roja y toxicidad de moluscos bivalvos en el Mar Argentino. Contrib Inst Nac Invest Des Pesq 399, Mar del Plata, Argentina, 55 pp

Carreto JI, Negri RM, Benavides HR et al (1985) Toxic dinoflagellates blooms in the Argentine sea. In: Anderson DM, White AW, Baden DG (eds) Toxic Dinoflagellates. Elsevier, New York, pp 147–152

Carreto JI, Benavides HR, Negri RM et al (1986) Toxic red–tide in the argentine sea. phytoplankton distribution and survival of the toxic dinoflagellate *Gonyaulax excavata* in a frontal area. J Plankton Res 8:15–28

Carreto JL, Elbusto C, Sancho B et al (1993) An exploratory analysis of the Mar del Plata shellfish toxicity area (1980–1990) In: Smayda TJ, Shimizu S (eds) Toxic phytoplankton blooms in the sea. Proceedings of the Fifth International Conference on Toxic Marine Phytoplankton, Newport, Rhode Island, USA, Elsevier Amsterdam, pp 229–233

Carreto JI, Elbusto C, Sancho H et al (1996) Comparative studies on paralytic shellfish toxin profiles of marine snails, mussels and an *Alexandrium tamarense* isolate from the mar del Plata coast (Argentina). Rev Invest Des Pesq 10:101–107

Carreto JI, Akselman R, Montoya NG et al (1998a) *Alexandrium tamarense* blooms and shellfish toxicity in the Argentine Sea: a retrospective view. In: Reguera B, Blanco J, Fernández ML, Wyatt T (eds) Harmful microalgae IOC–UNESCO, pp 131–134

Carreto JI, Montoya, NG, Cucchi Colleoni AD et al (1998b) *Alexandrium tamarense* bloom dynamics and *Mytilus edulis* toxicity in the coastal waters off Mar del Plata (Argentina). In: Reguera B, Blanco J, Fernández ML, Wyatt T (eds) Harmful microalgae IOC–UNESCO, pp 135–138

Carreto JI, Carignan MO, Montoya NG (2001) Comparative studies on mycosporine–like amino acids, paralytic shellfish toxins and pigment profiles of the toxic dinoflagellates *Alexandrium tamarense, Alexandrium catenella* and *Alexandrium minutum*. Mar Ecol Prog Ser 223:49–60

Carreto JI, Montoya NG, Akselman R et al (2004) Differences in the PSP toxin profiles of *Mytilus edulis* during spring and autumn blooms of *Alexandrium tamarense* off Mar del Plata coast. In: Steidinger KA Landsberg J HC, Tomas R, Vargo GA (eds) Harmful Algae 2002, Florida Fish and Wildlife Conservation Commission Florida Institute of Oceanography and Intergovernmental Oceanographic Commission of UNESCO, pp 100–102

Carreto JI, Carignan MO, Montoya NG (2007) Florecimientos de algas nocivas Atlas de Sensibilidad Ambiental del Mar y la Costa Ed: D Boltovskoy, Secretaría de Ambiente y Desarrollo Sustentable. Jefatura de Gabinete de Ministros, Argentina. CDRom

Córdova JL, Müller I (2002) Use of PCR and partial sequencing of the large–subunit rRNA gene to identify *Alexandrium catenella* (Dinophyceae) from the south of Chile. Harmful Algae 1:343–350

D'Agostino VC (2017) Dinámica planctónica y transferencia de biotoxinas marinas a la Ballena Franca Austral (*Eubalaena australis*) a través de su alimentación en los golfos Nuevo y San José, Península Valdés, Chubut. Doctoral thesis. Universidad Nacional del Sur, Argentina

Dale B (1983) Dinoflagellate resting cysts: "benthic plankton". In: Fryxell GA (ed) Survival strategies of the algae. Univ Press, Cambridge, pp 69–136

El Busto C, Carreto JI, Benavides HR et al (1992) Paralytic shellfish toxicity in the Argentine sea 1990: an extraordinary year. In: Smayda J, Shimizu Y (eds) Toxic Phytoplankton Blooms in the Sea. Proceedings of the Fith International Conference on Toxic Marine Phytoplankton, Newport, Rhode Island, USA Elsevier, Amsterdam, p 952

Esteves JL, Santinelli N, Sastre V (1992) A toxic dinoflagellate bloom and PSP production associated with upwelling in Golfo Nuevo, Patagonia, Argentina. Hydrobiologia 242:115–122

Etheridge SM (2010) Paralytic shellfish poisoning: sea food safety and human health perspectives. Toxicon 56:108–122

Fabro E, Almandoz GO, Ferrario ME et al (2017) Alexandrium species from the Argentine sea: diversity, distribution and associated toxins. J Phycol. https://doi.org/10.1111/jpy.12574

Fraga S, Sampedro N, Larsen J et al (2015) Arguments against the proposal 2302 by John U al to reject the name *Gonyaulax catenella* (*Alexandrium catenella*). Taxon 64(3):634–635

Fukuyo Y (1985) Morphology of *Protogonyaulax tamarensis* (Lebour) Taylor and *Protogonyaulax catenella* (Whedon and Kofoid) Taylor from Japanese coastal waters. Bull Mar Sci 37:529–537
Gayoso AM (2001) Observations on *Alexandrium tamarense* (Lebour) Balech and other dinoflagellate populations in Golfo Nuevo, Patagonia (Argentina). J Plankton Res 23:463–468
Gayoso AM, Fulco K (2006) Occurrence patterns of *Alexandrium tamarense* (Lebour) Balech populations in the Golfo Nuevo (Patagonia, Argentina) with observations on ventral pore occurrence in natural and cultured cells. Harmful Algae 5:233–241
GEOHAB (2001) Global Ecology and Oceanography of Harmful Algal Blooms. In: P Glibert and G Pitcher (eds) Science Plan SCOR and IOC, Baltimore and Paris, p 87
Goya A, Maldonado S (2014) Evolution of PSP Toxicity in Shellfish from the Beagle Channel (Tierra del Fuego, Argentina): An Overview. Chapter II, Jan 2014, Molluscan Shellfish Safety
Guzmán L, Campodonico I, Antunovic M (1975) Estudios sobre un florecimiento tóxico causado por *Gonyaulax catenella* en Magallanes IV Distribución y niveles de toxicidad del veneno paralítico de los mariscos. Anales Instituo Patagonia, Chile 6:229–233
Guzmán L, Pacheco H, Pizarro G, Alarcón C (2002) *Alexandrium catenella* y veneno paralizante de los mariscos en Chile. In: Sar EA, Ferrario NE, Reguera B (eds) Floraciones algales nocivas en el cono sur americano. Instituto Español de Oceanografía Vigo, Pontevedra, pp 235–255
Hattenrath-Lehmann TK, Ossiboff RJ, Craig A et al (2017) The role of a PSP–producing *Alexandrium* bloom in an unprecedented diamond back terrapin (*Malaclemys terrapin*) mortality event in Flanders Bay, New York, USA. Toxicon 129:36–43
Ho KC, Kang SH, Lam IY et al (2003) Distribution of *Alexandrium tamarense* in drake passage and the threat of harmful algal blooms in the Antarctic Ocean. Ocean and Polar Res 25(4):625–631
John U, Litaker RW, Montresor M et al (2014) Formal revision of the *Alexandrium tamarense* species complex (Dinophyceae) taxonomy: the introduction of five species with emphasis on molecular–based (rDNA) classification. Protist 165:779–804
Krock B, Seguel CG, Cembella AD (2007) Toxin profile of *Alexandrium catenella* from the Chilean coast as determined by liquid chromatography with fluorescence detection and liquid chromatography coupled with tandem mass spectrometry. Harmful Algae 6:734–744
Krock B, Borel M, Barrera F et al (2015) Analysis of the hydrographic conditions and cyst beds in the San Jorge Gulf, Argentina, that favor dinoflagellate population development including toxigenic species and their toxins. J Mar Syst 148:86–100
Landsberg JH, Lefebvre KA, Flewelling A (2014) Effects of toxic microalgae on marine organisms. In: Rossini GP (ed) Toxins and biologically active compounds from microalgae, vol 2. Biological Effects and Risk Management CRC Press, Boca Raton, pp 379–449
Lembeye G (2004) Distribución de quistes de *Alexandrium catenella* y otros dinoflagelados en sedimentos de la zona sur–austral de Chile. Cienc Tecnol Mar 27(2):21–31
Lilly EL, Halanych KM, Anderson DM (2007) Specie boundaries and global biogeography of the *Alexandrium tamarense* complex (Dinophyceae). J Phycol 43:1329–1338
MacKenzie L, de Salas M, Adamson J et al (2004) The dinoflagellate genus *Alexandrium* (Halim) in New Zealand coastal waters: comparative morphology, toxicity and molecular genetics. Harmful Algae 3:71–92
Mardones JI, Bolch C, Guzman L et al (2016) Role of resting cysts in Chilean Alexandrium catenella dinoflagellate blooms revisited. Harmful Algae 55:238–249
Méndez S, Kulis D, Anderson DM (2001) PSP toxin production of Uruguayan isolates of *Gymnodinium catenatum* and *Alexandrium tamarense*. In: Hallegraeff G, Blackburn S, Lewis R, Bolch C (eds.) Harmful Algal Blooms. IOC–UNESCO, pp. 352–355
Menezes M, Varela DM, Proença LAO et al (2008) Characterization of *Alexandrium tamiyavanichi* (Dinophyceae) in tropical Brazilian coast. In: The 13th International Conference on Harmful Algae, Hong Kong, 3–7 November 2008
Montoya NG, Carignan M (2011) Análisis de toxinas paralizantes de moluscos (TPM) de algunos organismos de la comunidad biológica asociada al frente de Península de Valdés durante la campaña CC01/08. Inf. Invest. N 34 INIDEP, Mar del Plata Argentina, p 10

Montoya NG, Carreto JI (2007) Informe sobre mortandad de aves marinas ocurrida en las costas de Chubut (noviembre de 2006) asociada a la presencia de toxinas paralizantes de moluscos. Inf Técn INIDEP N° 34/2007, INIDEP, Mar del Plata, Argentina

Montoya NG, Akselman R, Franco J et al (1996) Paralytic shellfish toxins and mackerel (Scomber Japonicus) mortality in the Argentine sea. In: Yasumoto T, Y Oshima, Fukuyo Y (eds) Harmful and Toxic Algal Blooms Intergovernmental Oceanographic Commission (IOC) of UNESCO, pp 417–420

Montoya NG, Reyero MI, Akselman R et al (1998) Paralytic shellfish toxins in the anchovy *Engraulis anchoita* from Argentinian coasts. In: Reguera B, Blanco ML, Fernández J, Wyatt T (eds) Harmful microalgae IOC–UNESCO, pp 72–73

Montoya NG, Fulco KV, Carignan MO et al (2010) Toxin variability in cultured and natural populations of *Alexandrium tamarense* from southern South America – evidences of diversity and environmental regulation. Toxicon 56:1408–1418

Mouzo F, Garza ML, Izquierdo JF et al (1978) Rasgos de la Geología del Golfo Nuevo (Chubut). Acta Oceanog Argentina 2:69–91

Odebrecht C, Mendez S, Garcia VMT (1997) Oceanographic processes and harmful algae blooms in the southwestern Atlantic. In: VIII International Conference on Harmful Algae, Vigo, Abstract

Orlova TY, Selina MS, Lilly EL et al (2007) Morphogenetic and toxin composition variability of *Alexandrium tamarense* (Dinophyceae) from the east coast of Russia. Phycologia 46:534–548

Orozco FE, Carreto JI (1989) Distribution of *Alexandrium excavatum* resting cysts in a patagonic shelf area (Argentina). In: Okumura Y, Anderson DM, Nemoto T (eds) Red tides: biology, environmental sciences and toxicology. Elsevier, London, pp 309–312

Penna A, Fraga S, Maso M, Giacobbe MG, Bravo I, Garces E, Vila M, Bertozzini E, Andreoni F, Luglie A, Vernes C (2008) Phylogenetic relationships among the Mediterranean *Alexandrium* (Dinophyceae) species based on sequences of 5.8S gene and internal transcript spacers of the rRNA operon. Eur J Phycol 43:163–178

Pérez LB, Santinelli N, Sastre V et al (2012) Morphology and seasonal presence of *Alexandrium Tamarense* (Lebour) Balech in coastal stations of San Jorge Gulf (Patagonia, Argentina). Nat Patag 6:50–60

Persich GR, Kulisb DM, Lilly EL et al (2006) Probable origin and toxin profile of *Alexandrium tamarense* (Lebour) Balech from southern Brazil. Harmful Algae 5:36–44

Pizarro G, Garrido C, Zamora C et al (2015) Distribución de *Alexandrium catenella* y de toxinas paralizantes en el fitoplancton y mariscos entre el Estrecho de Magallanes y Tierra del Fuego (primavera 2010). Cienc Tecnol Mar 36, años 2013–2015

Prud'homme VR, Willem F (2017) Report of the nomenclature Committee for Algae: 15. Taxon 66:192–192

Reyero MN, Santinelli N, Otaño S et al (1998) Toxin profiles of PSP from molluscs and phytoplankton containing *Alexandrium tamarense* (Lebour) Balech in two patagonian gulfs (Argentina). In: Reguera B, Blanco J, Fernández ML, Wyatt T (eds.) Harmful Microalgae, IOC–UNESCO, pp. 507–508

Rivas AL, Beier EJ (1990) Temperature and salinity fields in the north patagonic gulf. Oceanog Acta 13:15–20

Sako Y, Kim H, Ninomiya H et al (1990) Isozyme and cross analysis of mating populations in the *Alexandrium catenella/tamarense* species complex. In: Graneli E, Sunderstrom E, Edler L, Anderson DM (eds) Toxic marine phytoplankton. Elsevier, New York, pp 320–323

Santinelli N (2013) Monitoreo de FAN en la costa de la Provincia de Chubut http://wwwagroindustriagobar/sitio/areas/acuicultura/zonificacion. Accessed 15 Oct 2013

Santinelli N, Sastre V, Esteves JL (2002) Episodios de Algas Nocivas en La Patagonia Argentina. In: Sar EA, Ferrario NE, Reguera B (eds) Floraciones algales nocivas en el cono sur americano. Instituto Español de Oceanografía Vigo, Spain, pp 199–208

Sastre V, Santinelli N, Esteves JL (1997) Fitoplancton del Golfo San Matias de tres campañas de muestreo (noviembre 92, septiembre 93 y marzo 94). Physis A 53:124–125

Segers PA (1908) Primera observación de una causa nueva de enfermedad del hígado causando una hipertrofia y cirrosis atrófica consecutivas por excesividad funcional, debido a absorción de toxinas y primera observación de esplenomegalia concomitante con hipertrofia de bazo en estas afecciones. 1° Semana Médica (Buenos Aires) 20:117–119

Sunesen I, Lavigne A, Goya A et al (2014) Episodios de toxicidad en moluscos de aguas marinas costeras de la Provincia de Buenos Aires (Argentina) asociados a algas toxígenas (marzo de 2008–marzo de 2013) Bol Soc Argent Bot 49: 327–339

Turner AD, Tarnovius S, Goya A (2014) Paralytic shellfish toxins in the marine gastropods *Zidona dufresnei* and *Adelomelon beckii* from Argentina: toxicity and toxin profiles. J Shellfish Res 33(2):519–530

Uhart M, Karesh W, Cook R et al (2004) Paralytic shellfish poisoning in gentoo penguins (*Pygoscelis papua*) from the Falkland (Malvinas) Islands. Proceedings AAZV/AAWV/WDA Joint Conference San Diego, California, pp 481–486

Varela D, Paredes J, Alves–de–Souza C et al (2012) Intraregional variation among *Alexandrium catenella* (Dinophyceae) strains from southern Chile: morphological, toxicological and genetic diversity. Harmful Algae 15: 8–18

Vinuesa J (1993) Marea Roja en el Canal de Beagle. Centro Austral de Investigaciones Científicas Contribución N° 16 Tierra del Fuego, Argentina pp 32

Wiese M, D'Agostino PM, Mihali TK et al (2010) Neurotoxic alkaloids: saxitoxin and its analogs. Mar Drugs 8:2185–2211

Williams GN, Dogliotti AI Zaidman P et al (2013) Assessment of remotely–sensed sea–surface temperature and chlorophyll–a concentration in San Matias Gulf (Patagonia, Argentina). Cont Shelf Res 52:159–171

Wilson C, Sastre AV, Hoffmeyer M et al (2016) Southern right whale (*Eubalaena australis*) calf mortality at Península Valdés, Argentina: are harmful algal blooms to blame? Mar Mam Sci 32:423–451

Harmful Algal Blooms in the Río de la Plata Region

Silvia M. Méndez and José I. Carreto

Abstract The Río de la Plata oceanic front has recurrent blooms of species that cause harm to humans and marine ecosystems, especially toxic species that produce paralytic shellfish toxins (PSTs), amnesic shellfish toxins (AST), diarrhetic shellfish toxins (DST), yessotoxins (YTXs), azaspiracid toxins (AZAs), and cyanotoxins (CTXs). Up to now, there has been a growth in the number of reports of human health and economic impacts caused by the increase in intensity and geographic spread of blooms of toxic species like *Alexandrium tamarense/catenella*, *Gymnodinium catenatum*, and *Dinophysis acuminata* complex. This review seeks to provide a broad overview of the harmful species recorded for this region. Another key issue is the relationship between the observed spread of *A. tamarense/catenella* and *G. catenatum* blooms and the oceanographic patterns that appear to control the occurrence, distribution, and toxicity of these toxic dinoflagellates along the Río de la Plata oceanic front.

Keywords Harmful algae · Río de la Plata · Toxic phytoplankton

1 Introduction

Among the almost 4000 marine phytoplankton species at the base of the marine food chain, a small percentage can cause harm to humans and the marine ecosystem (Sournia 1995). These species show their presence in many ways, ranging from massive blooms of cells that discolor the water to low concentrations of cells detected only from the negative effects of their highly potent toxins. Toxins from blooms of some phytoplankton species can result in high concentrations of those toxins. Fortunately, the capacity to produce toxins is limited to approximately 80

S. M. Méndez (✉)
Dirección Nacional de Recursos Acuáticos (DINARA), Montevideo, Uruguay
e-mail: smendez@dinara.gub.uy

J. I. Carreto
Instituto Nacional de Investigación y Desarrollo Pesquero (INIDEP), Buenos Aires, Argentina
e-mail: jcarreto@inidep.edu.ar

M. S. Hoffmeyer et al. (eds.), *Plankton Ecology of the Southwestern Atlantic*,
https://doi.org/10.1007/978-3-319-77869-3_21

phytoplankton species (Sournia 1995). Bloom events are commonly called red tides but more formally termed "harmful algal blooms" (HABs). The impacts of these phenomena include mass mortalities of wild and farmed fish and shellfish, human intoxication and death, alterations of a marine ecosystem's structure through adverse effects on larvae and other life history stages of commercial fisheries species, and deaths of marine mammals, seabirds, and other animals (GEOHAB 2001). Of major concern are the marine and brackish water HAB species that impact public health due to the production of toxins. Human intoxication occurs from consumption of shellfish that have filtered toxic phytoplankton from the water as food, causing bio-accumulation and thus raising the algal toxins to levels which can be lethal.

The Río de la Plata maritime front has recurrent blooms of some HAB species, especially toxic species that result in paralytic shellfish poisoning (PSP) and diarrhetic shellfish poisoning (DSP) outbreaks. In this review, using historical data of plankton from references, harmful algae monitoring programs, and toxicity records, we present an analysis of the patterns of some recurrent toxic algae blooms in this region. These results are correlated with oceanographic patterns to elucidate the factors controlling occurrence and distribution of toxic species within the Río de la Plata region.

2 Toxic Microalgae in the Region: Taxonomy, Ecology, and Toxicology

2.1 Dinoflagellates Producing Paralytic Shellfish Toxins (PSTs)

PSP is a syndrome caused by ingesting shellfish and other organisms that contain high levels of the potent neurotoxins called paralytic shellfish toxin (PSTs), which are produced by several species of dinoflagellates.

In the Río de la Plata maritime front, PSP have been caused by blooms of the toxic dinoflagellates *Alexandrium tamarense/catenella/catenella* and *Gymnodinium catenatum* (Carreto et al. 1996; Méndez and Medina 2004). It was suggested that *Alexandrium fraterculus* from northern Argentina and Uruguay could produce PST. However, the results of Proença and Resgalla (2004), in concordance with previous studies from other regions, indicated that *A. fraterculus* is not a PST producer.

2.1.1 Occurrence and Distribution of PST Producers

The presence of *G. catenatum* in the region was first reported at the Argentinian coastal waters in front of Mar del Plata City in 1964 (Balech 1964). However, the first toxic episode associated with this species in the Southwestern Atlantic Ocean

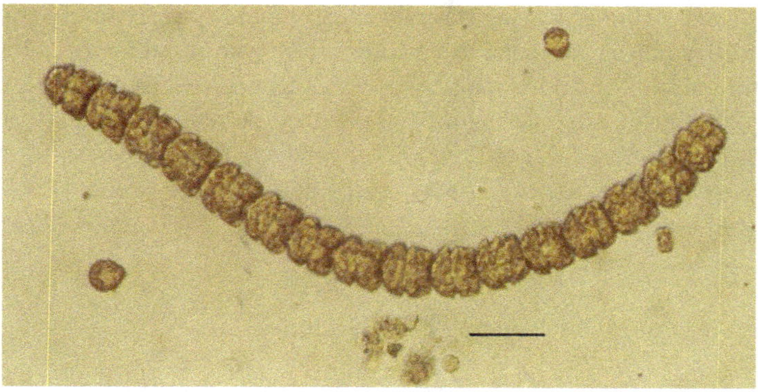

Fig. 1 *Gymnodinium catenatum* a live five-cell chain from Punta del Este March 2011 from a toxic PST bloom, microphotography at LM (S. Méndez/DINARA). Scale bar = 50 μm

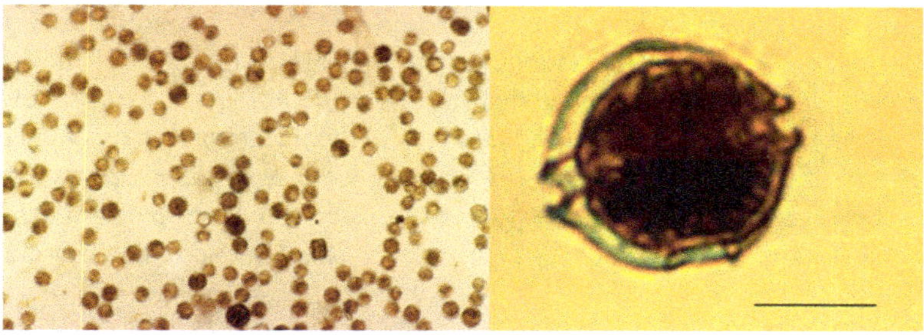

Fig. 2 *Alexandrium tamarense/catenella* bloom microphotography at LM, two-cell chain fixed with Lugol acid from a PST toxin bloom in Uruguay (S. Méndez/DINARA). Scale bar = 20 μm

(Fig. 1) was reported in 1992 along the Uruguayan coast, when the presence of *G. catenatum* was associated with a PSP in mussels (Méndez 1993). Later, the occurrence of *G. catenatum* was also documented in the northern coastal waters of southern Brazil (Proença et al. 2001), although with less frequent and intense blooms than those observed along the Uruguayan coast (Méndez and Ferrari 2003, Méndez and Gallit 2008). Since 1992, periods with PSP toxins in mussels associated with *G. catenatum* blooms were recorded in Uruguay in the warm seasons of 1993, 1994, 1995, 2001, 2002, 2004, 2005, 2010, 2013, and 2014.

The first documentation of *Alexandrium tamarense/catenella/catenella* in South America was related with a PSP outbreak detected in 1980 in the tidal front of the Valdes Peninsula, Argentina (Carreto et al. 1981). During this event, fatal intoxication of two fisherman occurred due to *Mytilus edulis* mussel consumption with extremely high toxicity levels (173.000 UR/l00gr of shellfish meat) (Carreto et al. 1981). In the same year, during summer, organisms along the coast of Uruguay were affected by paralytic shellfish toxins. That incident was the first toxic event in Uruguay, which lasted 10 months and reached 1248 μg STX eq/100 g (Davison and Medina 1982), when the associated species was *Gonyaulax* sp. Since then, the toxic area expanded to the north, periodically affecting the Argentinian (Carreto et al. 1998) and Uruguayan (Méndez and Ferrari 2002) coastal region of the Río de la Plata maritime front and occasionally up to southern Brazil (Odebrecht et al. 1997) (Fig. 2). Experimental data on cultured strains from Argentina, Uruguay, and Brazil have showed identical D1–D2 LSU rRNA gene sequences, indicating that these strains of *A. tamarense/catenella* may represent a continuous population (Persich et al. 2006).

2.1.2 Ecology of PSP-Producing Dinoflagellates

Alexandrium Tamarense/Catenella On the Atlantic coast of South America, the highest *A. tamarense/catenella* concentrations (1.6×10^6 cell L^{-1}) were associated with the tidal front of the Valdés Península (42°S) (Carreto et al. 1986). Currents link the Patagonian tidal fronts with northern areas, but alongshore transport of vegetative cells cannot explain the synchronicity in bloom initiation observed over such a wide area (35–45°S). In contrast, several studies (Carreto et al. 1998; Brazeiro et al. 1997) would support the idea that in the Río de la Plata maritime front, offshore growth of *A. tamarense/catenella* begins in the frontal region associated with the Subantarctic Shelf Water (SSW). Benthic *A. tamarense/catenella* cyst concentrations observed along Uruguayan (Méndez 2006) and Brazilian (Persich et al. 2006) shores are substantially lower than concentrations found in the Mar del Plata frontal area (Carreto et al. 1998).

In this area, the first planktonic forms are detected in late winter, before the beginning of the spring algal bloom dominated by diatoms (Carreto et al. 2004). The observed cyclical trend agrees with the hypothesis that *A. tamarense/catenella* blooms are initiated from germination of benthic cysts, and that this process is controlled by an endogenous biological mechanism synchronized by certain environmental factors. *A. tamarense/catenella* blooms are subject to environmental disturbances; maximum levels of toxicity are the result of prolonged periods of sustained growth (Carreto et al. 1996). Solar radiation and wind velocity prevailing during spring may regulate the phytoplankton composition of the bloom; *A. tamarense/catenella* growth is favored by calm periods and high solar radiation (Carreto et al. 1993).

The occasional and relatively minor toxicity episodes detected in the yellow clam (*Mesodesma mactroides*) and in the cockles (*Donax hanleyanus*) from the northern sandy shores of Argentina suggest that *A. tamarense/catenella* populations already established in the SSW were eventually advected onshore. A similar but more complex mechanism has been suggested to explain toxicity phenomena recorded on the Uruguayan coast (Méndez et al. 1996; Carreto et al. 1998). During winter, offshore winds and high rainfall predominate in the Río de la Plata front, resulting in an important northward drift of low-salinity water along the Uruguayan coast (Piola et al. 2005) that limited the onshore transport of *A. tamarense/catenella* populations. However, during anomalous periods (Acha et al. 2008) of strong

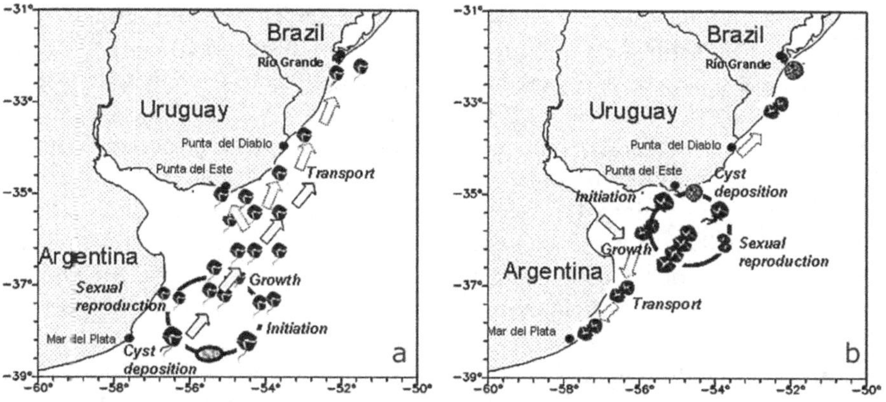

Fig. 3 Hypothesis of mechanisms of in situ growth and transport of established populations of (**a**) *A. tamarense/catenella* during the spring and (**b**) *G. catenatum* during the summer-autumn in the Río de La Plata Maritime front

southwesterly winds and low river discharge, *A. tamarense/catenella* populations can be transported, onshore and along the Uruguayan coast (Fig. 3) by the onshore SSW drift (Méndez et al. 1996; Méndez and Ferrari 2002; Carreto et al. 1998). Exceptionally, *A. tamarense/catenella* populations can be transported up to southern Brazil (~32 °S) as appears to have occurred during the late spring of 1996 (Odebrecht et al. 1997). During that episode, *A. tamarense/catenella* was recorded at concentrations up to 2 × 10^5 cell L^{-1}, indicating that whereas transport may be the mechanism for inoculation, local growth or cell accumulation also may be occurring. For instance, *A. tamarense/catenella* was also observed (8 × 10^4 cell L^{-1}) at the same time in Uruguayan waters (Méndez and Ferrari 2002). Therefore, it is possible that the southern Brazilian coastline may experience additional outbreaks of PSP, originating from localized cyst germination (Persich et al. 2006). However, the size of the inoculum is an important regulatory factor for bloom success (Eilertsen and Wyatt 2000) that can explain why the 1996 event recorded in Brazil (Persich et al. 2006) has not been repeated (Odebrech pers.com.), and that after 1997, *A. tamarense/catenella* has not produced any other toxic events in Uruguay.

Gymnodinium Catenatum Blooms of *G. catenatum* are involved in recurrent episodes of PSP in Uruguayan coastal waters during late summer/early autumn (Méndez and Ferrari 2003; Méndez and Galli 2008). Studies in Northern Argentinean shelf waters have also shown the presence of this species during autumn, although in low numbers (Balech 1964; Akselman et al. 1998). Blooms of *G. catenatum* in the Río de la Plata maritime front are considered localized events; seedbeds with more than 800 resting cysts cc^{-1} wet sediment have been discovered in the Uruguayan nearshore area (Méndez and Ferrari 2003; Méndez 2006).

Nevertheless, in the Argentinean coast, resting cyst of *G. catenatum* was found only exceptionally and in low abundance (18 resting cysts cc^{-1}) (Akselman et al. 1998). Blooms of *G. catenatum* along the Uruguayan coast have reached cell densities of over 9.3×10^4 cells L^{-1} when warmer water temperatures prevailed during the late summer and early fall months (Méndez and Ferrari 2003). During this time, onshore winds generally predominate, and the flow of the Río de la Plata is minimal, resulting in Ekman surface advection of low-salinity water southward (Acha et al. 2008). These facts and the results of several studies support the idea that *G. catenatum* populations from the estuarine region of the Río de la Plata are occasionally transported along the coast of Argentina (Fig. 3) during early autumn (Carreto et al. 1998). Even though a seasonal (summer and winter) circulation pattern has been reported, recent model studies suggest that the dynamic condition of "summer" and "winter" (Piola et al. 2005) are likely to occur during any season (Acha et al. 2008).

Long-term studies showed that *G. catenatum* blooms (>10^4 cells L^{-1}) had occurred on the Uruguayan coast in a narrow temperature range (21.8–24 °C) with a wide range of salinities (18.4–32.6 psu) during periods of positive surface seawater temperature anomalies (Méndez and Galli 2008).

2.1.3 Toxicology of PSP-Producing Dinoflagellates

Alexandrium tamarense/catenella Among cultured strain isolates from Argentina (Carreto et al. 1996), Uruguay (Méndez et al. 2001), and Brazil (Persich et al. 2006), cell toxicity ranged from 2.3 to 65.6 pg STX eq. cell^{-1}. Therefore, the observed differences in regional toxicity (Carreto et al. 1998; Méndez and Ferrari 2002) may be due to bloom intensity but also to differences in spatial and temporal distribution of *A. tamarense/catenella* subpopulations. In addition, environmental parameters such as irradiance, temperature, salinity, or inorganic nutrients have been shown to affect toxin content and composition for several *Alexandrium* strains.

Although a comprehensive view of the toxin composition of *A. tamarense/catenella* from the Río de la Plata maritime front is lacking, preliminary results showed that under nutrient-replete conditions, the low-potency N-sulfocarbamoil toxins C1,2 were predominant in almost all strains analyzed (Carreto et al. 1996, 2001; Méndez et al. 2001; Persich et al. 2006; Montoya et al. 2010). One notable exception was a rare highly toxic strain isolated from Los Patos lagoon (Brazil) (Persich et al. 2006).

Gymnodinium catenatum Analysis of toxin composition of two clonal culture (GCURD5 and GCURF11) isolates from the Uruguayan coast showed that, as in most *G. catenatum* strains from different geographical origins, the toxin profile was dominated (96%) by the less potent N-sulfocarbamoyl toxins C1,4, GTX5, and GTX6, predominantly C1 and C2 (Méndez et al. 2001; Negri et al. 2001). In a comprehensive view of the toxin composition of *G. catenatum,* from different regions of the world, the Uruguayan strains (GCUR52 and GCUR45) are grouped separately due to the high proportions of C3,4 and the low proportions of the recently described hydroxybenzoate saxitoxin analogues GC1,3 (Negri et al. 2007). However, *G.*

catenatum field populations from Mar del Plata exhibited a different profile; they contained the toxins C1,2, GTX4, GTX2,3, and dcGTX2,3 while lacking GTX5, GTX6, C3, C4, and GC1,3 toxins (Montoya et al. 2010). These results indicate a high heterogeneity in the toxin profile of *G. catenatum* populations from the Río de la Plata maritime front.

2.1.4 Detoxification and Toxin Transformations in Mollusk Tissues

The detoxification rates of *Mytilus edulis* following an *A. tamarense/catenella* bloom event decay rapidly, following an exponential function, until reaching the accepted PST regulatory levels for human consumption (<80 μgSTX 100 g meat). The half-life for PSP toxicity was estimated to be about 8 days (Carreto et al. 2004). In contrast, the detoxification of the snail *Zidona angulata* is slower, remaining above the mentioned regulatory levels for extended periods of time from several months to over a year (Carreto et al. 1996).

During uptake and detoxification, the toxin composition of dinoflagellate cells may undergo changes in the tissues of their primary and secondary consumers due to metabolic transformation and/or selective retention of toxins (Oshima 1995). For example, in the mussel *M. edulis*, GTXs comprise more than 95% of total toxins (Carreto et al. 1996; Méndez et al. 2001), whereas a scarcity of GTXs in comparison with high SX abundance (99%) is characteristic of their secondary consumers, the marine snails *Zidina angulate, Zidona dufresnei*, and *Adelomelon brasiliana* (Carreto et al. 1996).

2.2 *Dinoflagellates Producing Diarrhetic Shellfish Poisoning (DSP)*

2.2.1 Okadaic Acid Producers

Diarrhetic shellfish poisoning (DSP) is a potential intoxication caused by ingesting shellfish that contain high levels of several lipophilic toxins called diarrhetic shellfish toxins (DST), which are produced by some species of dinoflagellates.

A recent report of a massive and unusually long-duration (8 months) DSP event in 2015 was associated with high densities of *Dinophysis ovum* (30.10^4 cell L^{-1}), a species never reported before in Uruguay (Fig. 4), that produces okadaic acid (Méndez et al. 2016b). A narrow range of salinity (31.4–32) and temperature (20–24 °C) were associated with the peak of the bloom (Méndez et al. 2016). Since 2006, lipophilic toxins have been reported annually in Uruguayan wild mollusks, showing an increase in duration annually in this region (Martínez et al. in press). On the Argentinean maritime front coast, the first recorded detection of a DSP outbreak was in 2010, near Villa Gesell and Mar Azul, associated with the presence of

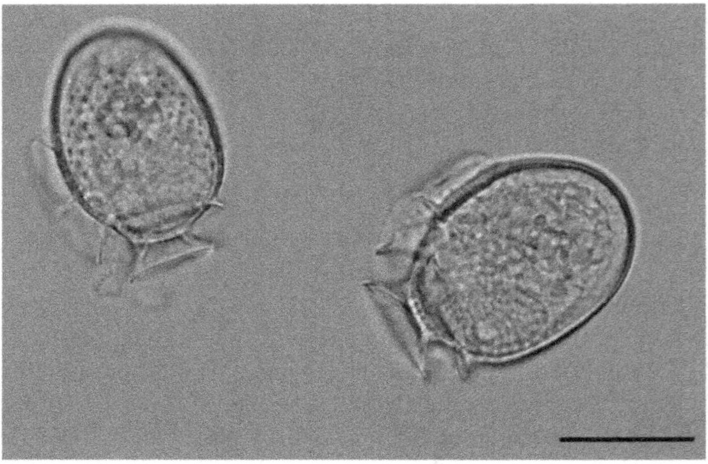

Fig. 4 *Dinophysis ovum* from a Uruguayan toxic bloom 2015 (Méndez et al. 2016a). Scale bar = 20 μm

Dinophysis acuminata and *D. caudata*. The highest density recorded was 4.10[4] cells L^{-1} of *D. acuminata* in Villa Gesell (Sar et al. 2010).

In the Río de la Plata maritime front, several potentially toxic species of the genus *Dinophysis* have been recorded: *D. acuminata, D. caudata, D. rotundata (Phalacroma rotundatum), D. sacculus*, and *D. acuta* (Ferrari et al. 2000). Some DSP-positive results reported on the Uruguayan coast using biological assay were associated with the presence of high abundances of *D. caudata* and *D. acuminata* "complex." DSP was recorded in clams (*Mesodesma mactroides*), cockles (*Donax hanleyanus*), and mussels (*Mytilus edulis*) (Méndez and Medina 2004).

Phalacroma rotundatum has been often observed in Uruguayan oceanic coastal waters (Fig. 5a). It is the second most abundant species in the coastal waters of Uruguay. Concentrations >500 cells L^{-1} were reported at a mean temperature of 22 °C and salinity between 13 and 27 psu. It is not suspected to be a toxin producer itself but may contain DSP toxins and act as a vector after feeding on tintinnids who preyed on toxic *Dinophysis* (reviewed in González-Gil et al. 2011). *Dinophysis sacculus* had been reported in Spain as an okadaic acid producer (Delgado et al. 1996) but has been observed in Uruguayan oceanic coastal waters only in low densities (Fig. 5b). *Dinophysis tripos* produced high levels of pectenotoxins PTX2 (179–232 pg.cell^{-1}) in cultures of Galician strains analyzed by LC-MS (Rodríguez et al. 2012). However, when *D. tripos* was the only species in *Dinophysis* blooms, no toxins were detected in shellfish, possibly because PTX2 is very quickly converted into nontoxic PTX2 SA by mussels (Rodríguez et al. 2012). This species had never been reported associated to a DSP event in this region (Fig. 5c). Other species like *D.* cf *parvula*, *D.* cf *truncata*, and *D.* cf *scrobiculata* had been recorded occasionally in Uruguayan waters in periods without DSP in mussels (Fig. 5g–i).

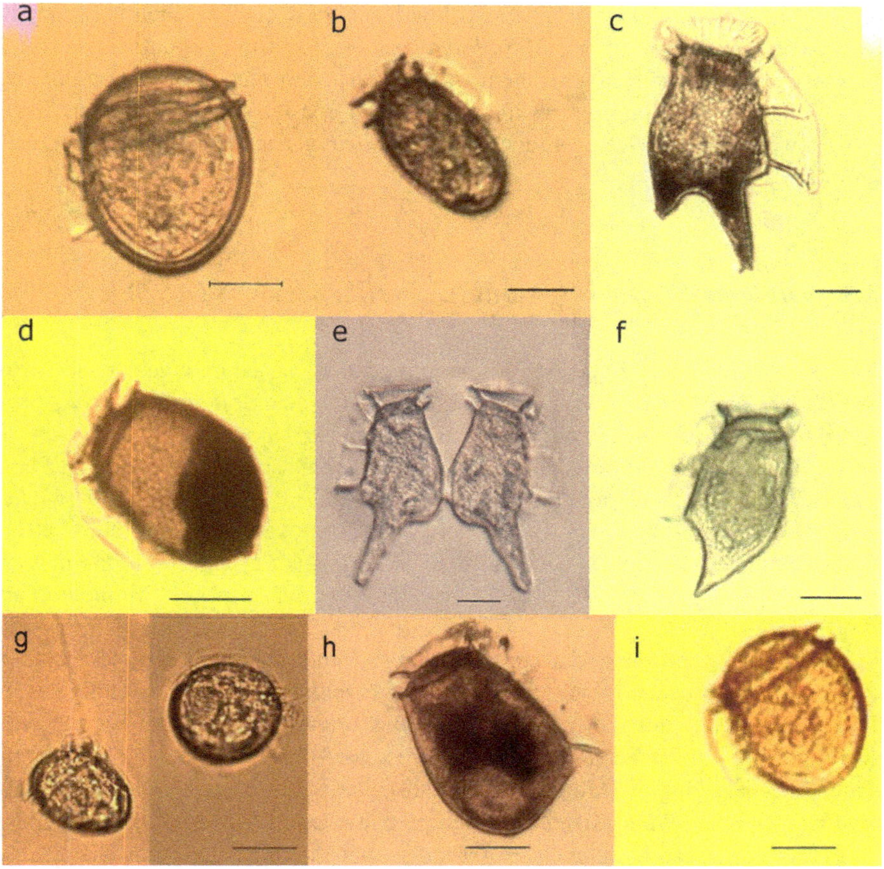

Fig. 5 *Dinophysis* spp. recorded in Uruguayan waters: (**a**) *Phalacroma rotundatum*, (**b**) *Dinophysis sacculus*, (**c**) *Dinophysis tripos*, (**d**) *D. acuminata* "complex," (**e**) *D. caudata* in division, (**f**) *D. diegensis* (early stage of *D. Caudata*), (**g**) *D.* cf *parvula*, (**h**) *D.* cf *truncata*, (**i**) *D.* cf *scrobiculata*. Scale bar = 20 μm

2.2.2 Dinoflagellates Producing Azaspiracids (AZAs)

Recent studies of a bloom collected from Argentinean waters in 1991 (Tillmann and Akselman 2016) showed the diversity of *Amphidomataceae* (*Azadinium spinosum, Az. dalianense, Az. dexteroporum, Az. luciferelloides*, and *Amphidoma languida*). Azaspiracids are a group of lipophilic polyether toxins. Symptoms of azaspiracid poisoning (AZP) in humans include diarrhea, stomach cramps, nausea, and vomiting after consumption of bivalve shellfish. *Azadinium* from the South Atlantic showed a wide spatial distribution that encompassed Northern Argentine and Southern Uruguayan shelves including the mouth of Río de la Plata and presented a marked seasonality in spring and autumn. Nevertheless, AZA toxins in shellfish were just recently reported probably because it had not been specifically detected before. Turner and Goya (2015) found low levels of AZA-2 in shellfish samples (*Brachidontes rodriguezii* and *Mesodesma mactroides*) from Argentina, recorded for the first time. This is in agreement with the first confirmation of the presence of AZA-2 producing *Azadinium* in the Argentinean coastal area by Tillmann et al. (2016), who isolated AZA-2 producing *Azadinium poporum* from coastal sediment samples.

2.3 Diatoms Producing Amnesic Shellfish Toxins (AST)

Amnesic shellfish poisoning (ASP) is an intoxication caused by ingesting shellfish that contain high levels of the neuroexcitatory amino acid domoic acid that is produced by some toxic diatoms of the genus *Pseudo-nitzschia*.

The first documentation of the presence of domoic ascid in the Río de la Plata region was related to a bloom of the diatom *P. australis* that occurred during winter of 2000 in the coastal waters of Mar del Plata (Montoya et al. 2000; Negri et al. 2004). In this event, domoic acid was detected in plankton samples, in the mussel *Mytilus edulis*, and in the viscera of the anchovy *Engraulis anchoita* (Montoya et al. 2000).

In Uruguay, the first ASP event occurred in Punta del Este during December 2001 and was associated with the presence of *P. multiseries* (Medina et al. 2003). The capacity of regional *P. multiseries* populations to produce domoic acid was unequivocally confirmed in culture strains isolated from waters of the northern Argentina coastal region (Montoya et al. 2008). During summer 2009, another ASP-producing species, *P. multistriata*, was reported in Uruguayan waters, the first report for the Southwestern Atlantic Ocean (Méndez et al. 2012a).

A great diversity of potential producers of domoic acid, species of the genus *Pseudo-nitzschia,* have been recorded in the Río de la Plata maritime front: *P. australis*, *P. pseudodelicatissima*, *P. multiseries*, *P. turgidula*, and *P. pungens* (Ferrario et al. 2002), but little is known about their ecology and toxigenicity.

In Uruguay, the first event with detected levels of DA (9.9 mg DA g^{-1} shellfish tissue), as determined by the HPLC standard method, was caused by *P. multiseries* in December 2001 with concentrations reaching 1200 cell L^{-1}. This record was the first DA report associated with this species in Western Atlantic waters (Medina et al. 2003).

Six species of *Pseudo-nitzschia, P. australis, P. delicatissima, P. fraudulenta, P. multiseries, P. multistriata*, and *P. pungens*, were reported in Uruguayan waters.

In February 2009, *P. multistriata* was detected for the first time in Uruguayan waters (Méndez and Ferrario 2009) when the highest density observed was 2×10^6 cell.L^{-1}. That report suggests that Uruguay is the southern limit of its biogeographic distribution in the Southwest Atlantic (Méndez et al. 2012a).

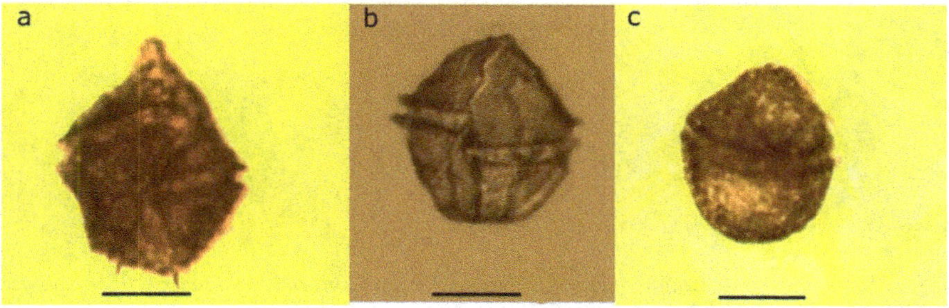

Fig. 6 (**a**) *Gonyaulax spinifera*, (**b**) *Lingulodinium polyedrum*, and (**c**) *Protoceratium reticulatum* from Uruguay. Scale bar = 20 μm

2.4 Dinoflagellates Potentially Producing Yessotoxins (YTXs)

Yessotoxins (YTXs) are produced by the dinoflagellates *Protoceratium reticulatum, Lingulodinium polyedrum*, and *Gonyaulax spinifera.* This complex group of toxins has been associated with DSP because they give positive results for the conventional mouse bioassay for DSP toxins. But unlike okadaic acid (OA) and dinophyisis-toxin-1 (DTX-1), YTXs do not cause either diarrhea or inhibition of protein phosphatases. Structural analysis of YTX analogues showed great variability in the abundance and profile of YTXs in both bivalves and dinoflagellates (Paz et al. 2008).

Protoceratiun reticulatum has been recorded in coastal Atlantic waters of Argentina and Brazil (Balech 1988; Odebrecht 2010). Recent studies confirmed the presence of YTXs in plankton field samples and in cultures of *P. reticulatum* from San Jorge Gulf (Argentina) (Akselman et al. 2015). *Gonyaulax spinifera, Lingulodinium polyedrum*, and *Protoceratium reticulatum* (Fig. 6) had been recorded on the Uruguayan Atlantic coast, but no information about YTX detection have been reported.

2.5 Raphydophyceans Producing Ichthyotoxins (ITX)

Raphydophyceans had been reported worldwide as associated with massive fish mortalities. The most dramatic massive fish mortality in South America occurred during March 2016 in the Los Lagos region in Chile due to the *Pseudochattonella* cf. *verruculosa* bloom. The estimated loss was around 100,000 tons of Atlantic and Coho salmon and trout. During this bloom, the peak cell density reached 7700 cell mL^{-1} (Clement et al. 2016). Their mechanism of toxicity is still unknown, but the most common mode of action is by acute gill irritation, reducing gas exchange efficiency (Eckford-Soper and Daugbjerg 2016). There is evidence of superoxide and hydroxyl radicals; some species may contain brevetoxin-like neurotoxins and free polysaturated fatty acids (Marshall et al. 2005). *Heterosigma akashiwo, Fibrocapsa*

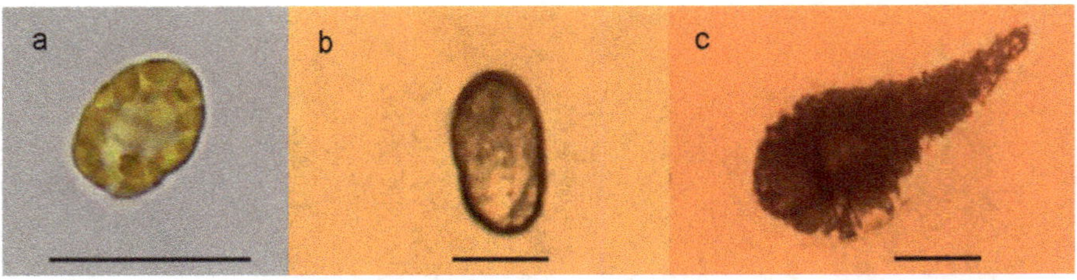

Fig. 7 Raphidophyceae from Uruguay: (**a**) *Heterosigma akashiwo*, (**b**) *Fibrocapsa japonica*, and (**c**) *Chattonella marina* var *antigua*. Scale bar = 20 μm

japonica, and *Chattonella marina* var *antigua* (Fig. 7a–c) were reported in Uruguayan coastal waters of the Río de la Plata during summer and fall periods (Méndez et al. 2012b). *H. akashiwo* was also observed as causing water discoloration on the Argentinean coast (Negri pers. Com) and in Brazil, where it was associated with benthic fish mortality (Proença and Fernandez 2004).

2.6 Cyanobacteria Producing Cyanotoxins (CTXs)

Cyanobacterial blooms are very frequent in the Río de la Plata, especially in protected areas, where a bright green scum of *Microcystis aeruginosa* can be seen in the summer (Ferrari and Méndez 2000; De León and Yunes 2001; Sienra and Ferrari 2006). This species produces a series of hepatotoxic peptides, called microcystins (MCs), whose potent action can trigger lethal effects or, in sublethal doses, induce the formation of tumors. The risk for human consumption of MCs present in drinking water is high because the Río de la Plata is the primary water supply for several Argentine cities such as Buenos Aires and La Plata.

Toxicity of this species was reported in the summer of 1997 and 1999 on the Rio de la Plata Uruguayan coast near the city of Colonia (De León and Yunes 2001; Méndez and Ferrari 2002).

Several species of cyanobacteria had been reported from Uruguayan freshwater and estuaries. Concentrations of MCs between 0.2 and 10.5 μg L^{-1} were found in rivers, reservoirs, and lakes of Uruguay associated to *Dolichospermum* sp. *Microcystis* sp., *Aphanizomenon gracile*, and *Cuspidothrix issatschenkoi*. The presence of saxitoxin was associated to *Aphanizomenon gracile, Cylindrospermopsis raciborskii*, and *Nostocalean* spp. Cylindrospermopsin was recorded in rivers containing populations of *Dolichospermum crassum and Aphanizomenon* sp. (Vidal and Britos 2012).

On the Argentinean coast, two blooms of *Microcystis aeuriginosa* were recorded in March 2005 and December 2006, respectively. Highest abundance was detected on December 2006 (7.8 10^4 cell. mL^{-1}). The MCs levels detected (0.93 and 78 μg. mg^{-1}) were similar to the values found in a toxin-containing bloom of *M. aeruginosa* on the Uruguayan side of the Río de la Plata (De León and Yunes 2001).

HPLC-MS analysis of samples revealed the presence of two major MCs: MC LR and another not yet identified MC (Andrinolo et al. 2007).

Some species of the genus *Trichodesmium* that produce extensive blooms in tropical marine waters can also produce microcystin-like compounds and saxitoxins (Long and Carmichael 2004). Respiratory problems, headaches, and vomiting (Fiebre de Tramandaré) have been reported in people who were close to *Trichodesmium* blooms in Northeast Brazil (Sato et al. 1963). *Trichodesmium erithraeum* had been reported in the Southeast region of Brazil and the Uruguayan oceanic coast (approx. 34°–35°S) (Rörig et al. 1998; Méndez and Ferrari 2002), producing blooms during spring/summer/autumn periods associated to the Brazil current. This species had produced dense blooms with red water discoloration on the Uruguayan marine coast (Ferrari and Méndez 2000). The Río de la Plata seems to be the salinity barrier for this species' distribution in the South.

3 Conclusions

There is a high diversity of harmful and toxic species of phytoplankton present at the Rio de la Plata, associated to production of different groups of toxins. The most relevant are those that produce toxic effects in human consumers of marine products. Several fatal intoxication victims have been reported associated to PST in Argentina (Montoya et al. in this volume), and 60 people were hospitalized due to PST intoxication in Uruguay in 1980 (Davison and Yentsch 1985). However, numerous sublethal intoxications during lipophilic toxin events had been reported officially or unofficially because of gastrointestinal disorders (Méndez 2006). Although there are several species potentially producing ASP, YTXs, AZAs, and CTXs, no human illnesses from marine product consumption were reported during their blooms. In addition, blooms of rhaphidophycean in the region are a potential risk of massive fish mortalities. Further research is needed to define the trophic transfer and accumulation of toxins produced by noxious species in fish, seabirds, and mammals and their effects on the ecosystem.

The natural fluctuations in the Río de la Plata discharge to the Atlantic Ocean cause a dynamic salinity front, with the phytoplankton community shifting from estuarine to marine (Brandini 1988; Móller et al. 2008). We noted that this front also seems to be a distribution barrier for some species extending further south, such as *T. erithraeum* and *P. multistriata*, probably due to the coastal incidence of the well-known thermal and haline front of the oceanic subtropical convergence.

The presented theoretical diagram for mechanisms of in situ growth and transport of PST-producing species, built from historical toxic event records, shows the association with and dependence on the hydrographic conditions for these toxic microalgae. Since 2014, an increase in DST events in the region has been noted, which shows the need of further research efforts regarding the dynamics of lipophilic toxins producer species.

There is a growing need for research on toxicology, molecular biology, and taxonomy applied to harmful algal species. The only way to mitigate the effects of harmful algae and decrease the risk of human intoxication is to continue improving the present monitoring programs of phytoplankton and biotoxins in this region.

References

Acha EM, Mianzan H, Guerrero R et al (2008) An overview of physical and ecological processes in the Río de la Plata estuary. Cont Shelf Res 28(13):1579–1588

Akselman R, Carreto JI, Montoya NG (1998) *Gymnodinium catenatum* and autumn toxicity in northern shelf waters of Argentina. In: Reguera B, Blanco J, Fernández ML Wyatt T (eds) Harmful Microalgae, IOC-UNESCO, pp 122–123

Akselman R, Krock B, Alpermann TJ et al (2015) *Protoceratium reticulatum* (Dinophyceae) in the austral southwestern Atlantic and the first report on YTX-production in shelf waters of Argentina. Harmful Algae 45:40–52

Andrinolo D, Pereira P, Giannuzzi L et al (2007) Occurrence of *Microcystis aeruginosa* and microcystins. In: Río de la Plata river (Argentina). Acta Toxicol Argent, 15(1):8–14

Balech E (1964) El plancton de Mar del Plata durante el período 1961–1962. Bol Inst Biol Marina (Mar del Plata) 4:1–49

Balech E (1988) Los dinoflagelados del Atlántico Sudoccidental. Publ. Espec., Inst. Esp. Oceanogr. Madrid, España 1:310

Brandini FP (1988) Composição e distribuição do fitoplâncton na região Sueste do Brasil e suas relações com as massas de água (Operação Sueste – julho/agosto 1982). Cienc Cult 40(4):334–341

Brazeiro A, Mendez S, Ferrari G (1997) The first toxic bloom of *Alexandrium tamarense* in Uruguay: associated environmental factors. Atlantica 19:19–29

Carreto JI, Lasta M, Negri RM, Benavides HR (1981) Los fenómenos de marea roja y toxicidad de moluscos bivalvos en el Mar Argentino. Contrib Inst Nac Invest Des Pesq Mar del Plata 399:181–201

Carreto JI, Benavides HR, Negri RM et al (1986) Toxic red tide in the argentine sea: phytoplankton distribution and survival of the toxic dinoflagellate *Gonyaulax excavata* in a frontal area. J Plankton Res 8:15–28

Carreto JI, Elbusto C, Sancho H et al (1993) An exploratory analysis of the Mar del Plata shellfish toxicity area (1980–1990). In: Smayda TJ, Shimizu Y (eds) Toxic phytoplankton blooms in the sea. Elsevier Sc. Publ, Amsterdam, pp 377–382

Carreto JI, Elbusto C, Sancho H et al (1996) Comparative studies on paralytic shellfish toxin profiles of marine snails, mussels and an *Alexandrium tamarense* isolate from the Mar del Plata coast (Argentina). Rev Invest Des Pesq 10:101–107

Carreto JI, Montoya NG, Cucchi Colleoni AD et al (1998) *Alexandrium tamarense* blooms and shellfish toxicity in the Argentine sea: a retrospective view. In: Reguera B, Blanco J, Fernández ML Wyatt T (eds) Harmful Microalgae, IOC-UNESCO, pp 131–134

Carreto JI, Montoya NG, Akselman R et al (2004) Differences in the PSP toxin profiles of *Mytilus edulis* during spring and autumn blooms of *Alexandrium tamarense* off Mar del Plata coast. In: Steidinger KA, Landsberg J HC, Tomas R, Vargo GA (eds.) Harmful Algae 2002, Florida Fish and Wildlife Conservation Commission, Florida Institute of Oceanography, and IOC-UNESCO, pp 100–102

Clement A, Lincoquero L, Saldivia M et al (2016) Exceptional summer conditions and HABs of *Pseudochattonella* in southern Chile create record impacts on salmon farms. Harmful Algal News 53:1–3

Davison P, Medina D (1982) Control de la toxina paralítica de los moluscos en el Uruguay. In: III Congreso Nacional de Veterinaria. Montevideo, pp 997–1007
Davison P, Yentsch CM (1985) Occurrence of toxic dinoflagellate and shellfish toxicity along the Uruguayan coast, South America. In: Anderson DM, White AW, Baden DG (eds) Toxic dinoflagellates. Elsevier, New York, pp 153–158
De León L, Yunes J (2001) First report of a microcystin-containing bloom of the Cyanobacterium *Microcystis aeruginosa* 2001 in the Río de la Plata river, South America. Environ Toxicol 16:110–112
Delgado M, Garcés E, Camp J (1996) Growth and behaviour of Dinophysis sacculus from NW Mediterranean Sea. In: Yasumoto T, Oshima Y, Fukuyo Y (eds) Harmful and Toxic Algal Blooms, Proceedings of the Seventh International Conference on Toxic Phytoplankton, Sendai, Japan. 12–16 July 1995, Paris, France: Intergovernmental Oceanographic Commission of UNESCO, pp 261–264
Eckford-Soper L, Daugbjerg N (2016) The Ichthyotoxic genus Pseudochattonella (Dictyochophyceae): distribution, toxicity, enumeration, ecological impact, succession and life history – a review. Harmful Algae 58:51–52
Eilertsen HC, Wyatt T (2000) Phytoplankton models and life history strategies. S Afr J Mar Sci 22:323–338
Ferrari G, Méndez SM (2000) Report of phytoplankton species producing coastal water discoloration in Uruguay. Iheringia Ser Bot 54:3–18
Ferrari G, Méndez SM, Brazeiro A (2000) *Dinophysis acuminata* associated to diarrehtic shellfish poisoning reports in Uruguay. Fr Marít 19:91–95
Ferrario ME, Sar AE, Salas E (2002) Diatomeas potencialmente tóxicas del Cono Sur Americano. In: Sar EA, Ferrario ME, Reguera B (eds) Floraciones Algales Nocivas en el Cono Sur Americano. Instituto Español de Oceanografía, Madrid, pp 167–194
GEOHAB (2001) In: Gibert P, Pitcher G (eds) Global ecology and oceanography of harmful algal blooms, science plan. SCOR/ IOC, Baltimore/Paris
González-Gil S, Pizarro G, Paz B et al (2011) Considerations on the toxigenic nature and prey sources of *Phalacroma rotundatum*. Aquat Microb Ecol 64:197–203
Long BM, Carmichael WW (2004) Marine cyanobacteria toxins. In: Hallegraeff GM, Anderson DM, Cembella AD (eds) Manual on harmful marine microalgae, Monographs on oceanographic methodology, vol 11. UNESCO Publishing, Paris, pp 279–296
Marshall JA, De Salas M, Odas T (2005) Superoxide production by marine microalgae. Mar Biol 147:533–540
Medina D, Méndez SM, Inocente G et al (2003) Shellfish monitoring programme in Uruguay. In: Villalba A, Reguera B, Romalde JL, Beiras R (eds) Molluscan Shellefish safety. Consellería de Pesca e Asuntos Maritimos da Xunta de Galicia and oceanographic commission of UNESCO, pp 197–202
Méndez SM (1993) Uruguayan red tide monitoring programme: preliminary results (1990–1991). In: Smayda TJ, Shimizu Y (eds) Toxic phytoplankton blooms in the sea. Elsevier Sc. Publ, Amsterdam, pp 287–289
Méndez SM (2006) El impacto de las floraciones algales nocivas, un tema de gestión costera: monitoreo, inicio y dispersión, control y mitigación. In: Menafra R, Rodriguez C, Scarabino F, Conde D (eds), Bases para la conservación y el manejo de la costa uruguaya. Vida Silvestre and Fac. de Ciencias, Montevideo, pp 57–71
Méndez S, Ferrari G (2002) Floraciones algales nocivas en Uruguay: antecedentes, proyectos en curso y revisión de resultados. In: Sar EA, Ferrario ME, Reguera B (eds) Floraciones Algales Nocivas en el Cono Sur Americano. Instituto Español de Oceanografía, Madrid, pp 271–288
Méndez S, Ferrari G (2003) Floraciones tóxicas de *Gymnodinium catenatum* en aguas uruguayas. Pub Com Tec Mix Fr Marít 19:97–102
Méndez S, Ferrario M (2009) First report of *Pseudo-nitzschia multistriata* in Uruguay. Harmful Algae News 40(4):5–6

Méndez SM, Galli MO (2008) Condiciones ambientales asociadas con floraciones de dinoflagelados productores de veneno paralizante. In: Actas del XI Congresso Brasileiro de Ficología y Simposio Latino-Americano sobre algas nocivas, pp 243–257

Méndez SM, Medina D (2004) Twenty-three years of red tide monitoring at fixed stations along the coast of Uruguay. In: Steidinger KA, Landsberg JHC, Tomas R, Vargo GA (eds) Harmful Algae 2002, Florida Fish and Wildlife Conservation Commission, Florida Institute of Oceanography, and IOC-UNESCO, pp 184–185

Méndez S, Severov D, Ferrari G et al (1996) Early spring *Alexandrium tamarense* toxic blooms in the Uruguayan waters. In: Yasumoto T, Oshima Y, Fukuyo Y (eds) Harmful and toxic algal blooms. IOC-UNESCO, Paris, pp 113–119

Méndez S, Kulis D, Anderson DM (2001) PSP toxin production of Uruguayan isolates of *Gymnodinium catenatum* and *Alexandrium tamarense*. In: Hallegraeff G, Blackburn S, Lewis R, Bolch C (eds), Harmful algal blooms, IOC-UNESCO, pp 352–355

Méndez SM, Ferrario M, Cefarelli AO (2012a) Description of toxigenic species of the genus *Pseudo-nitzschia* in coastal waters of Uruguay: morphology and distribution. Harmful Algae 19:53–60

Méndez SM, Akselman R, Thomas C (2012b) First report of *Heterosigma akashiwo*, *Fibrocapsa japonica* and *Chattonella marina* var. *antiqua* in Uruguay. In: C.R. Pagou, P. and Hallegraef, G. (eds). Proceedings of the 14th International Conference on Harmful Algae. Hersonissos, Creta, Grecia, 1-5 noviembre 2010. International Society for the Study of Harmful Algae and Intergovernmental Oceanographyc Commission of UNESCO, pp 134–136

Méndez SM, Martinez A, Fabre A (2016a) Extreme abundant bloom of *Dinophysis* of the *acuminata* complex associated to warm marine waters in Uruguay. In: Abstracts of the International Conference of Harmful Algae. Florianopolis, Brazil, 9–15 Oct 2016

Méndez SM, Rodriguez F, Reguera B et al (2016b) Characterization of a *Dinophysis* of the *acuminata* complex causing an exceptional DSP event in Uruguay during 2015. In: Abstracts of the International Conference of Harmful Algae. Florianopolis, Brazil, 9–15 Oct 2016

Möller Jr OO, Piola AR, Freitas AC et al (2008) The effects of river discharge and seasonal winds on the shelf off southeastern South America. Cont Shelf Res 28(13):1607–1624

Montoya NG, Negri RM, Carreto JI (2000) Primera detección de toxina amnésica de moluscos en el Mar Argentino asociada a un florecimiento de la diatomea *Pseudo-nitzschia australis* en la Zona Común de Pesca Argentino-Uruguaya. In: Abstracts of the XV Simposio Científico-Tecnológico de la Com Técn Mixta Frente Marítimo, Mar del Plata, p 21

Montoya NG, Negri RM, Carignan MO et al (2008) Algunas características bioquímicas de la diatomea tóxica *Pseudo-nitzschia multiseries* aislada en el mar Argentino. In: Actas del XI Congresso Brasileiro de Ficología y Simposio Latino-Americano sobre algas nocivas, pp 259–266

Montoya NG, Fulco VK, Carignan MO et al (2010) Toxin variability in cultured and natural populations of *Alexandrium tamarense* from southern South America – evidences of diversity and environmental regulation. Toxicon 56(8):1408–1418

Negri A, Bolch CJ, Lindon E et al (2001) Paralytic shellfish toxins in *Gymnodinium catenatum* strains from six countries. In: Hallegraeff GM, Blackburn S, Bolch CJ, Lewis R (eds) Harmful algal blooms 2000, IOC-UNESCO, pp 210–214

Negri RM, Montoya N, Carreto JI (2004) *Pseudo-nitzschia australis*, *Mytilus edulis*, *Engraulis anchoita* and domoic acid in the Argentine sea. In: Steidinger KA, Landsberg JHC, Tomas R, Vargo GA (eds) Harmful Algae 2002, Florida Fish and Wildlife Conservation Commission, Florida Institute of Oceanography, and IOC-UNESCO, pp 139–141

Negri AP, Bolch CJC, Geier S et al (2007) Widespread presence of hydrophobic paralytic shellfish toxins in *Gymnodinium catenatum*. Harmful Algae 6:774–780

Odebrecht C (2010) Dinophyceae. In: Forzza et al (Eds.), Catálogo de plantas e fungos do Brasil 1, pp 366–383

Odebrecht C, Méndez S, Garcia VMT (1997) Oceanographic processes and harmful algal blooms in the Southwestern Atlantic. In: Abstracts of VIIIth Intern Conference on Harmful Algae Vigo, Spain, p 152

Oshima Y (1995) Chemical and enzymatic transformation of paralytic shellfish toxins in marine organisms. In: Lassus P, Arzul G, Gentien P, Marcaillou N (eds) Harmful marine algal blooms. Lavoisier Intercet Ltd, Paris, pp 475–480

Paz B, Daranas AH, Norte M et al (2008) Yessotoxins, a group of marine polyether toxins: an overview. Mar Drugs 6:73–102

Persich GR, Kulis DM, Lilly EL et al (2006) Probable origin and toxin profile of *Alexandrium tamarense* (Lebour) Balech from southern Brazil. Harmful Algae 5:36–44

Piola AR, Matano RP, Palma ED et al (2005) The influence of the Plata River discharge on the western South Atlantic shelf. Geophys Res Lett 32:1603–1606

Proença LAO, Fernandez LF (2004) Introdução de microalgas no ambiente marinho: impactos negativos e fatores controladores. In: Silva JSV, Souza RCCL (Ed.). Água de lastro e bioinvasão. Porto Alegre: Interciência, p 77–97

Proença LAO, Resgalla C (2004) Toxicity of the aqueous extract of *Alexandrium fraterculus* (Balech) Balech. In: Steidinger KA, Landsberg J HC, Tomas R, Vargo GA (eds.) Harmful Algae 2002, Florida Fish and Wildlife Conservation Commission, Florida Institute of Oceanography, and IOC-UNESCO, pp 184–185

Proença LAO, Tamanaha MS, Souza NP (2001) The toxic dinoflagellate *Gymnodinium catenatum* Graham in southern Brazilian waters: ocurrence, pigments and toxins Atlântica. Río Grande 23:59–65

Rodríguez F, Escalera L, Reguera B et al (2012) Morphological variability, toxinology and genetics of the dinoflagellate *Dinophysis tripos* (Dinophysiaceae, Dinophysiales). Harmful Algae 13:26–33

Rörig LR, Yunes JS, Kuroshima KN et al (1998) Studies in the ecology and toxicity of *Trichodesmium* spp. blooms in southern Brazilian coastal waters. In: Reguera B, Blanco J, Fernandez ML, Wyatt T (eds), Harmful algae. Xunta de Galicia and IOC-UNESCO, pp 22–25

Sar EA, Sunensen I, Lavigne AS, Goya AB (2010) *Dinophysis* spp. associated to detection of diarrheic shellfish toxins (DSTs) in mollusks and to human diarrhoetic disease episode (Buenos Aires Province, Argentina). Rev Biol Mar Oceanogr 45(3):451–460

Satô S, Paranaguá MN, Eskinazi E (1963) On the mechanism of the red tide *Trichodesmium* in Recife north eastern Brazil, with some considerations of the relation to the human disease "Tamandaré fever". Trabhs Inst Oceanogr (Univ Recife) 5:7–49

Sienra D, Ferrari G (2006) Monitoreo de cianobacterias en la costa de Montevideo (Uruguay). En: Menafra, Rodriguez-Gallego L, Scarabino F, Conde D (eds.). Bases para la conservación y el manejo de la costa uruguaya. Montevideo: Vida Silvestre, pp 413–420

Sournia A (1995) Red tide and toxic marine phytoplankton of the world ocean: an inquiry into biodiversity. In: Lassus P, Arzul G, Gentien P, Marcaillou N (eds) Harmful marine algal blooms. Lavoisier Intercet Ltd, Paris, pp 103–112

Tillmann U, Akselman R (2016) Revisiting the 1991 algal bloom in shelf waters off Argentina: *Azadinium luciferelloides* sp. nov. (Amphidomataceae, Dinophyceae) as the causative species in a diverse community of other amphidomataceans. Phycol Res 64:160–175

Tillmann U, Borel M, Barrera F et al (2016) *Azadinium poporum* (Dinophyceae) from the South Atlantic off the Argentinean coast produce AZA-2. Harmful Algae 51:40–55

Turner AD, Goya AB (2015) Occurrence and profiles of lipophilic toxins in shellfish harvested from Argentina. Toxicon 102:32–42

Vidal L, Britos A (2012) Uruguay: occurrence, toxicity and regulation of Cyanobacteria. Ingrid Chorus (ed) Federal Envirnonment Agency, Germany, pp 130–136

Harmful Marine Microalgae in Coastal Waters of Chubut (Patagonia, Argentina)

Alicia V. Sastre, Norma H. Santinelli, Miriam E. Solís, Laura B. Pérez, Soledad Díaz Ovejero, Leilén Gracia Villalobos, Andreana Cadaillón, and Valeria C. D'Agostino

Abstract A Harmful Algal Blooms Regional Monitoring Program has been carried out in Chubut coastal waters (Patagonia, Argentina) since the year 2000. This program surveys an extended shoreline, with bays and gulfs with shellfish natural banks and farms. Paralytic shellfish poison (PSP)-toxin-producing species, *A. tamarense*, have been observed during the study period; in addition, species producing diarrheic shellfish poison (DSP)-toxins, such as *Dinophysis acuminata* and *D. tripos* and *Prorocentrum lima*, and amnesic shellfish poison (ASP)-toxins, as several species of genus *Pseudo-nitzschia*, have been identified. Moreover, the production of the three types of toxins has been proven. Other harmful but nontoxic species have been registered in the area. The aim of this review is to show the temporal and spatial distribution of harmful microalgae species, the environmental factors associated with their occurrence, and their relation to toxic outbreaks during more than 15 years of observations, with special attention focused on the episodes of human intoxications. In addition, we discussed the accumulation and transfer of some phycotoxins through pelagic food webs, from the first trophic levels to large marine mammals, such as whales.

Keywords Harmful marine microalgae · Phycotoxins · Coastal waters · Patagonia Argentina · Southwest Atlantic Ocean

A. V. Sastre (✉) · N. H. Santinelli · S. D. Ovejero · A. Cadaillón
Instituto de Investigación de Hidrobiología, Facultad de Ciencias Naturales y Ciencias de la Salud, Universidad Nacional de la Patagonia San Juan Bosco, Trelew, Argentina

M. E. Solís · L. G. Villalobos · V. C. D'Agostino
Centro para el Estudio de Sistemas Marinos (CESIMAR), Consejo Nacional de Investigaciones Científicas y Técnicas (CONICET), Puerto Madryn, Argentina
e-mail: miriam@cenpat-conicet.gob.ar; gracia@cenpat-conicet.gob.ar

L. B. Pérez
Laboratorio de Biología, Facultad de Ciencias Naturales y Ciencias de la Salud, Universidad Nacional de la Patagonia San Juan Bosco, Comodoro Rivadavia, Argentina

M. S. Hoffmeyer et al. (eds.), *Plankton Ecology of the Southwestern Atlantic*,
https://doi.org/10.1007/978-3-319-77869-3_22

1 Introduction

While harmful algal blooms, in a strict sense, are completely natural phenomena that have occurred throughout recorded history, in the past two decades, the impact of such events on public health and economy appears to have increased in frequency, intensity, and geographical distribution (Van Dolah 2000; Hallegraeff 2004). Harmful algal blooms (HABs), commonly called red tides, affect virtually every coastal region of the world. Since the latter term erroneously includes many blooms that discolor the water but cause no harm and also excludes blooms of highly toxic cells that cause problems at low (and essentially invisible) cell concentrations, scientists prefer the term HAB (Anderson et al. 2012).

In the broad sense, UNESCO's Intergovernmental Oceanographic Commission (IOC) coined the term "harmful algal blooms" (HABs) to designate the occurrence of a heterogeneous group of microorganisms that are perceived as harmful. Then, the HAB designation is a societal concept rather than a scientific definition. Blooms are considered to fit the HAB criterion if they cause injury to human health, to socioeconomic interests, or to components of aquatic ecosystems (Reguera 2002; Anderson et al. 2012).

Among the different types of HABs, some may be harmful non-toxigenic. They may cause serious damage to marine wildlife by clogging the fish gills, decreasing oxygen levels in the water column, triggering death by anoxia, etc. Some HABs may injure gill tissues and membranes in fish producing their death by a mechanical effect of their cellular spines and horns (Hasle and Fryxell 1995) or by producing hemolytic substances, as it is the case with some species of Raphydophyceae (Suárez-Isla and Guzmán 1999; Hallegraeff and Hara 2004). On the other hand, some HAB species are toxigenic and produce blooms that cause illness and death of fish, seabirds, mammals, and other marine life, often via toxin transfer through the food web by consumption of filter-feeding organisms that accumulate the toxin-producing species. Human consumers of seafood contaminated by these toxins may also be poisoned, suffering acute toxic symptoms and even fatalities in extreme cases (Anderson et al. 2012). Based on the symptomatology of the intoxication and on the vectors of transmission, different types of poisoning have been defined (Fernández et al. 2002; Wang 2008).

Zooplankton can be an intermediate link in the trophic transfer of phycotoxins to higher consumers, potentially intoxicating marine fish, birds, and mammals (Lefebvre et al. 1999; Scholin et al. 2000). The Valdés Peninsula, declared a World Heritage Site by UNESCO in 2000, and its surrounding gulfs have a rich biodiversity of marine birds and mammals of a great tourist importance; the most representative species is the southern right whale, *Eubalaena australis* UNESCO's Natural Monument, which finds in these waters an important calving ground. Doucette et al. (2006), Leandro et al. (2010), and Fire and Van Dolah (2012), among others, have clearly demonstrated that whales ingest the different toxins that pass through their digestive tract. Thus, harmful algal blooms pose a risk for this protected species as well as for other animals that feed on zooplankton (D'Agostino et al. 2015) and shellfish.

Red tides in the South American Cone, as in other regions of the world, have been recorded as early as the time of the great oceanographic expeditions of the nineteenth century (Reguera 2002). Carreto and Benavides (1993) reported the existence of historical archives in Argentina describing mortalities within the indigenous populations of Ushuaia, caused by shellfish consumption, since 1886. In the Argentine seashore, this phenomenon was first reported in shelf waters off the Valdés Peninsula in 1980, when two fishermen died after eating mussels with high concentrations of paralytic shellfish poison (PSP)-toxins, associated with *Gonyaulax excavata* (Braarud) Balech (Carreto et al. 1981). This species was later transferred to the genus *Alexandrium* Halim under the name *Alexandrium excavatum* (Braarud) Balech et Tangen and synonymized by Balech (1995) as *Alexandrium tamarense* (Lebour) Balech. Molecular studies revealed that *A. fundyense* from eastern USA, *A. catenella* from western USA and western South America, and *A. tamarense* from western South America are all the same species, the so-called *A. tamarense* complex Group I (Lilly et al. 2007), lately renamed *A. fundyense* (John et al. 2014). After strong controversy, the final name adopted for the *A. tamarense* Group I was *A. catenella* (Fraga et al. 2015), priority based on seniority (Prud'homme van Reine and Willem 2017). This study retains the name *A. tamarense* because of the previously published literature for this region and since the confirmation of the presence or absence of alternative ribotypes of this species complex in the eastern South America has not yet been communicated.

Monitoring of PSP-toxins in shellfish in Chubut coastal waters started after the 1980 event. Nevertheless, several toxic outbreaks, some of them including fatal cases, have occurred in different sites of the coast ever since (Vecchio et al. 1986; Esteves et al. 1992; Andrade 2001; Andrade 2002; Santinelli et al. 2002; Baulde 2010; Baulde 2011).

2 The Chubut Province Monitoring Program

Since the year 2000, a Harmful Algae Blooms and Shellfish Toxicity Monitoring Program has been carried out in Chubut coastal waters (Patagonia, Argentina) as part of the Provincial Plan for Prevention and Control of Red Tide. This area covers an extended shoreline of approximately 1600 km, with bays and gulfs with natural shellfish banks and farms. The program includes monitoring of harmful algal blooms and marine environmental conditions, shellfish toxicity control and harvesting closures, educational and training activities, dissemination of information, and detection of intoxicated consumers in public health centers. At the beginning, only the northern coastal zone (North Patagonian gulfs) was controlled, but the southern stations were added several years later. Moreover, amnesic shellfish poison (ASP) and diarrheic shellfish poison (DSP) have been measured since 2005 and 2008,

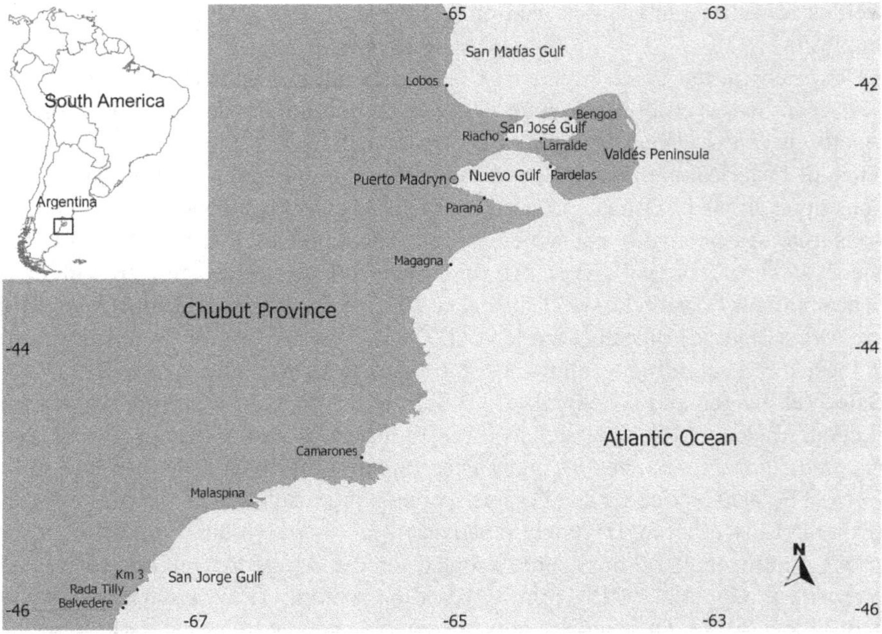

Fig. 1 Study area and location of the sampling stations

respectively. Nowadays, samples are taken monthly or bi-weekly from 12 sites of the Patagonian shoreline.

The study area is located approximately between 42°–46°S and 64°–67°30′W (Fig. 1). We will consider the North Patagonian gulfs of San Matias, San José, Nuevo, and Magagna as part of the northern zone and Camarones and the San Jorge Gulf as the southern zone. Samples were collected from integrated water column; 250 ml subsamples were preserved with Lugol's solution and stored for species identification and enumeration. Afterward, they were counted using the Utermöhl (1958) method with a Leica DMIL phase contrast inverted microscope. In addition, qualitative phytoplankton samples were taken using a 25 μm mesh net through oblique tows on a boat and fixed with formaldehyde at a final concentration of 4%. They were later analyzed at the Instituto de Investigación de Hidrobiología of the Universidad Nacional de la Patagonia.

Living and fixed net samples were observed with an Olympus CX31 phase contrast light microscope. Taxonomic identification of dinoflagellates was carried out following Balech (1995), and diatoms frustules were cleaned following the Hasle and Fryxell (1970) procedure. Scanning electron microscopy observations of the samples were made with a Jeol JSM-6360 LV scanning electron microscopy at the Facultad de Ciencias Naturales y Museo and Universidad Nacional de La Plata and with a Zeiss Supra 40 at the Universidad de Buenos Aires advanced microscopy center.

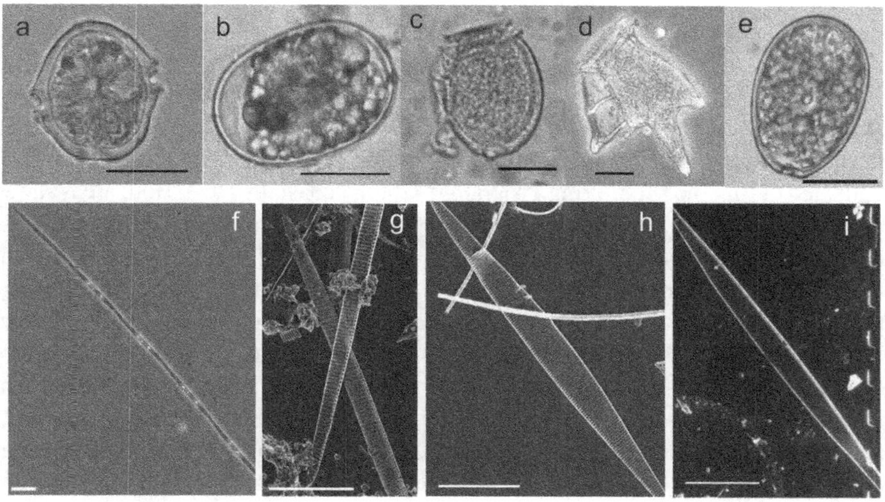

Fig. 2 Main toxin-producing species in Chubut province. (**a**) *Alexandrium tamarense* vegetative cell, (**b**) *A. tamarense* resting cist, (**c**) *Dinophysis acuminata*, (**d**) *D. tripos*, (**e**) *Prorocentrum lima*, (**f**) *Pseudo-nitzschia calliantha*, (**g**) *P. pungens*, (**h**) *P. australis*, (**i**) *P. fraudulenta* (**a–f** LM; **g–i** SEM) (Scale bars = 20 μm)

3 Harmful Toxin-Producing Species

Outbreaks of harmful algal blooms (HAB) have occurred in coastal waters of Chubut province (Patagonia, Argentina) since 1980. Different toxin-producing species associated with these events have been identified (Fig. 2): (i) paralytic shellfish poisoning (PSP), *Alexandrium tamarense*; (ii) diarrheic shellfish poisoning (DSP), *Dinophysis acuminata*, *D. tripos*, *D. caudata*, *D. fortii*, *D. acuta*, *Phalacroma rotundatum*, *Prorocentrum lima*, *P. cordatum*, and *Protoceratium reticulatum*; and (iii) amnesic shellfish poisoning (ASP), *Pseudo-nitzschia australis*, *P. multiseries*, *P. pungens*, *P. fraudulenta*, and *P. calliantha*.

4 Harmful Non-toxigenic Species

Among the harmful non-toxigenic species, the most frequent in the samples were the diatoms *Coscinodiscus wailesii*, *Thalassiosira mala*, *Leptocylindrus minimus*, *Chaetoceros socialis*, *C. concavicornis*, and *Asterionellopsis glacialis*, the dinoflagellates *Prorocentrum micans* and *Lepidodinium* sp., and the silicoflagellates *Dictyocha fibula*, *D. speculum*, and *D. octonaria*. Although they have not caused any proven damage to the fauna or flora in this region, they are responsible for harmful events in other geographic areas (Clément and Lembeye 1993; Hallegraeff 2004; Hasle and Fryxell 1995; Andersen et al. 1995; Hargraves and Maranda 2002; Fryxell and Hasle 2004; Smayda 2006).

5 Environmental Features

The seasonal variability of the hydrographic conditions of both the northern and southern areas is summarized in Fig. 3 and is based on monthly mean temperature, nitrate + nitrite (hereinafter identified as nitrate), phosphate, and silicic acid data collected during the Monitoring Program in Chubut coastal waters. Both areas show the typical seasonal variability of temperate-cold regions. The northern area was characterized by an annual average salinity of 34.06 ± 0.22 ($n = 818$) (not shown in the figure), temperature in the range of 10.7 °C–17.5 °C (Fig. 3a). Nutrient concentrations (Fig. 3b–d) were highest in winter (nitrate ~ 5 μM, phosphate ~ 1.4 μM, and silicic acid ~ 5.7 μM) and lowest from spring to early autumn (nitrate <1 μM or no detected, phosphate <1.1 μM, and silicic acid mainly <2.5 μM). In the southern area, annual average salinity was 33.46 ± 0.41 ($n = 336$) (not shown in the figure), and temperature was between 8.4 °C and 15.5 °C (Fig. 3e). The highest nutrient concentrations (Fig. 3f–h) were also observed in winter, but values were higher than in the northern area (nitrate ~ 10.5 μM, phosphate ~ 1.6 μM, and silicic acid ~ 6.1 μM). The lowest nutrient values were recorded from spring to the end of summer; concentrations of phosphate and silicic acid concentrations were similar to the northern area, but nitrate did not reach undetected levels.

6 Spatial and Temporal Dynamics of Harmful Species

Only the more frequent species in the phytoplankton samples were considered in this analysis. The harmful species showed a large interannual variability in their occurrence and density and differences between sampling sites (Fig. 4). There were years, such as 2001, with great species diversity but others with a strong dominance of a single species, such as 2012 and 2013. *A. tamarense* was an important component within the harmful species in the years 2000, 2004, and 2005 in some stations of the north zone (Magagna, Nuevo, and San José gulfs) and in the 2010 and 2015 in the stations of the south zone (San Jorge Gulf). Among *Pseudo-nitzschia* species, *P. pungens* was well represented until 2010 in the northern area (except 2006 and 2008); *P. australis* until 2006; *P. calliantha* during 2007, 2008, 2009, 2010, and 2015; and *P. fraudulenta* from 2006 onward.

7 Spatial and Temporal Dynamics of *Alexandrium tamarense* and PSP

Alexandrium tamarense was present throughout Chubut province coasts, with maximum cell densities at the end of winter and during spring in the northern zone (Santinelli et al. 2002) and during spring and summer in the south (Pérez et al. 2013)

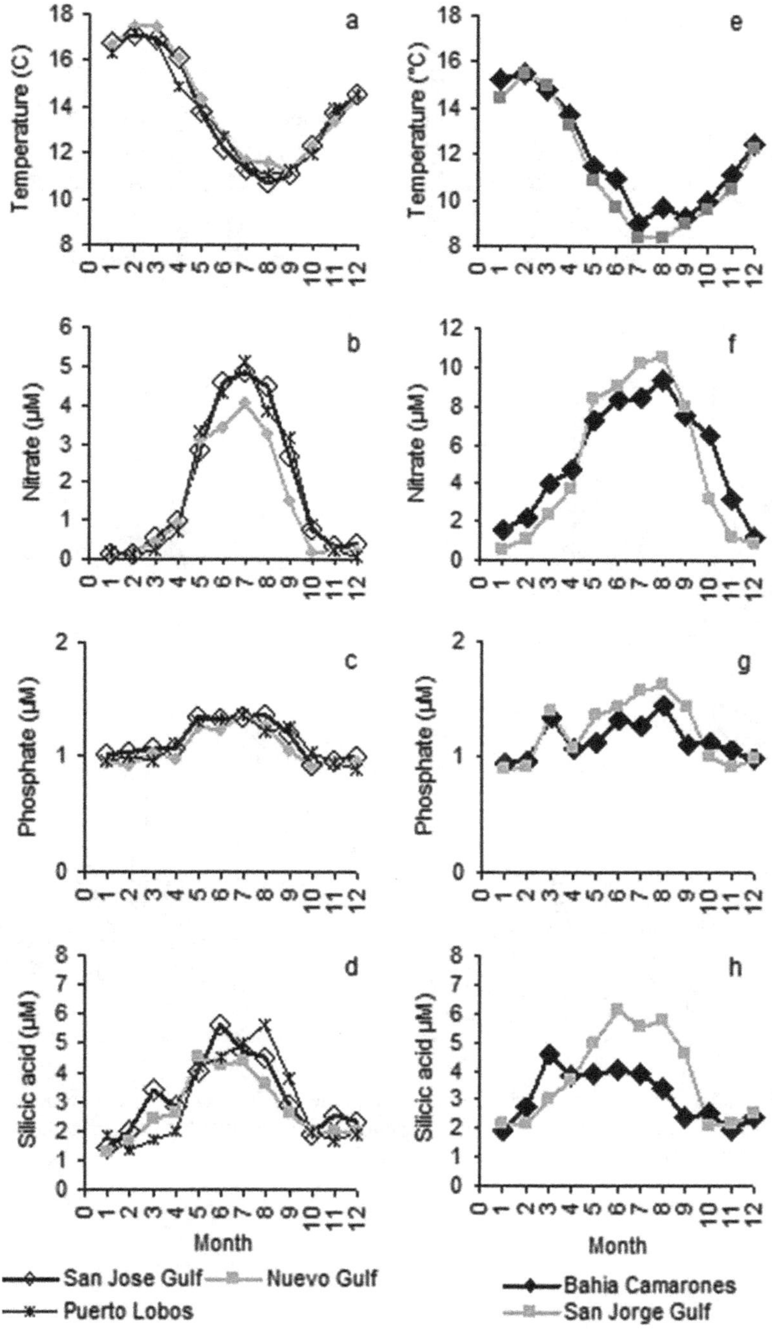

Fig. 3 Monthly mean temperature and seasonal variation of nitrate, phosphate, and silicate in northern (**a–d**) and southern (**e–h**) areas

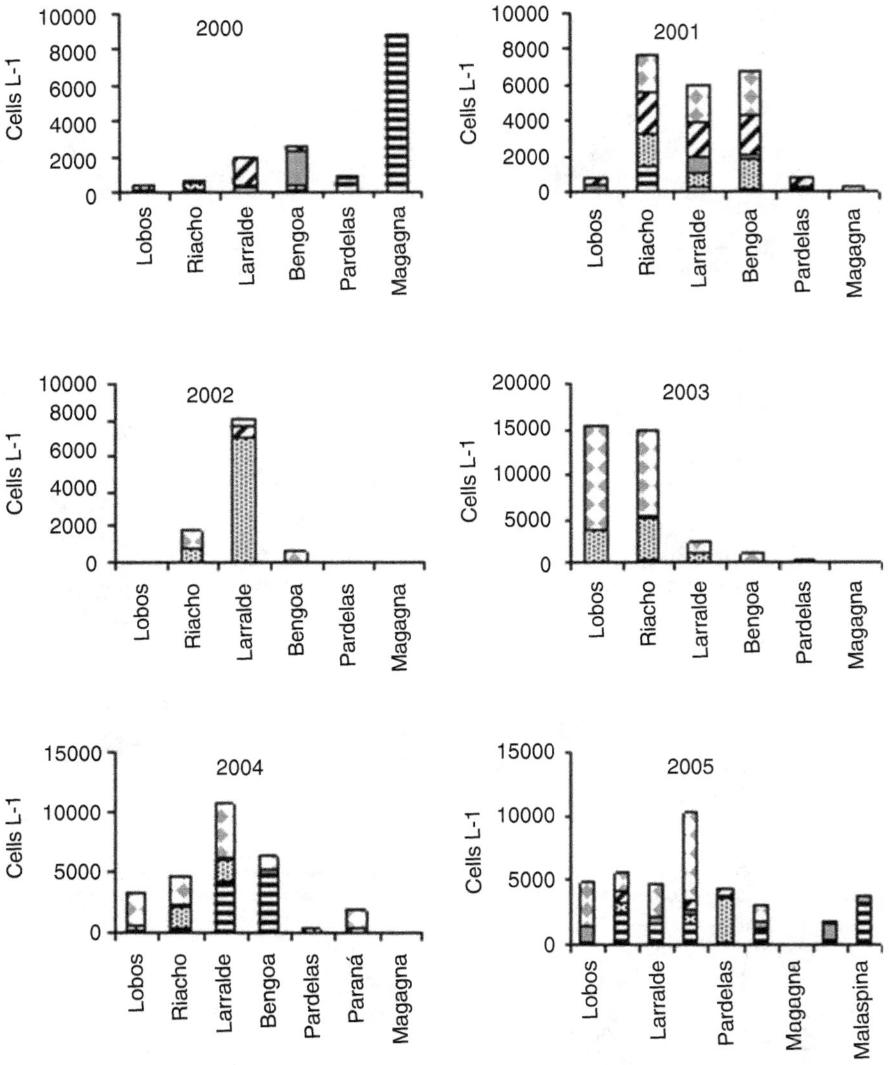

Fig. 4 Spatial and temporal distribution of toxin-producing species.(**a**) Annual average of cell densities between 2000 and 2005, (**b**) Annual average of cell densities between 2006 and 2011, (**c**) Annual average of cell densities between 2012 and 2016

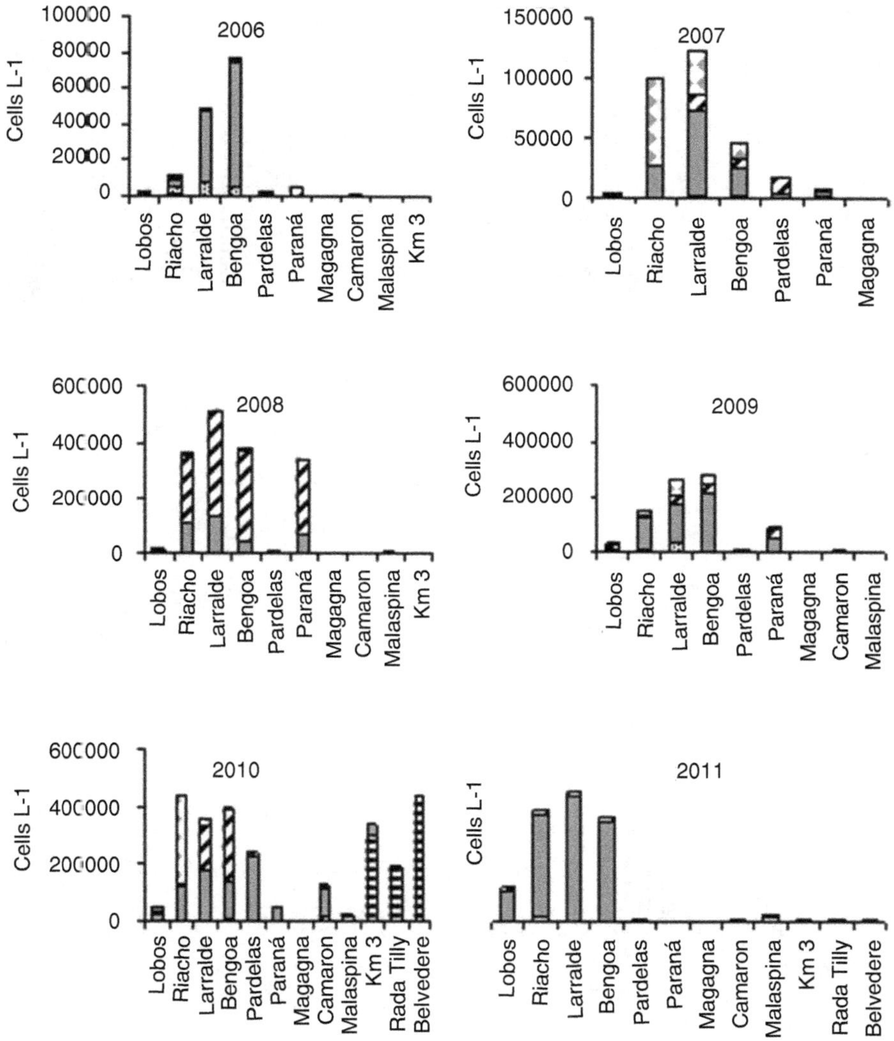

Fig. 4 (continued)

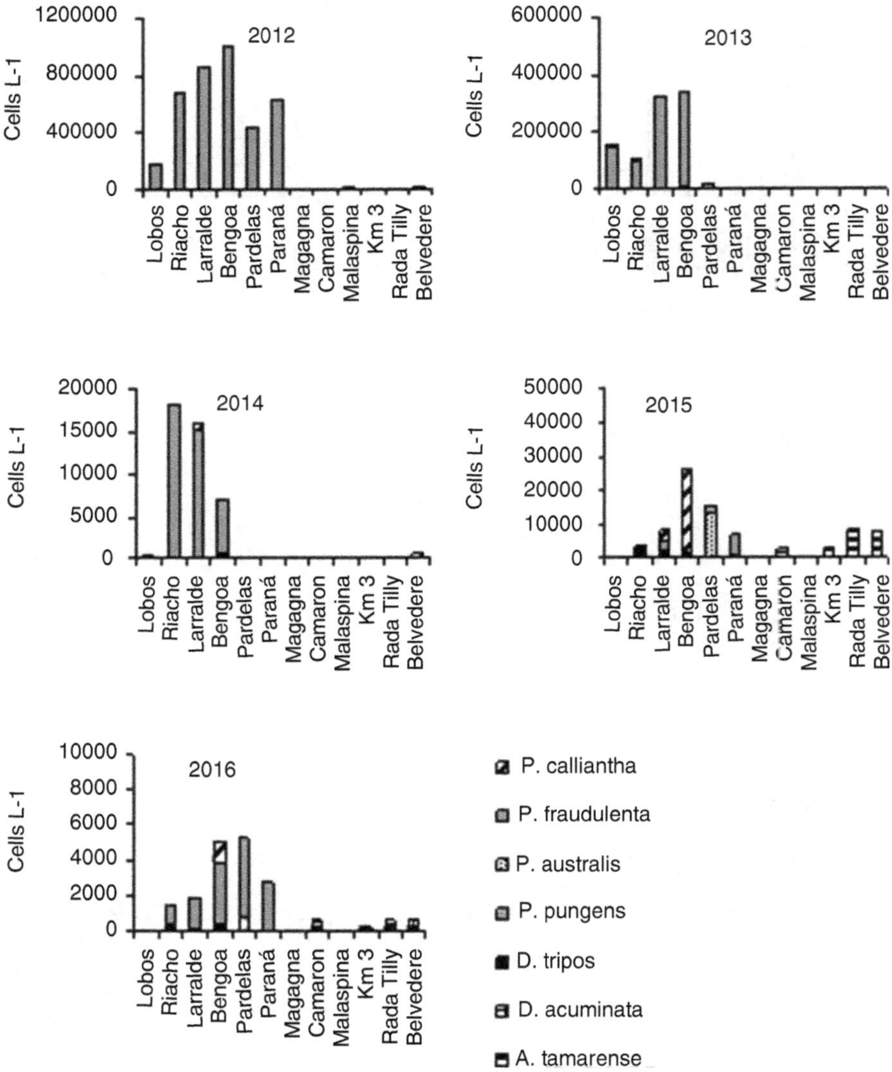

Fig. 4 (continued)

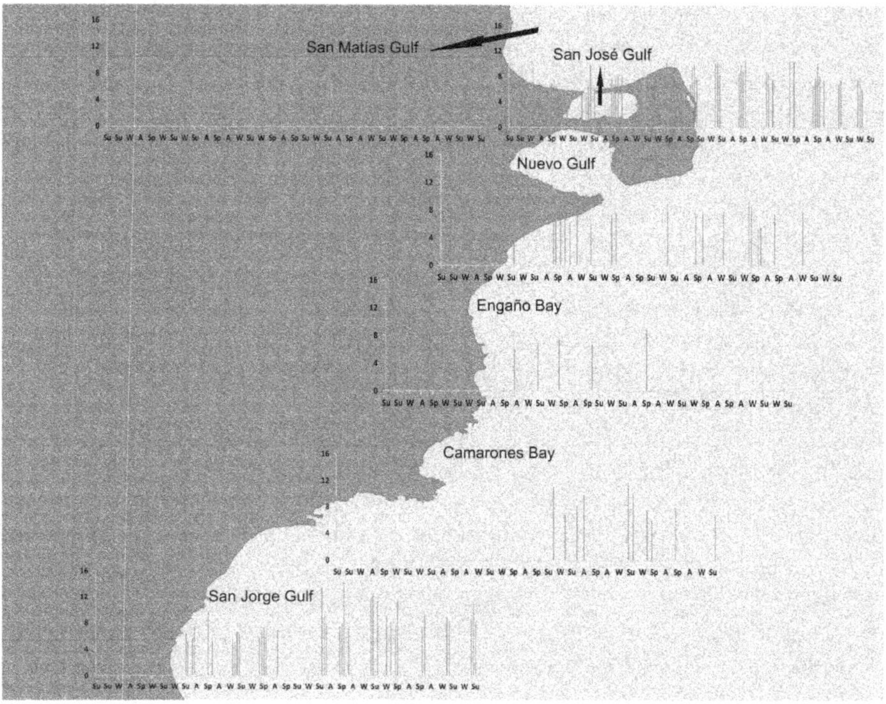

Fig. 5 Seasonal dynamics of *Alexandrium tamarense* density (as ln cell L^{-1}) in Chubut coastal waters between 2000 and 2015

(Fig. 5). This species produced recurrent toxic events during spring and summer associated with a wide interannual variability in toxin concentrations (Andrinolo et al. 1999; Carreto et al. 1996; 1998a, b; Esteves et al. 1992; Reyero et al. 1998; Santinelli et al. 2002).

In the northern zone, cell densities of this species were below 10^4 cells L^{-1} although they led to toxin levels in shellfish, frequently exceeding the regulatory levels established by the World Health Organization (WHO) and by the Argentinian Food Code (AFC) (800 μg STX eq./kg of tissue). The maximum toxin level recorded was 41,114 μg STX eq./kg of tissue in *Aequipecten tehuelchus* (shellfish) in Bengoa station (San José Gulf) in November 2005, following a peak of 6.6×10^3 cells L^{-1} of *A. tamarense* in October. In contrast, high densities of this species were recorded in the Nuevo Gulf in January 1988, (75×10^4 cells L^{-1}), associated with 13,230 μg STX eq./Kg of tissue in *Aulacomya atra* (Esteves et al. 1992) and in December 1993 (2.3×10^6 cells L^{-1}) without toxicity results (Santinelli 2008).

In the southern zone, the highest cell density (7×10^5 cells L^{-1}) and the highest percentage of this dinoflagellate (52%) in relation to the whole phytoplankton community were recorded in January 2010 associated with a toxicity peak of more than 22,000 μg STX eq./kg of shellfish tissue. At this time, five cases of intoxication with a half-hour incubation period and one fatal case occurred in Rada Tilly (Baulde

2010). By late spring (November), the annual maximum of this species (1.9×10^6 cells L^{-1}, 97% of total phytoplankton) was recorded. It should be noted that this has been the record value for all the sampling stations that are part of the monitoring program since 2000. At Belvedere station, the net sample was formed exclusively by *A. tamarense* resting cysts, indicating that the populations were going through an encystment phase. Toxin levels in *Mytilus edulis platensis* reached 88,596 μg STX eq Kg^{-1} of tissue a month later (December).

Blooms occurred in summer with temperatures between 15 and 17 °C and nitrate + nitrite practically depleted. Those from the spring were observed with temperatures between 8 and 10 °C and with undetectable nitrate + nitrite concentrations. One year later (December 2011), a new toxic bloom occurred north of the San Jorge Gulf associated with more than 100,000 μg STX eq Kg^{-1} of shellfish tissue, causing two severe cases of intoxicated people, one of them fatal (Baulde 2011). HPLC analyses of phytoplankton and filter-feeding bivalves, both in the north (Nuevo and San José gulfs) and in the south (San Jorge Gulf), showed that gonyautoxins (GTXs) were the most abundant PSP-toxins. Among these, GTX 1–4 epimers were predominant, and GTX 2 and GTX 3 were present in low concentrations. Small amounts of N-sulfocarbamoyl (C1–C4) and (C1–C2) were detected in mollusk samples but not in phytoplankton and only traces of STX and dc STX in both phytoplankton and shellfish (Reyero et al. 1998; Andrinolo et al. 1999; Sastre et al. 2013).

8 Spatial and Temporal Dynamics of DSP-Toxin-, Pectenotoxin-, and Yessotoxin-Producing Species

Table 1 shows the sites, dates, and environmental conditions in which the highest cell densities of the DSP-toxin-, pectenotoxin-, and yessotoxin-producing species were recorded. The presence of several *Dinophysis* species has been reported since the start of the monitoring program in 2000. *Dinophysis tripos* has also been detected along the coast throughout all the seasons, but peaks of abundance were observed mainly in autumn and winter. It was recorded for the first time in May 2007 with a cell density of 2×10^3 cells L^{-1} in Lobos (San Matías Gulf) and reached a maximum of 26×10^3 cells L^{-1} in Riacho (San José Gulf) in March 2015 (Fig. 6a). *Dinophysis acuminata* was registered along the whole coast (San Matías, San José, Nuevo, and San Jorge gulfs and in Magagna and Camarones) mainly in spring and summer. In November 2007, it reached a maximum density of 5×10^3 cells L^{-1} in the San José Gulf (Larralde) (Fig. 6b). *D. acuminata* is the main agent of chronic and persistent episodes (spring-autumn) of DSP (part of the lipophilic toxin complex), mainly okadaic acid, in bivalves from the Galician coast (Reguera and Pizarro 2008). *Dinophysis fortii* and *D. acuta* were registered in the Nuevo Gulf in January 2006, with a maximum density of 4×10^2 cells L^{-1}. Two species, *Dinophysis caudata* and *Phalacroma rotundatum* (*Dinophysis rotundata*), were

Table 1 Maximum cell density, sampling site and date, and hydrographic data of lipophilic shellfish toxin (LST)-producing species (nutrients in μM)

Species	Cells L^{-1}	Site	Date	T°C	Salinity	NO_3-N	PO_4-P	SiO_2-Si
D. tripos	8400	Bengoa	Aug 2010	11	34.5	2.53	1.20	3.79
D. acuminata	5400	Larralde	Nov 2007	12	34.0	0.03	0.83	0.20
D. caudata	Scarce in net samples	Lobos	Feb 2009	17	34.4	0.16	1.59	0.03
D. fortii	400	Paraná	Jan 2006	17	33.8	0.25	0.78	1.19
D. acuta	440	Paraná	Jan 2006	17	33.8	0.06	0.89	1.53
P. rotundatum	Scarce in net samples	Malaspina	Oct 2005	9	33,7	0,03	0,70	1,07
P. lima	1320	Riacho	May 2004	14	33.7	2	4.17	1.32
P. cordatum	6200	Camarones	Nov 2005	12.5	33.6	0.21	0.95	2.48
P. reticulatum	Scarce in net samples	Pardelas	Oct 2010	12	34.3	0.29	0.73	1.15

not detected in samples for quantitative analysis; they occurred sporadically and were detected in qualitative analysis. *D. caudata* was only found in the San Matias Gulf in summer. These results show that *Dinophysis* species only appeared sporadically in quantitative samples. There are two possible explanations for this finding: on the one hand, the sporadic occurrence in quantitative plankton samples may be due to low cell densities, a common feature among *Dinophysis* spp., which makes it difficult to acquire accurate quantitative information and will often be associated with high counting errors (Reguera et al. 2012). On the other hand, it is known that populations of *Dinophysis* are aggregated in patches or in thin layers of the water column and thus they may escape observation with conventional sampling methods (Escalera et al. 2012).

Other producers of lipophilic toxins present in the study zone were *Prorocentrum lima*, first reported by Santinelli et al. (1995) and related to human intoxications in Puerto Madryn city (Gayoso and Ciocco 2001; Gayoso et al. 2002), *P. cordatum* and *Protoceratium reticulatum*. *P. lima* was most abundant in the northern zone in autumn. *P. cordatum* in the southern coast in summer and *P. reticulatum* was only registered in net samples in the Nuevo Gulf in spring. *D. tripos* could be identified as a pectenotoxin (PTX)-producing species in North Patagonian gulfs and thus most likely responsible for positive DSP mouse bioassays in the region. The PTX-2 production associated with *D. tripos* along the Chubut coast is in accordance with other observations in the Argentine Sea (Fabro et al. 2015) as well as in other regions (Rodríguez et al. 2012). Our study suggests that *D. tripos* blooms associated with the presence of DSP in shellfish are becoming a recurrent phenomenon in the North

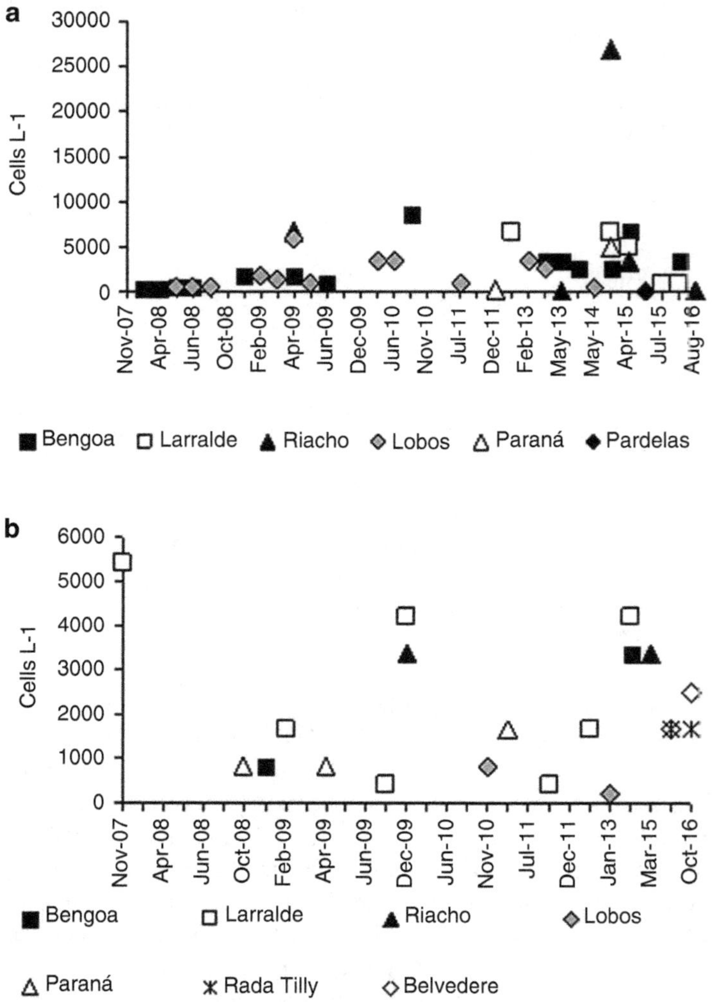

Fig. 6 Cell densities of *Dinophysis tripos* (**a**) and *D. acuminata* (**b**) in different locations of the Chubut coastal area between November 2007 and August/October 2016

Patagonian gulfs (Gracia Villalobos et al. 2015). However, *D. tripos* has never been cited as the causative agent of DSP events when it was the only or the overwhelmingly dominant species of *Dinophysis* in the micro-sized phytoplankton (Reguera et al. 2014). In contrast, *D. acuminata* has been identified as the causative agent of DSP in Southern Brazil (Proença et al. 2007) and, combined with *D. caudata*, in Uruguay (Méndez and Ferrari 2002) and Argentina (Sar et al. 2010, 2012; Sunesen et al. 2014). A coastal species, *D. acuminata*, has a strong negative impact on shellfisheries, because it is an early blooming species with a very long growing season (spring to autumn). This is the most cosmopolitan *Dinophysis* species associated with DSP events (Reguera et al. 2014). Between 2009 and 2016, fishing

Table 2 Maximum cell density, sampling site and date, and hydrographic data of amnesic shellfish toxin (AST)-producing species (nutrients in μM)

Species	Cells L^{-1}	Site	Date	T°C	Salinity	NO_3-N	PO_4-P	SiO_2-Si
P. australis	394,800	Larralde	Nov 2009	12	33.9	0.00	0.78	0.26
P. pungens	2,197,440	Riacho	Nov 2010	12	34.2	0.55	0.86	1.19
P. fraudulenta	4,174,400	Larralde	Dec 2012	13	–	0.00	0.97	0.34
P. calliantha	3,292,800	Bengoa	Dec 2010	13	34.0	0.00	0.96	0.37

authorities had to implement 15 closures for commercial extraction due to DSP-toxins in the Chubut coastal zone.

9 Spatial and Temporal Dynamics of *Pseudo-nitzschia* spp.

P. australis, *P. pungens*, *P. multiseries*, and *P. pseudodelicatissima* have been cited in North Patagonian gulfs (San Matías, San José, and Nuevo Gulfs) (Sastre et al. 1995, 2001). Later, following the taxonomic review by Lundholm et al. (2003), it was verified that according to the ultrastructure of the areolae, the latter species corresponded to *P. calliantha.* Table 2 shows the sites, dates, and environmental conditions in which the highest cell densities of the amnesic toxin-producing species were recorded.

P. fraudulenta has been the most common species in Chubut coastal waters. It has been present in all the sampling stations and showing the highest cell densities. This species, together with *P. calliantha* and *P. pungens*, has several times exceeded densities of 10^6 cells L^{-1}, in the northern gulfs (San Matías, San José, and Nuevo) reaching levels between 2 and 4 × 10^6 cells L^{-1} in the San José Gulf (Table 2).

The highest densities of *Pseudo-nitzschia* spp. were observed between 2008 and 2012. The ASP-toxin, domoic acid (DA), was detected in continental shelf waters of the Argentine Sea (Southwestern Atlantic Ocean) in July 2000 associated with *P. australis* (Negri et al. 2004). In Chubut coastal waters, this ASP-toxin was measured in phytoplankton concentrates from the Nuevo Gulf (Pardelas station) and Camarones in October 2005, rich in *P. fraudulenta* and *P. pungens* (Sastre et al. 2007). Both species co-occurred during the spring bloom at the Pardelas and Camarones stations. Blooms coincided with nutrient decline and increased values of temperature and Secchi disk depth reads. The species *P. australis*, *P. fraudulenta, and P. pungens* are worldwide known as DA producers (Lelong et al. 2012).

10 Accumulation and Transport of Some Phycotoxins Through Pelagic Food Webs

Phycotoxins in Chubut coastal waters are concentrated by primary consumers as filter-feeding bivalve mollusks, such as *Mytilus edulis platensis*, *Aulacomya atra*, *Aequipecten tehuelchus*, *Ameghinomya antiqua*, *Ensis macha*, and *Panopea abbreviata*, and by secondary consumers, such as gastropods *Odontocymbiola magellanica*, *Buccinanops cochlidium*, *Buccinanops globulosus*, and *Trophon geversianus*. When PSP-toxin levels in these species exceed the regulatory limits (800 μg Kg^{-1} of shellfish tissue) established by the World Health Organization (WHO), the fishing authority implements closures for commercial extraction. These closures mainly affect the artisanal fisheries and mariculture between September and March.

In addition, Cadaillón (2012) demonstrated the transfer of these phycotoxins (PSP and DA) to zooplankton, the next trophic link in the planktonic food chain, in the study area. This author showed that during spring 2010, ASP-toxins reached a maximum concentration of 42.78 μg DA g $tissue^{-1}$ in zooplankton from Riacho, associated with a sharp increase of *P. pungens* and *P. fraudulenta* populations. In addition, *P. australis* and *P. calliantha* were detected in net samples. Overall, species of the genus *Pseudo-nitzschia* represented 81% of the total phytoplankton community. *A. tamarense* was present in this gulf since the end of winter (September), and STX levels in phytoplankton reached a maximum of 1.866 μg STX eq g $cell^{-1}$ in Bengoa and 3.380 μg STX eq g $tissue^{-1}$ in zooplankton in Larralde (November). In the Nuevo Gulf, the maximum toxin level in zooplankton, 25.75 μg DA g $tissue^{-1}$, was associated with a bloom of *P. fraudulenta*, when this species represented 90% of the total phytoplankton.

Undoubtedly, the presence of toxin-producing phytoplankton species and the possibility of transfer of their toxins to the zooplankton pose a risk to marine organisms, such as fish, birds, and mammals, which are zooplankton consumers belonging to a unique ecosystem of great tourist interest. D'Agostino et al. (2015) found fragments of *Pseudo-nitzschia* spp. frustules in all the southern right whale *Eubalaena australis* fecal samples analyzed. At least four taxa, *P. australis*, *P. fraudulenta*, *P. pungens*, and the *P. pseudodelicatissima* complex, were identified in the feces and water samples analyzed. Micro-sized crustacean remains, mainly copepodite five mandibular gnathobases of *Calanus australis* (a common species in the Argentine Sea), were also found in fecal samples of live and dead whales from the San José Gulf (D'Agostino et al. 2016). These findings indicate that southern right whales may have been exposed to DA, while feeding in this area and copepods could have acted as the main vector of this neurotoxin.

In recent years, there have been an increased number of dead whales found in the Valdés Peninsula area (753 dead whales from 2003 to 2016, Southern Right Whale Health Monitoring Program), and 4 main hypotheses were proposed to explain this phenomenon (Rowntree et al. 2013). One of them, addressed in this study, is the transfer of marine biotoxins along the food web from producing organisms to southern right whales through mesozooplanktonic vectors. Trace levels of STXs and DA were detected in samples of feces, urine, and tissues collected from dead individuals of *E. australis* in the Valdés Peninsula (Uhart et al. 2009; Rowntree et al. 2013). Southern right whale mothers and their calves are exposed to biotoxin-producing algae in the Valdés Peninsula area. High-risk levels of *Pseudo-nitzschia* spp. and *A. tamarense* occurred between 2007 and 2013, the period with the highest number of dead whales (>50 deaths/year) (Wilson et al. 2016). The occurrence of toxic *Pseudo-nitzschia* spp. blooms concurrently with the whale season and the detection of high levels of DA (710 μg DA g^{-1} dry weight) in southern right whale's feces, in some cases higher than those reported during marine mammal mortality events, demonstrate the natural risk to which the whales are exposed during their stay in this area (D'Agostino et al. in press).

11 Conclusions

Distinct environmental conditions in the northern and southern areas of Chubut coastal waters and large differences between seasons have been observed. The composition and abundance of harmful microalgae species showed a large site-specific interannual variability. Some years presented a good diversity of harmful species, whereas others had an almost exclusive dominance of a single species. The PSP-toxin producer *Alexandrium tamarense* was present throughout the entire coast of the Chubut province. In the north, maximal cell densities appeared by the end of winter and during spring; however, it was not necessary to reach high densities to cause toxic events (low biomass toxic HABs). In contrast, cell maxima occurred in spring-summer in the south, and densities exceeding one million cells per liter were necessary to produce a high accumulation of toxins. The similarities between the PSP-toxin profiles in the filter-feeding bivalves and in the co-occurring phytoplankton strongly support the view that the dinoflagellate *A. tamarense* is the source of PSP-toxin contamination in the Valdés Peninsula. GTX1 and STX were the most powerful PSP-toxins in the toxin profile. The predominance of GTX1 may have been decisive in the intoxication episodes registered in the San Jorge Gulf in 2010 and 2011. In addition to human illness, PSP-toxins are cause of major economic losses due to the commercial shellfish harvesting closures.

Among the producers of DSP-toxins and pectenotoxins, high cell densities of *D. tripos* were found mainly in autumn and winter but restricted to the San Matías and San José gulfs, whereas *D. acuminata* peaked in December and was not frequent in all the stations. Closures to commercial shellfish extraction due to DSP-toxins associated with *D. tripos* are becoming a recurrent phenomenon in the North Patagonian gulfs. The other lipophilic toxin-producing species found in the Chubut province coastal zone occurred only sporadically and in low densities.

Although several species of the genus *Pseudo-nitzschia* cited as toxic in other parts of the world occur in high cell densities in Chubut, no human poisoning by

DA has been reported in the region. It has not been confirmed if toxin accumulation has been the probable cause of death in stranded whales. However, it has been demonstrated that whales are exposed to toxins (PSP-toxins and DA) in the Valdés Peninsula since their stay coincides with the toxic phytoplankton bloom season, and it has been confirmed that they feed on contaminated zooplankton in this breeding area.

Chubut coastal waters are an adequate environment for the development of Harmful Algal Blooms and the production of phycotoxins. Therefore, it is necessary to maintain a permanent monitoring to take measures to prevent human poisoning and to understand possible effects on the marine fauna as well as to initiate new lines of research including molecular analysis and the use of satellite images for the early detection of the blooms.

Acknowledgments We are grateful to Dr Martha Ferrario for her great help in identifying species of the genus *Pseudo-nitzschia* and the divers and helmsmen of the Provincial Plan for Prevention and Control of Red Tide (Chubut province) for the collection of phytoplankton samples and to the Ministry of Fisheries of Chubut province for providing data on timing of shellfish harvesting closures. This work was supported by Project 10/c312, "Atlas de Microalgas Marinas Nocivas de la Costa de la Provincia de Chubut" from Universidad Nacional de la Patagonia San Juan Bosco.

References

Andersen RJ, Blackburn SI, Taylor FJR et al (1995) Algal culture collections and toxic algal strains. In: Hallegraeff GM, Anderson DM, Cembella AD (eds) Manual on harmful marine microalgae. IOC, UNESCO, Paris, pp 489–531

Anderson DM, Cembella AD, Hallegraeff GM (2012) Progress in understanding harmful algal blooms: paradigm shifts and new technologies for research, monitoring, and management. Annu Rev Mar Sci 4:143–176

Andrade A (2001) Brote de Intoxicación por Veneno Paralizante de Molusco. Informe de la Dirección de Salud Ambiental, Chubut, diciembre 2000, p 6

Andrade A (2002) Brote de Intoxicación por Veneno Paralizante de Molusco. Informe de la Dirección de Salud Ambiental, Chubut, octubre 2002, p 4

Andrinolo D, Santinelli N, Otaño S et al (1999) Paralytic shellfish toxins in mussel and *Alexandrium tamarense* at Valdés Península, Chubut, Patagonia, Argentina: kinetics of a natural depuration. J Shellfish Res 18(1):1–17

Balech E (1995) The genus *Alexandrium* Halim (Dinoflagellata). Sherkin Island Marine Station, Sherkin Island, p 151

Baulde D (2010) Brote de Intoxicación por Veneno Paralizante de Molusco. Informe de la Dirección de Salud Ambiental, Chubut, enero 2010, p 7

Baulde D (2011) Brote de Intoxicación por Veneno Paralizante de Molusco. Informe de la Dirección de Salud Ambiental, Chubut, diciembre 2011, p 6

Cadaillón AM (2012) Floraciones algales nocivas y ficotoxinas en fitoplancton y zooplancton de los Golfos Nuevo y San José. Tesis de grado, Facultad de Ciencias Naturales, Sede Puerto Madryn, Universidad Nacional de la Patagonia San Juan Bosco, Chubut, Argentina, p 168

Carreto JI, Benavides HR (1993) World record of PSP in Southern Argentina. Harm Alg News, IOC, UNESCO 5:2

Carreto JI, Lasta M, Negri RM et al (1981) Los fenómenos de Marea Roja y toxicidad de moluscos bivalvos en el Mar Argentino. Contr INIDEP 399:55

Carreto JI, Elbusto C, Sancho H et al (1996) Comparative studies on paralytic shellfish toxins profiles of marine snails, mussels and an Alexandrium tamarense isolated from the Mar del Plata coast (Argentina). Rev Invest Desarr Pesq 10:101–107

Carreto JI, Akselman R, Montoya NG et al (1998a) *Alexandrium tamarense* blooms dynamics and *Mitylus edulis* toxicity in coastal waters of mar del Plata (Argentina). In: Reguera B, Blanco J, Fernández ML, Wyatt T (eds.) Harmful Microalgae, IOC, UNESCO, Santiago de Compostela, pp 135–138

Carreto JI, Montoya NG, Cucchi Colleoni AD et al (1998b) Alexandrium tamarense blooms and shellfish toxicity in the Argentine Sea: a retrospective view. In: Reguera B, Blanco J, Fernández ML, Wyatt T (eds.) Harmful Microalgae, IOC, UNESCO, Santiago de Compostela, pp 131–134

Clément A, Lembeye G (1993) Phytoplankton monitoring programme in the fish farm region of the South of Chile. In: Smayda TJ, Shimizu Y (eds) Toxic phytoplankton blooms in the sea. Elsevier, Amsterdam, pp 223–228

D'Agostino VC, Hoffmeyer MS, Almandoz GO et al (2015) Potentially toxic pseudo-nitzschia species in plankton and fecal samples of *Eubalaena australis* from Península Valdés calving ground, Argentina. J Sea Res 106:39–43

D'Agostino VC, Hoffmeyer MS, Degrati M (2016) Fecal analysis of southern right whales (*Eubalaena australis*) in Península Valdés calving ground, Argentina: *Calanus australis*, a key prey species. J Mar Biol Assoc UK 96(4):859–868

D'Agostino VC, Degrati M, Sastre V et al (in press) Domoic acid in marine pelagic food web: Exposure of southern right whale *Eubalaena australis* in Península Valdés calving ground, Argentina. Harm Al

Doucette GJ, Cembella AD, Martin JL et al (2006) Paralytic shellfish poisoning (PSP) toxins in North Atlantic right whales *Eubalaena glacialis* and their zooplankton prey in the Bay of Fundy, Canada. Mar Ecol Prog Ser 306:303–313

Escalera L, Pazos Y, Doval MD et al (2012) A comparison of integrated and discrete depth sampling for monitoring *Dinophysis*. Pollut Bull 64:106–113

Esteves JL, Santinelli N, Sastre V et al (1992) A toxic dinoflagellate bloom and PSP production associated with upwelling in Golfo Nuevo, Patagonia Argentina. Hydrobiology 242:115–122

Fabro E, Almandoz GO, Ferrario ME et al (2015) Co-occurrence of *Dinophysis tripos* and pectenotoxins in Argentinean shelf waters. Harmful Algae 42:25–33

Fernández ML, Miguez A, Cacho E et al (2002) Bioensayos con mamíferos y ensayos bioquímicos y celulares para la detección de toxinas. En: Sar EA, Ferrario ME, Reguera B (eds.) Floraciones Algales Nocivas en el Cono Sur Americano, Instituto Español de Oceanografía, Madrid, pp 77–120

Fire SE, Van Dolah FM (2012) Marine toxins emergence of harmful algal blooms as health threats to marine wildlife. In: Aguirre AA, Ostfeld R, Daszak P (eds) New directions in conservation medicine: applied cases of ecological health. Oxford University Press, New York, pp 374–389

Fraga S, Sampedro N, Larsen J et al (2015) Arguments against the proposal 2302 by John et al to reject the name *Gonyaulax catenella* (*Alexandrium catenella*). Taxon 64(3):634–635

Fryxell GA, Hasle GR (2004) Taxonomy of harmful diatoms. In: Hallegraeff GM, Anderson DM, Cembella AD (eds) Manual on harmful marine microalgae, IOC. UNESCO, Paris, pp 465–509

Gayoso AM, Ciocco NF (2001) Observations on *Prorocentrum lima* of North-Patagonian coastal waters (Argentina) associated with diarrhoeic disease episode. Harm Alg News, IOC, UNESCO 22: 4

Gayoso AM, Dover S, Morton SL et al (2002) Possibility of diarrheic shellfish poisoning associated with *Prorocentrum lima* (Dinophyceae) in Patagonian Gulfs (Argentina). J Shellfish Res 21:461–463

Gracia Villalobos L, Santinelli N, Sastre V et al (2015) *Dinophysis* species associated with Diarrhetic Shellfish Poisoning (DSP) episodes in North Patagonian gulfs (Chubut, Argentina). J Shellfish Res 34(3):1141–1149

Hallegraeff GM (2004) Harmful algal blooms: a global overview. In: Hallegraeff GM, Anderson DM, Cembella AD (eds) Manual of harmful marine microalgae, IOC. UNESCO, Paris, pp 25–49

Hallegraeff GM, Hara Y (2004) Taxonomy of harmful marine raphidophytes. In: Hallegraeff GM, Anderson DM, Cembella AD (eds) Manual of harmful marine microalgae. IOC, UNESCO, Paris, pp 511–522

Hargraves PE, Maranda L (2002) Potentially toxic or harmful microalgae from the northeast coast. Northeast Nat 9:81–120

Hasle GR, Fryxell GA (1970) Diatoms: cleaning and mounting for light and electron microscopy. Trans Am Microsc Soc 89(4):469–474

Hasle GR, Fryxell GA (1995) Taxonomy of diatoms. In: Hallegraeff GM, Anderson DM, Cembella AD (eds) Manual of harmful marine microalgae. IOC, UNESCO, Paris, pp 339–364

John U, Litaker RW, Montresor M (2014) Formal revision of the *Alexandrium tamarense* species complex (Dinophyceae) taxonomy: the introduction of five species with emphasis on molecular-based (rDNA) classification. Protist 165:779–804

Leandro LF, Rolland RM, Roth PB et al (2010) Exposure of the North Atlantic right whale *Eubalaena glacialis* to the marine algal biotoxin, domoic acid. Mar Ecol Prog Ser 398:287–303

Lefebvre KA, Powell CL, Busman M et al (1999) Detection of domoic acid in northern anchovies and California Sea lions associated with an unusual mortality event. Nat Toxins 7 85–92

Lelong A, Hégaret H, Soudant P et al (2012) *Pseudo-nitzschia* (Bacillariophyceae) species, domoic acid and amnesic shellfish poisoning: revisiting previous paradigms. Phycologia 51:168–216

Lilly EL, Halanych KM, Anderson DM (2007) Species boundaries and global biogeography of the *Alexandrium tamarense* complex (Dinophyceae). J Phycol 43:1329–1338

Lundholm N, Moestrup O, Hasle GR et al (2003) A study of the *Pseudo-nitzschia pseudodelicatissima/cuspidata* complex (Bacillariophyceae): what is *P. pseudodelicatissima*? J Phycol 39:797–813

Méndez, S, Ferrari G (2002) Floraciones algales nocivas en Uruguay: Antecedentes, proyectos en curso y revisión de resultados. En: Sar EA, Ferrario ME, Reguera B (eds.) Floraciones algales nocivas en el Cono Sur Americano, Instituto Español de Oceanografía, Madrid, pp 271–288

Negri RM, Montoya NG, Carreto JI et al (2004) *Pseudo-nitzschia australis, Mytilus edulis, Engraulis anchoita*, and domoic acid in the argentine sea. In: Steidinger K, Landsberg JH, Tomas CR, Vargo GA (eds.). Harmful and toxic algal blooms, IOC, UNESCO, pp 139–141

Pérez LB, Santinelli N, Sastre V et al (2013) Morfología y presencia estacional de Alexandrium tamarense (Lebour) Balech en estaciones costeras del Golfo San Jorge (Patagonia, Argentina). Rev Natur Patag 6:50–60

Proença LAO, Schramm MA, Tamanaha MS et al (2007) Diarrhetic shellfish poisoning (DSP) outbreak in subtropical Southwest Atlantic. Harm Alg News 33:19–20

Prud'homme van Reine W, Willem F (2017) Report of the Nomenclature Committee for Algae: 15. Taxon 66:191–192

Reguera B (2002) Establecimiento de un programa de seguimiento de microalgas tóxicas. En: Sar EA, Ferrario ME, Reguera B (eds.) Floraciones Algales Nocivas en el Cono Sur Americano. Instituto Español de Oceanografía, Madrid, pp 19–54

Reguera B, Pizarro G (2008) Planktonic dinoflagellates which produce polyether toxins of the old DSP complex. In: Botana LM (ed) Seafood and freshwater toxins: pharmacology, physiology and detection, 2nd edn. CRC Press, London, pp 257–284

Reguera B, Velo-Suárez L, Raine R et al (2012) Harmful *Dinophysis* species: a review. Harmful Algae 14:87–106

Reguera B, Riobó P, Rodríguez F et al (2014) *Dinophysis* toxins: causative organisms, distribution and fate in shellfish. Mar Drugs 12:394–461

Reyero M, Santinelli N, Otaño S et al (1998) Toxins profiles of PSP from molluscs and phytoplankton containing *Alexandrium tamarense* (Lebour) Balech in two patagonian gulfs (Argentina). In: Reguera B, Fernández ML, Wyatt T (eds) Harmful Algae. Xunta de Galicia and IOC–UNESCO, pp 507–508

Rodríguez F, Escalera L, Reguera B et al (2012) Morphological variability, toxicology and genetics of the dinoflagellate *Dinophysis tripos* (Dinophysiaceae, Dinophysiales). Harmful Algae 13:26–33

Rowntree VJ, Uhart MM, Sironi M et al (2013) Unexplained recurring high mortality of southern right whale *Eubalaena australis* calves at Península Valdés, Argentina. Mar Ecol Prog Ser 493:275–289

Santinelli N (2008) Fitoplancton de un ambiente costero sometido a perturbación antrópica: Bahía Nueva, Provincia del Chubut. Tesis Doctoral, Facultad de Ciencias Naturales, Sede Trelew, Universidad Nacional de la Patagonia San Juan Bosco, Chubut, Argentina, p 176

Santinelli N, Caille G, Lettieri A (1995) Harmful algae and PSP toxicity along the northern Patagonian coast. Harmful algae News, IOC, UNESCO 9:6

Santinelli N, Sastre V, Esteves JL (2002) Episodios de algas nocivas en la Patagonia Argentina. En: Sar EA, Ferrario ME, Reguera B (eds.) Floraciones Algales Nocivas en el Cono Sur Americano. Instituto Español de Oceanografía, Madrid, pp 197–208

Sar EA, Sunesen I, Lavigne A et al (2010) Dinophysis spp. asociadas a detección de toxinas diarreicas (DSTs) en moluscos y a intoxicación diarreica en humanos (Provincia de Buenos Aires, Argentina). Rev Biol Mar Oceanogr 45:451–460

Sar EA, Sunesen I, Goya AB et al (2012) First report of diarrheic shellfish toxins in mollusks from Buenos Aires Province (Argentina) associated with *Dinophysis* spp.: evidence of okadaic acid, dinophysis toxin-1 and their acyl derivatives. Bol Soc Argent Bot 47:5–14

Sastre AV, Santinelli NH, Esteves JL (1995) Fitoplancton del Golfo San Matías de tres campañas de muestreo (Noviembre'92, Septiembre'93 y Marzo'94). Physis Sec A 53(124–125):7–12

Sastre AV, Santinelli NH, Esteves JL et al (2001) Aspectos ecológicos de especies de *Pseudo-nitzschia* en aguas costeras patagónicas. En: Alveal K, Antezana T (eds.) Sustentabilidad de la biodiversidad. Univ de Concepción, Concepción, pp 217–235

Sastre V, Santinelli N, Marino G et al (2007) First Domoic acid detection produced by *Pseudo-nitzschia* species, Chubut coastal waters, Patagonia, Argentina. Harm Alg News, IOC, UNESCO 34:12–14

Sastre AV, Santinelli NH, Willers V et al (2013) Floraciones de *Alexandrium tamarense* y TPM en el Golfo San Jorge. XXXIV Jornadas Argentinas de Botánica. Bol Soc Arg Bot 48(Supl):175–176

Scholin CA, Gulland F, Doucette GJ, Benson S, Busman M, Chavez FP, Cordaro J, DeLong R, De Vogelaere A (2000) Mortality of sea lions along the Central California coast linked to a toxic diatom bloom. Nature 403:80–83

Smayda TJ (2006) Harmful algal bloom communities in Scottish coastal waters: Relationship to fish farming and regional comparisons –A Review. Scottish Executive Environment Group. [online]. http://www.scotland.gov.uk/publications

Suárez-Isla B, Guzmán L (1999) Mareas rojas y toxinas marinas. Serie Orientaciones en Ciencias, Tecnología y Cultura, Editorial Universitaria, Santiago, Chile, p 77

Sunesen I, Lavigne AS, Goya AB et al (2014) Episodios de toxicidad en moluscos de aguas marinas costeras de la provincia de Buenos Aires (Argentina) asociados a algas toxígenas (marzo de 2008–marzo de 2013). Bol Soc Argent Bot 49(3):327–339

Uhart M, Rowntree VJ, Sironi M et al (2009) Continuing southern right whale mortality events at Península Valdés, Argentina. Report IWC SRW Assessment Workshop SC/61/BRG18

Van Dolah FM (2000) Marine algal toxins: origins, health effects, and their increased occurrence. Environ Health Perspect 108:133–141

Vecchio J, Tartaglione J, Orozco J et al (1986) Intoxicación por VPM (Marea Roja). Medicina 6:705–707

Wang DZ (2008) Neurotoxins from marine dinoflagellates: a brief review. Mar Drugs 6:349–371

Wilson C, Sastre AV, Hoffmeyer M et al (2016) Southern right whale (Eubalaena australis) calf mortality at peninsula Valdes, Argentina: are harmful algal blooms to blame? Mar Mam Sci 32(2):423–451

Part VII
Plankton, Climate Change and Human-Induced Changes

Composition and Structure of Phytoplankton Communities in Coastal Environments with Anthropogenic Disturbance (Patagonia, Argentina)

Norma H. Santinelli, Alicia V. Sastre, Mónica N. Gil, and José L. Esteves

Abstract The purpose of this study is to review as well as to compare environmental and phytoplankton data collected from coastal Patagonian ecosystems (Argentina) located at different latitudes and exposed not only to freshwater discharges (Chubut river in Chubut province, Deseado river in Santa Cruz province, Pipo and Olivia rivers in Tierra del Fuego province) but also to urban, industrial, and port activities related to the main cities on the Patagonian coast. The data analyzed were collected during the spring, summer, and fall of 1993, 1994, and 2001. The presence of Bacillariophyta, the dominant phylum in most environments, is the main characteristic of the composition and structure of phytoplankton community in the study area. A total of 127 taxa were identified, 9 of which are known to be potentially toxic species. Taxa richness was found to be highest in Nueva and Engaño Bays, the order of abundance being 10^6 cells L^{-1} in the majority of the environments of our study area. Blooms of *Alexandrium tamarense* and *Pseudo-nitzschia* cf. *pseudodelicatissima* were recorded in Nueva Bay in the summer of 1993, whereas blooms of *Skeletonema costatum* were recorded in Deseado Ría in the fall of 1994. Maximum biomass was recorded in Ushuaia Bay with 10 μg/L Chl *a*. N:P ratios indicated nitrogen limitation in the majority of the environments analyzed. Changes in the structure of the microalgal community, bloom occurrence, and the presence of organisms indicating environmental alterations associated with eutrophication

N. H. Santinelli (✉) · A. V. Sastre
Instituto de Investigación de Hidrobiología, Facultad de Ciencias Naturales y Ciencias de la Salud, Universidad Nacional de la Patagonia San Juan Bosco, Trelew, Argentina

M. N. Gil
Centro para el Estudio de Sistemas Marinos (CESIMAR), Consejo Nacional de Investigaciones Científicas y Técnicas (CONICET), Puerto Madryn, Argentina
e-mail: monicagil@cenpat-conicet.gob.ar

J. L. Esteves
Fundación Patagonia Natural, Puerto Madryn, Argentina

M. S. Hoffmeyer et al. (eds.), *Plankton Ecology of the Southwestern Atlantic*,
https://doi.org/10.1007/978-3-319-77869-3_23

events in the coastal zones could all be bioindicators of changes in water quality and incipient eutrophication processes in the studied environments.

Keywords Phytoplankton · Nutrients · Eutrophication · Toxic species · Argentine coast

1 Introduction

The Patagonian Sea in Argentina is known as an area with high primary production and biodiversity. The complex hydrodynamic environment resulting from the interaction between the Malvinas-Brazil convergence and the waters over the continental shelf and slope enhances the development of high chlorophyll concentrations, especially in frontal and coastal areas (Vega Moreno et al. 2012).

The Argentine coast has an estimated length of more than 6880 km, of which over 70% belongs to the Patagonian coast.: 334 km for the Río Negro province, 1634 km for Chubut province, 1749 km for Santa Cruz province, and 1105 km for the Tierra del Fuego province. The coast is characterized by a wide variety of environments and productive zones; its continental shelf is known to be rich in marine resources. Both the Argentine Sea and coast are therefore of great interest for conservation as well as for local and regional economies. However, pollution, a direct consequence of human activities, is one of the main threats to the marine biological diversity that characterizes this area (González and Esteves 2008).

Between the Río Negro (41 °S) and Tierra del Fuego (55 °S) provinces (1,548,922 inhabitants, Indec 2012), 21 populated areas with 800,000 inhabitants are either on or near the coast. Among them, the most important cities are San Antonio Oeste, Puerto Madryn, Rawson (and its zone of influence), Comodoro Rivadavia, Caleta Olivia, Puerto Deseado, Río Gallegos, and Ushuaia. They are located in sheltered or exposed zones, which either broaden or mitigate, respectively, the environmental impact of industrial, oil, and urban activities.

Nutrients, which are considered to be one of the most important variables in marine environments, exert a strong influence on the growth, reproduction, and metabolic activities of living beings. The distribution of nutrients depends on several factors, namely, the season, the tidal conditions, the water flow that reaches the coastal sector from the continent (Achary et al. 2014), and the oceanographic conditions of adjacent water masses.

The continuous discharge of organic material into coastal environments may lead to nutrient enrichment, namely, eutrophication (Nixon 2009). However, when in excess, nutrients give rise to what is known as blooms, i.e., an excessive growth of phytoplankton populations. Redfield's N:P ratio in estuaries or coastal waters indicates phytoplankton growth patterns: an N:P ratio ≤ 5 indicates nitrogen-limited growth, an N:P ratio ≥ 10 indicates phosphate-limited growth, while an N:P ratio between 5 and 10 is an intermediate state (Duarte 2009).

Marine eutrophication stimulates primary production and, in its most serious manifestations, leads to visible algal blooms, algal foams, excessive growth of benthic algae, and massive growths of submerged and floating macrophytes. There is a general consensus that algal blooms and toxicity, which affect marine biota and humans, are present in eutrophicated coastal areas (Hallegraeff 2004, Lassus et al. 2016).

Among the several biological indicators of possible changes in the trophic state of coastal and marine ecosystems, phytoplankton community analysis is particularly useful thanks to the speed and magnitude of its response to environmental disturbances (De la Lanza-Espino et al. 2000, Moreira et al. 2007).

In this chapter, we review and compare information from the following Argentine coastal ecosystems: San Antonio Bay (San Matías Gulf), Nueva Bay (Nuevo Gulf, which forms the southwestern and southern boundaries of Valdés Península), Engaño Bay (which includes freshwater discharge from Chubut river), Deseado Ría, and Ushuaia and Golondrina Bays in the Beagle Channel. All of them are influenced by the impact of coastal cities and ports, such as San Antonio Oeste, Puerto Madryn, Puerto Rawson, Puerto Deseado, and Ushuaia. Although several studies have been performed in these areas (Santinelli and Esteves 1993, Ferrario and Sastre 1990, Santinelli et al. 1990, Sastre et al. 1990, Esteves et al. 1992, 1996a, b, 1997; Benavides et al. 1995, Esteves and Amín 2005), no comparative analyses of their phytoplankton structures have been performed to date.

In view of this, the purpose of the present study was to analyze phytoplankton structure and dynamics, giving particular attention to noxious species living in Patagonian coastal environments located at different latitudes and exposed to anthropogenic disturbances. To this, the specific environmental conditions under which the studied communities developed during the spring, summer, and fall of 1993, 1994, and 2001 were taken into account.

2 Study Area

Figure 1 shows the location of the study area in the Patagonian coastal zone, Argentina, and Fig. 2 shows details of each of the sampling stations. The sampling strategy followed has been described in detail in Esteves et al. (1996a, b, 1997), Esteves and Amín (2005), and Santinelli (2008).

San Antonio Bay (40° 42′ and 40° 49′ S and 64° 45′ and 65° 00′ W) is situated in the northwestern sector of San Matías Gulf (Fig. 2a). San Antonio Oeste city, where fishing activities are one of the main sources of livelihood, and San Antonio Este port are both located in this gulf. Las Grutas, a well-known seaside resort, is nearby but outside San Antonio Bay. The 1991 official census reported that the population of these cities reached a total of 24,216 inhabitants (Indec 1991). San Antonio Oeste and San Antonio have neither industrial nor domestic sewage systems. Lead and other metal pollutants coming from mining residues that have begun to be cleaned have been recorded in the inner sector of San Antonio Bay. Las Grutas

Fig. 1 Map of the study area in the coastal zone of Patagonia (Argentina)

is known to have a secondary treatment system consisting in open-air pools, and the remaining effluents go to San Matías Gulf. In view of this, water quality evolution was studied from the physical, chemical, and phytoplanktonic points of view in Villarino beach (northern coast) and Las Grutas (western coast) in December 1994.

Nueva Bay (42° 46′ S, 65° 02′ W) is located within Nuevo Gulf (Fig. 2b). When the present study was conducted, the effluents derived from urban, industrial, and port activities in Puerto Madryn (50,000 inhabitants in the 1990s) were discharged with or without previous treatment. Samplings were carried out at the stations located in the bay and at two reference stations. Two field trips were carried out in December 1993 and November 1994. The remarkable development of different

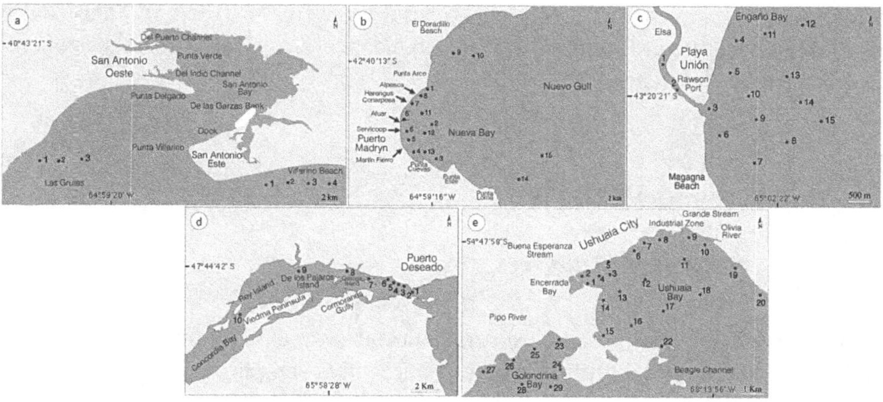

Fig. 2 Location of the sampling stations at each of the sites analyzed. (**a**) San Antonio Bay (summer 1994), (**b**) Nueva Bay (spring and summer 1993–1994), (**c**) Engaño Bay (spring and summer 1993–1994), (**d**) Deseado Ría (autumn 1994), (**e**) Golondrina and Ushuaia Bays (summer 2001)

macroalgal species, particularly of *Ulva* sp., in Nueva Bay could be considered an indicator of urban pollution. Furthermore, blooms of the invasive species *Undaria pinnatifida* were reported to have changed the biotic characteristics of this system (Piriz and Casas 1994, Casas and Piriz 1996).

In Engaño Bay (43° 18′ S, 65° 06′ W, Fig. 2c), Chubut river flows out forming an estuary that receives anthropogenic contributions not only from port and fishing activities but also from urban sewage of upstream localities, namely, Rawson, Trelew, Gaiman, and Dolavon. Freshwater contribution to Engaño Bay comes mainly from Chubut river and also brings nutrients and pollutants that are likely to alter the ecosystem. It represents an average flow of 49 m^3/s with minimum and maximum values of 4 and 261 m^3/s, respectively. The following can be mentioned as different contributions that may affect water quality: irrigation system drainage, Rawson sewage system, Rawson port, and bycatch discards from ships belonging to the so-called yellow fleet. Two field trips were carried out between 1993 and 1994 in Engaño Bay to analyze the impact of nutrients coming from Chubut river, their distribution, and phytoplankton abundance and structure.

Puerto Deseado city is located in Deseado Ría (51° 37′ S, 69° 13′ W, Fig. 2d). It has undergone an important increase in industry, particularly in fishing and port activities, both of which add their influence to that exerted by the effluents and solid wastes from the urban sector. A field trip was carried out in this coastal system in May 1994 in order to study the influence of human activities on the spatial distribution of nutrients and on the structure of phytoplankton population.

Ushuaia city is located on the coastal area of Ushuaia and Golondrina Bays (54° 49′ S, 68° W, Fig. 2e). The impact of urban, industrial, and port activities is a matter of concern in these ecosystems. When our study was carried out, part of the sewage coming from these activities was collected and discharged into Golondrina Bay after a primary treatment, via a submarine pipeline. The remaining fraction was introduced into rainwater drains and finally reached the coastal zone of Ushuaia Bay without any treatment. The following variables were analyzed in these bays: phytoplankton structure, composition and abundance, as well as physical and chemical variables of seawater.

3 Material and Methods

3.1 Phytoplankton Analyses

Qualitative phytoplankton samples were collected with a 30 μm mesh plankton net and were fixed with 4% formaldehyde. Quantitative samples were collected with Van Dorn bottles and preserved with Lugol solution. The taxonomic identification of dinoflagellates was carried out following Balech's methodology (Balech 1995).

Cleaning of *Pseudo-nitzschia* spp. frustules was performed according to the method of Hasle and Fryxell (1970). Naphrax-mounted slides were observed with an optical microscope equipped with phase contrast, and selected samples were observed by SEM (Jeol JSM-6360 LV) for species identification. Microplankton was quantified with an inverted microscope using 200–400× magnification according to Lund et al. (1958). Minimum size of organisms identified and counted was approximately 20 μm, i.e., counting included larger nanoplankton and the microphytoplankton between 20 and 200 μm. Most taxa were identified and counted using 200× magnification, except for nanoplankton (small flagellates and small centric diatoms <20 μm) where 400× magnification was used.

3.2 Chemical Analyses

At all stations, water temperature was measured, and samples were collected for the analysis of dissolved oxygen, salinity, nutrients (nitrate, nitrite, ammonium, phosphate, silicate), and chlorophyll *a*. The specific techniques used to measure the physicochemical variables are described in the studies cited in the Study Area section of this chapter.

4 Results

4.1 Phytoplankton

The following phyla were identified at the six sites analyzed: Bacillariophyta, Dinophyta, Chlorophyta, Ochrophyta, Cryptophyta, Euglenophyta, *Cyanobacteria*, and Classe Ebriophyceae (Table 1). The taxonomic criterion adopted was based on AlgaeBase (Guiry and Guiry 2017). Total phytoplankton abundance varied between

Table 1 List of taxa identified during the sampling period

	SAB	*NB*	*EB*	*DR*	*GB*	*UB*		*SAB*	*NB*	*EB*	*DR*	*GB*	*UB*
Bacillariophyta							**Dinophyta**						
Actinoptychus vulgaris		X		X			*Alexandrium* cf. *acatenella*						X
Amphiprora alata		X					*Alexandrium catenella*						X
Amphiprora marina	X						*Alexandrium tamarense*		X	X			
Aulacoseira granulata			X				*Amphidinium* spp.		X				
Bacteriastrum furcatum		X					*Ceratium fusus*	X	X				
Biddulphia alternans		X					*Ceratium lineatum*		X		X	X	X
Campylodiscus sp.		X					*Ceratium longipes*		X				
Ceratoneis arcus					X	X	*Ceratium tripos*	X	X				
Ceratoneis closterium	X	X	X		X	X	*Ceratium tripos tripos*	X					
Chaetoceros cf.*ine*	X						*Dinophysis acuminata*		X	X		X	
Chaetoceros decipiens		X					*Dinophysis punctata*		X			X	X
Chaetoceros didymus	X						*Diplopelta* sp.						X
Chaetoceros eibenii		X					*Ensiculifera angulata*	X					
Chaetoceros spp.	X					X	*Gonyaulax* sp.	X				X	
Cyclostephanos sp.			X				*Gyrodinium* spp.		X				X
Cocconeis placentula			X	X			*Oblea baculifera*	X					
Cocconeis scutellum					X	X	*Prorocentrum micans*	X	X	X			
Cocconeis sp.	X	X		X			*Prorocentrum lima*						X
Coscinodiscus curvatulus		X			X	X	*Prorocentrum* sp.			X			
Coscinodiscus granii			X				*Protoperidinium* cf. *aerolatum*	X					
Coscinodiscus sp.			X				*Protoperidinium* cf. *bispinium*	X	X				
Cyclotella sp.			X				*Protoperidinium* cf. *capurroi*					X	X
Cymatopleura solea			X				*Protoperidinium balechii*			X	X		
Cymbella cf.*inis*			X				*Protoperidinium conicoides*	X					

(continued)

Table 1 (continued)

	SAB	*NB*	*EB*	*DR*	*GB*	*UB*		*SAB*	*NB*	*EB*	*DR*	*GB*	*UB*
Dactyliosolen fragilissimus	X						*Protoperidinium grani*				X		
Delphineis surirella	X						*Protoperidinium punctulatum*	X					
Diatoma vulgare			X				*Protoperidinium pyrum*		X				
Epithemia sorex			X				*Protoperidinium solitarium*		X				
Fragilaria spp.		X		X			*Protoperidinium* spp.		X	X	X	X	X
Diploneis cabro		X					*Scrippsiella trochoidea*		X	X			
Gomphoneis minuta			X				*Scrippsiella* sp.	X				X	X
Gomphoneis sp.					X	X	*Tripos furca*					X	X
Gomphonem pseudotenellum			X				*Zigabikodinium lenticulatum*		X				
Grammatophora marina	X	X	X	X			**Ochrophyta**						
Gyrosigma acuminate			X				*Dictyocha fibula*	X	X				
Gyrosigma fasciola		X		X			*Dictyocha speculum*	X	X	X			
Gyrosigma spencerii		X					**Chlorophyta**						
Gyrosigma sp.		X		X	X	X	*Chlorella* sp.			X			
Helicotheca tamesis		X	X				*Schroederia setigera*			X			
Leptocylindrus danicus		X					Clorofícea filamentosa				X		
Leptocylindrus minimus	X	X					***Cyanobacteria***						
Licmophora abbreviata		X				X	*Anabaena* sp.		X				
Licmophora flabellata		X					*Oscillatoria* sp.				X	X	
Lithodesmium undulatum			X				*Lyngbya* sp.						X
Melosira moniliformis		X					**Euglenophytes**						
Melosira nummuloides	X						*Trachelomonas* sp.				X		
Melosira varians			X				*Euglena* sp.				X	X	
Melosira sp.		X		X	X		*Eutreptiella* sp.					X	
Navicula capitatoradiata			X				**Cryptophyta**						
Navicula radiosa			X				*Rhodomonas* sp.	X					

	SAB	*NB*	*EB*	*DR*	*GB*	*UB*		*SAB*	*NB*	*EB*	*DR*	*GB*	*UB*
Navicula spp.		X	X	X	X		Cryptophyceae s/i	X					
Nitzschia angularis		X					**Ebriaceae**						
Nitzschia frustulum		X					*Ebria* sp.						X
Nitzschia linearis			X				**Nanoplankton**						
Nitzschia longissima	X	X		X	X		Nanoflagellates <10 μm	X	X			X	X
Nitzschia vermicularis													
Nitzschia spp.													
Odontella aurita	X	X	X	X	X								
Odontella sinensis		X											
Paralia sulcata			X	X									
Pleurosigma formosum		X											
Pleurosigma normanii		X											
Pleurosigma spp.	X		X	X									
Pleurosira laevis			X										
Pseudo-nitzschia cf.	X	X											
Pseudodelicatissima													
Pseudo-nitzschia australis		X											
Pseudo-nitzschia seriata						X							
Raphoneis amphiceros		X											
Rhabdonema adriaticum													
Rhizosolenia styliformis	X	X											
Rhizosolenia setigera	X	X		X		X							
Skeletonema costatum		X		X									
Stephanodiscus sp.			X		X								
Striatella unipunctata		X	X	X									
Synedra ulna				X									

(continued)

Table 1 (continued)

	SAB	*NB*	*EB*	*DR*	*GB*	*UB*		*SAB*	*NB*	*EB*	*DR*	*GB*	*UB*
Synedra sp.		X	X		X	X							
Thalassionema nitzschioides													
Thalassiosira <10 μ	X	X											
Thalassiosira 20 μ	X	X											
Thalassiosira eccentrica		X		X									
Thalassiosira spp.	X	X	X	X	X	X							
Toxarium undulatum													

SAB San Antonio Bay, *NB* Nueva Bay, *EB* Engaño Bay, *DR* Deseado Ría, *GB* Golondrina Bay, *UB* Ushuaia Bay

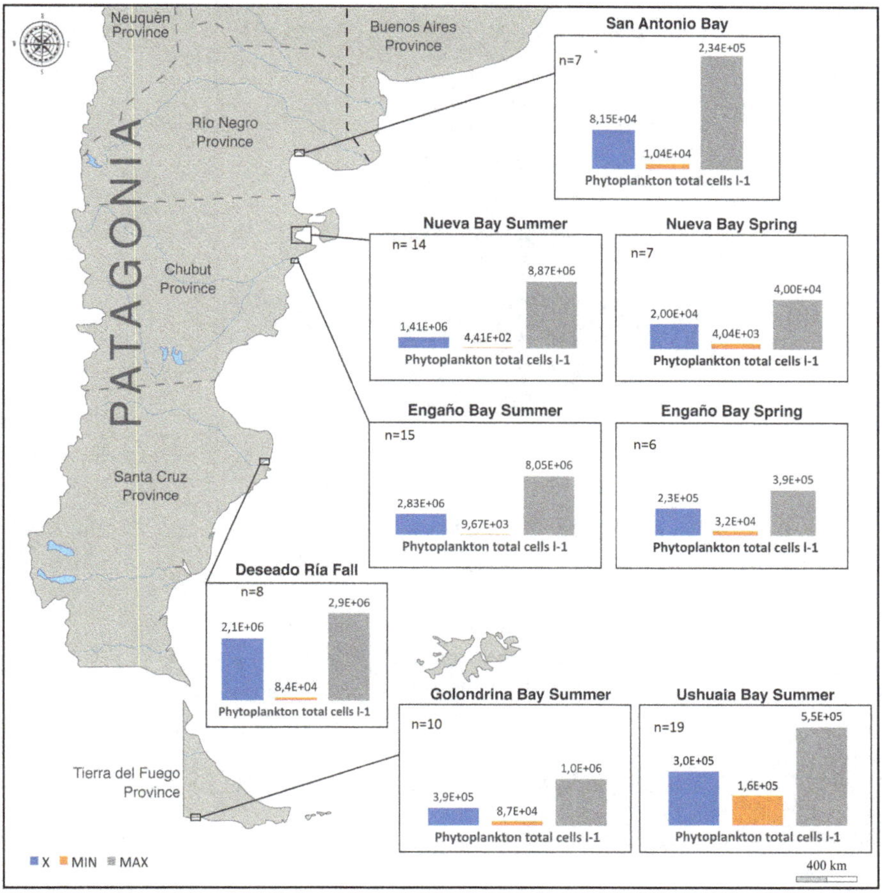

Fig. 3 Mean (X), minimum (Min), and maximum (Max) values of total phytoplankton abundance at each site analyzed. Number of cases (n)

10^2 (Nueva Bay) and 10^6 cells L^{-1} (Nueva, Engaño, Golondrina Bay, and Deseado Ría, Fig. 3).

Figure 4 shows the taxa richness corresponding to each of the phytoplankton groups identified. Diatoms (83 taxa) were the most represented group in each environment analyzed; the maximum species richness values were recorded in Nueva Bay. Dinoflagellates (32 taxa) were also observed to evidence maximum species richness in Nueva Bay, but they were scarcely represented in Engaño Bay and Deseado Ría. Chlorophyta (three taxa) were found in Engaño Bay and Deseado Ría, and Euglenophyta (three taxa) were recorded in Engaño Bay and Deseado Ría. Cryptophyta (two taxa) were found in San Antonio Bay. Ebriceae (one taxon) were characteristic in Ushuaia Bay, and *Cyanobacteria* (three taxa) were recorded in Nueva Golondrina Bay and Deseado Ría. Nanoplankton was observed in San Antonio, Nueva, Golondrina, and Ushuaia Bays.

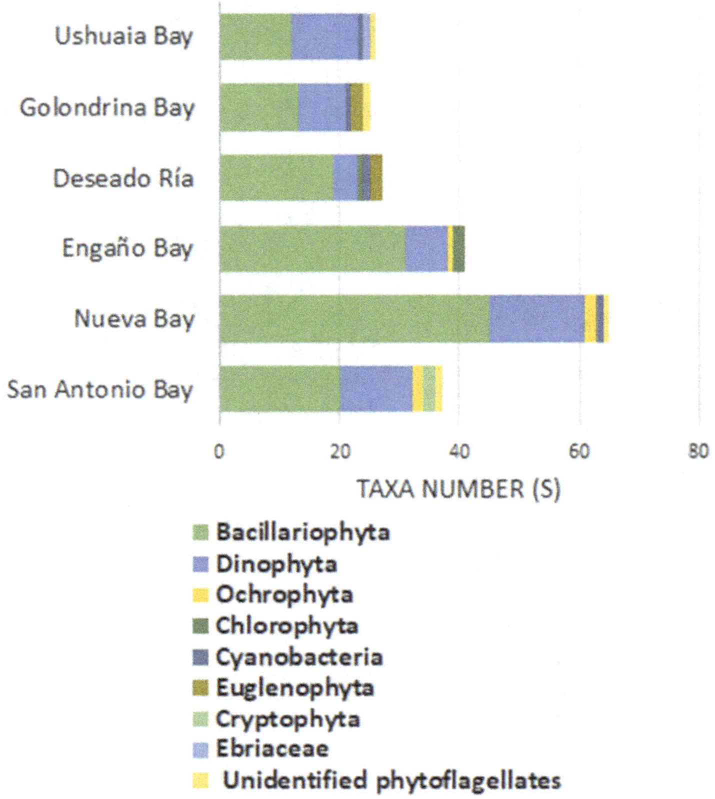

Fig. 4 Phytoplankton groups identified and number of taxa recorded at each site

4.2 *Anthropogenically Affected Coastal Water Areas: San Antonio and Nueva Bays*

In San Antonio Bay, phytoplankton concentrations oscillated between a minimum value of 10^4 cells L^{-1} in Las Grutas and a maximum value of 10^5 cells L^{-1} in Villarino beach (Fig. 3). Diatoms were dominant at all stations analyzed, representing more than 90% of total organisms in all samples, except for one station in Villarino beach where dinoflagellates represented 18%. The most abundant species in all the samples collected from San Antonio Bay was *Rhizosolenia setigera*. Villarino beach evidenced the highest cellular abundance and the highest number of species, mainly dinoflagellates (*Ensiculifera angulata*). Among diatoms, species of the genera *Nitzschia* and *Pseudo-nitzschia* and the species *Dactyliosolen fragilissimus* were dominant. In Las Grutas, the dominant species was the diatom *R. setigera*. Other algal groups that were scarcely present were silicoflagellates, such as *Dictyocha speculum* and *Dictyocha fibula*, and unidentified nanoplankton.

Nueva Bay (Fig. 2b) showed great variability in relation to phytoplankton taxa distribution, which could be indicative of a heterogeneous distribution within which water quality favors the development of different species. Phytoplankton abundance was found to be highest inside the bay, with a variability ranging between 10^2 and 10^6 cells L^{-1} (Fig. 3). Both the reference stations (9 and 10 north and 14 and 15 south) and those located far from anthropogenic impact were found to be characterized by the presence of nanoplankton. The density of the paralytic shellfish toxin-producing species *Alexandrium tamarense* reached maximum values (2.10^6 cells L^{-1}) near the coast, at stations 11 and 12, during summer, leading to a bloom (Fig. 2b). The toxigenic diatoms *P.* cf. *pseudodelicatissima* (station 12, 11) contributing with 75% of total phytoplankton and *P. australis* (Sastre et al. 2001) and of the diarrhetic shellfish poison (DSP)-producing species *D. acuminata* were also recorded in Nueva Bay.

4.3 Anthropogenically Induced Disturbance in Areas with Freshwater Discharges: Engaño, Golondrina, Ushuaia, and Río Deseado Bays

In Engaño Bay, phytoplankton concentration at the 15 stations selected for the present study presented a minimum value of 10^3 cells L^{-1} and a maximum of 8×10^6 cells L^{-1} (Fig.3). The stations near the coast showed low cellular concentrations which increased considerably at the offshore stations (Fig. 2c). Diatoms were found to be the dominant algal type at all stations analyzed. *Aulacoseira granulata* and *Stephanodiscus* sp., the latter with abundances higher than 50% in some of the samples collected, were both contributions coming from Chubut river. Of the two marine species *Rhizosolenia setigera* and *Helicotheca tamesis* found in the study area, the former reached an abundance percentage higher than 50% in some of the samples collected. Dinoflagellates were not found at the coastal stations probably because of the high turbulence in the breaking area. In contrast, their presence was recorded at the offshore stations, *Protoperidinium balechii* being the most important. *Alexandrium tamarense* and *Dinophysis acuminata* were found with maximum values of 9×10^4 and 4×10^4 cells L^{-1}, respectively. Nanoplankton was observed to be an important component at the coastal stations, particularly at station 5, where it represented a 37.5% of the organisms in the samples collected. Chrysophytes, Chlorophyte, and *Cyanobacteria* were found in low percentages and only occasionally.

In Deseado Ría, phytoplankton concentration showed a minimum value of 10^5 cells L^{-1} under rising tide conditions and a maximum value of 10^6 cells L^{-1} in del Rey Island (Fig. 2d). Phytoplankton abundance was observed to increase at the inner stations of Deseado Ría. Algal class composition was variable, diatoms represented 100% of the total samples collected in Quiroga Island, whereas nanoplankton represented an important portion that ranged from 42% to 74% at the outermost stations. Euglenophyceans were also present but in low values at these stations.

Dinoflagellates were detected in quantitative samples from station 9. Among the dominant species, the abundance of *Skeletonema costatum* was remarkable, the highest values being recorded in del Rey Island (Fig.2d) where it exceeded 80% of total organisms in the samples collected (1.9×10^6 cells L^{-1}). The outermost stations showed a dominance of nanoflagellates with percentages ranging from 33% to 41%.

In Ushuaia Bay, the order of magnitude of phytoplankton density was of 10^5 cells L^{-1}. Diatoms, cryptophyceae, and dinoflagellates were the most represented algal classes, and *Thalassiosira*, *Rhodomonas*, and *Protoperidinium* were the most abundant genera.

In Golondrina Bay, the order of magnitude of phytoplankton density ranged from 10^5 to 10^6 cells L^{-1}. Diatoms and cyanobacteria were the best represented algal classes, and *Thalassiosira* and *Oscillatoria* were the dominant genera in all samples and maximum abundances recorded.

4.4 Relationships Between Phytoplankton and Environmental Variables

Table 2 lists the values corresponding to the following environmental variables: temperature, salinity, oxygen, nitrate, nitrite, ammonium, phosphate, silicate, and chlorophyll a, all of which were measured in each ecosystem analyzed. The dissolved oxygen concentrations recorded – always above saturation level could be interpreted as indicative of a proper functioning of the overall autotrophic community. Each environment had particular features in terms of dissolved inorganic nutrients (Table 2), ammonium concentrations were higher in Engaño Bay in spring and Deseado Ría (ammonium mean 3,16 ± 2.78 μM and 3.13 ± 1.52 μM, respectively), while the silicate was elevated in those same environments (Engaño Bay silicate mean 60 ± 91.5 μM, Deseado Ría 6.67 ± 2.45 μM). The silicate was also elevated in the Ushuaia and Golondrina Bays, showing the influence of the Pipo River in the latter and the Olivia River in the Ushuaia Bay.

Table 3 shows the mean and median N:P ratios and their corresponding limiting-nutrient status, nitrogen being the limiting nutrient in all the ecosystems studied, except in Deseado Ría where an intermediate condition was observed. Summarized information on phytoplankton community is also included in Table 3.

No correlation was observed between chlorophyll *a*, total phytoplankton, and N:P (molar ratio). There was also no correlation between nutrient and total phytoplankton either through cell number or chlorophyll *a* concentration. Nonetheless, the only negative correlation observed was between the median values of taxa richness (S) and N:P (molar ratio) (Fig. 5).

The summer field trip carried out in Engaño Bay (Fig. 2c) showed that nutrient concentrations, particularly those of phosphate and nitrate, which appeared to be contributions coming from Chubut river and from the sea, were higher at the offshore stations in the southern sector than those in the northern sector of this area. On

Table 2 Values for the environmental parameters measured in the seawater samples ($n = 85$) in Bahía San Antonio, data collected by Esteves et al. (1996a); in Nueva Bay, data collected by Esteves et al. (1996b); in Engaño Bay, data collected by Esteves et al. (1996b); in Deseado Ría, data collected by Esteves et al. (1997); in Ushuaia Golondrina Bays, data collected by Esteves and Amín (2005)

	San Antonio Bay				Nueva Bay				Nueva Bay				Engaño Bay			
	Summer				Summer				Spring				Summer			
	Min	Max	X	Sd	Min	Max	X	Sd	Min	Max	X	Sd	Min	Max	X	Sd
T (°)	19.80	19.9	19.8	0.03	14.1	15	15.06	0.49	14.1	16.5	16.5	14.1	16.6	24.0	19.1	2.09
S	33.83	34.0	33.9	0.01	33.5	33	33.65	0.06	33.7	34.0	33.9	0.12	0.00	33.4	25.5	13.4
O (%)	92.78	101	99.6	2.36	93.8	124	105.5	7.57	90.0	109	103	6.4	93.0	132	105	11.6
N-NO3 (μM)	0.39	1.22	0.7	0.28	0.00	0.3	0.14	0.10	0.36	1.24	0.6	0.31	0.00	2.25	0.66	0.68
N-NO2 (μM)	0.00	0.04	0.01	0.02	0.00	0.1	0.03	0.04	0.00	0.00	0.00	0.00	0.00	0.38	0.08	0.11
N-NH4 (μM)	1.07	4.49	1.85	1.24	0.00	1.3	0.48	0.52	0.00	5.17	1.01	1.80	0.25	6.75	2.06	1.92
P-PO4 (μM)	0.59	0.79	0.69	0.06	0.02	1.2	0.92	0.31	1.06	1.56	1.06	0.23	0.73	2.10	1.08	0.38
Si (μM)	0.00	11.0	2.76	1.35	2.06	100	3.72	1.78	4.19	12.6	12.6	4.19	3.73	243	30.7	65.0
Chl a (μg/l)	0.54	1.49	0.89	0.34	0.17	0. 5	0.59	0.20	0.17	0.69	1.34	0.17	0.16	21.0	2.44	5.12

	Engaño Bay				Deseado Ría				Ushuaia Bay				Golondrina Bay			
	Spring				Fall				Summer				Summer			
	Min	Max	X	Sd	Min	Max	X	Sd	Min	Max	X	Sd	Min	Max	X	Sd
T (°)	13.0	16.0	14.4	2.00	14.6	8.20	2.40	0.13	2.00	14.6	8.20	2.40	2.00	14.6	8.20	2.40
S	31.9	33.5	32.87	5.00	32.8	33.3	33.1	0.12	5.00	34.2	29.7	3.62	nd	nd	nd	nd
O (%)	87.2	98.1	91.19	95.0	nd	nd	nd	nd	115	135	115	12.6	95.0	95.0	95.0	0.00
N-NO3 (μM)	0.80	10.0	4.57	0.50	0.71	1.06	0.86	0.12	0.00	1.00	0.49	0.33	3.00	4.00	3.50	0.5
N-NO2 (μM)	0.00	0.10	0.07	0.00	7.07	7.56	7.31	0.18	0.03	0.08	0.04	0.01	0.08	0.20	0.16	0.05
N-NH4 (μM)	0.55	6.69	3.16	2.78	1.36	6.00	3.13	1.52	0.05	2.00	0.89	0.89	0.5	1.00	0.84	0.25
P-PO4 (μM)	0.74	2.60	1.36	0.04	0.92	1.31	1.15	0.11	0.20	0.60	0.43	0.10	0.25	1.20	0.84	0.25
Si (μM)	2.38	210	60.7	91.5	3.56	12.2	6.67	2.45	0.57	10.0	5.29	3.03	4.00	20.0	7.8	12.3
Chl *a* (μg/l)	1.07	5.06	3.08	1.65	0.70	1.29	0.98	0.28	10.0	18.0	10.6	4.71	1.00	2.00	1.10	0.73

Min minimum, *Max* maximum, *X* mean, *SD* standard deviation, *nd* no data

Table 3 N:P ratio, biomass (Chl *a*), cell density (cells L^{-1}), and richness of taxa (S): mean and median at each site analyzed

Sites	Ratio Total N:P Mean	Median	Status	Chlorophyll Chl *a* (µg/L) Mean	Median	Density Cells L^{-1} Mean	Median	Richness S Mean	Median
San Antonio Bay Summer	3.72	3.36	N-limiting	0.89	0.76	81,500	66,000	12.8	11
Nueva Bay Summer	0.97	0.94	N-limiting	0.59	0.59	14,09,815	5,41,500	12	12.5
Nueva Bay Spring	1.34	0.8	N-limiting	1.34	0.29	20,036	18,396	14.4	15
Engaño Bay Summer	2.33	2.47	N-limiting	2.44	2.33	28,31,544	29,84,500	15.6	14
Engaño Bay Spring	5.11	4.15	N-limiting	3.08	3.13	2,30,978	2,68,764	12	12.5
Deseado Ría Fall	9.8	9.4	Intermediate	0.98	0.96	20,61,142	23,70,000	6	4.78
Ushuaia Bay Summer	3.23	2.75	N-limiting	10.63	10	3,89,947	3,27,600	12.5	9
Golondrina Bay Summer	6	2.75	N-limiting	1.1	2	3,03,291	3,92,500	9	9

the coast, the high values of chlorophyll *a* were found to be coincident with an oversaturation of dissolved oxygen indicating active primary productivity.

In Deseado Ría (Fig. 2d), salinity and silicate values showed a minimal influence of either Deseado river or freshwater contribution from the Ría to the sea.

Ushuaia and Golondrina Bays (Fig. 2e) evidenced a different behavior from the chemical point of view and in terms of phytoplankton. Temperature and the concentrations of nitrate, phosphate, and silicate were found to be higher in Golondrina Bay than in Ushuaia Bay (Table 2). In the latter, a clear influence from Pipo River was observed with chlorophyll *a* concentrations remarkably lower than in Ushuaia Bay.

In Ushuaia Bay, nutrient concentrations were higher in the coastline than at the offshore station. The same pattern was observed for chlorophyll *a* concentration, which was higher than 10 µg/l in some of the surface water samples collected, evidencing an efficient uptake of anthropogenic nutrients by primary producers.

5 Discussion

Temperate shelf ecosystems are subject to a strong seasonality as a result of processes of mixing and stratification of the water column and the availability of macro- and micronutrients, mainly from the forms of nitrogen. These two processes

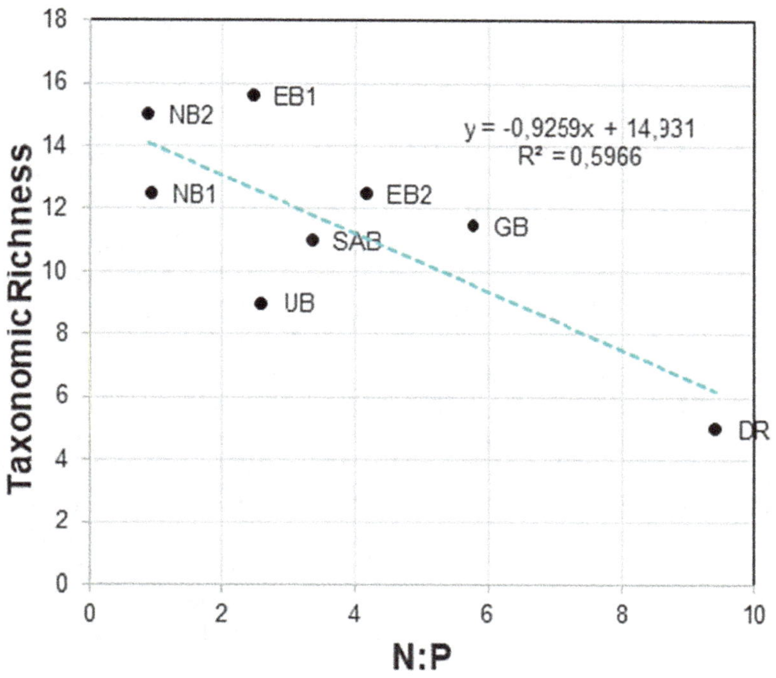

Fig. 5 Scatterplot of taxonomic richness vs. N:P ratio at each site analyzed: SA, San Antonio Bay; NB, Nueva Bay; EB, Engaño Bay; DR, Deseado Ría; UB, Ushuaia Bay; GB, Golondrina Bay

as well as the availability of macro- and micronutrients determine not only the phytoplankton biomass levels but also the dominance and replacement of the different taxa during phytoplankton succession (Fu et al. 2009). During winter, intense vertical mixing maintains high levels of nutrients at the superficial layers. However, phytoplankton growth is limited mainly as a result of low light intensity (Gargett and Marra 2002). During spring, the increase in solar radiation and, consequently, both the temperature of the seawater and the degree of stratification of the water column promote the development of "new phytoplankton production" based mainly on "new nitrogen" (nitrate) from the deep strata (e.g., Dugdale and Goering 1967, Rodrigues et al. 2002).

Previous research on nutrient distribution in the different latitudinal gradients on the Patagonian Continental Shelf of Argentina (PCSA) showed an overall south-north decrease in nitrate along the PCSA (Paparazzo et al. 2010). This could be due to the subantarctic origin of the waters in this region, which flow from the northern boundary of Drake Passage onto the Patagonian shelf and may provide nutrients to the coastal zone. The coastal zone may exhibit different nutrient concentrations from those of the continental shelf as it is also subject to the effect of discharges from different sources (cities and rivers) (Paparazzo 2011). In line with this, our findings reveal variations in the nutrient concentrations recorded which are due to the different types of discharges and freshwater contributions coming from the rivers that flow into the coastal zone.

Various coastal environments which experience increased nutrient loading were observed to exhibit general trends of shifts in phytoplankton community dominance from larger to smaller cells (Cederwall and Elmgren 1990; Cadee 1992). In addition, Finni et al. (2001) found that small or fragile forms of diatoms (*Skeletonema* spp., *Chaetoceros minimus*, small centrals, etc.) as well as small flagellated forms (Cryptophyceae) could be well represented in areas with considerable eutrophication. A pronounced decrease in the size structure of phytoplankton from pico- to nanoplankton is considered a sign of stress (Relevante and Gilmartin 1988).

As to the composition of algal groups, a dominance of diatoms was observed in general, followed by dinoflagellates. However, many of the variations in the size of phytoplankton community recorded in some of the stations analyzed could be indicative of eutrophication symptoms with predominance of flagellates and cryptophyceae. Several species, such as *P.* cf. *pseudodelicatissima*, *Nitzschia longissima*, *Leptocylindrus minimus*, *Ceratium tripos*, and *Prorocentrum micans*, are cited as characteristic of eutrophic waters (Bodeanu 1992).

Jaanus et al. (2009) determined phytoplankton species indicating eutrophication to monitor the coastal waters of the Baltic, among them, oscillatory cyanobacteria, diatoms, such as *Cyclotella choctawhatcheeana*, *Ceratoneis closterium*, and *Skeletonema costatum*, and dinoflagellates of the genera *Prorocentrum* and *Heterocapsa*. These taxa are also characteristic components of the coastal environments of Patagonia.

The absence of significant correlations between phytoplankton density and chlorophyll *a* agrees with the fact that photosynthetic pigment concentrations can vary both among species and inside the cell itself, depending on salinity, incident light intensity, temperature, the physiological state of cells, and the size of organisms (Regadera 2006). The relationship between nutrients (limiting or not) and their chemical form influences the structure and functioning of the ecosystem and the development of harmful algal blooms (Glibert 2017).

Laboratory bioassays carried out to test the effects of nutrient enrichment on the growth of isolated native species from Nueva Bay showed that fishery industry and domestic effluents stimulate algal growth and biomass development but decrease algal biodiversity (Santinelli 2008).

Gil (2001) suggested that nitrogen is key to triggering eutrophication processes in Nueva Bay's ecosystem. She demonstrated that allochthonous ammonium derived from residual waters could have been the main cause of water and sediment quality deterioration during the 1990s, which could, in turn, be considered an incipient eutrophication process. Macroalgae could have been an essential factor in regulating nitrogen concentration in the seawater of this Nueva Bay by that time (Gil et al. 2005). The fact that *Alexandrium tamarense* bloom recorded in Nueva Bay coincided with low inorganic nitrogen concentrations in seawater was in agreement with Anderson et al. (2002) according to which the low availability of inorganic nitrogen stimulates the development of opportunistic species, such as cyanobacteria and dinoflagellates.

In the environments with significant contribution of freshwater, for example, the mouth of Chubut river, the high concentrations of chlorophyll *a* were found to be coincident with an oversaturation of dissolved oxygen indicating active primary productivity (Esteves et al. 1996b).

In the case of Deseado Ría, a great homogenization of all chemical variables between surface and depth was observed. Nitrate showed neither significant variability inside the Ría nor in the open sea, and its low concentrations inside the Ría were not only indicative of consumption by primary producers but also coincident with high densities of phytoplankton (Esteves et al. 1997).

The analysis of water contribution to Ushuaia and Golondrina Bays showed that the sewage discharged to these bays had low concentrations of dissolved oxygen and high concentrations of nutrients, such as ammonium, nitrite, and phosphate (Esteves and Amín 2005). Gil et al. (2011) showed that the northwestern area of Ushuaia Bay is the most impacted by organic matter contribution, nutrient enrichment, and oxygen demand.

According to chlorophyll *a* indexes of trophic state (Contreras et al. 1994), the ecosystems studied in the present work could be classified as oligotrophic (0.58 to 2.88 μg/L Chl *a*) except for Ushuaia Bay, whose maximum value of 10 μg/L chlorophyll *a* suggests a mesotrophic category.

6 Conclusions

Diatoms are the dominant group in all the environments followed by dinoflagellates, cryptophytes, and silicoflagellates. However, the presence of other groups, such as Euglenophyceae and *Cyanobacteria*, particularly in the case of Golondrina Bay with the genus *Oscillatoria*, could be indicative of eutrophication processes.

Blooms of toxic species, such as the dinoflagellate *Alexandrium tamarense*, were detected in Nueva Bay, and blooms of the diatoms *Pseudo-nitzchia* cf. *pseudodelicatissima* and *Skeletonema costatum* were recorded in Deseado Ría.

The phytoplankton community composition was characterized by the presence of species that are typical of coastal environments, many of which have been reported to belong to eutrophic coastal environments, whereas others have been reported to be characteristic of oceanic habitats.

In the ecosystems characterized by the mouth of Chubut river, such as Engaño Bay, phytoplankton composition was observed to be dominated by freshwater species in spite of the salinity values detected at the stations located in this bay. This could be due to the fact that these species are euryhaline and can be found active in these environments. In Deseado Ría, the highest concentrations of phytoplankton were detected upstream, and the species identified were mostly euplanktonic, euryhaline, and eurythermic and some ticoplanktonic or meroplanktonic. No species cited as potentially noxious for the Argentine Sea were detected in this sector.

In Ushuaia Bay, the paralytic shellfish toxin-producing species *Alexandrium catenella*, *A* cf. *acatenella*, the potential lipophilic toxin producer of the genus *Dinophysis*, and species of the genus *Pseudo-nitzchia* were detected.

As to the nutrients observed in the study area, the majority of the ecosystems analyzed showed that nitrogen is the limiting nutrient. The exception was Deseado Ría which showed an intermediate status. Phytoplankton abundance was observed to increase significantly, whereas the number of species did not increase, this being indicative of the dominance of certain species.

A negative relationship between the increase in N:P ratio and number of taxa was observed. Cellular concentrations ranging in the order of 10^6 cells L-1 were recorded in Nueva Bay, Engaño Bay, and Deseado Ría as observed in other anthropized ecosystems. The trophic index based on chlorophyll *a* showed that these environments can be mostly classified as oligotrophic with a tendency toward mesotrophia in the case of Ushuaia Bay.

In all the environments analyzed, differences in composition, structure, and abundance of phytoplankton communities were observed, indicating different responses to the water quality changes resulting from the contributions produced by human activities.

Acknowledgments This work has been carried out within the framework of the Program for the Implementation and Consolidation of the Integrated Patagonian Coastal Zone Management Plan (PMZCP), a project financed by the Global Environment Facility (GEF) through the United Nations Program for Development (UNDP) and implemented by the United Nations Office for Project Services (UNOPS) and Patagonia Natural Foundation (FPN) (Project ARG/02/G3).

References

Achary SM, Panigrahi S, Satpathy K et al (2014) Nutrient dynamics and seasonal variation of phytoplankton assemblages in the coastal waters of southwest Bay of Bengal. Environ Monit Assess. https://doi.org/10.1007/s10661-014-3812-8

Anderson DM, Glibert PM, Burkholder JM (2002) Harmful algal blooms and eutrophication: nutrients sources. Composition. And consequences. Estuaries 25(4b):707–726

Balech E (1995) The genus *Alexandrium* Halim (Dinoflagellate). Sherkin Island Marine Station, Cork, p 151

Benavides HL, Prado S, Diaz F et al (1995) An exceptional bloom of *Alexandrium catenella* in the Beagle Channel. Argentina. In: Harmful Marine Algal Bloms. Proceedings of the Sixth International Conference on toxic Marine Phytoplankton, Nantes France, pp 113–119

Bodeanu N (1992) Algal blooms and development of the main phytoplanktonic species at the Romanian Black Sea littoral in conditions of intensification of the eutrophication process. Eds Vollenweider RA, Marchetti R, Viviani R. Marine Coastal Eutrophication, pp 891–906

Cadee GC (1992) Trends in Marsdiep phytoplankton. Netherlands. J Sea Res 20:143–149

Casas G, Piriz ML (1996) Surveys of *Undaria pinnatifida* (Laminariales. Phaeophyta) in Golfo Nuevo. Argentina. Hydrobiologia 326/327:213–215

Cederwall H, Elmgren R (1990) Biological effects of eutrophication in the Baltic Sea, particularly in the coastal zone. Ambio 19:109–112

Contreras F, Castañeda O, García A (1994) La clorofila-a como base para un índice trófico en lagunas costeras mexicanas. Anales Instituto. Ciencias del Mar y Limnología México UNAM, pp 55–66

De la Lanza-Espino G, Hernández-Pulido S, Carvajal-Pérez JL (2000) Organismos indicadores de la calidad del agua y de la contaminación (bioindicadores). Plaza y Valdés, México

Duarte CM (2009) Coastal eutrophication research: a new awareness. Hydrobiologia 629:263–269

Dugdale RC, Goering JJ (1967) Uptake of new and regenerated forms of nitrogen in primary productivity. Contribution no. 6 from the Institute of Marine. Science 12:196–206

Esteves JL, Amín O (2005) Evaluación de la contaminación urbana de las bahías de Ushuaia. Encerrada y Golondrina. Informe Técnico del Proyecto Consolidación e Implementación del Plan de Manejo de la zona costera patagónica para la conservación de la biodiversidad, pp 1–69

Esteves JL, Santinelli N, Sastre V (1992) A dinoflagellate bloom and P.S.P. Production associated with upwelling in Golfo Nuevo. Patagonia. Argentina. Hydrobiologia 242:115–122

Esteves JL, Solís M, Sastre V et al (1996a) Evaluación de la contaminación urbana de la bahía de San Antonio (Provincia del Río Negro). Informes Técnicos del Plan de Manejo Integrado de la Zona Costera Patagónica (Puerto Madryn. Argentina). N° 20: 1–26

Esteves JL, Solis M, Gil M et al (1996b) Dos Bahías Patagónicas: Bahía Nueva y Bahía Engaño. En Jorge Marcovecchio (Ed.): Pollution Processes in coastal environments. Chapter I General Surveys, pp 64–70

Esteves JL, Gil MN, Commendatore M et al (1997) Evaluación de la contaminación urbana de la Ría de Deseado (provincia de santa cruz) Informes Técnicos del Plan de Manejo Integrado de la Zona Costera Patagónica (Puerto Madryn. Argentina). N° 36: 1–59

Ferrario M, Sastre V (1990) Ultraestructura. polimorfismo y ecología de *Odontella aurita* (Lyng.) Agardh (Bacillariophyceae) en el estuario del río Chubut. Argentina. En Anales II Congreso Latinoamericano de Ficología Marina. Revista de la Facultad de Oceanografía. Pesquería y Ciencias Alimentarias. Universidad Nacional Federico Villarreal. N° 2: 98–106 (Publicado en 1992)

Finni TK, Kononen R, Olsonen et al (2001) The history of cyanobacterial blooms in the Baltic Sea. Ambio 30: 172–178

Fu M, Wang Z, Li Y et al (2009) Phytoplankton biomass size structure and its regulation in the Southern Yellow Sea (China): seasonal variability. Cont Shelf Res 29:2178–2194

Gargett AE, Marra J (2002) Effects of upper ocean physical processes (turbulence, advection and air-sea interaction) on oceanic primary production. In: Robinson ARM, Rothschild BJ (eds) The sea, vol 12. John Wiley & Sons Inc., New York, pp 19–49

Gil MN (2001) Eutroficación: el rol del nitrógeno en ecosistemas marinos costeros. Doctorado en Ingeniería Química Universidad Nacional del Sur

Gil MN, Torres A, Esteves JL (2005) Uptake of nitrogen by *Ulva rigida* (Chlorophyceae) when exposed to treated sewage effluent in culture. Hydrobiologia 532:39–43

Gil MN, Torres A, Amín O et al (2011) Assessment of recent sediment influence in an urban polluted subantarctic coastal ecosystem. Beagle Channel (Southern Argentina). Mar Pollut Bull 62(1):201–207

Glibert GM (2017) Eutrophication. Harmful algae and biodiversity — challenging paradigms in a world of complex nutrient changes. Mar Pollut Bull. https://doi.org/10.1016/j.marpolbul.2017.04

González P, Esteves JL (2008) Relevamiento de la situación ambiental urbana en la zona costera patagónica. Informe Técnico N° 3 – 1a ed. – Puerto Madryn: Fundación Patagonia Natural, pp 1–57

Guiry MD and Guiry GM (2017) *AlgaeBase*. In: World-wide electronic publication. National University of Ireland. Galway. http://www.algaebase.org. Accessed 30 July 2017

Hallegraeff GM (2004) Harmful algae blooms: a global overview. In: Anderson H, Cembella (eds) Manual on harmful marine microalgae. Monographs on oceanographic methodology 11. UNESCO, Paris, pp 25–49

Hasle GR, Fryxell GA (1970) Diatoms cleaning and mounting for light and electron microscopy. Trans Am Microsc Soc 89:468–474

INDEC (1991) Instituto Nacional de Estadísticas y Censo. Censo Nacional de Población 1991. http://www.indec.gob.ar. Accessed 15 July 2017

INDEC (2012) Instituto Nacional de Estadísticas y Censo. Censo Nacional de Población 2012. http://www.sig.indec.gov.ar/censo2012/. Accessed 15 July 2017

Jaanus A, Toming K, Hallfors S et al (2009) Potential phytoplankton indicator species for monitoring Baltic coastal waters in the summer period. Eutrophication in coastal ecosystem. Hydrobiologia. https://doi.org/10.1007/s10750-009-9768

Lassus P, Chromérat N, Hess P et al (2016) Toxic and harmful microalgae of the world ocean / Micro-algues toxiques et nuisibles de l' océan mundial. Denmark. International Society for the Study of Harmful Algae/Intergovernmental Oceanographic Commision of UNESCO. IOC Manuals and Guides 68

Lund JWG, Kilpling C, Le Cren ED (1958) The inverted microscope method of estimating algal numbers and the statistical basis of estimation by counting. Hidrobiologia 11(2):143–170

Moreira A, Seisdedo M, Leal S et al (2007) Composición y abundancia del fitoplancton de la bahía de Cienfuegos. Cuba. Invest Mars 28(2):97–109

Nixon SW (2009) Eutrophication and macrosscope. Hydrobiologia 629:5–19. https://doi.org/10.1007/s10750-009-9759-z

Paparazzo FE (2011) Distribución espacio-temporal de nutrientes en el Mar Argentino, Pasaje Drake y Península Antártica. Tasa de incorporación por fitoplancton Tesis Doctoral Facultad de Ciencias Exactas y Naturales Universidad de Buenos Aires 187. www.digital bl.fcen.uba.ar

Paparazzo FE, Bianucci L, Schloss IR et al (2010) Cross-frontal distribution of inorganic nutrients and chlorophyll-a on the Patagonian continental shelf of Argentina during summer and fall. Rev Biol Mar Oceanogr 45(1):107–119

Piriz ML, Casas GM (1994) Occurrence of *Undaria pinnatifida* in Golfo Nuevo. Argentina. Applied Phycology Forum 10.4

Relevante N, Gilmartin M (1988) Some observations on eutrophication associated changes in phototrophic and heterotrophic pico and nanoplankton assemblages in the northen Adriatic Sea. Rapp. Comm. Int. Mer Médit 32(2):49–66

Regadera R (2006). Fitoplancton y clorofila-a como indicadores de eutrofización en la bahía de La Habana. CIM. La Habana. Cuba, pp 1–100

Rodrigues RM, Williams NV, Le BP (2002) Inorganic nitrogen assimilation by picoplankton and whole plankton in a coastal ecosystem. Limnol Oceanogr 47:1608–1616

Santinelli NH (2008) Fitoplancton de un ambiente costero sometido a perturbación antrópica: Bahia Nueva, Pcia del Chubut. Tesis Doctoral, Facultad de Ciencias Naturales, Sede Trelew, Universidad Nacional de la Patagonia San Juan Bosco, Chubut, Argentina

Santinelli NH, Esteves JL (1993) Características químicas y fitoplanctónicas del Estuario del río Chubut, Patagonia, Argentina. Natur Patag 1(1):22–34

Santinelli N, Sastre V, Caille G (1990) Fitoplancton del Estuario Inferior del río Chubut (Patagonia Argentina) y su relación con la Salinidad y la Temperatura. Rev Asoc Cien Nat Lit 21(1):69–79

Sastre V, Santinelli N, Caille G (1990) Diatomeas y Dinoflagelados del Estuario del río Chubut (Patagonia. Argentina). II Estructura de las Comunidades. Anales 2o Congreso Latinoamericano de Ficología Marina. Rev Fac Oceanogr Pesq Cien Alim 2:181–192

Sastre V, Santinelli NH, Esteves JL et al (2001) Aspectos ecológicos de especies de *Pseudo-nitzschia* en aguas costeras patagónicas (Argentina): 217–235. En: Alveal K, Antezana T (eds) Sustentabilidad de la biodiversidad: un problema actual bases científico-técnicas. teorizaciones y proyecciones. ISBN: 956-227-257-5. Universidad de Concepción. Concepción. Chile. 896

Vega Moreno D, Pérez-Marrero J, Morales J et al (2012) Phytoplankton functional community structure in Argentinian continental shelf determined by HPLC pigment signatures. Estuar Coast Shelf Sci 100:72–81

Responses of Subantarctic Marine Phytoplankton to Ozone Decrease and Increased Temperature

Marcelo P. Hernando, Gabriela Malanga, Gastón O. Almandoz, Irene R. Schloss, and Gustavo A. Ferreyra

Abstract Temperature and ultraviolet B radiation (UVB, 280–315 nm) are external stressors that affect organisms in mid and high latitudes in a combined way. The combined effects of both variables on natural marine phytoplankton from the Beagle Channel (Argentina) were examined during a 7-day mesocosm experiment. We tested the hypothesis that increased temperature (HT, +3 °C) will offset negative effects on phytoplankton by UVB (natural, NUVB, and high, HUVB, simulating a 60% decrease in stratospheric ozone layer thickness). The response of the entire phytoplankton assemblage, in terms of phytoplankton biomass, community composition,

M. P. Hernando (✉)
Departamento Radiobiología CAC, Comisión Nacional de Energía Atómica, Buenos Aires, Argentina
e-mail: mhernando@cnea.gov.ar

G. Malanga
Facultad de Farmacia y Bioquímica, Fisicoquímica, Universidad de Buenos Aires Ciudad Autónoma de Buenos Aires, Buenos Aires, Argentina

Instituto de Bioquímica y Medicina Molecular (IBIMOL), Consejo Nacional de Investigaciones científicas – Universidad de Buenos Aires (CONICET–UBA), Buenos Aires, Argentina

G. O. Almandoz
División Ficología, Facultad de Ciencias Naturales y Museo, Universidad Nacional de La Plata, La Plata, Argentina

I. R. Schloss
Instituto Antártico Argentino, Buenos Aires, Argentina

Centro Austral de Investigaciones Científicas (CADIC), Consejo Nacional de Investigaciones Científicas (CONICET), Ushuaia, Tierra del Fuego, Argentina

Universidad Nacional de Tierra del Fuego, Ushuaia, Argentina

G. A. Ferreyra
Centro Austral de Investigaciones Científicas (CADIC), Consejo Nacional de Investigaciones Científicas (CONICET), Ushuaia, Tierra del Fuego, Argentina
e-mail: gferreyra@cadic-conicet.gob.ar

M. S. Hoffmeyer et al. (eds.), *Plankton Ecology of the Southwestern Atlantic*,
https://doi.org/10.1007/978-3-319-77869-3_24

reactive oxygen species (ROS), lipid damage (TBARS), nonenzymatic antioxidants (α-tocopherol (αT) and β-carotene (βC)), and mycosporine-like amino acids (MAAs), was evaluated. On the first exposure day, assemblages exposed to HUVB showed a significant increase in ROS content, regardless of the temperature, while lipid damage was significantly higher at HT and HUVB. However, on day 2, lipid damage was significantly lower possibly due to the consumption of the nonenzymatic antioxidants that protected the membranes from further damage. Under normal temperature (NT) conditions, ROS concentrations were significantly lower compared with day 1, and nonenzymatic antioxidant concentrations remained high (0.025 nmol C^{-1} compared with 0.05 nmol C^{-1} at initial time). ROS increased again in HT-HUVB and in control (NT-NUVB), in coincidence with a significant increase in UVB radiation on day 4. However, the lipid damage was significantly lower in HT-HUVB than in control conditions possibly due to a higher consumption of nonenzymatic antioxidants and probably also to a higher activity of enzymatic antioxidants by the effect of the higher temperature. The same results were observed for HT-NUVB, with low lipid damage. During all experiment no significant differences were observed in carbon-normalized MAAs. After day 4, when nutrients became limiting, high temperature significantly influenced community structure, with a negative impact on diatoms and positive on phytoflagellates, independently of the UVB doses. Our results show that subantarctic phytoplankton is able to respond to a ROS increase via antioxidant response in high irradiance conditions. In addition, increased temperature and phytoplankton community composition play a central role in this response. At lower UVB doses, diatoms were able to avoid UVB lipid damage by αT and βC synthesis. However, with maximum doses, phytoflagellates showed a best UVB adaptation to high temperature conditions.

Keywords UBVR · Increased temperature · Beagle Channel · Phytoplankton assemblage · ROS · Nonenzymatic antioxidants

1 Introduction

Climate change over the next century is expected to modify ocean ice cover in polar regions, water column stratification via temperature and rainfall increases, and circulation (Meehl et al. 2007), altering the environmental conditions that influence phytoplankton standing stock and primary production (Sarmiento et al. 2004; Irwin and Finkel 2008). A global warming induced shallower mixed layer, and strengthened water column stratification may lead to a higher exposure of phytoplankton to harmful surface ultraviolet B radiation (UVB, 280–315 nm) (Beardall and Stojkovic 2006; Häder et al. 2007). Within this context, phytoplankton assemblage structure changes are expected, which can have cascading effects on marine food web dynamics and elemental cycling (Beardall and Stojkovic 2006). An increase in phytoplankton assemblages dominated by small picoplankton species has been associated with nutrient limitation due to increases in ocean stratification and the expansion of the ocean gyres as a consequence of increased temperature (Behrenfeld et al. 2006; Irwin and Oliver 2009). Mesocosm experiments, exposing natural plankton communities to increased UVB, provide information in relation to modifications in their structure and trophic interactions (Mostajir et al. 1999, 2000; Ferreyra et al. 2006). Such studies showed that the planktonic system develops toward a microbial food web potentially weakening trophic transfer of matter and energy toward higher trophic levels. Thus, UVB potentially influences aquatic and coastal bio-resources, and deleterious effects from this perspective could impact not only on primary producers but also other components of the system. If the observed experimental responses of plankton communities to UVB are extrapolated to natural conditions, then the efficiency of the biological pump, the energy transfer toward top predators, as well as the dynamics of CO_2 atmosphere-ocean fluxes could be significantly altered.

Plant and animal cells are capable of generating, via multiple pathways, a number of different reactive oxygen species (ROS), including the superoxide anion (O_2^-), hydrogen peroxide (H_2O_2), singlet oxygen ($^1O_2$), and, by Fenton reaction, the hydroxyl radical (OH^-) (Halliwell and Gutteridge 2007). These ROS species occur transiently and are regular products of oxidative metabolism, results of, for example, increasing temperature or UVB doses (Hernando et al. 2011, 2006; González et al. 2015). Although some ROS may function as important signaling molecules that alter gene expression and modulate the activity of specific defense proteins, all ROS are harmful to organisms at high concentrations (Apel and Hirt 2004). UVB has been shown to be very effective in inducing lipid peroxidation of biological membranes (Hideg and Vass 1996; Hernando et al. 2005, 2012) and polyunsaturated fatty acids (Won Eun-Ji et al. 2014). Moreover, UVB can destroy the natural lipid-soluble antioxidants and promote the formation of lipid peroxidation products (Malanga and Puntarulo 1995; Hernando et al. 2011).

One of the possible mechanisms that could counteract the damage of oxidative stress generated by UVB radiation is the synthesis of both enzymatic and nonenzymatic antioxidants (Niyogi 1999; Hernando 2011; Hernando et al. 2015). The nonenzymatic antioxidants are generally small molecules, such as ascorbate and glutathione, acting in the aqueous phase, whereas the lipophilic antioxidants (such as α-tocopherol (αT) and β-carotene (βC)) are active in the membrane environment. Especially αT is known for its protective effect against lipid peroxidation of biological membranes via peroxyl and alcoxyl radical scavenging (Häubner et al. 2014). In contrast, the main function of βC is photoreceptive, because it acts as a pigment antenna in the photosynthesis reaction center complex (Hernando et al. 2015).

In a scenario of global change, with predictions of increased temperatures of up to 3–5 °C by the end of the century (Alley et al. 2007), it is highly plausible that the impact of UVR on metabolic processes will differ in comparison to present-day conditions. High temperature could increase the activity of antioxidant enzymes such as superoxide dismutase, ascorbate peroxidase, catalase, or glutathione reductase avoiding the negative effects of UVB by increasing metabolic activities and, therefore, by increasing the rates of enzymatic photorepair mechanisms (Bouchard et al. 2006). Halac et al. (2010) observed low UVB-induced photoinhibition in two diatom species exposed to a 5 °C increase in water temperature. However, the combined effects of UVR and increased temperature on phytoplankton are variable. In general, increased temperatures counteract the negative effects produced by UVR by reducing UVR-induced photoinhibition in marine diatoms (Sobrino and Neale 2007; Halac et al. 2010). However, in investigations carried out with tropical marine assemblages of the Caribbean Sea, Halac et al. (2013) also found that increased temperature negatively affected the growth of some diatoms. These variable responses occurred because of the differences in the taxonomic composition of the assemblages and due to their differential capacity to acclimate to solar radiation. In addition, due to the antagonistic or synergistic nature of the interactions among stressors (Folt et al. 1999) together with species-specific responses, it is not possible to predict the phytoplankton response in a global perspective.

Another adaptive mechanism by which phytoplankton can reduce UVR damage is the synthesis of mycosporine-like amino acids (MAAs) (Carreto et al. 1990; Karsten et al. 1999; Hernando et al. 2012). Recent studies have revealed their radical scavenging activities in vitro (Naoki Wada et al. 2015). A series of results shows a direct relationship between MAA concentrations and photoprotection on photosynthesis, growth, and motility of algae (Klisch and Häder 2002; Hernando et al. 2012; Rastogi and Incharoensakdi 2014).

Within this context, the main goal of this work was to study the effects of several days' exposure to increased UVB and increased temperature on the relative abundance of phytoplankton groups, oxidative stress, and antioxidant responses in large-volume mesocosms of a subantarctic phytoplankton community. We tested the hypothesis that increased temperature (HT, +3 °C) will offset negative effects on phytoplankton by UVB. Such objectives are a complement of the previous analysis evaluated with the same experimental design but determining the plankton responses in relation to the interaction of phyto-bacterioplankton and particulate organic carbon (POC) and the nutrient consumption (Moreau et al. 2014).

2 Materials and Methods

2.1 Experimental Setup

The experiments were performed in Ushuaia (54° 52′S, 68° 18′W, Tierra del Fuego, Argentina) between February 10 (T0) and February 17, 2008. Seawater was collected at 5 m depth in a small bay in the Beagle Channel far from the influence of the city of Ushuaia (Fig. 1). Filling of the mesocosms was done as in Ferreyra et al. (2006). Briefly, the water was filtered through a 300 μm Nitex net placed inside a water distributor to avoid the passage of mesozooplankton. This distributor was a 150 L plastic container connected to eight tubes of the same diameter (1/2 po) and length to equally distribute the water in the different mesocosms. The filling of the tanks started on February 9 in late afternoon and was terminated on February 10 at 14 h. At that moment the temperature control system was set at 15 °C (+3 °C

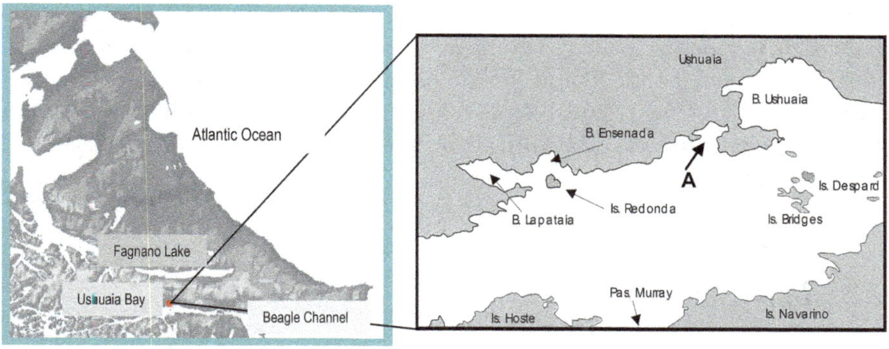

Fig. 1 Map with sample location (A) in the Beagle Channel, Ushuaia, Argentina

compared with natural conditions). Surface water was pumped within a mesocosm system constituted by four 2 m^3 stainless steel tanks. The water column in each mesocosm was completely mixed with pumps at a ~1000 L/h flux. Water was taken from the bottom of tanks and transported toward a temperature control system and then forced it to reenter the tanks from the surface. The first sampling of the water column in the mesocosms (T0) was done at 17 h on February 10. Following an acclimation period, 125 mL of a solution containing 8.088 gL^{-1} of KNO_3 and 0.6805 g L^{-1} of KH_2PO_4 and 20.35 mL of a solution of Na_2SiO_3 (30 g L^{-1}) were injected in each mesocosm on February 11. This procedure was followed to increase the initial concentrations of nitrate, phosphate, and silicate in 5.00, 0.31, and 0.39 μM, respectively (to approximate the Redfield ratio), without extra nutrient addition during the experiment, due to the fact that the Beagle Channel was in a post-bloom situation with low nutrient load (Hernando, unpublished data).

Given the presence of nearly constant winds (most frequent between 3 and 8 m s^{-1}, with maximum speeds >15 m s^{-1}) and the exposure to particles and organisms (particularly bugs during the night), the mesocosms were covered with plastic transparent sheets between 22 h and 7 h of the following day. Temperature was recorded every 60 s in a data logger. An hourly direct reading of the data was also performed.

The experimental design included two different light treatments (two replicates per treatment): natural ambient irradiance (NUVB) and natural irradiance plus increased UVB (HUVB) equivalent to 60% ozone depletion (Díaz et al. 2003). For this, four UVB fluorescent light tubes (TL40 W/12RS, Philips, Markham, Ontario, Canada) were installed over the mesocosms. Cells were exposed all day to natural UVB, and additionally, the exposure with UVB lamps was done for a 5-h daily period during the complete experiment, starting on February 12 between 12 h and 17 h. The lamps were covered with cellulose acetate film (SABIC Polymer Shapes, cat. Nr. 70,600,605) in order to filter UV-C radiation (< 280 nm) (Díaz et al. 2006). The filters were changed daily due to the changes in the spectral transmission after exposure to UVB irradiance. The reduction in transmittance between 294 and 320 nm was ~22% (SD ± 1.4).

Surface water salinity was determined daily using a Horiba U-10. Water samples from each mesocosm were taken every day at 12 h a.m. as follows: 500 ml for chlorophyll *a* measurements, 200 ml for MAA determinations, 200 ml for cell counts (see below), and 800 ml for dichlorofluorescein diacetate (DCFH-DA), thiobarbituric acid reactive substance (TBARS, an index of lipid peroxidation), and αT and βC measurements.

2.2 Light Measurements

During the experiments, incident solar radiation was recorded at 305, 313, 320, 340, and 380 nm and PAR (photosynthetic active radiation 400–700 nm) every 15 min using a GUV-R510 surface radiometer (Biospherical Instruments, Inc., San Diego, CA, USA). The equation from Orce et al. (1997) was used in order to calculate UVB and ultraviolet A radiation (UVA, 315–400 nm) doses in kJm^{-2}. PAR doses were calculated integrating the irradiance during sunlight period.

2.3 Chlorophyll a Analyses and Cell Counts

Chlorophyll *a* (Chl *a*) analyses were evaluated by high-performance liquid chromatography (HPLC) (Moreau et al. 2014). Analyses of the particulate organic carbon concentrations were done with a CHN elemental analyzer (Costech 4010) (Moreau et al. 2014).

In addition, for the identification of phytoplankton, subsamples were fixed with acidic Lugol solution in 250 ml bottles. Moreover, qualitative phytoplankton samples were taken at the end of the experiment using a 20 μm mesh net. For quantitative estimations, cells were enumerated with a phase contrast inverted microscope according to the procedures described by Utermöhl (1958). Subsamples were settled for 24 h in a composite sedimentation chamber. At least 100 cells of the dominant taxa were counted in one or more strips of the chamber or random fields at 250 or 400×, depending on their concentration and size. The whole chamber bottom was also scanned at 100× to count large and sparse species.

Cell biovolumes were measured using the geometric shapes proposed by Hillebrand et al. (1999) and corrected to account for cell shrinkage caused by fixation of samples (Montagnes et al. 1994). Cell carbon content was calculated with two different C to volume ratios: one for diatoms and one for all other groups (Menden-Deuer and Lessard 2000). In order to determine the variation of the relative abundance of small coccoid phytoflagellates and diatoms during the experiment, the ratio phytoflagellates (%)/diatoms (%), based on carbon biomass estimates, was used.

2.4 DCFH-DA Oxidation Rate

The membrane-permeable nonfluorescent DCFH-DA oxidation has been used for detecting several ROS in biological media (McDowell et al. 2013). Cells obtained from filtered samples (14 ml filtered on GF/F filters) were incubated in vivo in the dark for 30 min in 2 ml of 40 mM Tris-HCl buffer (pH 7.0), in the presence of 5 μM DCFH-DA at 27 °C (Malanga et al. 2001). Fluorescence in the supernatant (without cells) was monitored in a microplate reader (Beckman Coulter DTX 880, Multimode Detectors) with excitation (λ_{ex}) at 498 nm and emission (λ_{em}) at 525 nm. In all cases, parallel blank controls were included. Results were expressed as arbitrary units (AU) in 1-h exposure and normalized by total carbon.

2.5 Lipid Peroxidation

TBARS comprise a rough estimate of the presence of aldehydes, yet most reactivity originates from malondialdehyde (MDA), a product of lipid peroxidation, which can react with thiobarbituric acid under acidic and boiling conditions to allow a colorimetric assay (Janknegt et al. 2008). In consequence, cellular TBARS were used as an indicator of ROS-induced lipid peroxidation. The GF/F filters with cells were suspended in 2 ml of 120 mM KCl and 50 mM potassium phosphate buffer (pH 7.0), sonicated and centrifuged 10 min at 600 *g*. An aliquot (0.8 ml) of the supernatant was treated with 0.7 ml 30% (w/v) TCA and 50 mM potassium phosphate buffer (pH 7.0) and brought to a final volume of 2 ml before centrifugation. An aliquot of 0.2 ml of 3% (w/v) SDS and 0.05 ml of 4% (w/v) butylated hydroxytoluene (BHT) in ethanol were added to 1 ml of the supernatant. After mixing, 2 ml of 0.1 N HCl, 0.3 ml of 10% (w/v) phosphotungstic acid, and 1 ml of 0.7% (w/v) 2-thiobarbituric acid were added. The mixture was heated at 95 °C in a water bath for 45 min and 5 ml of n-butanol was added, and then the samples were vortexed and centrifuged at 600 *g* during 10 min. The fluorescence of the organic layer (upper layer) was measured at λ_{ex} = 515 nm and λ_{em} = 555 nm. The values were expressed as nmol TBARS (malondialdehyde equivalents) per carbon units. Malondialdehyde standards were prepared from 1,1,3,3-tetramethoxypropane (Malanga and Puntarulo 1995).

2.6 Lipid-Soluble Antioxidants

The content of αT and βC in the cell homogenates was quantified filtering 5 ml samples by reverse-phase HPLC with electrochemical detection using a Bioanalytical Systems LC-4C amperometric detector with a glassy carbon working electrode at an applied oxidation potential of 0.6 V (Desai 1984). Extraction from the samples was performed with 1 ml of methanol and 4 ml of hexane. After centrifugation at 600 *g* for 10 min, the hexane phase was removed and evaporated to dryness under N_2. Extracts were dissolved in methanol/ethanol (1:1 v/v) and injected for HPLC analysis; HPLC conditions were isocratic reversed-phase column, SUPELCOSIL LC-8, 3.3 cm × 4.6 cm × 3 cm; mobile phase, 20 mM lithium perchlorate in methanol/water 99/1 (v/v); flow rate, 8 ml/min; and retention time, αT = 0.8 min and βC = 1.6 min. d,l-αT from synthetic phytol (Sigma) and βC were used as standards.

2.7 MAA Measurements

For identification and quantification of MAAs, 10–30 ml subsamples were filtered onto Whatman GF/F filters and frozen (−20 °C) until the time of analysis. Filters were extracted with a Vibra-Cell probe sonicator (Sonics & Materials, 1 min, 100 W pulse mode, °C) into 4 ml 100% HPLC grade methanol. The extracts were filtered (Whatman GF/F) and dried using a centrifugal vacuum evaporator (Centrivap, Labconco, Co.). The residue was redissolved in 500 μl water and vortexed for 30 s. After passing through a 100 k Dalton ultrafilter (Ultraspin™), samples were analyzed by HPLC according to the method of Carreto et al. (2002). Briefly, the individual MAAs were separated by reverse-phase, gradient elution on Alltima (Alltech) C_{18}, 5 μm columns (4.6 mm i.d. × 150 mm length) protected with an Alltech guard column cartridge (4.6 mm i.d. × 20 mm length) filled with the same material. They were connected in series and thermostated at 30 °C. An initial isocratic hold up to 8 min with 0.2 acetic acid (Solvent A) was followed by a gradient with Solvent B (methanol/acetonitrile/0.2% acetic acid, 25:25:50) at a flow rate of 1.0 ml min^{-1}. Peak detection was carried out using a diode array detector. Individual peaks were identified by online absorption spectra, retention time, and cochromatography with authentic standards isolated from the red algae *Porphyra* sp. Quantification was made by peak-area integration at 330 nm. The results were presented normalized by total diatom carbon considering that in previous studies it was determined that the phytoflagellates are not able to produce MAAs due to the elevated energetic costs necessary for the small cells (Garcia-Pichel F. 1994) and it was also observed from phytoflagellate cultures and in situ analysis from the Beagle Channel (Hernando et al. 2011).

2.8 Statistical Analyses

Repeated measures ANOVA (RMANOVA) were performed (Statistica, version 9) to determine the significance of the differences observed for relative abundance of main phytoplankton groups, TBARS, αT, βC, and Chl *a* in each treatment. Normality was checked using a one-sample Kolmogorov-Smirnov test, whereas the sphericity assumption that concerns variance homogeneity was checked using Mauchly's test. The main factors considered in the analysis were exposure days and treatments.

Tukey test was additionally done to determine the differences between factors. When interaction was significant or the assumptions of sphericity were not satisfied, a one-factor ANOVA was performed (Scheiner 2001).

3 Results

3.1 Temperature

The water temperature inside the low-temperature treatments was, on average, 12.05 ± 0.26 °C from day 1 to day 9 of the experiment. In the elevated temperature treatments, water temperature was, on average, 14.92 ± 0.18 °C from day 1 to day 9.

3.2 Light Environment

PAR, UVA, and UVB natural daily doses presented high variability (UVB Max. = 64 kJ m^{-2}, Min. = 40 kJ m^{-2}) due to varying cloudy conditions. Maximal doses of UVB, UVA, and PAR were recorded on day 4 of the experiment (February 15) in coincidence with relatively low ozone concentrations (270 D.U., Fig. 2a, b). The incident UVB around local noon reached with supplemented UVB lamps averaged values of 377 ± 394 μW cm^{-2} s^{-1}, which correspond to ~510% of the UVB in the normal UVB treatments (Fig. 2c).

3.3 Phytoplankton Biomass and Composition

The time evolution of Chl *a* did not show significant differences between HUVB and NUVB. Chl *a* increased since the beginning of the experiment, reaching maximum values of ~9.5 μgL^{-1} at the fourth exposure day (D4) in all treatments. After that, Chl *a* decreased sharply until the end of the experiment, being significantly lower in both HT treatments in days 5, 6, and 7 ($p < 0.01$). Diatoms were composed of two main taxa, *Thalassiosira* sp. and *Asterionellopsis glacialis*, with an average contribution of 64 and 8% to total microphytoplankton carbon biomass, respectively. The rest of the phytoplankton assemblages were composed of euglenophytes and small phytoflagellates, mainly coccolithophorids resembling *Emiliania huxleyi* and other prymnesiophytes, as well as *Pyramimonas*-like prasinophytes and other unidentified flagellates. Over the course of the experiment, however, euglenophytes progressively disappeared in all treatments, while prasinophytes remained stable. According to the ratio small coccoid phytoflagellates (%)/diatoms (%), there was a gradual replacement of diatoms by small coccoid phytoflagellates in the high-temperature treatment, being significantly higher in both UV (NUVB and HUVB) in comparison with both normal temperature treatments at D5 and D7 ($p < 0.01$, Fig. 3).

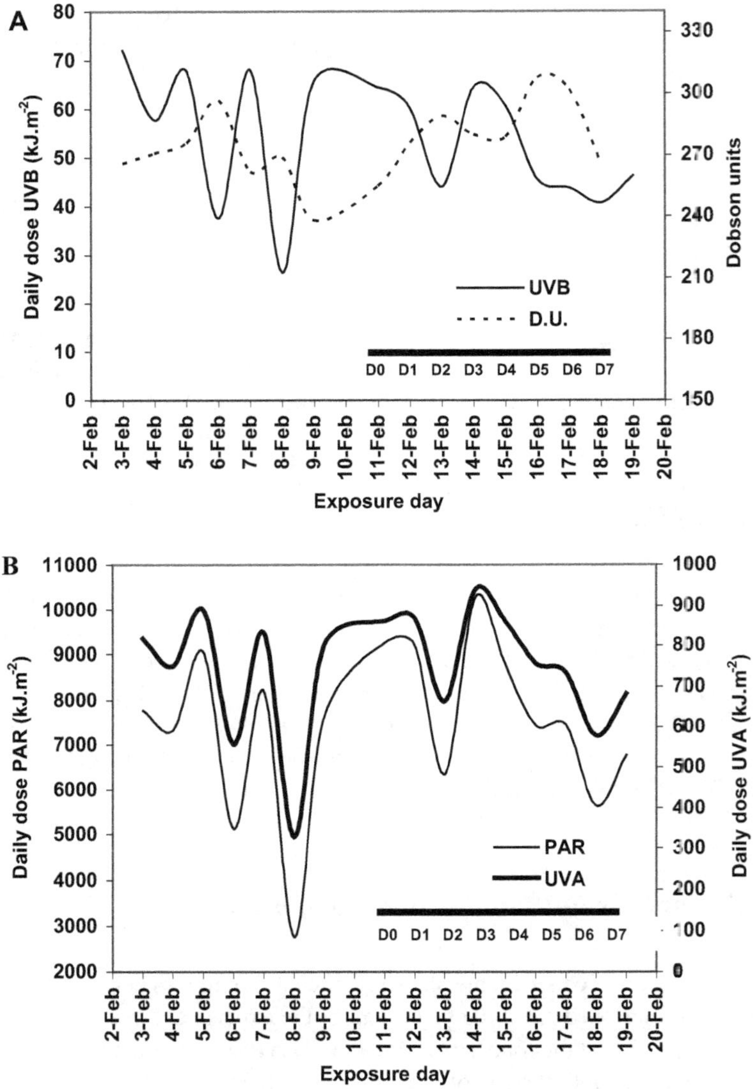

Fig. 2 Variations in UVR doses, PAR doses, and ozone concentrations (Dobson Units, D.U.) at the Beagle Channel from February 2–19, 2008. (**a**) UVB doses (280–315 nm) in kJ.m^{-2} and D.U., (**b**) UVA (315–400 nm) and PAR doses in kJ.m^{-2}, (**c**) incident UVB irradiance at the surface of the experimental mesocosms (average ± SD). The four treatments were the following: NT-NUVB, natural temperature and natural UVB; NT-HUVB, natural temperature and high UVB (simulating a 60% decrease in stratospheric ozone layer thickness); HT-NUVB, high temperature (+3 °C) and normal UVB; and HT-HUVB, high temperature and high UVB. Each point represents the integrated value doses. The experimental period is indicated with a black bar (D0 = February 10). Solar radiation data were obtained with a GUV radiometer. Ozone concentration data were obtained from NASA (http://jwocky.gsfc.nasa.gov). Note the low ozone values during February

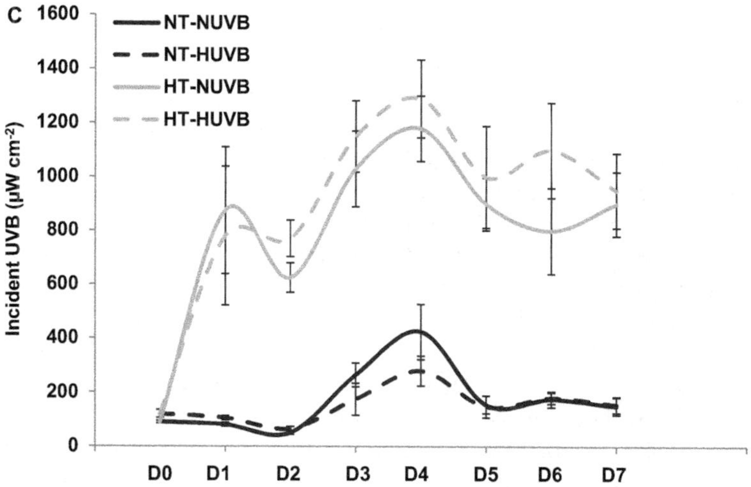

Fig. 2 (continued)

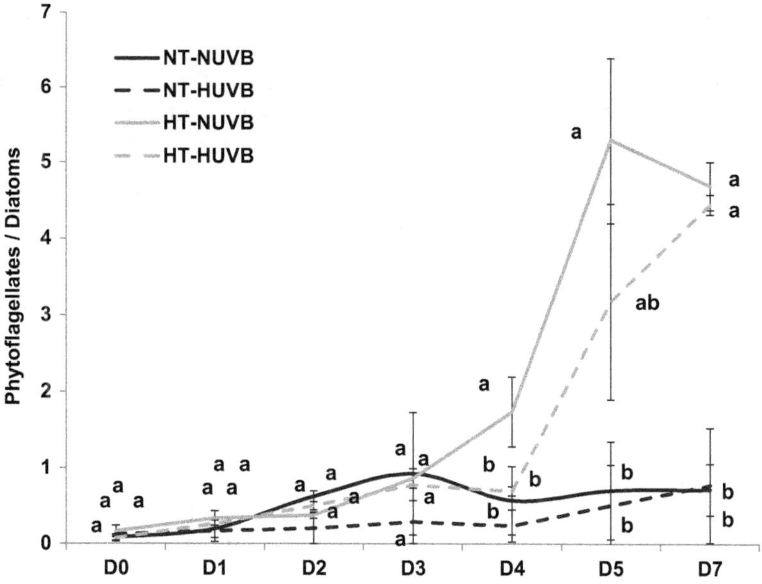

Fig. 3 Variation of the relative abundance (ratio phytoflagellates (%)/diatoms (%)) based on carbon biomass estimates as a function of the exposure time. Each point represents the mean ± SD. The same letter means no significant differences between treatments at the same exposure day, according to Tuckey test after RMANOVA analysis

3.4 Oxidative Stress

Here we will simultaneously analyze the damage as indicated by the presence of ROS (Fig. 4) and TBARS (Fig. 5) on a day-by-day comparison. ROS concentration was significantly higher in NT-HUVB treatment ($p < 0.01$) compared with the rest of treatments, reaching the maximum concentration of 4.6 AU h^{-1} C^{-1} during D1 in relation to the initial time. In HT treatments (HUVB and NUVB), ROS concentrations increased but were not significantly higher compared with the control ($p = 0.78$, Fig. 4). However, TBARS content was significantly higher in cells exposed to HT-HUVB ($p < 0.01$), while NT-NUVB and HT-NUVB TBARS content did not show any significant differences compared with the control ($p = 0.99$, Fig. 5). On D2, the ROS concentrations decreased in all treatments, reaching in average 0.2 AU h^{-1} C^{-1} without significant differences between treatments ($p = 0.3$, Fig. 4). In contrast, TBARS showed an increasing trend in all treatments except for HT-HUVB without significant differences between treatments ($p = 0.16$, Fig. 5). On D3, the ROS concentration showed a slight increase without significant differences between treatments ($p = 0.31$, Fig. 4), but TBARS showed neither significant differences with respect to the previous day nor between treatments ($p = 0.8$, Fig. 5). ROS concentrations on D4 remained without significant changes with respect to the previous day in HT-NUVB, being significantly lower ($p < 0.05$) than the control. In contrast, in the HT-HUVB treatment, ROS concentration was significantly higher compared with HT-NUVB ($p < 0.01$) but did not show significant differences with the control nor with NT-HUVB ($p = 0.98$ and 0.2, respectively, Fig. 4). In agreement

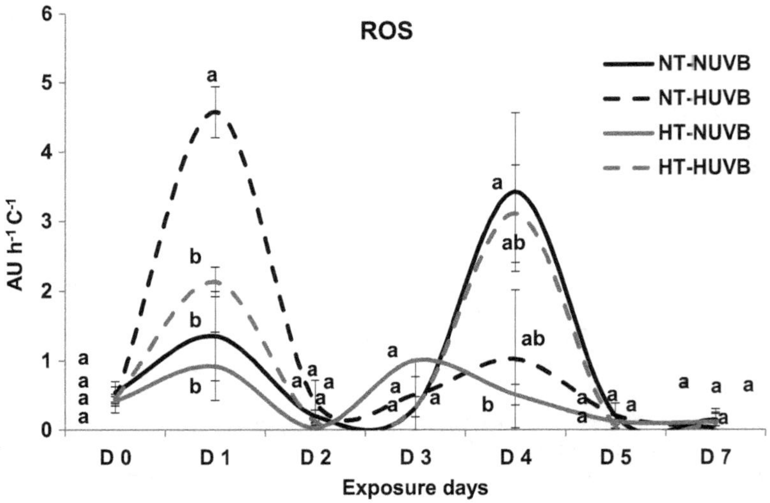

Fig. 4 DCF-DA oxidation rate (expressed as arbitrary units (AU) in 1-h exposure and normalized by total carbon) in plankton assemblages exposed to NT-HUVB (normal temperature and high UVB), HT-HUVB (high temperature and high UVB), HT-NUVB (high temperature and natural UVB), and control (normal temperature and natural UVB) as a function of exposure time. Each point represents the mean ± SD. The same letter means no significant differences between treatments at the same exposure day, according to Tuckey test after RMANOVA analysis

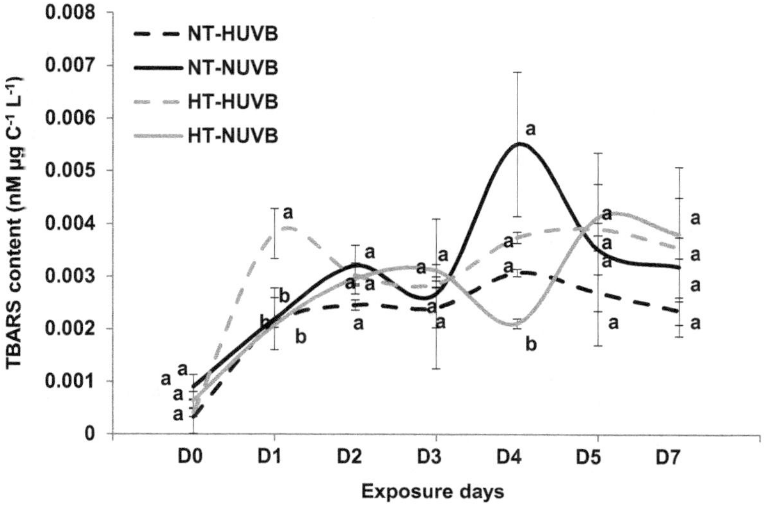

Fig. 5 Lipid peroxidation (TBARS nMug C^{-1} L^{-1}) in plankton assemblages exposed to NT-HUVB (normal temperature and high UVB), HT-HUVB (high temperature and high UVB), HT-NUVB (high temperature and natural UVB), and control (normal temperature and natural UVB) as a function of exposure time. Each point represents the mean ± SD. The same letter means no significant differences between treatments at the same exposure day, according to Tuckey test after RMANOVA analysis

with these results, TBARS content was significantly lower in cells exposed to HT-NUVB ($p < 0.01$) compared with the rest of the treatments, showing increasing values, reaching a maximum of 0.0055 nM µg C^{-1} L^{-1} in cells exposed to the control treatment (Fig. 5). For D5 and D7, the ROS concentration was minimum without significant differences between treatments ($p = 0.4$ and 0.8, respectively, Fig. 4), and TBARS content was at the same level than in the previous day without significant differences between treatments ($p = 0.54$ and 0.5, respectively, Fig. 5).

3.5 *Lipid-Soluble Antioxidants*

During D1, no significant differences were found between the cellular αT concentrations comparing neither with initial time nor between treatments. However, in both HUVB treatments (NT and HT), the values were higher compared with both NUVB treatments (NT and HT) (Fig. 6a). On D2, αT was maximum for HT-NUVB and control, reaching 0.025 nmol C^{-1} on average for both treatments. The concentration in both HUVB treatments (NT and HT) was significantly lower than in both NUVB treatments (NT and HT) ($p < 0.01$) (Fig. 6a). On D3, αT were significantly lower compared with D2, reaching minimum values of 0.003 nmol C^{-1} and without significant differences between treatments ($p = 0.29$, Fig. 6a).

The βC concentrations followed the same trend than αT, but on D2 a consumption of βC was evident in all treatments except for control in which the βC concentration

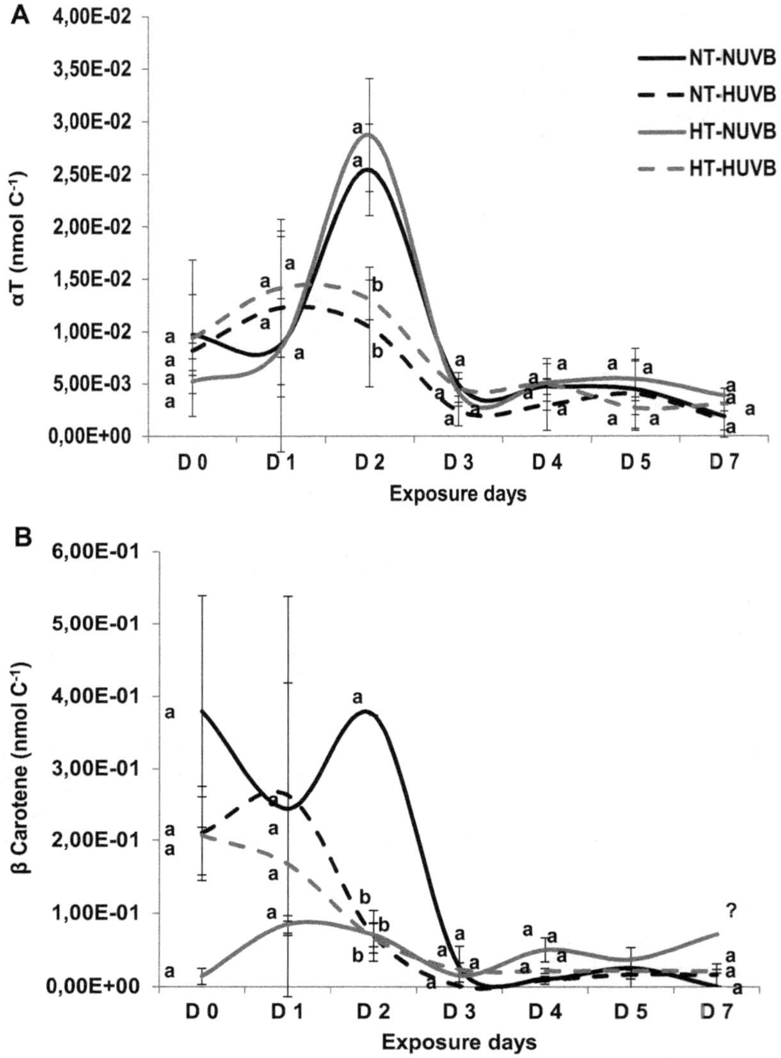

Fig. 6 Effect of temperature and UVB radiation on lipid-soluble antioxidant normalized by total carbon. (**a**) αT (nmol C^{-1}) and (**b**) βC (nmol C^{-1}) as a function of exposure time. Each point represents the mean ± SD. The same letter means no significant differences between treatments at the same exposure day, according to Tuckey test after RMANOVA analysis

was significantly higher than in the treatments (reaching 0.4 nmol C^{-1}, $p < 0.01$, Fig. 6b). Finally, MAAs did not show significant differences in the concentration normalized by total carbon diatoms along the experiment for any treatment and the control or between days. It was observed a high MAA concentration on D5 in HT-HUVB; however, due to the high variability of the values, there were no significant differences between treatments ($p > 0.05$). Additionally, on day 7 the MAA concentrations were undetectable (Carreto personal communication) (Fig. 7).

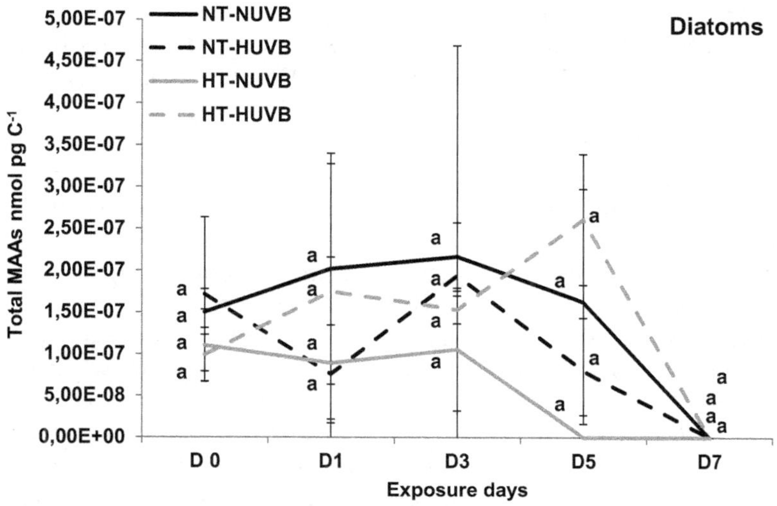

Fig. 7 Total MAA concentrations (shinorine+palythine+porphyra-334) normalized by diatom carbon in plankton assemblages exposed to NT-HUVB (normal temperature and high UVB), HT-HUVB (high temperature and high UVB), HT-NUVB (high temperature and natural UVB), and control (normal temperature and natural UVB) as a function of exposure time. Each point represents the mean ± SD. The same letter means no significant differences between treatments at the same exposure day, according to Tuckey test after RMANOVA analysis

4 Discussion

The purpose of this mesocosm study was to determine if increased temperature (HT, +3 °C) will offset negative effects on phytoplankton by UVB, considering the physiology and composition of natural subantarctic phytoplankton assemblages. Mesocosms allowed us to study the natural phytoplankton assemblages from the Beagle Channel under mixing conditions. Vertical mixing is one of the most important variables affecting the performance of phytoplankton within the water column (Litchman 2000) because it exposes organisms to fluctuating (i.e., variable) radiation regimes (Helbling et al. 1994). Several UV studies with natural phytoplankton communities have been performed using outdoor mesocosms (e.g., Wängberg et al. 1999; Bergmann et al. 2002; Hernando et al. 2006; Longhi et al. 2006) in which mixing is well simulated, overcoming this limitation.

4.1 The Importance of Phytoplankton Community Structure

Several days' exposure to high UVB may induce changes in the phytoplankton assemblages' dominant species, from less to more tolerant ones (Vernet 2000; Hernando et al. 2006). Warmer temperatures have shown to cause shifts in phytoplankton composition, favoring small phytoflagellates, in experimentally manipulated plankton communities from other high-latitude regimes such as the Bering Sea and the North Atlantic (Noiri et al. 2005; Hare et al. 2007; Rose et al. 2009).

Hernando et al. (2006) showed that the composition of phytoplankton assemblages seems to play a central role in the acclimation to high irradiance and temperature conditions for both growth rate and photosynthesis in the Beagle Channel. During the present experiment, the initial phytoplankton assemblage's composition was mainly composed of diatoms, genus *Thalassiosira*, which contributed to around 60%, and *Asterionellopsis glacialis* with 8% of the relative cell abundance. Small nanophytoplankton was mainly composed of coccolithophorids resembling *Emiliania huxleyi* as well as other prymnesiophytes, prasinophytes (*Pyramimonas-like*), cryptophytes, and other unidentified flagellates (Moreau et al. 2014). Temperature had a positive effect on the abundance of large nanophytoplankton (mainly prymnesiophytes), but no effects were observed in other phytoplankton groups independently of UVB doses (Moreau et al. 2014). No significant differences were observed in the relative abundance of small coccoid phytoflagellates and diatoms when the assemblages were exposed to normal temperature, both in natural and high UVB doses (NUVB and HUVB). However, under increased temperature conditions, the relative abundance of small coccoid phytoflagellates was significantly higher compared to normal temperature after 5 days' exposure, independently of the UVB doses received. In coincidence with our results, Hare et al. (2007) reported a dominance of nanophytoplankton after mesocosm incubations at high temperature in the Bering Sea. By contrast, Lionard et al. (2012) showed that large centric diatoms were favored in terms of cumulative biomass when natural marine phytoplankton was exposed to high temperature in temperate environments. The increase in the relative abundance of small coccoid phytoflagellates agrees with previous results from Hernando et al. (2006), who exposed to UVB other phytoplankton assemblages from the Beagle Channel in mesocosms with mixed incubation conditions with addition of nutrients only at the beginning of the experiment and normal temperature. After 10 days, the euglenoid algae (*Eutreptiella* sp.) had increased, and phytoflagellates were the dominant group (Hernando et al. 2006).

Different taxa of primary producers have different abilities to acclimate to changing temperature (Huertas et al. 2011). A possible explanation for the observed higher ratio of phytoflagellates/diatoms since D4 for high temperature compared with normal temperature could be related to the ratio of nutrient consumption (limitations for the experiment). In Moreau et al. (2014), we proposed that the competition for nutrients, particularly for nitrate, was the reason for such relative increase in small coccoid phytoflagellates in high-temperature treatments. However, the analysis of oxidative stress parameters provides new hypotheses to explain the change in relative phytoplankton abundances. Van Leeuwe et al. (2005) have determined differences in the photoacclimation of diatoms and small coccoid phytoflagellates. In this last group, these authors determined a strong regulation of the photosynthetic efficiency during the day. Diatoms adjust to average light intensity rather than to maximum irradiance (Ibelings et al. 1994; Dijkman 2001). This strategy renders them vulnerable to photoinhibition. During exposure to high irradiance, downregulation of photosynthetic efficiency is required to prevent the generation of an energy excess that can result in radical formation and consequent photodamage (van Leeuwe et al. 2005). On the other hand, phytoflagellates readily acclimated to dynamic light conditions (van Leeuwe et al. 2005). Several years of seasonal exposure of phytoplankton assemblages from the Beagle Channel to solar UVR showed that photosynthesis inhibition was significantly lower in assemblages dominated by small coccoid phytoflagellates compared with those dominated by diatoms (Hernando 2011). The repair of UV-damaged structures requires a healthy cell metabolism. A cell that is replete with respect to its nutritional demands is best equipped to both continue fixing carbon and to repair UV-induced damage with a higher activity of enzymatic antioxidants, for example. In our experiments, it was observed a significant increase in the abundance of phytoflagellates in HT treatments after the maximum UVB doses received for the phytoplankton assemblages. Complementing the hypothesis presented in Moreau et al. (2014) regarding the influence of nutrients on the increase of the relative abundance of small coccoid phytoflagellates in high temperature conditions, the high nonenzymatic antioxidant consumption presented in the present work allows to propose that phytoflagellates were probably better adapted to photoinhibition. Phytoflagellates could have a higher photosynthesis rate to support the extra metabolic available energy necessary for the synthesis of nonenzymatic antioxidants (see above) as well as for the activity of enzymatic antioxidants (not measured) that could have additionally helped in avoiding the oxidative stress.

4.2 Damage/Protection Mechanisms

Oxidative stress has been linked to a number of cellular toxic processes, including damages to proteins, membrane lipid peroxidation, enzyme inactivation, and DNA breakage (Halliwell and Gutteridge 2007). In addition, the increased ROS levels could damage other target cells (like proteins or DNA) that we did not measure in our experiment. Although UVB had no significant effects on the observed changes in phytoplankton assemblages' composition (Moreau et al. 2014), when adding the temperature factor, HUVB conditions produced the maximum ROS concentrations under NT, especially during D1. Such high ROS concentrations did not produce significant lipid damage in NT-HUVB. However, in the HT-HUVB treatment, lipid damage was significantly higher compared with the rest of the treatments. In cells exposed to NT-HUVB, protection was present considering the production of nonenzymatic antioxidant molecules. The fact that the cellular ROS concentrations decreased since D2 and during the subsequent days could be related to consumption of antioxidants (Mittler 2002). In fact, a decreased αT concentration was observed in cells exposed to both HUVB treatments (HT and NT) on D2 in comparison with both NUVB treatments. Lipophilic molecules such as αT and βC are able to deactivate $^1O_2$, reduce O_2, and terminate lipid radical chain reactions (Polle and Rennenberg

1994). It was previously demonstrated that αT and βC avoid damage in unialgal phytoflagellate cultures isolated from the Beagle Channel (Hernando et al. 2005) and provide protection to diatoms (*Thalassiosira* sp.) and phytoplankton assemblages from coastal Antarctic waters (Hernando et al. 2011, 2015, respectively). In addition, βC concentration was significantly lower in all experimental treatments compared with control at D2. For D3 and until the end of the experiment, αT and βC concentrations were maintained at low levels in all treatments including control, due to consumption. The connection between oxidative damage, on the one hand, and antioxidant defense mechanisms, on the other hand, has been postulated in both animal and plant cells (Kingston-Smith and Foyer 2000).

A significantly higher ROS concentration was produced in cells exposed to maximum UVB doses. An additional increase in the rates of ROS production, due to temperature-enhanced metabolic rates, occurs primarily in the mitochondria (González et al. 2015). No lipid damage, however, was registered on any of both HT treatments (HUVB and NUVB) probably as a consequence of consumption of αT and βC. The lipid damage observed in cells exposed to HT-NUVB was significantly lower compared with the rest of the treatments probably due to the higher activity of enzymatic antioxidants such as catalase (CAT) or superoxide dismutase (SOD), not measured in our experiment. Rijstenbil (2002) showed an increased activity of SOD on diatoms exposed to UVB radiation.

Upper ocean temperature plays a fundamental role in controlling phytoplankton metabolic processes (Moisan et al. 2002). Experiments in which multiple environmental variables were manipulated reveal that temperature has significant interactive (i.e., synergistic or antagonistic) effects on growth and photo-physiology (Rose et al. 2009). Temperature determines enzymatic activity and metabolic processes by increasing reaction rates with increasing temperature as was seen in early laboratory studies (Eppley 1972) and in the field (Montagnes and Franklin 2001). Supporting the hypothesis of increased enzymatic antioxidants activity with temperature, Giannuzzi et al. (2016) showed an increased CAT and SOD activity on *M. aeruginosa* cultures as a consequence of 3 °C of temperature increase. Opposite results were observed in cells exposed to control treatment on maximum UVB doses with maximum TBARS concentrations observed. Such results reinforce the hypothesis of a lower activity of enzymatic antioxidants under normal temperature conditions compared with 3 °C of increase. If photoprotective and antioxidant mechanisms are overwhelmed, increased production of ROS leads to oxidative damage to lipids, proteins, and other macromolecules (Foyer et al. 1994) as it was observed in control cells exposed to maximum UVB doses.

Another process by which phytoplankton can prevent damage produced by UVR is through the synthesis of UV-absorbing compounds (e.g., MAAs), a family of compounds that absorb radiation in the range of 310–360 nm (Shick et al. 1996). In addition, it was recently postulated that MAA play a role in scavenging of hydrosoluble radicals. Thus, MAAs may act as antioxidants to prevent cellular damage resulting from UV-induced production of ROS (Rastogi and Incharoensakdi 2014). However, we did not observe any significant differences between days nor treatments in MAA concentration in our experiment. From previous mesocosm experiments with phytoplankton assemblages from the Beagle Channel, Hernando et al. (2006) showed that the MAA synthesis was the most effective photoprotective mechanism only under fixed incubations simulating surface exposure with maximum UVR doses received. In coincidence with these results, Hernando et al. (2012) showed a higher synthesis of porphyra-334 and shinorine in simulated fixed conditions of *Thalassiosira* sp. cultures compared with mixing exposure conditions from coastal Antarctic waters. Contrasting with the fixed exposure in which cells are exposed to maximum UVB doses (in our experiment the normal UVB irradiance in the surface was $172 \pm 105\ \mu W\ cm^{-2}\ s^{-1}$ in average for all the exposure period, and the increased UVB irradiance was $877 \pm 394\ \mu W\ cm^{-2}\ s^{-1}$), under mixed conditions, and low UVB doses (in present experiment, the 10% of the surface irradiance was $5 \pm 0.8\ \mu W\ cm^{-2}\ s^{-1}$ for HUVB treatments and $0.8 \pm 0.4\ \mu W\ cm^{-2}\ s^{-1}$ for the NUVB treatments, in average for the complete exposure period), photoprotective mechanisms, such as non-photochemical heat dissipation of energy (i.e., non-photochemical quenching, Hernando et al. 2012), may contribute to the defense against UVB photodamage.

5 Conclusion

Although different from research in the natural environment, studying plankton in large volume mesocosms has the advantage of working with controlled variables. In addition, compared with monospecific cultures, the use of natural planktonic communities gives a better representation of real conditions. However, one should be cautious when interpreting results obtained in this kind of experiments, considering that other factors (e.g., migration, grazing pressure, nutrient renewal) are important components when addressing UVR effects from an ecological point of view (Neale et al. 2003; Villafañe et al. 2003).

Our results show that subantarctic phytoplankton can adapt to high UVB irradiance conditions (e.g., by nonenzymatic antioxidant consumption), in agreement with other studies in the area (Villafañe et al. 1995; Davidson et al. 1996; Hernando et al. 2006; Hernando 2011), as well as to high temperature (Helbling et al. 2015). The structure of phytoplankton community seems to play a central role in this response (being the small coccoid phytoflagellates better adapted to high temperature conditions independently of UVB doses after several days of exposure). In environments like the Beagle Channel, seasonally exposed to high UVB radiation in surface layers due to the springtime stratospheric ozone layer thinning, phytoplankton can cope with UVB by means of αT and βC synthesis that helps to avoid lipid damage in diatom-dominated assemblages. Overall, the present biochemical analysis let us to reject the hypothesis that increased temperature offset negative effects on phytoplankton by UVB. An increased ROS concentration was observed as a consequence of high UVB doses, independently of temperature, but the cells were able to avoid the oxidative stress via the nonenzymatic antioxidant consumption and a possible increased activity of enzymatic antioxidants after 7 days of exposure. However, there was a clear effect of temperature on phytoplankton assemblage composition, being the small coccoid phytoflagellates the most adapted to high temperature on UVB exposure and nutrient-limiting conditions.

Acknowledgments This research is part of the project "Combined Effects of Ultraviolet-B Radiation, Increased CO_2 and Climate Warming on the Biological Pump: A Temporal and Latitudinal Study," led by S.D. and supported by the Natural Sciences and Engineering Research Council of Canada (NSERC, SRO Grant# 334876-2005) and by a grant from the Instituto Antártico Argentino (IAA) to G.A.F. in the frame of the project "Research on Ultraviolet and Global warming effects on Biological pump Yields" (RUGBY). We want to thank Dr. Jose Carreto and Mario Carignan for providing the MAA data. We also want to thank the numerous people that helped us with the setup of the mesocosms and the experiment in Ushuaia: Sylvain Leblanc, Patrick Poulin, Alejandro Olariaga, Alejandro Ulrich, and Raúl Codina.

References

Alley R, Berntsen T, Bindoff NL et al (2007) Climate change 2007: the physical science basis. Summary for policymakers, Intergovernmental Panel on Climate Change, Geneva

Apel K, Hirt H (2004) Reactive oxygen species: metabolism, oxidative stress, and signal transduction. Annu Rev Plant Biol 55:373–379

Beardall J, Stojkovic S (2006) Microalgae under global environmental change: implications for growth and productivity, populations and trophic flow. Sci Asia 32:1–10

Behrenfeld MJ, O'Malley R, Siegel D et al (2006) Climate-driven trends in contemporary ocean productivity. Nature 444:752–755

Bergmann T, Richardson TL, Paerl HW et al (2002) Synergy of light and nutrients on the photosynthetic efficiency of phytoplankton populations from the Neuse River Estuary, North Carolina. J Plankton Res 24:923–933

Bouchard JN, Roy S, Campbell DA (2006) UVB effects on the photosystem II-D1 protein of phytoplankton and natural phytoplankton communities. Photochem Photobiol 82:936–951

Carreto JI, Carignan MO, Daleo G et al (1990) Occurrence of Mycosporine-like Aminoacids in the red-tide dinoflagellate *Alexandrium excavatum*: UV-Photoprotective compounds? J Plankton Res 12:909–921

Carreto JI, Carignan MO, Montoya NG (2002) Short-term effects of ultraviolet radiation on the dinofagellate *Alexandrium catenella*. Pigment bleaching and MAAs synthesis inhibition. In: G. Arzul (ed). Aquaculture, environment and marine phytoplankton. IFREMER, Actes colloq 34:173–190

Davidson AT, Marchant HJ, de la Mare WK (1996) Natural UVB exposure changes the species composition of Antarctic phytoplankton in mixed cultures. Aquat Microb Ecol 10:299–305

Desai ID (1984) Vitamin E analysis methods for animal tissues. Methods Enzymol 105:138–147

Díaz S, Nelson D, Deferrari G et al (2003) A model to extend spectral and multi-wavelength UV irradiances time series. Model development and validation. J Geophys Res. https://doi.org/10.1029/2002JD002134

Diaz S, Camilion C, Cassiccia C et al (2006) Symposium-in-print: UV effects on aquatic and coastal ecosystems simulation of ozone depletion using ambient irradiance supplemented with UV lamps. Photochem Photobiol 82:857–864

Dijkman NA (2001) The regulation of photosynthesis in diatoms under dynamic irradiance. PhD thesis. Universität Bremen

Eppley RW (1972) Temperature and phytoplankton growth in the sea. Fish Bull Nat Ocean Atmos Adm 70:1063–1085

Eun-Ji W, Lee Y, Han J et al (2014) Effects of UV radiation on hatching, lipid peroxidation, and fatty acid composition in the copepod *Paracyclopina nana*. Comp Biochem Physiol C Toxicol Pharmacol 165:60–66

Ferreyra GA, Mostajir B, Schloss IR et al (2006) Ultraviolet-B radiation effects on the structure and function of lower trophic levels of the marine planktonic food web. Photochem Photobiol 82:887–897

Folt CL, Chen CY, Moore MV et al (1999) Synergism and antagonism among multiple stressors. Limnol Oceanogr 44:864–877

Foyer CH, Lelandais M, Kunert KJ (1994) Photooxidative stress in plants. Physiol Plant 92:696–717

Garcia-Pichel F (1994) A model for shelf-shading in planktonic organisms and its implications for the usefulness of ultraviolet sunscreens. Limnol Oceanogr 39:1704–1717

Giannuzzi L, Krock B, Crettaz MC et al (2016) Growth, toxin production, active oxygen species and antioxidants responses of *Microcystis aeruginosa* (Cyanophyceae) to temperature stress. Comp Biochem Physiol C189:22–30

González PM, Malanga G, Puntarulo S (2015) Cellular oxidant/antioxidant network: update on the environmental effects over marine organisms. Open Mar Biol J 9:1–13

Häder DP, Kumar HD, Smith RC et al (2007) Effects of solar UV radiation on aquatic ecosystems and interactions with climate change. Photochem Photobiol Sci 6:267–285

Halac SR, Villafañe VE, Helbling EW (2010) Temperature benefits the photosynthetic performance of the diatoms *Chaetoceros gracilis* and *Thalassiosira weissflogii* when exposed to UVR. J Photochem Photobiol B Biol 101:196–205

Halac SR, Guendulain-García SD, Villafañe VE et al (2013) Responses of tropical plankton communities from the Mexican Caribbean to solar ultraviolet radiation exposure and increased temperature. J Exp Mar Biol Ecol 445:99–107

Halliwell B, Gutteridge JMC (2007) Free radicals in biology and medicine, 4th edn. Clarendon, Oxford

Hare CE, DiTullio GR, Popels LC et al (2007) Effects of changing continuous iron input rates on a Southern Ocean algal assemblage. Deep-Sea Res I 54:732–746

Häubner N, Sylvander P, Vuori K et al (2014) Abiotic stress modifies the synthesis of alpha-tocopherol and beta-carotene in phytoplankton species. J Phycol 50:753–759

Helbling EW, Villafañe VE, Holm-Hansen O (1994) Effects of Ultraviolet Radiation on Antarctic Marine Phytoplankton Photosynthesis with Particular Attention to the Influence of Mixing. In: Weiler CS, Penhale P (eds) Ultraviolet radiation in Antarctica: measurements and biological effects. American Geophysical Union, Antarct Res Ser, 62:207–227

Helbling EW, Banaszak AT, Villafañe VE (2015) Differential responses to the combination UVR and elevated temperature of two phytoplankton communities of the Chubut River estuary (Patagonia, Argentina). Estuar Coasts 38:1134–1146

Hernando M (2011) Fitoplancton de altas latitudes en condiciones de ozono disminuido. 1 ed. Editorial Académica Española, Reino Unido, p300. ISBN-10: 3846560545

Hernando M, Malanga G, Ferreyra GA (2005) Oxidative stress and antioxidant defenses generated by solar UV in a subantarctic marine phytoflagellate. Sci Mar 69:287–295

Hernando M, Schloss I, Roy S et al (2006) Photoacclimation to long-term UVR exposure of natural Subantarctic phytoplankton communities: fixed surface incubations versus mixed mesocosms. Photochem Photobiol 82:923–935

Hernando M, Malanga G, Puntarulo S et al (2011) Non-enzymatic antioxidant photoprotection against potential UVBR-induced damage in an Antarctic diatom (*Thalassiosira* sp). Lat Am J Aquat Res 39(3):397–408

Hernando M, Carreto JI, Carignan MO et al (2012) Effect of vertical mixing on short-term mycosporine-like amino acids (MAAs) synthesis in the Antarctic diatom, *Thalasiossira* sp. Sci Mar 76(1):49–57

Hernando M, Schloss IR, Malanga G et al (2015) Effects of salinity changes on coastal antarctic phytoplankton physiology and assemblage composition. J Exp Mar Biol Ecol 466:110–119

Hideg E, Vass I (1996) UV-B induced free radical production in plant leaves and isolated thylakoid membranes. Plant Sci 115:251–260

Hillebrand H, Dürselen C-D, Kirschtel D et al (1999) Biovolume calculation for pelagic and benthic microalgae. J Phycol 35:403–424

Huertas IE, Rouco M, López-Rodas V et al (2011) Warming will affect phytoplankton differently: evidence through a mechanistic approach. Proc R Soc B 278:3534–3543

Ibelings BW, Kroon BMA, Mur LR (1994) Acclimation of photosystem II in a cyanobacterium and a eukaryotic green alga to high and fluctuating photosynthetic flux densities, simulating light regimes induced by mixing lakes. New Phytol 128:407–424

Irwin AJ, Finkel ZV (2008) Mining a sea of data: determining controls of ocean chlorophyll. PLoS One 3:e3836. https://doi.org/10.1371/Journal.pone.0003836

Irwin AJ, Oliver MJ (2009) Are ocean deserts getting larger? Geophys Res Lett. https://doi.org/10.1029/2009GL039883

Janknegt PJ, van de Poll WH, Visser RJW et al (2008) Oxidative stress responses in the marine antarctic diatom *Chaetoceros brevis* (bacillariophyceae) during photoacclimation. J Phycol 44:957–966

Karsten U, Bischof K, Hanelt D et al (1999) The effect of UV radiation on photosynthesis and UV-absorbing substances in the endemic Arctic macroalga *Develaraea ramentacea* (Rhodophyta). Physiol Plant 105:58–66

Kingston-Smith AH, Foyer CH (2000) Over expression of Mn-superoxide dismutase in maize leaves leads to increased monodehydroascorbate reductase, dehydroascorbate reductase and glutathione reductase activities. J Exp Bot 51:1867–1877

Klisch M, Häder DP (2002) Wavelength dependence of mycosporine-like amino acid synthesis in *Gyrodinium dorsum*. J Photochem Photobiol B 66(1):60–66

Lionard M, Roy S, Tremblay-Létourneau M et al (2012) Combined effects of increased UV-B and temperature on the pigment-determined marine phytoplankton community of the St. Lawrence Estuary. Mar Ecol Prog Ser 445:219–234

Litchman E (2000) Growth rates of phytoplankton under fluctuating light. Freshw Biol 44:223–235

Longhi ML, Ferreyra G, Schloss I et al (2006) Variable phytoplankton response to enhanced UV-B and nitrate addition in mesocosm experiments at three latitudes (Canada, Brazil and Argentina). Mar Ecol Prog Ser 313:57–72

Malanga G, Puntarulo S (1995) Oxidative stress and antioxidant content in *Chlorella vulgaris* after exposure to ultraviolet-B radiation. Physiol Plant 94(4):672–679

Malanga G, Juarez AB, Albergheria JS et al (2001) Efecto de la radiación UVB sobre el contenido de ascorbato y radical ascorbilo en algas verdes. In: Alveal K, Antezana T (eds) Sustentabilidad de la Biodiversidad, un problema actual, bases científico-técnicas, teorizaciones y proyecciones. Universidad de Concepción, Concepción, pp p389–p398

McDowell RE, Amsler CD, Dickinson DA et al (2013) Reactive oxygen species and the Antarctic macroalgal wound response. J Phycol 50:71–80

Meehl GA, Stocker TF, Collins WD et al (2007) Global climate projections. In: Solomon S, Qin D, Manning M et al (eds) Climate change 2007: the physical science basis. Contribution of working group I to the fourth assessment report of the intergovernmental panel on climate change. Cambridge University Press, Cambridge, UK/New York

Menden-Deuer S, Lessard EJ (2000) Carbon to volume relationships for dinoflagellates, diatoms, and other protist plankton. Limnol Oceanogr 45(3):569–579

Mittler R (2002) Oxidative stress, antioxidants, and stress tolerance. Trends Plant Sci 7:405–410

Moisan JR, Moisan TA, Abbott MR (2002) Modelling the effect of temperature on the maximum growth rates of phytoplankton populations. Ecol Model 153:197–215

Montagnes DJS, Franklin DJ (2001) Effect of temperature on diatom volume, growth rate, and carbon and nitrogen content: reconsidering some paradigms. Limnol Oceanogr 46(8):2008–2018

Montagnes DJS, Berges JA, Harrison PJ et al (1994) Estimating carbon, nitrogen, protein, and chlorophyll-*a* from volume in marine phytoplankton. Limnol Oceanogr 39(5):1044–1060

Moreau S, Mostajir B, Almandoz GO et al (2014) Effects of enhanced temperature and ultraviolet B radiation on a natural plankton community of the Beagle Channel (Southern Argentina): a mesocosm study. Aquat Microb Ecol 72:155–173

Mostajir B, Demers S, de Mora S et al (1999) Experimental test of the effect of ultraviolet-B radiation in a planktonic community. Limnol Oceanogr 44:586–596

Mostajir B, Demers S, de Mora SJ et al (2000) Implications of UV radiation on the food web structure and consequences on the carbon flow. In: de Mora SJ, Demers S, Vernet M (eds) The effects of UV radiation in the marine environment. Cambridge University Press, Cambridge, pp 310–320

Neale PJ, Helbling EW, Zagarese HE (2003) Modulation of UVR exposure and effects by vertical mixing and advection. In: Helbling EW, Zagarese HE (eds) UV effects in aquatic organisms and ecosystems, Comprehensive series in Photochemical and Photobiological Sciences. The Royal Society of Chemistry, Cambridge, pp 107–134

Niyogi KK (1999) Photoprotection revisited: genetic and molecular approaches. Annu Res Plant Physiol Plant Mol Biol 50:333–359

Noiri Y, Kudo I, Kiyosawa H et al (2005) Influence of iron and temperature on growth, nutrient utilization ratios and phytoplankton species composition in the western subarctic Pacific Ocean during the SEEDS experiment. Prog Oceanogr 64:149–166

Orce LV, San Román NA, Paladini A et al (1997) Multiple regression fit. Latitudinal UVR-PAR measurements in Argentina: extent of the 'ozone hole'. Glob Planet Chang 15:113–121

Polle A, Rennenberg H (1994) Photooxidative stress in trees. In: Foyer CH, Mullineaux PM (eds) Causes of photooxidative stress and ameloration of defence systems in plants. CRC Press, London, pp 199–218

Rastogi RP, Incharoensakdi A (2014) UV radiation-induced biosynthesis, stability and antioxidant activity of mycosporine-like amino acids (MAAs) in a unicellular cyanobacterium Gloeocapsa sp. CU2556. J Photochem Photobiol B 130:287–292

Rijstenbil JW (2002) Assessment of oxidative stress in the planktonic diatom Thalassiosira pseudonana in response to UVA and UVB radiation. J Plankton Res 24:1277–1288

Rose JM, Feng Y, Gobler CJ et al (2009) The effects of increased pCO2 and temperature on the North Atlantic Spring Bloom. II. Microzooplankton abundance and grazing. Mar Ecol Prog Ser 388:27–40

Sarmiento J, Slater RD, Barber R et al (2004) Response of ocean ecosystems to climate warming. Global Biochem Cycles 18:1–23

Scheiner SM (2001) MANOVA: multiple response variables and multispecies interactions. In: Scheiner G (ed) Design and analysis of ecological experiments, 2nd edn. Oxford University Press, Oxford

Shick JM, Lesser MP, Jokiel PL (1996) Effects of ultraviolet radiation on corals and other coral reef organisms. Glob Chang Biol 2:527–545

Sobrino C, Neale PJ (2007) Short-term and long-term effects of temperature on photosynthesis in the diatom *Thalassiosira pseudonana* under UVR exposures. J Phycol 43:426–436

Utermöhl H (1958) Zur vervollkommnung der quantitativen phytoplankton-methodik. Mitt Int Ver Theor Angew Limnol 9:1–38

Van Leeuwe MA, van Sikkelerus B, Gieskes WWC et al (2005) Taxon-specific differences in photoacclimation to fluctuating irradiance in an Antarctic diatom and a green flagellate. Mar Ecol Prog Ser 288:9–19

Vernet M (2000) Effects of UV radiation on the physiology and ecology of marine phytoplankton. In: De Mora S, Demers S, Vernet M (eds) The effects of UV radiation in the marine environment. Cambridge Environ Chem Ser, 10, Cambridge University Press, Cambridge, pp 237–278

Villafañe VE, Helbling EW, Holm-Hansen O et al (1995) Acclimatization of Antarctic natural phytoplankton assemblages when exposed to solar ultraviolet radiation. J Plankton Res 17:2295–2306

Villafañe VE, Sundback K, Figueroa FL et al (2003) Photosynthesis in the aquatic organisms and ecosystems. The Royal Society of Chemistry, Cambridge, pp 357–397 (Chapter 2)

Wada N, Sakamoto T, Matsugo S (2015) Mycosporine-like amino acids and their derivatives as natural antioxidants. Antioxidants 4(3):603–646

Wängberg SÅ, Garde K, Gustavson K et al (1999) Effects of UV-B radiation on marine phytoplankton communities. J Plankton Res 21:147–166

Global Change and Plankton Ecology in the Southwestern Atlantic

Elena S. Barbieri, María A. Marcoval, Rodrigo D. Hernández-Moresino, Mariela L. Spinelli, and Rodrigo J. Gonçalves

Abstract In aquatic ecosystems, global change encompasses human-induced variations in the upper layer of the water column, where most of the carbon fixation takes place. These changes include increments of exposure to solar UV radiation, mean temperature of surface seawater, ocean acidification and variations in nutrient concentrations. These factors are well known to affect plankton ecosystems, although not all organisms are equally affected. In addition, many times these factors interact so that the final results are not always additive or synergistic as expected. Finally, indirect effects (e.g. through trophic interactions) are often more pronounced than direct ones (e.g. survival or growth), which in some cases mask overall and long-term expected ecological shifts in plankton communities. We review the main contributions in this field from the Atlantic coast of South America and discuss in a final section what have we learned and what is still unknown as global changes seem to be here to stay.

Keywords Global change · Planktonic communities · UV radiation · Human impact

---

E. S. Barbieri · R. D. Hernández-Moresino · R. J. Gonçalves (✉)
Laboratorio de Oceanografía Biológica (LOBio), Centro para el Estudio de Sistemas Marinos (CESIMAR), Consejo Nacional de Investigaciones Científicas y Técnicas (CONICET), Puerto Madryn, Chubut, Argentina
e-mail: barbieri@cenpat-conicet.gob.ar; rodrigo@cenpat-conicet.gob.ar; patagoniaplankton@gmail.com

M. A. Marcoval
Instituto de Investigaciones Marinas y Costeras (IIMyC), Consejo Nacional de Investigaciones Científicas y Técnicas–Universidad Nacional de Mar del Plata (CONICET–UMNdP), Buenos Aires, Argentina
e-mail: marcoval@mdp.edu.ar

M. L. Spinelli
Instituto de Biodiversidad y Biología Experimental y Aplicada (IBBEA), Consejo Nacional de Investigaciones Científicas y Técnicas–Universidad de Buenos Aires (CONICET – UBA), Buenos Aires, Argentina
e-mail: marielaspinelli@bg.fcen.uba.ar

M. S. Hoffmeyer et al. (eds.), *Plankton Ecology of the Southwestern Atlantic*,
https://doi.org/10.1007/978-3-319-77869-3_25

1 Introduction

Given the ubiquitous and persistent activities by human population, as of today, all major oceans and coastal ecosystems can be considered affected to a certain degree by cumulative action of one or more factors associated with "global change". These may include higher exposure to solar UV radiation (UVR), increased ocean acidification (OA), rising surface seawater temperature (SST) and higher concentration of nutrients (and in many cases, contaminants). Those changes have been documented in different sites around the world (Bates et al. 2014) and occur most notably in the upper layer of the water column, where most of the carbon fixation takes place. These factors are well known to affect several plankton communities, although not all organisms are equally affected. Since South America includes a variety of marine habitats (completely or partially including up to seven biogeographical provinces, at least ~3 million km^2), the material considered in this review covers the coast of South America and in some cases also studies from the wider South Atlantic basin, since they share many biological and physical characteristics.

Some of the documented factors affecting South Atlantic waters can be briefly mentioned. Kitidis et al. (2016) found clear evidence of OA across a wide range of latitudes in the Atlantic Ocean (including those encompassing a significant part of South America), and their results suggest that the fugacity of CO_2(fCO_2) in seawater increases at the same rate as atmospheric CO_2. Increases of solar UVR are largest (zonal average) in the Southern Hemisphere and have been well documented in South America in the last decades (Jaque et al. 1994; Herman 2010), and interestingly, compounds which act as natural "sunscreens" (e.g. mycosporine-like amino acids, Karentz et al. 1991; Dunlap and Yamamoto 1995) are found in high concentrations in the South Atlantic from 40°S poleward (Fileman et al. 2017). In this review we consider the term "global change" in a broad sense, i.e. including not only UVR (although the review itself is somewhat skewed towards UVR due to our background), temperature and OA but also some selected studies on human impact (e.g. pollution) on marine organisms. Aiming to get an integral overview, we emphasize studies dealing with multifactorial or interactive effects whenever possible.

2 Bacterioplankton

Given the importance of heterotrophic bacterioplankton to biogeochemical cycles, there seems to be very little information in the context of global change in Atlantic South America. Mesocosm studies with water from the Atlantic coast of Patagonia show that UVR can affect the taxonomic composition of marine bacterial populations (Manrique et al. 2012). When nutrients and OA are added to UVR to consider a hypothetical future scenario of global change, sensitivity of heterotrophic bacterioplankton to UVR may change from inhibition to enhancement, presumably an combination of direct effects on bacterial groups with differential sensitivities, and indirect effects through the responses of other plankton components (e.g. phytoplankton) which may act as sources of dissolved organic carbon (DOC) for heterotrophic bacterioplankton (Durán-Romero et al. 2017).

3 Phytoplankton

Being primary producers, phytoplankton organisms are of utmost importance in marine food webs, and thus, considerable effort has been put into elucidating its responses to global change. As autotrophic organisms, they are exposed to sunlight and therefore to the potentially damaging levels of solar UVR. Since many responses to UVR (as well as to other factors) are species-specific, it follows that in order to evaluate potential responses under different environmental conditions, it is necessary to consider other factors such as changes in cloudiness, ozone concentrations, differential sensitivity of organisms, stratification of the water column, etc. (Villafañe et al. 2001; Vernet et al. 2009). At least five interrelated aspects play important roles when evaluating responses of phytoplankton to global change variables: cell size, bloom phase (i.e. pre-, during and post-bloom assemblages), mixing/light acclimation, taxonomic composition and available defence mechanisms. Similarly to studies from elsewhere, it has been shown that smaller cells in phytoplankton samples from Patagonia can be more sensitive to UVR in terms of DNA damage, whereas larger cells tend to be more sensitive regarding photosynthesis inhibition (Helbling et al. 2005). In incubation experiments in coastal waters of Bahía Bustamante (Patagonia Argentina), picoplankton cells (0.2–2 μm) appeared as very sensitive to UVR-induced damage to DNA, while photosynthetic inhibition showed a different pattern, suggesting photosystems and DNA are independent targets for UVR (Buma et al. 2001; Helbling et al. 2001). Also in the picoplanktonic size category, an example can be found in waters off Brazil and Uruguay (as part of a larger sampling area), where the cyanobacteria *Prochlorococcus* and *Synechococcus* as well as picoeukaryotes showed differential vulnerability to solar radiation exposure (with *Prochlorococcus*, among the smallest oxygenic photosynthetic organisms, with the smallest known genome – being more sensitive) (Agusti and Llabrés 2007). The reason for smaller cells being more vulnerable to UVR seems to be related to the lower theoretical efficiency of very small cells to accommodate protective compounds[1] within such limited volume (Garcia-Pichel 1994). However this cell size-related pattern may be completely reversed when microalgae are exposed to fluctuating light levels (i.e. simulating the upper mixed layer). In this regard, the photosynthetic performance of *Prorocentrum micans* (50 μm) was less affected than that of *Gymnodinium chlorophorum* (5 μm), probably due to a lower capacity of the latter to cope with UVR under fast mixing and higher "sunscreen" compounds in

[1] We will use collective terms for these compounds ("sunscreens", photoprotective compounds PPC, protective compounds) interchangeably. They are in general mostly catorenoids and mycosporine-like amino acids (MAAs).

the former species. This led to the suggestion that, under mixing conditions, photosynthesis may be a more important target for UVR (Helbling et al. 2008).

As mentioned above, cell size alone is not sufficient to explain nor to predict responses of phytoplankton, as different microalgal groups may respond differently and thus taxonomic composition of plankton community becomes critical in shaping the overall effect of global change. For example, when considering interactive effects of UVR and nutrients addition on microalgal cultures, dinoflagellates may show higher growth and photosynthesis than diatoms (Marcoval et al. 2007), probably due to different protective mechanisms of these groups. However, when interaction between nutrients and UVR was considered upon natural plankton assemblages, diatoms ended up dominating biomass (versus small flagellates and monads), being greater the effect of nutrients availability than solar UVR exposure (Marcoval et al. 2008). Thus, due to species- or group-specific effects of UVR, OA and SST, overall data show that most likely result would be a shift in dominant groups, but it is not clear how well this can be predicted. As pointed out it has been found often that combined global change factors favour the dominance of diatoms and relatively larger micro-phytoplankton-like diatoms (e.g. Marcoval et al. 2008; Villafañe et al. 2014, 2015), although in some cases the dominant group remains the same (e.g. Villafañe et al. 2017).

In natural phytoplankton communities, the taxonomic composition will vary seasonally in areas with a characteristic bloom (with several examples in South America coast, e.g. Villafañe et al. 2004a, Guinder et al. 2015); thus, the timing of the experiments/sampling with respect to the bloom development can result in different photosynthetic responses of phytoplankton assemblages. In species from coastal Patagonia, only post-bloom assemblages were affected by UVR, while mixing (i.e. varying light regime) enhanced carbon fixation in nanoplanktonic flagellates and diatom-dominated communities. Different responses before, during and after bloom may be related to light history, cell size and taxonomic composition of phytoplankton assemblages (Barbieri et al. 2002). Mixing of the water column (i.e. a shorter exposure to surface solar radiation) may also help explaining the moderate/null UVR effects in a mesocosm study which included Ubatuba (Brazil) and Ushuaia (Argentina), thus suggesting that light acclimation, temperature and community composition all play a role (Bouchard et al. 2005; Longhi et al. 2006; Roy et al. 2006). In experiments on coastal phytoplankton under increased UVR, nutrients and temperature, it has been observed that net community production (NCP) can be enhanced by interaction of those factors during pre-bloom conditions (warmer in situ temperatures, > 15 °C) but was inhibited during bloom (temperatures < 13 °C). Thus, the effects of global change on communities needs to consider not only interaction of factors and diverse biological targets but also the timing and length of the experimentation (Cabrerizo et al. 2017).

Overall, throughout the years, studies in this and other areas have found repeatedly that the effects of UVR and other global change factors depend markedly on the composition and physiological state of the phytoplankton community in terms of pre-, during and post-bloom assemblages; this in turn highlights the importance of acclimation as a key role in shaping the responses of primary producers (e.g.

Villafañe et al. 2004a, b, 2008, 2013, 2017; Helbling et al. 2015). Arguably, species-specific responses may hinder broad generalizations from experiments with single or few species, as some organisms may have different ways of counteracting negative impacts – e.g. of UVR – by using protective effect of pigments/"sunscreen" compounds (Marcoval et al. 2007; Carreto and Carignan 2011) or even escape from damaging UVR resorting to motility (Richter et al. 2007).

4 Zooplankton

Human-induced environmental changes include the facilitation of biological invasions and shifting in species dominance. Hoffmeyer (2004) documented the appearance of holo- and meroplankton species in the Bahía Blanca Estuary (the copepod *Eurytemora americana* and larvae of the cyrripede *Balanus glandula*, respectively). In this context, it would be interesting to further study how this affects benthic-pelagic coupling (via meroplanktonic larvae). Regarding UVR, it has been found that planktonic larvae of the decapod *Cyrtograpsus altimanus* is affected in terms of delayed development (moulting) and enhanced swimming activity (Hernández-Moresino et al. 2011). More actively-swimming larvae may reflect that the individuals are avoiding the UVR-rich water surface. This was specifically tested in a subsequent study on free-swimming larvae of *C. altimanus*, in which it was found that UVR may repeal individuals away from surface waters (Gonçalves et al. 2014); thus, organisms would need to quickly adjust their vertical position to cope with the high solar radiation which is present during their reproductive season and further development. Additionally, UVR tolerance of each species can also depend on temperature: in a study with simulated solar ultraviolet radiation comparing planktonic larvae of *C. altimanus*, *C. angulatus* and *Leucippa pentagona*, it was found that *C. altimanus* was the most resistant and *L. pentagona* the most sensitive to ultraviolet radiation; however, both *C. altimanus* and *C. angulatus* showed lower mortality at 20 °C than at 15 °C, although no difference was found for *L. pentagona* (Hernández-Moresino and Helbling 2010). As mentioned above, some organisms may use photoprotective compounds (PPC) against damaging solar radiation. That might explain, at least partially, the resistance of *C. altimanus* and *C. angulatus* to UVR, as larvae of *C. altimanus* may acquire PPC directly from the diet or indirectly through maternal feeding (Hernández-Moresino et al. 2014). Thus, not only there seems to be more than one way to obtain protection from PPC-producing organisms – such as phytoplankton – but also trophic relationships are again very important when considering effects of global change for a given taxonomic group.

Environmental stress such as UVR or pollution may trigger antioxidant defence in different aquatic organisms (Abele et al. 2011). For example, in coastal waters of Uruguay, the copepod *Acartia tonsa* shows signs of enhanced oxidative stress in sites with more human impact, although it is not entirely clear how this translates into changes in fitness (Martínez et al. 2017). This may worsen in the next decades, as eutrophication is expected to increase in Río de la Plata estuary (Nagy et al.

2002). In addition, UVR may pose additional oxidative stress on zooplankton (Lesser et al. 2001; Abele et al. 2011), which should be incorporated as one more factor that interacts with OA, pollution and warming.

5 Conclusions

Studies on the effects of one or more of the 'global change factors' have been growing rapidly worldwide, being SST and UVR the most frequent factors evaluated and organism and/or population levels the most frequent approaches (Harley et al. 2006). An ecologist may be tempted to think that more integral approaches are needed, but our knowledge of the study area of this book (with few exceptions) is so scarce that we are only beginning to understand its functioning, let alone its response to global change in a more ecosystemic approach. One characteristic of global change is that several correlated factors interact simultaneously, and thus, the inferences from single-factor studies should be taken cautiously (Hoffman et al. 2003). However, considering several factors simultaneously (e.g. UVR, temperature, OA) greatly increases the complexity of experimental studies, while it is not always clear how much this actually enhances our explanatory ability to assess changes in abundance/distribution of organisms, or our predictive power to mitigate the effects. Additionally, poor initial understanding of, e.g. biogeochemical cycles may prevent us from predicting changes in particular regions (e.g., some coastal areas) which are not well resolved in larger, global models. Laboratory and enclosed-simulated studies are a first step, although extrapolating those results to the field (i.e. much larger time, space and trophic scales) poses challenging questions. Even when somewhat "clear" trends are detected in experimental studies, one must be cautious when extrapolating or making predictions based on these responses, even under similar conditions (Villafañe et al. 2017).

There is very little information on several important key trophic groups such as bacterioplankton, virus and zooplankton. Most studies are focused on phytoplankton, with a remarkable amount of experimental results obtained in coastal waters of Chubut (Argentina). Integrating studies from different environments (river-dominated estuaries, gulfs, continental shelf waters, etc.) should give a broader picture to identify key processes. Also, much work remains to be done about marine viruses, which are still virtually unknown in spite of growing awareness of their importance in pelagic ecosystems (Jacquet and Bratbak 2003) as key players in nutrient cycling and trophic networks (Roux et al. 2016). Time series of in situ data are much needed to detect trends beyond seasonal variations of the physical (temperature, mixing depth, etc.) and biological environment (presence of new species, changes in abundances, etc.). One case worth mentioning is the 15-year time series in Cabo Frio (Brazil), in which meroplankton abundance was positively correlated to sea surface temperature and negatively related to phytoplankton biomass (Fernandes et al. 2012). Another remarkable case is from Bahía Blanca estuary (Guinder et al. 2013) where a trend of increasing minimal water temperature in winter, together with more turbidity and dissolved nutrients in recent summers has been detected. There are examples of long-term data in other regions of the world, regarding ocean acidification (Bates et al. 2014), sea surface warming and zooplankton decline (Roemmich and McGowan 1995) and expansion of low-chlorophyll water areas (Polovina et al. 2008). In this review, studies related to OA have been mainly focused on phytoplankton, and although there are studies elsewhere on species of zooplankton which are also found in western South Atlantic (e.g. the copepod *Acartia* sp.), more zooplankton studies are lacking. In addition, global change effects may sometimes be more important in indirect (e.g. trophic interactions, maternal effects, etc.; Gonçalves and Hylander 2014; Hernández-Moresino et al. 2014) than direct fashion; thus, more studies considering several trophic levels would help to get a better understanding of potential, ecosystem-level effects. Even if single-species are to be considered, perhaps key species could be chosen if possible. For example, holoplanktonic gastropods like pteropods are the major metazoan calcifiers in the water column and are most abundant at South America latitudes ~40°S (Burridge et al. 2016); thus, this group can be good indicator of the effects of OA.

Finally, key factors associated to climate change modify the physical and chemical environment of planktonic organisms, so they are expected to affect the bottom-up transfer of energy in the upper ocean. However, human activities such as fishing are likely exerting additional pressure over the pelagic realm, affecting top-down processes (by removing planktivorous predators). Thus, human activities are changing energy flow and biogeochemical cycling in marine ecosystems of South America, and one of the main challenges for the next decades is to estimate the magnitude and characteristic time scales of these interactions.

References

Abele D, Vazquez-Medina JP, Zenteno-Savin T (eds) (2011) Oxidative stress in aquatic ecosystems. Wiley-Blackwell, Chichester, West Sussex (UK)

Agusti S, Llabrés M (2007) Solar radiation-induced mortality of marine pico-phytoplankton in the oligotrophic ocean. Photochem Photobiol 83:793–801

Barbieri ES, Villafañe VE, Helbling EW (2002) Experimental assessment of UV effects upon temperate marine phytoplankton when exposed to variable radiation regimes. Limnol Oceanogr 47:1648–1655

Bates N, Astor Y, Church M et al (2014) A time-series view of changing ocean chemistry due to ocean uptake of anthropogenic CO2 and ocean acidification. Oceanography. https://doi.org/10.5670/oceanog.2014.16

Bouchard JN, Campbell DA, Roy S (2005) Effects of UV-B radiation on the D1 protein repair cycle of natural phytoplankton communities from three latitudes (Canada, Brazil, and Argentina). J Phycol. https://doi.org/10.1111/j.1529-8817.2005.04126.x

Buma AGJ, Helbling EW, de Boer MK et al (2001) Patterns of DNA damage and photoinhibition in temperate South-Atlantic picophytoplankton exposed to solar ultraviolet radiation. J Photochem Photobiol B 62:9–18

Burridge AK, Goetze E, Wall-Palmer D et al (2016) Diversity and abundance of pteropods and heteropods along a latitudinal gradient across the Atlantic Ocean. Prog Oceanogr. https://doi.org/10.1016/j.pocean.2016.10.001

Cabrerizo MJ, Carrillo P, Villafañe VE et al (2017) Differential impacts of global change variables on coastal South Atlantic phytoplankton: role of seasonal variations. Mar Environ Res. https://doi.org/10.1016/j.marenvres.2017.01.005

Carreto J, Carignan M (2011) Mycosporine-like amino acids: relevant secondary metabolites. Chemical and ecological aspects. Mar Drugs 9:387–446

Dunlap WC, Yamamoto Y (1995) Small-molecule antioxidants in marine organisms: antioxidant activity of mycosporine-glycine. Comp Biochem Physiol 112B:105–114

Durán-Romero C, Villafañe VE, Valiñas MS, et al (2017) Solar UVR sensitivity of phyto- and bacterioplankton communities from Patagonian coastal waters under increased nutrients and acidification. ICES J Mar Sci. https://academic.oup.com/icesjms

de Fernandes LDA, Quintanilha J, Monteiro-Ribas W et al (2012) Seasonal and interannual coupling between sea surface temperature, phytoplankton and meroplankton in the subtropical South-Western Atlantic Ocean. J Plankton Res. https://doi.org/10.1093/plankt/fbr106

Fileman ES, White DA, Harmer RA et al (2017) Stress of life at the ocean's surface: latitudinal patterns of UV sunscreens in plankton across the Atlantic. Prog Oceanogr. https://doi.org/10.1016/j.pocean.2017.01.001

Garcia-Pichel F (1994) A model for internal self-shading in planktonic organisms and its implications for the usefulness of ultraviolet sunscreens. Limnol Oceanogr 39:1704–1717

Gonçalves RJ, Hylander S (2014) Marine copepods and solar radiation. In: Seuront L (ed) Copepods: diversity, habitat and behavior. Nova Science Publishers, Inc., Hauppauge, pp 101–120

Gonçalves RJ, Hernández - Moresino RD, Spinelli ML (2014) Short-term effect of UVR on vertical distribution of *Cyrtograpsus altimanus* and *Alexandrium tamarense* from Atlantic Patagonia. Lat Am J Aquat Res 42:963–970

Guinder VA, Popovich CA, Molinero JC et al (2013) Phytoplankton summer bloom dynamics in the Bahía Blanca Estuary in relation to changing environmental conditions. Cont Shelf Res. https://doi.org/10.1016/j.csr.2012.11.010

Guinder VA, López-Abbate MC, Berasategui AA et al (2015) Influence of the winter phytoplankton bloom on the settled material in a temperate shallow estuary. Oceanologia. https://doi.org/10.1016/j.oceano.2014.10.002

Harley CDG, Hughes AR, Hultgre KM et al (2006) The impacts of climate change in coastal marine systems. Ecol Lett 9:228–241

Helbling EW, Buma AGJ, de Boer MK et al (2001) In situ impact of solar ultraviolet radiation on photosynthesis and DNA in temperate marine phytoplankton. Mar Ecol Prog Ser 211:43–49

Helbling EW, Barbieri ES, Marcoval MA et al (2005) Impact of solar ultraviolet radiation on marine phytoplankton of Patagonia, Argentina. Photochem Photobiol 81:807–818

Helbling EW, Buma AGJ, Van de Poll W et al (2008) UVR-induced photosynthetic inhibition dominates over DNA damage in marine dinoflagellates exposed to fluctuating solar radiation regimes. J Exp Mar Biol Ecol 365:96–102

Helbling EW, Banaszak AT, Villafañe VE (2015) Differential responses of two phytoplankton communities from the Chubut River estuary (Patagonia, Argentina) to the combination of UVR and elevated temperature. Estuar Coasts. https://doi.org/10.1007/s12237-013-9752-7

Herman JR (2010) Global increase in UV irradiance during the past 30 years (1979–2008) estimated from satellite data. J Geophys Res-Atmos. https://doi.org/10.1029/2009JD012219

Hernández-Moresino RD, Helbling EW (2010) Combined effects of UVR and temperature on the survival of crab larvae (Zoea I) from Patagonia: the role of UV-absorbing compounds. Mar Drugs 8:1681–1698

Hernández-Moresino RD, Gonçalves RJ, Helbling EW (2011) Sublethal effects of ultraviolet radiation on crab larvae of *Cyrtograpsus altimanus*. J Exp Mar Biol Ecol. https://doi.org/10.1016/j.jembe.2011.07.019

Hernández-Moresino RD, Gonçalves RJ, Helbling EW (2014) Direct and indirect acquisition of photoprotective compounds in crab larvae of coastal Patagonia (Argentina). J Plankton Res. https://doi.org/10.1093/plankt/fbu010

Hoffman JR, Hansen LJ, Klinger T (2003) Interactions between UV radiation and temperature limit inferences from single-factor experiments. J Phycol 39:268–272

Hoffmeyer M (2004) Decadal change in zooplankton seasonal succession in the Bahía Blanca estuary, Argentina, following introduction of two zooplankton species. J Plankton Res 26:181–189

Jacquet S, Bratbak G (2003) Effects of ultraviolet radiation on marine virus-phytoplankton interactions. FEMS Microbiol Ecol. https://doi.org/10.1016/S0168-6496(03)00075-8

Jaque F, Tocho JO, Silva LFD et al (1994) Ground-based ultraviolet-radiation measurements during springtime in the southern hemisphere. Europhys Lett EPL. https://doi.org/10.1209/0295-5075/28/4/012

Karentz D, McEuen FS, Land MC, Dunlap WC (1991) Survey of mycosporine-like amino acid compounds in Antarctic marine organisms: potential protection from ultraviolet exposure. Mar Biol 108:157–166

Kitidis V, Brown I, Hardman-Mountford N et al (2016) Surface ocean carbon dioxide during the Atlantic Meridional transect (1995–2013); evidence of ocean acidification. Prog Oceanogr. https://doi.org/10.1016/j.pocean.2016.08.005

Lesser MP, Farrell JH, Walker CW (2001) Oxidative stress, DNA damage and p53 expression in the larvae of Atlantic cod (*Gadus morhua*) exposed to ultraviolet (290–400 nm) radiation. J Exp Biol 204:157–164

Longhi ML, Ferreyra G, Schloss I et al (2006) Variable phytoplankton response to enhanced UV-B and nitrate addition in mesocosm experiments at three latitudes (Canada, Brazil and Argentina). Mar Ecol Prog Ser 313:57–72

Manrique, J. M., Calvo, A. Y., Halac, S. R., Villafañe, V. E., Jones, L. R., & Helbling, E. W. (2012). Effects of UV radiation on the taxonomic composition of natural bacterioplankton communities from Bahía Engaño (Patagonia, Argentina). J Photochem Photobiol B 117: 171–178

Marcoval MA, Villafañe VE, Helbling EW (2007) Interactive effects of ultraviolet radiation and nutrient addition on growth and photosynthesis performance of four species of marine phytoplankton. J Photochem Photobiol B 89:78–87

Marcoval MA, Villafañe VE, Helbling EW (2008) Combined effects of solar ultraviolet radiation and nutrients addition on growth, biomass and taxonomic composition of coastal marine phytoplankton communities of Patagonia. J Photochem Photobiol B 91:157–166

Martínez M, Rodríguez-Graña L, Santos L et al (2017) Oxidative damage and vital rates in the copepod *Acartia tonsa* in subtropical estuaries with contrasting anthropogenic impact. J Exp Mar Biol Ecol. https://doi.org/10.1016/j.jembe.2016.11.016

Nagy GJ, Gómez-Erache M, López CH et al (2002) Distribution patterns of nutrients and symptoms of eutrophication in the Rio de la Plata River estuary system. In: Orive E, Elliott M, de Jonge VN (eds) Nutrients and eutrophication in estuaries and coastal waters: proceedings of the 31st symposium of the Estuarine and coastal sciences association (ECSA), held in Bilbao, Spain, 3–7 July 2000. Springer Netherlands, Dordrecht, pp 125–139

Polovina JJ, Howell EA, Abecassis M (2008) Ocean's least productive waters are expanding. Geophys Res Lett. https://doi.org/10.1029/2007GL031745

Richter PR, Häder D-P, Goncalves RJ et al (2007) Vertical migration and motility responses in three marine phytoplankton species exposed to solar radiation. Photochem Photobiol 83:810–817

Roemmich D, McGowan J (1995) Climatic warming and the decline of zooplankton in the California current. Science. https://doi.org/10.1126/science.267.5202.1324

Roux S, Brum JR, Dutilh BE et al (2016) Ecogenomics and potential biogeochemical impacts of globally abundant ocean viruses. Nature. https://doi.org/10.1038/nature19366

Roy S, Mohovic B, Gianesella SMF et al (2006) Effects of enhanced UV-B on pigment-based phytoplankton biomass and composition of mesocosm-enclosed natural marine communities from three latitudes. Photochem Photobiol 82:909–922

Vernet M, Diaz SB, Fuenzalida HA et al (2009) Quality of UVR exposure for different biological systems along a latitudinal gradient. Photochem Photobiol Sci. https://doi.org/10.1039/B904540F
Villafañe VE, Helbling EW, Zagarese HE (2001) Solar ultraviolet radiation and its impact on aquatic systems of Patagonia, South America. Ambio 30:112–117
Villafañe VE, Barbieri ES, Helbling EW (2004a) Annual patterns of ultraviolet radiation effects on temperate marine phytoplankton off Patagonia, Argentina. J Plankton Res 26:167–174
Villafañe VE, Marcoval MA, Helbling EW (2004b) Photosynthesis versus irradiance characteristics in phytoplankton assemblages off Patagonia (Argentina): temporal variability and solar UVR effects. Mar Ecol Prog Ser 284:23–34
Villafañe VE, Janknegt PJ, de Graaff M et al (2008) UVR-induced photoinhibition of summer marine phytoplankton communities from Patagonia. Mar Biol 154:1021–1029
Villafañe VE, Banaszak AT, Guendulain-García SD et al (2013) Influence of seasonal variables associated with climate change on photochemical diurnal cycles of marine phytoplankton from Patagonia (Argentina). Limnol Oceanogr. https://doi.org/10.4319/lo.2013.58.1.0203
Villafañe VE, Erzinger GS, Strauch SM et al (2014) Photochemical activity of PSII of tropical phytoplankton communities of Southern Brazil exposed to solar radiation and nutrient addition. J Exp Mar Biol Ecol. https://doi.org/10.1016/j.jembe.2014.05.027
Villafañe VE, Valiñas MS, Cabrerizo MJ et al (2015) Physio-ecological responses of Patagonian coastal marine phytoplankton in a scenario of global change: role of acidification, nutrients and solar UVR. Mar Chem. https://doi.org/10.1016/j.marchem.2015.02.012
Villafañe VE, Cabrerizo MJ, Erzinger GS et al (2017) Photosynthesis and growth of temperate and sub-tropical estuarine phytoplankton in a scenario of nutrient enrichment under solar ultraviolet radiation exposure. Estuar Coasts. https://doi.org/10.1007/s12237-016-0176-z

Index

M. S. Hoffmeyer et al. (eds.), *Plankton Ecology of the Southwestern Atlantic*,
https://doi.org/10.1007/978-3-319-77869-3

N

CPSIA information can be obtained
at www.ICGtesting.com
Printed in the USA
LVHW081943290519
619462LV00005B/28/P

9 783030 085605